# Nuclear and Particle Physics with Cosmology, Volume 1

## Nuclear physics

Online at: https://doi.org/10.1088/978-0-7503-5027-3

# Nuclear and Particle Physics with Cosmology, Volume 1

Nuclear physics

**Jyotirmoy Guha**

*Department of Physics, Santipur College, Santipur, Nadia, West Bengal, India*

**IOP** Publishing, Bristol, UK

ISBN    978-0-7503-5027-3 (ebook)
ISBN    978-0-7503-5025-9 (print)
ISBN    978-0-7503-5028-0 (myPrint)
ISBN    978-0-7503-5026-6 (mobi)

DOI    10.1088/978-0-7503-5027-3

Version: 20240501

IOP ebooks

British Library Cataloguing-in-Publication Data: A catalogue record for this book is available from the British Library.

Published by IOP Publishing, wholly owned by The Institute of Physics, London

IOP Publishing, No.2 The Distillery, Glassfields, Avon Street, Bristol, BS2 0GR, UK

US Office: IOP Publishing, Inc., 190 North Independence Mall West, Suite 601, Philadelphia, PA 19106, USA

*Dedicated to Students and Colleagues of Santipur College, West Bengal, India.*

*That completed 75 years in 2023 of training young minds —to think, to push the limits and to rise every time they fail.*

# Contents

# Foreword

Dr Jyotirmoy Guha is an asset of Santipur College and has compiled the first volume of *Nuclear and Particle Physics with Cosmology* based on his lectures to students in different academic sessions. This volume can be treated as a course of a semester that includes nuclear physics. It is indeed remarkable that this book is getting published by the famed IOP Publishing Ltd at a time when Santipur College celebrates and commemorates its **75 years** of service of grooming students in the field of higher learning by strengthening innovation and research and stretching the frontiers of knowledge. The detailed analysis of topics performed in a simple style and lucid approach as well as several solved problems are highlights of this volume and are sure to attract and benefit learners in the relevant arena of science and help generate love of the subject.

**Dr Chandrima Bhattacharya**
Principal
Santipur College
West Bengal, India
22 July 2023
Santipur College
Santipur, Nadia
West Bengal, India

# Preface

There are so many good books in the field of nuclear physics written by eminent authors, and these books contain several suggested problems that are intended to generate passion and the desire to learn more. However, in the semester system that is prevalent nowadays in the education system, very little time can be invested to think over a problem and there is virtually no scope to delve deep into a subject, given its vastness, in a limited period of time. One has thus to remain contented with a superficial shallow knowledge about the subject.

This book is an effort to address this issue. Following a simple style, lucid approach and step-by-step explanation efforts have been made to describe things in a manner that is understandable without any trouble. Care has been taken to give illustrations whenever applicable. The characteristic of this book is not only to ask questions but to answer them also. This would definitely satisfy a learner and help generate more queries among the inquisitive. They would then search for alternative ways to think over the same problem and to create avenues where they can give their thoughts a free run.

This volume is the first of two; the second volume will address particle physics and cosmology.

This volume on nuclear physics can be treated as a complete course on nuclear physics. It has been divided into a set of eight modules, or chapters. Each module or chapter has been subdivided into sections. Two or three sections can be looked upon as a single lecture.

The majority of the topics referred to in any undergraduate and postgraduate course on nuclear physics have been covered in the book in this volume.

Any suggestion, constructive criticism emailed to me would help improve the book further.

Dr Jyotirmoy Guha
jgsantipurcollege@gmail.com
https://youtube.com/@jgphysics
22 July 2023
Santipur College
Santipur, Nadia
West Bengal, India

# Acknowledgments

I thank IOP Publishing Ltd for providing me with the wonderful opportunity to stitch the lectures delivered in various semester courses that included nuclear physics as a component and give it the shape of a book. I express my sincere gratitude to Robert Trevelyan, Commissioning Editor for the IOP book series for being in constant touch with me and discussing the progress of the book periodically.

Special thanks are due to Dr Dipankar Bhattacharyya, Associate Professor of Physics of Santipur College with whom I co-authored the book entitled *Quantum Optics and Quantum Computing: An introduction* in 2022, also published by IOP. It was through him that the contact with IOP took a concrete shape.

I gratefully acknowledge the words of encouragement and constant support of my departmental colleagues Dr Atreyi Paul, Dr Anita Gangopadhyay, Dr Palash Das, Professor Chintaharan Majumder, Dr Dibyendu Biswas and Professor Subhau Choudhury.

The appreciation, love and words of encouragement that I received from Principal Dr Chandrima Bhattacharya of Santipur College and from all my colleagues of Santipur College were incredibly gratifying.

# Author biography

**Jyotirmoy Guha**

**Dr Jyotirmoy Guha** is a faculty of Santipur College, West Bengal, India in the Department of Physics. He is the Head of the department and serves as Associate Professor.

Dr Guha has a brilliant academic record, having secured first class first both in undergraduate and post graduate examinations. He did his doctoral work on quantum cosmology and has published works in international journals of repute.

Dr Guha has authored several books, of which those that need special mention are the following: *Solid state Physics: Theory, Problems and Solutions* published by Books and Allied, Kolkata; *Quantum Mechanics: Theory, Problems and Solutions* published by Books and Allied, Kolkata; *Modern Physics Volumes I and II* published by Techno World, Kolkata. Dr Guha is a co-author of the book *Quantum Optics and Quantum Computation: An introduction* published by IOP Publishing Ltd.

Several of his students are well established in various institutes of international repute and carry out research around the globe. Dr Guha regularly contributes in his you tube channel https://youtube.com/@jgphysics.

**IOP** Publishing

Nuclear and Particle Physics with Cosmology, Volume 1
Nuclear physics
**Jyotirmoy Guha**

# Chapter 1

## General properties of the nucleus

## 1.1 Introduction: motivation of studying nuclear physics

Nuclear physics began its journey with Rutherford's experiment that revealed the presence of a tiny dense nucleus at the centre of atoms. In nuclear physics we study properties and interactions of the nucleus, while in atomic physics we study the properties and interactions of the atom.

Nuclear physics is the study of various aspects of nuclei—their formation, structure, stability and decay, as well as interaction among the constituents.

Important advancements in medicine, materials, energy, security, climatology etc are because of research and development in nuclear science. In fact, nuclear physics has impacted our lives, especially in the field of safety, health and security.

Nuclear physics is essential for understanding the basics of radioactivity as well as nuclear reactions; gaining knowledge about industrial and agricultural isotopes, radio carbon dating, ion implantation in materials etc.

Nuclear physics finds application in many fields.

Nuclear power generation is a major area. Safe and reliable use of reactors with advanced error-free design is a necessity.

Nuclear imaging technology is used in healthcare to diagnose and treat different types of cancers, cardiovascular disease, some neurological disorders in their early stages. Nuclear medicine procedures are used to diagnose Alzheimer's disease, treat hyperthyroidism, assess coronary artery disease, localize tumors etc. Nuclear medicine can provide vital information about the function of major organs within the body. Advances in nuclear medicine are closely connected with advances in nuclear techniques. Further progress in the field of healthcare depends on: basic research in nuclear science; more use of accelerators, detectors; understanding interaction of radiation with matter and improvement in data analysis techniques. Nuclear physics facilities will broaden the range of isotopes for medical applications, i.e. bring more isotopes into routine use for human benefit.

Nuclear science has an important and sensitive role in national security of a country. Nuclear weapons are stockpiled for defense preparedness of a country to act as deterrent. Nuclear devices determine the outcome of a war and can alter political boundaries. One major concern of nuclear scientists is the possibility of nuclear arms getting into the hands of a rogue state or terrorists, which could bring about disastrous consequences. Also, more progress is needed in the proper handling, treatment and storage of nuclear waste.

Nuclear science plays a critical role in global politics, protecting international borders, safeguarding nuclear materials, preventing nuclear terrorism and restricting proliferation of nuclear weapons.

Another motivation for studying nuclear physics is its all-round presence in astrophysics. Nuclear astrophysics has developed rapidly and has an important role in understanding stellar matters, origin of chemical elements, life cycle of stars, death of a star and how the elements thrown into space get recycled to produce new stars. This also helps in our understanding of the universe.

## 1.2 Constituents of a nucleus

A molecule is made up of atoms. An atom is made up of one nucleus plus electrons.

The nucleus or nuclide refers to one or a single nucleus. Nuclei or nuclides refer to a multi-nucleus system, e.g. there are two nuclei in a diatomic molecule (which is a two atom–two nucleus system). The nucleus is made up of $A$ number of constituents called nucleons where $A$ is called mass number. In other words the nucleus contains $A$ nucleons.

Nucleons are of two types: proton (symbol is $p$) neutron (symbol is $n$).

☐ Proton

✓ Proton is one constituent of a nucleus.

✓ Number of protons is called atomic number and is denoted by $Z$. Clearly

$$A = Z + N \tag{1.1}$$

✓ Mass of proton is

$$m_p = 1.67 \times 10^{-27} \, kg \tag{1.2}$$

It is slightly less than the mass of a neutron $m_n$.

✓ Charge of proton is

$$Q_p = |\,e\,| = 1.6 \times 10^{-19} C \tag{1.3}$$

i.e. a proton is positively charged. Here $|\,e\,|$ = magnitude of electronic charge.

✓ Rest energy of a proton is

$$m_p c^2 = (1.67 \times 10^{-27} \, kg)(3 \times 10^8 ms^{-1})^2$$

$$\approx 1.5 \times 10^{-10} \, kg. \, m^2 s^{-2} = \frac{1.5 \times 10^{-10}}{1.6 \times 10^{-19}} \, eV$$

$$m_p c^2 \approx 938 \times 10^6 \, eV = 938 \, MeV \tag{1.4}$$

Here $c$ denotes speed of light in free space and we take its value to be

$$c = 3 \times 10^8 ms^{-1}. \tag{1.5}$$

Energy unit is joule denoted by

$$J \rightarrow \text{force. distance} = MLT^{-2}. \, L = ML^2 T^{-2} \overset{\text{in S.I}}{\underset{\text{units}}{\rightarrow}} \, kg. \, m^2 s^{-2}. \text{ Also}$$

$$1 \, eV = 1.6022 \times 10^{-19} \, J \cong 1.6 \times 10^{-19} \, J. \tag{1.6}$$

☐ Neutron
  ✓ The neutron is one constituent of nucleus.
  ✓ Number of neutrons is $A - Z = N$ where $A$ is mass number = total number of nucleons.
  ✓ We take mass of a neutron to be

$$m_n = 1.675 \times 10^{-27} \, kg \tag{1.7}$$

This is slightly more than the mass of a proton $m_p$.
  ✓ Charge of a neutron is

$$Q_n = 0 \tag{1.8}$$

i.e. the neutron is electrically neutral.
  ✓ Rest energy of a neutron is

$$m_n c^2 = (1.675 \times 10^{-27} \, kg)(3 \times 10^8 ms^{-1})^2$$

$$\approx 1.507 \times 10^{-10} \, kg. \, m^2 \, s^{-2} \rightarrow \frac{1.507 \times 10^{-10}}{1.6022 \times 10^{-19}} \, eV \overset{\text{we take}}{\rightarrow} 939 \times 10^6 \, eV$$

$$m_n c^2 = 939 \, MeV \tag{1.9}$$

Proton and neutron are particles that almost have the same mass and hence are not distinguishable so far as their masses are concerned. They are disinguishable only through their electromagnetic interaction as their charges are different—one is charged and one is neutral. So if we ignore their electromagnetic property then they can virtually be treated as identical particles. The proton can then be called a charged neutron and the neutron can then be called a chargeless proton.

The symbol of nucleus is

$$_Z X_N^A \equiv \, _Z X^A$$

where the name of the chemical element is denoted by $X$. Mass number is $A$ (which is nucleon number), $Z$ is atomic number (i.e. proton number), $N$ is neutron number. For instance

✓ $_{17}\text{Cl}_{18}^{35} \equiv _{17}\text{Cl}^{35}$ is the chemical element chlorine having $A = 35$, $Z = 17$, $N = A - Z = 35 - 17 = 18$.

✓ $_{1}\text{H}_{0}^{1} \equiv _{1}\text{H}^{1}$ is the chemical element hydrogen having $A = 1$, $Z = 1$, $N = A - Z = 1 - 1 = 0$. Hydrogen nucleus has one proton, no neutron. So $_{1}\text{H}^{1} \equiv p$ is a one-proton system.

✓ $_{1}\text{H}_{1}^{2} \equiv _{1}\text{H}^{2} \equiv d$ is the chemical element deuterium having $A = 2$, $Z = 1$, $N = A - Z = 2 - 1 = 1$. So $_{1}\text{H}^{2} = \text{np} = d$ is a two-nucleon system.

✓ $_{1}\text{H}_{2}^{3} \equiv _{1}\text{H}^{3}$ is the chemical element called tritium having $A = 3$, $Z = 1$, $N = A - Z = 3 - 1 = 2$.

☐ Hydrogen has two isotopes

    ① Deuterium nucleus $_{1}\text{H}^{2} \equiv d$

    It is a bound stable $np$ system. It is the only stable two-nucleon system as $pp, nn$ systems do not exist in nature.

    ② Tritium nucleus $_{1}\text{H}^{3}$

    It is $\beta^-$ radioactive

$$_{1}\text{H}^{3} \rightarrow _{2}\text{He}^{3} + e^- + \bar{\nu}_e \text{ (half-life 12 years)}.$$

Here the symbol $e^- \equiv _{-1}e^0$ stands for electron and $\bar{\nu}_e$ is a particle called the electron anti-neutrino.

☐ Electron properties

The electron has mass

$$m_e = 9.1 \times 10^{-31} \, kg$$

$$= 9.1 \times 10^{-31} \, kg \frac{m_p}{m_p} = \frac{9.1 \times 10^{-31} \, kg}{1.67 \times 10^{-27} \, kg} m_p \tag{1.10}$$

$$= \frac{m_p}{1835} \tag{1.11}$$

Clearly $e^-$ is 1835 times less massive than a proton or neutron.

Electronic charge is

$$Q_e = -1.6 \times 10^{-19} \, C = -|e| = -Q_p \text{ (negative charge)} \tag{1.12}$$

The electron has rest energy

$$m_e c^2 = \left(9.1 \times 10^{-31} \, kg\right) \times \left(3 \times 10^8 ms^{-1}\right)^2 = 8.19 \times 10^{-14} \, J$$

$$= \frac{8.19 \times 10^{-14}}{1.6 \times 10^{-19}} \, eV = 0.511 \times 10^6 \, eV$$

$$m_e c^2 = 0.511 \, MeV \tag{1.13}$$

The atom in ground state is neutral. This is possible since the number of electrons carrying negative charge $-Z|e|$ and $Z$ number of protons carrying positive charge are equal in magnitude. So nuclear charge has to be $+Z|e|$.

In *exercise 1.3* we discuss whether it is possible for an electron to reside in a nucleus based on Heisenberg's uncertainty principle.

In *exercise 1.4* we show that the existence of an electron inside a nucleus is not consistent with the experimental observation of spin of deuteron to be 1.

In *exercise 1.5* we prove that an electron cannot occur in a nucleus from magnetic moment consideration.

## 1.3 Nuclear parameters

The nucleus is described by the following properties or parameters.

✓ Static properties

Electric charge, radius, mass, binding energy, angular momentum, parity, magnetic dipole moment, electric quadrupole moment, energy of excited state etc are referred to as static properties. These properties are not a function of time and hence are called static.

✓ Dynamic properties

Decay probabilities, reaction probabilities etc are referred to as dynamic properties. These properties are a function of time and hence called dynamic.

## 1.4 Unified atomic mass unit or amu or u

- Definition of *amu*

We define

$$1 \ amu = \frac{1}{12} \ \text{mass of } _6\text{C}^{12} \text{ atom} \tag{1.14}$$

This means that the mass of an atom of $_6\text{C}^{12}$ is 12 amu or 12 u. This is the unit of atomic mass as accepted in 1961 by the International Union of Pure and Applied Chemistry (IUPAC). (Before 1961 there was no unity between physicists and chemists regarding units.)

- *amu* in terms of proton mass

Avogadro's number is $N_A = 6.023 \times 10^{23}$ per mole. In other words in $_6\text{C}^{12}$ there are $N_A$ atoms in 1 mole, mass of which is 12 *gm*. So

1 atom of $_6\text{C}^{12}$ has mass $\frac{12 \ gm}{N_A}$. Hence we can write, using equation (1.14)

$$1 \ amu = \frac{1}{12} \ \text{mass of } _6\text{C}^{12} \text{ atom} = \frac{1}{12} \times \frac{12 \ gm}{N_A} = \frac{1 \ gm}{N_A} = \frac{10^{-3} \ kg}{N_A} = \frac{10^{-3} \ kg}{6.023 \times 10^{23}}$$

$$1 \ amu = 1.66 \times 10^{-27} \ kg \approx m_p \ \text{(proton mass)} \tag{1.15}$$

- *amu* in terms of *MeV*

$$1 \, amu = 1.66 \times 10^{-27} \, kg \xrightarrow[\text{multiply by } c^2]{\text{converting to energy}} \left(1.66 \times 10^{-27} \, kg\right)c^2$$

$$= 1.66 \times 10^{-27} \, kg \times (3 \times 10^8 ms^{-1})^2 = 1.49 \times 10^{-10} J$$

$$= \frac{1.49 \times 10^{-10} \, \text{eV}}{1.6 \times 10^{-19}} = 931 \times 10^6 \, eV$$

$$1 \, amu = 931 \, MeV \tag{1.16}$$

- Proton mass in terms of *amu*

  Proton rest mass energy is given by equation (1.4) viz $m_p c^2 = 938 \, MeV$. Using equation (1.16) we have

$$1 \, MeV = \frac{1}{931} \, amu \tag{1.17}$$

$938 \, MeV = \frac{938}{931} \, amu \approx 1.007\,276 \, amu$. Hence

$$m_p = 1.007\,276 \, amu \tag{1.18}$$

Thus from equations (1.2), (1.4), and (1.18) we can write

$$m_p = 1.67 \times 10^{-27} \, kg = 1.007\,276 \, amu = 938 \, \frac{MeV}{c^2} \tag{1.19}$$

- Neutron mass $m_n$ and electron mass $m_e$ in terms of *amu* and *MeV* have been dealt with in *exercise 1.16*.

## 1.5 Types of nuclei

There are 272 stable nuclei in nature. Depending on whether they possess odd or even number of protons ($Z$) or odd or even number of neutrons ($N$) we have four types of nuclei. We make a list of the possibilities.

| $Z$ value, $N$ value | Number of nuclei found in nature | Type of nuclei | $Z + N = A$ value | Number of nuclei found in nature |
|---|---|---|---|---|
| $Z$ = even, $N$ = even | 160 | even-even | even + even = even $A$ | 160 + |
| $Z$ = odd, $N$ = odd | 04 | odd-odd | odd + odd = even $A$ | 04 = 164 |
| | | | | (even $A$) |
| $Z$ = even, $N$ = odd | 56 | even-odd | even + odd = odd $A$ | 56 + 52 |
| $Z$ = odd, $N$ = even | 52 | odd-even | odd + even = odd $A$ | = 108 |
| | | | | (odd $A$) |
| | 272 | | | 272 |

We give example of such nuclei.

☐ $Z =$ even, $N =$ even, $A =$ even i.e. even–even nucleus

$$_2\text{He}_2^4, \ _6\text{C}_6^{12}, \ _8\text{O}_8^{16}, \ _{10}\text{Ne}_{10}^{20} \ \text{etc}$$

☐ $Z =$ odd, $N =$ odd, $A =$ even i.e. odd–odd nucleus

$$_1\text{H}_1^2, \ _3\text{Li}_3^6, \ _5\text{B}_5^{10}, \ _7\text{N}_7^{14}$$

☐ $Z =$ even, $N =$ odd, $A =$ odd i.e. even–odd nucleus

$$_2\text{He}_1^3, \ _4\text{Be}_5^9, \ _6\text{C}_7^{13}, \ _8\text{O}_9^{17}, \ _{10}\text{Ne}_{11}^{21}$$

☐ $Z =$ odd, $N =$ even, $A =$ odd i.e. odd–even nucleus

$$_1\text{H}_0^1, \ _{11}\text{Na}_{12}^{23}, \ _{17}\text{Cl}_{18}^{35}, \ _{29}\text{Cu}_{35}^{64}$$

Nuclei can be clubbed into various groups depending on various similarities they exhibit. We discuss them now.

## 1.6 Isotope

Nuclei with the same proton number, i.e. nuclei having the same atomic number $Z$ are called isotopes. Let us furnish examples of such nuclei.

✓ Hydrogen $_1\text{H}$ isotopes (with $Z = 1$)

$_1\text{H}^1 \equiv p$ is ordinary hydrogen. It is stable.

$_1\text{H}^2 \equiv np = d$ is deuterium or heavy hydrogen. It is just stable.

$_1\text{H}^3 \equiv 2n1p$ is tritium. It is unstable and radioactive.

✓ Carbon $_6\text{C}$ isotopes (with $Z = 6$)

$$_6\text{C}^{12}, \ _6\text{C}^{13}, \ _6\text{C}^{14}$$

✓ Oxygen $_8\text{O}$ isotopes (with $Z = 8$)

$$_8\text{O}^{16}, \ _8\text{O}^{17}, \ _8\text{O}^{18}$$

✓ Silicon $_{14}\text{Si}$ isotopes (with $Z = 14$)

$$_{14}\text{Si}^{28}, \ _{14}\text{Si}^{29}, \ _{14}\text{Si}^{30}, \ _{14}\text{Si}^{32}$$

✓ Iron $_{26}\text{Fe}$ isotopes (with $Z = 26$)

$$_{26}\text{Fe}^{54}, \ _{26}\text{Fe}^{56}, \ _{26}\text{Fe}^{57}, \ _{26}\text{Fe}^{58}$$

- Atomic number $Z$ or nuclear charge is responsible for the characteristic property of an atom.
- All isotopes of an element have identical chemical behavior—they differ only in mass.

☐ Radio isotope

If the isotope of an unstable nucleus is radioactive it is called a radio isotope. For instance

✓ $_1H^3$ is the radio isotope of hydrogen $_1H$ since it undergoes radioactive decay as

$$_1H^3 \rightarrow {}_2He^3 + e^- + \bar{\nu}_e$$

with a half-life of 12 years.

✓ $_6C^{14}$ is the radio isotope of carbon $_6C$ and is called radio carbon since it undergoes radioactive decay as

$$_6C^{14} \rightarrow {}_7N^{14} + e^- + \bar{\nu}_e$$

with a half-life of 5730 years.

## 1.7 Isotone

Nuclei with the same neutron number $N$ are called isotones. Let us furnish examples of such nuclei.

✓ Isotones with $N = 1$
$$_1H_1^2, \ _2He_1^3$$

✓ Isotones with $N = 2$
$$_2He_2^4, \ _3Li_2^5, \ _1H_2^3$$

✓ Isotones with $N = 8$
$$_6C_8^{14}, \ _7N_8^{15}, \ _8O_8^{16}.$$

## 1.8 Isobar

Nuclei with the same nucleon number, i.e. nuclei having the same mass number $A$ are called isobars. Let us furnish examples of such nuclei.

✓ Isobars with $A = 5$

$$_2He^5, \ _3Li^5$$

✓ Isobars with $A = 16$

$$_8O^{16}, \ _7N^{16}$$

## 1.9 Isomer or isomeric nuclei

Nuclei with the same proton number (i.e. same atomic number $Z$), same neutron number $N$ and hence same mass number $A = Z + N$, but different nuclear energy states are called isomers. These nuclei exhibit differences in internal structures.

Let us furnish an example of such a nuclear isomer.

✓ $_{90}Th^{234}$ (also called $UX_1$) suffers $\beta^-$ decay to give $_{91}Pa^{234}_{143}$ (protactinium) which can be in two energy states as shown in figure 1.1. The two energy states are denoted by the symbols namely

$UX_2$ which is in 0.394 MeV excited metastable state and
UZ which is in 0 MeV ground state.
The pair ( $UX_2$, $UZ$) are nuclear isomers.
Both $UX_2$, $UZ$ suffer $\beta^-$ decay with different half lives.

✓ $UX_2 \xrightarrow{\beta^-} {}_{92}U^{234}$ with half-life $T_{1/2} = 1.18$ min

✓ $UZ \xrightarrow{\beta^-} {}_{92}U^{234}$ with half-life $T_{1/2} = 6.66$ hrs.

As one state lives longer than the other the nuclear isomers $UX_2$, $UZ$ can be distinguished by their different lifetimes.

$UX_2$, $UZ$ are essentially the same nuclei but in different energy states and hence called nuclear isomers.

$UX_2$ can also shed off its extra energy through $\gamma$ emission as

$UX_2 \rightarrow UZ + \gamma \, (0.394 \, MeV)$

• An isomeric excited state differs from an ordinary excited state of a nucleus.

An isomeric excited state is a metastable state and lives for a measurably long time (e.g the isomeric excited state $UX_2$ of $_{91}Pa^{234}_{143}$ has half-life $T_{1/2} = 1.18$ min.

On the other hand an ordinary excited state of a nucleus decays instantaneously.

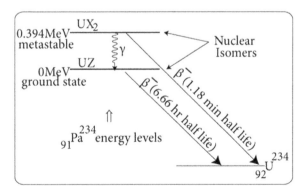

**Figure 1.1.** $_{91}Pa^{234}$ nuclear energy levels have been shown. ($UX_2$, $UZ$) form a pair of nuclear isomers. Both $UX_2$, $UZ$ suffer $\beta^-$ decay with different half lives. $UX_2$ can also shed off its extra energy through $\gamma$ emission.

## 1.10 Mirror nuclei

The elements $_ZX_N^A$ and $_NY_Z^A$ are called mirror nuclei.

Nuclei having the same mass number $A$, i.e. two isobaric nuclei with proton and neutron number interchanged are called mirror nuclei.

It is as if we are imagining a mirror that changes $p \to n$ and so $Z$ becomes $N$ and $n \to p$ so that $N$ becomes $Z$. In other words, the imagined mirror converts $_ZX^A$ to $_NY^A$ and vice versa. Obviously the mass number does not change since $A = Z + N = N + Z$ but the nucleus changes as atomic number changes from $Z$ to $N$. We furnish some examples of mirror nuclei.

✓ $_4Be_3^7$ and $_3Li_4^7$.

Here $(Z = 4, N = 3) \xrightarrow{p \leftrightarrows n \text{ transition}} (Z = 3, N = 4)$.

✓ $_5B_4^9$ and $_4Be_5^9$

Here $(Z = 5, N = 4) \xrightarrow{p \leftrightarrows n \text{ transition}} (Z = 4, N = 5)$.

✓ $_7N_8^{15}$ and $_8O_7^{15}$

✓ $_7N_6^{13}$ and $_6C_7^{13}$

✓ $_{20}Ca_{19}^{39}$ and $_{19}K_{20}^{39}$ etc

• Significance of mirror nuclei: charge symmetry hypothesis.

Let us note that one of the pairs of mirror nuclei contains an extra $nn$ bond while the other contains an extra $pp$ bond. This is evident in the following illustrations.

| $_4Be_3^7, \; _3Li_4^7$ | $_5B_4^9, \; _4Be_5^9$ | $_8O_7^{15}, \; _7N_8^{15}$ |
|---|---|---|
| • $_4Be_3^7 = 4p3n = \boxed{2p3n} + pp$ = core + $pp$ bond | • $_5B_4^9 = 5p4n = \boxed{3p4n} + pp$ = core + $pp$ bond | • $_8O_7^{15} = 8p7n = \boxed{6p7n} + pp$ = core + $pp$ bond |
| • $_3Li_4^7 = 3p4n \equiv 4n3p$ = $\boxed{2n3p}$ + $nn$ bond = core + $nn$ bond | • $_4Be_5^9 = 4p5n \equiv 5n4p$ = $\boxed{3n4p}$ + $nn$ bond = core + $nn$ bond | • $_7N_8^{15} = 7p8n \equiv 8n7p$ = $\boxed{6n7p}$ + $nn$ bond = core + $nn$ bond |
| The cores $2p3n$ and $2n3p$ are mirror images w.r.t $p \leftrightarrows n$ operation. Clearly $_4Be_3^7$ has one extra $pp$ bond while $_3Li_4^7$ has one extra $nn$ bond. | The cores $3p4n$ and $3n4p$ are mirror images w.r.t $p \leftrightarrows n$ operation. Clearly $_5B_4^9$ has one extra $pp$ bond while $_4Be_5^9$ has an extra $nn$ bond. | The cores $6p7n$ and $6n7p$ are mirror images w.r.t $p \leftrightarrows n$ operation. Clearly $_8O_7^{15}$ has one extra $pp$ bond while $_7N_8^{15}$ has an extra $nn$ bond. |

Experimentally it is observed that mirror nuclei have similar properties, e.g. same spin parity, same nuclear binding energy, similar energy level structure etc. This means that strong interactions that happen within mirror nuclei are identical. This suggests that the extra $pp$ bond and $nn$ bond should have the same strengths, i.e. they are identical in their strong interaction and cannot be distinguished on the basis of their strong interaction.

We thus can write strength of $pp$ bond = strength of $nn$ bond, i.e.

$$V_{pp} = V_{nn} \tag{1.20}$$

where $V$ stands for potential. So the interaction between nucleons is symmetric w.r.t interchange of $n$ by $p$ and $p$ by $n$, i.e. $pp$ and $nn$ interactions are identical.

This is charge symmetry hypothesis.

## 1.11 Charge independence hypothesis

Strong interaction between two protons and strong interaction between two neutrons are identical as per charge symmetry hypothesis. The next thought is about the strong interaction between a proton and a neutron.

Let us extend the charge symmetry hypothesis to deal with it. This leads us to the charge independence hypothesis that we explain now.

The ground state of $_6C_8^{14}$, the ground state of $_8O_6^{14}$ and the first excited state of $_7N_7^{14}$ have same spin, parity, binding energy. This means that the strong interaction prevalent in these nuclei are identical. Let us investigate the internal bond structure in these nuclei.

| | | |
|---|---|---|
| $_6C_8^{14} = 6p8n$ | $_8O_6^{14} = 8p6n$ | $_7N_7^{14} = 7p7n$ |
| $= 6p6n + nn$ | $= 6p6n + pp$ | $= 6p6n + np$ |
| $= {}_6C_6^{12}$ core + $nn$ | $= {}_6C_6^{12}$ core + $pp$ | $= {}_6C_6^{12}$ core + $np$ |
| $_6C_8^{14}$ has an extra $nn$ | $_8O_6^{14}$ has an extra $pp$ | $_7N_7^{14}$ has an extra $np$ |
| bond | bond | bond |

This suggests that the $nn$ bond, $pp$ bond and $np$ bond should have same strengths, i.e. they are identical in their strong interactions and cannot be distinguished on the basis of strong interaction. So nuclear interaction is charge independent. This means the strong interactions that happen between nucleons within the nuclei are identical. We can thus write strength of $pp$ bond = strength of $nn$ bond = strength of $np$ bond, i.e.

$$V_{pp} = V_{nn} = V_{np} \tag{1.21}$$

where $V$ stands for interaction potential.

Nuclear interaction is similar between any two sets of nucleons, i.e. does not distinguish between a neutron $n$ and a proton $p$.

So far as strong interaction is concerned $n$ and $p$ are identical and are called nucleons. Denoting nucleon by the symbol $N$ let us write

$$N = \begin{pmatrix} \text{proton} \\ \text{neutron} \end{pmatrix} = \begin{pmatrix} p \\ n \end{pmatrix} = \begin{pmatrix} \text{charged neutron} \\ \text{chargeless proton} \end{pmatrix} \tag{1.22}$$

This is the charge independence hypothesis.

Accordingly we identify the proton and neutron as two different charge states of the same particle called the nucleon.

- Consequence of the charge independence hypothesis: isospin quantum number.

The charge independence hypothesis tells us that when we are concerned with the strong interaction between nucleons it is not necessary to keep in mind the charges they carry. The fact that $p$ and $n$ cannot be distinguished on the basis of their strong interaction is described by saying that they have a conserved quantum number which we define as isospin denoted by $I$.

Isospin $I$ is the quantum number associated with an angular momentum called isospin angular momentum $\vec{I}$ defined in an internal spin space called isospace. With this we can characterize $p$ and $n$ by attaching quantum numbers associated with isospin angular momentum $\vec{I}$.

Since the nucleon has two charge states $p$ and $n$, it follows that there are two isospin states of nucleon $p$ and $n$. So the number of available states called multiplicity of $I$ is two, i.e.

$$2I + 1 = 2$$

$$I = \frac{1}{2}$$

Hence the nucleon is called iso-doublet (having two states $p$ and $n$).

Clearly $p$ and $n$ both have $I = \frac{1}{2}$. So $I$ is conserved for nucleon state. And let us designate the isospin projections as
$I_z = \frac{1}{2}$ for proton and $I_z = -\frac{1}{2}$ for neutron. Hence

$$| p > = | I = \frac{1}{2} \ I_z = \frac{1}{2} > \equiv | \frac{1}{2} \ \frac{1}{2} > \qquad (1.23)$$

$$| n > = | I = \frac{1}{2} \ I_z = -\frac{1}{2} > \equiv | \frac{1}{2} \ -\frac{1}{2} > \qquad (1.24)$$

☐ Actually $np$ interaction is a bit different from $pp$ and $nn$ interaction. This is evident from the fact that $pp$ and $nn$ bonds consist of identical particles, while $np$ bond is made of different particles. We discuss this in *exercise 1.17*.

The $np$ nuclear interaction can be mediated by all pions $\pi^0$, $\pi^\pm$ but $pp$ and $nn$ nuclear interaction can be mediated only by $\pi^0$. We discuss this in section 8.3.

## 1.12 Cross-section of an interaction

Cross-section of an interaction is a quantity:
- ✓ that specifies the amount of interaction. It indicates with how much vigour, strength or intensity the interaction takes place;
- ✓ that specifies the probability of interaction.

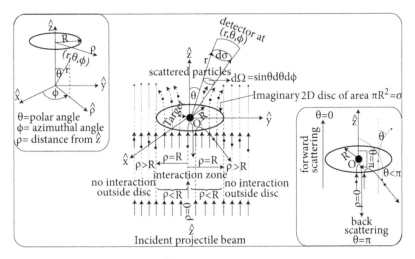

**Figure 1.2.** Cross-section of interaction is $\pi R^2 = \sigma$. Scattering occurs in 3D space. Disc radius $R$ is the impact parameter.

Let us consider the phenomenon of elastic scattering (in which there is no change in the entities before and after interaction). Consider the interaction

$$a + X \rightarrow X + a \tag{1.25}$$

$$\text{e.g. } \alpha + \text{Au} \rightarrow \alpha + \text{Au} \ (\alpha \equiv \text{alpha particle, Au} \equiv \text{gold nucleus}) \tag{1.26}$$

This is represented in figure 1.2.

The projectile $a$ (or $\alpha$) is thrown along $\hat{z}$ and strikes the target nucleus $X$ (or Au) sitting at the origin O where $\rho = 0$, the distance from $\hat{z}$ being $\rho$.

It is clear that the target has an influence over a region of space around it—say, influence over a region of radius $R$. In other words the target presents to the incident projectile, an effective area

$$\pi R^2 = \sigma \tag{1.27}$$

The distance $R$ is a measure of whether impact or interaction would occur or not and is called impact parameter.

Passage through this disc area $\sigma = \pi R^2$ ensures interaction and consequent deflection from the straight path of incidence.

Not all the incident projectile beam will hit and interact with the target. Only a few of those that try to fly sufficiently close to the target $(\rho \leqslant R)$ would feel the presence of the target and would get influenced (i.e. interact) by showing deflections from original path of projectile.

The angle of scattering is $\theta$ which also is the polar angle in this 3D scattering case. ($\theta$ is measured from $\hat{z}$, the direction of incidence of the projectile.)

For projectiles thrown at large distance from the $\rho = 0$, i.e. the large distance from the z-axis goes undeviated because of no interaction. This is forward scattering that corresponds to $\theta = 0$.

Obviously $R$ represents the distance from $\hat{z}$ that decides whether interaction occurs or not. And $\rho$ is any distance measured from $\hat{z}$. Clearly then:

- If $\rho \leqslant R$ then interaction occurs. The incident particle hits the target and interacts. There is a large probability of interaction for small $\rho$ and the probability decreases as $\rho$ increases, i.e. as the incident projectile passes by a large distance from the target.
- If $\rho > R$ then no interaction occurs.

The incident beam sees an effective target area of $\pi R^2 = \sigma$.

Consider alpha scattering by a gold nucleus. The projectile $\alpha$ is proceeding along $\hat{z}$ through a cylindrical region of space around $\rho = 0$ of cross-section $\pi R^2 = \sigma$ would hit the target nucleus Au and get deviated.

In particular, the beam along $\rho = 0$ will be back-scattered along $\theta = \pi$ and those in the range $\rho \leqslant R$ would get deviated from the straight path by angle $\theta < \pi$.

So the cross-section $\pi R^2$ of an imaginary 2D disc of radius $R$ defines the interaction zone. Outside this interaction region, i.e. in the region $\rho > R$ projectiles fly along $\hat{z}$ without any deflection as they are at too large a distance to get influenced by the target (in effect they are at $\rho \to \infty$).

The larger the area $\sigma$, the greater the amount of interaction.

Let us rewrite it from the perspective of probability of interaction.

If the projectile sees the target to have a small area ($\sigma$ small) only a few incident particles are likely to pass through it (disc) and so the probability of interaction will be small. On the other hand, if the projectile sees the target to have a large area ($\sigma$ large), a large number of incident particles is expected to pass through it (disc) and so the probability of interaction will be large. We note the following.

- $R$ represents the minimum distance between target nucleus ($X$ or Au say) and the undeviated path of projectile ($a$ or $\alpha$), i.e. the direction of the velocity of projection of projectile and is called impact parameter $R$ since it decides whether impact (i.e. collision) would occur or not.

    $\pi R^2$ is also called the impact cross-section.
- We note that the S.I unit of cross-section which is a disc area $\sigma = \pi R^2$ is m$^2$. The cross-section in various nuclear interactions turns out to be of the order of $10^{-28}$ m$^2$ and so we define a unit called barn as

$$1 \text{ barn} = 10^{-28} \, m^2. \tag{1.28}$$

- Cross-section $\sigma$ represents the total amount of interaction between the projectile and target.

## 1.13 Differential scattering cross-section

After interaction with target nucleus (located at origin O), projectiles incident along $\hat{z}$ are scattered in different directions—they come off the target with various angles $\theta$, $\phi$ ($\theta$ = polar angle, $\phi$ = azimuthal angle).

The differential scattering cross-section is a measure of the amount of interaction along a particular direction, and is measured by the number of scattered particles detected by the detector between $\theta$ and $\theta + d\theta$, $\phi$ and $\phi + d\phi$, i.e. within the solid angle

$$d\Omega = \sin\theta d\theta d\phi \qquad (1.29)$$

as shown in figure 1.2.

Now $d\sigma$ is the elemental cross-section along elemental solid angle $d\Omega$. The differential cross-section is cross-section per unit solid angle, i.e. $\frac{d\sigma}{d\Omega}$.

- Relation between total cross-section $\sigma$ and differential cross-section $\frac{d\sigma}{d\Omega}$ is

$$\sigma = \int \frac{d\sigma}{d\Omega} d\Omega = \int d\sigma \qquad (1.30)$$

Clearly total cross-section $\sigma$ is obtained by integrating over all solid angles.

## 1.14 Rutherford nuclear model of atom Geiger, Marsden and Rutherford α scattering experiment or gold foil experiment

In 1911, Geiger, Marsden and Rutherford performed an experiment (figure 1.3) of striking a metal foil having thickness $2 \times 10^{-7}$ m by $\alpha$ particles (from an $\alpha$ source, e.g. from $_{83}Bi^{214}$) having energy 5.5 $MeV$. They studied the pattern of scattering of the $\alpha$ particles from the foil and investigated the following three things in particular:

✓ Using foils of various metals like $_{79}Au$, $_{78}Pt$, $_{50}Sn$, $_{47}Ag$, $_{29}Cu$, $_{26}Fe$, $_{13}Al$ they studied the effect of atomic weight on scattering—whether the number of scattered $\alpha$ increased with atomic weight.

✓ Using gold foils of different thicknesses they studied the effect of thickness on scattering— whether the $\alpha$ particles scattered are coming more from the surface or more from interior of the metal.

✓ They investigated (i.e. counted the fraction of α particles) that are back-scattered.

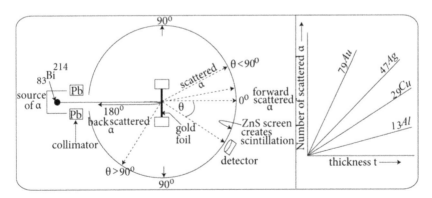

**Figure 1.3.** Geiger and Marsden experiment of scattering of $\alpha$ particles by metal foil. Experiment depends on foil thickness and materials (*exercise 1.24*)

Whenever a scattered $\alpha$ hits the ZnS screen, as shown in figure 1.3, scintillation is produced which is detected by the detector. The Pb block acts as collimator to collimate $\alpha$ particles from $\alpha$ source.

- Observation
    - ✓ Most of $\alpha$ went straight undeviated (i.e. were forward scattered) and emerged along the forward direction. Out of 20 000$\alpha$, 1999 $\alpha$ went forward (at angle $\theta < 90°$).
    - ✓ Very few $\alpha$ emerged in the backward direction (at angle $>90°$). Out of 20 000$\alpha$, one $\alpha$ was scattered back. This back-scattering of few $\alpha$ was the most significant observation in the Rutherford $\alpha$ scattering experiment.
    - ✓ The number of $\alpha$ scattered in the Rutherford scattering experiment depended on foil thickness and material. Figure 1.3 also shows the plot of number of $\alpha$ scattered against foil thickness for various materials.

In *exercise 1.18* we discuss why thin foil was chosen in the Rutherford $\alpha$ scattering experiment.

In *exercise 1.19* we discuss why thin gold foil is a better choice in the Rutherford $\alpha$ scattering experiment.

- Interpretation made by Rutherford:
    - ✓ Most of $\alpha$ went straight undeviated because they found an empty passage—nothing was there to obstruct them as evident from the paths of the $\alpha$ numbered 1, 2, 3, 4, 5 in figure 1.4—they were forward scattered ($\theta = 0$).

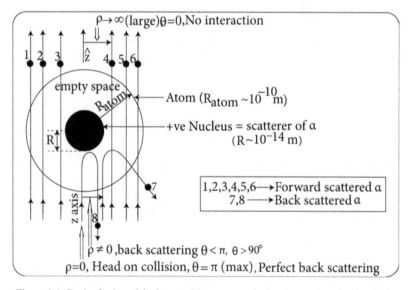

**Figure 1.4.** Rutherford model of atom. Most $\alpha$ go undeviated, very few ricochet back.

✓ Very few $\alpha$ emerged in the backward direction (at angle >90° since they found a concentrated lump of mass that obstructed their movement. So they suffered a direct collision, deviated from their course and came back. In figure 1.4 $\alpha$ numbered 7,8 are back-scattered. Also, the concentration of mass must be a concentration of positive charges so as to repel the $\alpha$ particles, which are also positively charged. This led to the nuclear model of the atom as proposed by Rutherford.

- Nuclear model of the atom:
    ✓ Most of the space in an atom is empty. The positively charged particles and most of the mass of the atom are concentrated in a small volume called nucleus with radius $\sim 10^{-14}\,m$. This can be compared with atomic radius $\sim 10^{-10}\,m$.
    ✓ The negative charges (electrons) surround the nucleus and revolve in fixed orbits.
    ✓ Negative electrons and positive nucleus are held together by electrostatic force of attraction.
- Limitation of nuclear model of atom of Rutherford:
    ✓ Electron arrangement outside the nucleus was not specified.
    ✓ An electron is a negatively charged particle. If a charged particle moves in an orbit (say in a circular path) it would accelerate at every instant. Again an accelerating charged particle radiates energy and therefore will lose energy continuously. So the electron will revolve in a circular orbit of gradually decreasing radius, i.e. will follow a spiral (figure 1.5) and collapse upon the nucleus (in time $\sim 10^{-8}\,s$) to make the atom shrink to a point. So the Rutherford nuclear model of atom does not predict a stable atom.

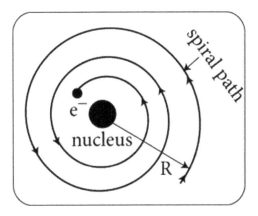

**Figure 1.5.** Rutherford nuclear model of atom is not stable since the electron, being a charged particle, will lose energy on acceleration in an orbit and would spiral down—the atom would thus collapse.

✓ In Rutherford theory it was assumed that Coulomb law of electrostatic repulsion between $\alpha$ and scatterer (say Au) holds good even at very small separation. But if:

(a) kinetic energy of $\alpha$ was very high;

(b) atomic number of scatterer was very small;
then there were deviations from predictions of Coulomb law.

If the distance between projectile $\alpha$ and target Au becomes very small, then a much stronger non-Coulomb force called strong force or nuclear force comes into play and the Coulomb interaction gets overshadowed or suppressed, i.e. fails to show up.

## 1.15 Proof of Rutherford scattering formula

To analyze Rutherford scattering of $\alpha \equiv {}_2\mathrm{He}^4$ by gold nucleus ${}_{79}\mathrm{Au}$ we consider the schematic diagram of figure 1.6. The target gold nucleus Au is at origin O and the $\alpha$ is being projected from a large distance (~ infinite distance) from point $M$ say. The projectile $\alpha$ falls within the target's influence—as it nears the target. The $\alpha$ is projected with velocity $v_\alpha$ and with a positive energy (i.e. positive kinetic energy)

$$E_\alpha = \frac{1}{2}m_\alpha v_\alpha^2 \tag{1.31}$$

As this $\alpha$ moves closer to target Au it feels the target potential and gets continuously deviated from its straight path $MNU$ and follows the deviated path $MSM'$. The angle of scattering is thus $\angle UNM' = \theta$.

The force of Coulomb interaction between $\alpha \equiv {}_2\mathrm{He}^4$ and ${}_{79}\mathrm{Au}$ is

$$\vec{F} = \frac{(Z \mid e \mid)(2 \mid e \mid)}{4\pi\varepsilon_0 r^2}\hat{r} = F(r)\hat{r} \tag{1.32}$$

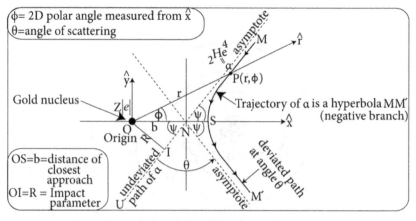

**Figure 1.6.** Schematic diagram of Rutherford experiment of $\alpha$ scattering by gold foil. Path of $\alpha$ is the negative branch of hyperbola (equations (1.204), (1.212), and (1.213)).

with

$$F(r) = \frac{2Z \mid e \mid^2}{4\pi\varepsilon_0 r^2} = \frac{\mid k \mid}{r^2} = +ve, \mid k \mid = \frac{2Z \mid e \mid^2}{4\pi\varepsilon_0} \tag{1.33}$$

The force is central (conservative), repulsive, inverse square.

As motion or scattering occurs under the central force $F(r)$ the angular momentum

$$\vec{L} = \vec{r} \times \vec{p} \tag{1.34}$$

is conserved. The motion of $\alpha$ particle will thus occur in a 2D plane that we take to be the $xy$ plane. We choose $OS$ as the $x$-axis.

At a time $t$ the projectile $\alpha$ is at position $P$ the coordinates of which are $(r, \phi)$, $r$ being the radial coordinate and $\phi$ being the polar coordinate in a 2D plane (= angle measured from $\hat{x}$). In other words, analysis of the 3D scattering problem reduces to a 2D problem. Here

$r$ = separation between $\alpha$ and gold nucleus

$Z$ = atomic number of gold nucleus = 79, charges are $Q_\alpha = 2 \mid e \mid$, $Q_{Au} = 79 \mid e \mid$ and the effective potential energy is

$$V_{eff} = \frac{\mid k \mid}{r} + \frac{L^2}{2\mu r^2} \tag{1.35}$$

Both terms on the RHS of equation (1.35) are positive, $\mu$ = reduced mass of $\alpha$ and Au system

$$\frac{1}{\mu} = \frac{1}{m_\alpha} + \frac{1}{m_{Au}} \tag{1.36}$$

- Assumptions
  - ✓ The gold nucleus is heavy and stationary (we take it to be infinitely heavy, so it won't move). We note that $m_{Au} \approx 50m_\alpha \approx \infty$ and so

$$\frac{1}{\mu} = \frac{1}{m_\alpha} + \frac{1}{m_{Au}} \approx \frac{1}{m_\alpha} + \frac{1}{\infty} \approx \frac{1}{m_\alpha} \Rightarrow \mu \approx m_\alpha \tag{1.37}$$

  Then we can develop the theory considering the mass of $\alpha$ only viz. $m_\alpha$ since then $\mu \approx m_\alpha$.
  - ✓ As the nucleus does not recoil (since it is too heavy to move) the initial and final kinetic energies of $\alpha$ particle are practically equal (—it is an elastic scattering).
  - ✓ The target is so thin that only a single scattering occurs.
  - ✓ The $\alpha$ particle and the target or scatterer are so small that they are treated as point masses and point charges.
  - ✓ Only Coulomb force is effective (which is repulsive).

✓ Relativistic effects are neglected since velocity of $\alpha$ is $\sim 0.1c$ ($c$ = speed of light in free space).

We have derived Rutherford scattering formula, i.e. the differential cross-section in *exercise 1.20* and here we give a brief outline.

The differential equation of the orbit followed by $\alpha$ in Rutherford scattering from gold nucleus is by *exercise 1.20(a)*

$$\frac{d^2u}{d\phi^2} + u = -\frac{m_\alpha}{L^2 u^2} F\left(\frac{1}{u}\right)$$
(1.38)

where

$$u = \frac{1}{r}, \; L = m_\alpha r^2 \dot{\phi}$$
(1.39)

This $L$ represents the magnitude of the angular momentum which is conserved.

The solution of this equation (1.38) is by *exercise 1.20(b)* given by

$$1 + \varepsilon \cos \phi = \frac{l}{r}$$
(1.40)

which represents a conic section—negative branch $MSM'$ of hyperbola to be precise, with gold nucleus $_{79}$Au (chosen as origin O) at the focus.

The eccentricity of trajectory or orbit will be by *exercise 1.20(c)*

$$\varepsilon = \sqrt{1 + \frac{2E_\alpha L^2}{m_\alpha \mid k \mid^2}} > 1$$
(1.41)

where

$$l = -\frac{L^2}{m_\alpha \mid k \mid}$$
(1.42)

is the semi-latus rectum.

The minimum distance between nucleus and the undeviated path of $\alpha$ is called the impact parameter $OI = R$. It is the perpendicular distance between the direction of initial velocity and the nucleus. The relation between impact parameter $R$ and angle of scattering $\theta$ is by *exercise 1.20(d)* given by

$$R = \frac{Z \mid e \mid^2}{4\pi\varepsilon_0 E_\alpha} \cot \frac{\theta}{2}$$
(1.43)

Using equation (1.33) viz. $\mid k \mid = \frac{2Z \mid e \mid^2}{4\pi\varepsilon_0}$ we rewrite

$$R = \frac{\mid k \mid}{2E_\alpha} \cot \frac{\theta}{2}$$
(1.44)

As the $\alpha$ particle approaches the nucleus its trajectory bends more and more due to increasing electrostatic repulsion. The $\alpha$ comes closest to the nucleus when it is at the vertex $S$ of the hyperbola where its velocity is minimum $v_\alpha^{\min}$.

For perfect back-scattering $\theta = \pi$, i.e. for head-on collision, the $\alpha$ particle will stop momentarily, velocity $v_\alpha^{\min} = 0$, kinetic energy $= 0$, and then it turns and moves away from the gold nucleus. The $\alpha$ cannot penetrate further towards the nucleus— it feels the nuclear boundary at this distance

$OS = b$ is called the distance of closest approach.

The relation between impact parameter $R$ and distance of closest approach $b$ is, by *exercise 1.20(e)*, given by

$$\frac{v_\alpha^{\min}}{v_\alpha} = \frac{R}{b} \tag{1.45}$$

The relation between distance of closest approach $b$, impact parameter $R$ and the angle of scattering $\theta$ is by *exercise 1.20(f)*

$$b = 2R \tan \frac{\theta}{2} \tag{1.46}$$

Total energy at a point of particle trajectory = kinetic energy + potential energy.

At point $M$ (where velocity of $\alpha$ is $v_\alpha$, figure 1.6)

$$\text{Total energy} = \text{K.E} \mid_M + \text{P.E} \mid_M = \frac{1}{2}m_\alpha v_\alpha^2 + 0 = \frac{1}{2}m_\alpha v_\alpha^2 \tag{1.47}$$

since point $M$ is at $r \to \infty$ the $\alpha$ particle cannot see the gold nucleus and so there is no interaction, hence $P.E \mid_M = 0$.

At point $P$ (where velocity of $\alpha$ is $v$, figure 1.6)

$$\text{Total energy} = \text{K.E} \mid_P + \text{P.E} \mid_P = \frac{1}{2}m_\alpha v^2 + \frac{(Z\mid e\mid)(2\mid e\mid)}{4\pi\epsilon_0 r} = \frac{1}{2}m_\alpha v^2 + \frac{2Z\mid e\mid^2}{4\pi\epsilon_0 r} \tag{1.48}$$

As energy is conserved in a central force field we can equate total energies at points $M$ (equation (1.47)) and at point $P$ (equation (1.48)) to get

$$\frac{1}{2}m_\alpha v_\alpha^2 = \frac{1}{2}m_\alpha v^2 + \frac{2Z\mid e\mid^2}{4\pi\epsilon_0 r}$$

$$\frac{1}{2}m_\alpha v_\alpha^2 = \frac{1}{2}m_\alpha v^2 + \frac{\mid k\mid}{r} \text{ (using equation 1.33)}$$

$$v_\alpha^2 = v^2 + \frac{2\mid k\mid}{m_\alpha r}. \tag{1.49}$$

At distance of closest approach (point $S$ of figure 1.6) $r = b$, $v = v_\alpha^{\min}$ and so from equation (1.49) we write

$$v_\alpha^2 = (v_\alpha^{\min})^2 + \frac{2\mid k\mid}{m_\alpha b} \tag{1.50}$$

- Let us get an estimate of nuclear radius as predicted by Rutherford scattering experiment.

For perfect back-scattering $\theta = \pi$, $v_\alpha^{min} = 0$ and so we have from equation (1.50)

$$v_\alpha^2 = 0 + \frac{2|k|}{m_\alpha b} = \frac{2|k|}{m_\alpha b}$$

$$b = \frac{2|k|}{m_\alpha v_\alpha^2} = \frac{|k|}{\frac{1}{2}m_\alpha v_\alpha^2}$$

$$b = \frac{|k|}{E_\alpha} \tag{1.51}$$

Using equation (1.33) viz. $|k| = \frac{2Z|e|^2}{4\pi\varepsilon_0}$ we rewrite

$$b = \frac{2Z|e|^2}{4\pi\varepsilon_0 E_\alpha} \tag{1.52}$$

For $E_\alpha = 5\ MeV$, the distance of closest approach is

$$b = \frac{2(79)(1.6 \times 10^{-19}C)^2}{4\pi\left(\frac{10^{-9}}{36\pi}Fm^{-1}\right)(5 \times 10^6 \times 1.6 \times 10^{-19}\ J)} = 4.6 \times 10^{-14}\ m \sim 10^{-14}\ m. \tag{1.53}$$

This $10^{-14}\ m$ is Rutherford's estimate of nuclear radius.(*exercise 1.29*)
The differential scattering cross-section is by *exercise 1.20(g)* given by

$$\frac{d\sigma}{d\Omega} = \left(\frac{Z|e|^2}{4\pi\varepsilon_0 2E_\alpha}\right)^2 \text{cosec}^4\frac{\theta}{2} \tag{1.54}$$

This is the Rutherford scattering formula. It gives the probability of scattering per unit solid angle.

✓ When incident $\alpha$ gets close enough to the target nucleus so that they can interact with nuclear force (over and above the Coulomb force) Rutherford scattering formula no longer holds. The point at which such breakdown occurs gives a measure of the size of the nucleus.

• The unit and dimension of $\frac{d\sigma}{d\Omega}$ can be found out from Rutherford scattering formula given by equation (1.54). The unit of $\frac{Z|e|^2}{4\pi\varepsilon_0 2E_\alpha}$ is $\frac{C^2}{\frac{F}{m}.J}$.

Using Farad $= \frac{coulomb}{volt}$, i.e. $F = \frac{C}{V}$ we have

$$\frac{Z|e|^2}{4\pi\varepsilon_0 2E_\alpha} \rightarrow \frac{C^2}{\frac{C}{Vm}.J} = \frac{CV.m}{.J}$$

Since $CV = J$ we have

$$\frac{Z|e|^2}{4\pi\varepsilon_0 2E_\alpha} \rightarrow m$$

Hence $\frac{d\sigma}{d\Omega} = \left(\frac{Z|e|^2}{4\pi\varepsilon_0 2E_\alpha}\right)^2 \text{cosec}^4\frac{\theta}{2}$ has unit $m^2$ and so dimension is $[L^2]$— that of area.

• The total scattering cross-section is obtained by integrating the differential scattering cross-section over all solid angles and since Coulomb potential has infinite range we get

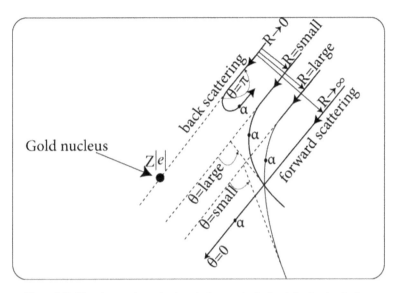

**Figure 1.7.** The closer $\alpha$ is to the target, the greater is the deflection/scattering.

$$\sigma = \int \frac{d\sigma}{d\Omega} d\Omega \rightarrow \infty. \tag{1.55}$$

Since the total cross-section $\sigma$ is infinity we cannot handle it directly. Instead we always refer to the differential cross-section $\frac{d\sigma}{d\Omega}$ in the case of Rutherford scattering.

The number of $\alpha$ particles recorded per unit area of detector is, by *exercise 1.20(h)*

$$N_\theta = \frac{Nnt}{r^2} \left( \frac{Z \mid e \mid^2}{4\pi\varepsilon_0 2 E_\alpha} \right)^2 \mathrm{cosec}^4 \frac{\theta}{2} \tag{1.56}$$

where $t=$ foil thickness, $A=$ foil area, $N=$ Number of $\alpha$ incident on the foil in a given time.

- If $\alpha$ is projected with more energy, i.e. for larger $E_\alpha$, it can penetrate further and come closer to the target nucleus. So the impact parameter $R$ is reduced. As $\alpha$-to-nucleus distance $r$ is small and the $\alpha$ now feels larger Coulomb potential $\frac{(Z \mid e \mid)(2 \mid e \mid)}{4\pi\varepsilon_0 r}$ hence $\alpha$ will be scattered more, i.e. angle of scattering $\theta$ will be large (path will be more bent, distorted). This is depicted in figure 1.7.

## 1.16 Nuclear radius

When $\alpha$ particles are projected towards gold foil with gradually increasing energies some are back-scattered due to Coulomb repulsion, implying existence of a concentrated positively charged lump of mass. And for $E_\alpha \sim 5\ MeV$, distance of closest approach as obtained from equation (1.53) is $b = \frac{2Z \mid e \mid^2}{4\pi\varepsilon_0 E_\alpha} \sim 10^{-14}\ m$.

This defines an approximate boundary within which the charged lump of mass, called nucleus is spread out.

But if $E_\alpha$ is large, $Z$ is small then Coulomb interaction, i.e. repulsion of $\alpha$ is no longer there and a non-Coulomb interaction takes over and Rutherford scattering formula fails (i.e. experimental data deviates from the Rutherford formula that was derived on the basis of Coulomb interaction. So we can think of an approximate boundary at which Coulomb interaction gets masked. Thus in the Rutherford $\alpha$ scattering experiment the critical energy $E_\alpha$ and the corresponding atomic number $Z$, at which the scattering law breaks down, clearly provides a rough estimate of nuclear radius of scatterer, namely $\sim 10^{-14}\,m$. This is Rutherford's estimate of nuclear radius (figure 1.8).

Deviation from the Rutherford formula (based upon alpha–nucleus Coulomb interaction) defines the nuclear boundary. If separation between $_2\mathrm{He}^4$ and $_{79}\mathrm{Au}^{197}$ nuclei is large, Coulomb repulsion is the only force of interaction and Rutherford scattering occurs. If $_2\mathrm{He}^4$ is sufficiently energetic to come closer to $_{79}\mathrm{Au}^{197}$ Coulomb repulsion is overcome as nuclear force dominates. Obviously, Rutherford scattering formula has no meaning in this region. This has been shown in figure 1.8.

The curve $abc$ is obtained from experiment for scattering angle 60°. Calculation using Rutherford formula that is based on Coulomb repulsion gives curve $abd$. Obviously, the portion $ab$ that corresponds to $\alpha$ energy $\leqslant 27\,MeV$ is explained by Rutherford Coulomb interaction. But if $\alpha$ energy is $\geqslant 27\,MeV$ nuclear interaction dominates and, Coulomb repulsion gets overshadowed and so Rutherford calculation fails. To explain the experimental plot of portion $bc$ one has to consider nuclear force.

In *exercise 1.21* we discuss whether nuclear boundary is precisely defined.

In *exercise 1.22* we calculate the size of wavelength and the amount of momentum or energy that a probing agency (say an electron) should possess to probe a nucleus.

In *exercise 1.23* we discuss whether the nucleus can be probed with electro-magnetic radiation.

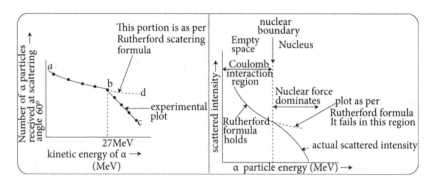

**Figure 1.8.** Rutherford's estimate of nuclear radius and departure from the Rutherford formula.

## 1.17 Radius of a spherical nucleus

Experiments indicate that the majority of nuclei are nearly spherical and some are perfectly spherical (called magic nuclei). It is thus a good approach to start with spherical nuclei (i.e. a spherically symmetric nuclei) and do calculations that will give a rough but good estimate.

The formula for spherical volume is

$$V = \frac{4}{3}\pi R^3$$

where $R$ is the radius of the sphere. Again nucleons (protons, neutrons) are packed over space to produce the nucleus. The greater the number of constituent nucleons the greater is the volume of the nucleus. So volume of nucleus is obviously proportional to the number of nucleons (which is called mass number $A$). Hence we expect that

$$V = \frac{4}{3}\pi R^3 \propto A \Rightarrow R^3 \propto A \Rightarrow R \propto A^{1/3} \tag{1.57}$$

This is how nuclear radius varies with mass number. So we write

$$R = R_0 A^{1/3}, \ R_0 = \text{constant} \tag{1.58}$$

It is clear that the nuclear radius has a constant part, the value of which has to be measured from experiment. The other part is $A^{1/3}$ and nuclear radius varies from element to element depending on mass number (i.e. number of nucleons).

For instance for $_3\text{Li}_4^7$, $R = R_0 7^{1/3} = 1.91 R_0$.

For $_{92}\text{U}_{143}^{235}$, $R = R_0 235^{1/3} = 6.17 R_0$ etc.

The value of $R_0$ can be determined from various methods.

## 1.18 Estimation of nuclear radius

Nuclear radius can be obtained by various methods.
- Estimation of nuclear radius can be done from scattering experiments.
  - ✓ Estimation of nuclear radius was done by Rutherford from his experiment that involved alpha scattering by metal (say gold) nucleus. It was $10^{-14}$ $m$.
  - ✓ Estimation of nuclear radius done from an experiment that involves neutron scattering by nucleus is referred to as nuclear force radius.
  - ✓ Estimation of nuclear radius done from an experiment that involves electron scattering by nucleus is referred to as charge radius.
- Estimation of nuclear radius can also be obtained from the study of atomic transition.
- Estimation of nuclear radius can also be obtained from the study of mirror nuclei.

We mention that the nucleus involves two types of interactions, namely:

   (i) charge interaction because protons carry positive charges. In other words there is a spread or distribution of positive charge (protons) in the nucleus;

  (ii) non-Coulomb interaction since nucleons (protons and neutrons) can exhibit charge independent interaction which is called nuclear force or nuclear interaction.

## 1.19 Nuclear force radius

Estimation of $R_0$ can be done from a study of data obtained from experiments of scattering of neutrons by the nucleus.

We note that the projectile neutrons carry no charge.

When a neutron (that is uncharged) strikes a nucleus it is clearly not interested in the charge distribution of the nucleus but will be interested in (i.e. influenced by) the distribution of nuclear fluid, i.e. distribution of all nucleons (protons + neutrons) in the nucleus. So neutron scattering data gives a value of $R_0$ which is an estimate of the influence of nuclear force leading to arrangement of nucleons, i.e. the spread out of the nuclear matter, and is called nuclear force radius.

Neutron scattering experiments give an estimate of

$$R_0 \sim 1.4 \, fm. \tag{1.59}$$

It is an experiment that measures strong nuclear interaction of nucleons and determines distribution of nucleons, i.e. distribution of nuclear matter or nuclear fluid.

## 1.20 Nuclear charge radius

An estimate of $R_0$ can be obtained from a study of data obtained from experiments of scattering of electrons by nucleus.

Projectile electrons carry negative charge.

When an electron strikes a nucleus it is clearly interested in the charge distribution (or proton distribution) of the nucleus but will not be interested in the distribution of neutrons because a proton, due to its charge can influence an electron but a neutron, due to its neutrality cannot detect an electron.

Electron scattering data thus gives a value of $R_0$ which is an estimate of the extension of charge distribution within the nucleus and is called charge radius or electromagnetic radius. Electron scattering experiments give an estimate of

$$R_0 \sim 1.2 \, fm \tag{1.60}$$

## 1.21 Experiment to measure charge radius

Beam of electrons of kinetic energy $420 \, MeV$ were scattered from an oxygen nucleus. The number of electrons scattered at various angles were noted by a detector, as shown in figure 1.9(a).

    ✓ The electrons recorded by the detector were scattered by the oxygen nucleus and were not its orbital electrons. This has been explained in *exercise 1.33(a)*.

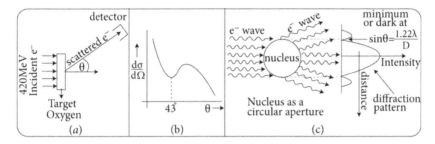

**Figure 1.9.** Experiment to measure charge radius of nucleus by electron scattering from oxygen nucleus and $\frac{d\sigma}{d\Omega}$ versus $\theta$ plot.

✓ The experiment did not find a second minimum. We address this in *exercise 1.33(b,c)*.

- Observation:

    The plot of $\frac{d\sigma}{d\Omega}$ (i.e. number of electrons per unit solid angle per second) against $\theta$ (angle of scattering) shows a minimum at $43°$, as shown in figure 1.9(b).

    Let us explain the $\frac{d\sigma}{d\Omega}$ versus $\theta$ plot and make a discussion.

    ✓ The plot shows a fall of $\frac{d\sigma}{d\Omega}$ with $\theta$. This is Rutherford scattering which predicts a $\theta$ dependence given by the relation

$$\text{Number of particles scattered} \propto \frac{1}{\sin^4 \frac{\theta}{2}} \qquad (1.61)$$

    As $\theta$ rises $\sin^4 \frac{\theta}{2}$ rises and so $\frac{d\sigma}{d\Omega}$ falls.

    ✓ As electrons are highly energetic there will be relativistic effects also.

    ✓ Rutherford's scattering formula was derived from classical considerations assuming it to be a particle–particle collision. But quantum effects manifest in this experiment. As the electron presents its wave nature, diffraction effects will be there. We can compare the nucleus as a circular aperture of diameter $D$. The de Broglie electron wave would create a diffraction pattern (called Airy's pattern from the circular aperture). The pattern consists of a central bright ring (or maximum due to constructive interference) surrounded by a dark ring (or minimum due to destructive interference) occurring at $\theta$ given by the relation

$$\sin \theta = \frac{1.22\lambda}{D} \qquad (1.62)$$

    The dark ring at this $\theta$ can be taken as an estimate of diameter $D$ of the nucleus (figure 1.9(c)). The de Broglie relation gives the wavelength of the electron wave to be

$$\lambda = \frac{h}{p} = \frac{hc}{pc} \tag{1.63}$$

With $E = pc$ we get, using *exercise 1.2*,

$$\lambda = \frac{hc}{E} = \frac{1242 \; MeV.fm}{420 \; MeV} = 2.96 \, fm \tag{1.64}$$

With $\lambda = 2.96 \, fm$ and $\theta = 43°$ (the first minimum or dark ring) we have from equation (1.62)

$$D = \frac{1.22\lambda}{\sin\theta} = \frac{1.22(2.96 \, fm)}{\sin 43°} = 5.3 \, fm \tag{1.65}$$

The radius of the oxygen nucleus is

$$R = \frac{D}{2} = \frac{5.3 \, fm}{2} = 2.65 \, fm. \tag{1.66}$$

This gives a rough estimate. We are getting a rough estimate because actual scattering is a 3D problem—only approximately related to diffraction by a 2D disc.

- From electron scattering experiment we get vital information about charge distribution (i.e. proton distribution) inside the nucleus. We discuss the results in the following.

    ✓ The plot of proton number density $\rho_{proton} = \frac{\text{number of protons}}{\text{volume}}$ versus $r$ (distance from centre of nucleus), based on experimental data, is shown in figure 1.10. The proton number density remains constant from the centre up to a certain distance and then falls slowly.

    ✓ In the electron scattering experiment the negatively charged electrons interact only with positively charged protons (i.e. scattering of electrons occurs from the protons). This means what we get in this experiment is proton density (how protons, i.e. the charges, are distributed within the nucleus).

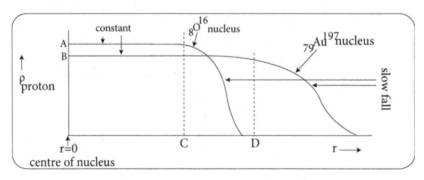

**Figure 1.10.** Proton distribution (or charge distribution) inside nucleus, from data of electron scattering by nucleus.

✓ Proton distribution or charge distribution differs in the case of lighter nuclei (say $_8\text{O}^{16}$) and heavy nuclei (say $_{79}\text{Au}^{197}$), as shown in figure 1.10.

- Lighter nuclei (say $_8\text{O}^{16}$):
  - ☐ Proton number density is larger at the centre of the nucleus, i.e. at $r = 0$ (point $A$). This means the centre is more charged.
  - ☐ Proton density (charge density) is constant up to a smaller distance from the centre (up to $C$).
- Heavy nuclei (say $_{79}\text{Au}^{197}$):
  - ☐ Proton number density is less at the centre of the nucleus, i.e. at $r = 0$ (point $B$). This means the centre is less charged.
  - ☐ Proton density (charge density) is constant up to a larger distance from the centre (up to $D$).

✓ Mass distribution, i.e. the nucleon density (proton density + neutron density):

Let us define the nucleon number density as

$$\rho = \frac{\text{number of nucleons}}{\text{volume}} = \frac{A}{V} \tag{1.67}$$

The proton number density is

$$\rho_{\text{proton}} = \frac{\text{number of protons}}{\text{volume}} = \frac{Z}{V}. \tag{1.68}$$

Assume that neutrons are similarly distributed as protons. Now from equations (1.67) and (1.68) we get

$$\frac{\rho}{\rho_{\text{proton}}} = \frac{A}{Z} \Rightarrow \rho = \frac{A}{Z}\rho_{\text{proton}} \tag{1.69}$$

The factor $\frac{A}{Z}$ acts as a scale factor.

For light nuclei like $_8\text{O}^{16}$ $\frac{A}{Z} = \frac{16}{8} = 2$, while for heavy nuclei like $_{79}\text{Au}^{197}$ $\frac{A}{Z} = \frac{197}{79} = 2.49$.

The plot of $\rho = \frac{A}{Z}\rho_{\text{proton}}$ versus $r$ gives a curve, as shown in figure 1.11. We mention the important features of the $\rho$ versus $r$ plot.

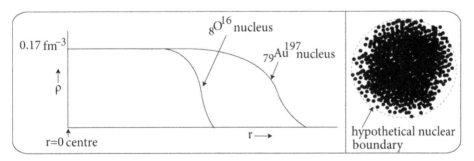

**Figure 1.11.** Nucleon distribution (or mass distribution) inside nucleus as per laboratory data.

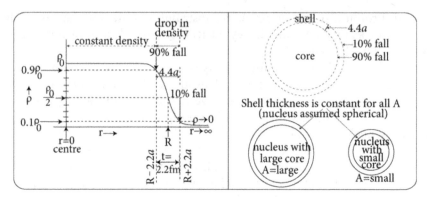

**Figure 1.12.** Woods–Saxon function describing nucleon distribution. Charge density is roughly constant up to a certain point and then drops relatively slowly to zero. Large $A$ nuclei have large core, small $A$ nuclei have small core. Shell thickness is constant for all $A$.

✓ At $r = 0$ (centre), $\rho = 0.17\,fm^{-3}$ for all nuclei. So the curve starts at the same point for both $_8O^{16}$, as well as for $_{79}Au^{197}$. This is independent of $A$.

✓ $\rho$ is practically constant throughout the nucleus, i.e. over various values of $r$. In other words the plot is independent of mass number $A$. The nucleus is equally dense throughout the central core, i.e. it contains a constant number of nucleons per unit volume.

✓ Nucleon density falls gradually near the surface. There is no sharp boundary at the surface, it is a diffused state outside the central uniformly dense core (figure 1.11).

✓ Clearly, experimental data suggests that the nucleon number density $\rho$ is constant in the region from $r = 0$ to some high value of $r$ and then falls off for all nuclei. In other words, nucleons do not gather near the centre of the nucleus, but spread out uniformly to the surface. This type of curve resembles the Woods–Saxon function (figure 1.12) described through the relation

$$\rho = \frac{\rho_0}{1 + e^{\frac{r-R}{a}}} \tag{1.70}$$

where $\rho_0$ and $a$ are constants.

We now describe nucleon distribution on the basis of Woods–Saxon function.

✓ At $r = 0$, $\rho = \dfrac{\rho_0}{1 + e^{-\frac{0-R}{a}}} = \dfrac{\rho_0}{1 + e^{-\frac{R}{a}}} \approx \rho_0$

✓ At $r = R$, $\rho = \dfrac{\rho_0}{1 + e^{-\frac{R-R}{a}}} = \dfrac{\rho_0}{1 + e^0} = \dfrac{\rho_0}{1 + 1} = \dfrac{\rho_0}{2} = 0.5\rho_0 \Rightarrow 50\%$ reduction in core density $\rho_0$.

✓ At $r \to \infty$, $\rho \to \dfrac{\rho_0}{1 + e^{-\frac{\infty-R}{a}}} = \dfrac{\rho_0}{1 + e^\infty} = \dfrac{\rho_0}{\infty} = 0$. This means nucleus does not have a sharp, well defined, precise boundary, rather the boundary is diffuse and with increase in $r$ the density gradually falls to zero.

✓ Let us find the distance $r$ at which density falls to 90% of $\rho_0$, i.e. to $\frac{90}{100}\rho_0 = 0.9\rho_0$. So from equation (1.70) we have

$$0.9\rho_0 = \frac{\rho_0}{1 + e^{\frac{r-R}{a}}}$$

$$1 + e^{\frac{r-R}{a}} = \frac{1}{0.9} = \frac{10}{9}$$

$$e^{\frac{r-R}{a}} = \frac{10}{9} - 1 = \frac{1}{9}$$

$$\frac{r-R}{a} = \ln\frac{1}{9} \Rightarrow r - R = -2.2a$$

$$r = R - 2.2a = r_{01} \tag{1.71}$$

✓ Let us find the distance $r$ at which density falls to 10% of $\rho_0$, i.e. to $\frac{10}{100}\rho_0 = 0.1\rho_0$. So from equation (1.70) we have

$$0.1\rho_0 = \frac{\rho_0}{1 + e^{\frac{r-R}{a}}}$$

$$1 + e^{\frac{r-R}{a}} = \frac{1}{0.1} = 10$$

$$e^{\frac{r-R}{a}} = 10 - 1 = 9$$

$$\frac{r-R}{a} = \ln 9 = 2.2 \Rightarrow r - R = 2.2a$$

$$r = R + 2.2a = r_{02} \tag{1.72}$$

✓ It follows that $a$ is a constant characterizing the rapidity or the rate of fall of core density. It is a measure of the surface thickness of the nucleus. If $a$ is large the fall is blunt and slow and if $a$ is small the fall is more steep and fast.

✓ The nucleus has thus two parts.

The nucleus has a central core which is almost uniformly populated by nucleons. This core is surrounded by a shell of width $\sim 4.4a$ between which the core density falls from 90% to 10% of the core density.

Nucleus → core + shell

The distance over which this drop occurs is nearly independent of the size of the nucleus and can be taken to be a constant.

✓ The value of $\rho_0$ as revealed from experimental data is

$$\rho = \rho_0 \approx 0.17 \text{ nucleons. } fm^{-3} \tag{1.73}$$

✓ The value of $a$ as revealed from experimental data is $a = 0.5\,fm$ for all nuclei. This means the shell thickness or surface width of the nucleus (or skin thickness) is

$$t = r_{02} - r_{01} = (R + 2.2a) - (R - 2.2a)$$

$$t = 4.4a = 4.4 \times 0.5\,fm = 2.2\,fm \tag{1.74}$$

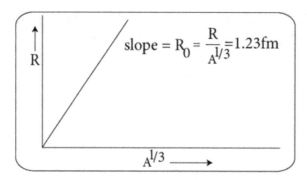

**Figure 1.13.** $R$ versus $A^{1/3}$ plot gives a straight line with slope $R_0 = 1.23\,fm$.

for all nuclei (figure 1.12). So the skin thickness parameter $t$ is the distance over which the charge density falls from 90% of its central value to 10%.

✓ The value of $R$ is different for different nuclei and we can take it as the mean radius. In other words the core size differs from nuclei to nuclei. A large nuclei has a large core while a small nuclei has a small core (figure 1.12).
• The value of $R$ as revealed from experimental data satisfies the relation

$$R = R_0 A^{1/3} \tag{1.75}$$

where $R_0 \approx 1.2\,fm$ and $A = Z + N =$ mass number.

In fact, this follows from the fact that the number of nucleons per unit volume is constant since

$$\rho = \text{constant}$$

$$\frac{A}{\frac{4}{3}\pi R^3} = \text{constant}$$

where $R$ is the mean nuclear radius. Thus
$$\frac{A}{R^3} = \text{constant} \Rightarrow R^3 \propto A$$

$$R \propto A^{1/3} \Rightarrow R = R_0 A^{\frac{1}{3}} \tag{1.76}$$

where $R_0$ is proportionality constant $=1.2\,fm$ (from electron scattering experiments) (figure 1.13).

• Parameters needed to define nuclear shape.

Nuclear shape is characterized by two parameters:
✓ Mean radius (at which density is half its central value);
✓ Skin thickness (over which the density drops from say 90% fall to 10% fall from central constant value $\rho_0$ of nuclear density).

## 1.22 Estimation of nuclear radius from study of atomic transition

The Coulomb interaction between an orbiting electron (charge $-|e|$) and nucleus (charge $Z|e|$) in an atom is given by the potential energy

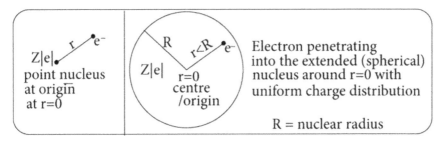

**Figure 1.14.** The change in potential energy of the electron due to extended spherical structure (i.e. finite size) of nucleus, by *exercise 1.49(b)* is $\Delta U = \frac{Z|e|^2}{4\pi\varepsilon_0}\left(\frac{1}{r} - \frac{3}{2R} + \frac{r^2}{2R^3}\right)$.

$$U = -\frac{Z|e|^2}{4\pi\varepsilon_0 r}. \tag{1.77}$$

The separation between electron and nucleus is $r$ with nucleus located at point $r = 0$ (origin) (figure 1.14).With this potential energy the time independent Schrödinger equation is

$$H_0\psi_0 = E\psi_0 \tag{1.78}$$

where $H_0$ is the Hamiltonian (assuming point-nucleus)

$$H_0 = -\frac{\hbar^2}{2m}\nabla^2 - \frac{Z|e|^2}{4\pi\varepsilon_0 r} \tag{1.79}$$

The equation can be solved and energy levels characterized by quantum number $n = 1,2,3,\ldots$ are obtained.

For a hydrogen-like atom the ground state energy is

$$E_{gs} = -13.6\ eV \tag{1.80}$$

and the ground state wave function is

$$\psi_{100} = \sqrt{\frac{Z^3}{\pi a_0^3}}\,e^{-\frac{Zr}{a_0}}. \tag{1.81}$$

We note that $\psi_{100}$ is not zero at $r = 0$, i.e. $\psi_{100}(r = 0) \neq 0$ implying that there is a probability that an electron can pass through the nucleus located at $r = 0$. This suggests that the orbiting electron can penetrate the nucleus. So if the nucleus is an extended object, i.e. if the nucleus has a structure, then it follows that the electron gets affected by such spread of the nucleus around $r = 0$ and this means the energy levels will change (they will no longer be $E$).

The amount of change in the electron energy depends on the amount of spread of nucleus. Assume nuclear spread to be spherical. If the change in energy due to spherical spread of the nucleus around $r = 0$ can be measured, then the radius of the nucleus can also be determined.

Due to nucleus occupying a spherical region around $r = 0$, the potential energy will change and so will the Hamiltonian. Assume that charge $Z|e|$ is uniformly distributed throughout sphere. The change in potential energy of the electron due to it seeing the extended spherical structure of the nucleus is by *exercise 1.49(b)* given by

$$\Delta U = \frac{Z \, | \, e \, |^2}{4\pi\varepsilon_0}\left(\frac{1}{r} - \frac{3}{2R} + \frac{r^2}{2R^3}\right) \tag{1.82}$$

So the Hamiltonian for the extended nucleus is given by

$$H = H_0 + \Delta U = H + H' \tag{1.83}$$

where $H' = \Delta U$ is the difference in the Hamiltonians with finite size of nucleus and point nucleus (it is a small term and can be treated as a perturbation term). This change in Hamiltonian will cause changes in energy from
$E \rightarrow E + \Delta E$
The corresponding Schrödinger equation is

$$H\psi = (H_0 + \Delta U)\psi = (E + \Delta E)\psi \tag{1.84}$$

According to quantum mechanics rules (first order perturbation theory) the process is to obtain $\psi_0$ from $H_0$. The process works if $H'$ is a small correction. Then the difference in energy (-due to the perturbation) is $\Delta E$ that is obtained from $\psi_0$ using the formula

$$\Delta E = <\psi_0 \, | \, H' | \, \psi_0 > \, = \int \psi_0^\star(\vec{r})\Delta U \, \psi_0(\vec{r})d\tau = \int | \, \psi_0(\vec{r}) \, |^2 \Delta U d\tau \tag{1.85}$$

Using equation (1.82) we have

$$\Delta E = \int | \, \psi_0(\vec{r}) \, |^2 \frac{Z \, | \, e \, |^2}{4\pi\varepsilon_0}\left(\frac{1}{r} - \frac{3}{2R} + \frac{r^2}{2R^3}\right)d\tau \tag{1.86}$$

Let us calculate the change in energy of $1s$ electron ($1s \Rightarrow n = 1$, $l = 0$, $m_l = 0$ i.e. $| \, 100 >$ state) which is closest to the nucleus having largest overlap with the nucleus and will thus involve largest energy shifts. We assume a point nucleus and ignore interaction of $1s$ electron with other electrons, i.e. assume that $1s$ electron interacts with nucleus only. In that case we can use $H$-like wave function denoted by

$$\psi_{1s}(\vec{r}) = \sqrt{\frac{Z^3}{\pi a_0^3}} \, e^{-\frac{Zr}{a_0}} \equiv \psi_0 \tag{1.87}$$

where $a_0 = 4\pi\varepsilon_0\frac{\hbar^2}{m_e \, | \, e \, |^2} = 0.52$ Å is the first Bohr radius. Hence from equation (1.86) we have

$$\Delta E = \int \left| \sqrt{\frac{Z^3}{\pi a_0^3}} \, e^{-\frac{Zr}{a_0}} \right|^2 \frac{Z \, | \, e \, |^2}{4\pi\varepsilon_0}\left(\frac{1}{r} - \frac{3}{2R} + \frac{r^2}{2R^3}\right)d\tau \tag{1.88}$$

With $d\tau = r^2 dr \sin\theta d\theta d\phi$ we get

$$\Delta E = \int_0^R \frac{Z^3}{\pi a_0^3} e^{-\frac{2Zr}{a_0}} \frac{Z\,|\,e\,|^2}{4\pi\varepsilon_0}\left(\frac{1}{r} - \frac{3}{2R} + \frac{r^2}{2R^3}\right) r^2 dr \int_0^\pi \sin\theta d\theta \int_0^{2\pi} d\phi$$

$$= \int_0^R \frac{Z^3}{\pi a_0^3} e^{-\frac{2Zr}{a_0}} \frac{Z\,|\,e\,|^2}{4\pi\varepsilon_0}\left(\frac{1}{r} - \frac{3}{2R} + \frac{r^2}{2R^3}\right) r^2 dr.\ 2.2\pi \qquad (1.89)$$

Consider the ratio

$\frac{r}{a_0} \xrightarrow{\text{upper limit is } R} \frac{R}{a_0} = \frac{10^{-15}\,m}{10^{-10}\,m} \sim 10^{-5}$ and $Z < 100$. It follows that $e^{-\frac{2Zr}{a_0}} \sim e^0 = 1$.

With this we get

$$\Delta E = 4\pi \frac{Z^3}{\pi a_0^3} \cdot \frac{Z\,|\,e\,|^2}{4\pi\varepsilon_0} \int_0^R \left(\frac{1}{r} - \frac{3}{2R} + \frac{r^2}{2R^3}\right) r^2 dr$$

$$= 4\pi \frac{Z^3}{\pi a_0^3} \cdot \frac{Z\,|\,e\,|^2}{4\pi\varepsilon_0} \int_0^R \left(r - \frac{3r^2}{2R} + \frac{r^4}{2R^3}\right) dr$$

$$= \frac{Z^4\,|\,e\,|^2}{\pi\varepsilon_0 a_0^3} \int_0^R \left(r - \frac{3r^2}{2R} + \frac{r^4}{2R^3}\right) dr = \frac{Z^4\,|\,e\,|^2}{\pi\varepsilon_0 a_0^3}\left(\frac{R^2}{2} - \frac{3}{2R}\cdot\frac{R^3}{3} + \frac{1}{2R^3}\cdot\frac{R^5}{5}\right)$$

$$\Delta E = \frac{Z^4\,|\,e\,|^2}{\pi\varepsilon_0 a_0^3}\left(\frac{R^2}{2} - \frac{R^2}{2} + \frac{R^2}{10}\right)$$

$$\Delta E = \frac{Z^4\,|\,e\,|^2\,R^2}{10\pi\varepsilon_0 a_0^3} \qquad (1.90)$$

This is the shift in energy of $1s$ electron considering finite size of nucleus.

The energy of $1s$ electron is thus

$$E_{1s} = \text{Energy of } 1s \text{ for a point nucleus} + \text{shift} = \text{E}_{1s}^{\text{point nucleus}} + \frac{Z^4\,|\,e\,|^2\,R^2}{10\pi\varepsilon_0 a_0^3} \qquad (1.91)$$

However, we cannot use this formula to find the nuclear radius. We have explained this in *exercise 1.50*.

To find the nuclear radius let us compare the $K_\alpha$ x-ray energies from $2p \to 1s$ electronic transition (figure 1.15) between two neighbouring isotopes.

If there is a vacancy in $1s$ ($K$ shell) it is filled up by electron of $2p$ ($L$ shell) level as it falls from $L$ shell to $K$ shell. The excess energy of $L$ shell is released in the form of an x-ray photon that is called $K_\alpha$ x-ray and this energy is measurable. The energy of $K_\alpha$ x-ray is

$$E_{K_\alpha} = E_L - E_K \equiv E_{2p} - E_{1s} \qquad (1.92)$$

The $L$ shell is distant from the nucleus and we can take the $2p$ wavefunction to go to zero at $r \to 0$ and so it has almost negligible overlap with the nucleus. So the energy of $2p$ state calculated with point nucleus (say $E_{2p}^{\text{point nucleus}}$) or with extended nucleus ($E_{2p}$) would be the same, i.e. $E_{2p} = E_{2p}^{\text{point nucleus}}$.

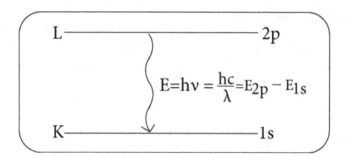

**Figure 1.15.** $K_\alpha$ x-ray from $2p \rightarrow 1s$ electronic transition.

Hence we can write with equation (1.91)

$$E_{K_\alpha} = E_{2p} - E_{1s} = E_{2p}^{\text{point nucleus}} - \left( E_{1s}^{\text{point nucleus}} + \frac{Z^4 |e|^2 R^2}{10\pi\varepsilon_0 a_0^3} \right) \qquad (1.93)$$

Let us calculate $E_{K_\alpha}$ for two different isotopes of an element that has several isotopes of mass numbers, say $A_1, A_2,...A,...$ ($A_1$ being the lowest, say). Isotope means same proton number ($Z$). Neutron number ($N$) differs, mass number ($A$) differs, size (radius $R$) differs. The terms $E_{2p}^{\text{point nucleus}}$, $E_{1s}^{\text{point nucleus}}$ can be taken to be effectively the same in different isotopes of an element since electric field produced by a point nucleus does not change if neutron number varies.

For mass number $A_1$ say, $R = R_1$ and we write from equation (1.93) the $K_\alpha$ x-ray energy to be

$$E_{K_\alpha}^{A_1} = E_{2p}^{\text{point nucleus}} - \left( E_{1s}^{\text{point nucleus}} + \frac{Z^4 |e|^2 R_1^2}{10\pi\varepsilon_0 a_0^3} \right) \qquad (1.94)$$

For another isotope with mass number $A$ and $R$ as radius

$$E_{K_\alpha}^{A} = E_{2p}^{\text{point nucleus}} - \left( E_{1s}^{\text{point nucleus}} + \frac{Z^4 |e|^2 R^2}{10\pi\varepsilon_0 a_0^3} \right) \qquad (1.95)$$

So the difference in the $K_\alpha$ x-ray energies coming from the two isotopes $A_1$ and $A$ is

$$\Delta E_{K_\alpha} = E_{K_\alpha}^{A} - E_{K_\alpha}^{A_1} = -\frac{Z^4 |e|^2 R^2}{10\pi\varepsilon_0 a_0^3} + \frac{Z^4 |e|^2 R_1^2}{10\pi\varepsilon_0 a_0^3}$$

$$\Delta E_{K_\alpha} = \frac{Z^4 |e|^2}{10\pi\varepsilon_0 a_0^3}(R_1^2 - R^2) \qquad (1.96)$$

This is called $K_\alpha$ x-ray isotope shift.

If we assume that $R_1 = R_0 A_1^{1/3}$, $R = R_0 A^{1/3}$ (we shall soon check whether assumption of this form is true), then

$$\Delta E_{K_\alpha} = \frac{Z^4 |e|^2}{10\pi\varepsilon_0 a_0^3}[(R_0 A_1^{1/3})^2 - (R_0 A^{1/3})^2] = \frac{Z^4 |e|^2 R_0^2}{10\pi\varepsilon_0 a_0^3}(A_1^{2/3} - A^{2/3})$$

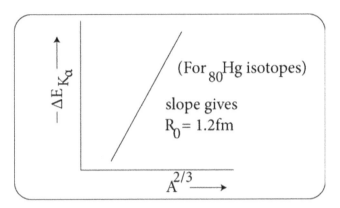

**Figure 1.16.** Plot of $\Delta E_{K_\alpha}$ versus $A^{2/3}$ for isotopes of $_{80}Hg^A$ where $A \to 198, 199, 200, 201, 202, 204$.

$$-\Delta E_{K_\alpha} = \frac{Z^4 \mid e \mid^2 R_0^2}{10\pi\varepsilon_0 a_0^3} A^{2/3} - \frac{Z^4 \mid e \mid^2 R_0^2}{10\pi\varepsilon_0 a_0^3} A_1^{2/3} \qquad (1.97)$$

$$-\Delta E_{K_\alpha} = m_0 A^{2/3} - c$$

with $m_0 = \frac{Z^4 \mid e \mid^2 R_0^2}{10\pi\varepsilon_0 a_0^3}$, $c = \frac{Z^4 \mid e \mid^2 R_0^2}{10\pi\varepsilon_0 a_0^3} A_1^{2/3}$.

This equation predicts a straight line in the plot of $-\Delta E_{K_\alpha}$ versus $A^{2/3}$ with slope $m_0 = \frac{Z^4 \mid e \mid^2 R_0^2}{10\pi\varepsilon_0 a_0^3}$ that contains the nuclear radius constant $R_0$. Now $\Delta E_{K_\alpha}$ was measured experimentally for different isotopes of $_{80}Hg^A$ ($A \to 198, 199, 200, 201, 202, 204$) and the plot against $A^{2/3}$ gives a straight line (figure 1.16) confirming that our assumption of $R = R_0 A^{1/3}$ was correct. The slope of the straight line gives the value $R_0 = 1.2\,fm$.

- Comments:
  - ✓ For $_{80}Hg$ the order of energy values are $E_{K_\alpha} \sim 100\,keV$ and $\Delta E \sim 0.1\,eV$. This means we have to measure changes $\frac{\Delta E}{E_{K_\alpha}} \sim \frac{0.1\,eV}{100\,keV} = 10^{-6}$ which is very small.
  - ✓ The situation can be improved and observations are easier if we employ muonic x-rays from a muonic atom. (*exercise 1.51*).
  - ✓ Charge radius (or proton radius) and nuclear force radius (or nuclear matter radius) of nucleus are nearly equal and both show $A^{1/3}$ dependence with $R_0 = 1.2\,fm$.

In *exercise 1.52* we explain the fact that proton radius and neutron radius are the same, although heavy nuclei have excess neutrons.

## 1.23 Estimate of nuclear radius from a study of mirror nuclei

The general formula of the electrostatic (Coulomb) energy of a charge $q$ which is uniformly distributed throughout a sphere of radius $R$ is

$$E_{\text{Coulomb}} = \frac{1}{4\pi\varepsilon_0}\frac{3q^2}{5R} \tag{1.98}$$

This is called self-energy of charge $q$. It represents the work done to build the charged sphere out of nothing.

Let us consider an assembly of $Z$ protons and that interaction occurs over a distance which is the spread or the size of nucleus, i.e. over $R = R_0 A^{1/3}$. The work done to build this assembly of $Z$ protons is

$$U = \begin{pmatrix}\text{Work to bring the protons from}\\ \text{infinity to their present location}\end{pmatrix} - \begin{pmatrix}\text{Number of}\\ \text{protons}\end{pmatrix}\begin{pmatrix}\text{Work to build an individual}\\ \text{proton out of vacuum}\end{pmatrix}$$

$$= \frac{1}{4\pi\varepsilon_0}\frac{3(Z|e|)^2}{5R} - Z\frac{1}{4\pi\varepsilon_0}\frac{3|e|^2}{5R}\frac{1}{4\pi\varepsilon_0}\frac{3|e|^2}{5R}(Z^2 - Z)$$

$$U = \frac{1}{4\pi\varepsilon_0}\frac{3Z(Z-1)|e|^2}{5R_0 A^{1/3}} \tag{1.99}$$

This is the electrostatic energy of nucleus with $Z$ number of protons.

Consider two mirror nuclei $_Z X^A_{Z+1}$ and $_{z+1} Y^A_Z$ (i.e. $_Z X^A_{Z+1}$ has $Z$ protons, $_{z+1} Y^A_Z$ has $Z+1$ protons.) The difference in their electrostatic energies is given by (assuming they have the same radius $R$)

$$\Delta E_C = \frac{1}{4\pi\varepsilon_0}\frac{3(Z+1)[(Z+1)-1]|e|^2}{5R_0 A^{1/3}} - \frac{1}{4\pi\varepsilon_0}\frac{3Z(Z-1)|e|^2}{5R_0 A^{1/3}}$$

$$= \frac{1}{4\pi\varepsilon_0}\frac{3|e|^2}{5R_0 A^{1/3}}[Z(Z+1) - Z(Z-1)]$$

$$\Delta E_C = \frac{1}{4\pi\varepsilon_0}\frac{3|e|^2}{5R_0 A^{1/3}}(2Z)$$

Use $A = (Z+1) + Z = 2Z + 1 \Rightarrow 2Z = A - 1$

$$\Delta E_C = \frac{1}{4\pi\varepsilon_0}\frac{3|e|^2}{5R_0 A^{1/3}}(A - 1) \tag{1.100}$$

Knowing the value of $\Delta E_C$, $A$ we can find the value of $R_0$. So $R = R_0 A^{1/3}$ is determined. We give an example in *exercise 1.53*. For $A - 1 \approx A$ we have from equation (1.100)

$$\Delta E_C \approx \frac{1}{4\pi\varepsilon_0}\frac{3|e|^2}{5R_0 A^{1/3}}A = \frac{1}{4\pi\varepsilon_0}\frac{3|e|^2}{5R_0}A^{2/3} \tag{1.101}$$

So the plot of $\Delta E_C$ versus $A^{2/3}$ gives a straight line (figure 1.17). From slope we have $R_0 = 1.22\,fm$.

- Measurement of $\Delta E_C$:

✓ One nuclei in the pair of mirror nuclei decays to another through nuclear $\beta^+$ decay. It is a $p \rightarrow n$ transformation with the emission of a $e^+$ (positron)

$$p \rightarrow n + e^+ + \nu_e$$

The maximum energy of the $e^+$ is a measure of $\Delta E_C$.

✓ The nuclear reaction

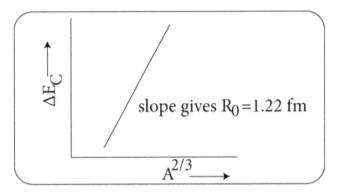

**Figure 1.17.** Measurement of $\Delta E_C$.

$$p + {}_5B^{11} \rightarrow {}_6C^{11} + n$$

can be studied. The minimum proton energy necessary to cause this reaction is a measure of $\Delta E_C$.

## 1.24 Nuclear mass density

Nuclear mass density is

$$\rho_m = \frac{\text{Nuclear mass}}{\text{Nuclear volume}} = \frac{Am_p}{\frac{4}{3}\pi R^3} \tag{1.102}$$

assuming a spherical nucleus and taking nuclear mass $=A$ times the mass of a proton.

Using $R = R_0 A^{1/3}$ and $R_0 = 1.2\,fm$ we have from equation (1.102)

$$\rho_m = \frac{Am_p}{\frac{4}{3}\pi (R_0 A^{1/3})^3} = \frac{Am_p}{\frac{4}{3}\pi R_0^3 A} = \frac{m_p}{\frac{4}{3}\pi R_0^3} = \frac{1.67 \times 10^{-27}\,kg}{\frac{4}{3}\pi (1.2 \times 10^{-15}\,m)^3}$$

$$\rho_m = 2.3 \times 10^{17}\,kg.\,m^{-3} \tag{1.103}$$

So nuclear mass density is very very large and constant.

## 1.25 Nucleon density

Nucleon density is

$$\rho = \frac{\text{Nucleon number}}{\text{Nuclear volume}}$$

$$= \frac{A}{\frac{4}{3}\pi R^3} = \frac{A}{\frac{4}{3}\pi (R_0 A^{1/3})^3} = \frac{A}{\frac{4}{3}\pi R_0^3 A} = \frac{1}{\frac{4}{3}\pi R_0^3} \tag{1.104}$$

$$= \frac{\text{nucleons}}{\frac{4}{3}\pi (1.2 \times 10^{-15}\,m)^3} = 1.38 \times 10^{44}\,\text{nucleons.}\,m^{-3}$$

So nucleon density is very very large and constant.

- Nuclear fluid:

Enormously large values of nuclear mass density $\rho_m \sim 10^{17}\, kg.\, m^{-3}$ and nucleon density $\rho \sim 10^{44}\frac{nucleons}{m^3}$ means that nuclear matter is in an extremely compressed or condensed state. Such highly compressed state of nucleus means that mass is essentially continuously distributed throughout the nuclear volume like a fluid. So we can look upon an assembly of nucleons in a nucleus as analogous to incompressible liquid or fluid. So we speak of nuclear fluid.

## 1.26 Binding energy of nucleus

Nucleus contains multiple nucleons (except hydrogen).

Energy that binds a set of nucleons into a nucleus is binding energy. And if this energy that binds is not available then the constituent nucleons would unbind. So binding energy is the energy needed to unbind or disintegrate a nucleus into its constituents.

Consider a nucleus with $Z$ protons, $N$ neutrons and so $A = Z + N$ nucleons.

When in infinite dilution, i.e. when nucleons (protons and neutrons) of a nucleus are free (i.e. separated by large distances), the energy is

$$\left(Zm_p + Nm_n\right)c^2.$$

Some energy (say $B$) is expended to bring the nucleons from infinity closer to each other so as to form the nucleus. When in bound state the nuclear mass is say $M$ and so energy is

$$Mc^2$$

and hence the energy balance or energy conservation suggests that

$$Mc^2 = \left(Zm_p + Nm_n\right)c^2 - B \tag{1.105}$$

$$B = \left(Zm_p + Nm_n\right)c^2 - Mc^2 \tag{1.106}$$

$$\frac{B}{c^2} = Zm_p + Nm_n - M. \tag{1.107}$$

This is the difference of energy of constituents and energy of nucleus.

Clearly energy in free state is more than the energy in bound state—the energy deficit being called binding energy $B$.

Note that in free state nucleons are independent of each other and have more energy; but, when within the nucleus, nucleons are not free but constantly interact, share and expend energy with the neighbours, which means that energy is reduced.

The binding energy holds nucleons together to maintain size of nucleus. It disallows nucleons to stray out of the nucleus to arbitrary distances, i.e. limits movement of nucleons within the nuclear boundary.

- Significance of binding energy:
  - ✓ If we supply the deficit energy $B$ to nucleus (mass $M$, energy $Mc^2$) then the nucleus just disintegrates— fragmentation just occurs, because now the nucleons get the needed energy to stay apart or free. Energy conservation gives

$$B + Mc^2 = \left(Zm_p + Nm_n\right)c^2 = \text{energy of free state} \qquad (1.108)$$

  - ✓ Infact binding energy is a measure of stability of nucleus against decay into fragments (or constituents). If energy supplied is greater than binding energy, fragments will move with some kinetic energy. And if energy supplied is less than binding energy, the nucleus will not disintegrate.
  - ✓ Supply of binding energy takes the nucleons from a bound nuclear state to free state. Here energy is converted to mass.
  - ✓ When free nucleons combine to form bound nucleus of lesser mass, the deficit in mass is converted to energy that holds nucleons together.
- The nature of binding between fragments or constituents is indicated by the value of binding energy $B$. A more useful measure of the binding between constituents of the nucleus is the average binding energy per nucleon or binding fraction

$$\frac{B}{A} = \frac{\text{Total binding energy}}{\text{Total number of nucleons}}. \qquad (1.109)$$

- Significance of binding fraction
  - ✓ If $B = \text{large} \gg 0$

$$\left(Zm_p + Nm_n\right)c^2 \gg Mc^2 \text{ [from equation (1.106)]} \qquad (1.110)$$

    The nucleus needs large energy to disintegrate and free its constituents. Normally the nucleus does not get such large energy and hence remains stable against decay (figure 1.18(a)).

    Example:
    $_2\text{He}_2^4$ has $\frac{B}{A} = 7.07\frac{MeV}{\text{nucleon}}$, $_{26}\text{Fe}_{30}^{56}$ has $\frac{B}{A} = 8.79\frac{MeV}{\text{nucleon}}$, $_{47}\text{Ag}_{60}^{107}$ has $\frac{B}{A} = 8.56\frac{MeV}{\text{nucleon}}$ (exercise 1.54).
  - ✓ If $B = \text{small} > 0$

$$\left(Zm_p + Nm_n\right)c^2 > Mc^2 \qquad (1.111)$$

    The nucleus needs small energy to disintegrate and free its constituents. There is a large probability that the nucleus might get that energy. And so the nucleus is not stable, but can decay easily (figure 1.18(b)).

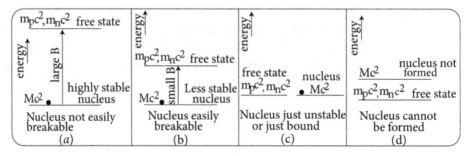

**Figure 1.18.** Significance of binding energy

Example:

$$_1H_1^2 \text{ has } \frac{B}{A} = 1.11\frac{MeV}{\text{nucleon}}, \quad _3Li_4^7 \text{ has } \frac{B}{A} = 5.74\frac{MeV}{\text{nucleon}}, \quad _4Be_5^9 \text{ has } \frac{B}{A} = 6.47\frac{MeV}{\text{nucleon}}$$

(*exercise 1.54*)

✓ If $B = 0$

$$\left(Zm_p + Nm_n\right)c^2 = Mc^2 \tag{1.112}$$

If such a nucleus is formed it is prone to decay by itself, the nucleus will just disintegrate on its own. It is just unstable or just bound (figure 1.18(c)).

✓ If $B < 0$

$$\left(Zm_p + Nm_n\right)c^2 < Mc^2 \tag{1.113}$$

The nucleus cannot be formed since energy is less in the free state, i.e. the constituents are already in the lower energy state. (figure 1.18(d)).

Binding energy $B$ is thus a measure of the stability of the nucleus. Binding energy tells us whether a nucleus is stable or not and if stable the amount of stability. Binding energy tells the amount of energy needed to dismantle it.

Binding energy per nucleon can be looked upon as the average energy per nucleon required to disintegrate a nucleus into its constituents (nucleons). The average value of the binding energy per nucleon reveals details of nuclear force (*exercise 1.58*).

Larger $\frac{B}{A}$ means the nucleus is more stable.

## 1.27 Binding energy curve

The plot of average binding energy per nucleon $\frac{B}{A}$ versus $A$ has been shown in figure 1.19(a).

The salient features are:

✓ Average binding energy per nucleon is $8\frac{MeV}{\text{nucleon}}$.

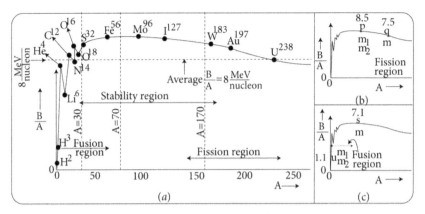

**Figure 1.19.** Plot of binding energy per nucleon $\frac{B}{A}$ versus mass number $A$.

✓ Nuclei which sit on the maxima of the curve are most tightly bound like $_{26}\text{Fe}_{30}^{56}$, $_{26}\text{Fe}_{32}^{58}$, $_{28}\text{Ni}_{32}^{60}$, $_{28}\text{Ni}_{34}^{62}$ with $\frac{B}{A} = 8.7904, 8.7922, 8.7808, 8.7946\frac{MeV}{\text{nucleon}}$. respectively.

✓ $\frac{B}{A}$ is practically independent of the mass number for nuclei in the middle region between $A = 30$ to $A = 170$ (with $\frac{B}{A} \sim 8\frac{MeV}{\text{nucleon}}$). This means that attractive force of strong interaction binds nucleons in a nucleus. Clearly $\frac{B}{A}$ fails to grow with size of nucleus and hence we say that $\frac{B}{A}$ saturates. This saturation property is a consequence of the short-range nature of nuclear force (*exercise 1.58*). It means that a nucleon influences only nucleons located close to it.

✓ For $A < 30$ and $A > 170$ the value of binding energy per nucleon is small (i.e. less than $8\frac{MeV}{\text{nucleon}}$).

✓ For $A < 30$ there is a cyclic recurrence of peaks for certain nuclei satisfying the formula $A = 4n$, $n = 1,2,3,4$. These are even–even nuclei such as:

$$_2\text{He}_2^4 \ (A = 4 = 4 \times 1)$$

$$_4\text{Be}_4^8 \ (A = 8 = 4 \times 2) \ (_4\text{Be}_4^8 \text{ is unstable and not shown in plot})$$

$$_6\text{C}_6^{12} \ (A = 12 = 4 \times 3)$$

$$_8\text{O}_8^{16} \ (A = 16 = 4 \times 4)$$

$$_{10}\text{Ne}_{10}^{20} \ (A = 20 = 4 \times 5)$$

having $\frac{B}{A}$ values 7.07, 6.8, 7.43, 7.73 and 7.78 $\frac{MeV}{\text{nucleon}}$, respectively (*exercises 1.54, 1.65*).

• The plot of binding energy per nucleon $\frac{B}{A}$ against $A$ has three regions having characteristics as mentioned in the following:

✓ Region of stability.

The curve of figure 1.19(a) shows a nearly flat region between $A = 30$ to $A = 70$ where $\frac{B}{A} = $ constant and represents the most stable elements like $_{26}Fe^{56}_{30}$, $_{26}Fe^{58}_{32}$, $_{28}Ni^{60}_{32}$, $_{28}Ni^{62}_{34}$.

✓ Region of fission reactions.

The curve of figure 1.19(b) shows that

$$\left(\frac{B}{A}\right)^{point\ q}_{high\ A} = \text{small as compared to} \left(\frac{B}{A}\right)^{point\ p}_{intermediate\ A} = \text{large}.$$

A high $A$ unstable nucleus can achieve more stability if it splits up to generate nuclei of intermediate $A$. So a high $A$ nucleus at point $q$ (mass $m$, $\frac{B}{A} = 7.5\frac{MeV}{nucleon} = $ small) can split to produce nuclei of intermediate masses at point $p$ ($m_1$, $m_2$, $\frac{B}{A} = 8.5\frac{MeV}{nucleon} = $ large) to achieve stability. The process thus liberates $(8.5 - 7.5)\frac{MeV}{nucleon} = 1\frac{MeV}{nucleon}$ energy and for nucleus with $A = 220$ nucleons, energy liberated will be

$$(220\ \text{nucleon})\left(1\frac{MeV}{nucleon}\right) = 220\ MeV$$

per fission (i.e. breaking) of the high $A$ nucleus. Fission of nucleus is described through the relation

$$m \rightarrow m_1 + m_2 + \text{energy}\ (MeV) \tag{1.114}$$

✓ Region of fusion reactions.

The curve of figure 1.19(c) shows that

$$\left(\frac{B}{A}\right)^{point\ u}_{low\ A} = \text{small as compared to} \left(\frac{B}{A}\right)^{point\ s}_{intermediate\ A} = \text{large}.$$

A low $A$ unstable nucleus can achieve more stability if it unites with another such nucleus to generate a nucleus of intermediate $A$. So two low $A$ nuclei at point $u$ (masses $m_1$, $m_2$, $\frac{B}{A} = 1.1\frac{MeV}{nucleon} = $ small) can fuse (i.e. join) to produce a nuclei of intermediate mass at point $s$ ($m$, $\frac{B}{A} = 7.1\frac{MeV}{nucleon} = $ large) to achieve stability. The process thus liberates

$$(7.1 - 1.1)\frac{MeV}{nucleon} = 6\frac{MeV}{nucleon}.$$

energy and for fusion of four nucleons energy liberated will be

$$(4\ \text{nucleons})\left(6\frac{MeV}{nucleon}\right) = 24\ MeV$$

per fusion. Fusion of nuclei is described through the relation

$$m_1 + m_2 \rightarrow m + \text{energy}\ (MeV) \tag{1.115}$$

## 1.28 Separation energy of a nucleon

Separation energy of a nucleon is the amount of energy needed to remove it from the nucleus. For a given $N$, $Z$ separation energy is larger for nuclei with even $Z$, even $N$ than for other combinations. Separation energy is analogous to the ionization energy of atomic electron.

✓ Separation energy $S_n$ for neutron $n$

Separating $n$ from nucleus $(Z, N, A)$ refers to the transformation

$$(Z, N, A) \to (Z, N - 1, A - 1) + n \qquad (1.116)$$

and hence

$$S_n = [M_{\text{final}} - M_{\text{initial}}]c^2 = [M(Z, N - 1, A - 1) + m_n - M(Z, N, A)]c^2 \quad (1.117)$$

$$= B(Z, N, A) - B(Z, N - 1, A - 1) \; (by \; exercise \; 1.56) \qquad (1.118)$$

✓ Separation energy $S_p$ for proton

Separating $p$ from nucleus $(Z, N, A)$ refers to the transformation

$$(Z, N, A) \to (Z - 1, N, A - 1) + p \qquad (1.119)$$

and hence

$$S_p = [M_{\text{final}} - M_{\text{initial}}]c^2 = [M(Z - 1, N, A - 1) + m_p - M(Z, N, A)]c^2 \qquad (1.120)$$

$$= B(Z, N, A) - B(Z - 1, N, A - 1) \; (exercise \; 1.56) \qquad (1.121)$$

## 1.29 Mass defect

• Definition 1:

Mass of the nucleus is less than the sum of the masses of its constituent nucleons. This missing mass $\Delta m$ is called mass defect

$$\Delta m = Zm_p + Nm_n - M \qquad (1.122)$$

The relation of mass defect $\Delta m$ with binding energy $B$ is

$$B = \left(Zm_p + Nm_n - M\right)c^2 = (\Delta m)c^2 \qquad (1.123)$$

We give an example. For $_2\text{He}^4 = \alpha = 2n2p$ system

$$\Delta m = 2m_p + 2m_n - m$$

$$\Delta m = 2(1.007\,276u) + 2(1.008\,665u) - 4.001\,506u = 0.030\,376u \qquad (1.124)$$

- Definition 2:

    Consider a nucleus of $Z$ protons and $N$ neutrons and hence $A = Z + N$ nucleons. We expect the mass of the nucleus to be

$$Zm_p + Nm_n \sim Am_p \sim A \; amu \qquad (1.125)$$

    since $m_p \sim 1 \; amu$. But the actual or measured mass $M = M(Z, N, A)$ in amu is slightly different (slightly greater or slightly less except for $_6C_6^{12}$) from $A \; amu$. The difference between measured mass $M$ and mass number $A$ (positive or negative, whatever) is referred to as mass defect, i.e. mass defect is

$$\Delta m = M - A \qquad (1.126)$$

- We note that while mass number $A$ is a whole number (it being the number of nucleons) the measured mass $M(Z, N, A) = M$ is not a whole number.
- Let us furnish some examples as in figure 1.20. It is clear that mass defect is
    ✓ zero for $_6C_6^{12}$;
    ✓ positive for very light and for very heavy atoms e.g. $_2He_2^4$, $_{88}Ra_{138}^{226}$;
    ✓ negative for atoms of intermediate mass, e.g. $_8O_8^{16}$.

- Atomic mass = Nuclear mass + Electronic mass $\qquad (1.127)$

$$\approx \text{Nuclear mass (since } m_e \approx 0) \qquad (1.128)$$

- Significance of mass defect:
✓ mass defect gives a measure of binding energy through the relation $B = \Delta mc^2$ (equation (1.123));

| Element | Au | M(Z,N,A) | Mass defect $M - A$ | Packing fraction $\dfrac{M - A}{A}$ |
|---|---|---|---|---|
| $_6C_6^{12}$ | 12u | 12u | 0 | 0 |
| $_1H_0^1$ | 1u | 1.007825u | +0.007825u | 0.007825 |
| $_1H_1^2$ | 2u | 2.014102u | +0.014102u | 0.007051 |
| $_2He_2^4$ | 4u | 4.0002603u | +0.002603u | 0.0006507 |
| $_8O_8^{16}$ | 16u | 15.994915u | −0.005085u | −0.0003178 |
| $_{79}Au_{118}^{197}$ | 197u | 196.96654u | −0.03346u | −0.0001698 |
| $_{88}Ra_{138}^{226}$ | 226u | 226.02543u | +0.02543u | 0.0001125 |

**Figure 1.20.** Mass defect and packing fraction for different elements.

✓ mass defect gives an indication of how the nucleons are packed in the nucleus. This is done through a study of mass defect per nucleon called packing fraction. So mass defect indicates stability also.

## 1.30 Packing fraction

Packing fraction $f$ is the mass defect per nucleon, i.e.

$$f = \frac{M - A}{A} = \frac{\Delta m}{A} \qquad (1.129)$$

$$fA = M - A$$

$$M = A(1 + f) \qquad (1.130)$$

Figure 1.21 shows a plot of packing fraction $f$ against mass number.
    For very light nuclei, $f$ is positive and decreases with increase in $A$.
    For $_6C_6^{12}$, $A = 12$, $f$ becomes zero.
    For $A > 20$, $f$ falls and attains minimum value for $A \sim 60$.
    For $A > 60$, $f$ rises slowly and becomes positive for $A > 180$.
    • Packing fraction is a measure of comparative stability of the nucleus:

✓ If $f$ is positive, $M > A$.
    The system has to handle more mass and this gives rise to instability. So $f$ positive indicates comparatively less stability.
    Example: very light and very heavy nuclei.
✓ If $f$ is negative $M < A$.
    The system has less mass to hold on. Actually, mass deficit has converted to binding energy that binds the constituents of the system to increase its stability. So $f$ negative indicates comparatively higher stability.
    Example: Nuclei with intermediate mass.

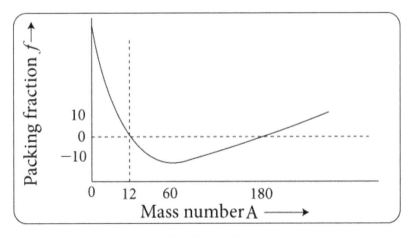

**Figure 1.21.** Plot of packing fraction $f$ against mass number $A$.

## 1.31 Nuclear stability, Segrè chart or stability curve

There are 272 stable nuclides out of which 160 nuclides are of even–even type. Existence of such a large number of stable nuclei suggests that even–even type nuclei is preferred in Nature.

On the other hand, there are only 4 odd–odd stable nuclei. Such rare occurrence of odd–odd nuclei shows that nature does not prefer odd–odd nuclei.

Figure 1.22 depicts the plot of proton number $Z$ versus neutron number $N$ in various nuclei.

- For $Z < 20$ the stability line is a 45° straight line with $N = Z$, which means that the neutron-to-proton ratio is unity as

$$\frac{N}{Z} = \frac{1}{1} = 1$$

Examples are $_6C_6^{12}(Z = 6, N = 6, \frac{N}{Z} = \frac{6}{6} = 1)$, $_8O_8^{16}(Z = 8, N = 8, \frac{N}{Z} = \frac{8}{8} = 1)$.

- For $Z > 20$, $N > 20$ the stability curve bends away from the $N = Z$ line towards the $N$-axis due to excess neutron population in the nucleus.

Examples are $_{20}Ca_{28}^{48}(Z = 20, N = 28, \frac{N}{Z} = \frac{28}{20} = 1.4)$, $_{91}Pa_{141}^{232}(Z = 91, N = 141, \frac{N}{Z} = \frac{141}{91} = 1.55)$.

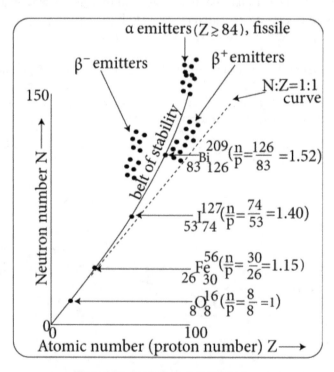

**Figure 1.22.** Segrè chart or stability curve.

As number of protons increase, Coulomb repulsion between them becomes more and more important. For stability of the nucleus (i.e. to prevent tearing apart of the nucleus by this repulsive Coulomb interaction), the neutron number has to be increased so as to screen or shield the protons from each other. In other words, the repulsive Coulomb interaction is checked, compensated or balanced by presence of excess neutrons in the nucleus. Actually, it is a quantum effect and the asymmetry in the number of neutrons and protons for high $Z$ nuclei is explained by the asymmetry energy correction term in the Bethe–Weizsacker semi-empirical mass formula (that we discuss in chapter 3).

- Belt of stability or stability curve.

  The belt of stability is defined by nuclei with a certain combination of protons and neutrons such that once a nucleus is formed it remains as it is for quite a long period of time, i.e. exists as a stable nucleus. Beyond the belt of stability the nuclei, after being formed changes itself in the course of time and so is unstable.

- Just below the stability curve, nuclei have too many protons and are unstable. They decay by emission of positron (as proton converts to neutron: $p \rightarrow n + e^+ + \nu_e$) to achieve stability through $\beta^+$ decay. For example

$$(Z, N, A) \rightarrow (Z - 1, \ N + 1, \ A) + e^+ + \nu_e$$

$$_{91}\text{Pa}_{139}^{230} \rightarrow {}_{90}\text{Th}_{140}^{230} + e^+ + \nu_e$$

- Just above the stability curve, nuclei have too many neutrons and are unstable. They decay by emission of electron (neutron converts to proton: $n \rightarrow p + e^- + \bar{\nu}_e$) to achieve stability through $\beta^-$ decay. For example

$$(Z, N, A) \rightarrow (Z + 1, \ N - 1, \ A) + e^- + \bar{\nu}_e$$

$$_{11}\text{Na}_{13}^{24} \rightarrow {}_{12}\text{Mg}_{12}^{24} + e^- + \bar{\nu}_e$$

- Very large nuclei have too many nucleons and decay by $\alpha$ emission that is often associated with $\gamma$ emission. For example

$$(Z, N, A) \rightarrow (Z - 2, \ N - 2, \ A - 4) + {}_2\text{He}_2^4$$

$$_{88}\text{Ra}_{134}^{222} \rightarrow {}_{86}\text{Rn}_{132}^{218} + {}_2\text{He}_2^4$$

Another process that they undertake is fission.

✓ Nuclei which are far away from the stability curve (i.e. far removed from the belt of stability) are unstable, radioactive and spontaneously decay successively such that eventually they reach a stable end product lying on the stability curve.

✓ Nuclei having $A > 200$, $Z > 81$, $N > 120$ are unstable.

✓ Clearly not all $Z$, $N$ combinations will form a stable nucleus. Only those that correspond to the belt of stability would be stable.

✓ Can we form a nucleus (stable or unstable) with any arbitrary value of $Z$, $N$?
The degree of instability is decided by life time of nucleus. If life time is long, the nucleus formed is less unstable. If life time is short, then the nucleus is more unstable.

✓ The total energy of a $Z$, $N$ nucleus decides whether it would be stable or unstable and if unstable how will it decay (whether by emitting $\alpha$, $\beta$ or by other means). If total energy is a minimum the nucleus will be stable. If otherwise (say if more minima are available), then it is likely to decay and move to a more stable state.

## 1.32 Nuclear angular momentum, nuclear spin

• Before going on to discuss nuclear angular momentum let us put our attention to electronic energy levels and fine structure.

Consider an atom which is 1 nucleus +1 electron (say a hydrogen atom).

We first consider the electronic energy levels. The first column on the left of figure 1.23 shows the Bohr levels $E_n$ corresponding to principal quantum number $n = 1,2,3,$ .. of the electron.

Next we consider the angular momentum of the electron.

✓ Orbital angular momentum

$$\vec{l} \xrightarrow{\text{eigenvalue}} \sqrt{l(l+1)}\,\hbar, \; l = 0, 1, 2, \ldots n-1 \tag{1.131}$$

Now the Bohr levels are denoted by two quantum numbers $n$, $l$, i.e. we replace $E_n$ by $E_{nl}$. So the Bohr levels are split up as shown in the second column from the left of figure 1.23. So $n = 1,2,3,\ldots$ will be replaced by $(n, l) = (n = 1, l = 0)$, $(n = 2, l = 0,1)$, $(n = 3, l = 0,1,2)$ etc.

✓ Intrinsic spin angular momentum

$$\vec{s} \xrightarrow{\text{eigenvalue}} \sqrt{s(s+1)} \xrightarrow{s=\frac{1}{2}\text{for electron}} \sqrt{\frac{1}{2}\left(\frac{1}{2}+1\right)}\,\hbar = \frac{\sqrt{3}}{2}\hbar \tag{1.132}$$

✓ Total angular momentum

$$\vec{j} \xrightarrow{\text{eigenvalue}} \sqrt{j(j+1)}\,\hbar \tag{1.133}$$

where $\vec{j} = \vec{l} + \vec{s}$ and $j = l + s$ to $|l - s|$ in unit steps $= l + \frac{1}{2}$ to $|l - \frac{1}{2}|$ in unit steps.

For $l = 0$, $j = \frac{1}{2}$ (in fact $0 + \frac{1}{2}$ and $|0 - \frac{1}{2}|$ both lead to the same value of $\frac{1}{2}$), while for $l \neq 0$, $j = l \pm \frac{1}{2}$ (in fact one value is $l + \frac{1}{2}$ and the next value is $l + \frac{1}{2} - 1 = l - \frac{1}{2}$). So we have two values corresponding

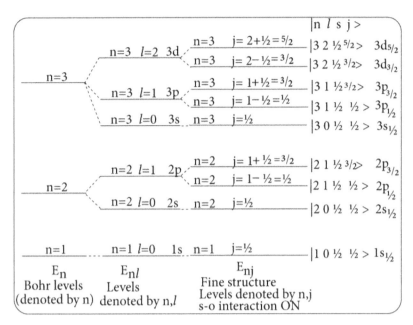

**Figure 1.23.** Fine structure of electronic levels of atom due to spin–orbit interaction. A $l = 0$ level is now denoted by $j = \frac{1}{2}$ (it is not split) while $l \neq 0$ level is doubly split into 2 levels denoted by $j = l + \frac{1}{2}$ (upper level) and $j = l - \frac{1}{2}$ (lower level).

to $l \neq 0$ and so $l \neq 0$ level splits up into two levels and it is said to show a fine structure (normal doublet in which $j = l + \frac{1}{2}$ is the upper level). Such splitting or fine structure arises because of the interaction of two magnetic dipole moments corresponding to two angular momenta viz. $\vec{\mu_l}$ (corresponding to $\vec{l}$) and $\vec{\mu_s}$ (corresponding to $\vec{s}$) since we have now considered both orbital and spin angular momentum. This is called spin–orbit or s–o, i.e. l–s interaction. Clearly, taking l–s interaction into account means we have to consider both types of angular momenta— orbital and spin (the resultant of which is $\vec{j}$) and so the fine structure levels are denoted by $j$ (instead of $l$).

So now $(n, l)$ will be replaced by $(n, j)$ as shown in the third column from the left of figure 1.23.

- Consider a multi-electron system (electron cloud):
  Atom = 1 nucleus + $Z$ electrons.
  We apply addition of angular momentum rules to get the following angular momenta of $Z$ electrons.

  ✓ Orbital angular momentum

$$\sum \vec{l} = \overrightarrow{L_e} \xrightarrow{\text{eigenvalue}} \sqrt{L_e(L_e + 1)}\,\hbar \tag{1.134}$$

✓ Intrinsic spin angular momentum

$$\sum \vec{s} = \overrightarrow{S_e} \xrightarrow{\text{eigenvalue}} \sqrt{S_e(S_e + 1)}\,\hbar \tag{1.135}$$

✓ Total angular momentum

$$\sum \vec{j} = \overrightarrow{J_e} \xrightarrow{\text{eigenvalue}} \sqrt{J_e(J_e + 1)}\,\hbar \tag{1.136}$$

where $\overrightarrow{J_e} = \overrightarrow{L_e} + \overrightarrow{S_e}$ and $J_e = L_e + S_e$ to $|L_e - S_e|$ in unit steps. There are $2L_e + 1$ values if $L_e < S_e$ and $2S_e + 1$ values if $S_e < L_e$.

- We now discuss angular momenta possessed by a nucleon.

  $\alpha, \beta$ spectra emitted by a nucleus suggest that the constituents of the nucleus, viz. nucleons (i.e. protons and neutrons), possess angular momenta. Let us use the symbol $n$ for nucleon (either a proton or a neutron) and write down nucleon angular momenta as in the following.

  ✓ Orbital angular momentum of nucleon

$$\vec{l}_n \xrightarrow{\text{eigenvalue}} \sqrt{l_n(l_n + 1)}\,\hbar \tag{1.137}$$

  ✓ Intrinsic spin angular momentum of nucleon

$$\vec{s}_n \xrightarrow{\text{eigenvalue}} \sqrt{s_n(s_n + 1)} \xrightarrow{s_n = \frac{1}{2} \text{for nucleon}} \sqrt{\frac{1}{2}\left(\frac{1}{2} + 1\right)}\,\hbar = \frac{\sqrt{3}}{2}\hbar \tag{1.138}$$

  ✓ Total angular momentum of nucleon

$$\vec{j}_n \xrightarrow{\text{eigenvalue}} \sqrt{j_n(j_n + 1)}\,\hbar \tag{1.139}$$

  where $\vec{j}_n = \vec{l}_n + \vec{s}_n$, and $j_n = l_n + s_n$ to $|l_n - s_n|$ in unit steps.

  For a nucleus having multiple nucleons, addition of angular momenta gives

  ✓ Orbital angular momentum

$$\sum \vec{l}_n = \overrightarrow{L_N} \xrightarrow{\text{eigenvalue}} \sqrt{L_n(L_n + 1)}\,\hbar \tag{1.140}$$

  ✓ Intrinsic spin angular momentum

$$\sum \vec{s_n} = \overrightarrow{S_N} \xrightarrow{\text{eigenvalue}} \sqrt{S_N(S_N + 1)}\,\hbar \qquad (1.141)$$

✓ Total angular momentum is also called nuclear spin in literature and we denote it by $J_N$. It is expressed as

$$\sum \vec{j_n} = \overrightarrow{J_N} \xrightarrow{\text{eigenvalue}} \sqrt{J_N(J_N + 1)}\,\hbar \qquad (1.142)$$

where $\overrightarrow{J_N} = \overrightarrow{L_N} + \overrightarrow{S_N}$ and $J_N = L_N + S_N$ to $|\,L_N - S_N\,|$ in unit steps. The value of $J_N$ can be computed as follows for odd $A$ and even $A$ nuclei.

| $Z + N \rightarrow A$ | $L_N$ | $S_N$ | $J_N$ |
|---|---|---|---|
| Odd + Even → Odd<br>Even + Odd → Odd | Integral | Half integral (upon adding odd number of $\frac{1}{2}s$) | $\frac{1}{2}, \frac{3}{2}, \frac{5}{2}, \frac{7}{2}\ldots$⇐half integral for odd $A$ nucleus |
| Even + Even → Even<br>Odd + Odd → Even | Integral | Integral (upon adding even number of $\frac{1}{2}s$) | $0,1,2,3\ldots$⇐integral for even $A$ nucleus |

The total angular momentum of atom is

$$\overrightarrow{J}_{\text{atom}} = \overrightarrow{J_e} + \overrightarrow{J_N} \qquad (1.143)$$

where $J_{\text{atom}} = J_e + J_N$ to $|\,J_e - J_N\,|$ in unit steps. There are $2J_e + 1$ hyperfine levels if $J_e < J_N$ and $2J_N + 1$ hyperfine levels if $J_N < J_e$.

- We now discuss hyperfine structure in atoms.

  We have discussed and shown in figure 1.23 fine structure of electronic levels. High resolution interferometry showed that electronic levels are further split, i.c. show a hyperfine (or finer than the fine) structure. This was explained by Pauli taking into consideration total angular momentum of nucleus $J_N$, also called nuclear spin.

  Figure 1.24 shows hyperfine structure corresponding to Bohr level $n = 2$. For $n = 2$ the $L_e$ values are 0, 1 leading to levels $2S$, $2P$ respectively.

  Now if we take spin $S_e = \frac{1}{2}$ (say) into account, i.e. if we switch on spin–orbit interaction

$$\overrightarrow{J_e} = \overrightarrow{L_e} + \overrightarrow{S_e}$$

then due to coupling between the magnetic dipole moments corresponding to the angular momenta $\overrightarrow{L_e}$ and $\overrightarrow{S_e}$ the energy levels will show fine structure and the levels have to be labeled by $J_e$ where $J_e = L_e + S_e$ to $|\,L_e - S_e\,|$ in unit steps. Since $S_e = \frac{1}{2}$ we have
$J_e = \frac{1}{2}$ for $L_e = 0$ and $J_e = L_e \pm \frac{1}{2}$ (doublet) for $L_e \neq 0$

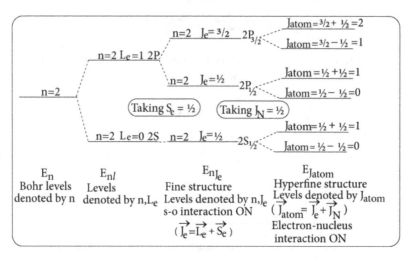

**Figure 1.24.** Hyperfine splitting due to electron–nucleus interaction. Hyperfine levels are labeled by $J_{\text{atom}}$.

This fine structure has been shown in the third column from the left in figure 1.24.

Now let us take total angular momentum of the nucleus (called nuclear spin) $\vec{J}_N$ into account

$$\vec{J}_{\text{atom}} = \vec{J}_e + \vec{J}_N$$

Then due to coupling between the magnetic dipole moments corresponding to the angular momenta $\vec{J}_e$ and $\vec{J}_N$ the energy levels will further split and show hyperfine structure. Levels are now labelled by $J_{\text{atom}}$ where $J_{\text{atom}} = J_e + J_N$ to $| J_e - J_N |$ in unit steps. For $J_N = \frac{1}{2}$ we have

$$J_{\text{atom}} = \frac{1}{2} \text{ for } J_e = 0 \text{ and } J_{\text{atom}} = J_e \pm \frac{1}{2} \text{ for } J_e \neq 0.$$

We thus get a hyperfine multiplet. The selection rule for transition between hyperfine levels is $\Delta J_{\text{atom}} = 0,1$ and $0 \rightarrow 0$ transition is forbidden.

## 1.33 Parity

We now briefly discuss what we mean by parity operation.

- Nuclear parity:

    Parity of nucleus is a purely quantum mechanical concept having no classical analogue.

    Parity operation (or parity inversion) means reflection of coordinates through the origin (0,0,0). It is a discrete orthogonal, non-rotational transformation that flips spatial coordinates so as to change a right-handed (RH) system to a left-handed (LH) system and vice versa (figure 1.25).

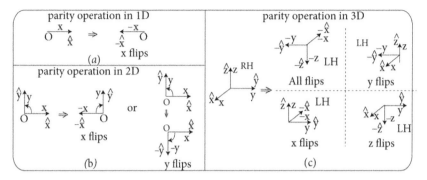

**Figure 1.25.** Parity operation in 1D, 2D and 3D.

Let $\hat{\pi}$ be the parity operator defined through the following operations.

$$x, y, z \xrightarrow{\hat{\pi}} -x, -y, -z \tag{1.144}$$

$$r, \theta, \phi \xrightarrow{\hat{\pi}} r, \pi - \theta, \pi + \phi \tag{1.145}$$

$$\rho, \phi, z \xrightarrow{\hat{\pi}} \rho, \pi + \phi, -z \tag{1.146}$$

$$\vec{r} \xrightarrow{\hat{\pi}} -\vec{r} \tag{1.147}$$

From figure 1.25 the following are evident
✓ Parity operation corresponds to a flip in the spatial coordinate in 1D (figure 1.25(a))

$$x \xrightarrow{\hat{\pi}} -x \tag{1.148}$$

✓ Parity operation in 2D corresponds to a flip in any of the two spatial coordinates

$$\text{either } x \xrightarrow{\hat{\pi}} -x \text{ or } y \xrightarrow{\hat{\pi}} -y \tag{1.149}$$

✓ Parity operation in 3D corresponds to a flip in any of the three spatial coordinates

$$\text{either } x \xrightarrow{\hat{\pi}} -x \text{ or } y \xrightarrow{\hat{\pi}} -y \text{ or } z \xrightarrow{\hat{\pi}} -z \tag{1.150}$$

or a flip in all the three spatial coordinates, i.e.

$$x, y, z \xrightarrow{\hat{\pi}} -x, -y, -z \text{ or equivalently } \vec{r} \rightarrow -\vec{r}. \tag{1.151}$$

In *exercise 1.66* we discuss what if we simultaneously flip two spatial coordinates in 2D.

Let $\vec{r}(x, y, z)$ be the coordinate of a nucleus. All information regarding the nucleus is contained in the state vector $| \psi >$

The nuclear wave function is

$$<\vec{r} \mid \psi > = \psi(\vec{r}) = \psi(x, y, z) \tag{1.152}$$

Parity of a nucleus is related to the behavior of nuclear wave function as a result of reflection, i.e. as a consequence of change of sign of spatial coordinates.

$$\text{If } \psi(-x) = \psi(x) \text{ (1D)}; \psi(-\vec{r}) = \psi(\vec{r}) \text{ (3D)} \tag{1.153}$$

then there is no change of sign, wave function is symmetric and has even parity.

$$\text{If } \psi(-x) = -\psi(x) \text{ (1D)}; \psi\left(-\vec{r}\right) = -\psi(\vec{r}) \text{ (3D)} \tag{1.154}$$

then there is change of sign, wave function is anti-symmetric and has odd parity.

- Incorporating spin, the wave function can be rewritten as
  $$\psi(\vec{r}, \vec{\sigma}) = \psi(\vec{r})\chi_{s, m_s}$$
  where $\psi(\vec{r})$ is the space part and $\chi_{s, m_s}$ is the spin part.
- Space parity or orbital parity (related to orbital part $l$)
  For a particle under central force (e.g. an H-like atom) the parity is decided by orbital angular momentum quantum number, the space parity eigenvalue being $(-1)^l$. So space parity eigenvalue is

$$(-1)^l = +1 \text{ for } l = \text{even and } (-1)^l = -1 \text{ for } l = \text{odd} \tag{1.155}$$

Clearly space parity goes with $l$, i.e. quantum state with definite $l$ has definite or fixed parity.

For a system of particles parity is $(-1)^L$ where $L = \sum l$.
- Intrinsic parity (related to spin part $s$)
  Spin part of wave function of a particle has definite parity called intrinsic parity of the particle. The intrinsic parity of a particle is determined relative to a particle whose intrinsic parity is already fixed by convention. Intrinsic parity of $n$, $p$, $e^-$ is taken to be even and w.r.t them pion parity is determined to be odd.
- Total parity = product of orbital parity and intrinsic parity (as parity is a multiplicative property).

## 1.34 Magnetic dipole moment

- Magnetic dipole moment of an electron:
  ✓ Orbital magnetic dipole moment of an electron is (equation 6.74)

$$\vec{\mu}_{le} = -g_{le}\mu_B \frac{\vec{l}_e}{\hbar} \tag{1.156}$$

where

$$g_{le} = 1 = \text{orbital gyromagnetic ratio or orbital } g \text{ factor for electron} \tag{1.157}$$

$$\mu_B = \frac{|e|\hbar}{2m_e} = \frac{(1.6 \times 10^{-19}\ C)\frac{1}{2\pi}(6.626 \times 10^{-34}\ J.\ s)}{2(9.1 \times 10^{-31}\ kg)} \tag{1.158}$$

$$\mu_B = 9.27 \times 10^{-24}\frac{J}{T} \text{ or } Am^2 = \text{Bohr magneton} \tag{1.159}$$

✓ Intrinsic spin magnetic dipole moment of electron is

$$\vec{\mu}_{se} = -g_{se}\mu_B \frac{\vec{s}_e}{\hbar} \tag{1.160}$$

where

$$g_{se} = 2 = \text{ gyromagnetic ratio or spin } g \text{ factor for electron (equation (6.79))} \tag{1.161}$$

For electron $s_e = \frac{1}{2}$ and taking

$$\vec{s}_e \xrightarrow{\text{eigenvalue}} s\hbar = \frac{1}{2}\hbar \text{ we have from equation (1.160)} \tag{1.162}$$

$$\mu_{se} = -2\mu_B\frac{\hbar/2}{\hbar} = -1\mu_B \text{ (experimental value is } \mu_{se}\left.\right|_{\text{expt}} = 1.001\ 159\mu_B) \tag{1.163}$$

- Nuclear magneton

    Atomic magnetic dipole moments are of the order of Bohr magneton $\mu_B = \frac{|e|\hbar}{2m_e}$ that involves electron mass $m_e$ in the denominator. In nuclear physics we have to replace $m_e$ by protonic mass $m_p$. So in nuclear physics instead of $\mu_B$ we use

$$\mu_N = \frac{|e|\hbar}{2m_p} = \frac{m_e}{m_p}\cdot\frac{|e|\hbar}{2m_e} = \frac{m_e}{m_p}\mu_B$$

$$= \frac{1}{1836}\left(9.27 \times 10^{-24}\frac{J}{T}\right) = 5.05 \times 10^{-27}\frac{J}{T} \text{ or } Am^2 \tag{1.164}$$

called nuclear magneton or Rabi magneton.

Nuclear magnetic dipole moments are of the order of $\mu_N = 5.05 \times 10^{-27}\frac{J}{T}$ or $Am^2$. Clearly $\frac{\mu_N}{\mu_B} = \frac{m_e}{m_p} = \frac{1}{1836}$. So $\mu_N$ is 1836 times smaller than $\mu_B$.

- Magnetic dipole moment of proton
  - ✓ Orbital magnetic dipole moment of proton is (equation (6.75))

$$\vec{\mu}_{lp} = g_{lp}\mu_N \frac{\vec{l}_p}{\hbar} \tag{1.165}$$

where

$$g_{lp} = 1 = \text{gyromagnetic ratio or orbital } g \text{ factor for proton} \tag{1.166}$$

  - ✓ Intrinsic spin magnetic dipole moment of proton is

$$\vec{\mu}_{sp} = g_{sp}\mu_N \frac{\vec{s}_p}{\hbar} \tag{1.167}$$

where

$$g_{sp} = 5.6 = \text{gyromagnetic ratio or spin } g \text{ factor for proton [equation (6.82)]} \tag{1.168}$$

For a proton

$$s_p = \frac{1}{2} \text{ and taking } \vec{s}_p \xrightarrow{\text{eigenvalue}} s\hbar = \frac{1}{2}\hbar \tag{1.169}$$

we have from equation (1.167)

$$\mu_{sp} = 5.6\,\mu_N \frac{\hbar/2}{\hbar} = 2.8\mu_N \text{ (experimental value is } \mu_{sp}\big|_{\text{expt}} = 2.7925\,\mu_N) \tag{1.170}$$

Hence the total magnetic dipole moment of proton is

$$\vec{\mu}_p = \vec{\mu}_{lp} + \vec{\mu}_{sp} \tag{1.171}$$

Comment:

Electron and proton both have spin $\frac{1}{2}$, the same charge magnitude $|e|$. Since $\mu_{se} = -1\mu_B$, we expect $\mu_{sp} = +1\mu_N$. But actually, $\mu_{sp} = 2.7925\mu_N \approx 2.8\mu_N$ (equation 6.104). This value of $\mu_{sp} = 2.8\mu_N$ could not be explained initially and is referred to as anomalous intrinsic spin magnetic dipole moment of proton.

- Magnetic dipole moment of neutron
  - ✓ Orbital magnetic dipole moment of neutron is (equation (6.77))

$$\vec{\mu}_{ln} = 0 \tag{1.172}$$

$$g_{ln} = 0 = \text{orbital gyromagnetic ratio or orbital } g \text{ factor for neutron} \tag{1.173}$$

  - ✓ Intrinsic spin magnetic dipole moment of neutron is

$$\vec{\mu}_{sn} = g_{sn}\mu_N \frac{\vec{s}_n}{\hbar} \tag{1.174}$$

where

$$g_{sn} = -3.8 = \text{spin gyromagnetic ratio or spin } g \text{ factor for neutron} \tag{1.175}$$

For neutron

$$s_n = \frac{1}{2} \text{ and taking } \vec{s}_n \xrightarrow{\text{eigenvalue}} s\hbar = \frac{1}{2}\hbar \tag{1.176}$$

we have

$$\mu_{sn} = (-3.8)\mu_N \frac{\hbar/2}{\hbar} = -1.9\mu_N \text{ (experimental value is } \mu_{sn}\big|_{\text{expt}} = -1.9128\mu_N) \tag{1.177}$$

Hence the total magnetic dipole moment of neutron is

$$\vec{\mu}_n = \vec{\mu}_{ln} + \vec{\mu}_{sn} \tag{1.178}$$

Comment:

Neutron has no charge. Again source of magnetic dipole moment is current or charge (since $\vec{\mu} = i\vec{A}$ and $i = \frac{dq}{dt}$). So the neutron should not have any magnetic dipole moment. But actually $\mu_{sn} = -1.9\mu_N$ (equation 6.105). This value could not be explained initially and is referred to as anomalous intrinsic spin magnetic dipole moment of neutron.

☐ The anomalous values of magnetic dipole moment of proton ($\mu_{sp} = 2.8\mu_N$, $g_{sp} = 5.6$) and magnetic dipole moment of neutron ($\mu_{sn} = -1.9\mu_N$, $g_{sn} = -3.8$) will be explained from the quark model, i.e. quark structure of proton and neutron.

Total magnetic dipole moment of nucleus will be obtained by summing over all nucleons in the nucleus, i.e.

$$\vec{\mu}^{\text{nucleus}} = \underset{\text{nucleons}}{\sum} \vec{\mu}_{\text{nucleon}} = \underset{\text{protons}}{\sum} \vec{\mu}_p + \underset{\text{neutrons}}{\sum} \vec{\mu}_n \tag{1.179}$$

- Estimation of nuclear magnetic dipole moment.

    Values of nuclear magnetic dipole moments can be estimated by nuclear magnetic resonance (NMR) spectrometers, microwave spectrometers, molecular beam deflection method etc.
- Example of magnetic dipole moment of some nuclei are

$\mu(_1\text{H}_1^2) = 0.8574\mu_N$ (compare with equation 6.99), $(_1\text{H}_2^3) = 2.9789\mu_N$, $\mu(_3\text{Li}_3^6) = 0.822\mu_N$, $\mu(_2\text{He}_2^4) = 0$, $\mu(_2\text{He}_1^3) = -2.1275\mu_N$, $\mu(_4\text{Be}_5^9) = -1.177\mu_N$.

## 1.35 Nuclear quadrupole moment

The electric quadrupole moment is a measure of the deviation of charge distribution of nucleus from spherical symmetry. Electrical quadrupole moment $Q$ is defined as

$$Q = \frac{1}{|e|} \int_{\tau} (3z^2 - r^2)\rho d\tau, \, \rho = \text{charge density} \tag{1.180}$$

Electrical quadrupole moment of a nucleus tells us whether it is deformed or not, i.e. gives information regarding the shape of the nucleus.

The S.I unit of electric quadrupole moment is $m^2$. In practice it is expressed in terms of $10^{-28} \, m^2$ called barn.

Nuclear quadrupole moments for different types of charge distributions have been shown in figure 1.26.

$a, b, c$ represent amount of extension of charge along $x, y, z$ axes, respectively. $a, b$ are equatorial radii, $c$ is polar radius.

- Significance of nuclear quadrupole moment:

  ✓ $Q = 0$ refers to spherical nuclei, a spherically symmetric charge distribution. These are called magic nuclei. For them $a = b = c$, i.e. stretching is same in all directions $x, y, z$. Figure 1.26(a) shows a spherical nucleus described by equation

  $$\frac{x^2 + y^2 + z^2}{a^2} = 1 \tag{1.181}$$

  Example:

  $$_2\text{He}_2^4, \, _{20}\text{Ca}_{20}^{40}$$

  ✓ $Q > 0$ refers to a charge distribution that is more stretched or elongated in one direction than the other two directions. In other words for them

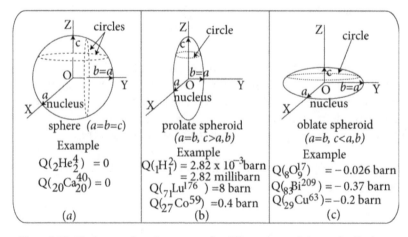

**Figure 1.26.** Nuclear quadrupole moments for different types of charge distributions.

stretching is the same in two directions $x$, $y$ ($a = b$) but stretching along $z$ direction is larger ($c > a, b$). Such distribution is pointy, cigar shaped and called prolate spheroid. Figure 1.26(b) shows a nucleus with the shape of a prolate spheroid described by the equation

$$\frac{x^2 + y^2}{a^2} + \frac{z^2}{c^2} = 1 \text{ with } c > a \qquad (1.182)$$

Example:

$$_1H_1^2, \ _{71}Lu^{176}, \ _{27}Co^{59}.$$

✓ $Q < 0$ refers to a charge distribution that is less stretched or elongated in one direction than the other two directions. In other words, for them stretching is the same in two directions $x$, $y$ ($a = b$) but stretching along $z$ direction is smaller ($c < a, b$). Such distribution is flattened, pancake shaped and called oblate spheroid. Figure 1.26(c) shows a nucleus with the shape of oblate spheroid described by equation

$$\frac{x^2 + y^2}{a^2} + \frac{z^2}{c^2} = 1 \text{ with } c < a \qquad (1.183)$$

Example:

$$_8O_9^{17}, \ _{83}Bi^{209}, \ _{29}Cu^{63}$$

## 1.36 Exercises

**Exercise 1.1** *Which of the following is the correct value of $\frac{|e|^2}{4\pi\varepsilon_0}$?*

  *(a)* 197 *MeV. fm (b)* 144 *MeV. fm (c)* 1.44 *MeV. fm (d)* 1.44 *GeV. fm*

Hint. $\frac{|e|^2}{4\pi\varepsilon_0} = \frac{\left(1.6 \times 10^{-19} C\right)^2}{4\pi\left(\frac{10^{-9}}{36\pi}\frac{F}{m}\right)} = 9 \times 10^9 \times (1.6 \times 10^{-19})^2 \frac{C^2 m}{C/V}$

Since $CVm = Jm$ we have upon using equation (1.6)

$\frac{|e|^2}{4\pi\varepsilon_0} = \frac{9 \times 10^9 \times \left(1.6 \times 10^{-19}\right)^2}{1.6 \times 10^{-19}} \ eV. \ m = 1.44 \times 10^{-9} \ eV. \ m = 1.44 \ (10^6 \ eV) \ (10^{-15} \ m)$

$= 1.44 \ MeV. \ fm.$

**Exercise 1.2** *Which of the following is the correct value of hc, ℏc?*
  *(a)* 1242 *eV. nm*, 0.197 *GeV. fm*   *(b)* 1242 *eV.* Å, 197 *MeV. fm*
  *(c)* 1242 *MeV. fm*, 197 *MeV. fm*   *(d)* 1242 *eV. nm*, 197 *MeV.* Å
  Hint. ✓ $hc = (6.626 \times 10^{-34} \ J. \ s)(3 \times 10^8 \ m \ s^{-1}) = 1.9878 \times 10^{-25} \ J. \ m$

$$= \frac{1.9878 \times 10^{-25} \, eV}{1.6 \times 10^{-19}} m = 1242 \times 10^{-9} \, eV. \, m = 1242 \, eV. \, 10^{-9} m$$

$$hc = 1242 \, eV. \, nm$$

$$hc = 1242 \times 10^{6} \, eV. \, 10^{-15} \, m$$

$$hc = 1242 \, MeV. \, fm$$

✓ $\hbar c = \left(\frac{1}{2\pi}6.626 \times 10^{-34} \, J. \, s\right)(3 \times 10^{8} \, ms^{-1}) = 3.164 \times 10^{-26} \, J. \, m$

$$= \frac{3.164 \times 10^{-26} \, eV}{1.6 \times 10^{-19}} m = 197 \times 10^{-9} \, eV. \, m = 197 \times 10^{6} \, eV. \, 10^{-15} \, m$$

$$\hbar c = 197 \, MeV. \, fm$$

$$\hbar c = 0.197 \times 10^{9} \, eV. \, 10^{-15} \, m$$

$$\hbar c = 0.197 \, GeV. \, fm$$

**Exercise 1.3** *Argue using Heisenberg's uncertainty principle that electrons cannot stay inside the nucleus.*

[Ans.] The size of the nucleus is $\sim 10^{-14}$ m. If an electron exists within the nucleus it should be able to stay anywhere within the nuclear extension $10^{-14}$ m. This means the position uncertainty is

$\Delta x \sim 10^{-14}$ m.

The corresponding uncertainty in linear momentum $\Delta p$ according to Heisenberg uncertainty principle $\Delta x \Delta p \sim \hbar$ will be

$\Delta p = \frac{\hbar}{\Delta x} = \frac{h}{2\pi \Delta x} = \frac{6.626 \times 10^{-34} \, J.s}{2\pi(10^{-14} \, m)}$

$\Delta p = 1.055 \times 10^{-20} \, kg. \, m. \, s^{-1}$.

If electron linear momentum is $p_e$ then, to stay within the nucleus, the electron should have linear momentum given by

$$p_e + \text{momentum uncertainty} = p_e + \Delta p \xrightarrow{\text{taking } p_e = 0} \Delta p.$$

Again, total energy $E$, kinetic energy $T$, linear momentum $p_e$ and rest mass $m_e$ are related through the relativistic formula

$$E = \sqrt{p_e^2 c^2 + m_e^2 c^4} = T + m_e c^2.$$

$$T = \sqrt{p_e^2 c^2 + m_e^2 c^4} - m_e c^2.$$

With $p_e \sim \Delta p$ we have

$$T = \sqrt{(\Delta p)^2 c^2 + m_e^2 c^4} - m_e c^2$$

Using $m_e c^2 = 0.511 \ MeV$ and

$$(\Delta p)c = (1.055 \times 10^{-20} \ kg. \ m. \ s^{-1})(3 \times 10^8 \ m \ s^{-1}) = 3.165 \times 10^{-12} \ J$$

$$= \frac{3.165 \times 10^{-12}}{1.6 \times 10^{-19}} \ eV$$

$$= 19.78 \times 10^6 \ eV = 19.78 \ MeV$$

we get

$$T = \sqrt{(\Delta p)^2 c^2 + m_e^2 c^4} - m_e c^2 = \sqrt{(19.78 \ MeV)^2 + (0.511 \ MeV)^2} - 0.511 \ MeV$$

$$= 19.28 \ MeV.$$

So electrons should at least possess a kinetic energy $T = 19.28 \ MeV$ to stay within the nucleus. But it is experimentally observed that electrons emitted from nucleus (during $\beta^-$ decay of nucleus) have a maximum kinetic energy $T_{max} \sim 4 \ MeV$ that is far less than the amount needed (i.e. $19.78 \ MeV$). This proves that electrons emitted from the nucleus, at the time of $\beta^-$ decay, were not staying within the nucleus. It thus follows that electrons do not reside within nucleus.

**Exercise 1.4** *The experimental observation of spin of deuteron is $S_d = 1$. Argue from this result if electrons can stay inside the nucleus.*

[Ans.] Consider deuterium nucleus $_1H_1^2 \equiv _1H^2 \equiv d$. Its observed spin is $S_d = 1$.

If a deuterium nucleus contained one proton ($s_p = \frac{1}{2}$), one neutron ($s_n = \frac{1}{2}$) and one electron ($s_e = \frac{1}{2}$) then according to addition of the angular momentum rule we would expect

$$\vec{S} = \vec{s}_p + \vec{s}_n + \vec{s}_e = \frac{\overline{1}}{2} + \frac{\overline{1}}{2} + \frac{\overline{1}}{2} = \frac{\overline{1}}{2} + \begin{cases} \vec{0} \\ \vec{1} \end{cases} \to \begin{cases} \frac{1}{2} \\ \frac{3}{2}, \frac{1}{2} \end{cases}$$

Clearly, the resultant of three spin-halves is $S = \frac{3}{2}$ or $S = \frac{1}{2}$ which contradicts the experimentally observed result of $S_d = 1$. This means that accommodating an electron in a nucleus (deuterium) is not consistent with experimental observation.

**Exercise 1.5** *Argue from magnetic moment consideration that electrons cannot stay inside the nucleus.*

[Ans.] If an electron is present in a deuterium nucleus then the magnetic moment of the nucleus should have been of the order of the magnetic moment of the electron, which is

$$\mu_e = 9.27 \times 10^{-24} \ JT^{-1} = \mu_B = \text{Bohr magneton [equation (1.159)]}$$

But the observed magnetic moment of deuterium is

$$\mu_d = 0.86\mu_N$$

where $\mu_N = 5.05 \times 10^{-27} \, JT^{-1}$ = nuclear magneton [equation (1.164)]. Hence

$$\mu_d = 0.86 \times (5.05 \times 10^{-27} \, JT^{-1}) = 4.343 \times 10^{-27} JT^{-1}$$

$$= (4.343 \times 10^{-27} \, JT^{-1})\frac{\mu_e}{\mu_e} = \frac{4.343 \times 10^{-27} \, JT^{-1}}{9.27 \times 10^{-24} \, JT^{-1}}\mu_e = 4.685 \times 10^{-4}\mu_e$$

$$\mu_d = \frac{1}{2135}\mu_e$$

This shows that electrons do not fit within the nucleus.


**Exercise 1.6** *Show from the de Broglie relation that a nucleus cannot accommodate a beta particle emitted during beta decay.*

[Ans.] It is experimentally observed that the beta particle ($\beta^-$ or $e^-$) emitted during $\beta^-$ decay from the nucleus ($_ZX^A \rightarrow \ _{Z+1}Y^A + e^- + \bar{\nu}_e$) has a maximum kinetic energy

$$T \sim 4 \, MeV$$

The corresponding de Broglie wavelength is

$$\lambda = \frac{h}{p}$$

From Einstein's special theory of relativity, the relation between total energy $E$, kinetic enrgy $T$, rest mass $m_e$ and momentum $p$ of an electron e$^-$ is given by

$$E = \sqrt{p^2 c^2 + m_e^2 c^4} = T + m_e c^2$$

Squaring, we have

$$E^2 = p^2 c^2 + m_e^2 c^4 = (T + m_e c^2)^2$$

$$p^2 c^2 + m_e^2 c^4 = T^2 + m_e^2 c^4 + 2 T m_e c^2$$

$$p^2 c^2 = T(T + 2m_e c^2)$$

$$p = \frac{1}{c}\sqrt{T(T + 2m_e c^2)}$$

Hence

$$\lambda = \frac{h}{p} = \frac{h}{\frac{1}{c}\sqrt{T(T + 2m_e c^2)}} = \frac{hc}{\sqrt{T(T + 2m_e c^2)}}.$$

Using $hc = 1242 \, eV. \, nm$ (*exercise 1.2*) and $m_e c^2 = 0.511 \, MeV$ [equation (1.13)] we get

$$\lambda = \frac{1242 \, eV. \, nm}{\sqrt{4 \, MeV(4 \, MeV + 2 \times 0.511 \, MeV)}} = 277 \times 10^{-15} \, m$$

$$\lambda = 277 \, fm.$$

This is the shortest wavelength associated with the beta particle emitted from the nucleus.

The nuclear radius of the large uranium nucleus $_{92}U_{146}^{238} \equiv \, _{92}U^{238}$(having mass number $A = 238$) is given by

$$R = R_0 A^{1/3} = (1.3 \, fm)(238)^{1/3} = 8 \, fm. \quad \text{(taking } R_0 = 1.3 \, fm)$$

So the diameter of $_{92}U^{238}$ is

$$D = 2R = 2(8 \, fm) \boxed{16 \, fm}.$$

We note that $D \ll \lambda$ implying that a nucleus cannot accommodate a beta particle ($\beta^-$ or $e^-$).

**Exercise 1.7** *Using the de Broglie relation, find the energy with which a beta particle must be emitted during beta decay to get accommodated within the nucleus.*

Ans. To get accommodated within the nucleus the beta particle ($\beta^-$ or $e^-$) with momentum $p$ should have a de Broglie wavelength $\lambda$ comparable to nuclear size, which is $\sim 10^{-14} \, m$. So

$$\lambda = \frac{h}{p} \sim 10^{-14} \, m$$

$$p = \frac{h}{\lambda} \Rightarrow pc = \frac{hc}{\lambda}.$$

Using $hc = 1242 \, eV. \, nm$ by *exercise 1.2* we have

$$pc = \frac{hc}{\lambda} = \frac{1242 \, eV. \, nm}{10^{-14} \, m} = \frac{1242 \, eV \times 10^{-9} \, m}{10^{-14} \, m} = 1242 \times 10^5 \, eV$$

$$pc = 124.2 \times 10^6 \, eV = 124.2 \, MeV.$$

Hence total energy that the beta should possess is

$$E = \sqrt{p^2 c^2 + m_e^2 c^4} = \sqrt{(pc)^2 + (m_e c^2)^2}$$

$$E = \sqrt{(124.2 \, MeV)^2 + (0.511 \, MeV)^2}$$

$$E = 124.2 \, MeV.$$

However, the beta particles emitted during beta decay are not that energetic but possess small energy $\sim 4\ MeV$. This implies that the beta particle has insufficient energy and cannot reside within the confines of the nucleus.

**Exercise 1.8** *If an electron cannot be accommodated in the nucleus then how do you explain emission of electrons during $\beta^-$ decay?*

Ans. $\beta^-$ is an electron, i.e. $e^- \equiv _{-1}e^0$. Consider the $\beta^-$ decay

$$_ZX^A \rightarrow _{Z+1}Y^A + _{-1}e^0 + \bar{\nu}_e$$

where $X \rightarrow$ parent nucleus, $Y \rightarrow$ daughter nucleus, $\bar{\nu}_e \rightarrow$ electron antineutrino. For instance

$$_{16}S^{35} \rightarrow _{17}Cl^{35} + _{-1}e^0 + \bar{\nu}_e$$

Here the parent nucleus is $_{16}S^{35} \equiv 16p19n = 16p18n + n$. At the time of $\beta^-$ decay weak interaction occurs that changes neutron to proton in this case, i.e. $n$ is converted to $p$ as

$$n \rightarrow p + e^- + \bar{\nu}_e$$

And an electron $e^-$ is produced. The daughter nucleus becomes $_{17}Cl^{35} \equiv 16p18n + p = 17p18n$. Since the electron $e^-$ cannot remain within the nucleus, it is emitted during $\beta^-$ decay.

**Exercise 1.9** *Show that the existence of a free electron having energy 1 MeV inside a nucleus is not consistent with Heisenberg uncertainty principle.*

Ans. Consider existence of a 1 $MeV$ electron inside the nucleus. Then

$$E = \sqrt{p^2 c^2 + m_e^2 c^4} = \sqrt{(pc)^2 + (m_e c^2)^2}$$

Neglecting $m_e c^2$ we can write

$$E = pc$$

$$p = \frac{E}{c} = \frac{1\ MeV}{c} = \frac{10^6\ eV}{c}$$

$$p = \frac{10^6 \times (1.6 \times 10^{-19}\ J)}{3 \times 10^8\ m\ s^{-1}} = 5.33 \times 10^{-22}\ J\ s\ m^{-1}$$

We take $p$ to be the uncertainty in momentum, i.e.

$$p \approx \Delta p \sim 5.33 \times 10^{-22}\ J\ s\ m^{-1}$$

Also, for the uranium nucleus $_{92}U_{146}^{238} \equiv _{92}U^{238}$ the position uncertainty is

$$\Delta x \sim D_{\text{uranium}} = 2R_{\text{uranium}} = 2R_0A^{1/3} = 2(1.3\,fm)(238)^{1/3} = 16\,fm.$$

$$\Delta x = 16 \times 10^{-15}\,m$$

Clearly then the uncertainty product is given by

$$\Delta x.\ \Delta p = (16 \times 10^{-15}\,m)(5.33 \times 10^{-22}\,J\,s\,m^{-1})$$

$$\Delta x.\ \Delta p = 8.528 \times 10^{-36}\,J.\,s.$$

Clearly

$$\Delta x.\ \Delta p = 8.528 \times 10^{-36}\,J\,s\frac{\hbar/2}{\hbar/2} = \frac{8.528 \times 10^{-36}\,J.\,s}{\frac{1}{2}.\,\frac{1}{2\pi}(6.626 \times 10^{-34}\,J.\,s)}\frac{\hbar}{2}.$$

$$\Delta x.\ \Delta p \approx 0.162\frac{\hbar}{2} < \frac{\hbar}{2}.$$

Clearly the Heisenberg uncertainty principle $\Delta x.\ \Delta p > \frac{\hbar}{2}$ is violated. So the electron cannot stay inside the nucleus.

**Exercise 1.10** *Electron is confined within an atom because of which of the following:*
 *(a) Coulomb force; (b) nuclear force; (c) atomic force; (d) electronic force?*

**Exercise 1.11** *Which of the following principles suggest that electrons cannot stay inside nucleus:*
*(a) Pauli exclusion principle; (b) Born intrerpretation;*
*(c) Heisenberg uncertainty principle; (d) de Broglie hypothesis;*
*(e) repulsion of nuclear force; (f) pull of atomic cloud?*

**Exercise 1.12** *Should we treat nucleon movement inside the nucleus relativistically?*
 Ans. Experiments on scattering of nuclei suggest that nucleons in a nucleus move with kinetic energy $\sim 10\,MeV$ which is very small compared to nucleon rest energies $m_pc^2 = 938\,MeV$, $m_nc^2 = 939\,MeV$.
 So we can use non-relativistic quantum mechanics to analyse nuclei.

**Exercise 1.13** *Justify the existence of protons inside a nucleus.*
 Ans. Nuclear dimension is $10^{-14}\,m$. As a proton should be able to stay anywhere within the nucleus, the maximum uncertainty in position is

$$\Delta x \sim 10^{-14}\,m$$

The corresponding minimum uncertainty in momentum, according to the Heisenberg uncertainty principle is

$$\Delta x. \, \Delta p \sim \hbar$$

$$\Delta p \sim \frac{\hbar}{10^{-14} \, m} = \frac{\frac{1}{2\pi} 6.626 \times 10^{-34} \, J.s}{10^{-14} \, m}$$

$$\Delta p = 1.05 \times 10^{-20} \, kg. \, ms^{-1}$$

If linear momentum of proton is $p$ then to stay within nucleus proton should have linear momentum given by

$$p + \text{uncertainty} = p + \Delta p \xrightarrow{\text{taking } p=0} \Delta p$$

This is the minimum linear momentum of proton. The corresponding kinetic energy is

$$T_{min} = \frac{(\Delta p)^2}{2m_p}$$

$$= \frac{\left(1.05 \times 10^{-20} \, kg.ms^{-1}\right)^2}{2\left(1.67 \times 10^{-27} \, kg\right)} = 3.3 \times 10^{-14} \, J$$

$$= \frac{3.3 \times 10^{-14} \, eV}{1.6 \times 10^{-19}} = 0.2 \times 10^6 \, eV = 0.2 \, MeV.$$

$$T_{min} = 0.2 \, MeV$$

This is the minimum energy needed for a proton to exist in the nucleus. Experimentally, it is found that energy of a proton is much larger than 0.2 $MeV$ (-the rest energy of a proton is 938 $MeV$). Hence a proton can exist in the nucleus (and this we have shown from the Heisenberg uncertainty principle).

**Exercise 1.14** *Can we consider a nucleon (neutron or proton) as a classical system? Is quantum mechanics necessary for studying a nucleon?*

Ans. Consider a proton $p$ in a nucleus having energy $\sim 1 \, MeV$. The de Broglie wavelength of a 1 $MeV$ proton is

$$\lambda_p = \frac{h}{p}$$

Using $p = \frac{1}{c} \sqrt{T(T + 2m_p c^2)}$ we can write

$$\lambda_p = \frac{hc}{pc} = \frac{hc}{\sqrt{T\left(T + 2m_pc^2\right)}}$$

$$= \frac{1242\ eV.nm}{\sqrt{1\ MeV(1\ MeV + 2 \times 938\ MeV)}} = 29 \times 10^{-15}\ m.$$

$$\lambda_p = 29\ fm$$

Again the size of a proton is given by its radius $r_p \sim 0.84\ fm$. Clearly

$$\lambda_p > r_p$$

Evidently we cannot ignore wave-like behavior of nucleons. We cannot treat them as classical particles. Hence to analyse their behavior we have to use quantum mechanics.

**Exercise 1.15** *The nuclear radius is taken to be $R = R_0A^{1/3}$ with $R_0 = 1.3\ fm$. Use this formula to calculate the proton radius.*

Ans. For hydrogen nucleus $_1H^1$ that contains a single proton $Z = 1$, $N = 0$, $A = 1$ and with $R_0 = 1.2\ fm$ we get proton radius, as predicted by the given formula to be

$$R = R_0A^{1/3} = (1.2\ fm)(1)^{1/3} = 1.2\ fm.$$

However, experimental data suggests proton radius to be around $0.84\ fm$.

**Exercise 1.16** *Express (a) neutron mass, and (b) electron mass in amu and MeV .*

Ans. *(a)* Neutron mass is by equation (1.7)

$$m_n = 1.675 \times 10^{-27}\ kg$$

Neutron rest mass energy is given by equation (1.9)

$$m_nc^2 = 939\ MeV$$

Using equation (1.16) we have

$$1\ MeV = \frac{1}{931}\ amu$$

$$939\ MeV = \frac{939}{931}\ amu = 1.0086\ amu \xrightarrow{\text{we take}} 1.008\ 665\ amu.\ \text{Hence}$$

$$m_n = 1.008\ 665\ amu$$

Thus with equations (1.7) and (1.9) we can write
$$m_n = 1.675 \times 10^{-27}\ kg = 1.008\ 665\ amu = 939\ MeVc^{-2}$$
*(b)* Electron mass by equation (1.10) is

$$m_e = 9.1 \times 10^{-31}\ kg$$

Electron rest mass energy is by equation (1.13)

$$m_e c^2 = 0.511 \ MeV.$$

Using equation (1.16) we have

$$1 \ MeV = \frac{1}{931} \ amu$$

$$0.511 \ MeV = \frac{0.511}{931} \ amu = 0.000 \ 54 \ amu \xrightarrow{\text{we take}} 0.000 \ 547 \ 21 \ amu. \ \text{Hence}$$

$$m_e = 0.000 \ 547 \ 21 \ amu$$

Thus with equations (1.10) and (1.13) we can write

$$m_e = 9.1 \times 10^{-31} \ kg = 0.000 \ 547 \ 21 \ amu = 0.511 \ MeVc^{-2}$$

**Exercise 1.17** *(a) Show that pp and nn interactions are identical.*
*(b) What about the np interaction?*
Ans. (a) Let us use the representation $|II_z>$.
The isospin quantum numbers associated with nucleons are
$I = \frac{1}{2}, I_z = \frac{1}{2}$ for $p$ written as $| p > = | \frac{1}{2} \frac{1}{2} > = | \uparrow >$ and

$$I = \frac{1}{2}, I_z = -\frac{1}{2} \ \text{for } n \text{ written as } | n > = | \frac{1}{2} - \frac{1}{2} > = | \downarrow > .$$

That both $p$ and $n$ have the same $I = \frac{1}{2}$ is indicative of identical strong interaction that they exhibit.

Let us now find the $I$ values of $pp$, $nn$ and $np$ interactions to investgate the nature of the strong interaction they exhibit.

For the $pp$ system
$I_z = I_z \mid_p + I_z \mid_p = \frac{1}{2} + \frac{1}{2} = 1$ (here both $p$ are up as $I_z = \frac{1}{2}$ for both $p$) and we can denote $| pp > = | \uparrow \uparrow >$.

For the $nn$ system
$I_z = I_z \mid_n + I_z \mid_n = -\frac{1}{2} - \frac{1}{2} = -1$ (here both $n$ are down as $I_z = -\frac{1}{2}$ for both $n$ and we can denote $| nn > = | \downarrow \downarrow >$.

Clearly $I_z = \pm 1$ projections can occur from $I = 1$. So $pp$ and $nn$ both have $I = 1$. We thus identify
the $pp$ system to have $I = 1, I_z = 1$ denoted by $| pp > = |11>$ and
the $nn$ system to have $I = 1, I_z = -1$ denoted by $| nn > = |1 - 1>$.
Since both $pp$ and $nn$ systems have $I = 1$ they are equivalent or identical so far as their strong interaction is concerned.

(b) The np system corresponds to

$$I_z = I_z \big|_n + I_z \big|_p = -\frac{1}{2} + \frac{1}{2} = 0 \ (np \to | \downarrow \uparrow > , |pn > \to | \uparrow \downarrow >).$$

The projection of $I = 0$ is $I_z = 0$ and the projection of $I = 1$ is $I_z \to 1, 0, -1$. Clearly the projection $I_z = 0$ can come from both $I = 0$ and $I = 1$.

So the np system has both $I = 0$ part mixed with $I = 1$ part. In other words, the np system is an admixture (linear combination) of $I = 0$ and $I = 1$ parts, and we can write this as

$$|np > = \frac{1}{\sqrt{2}} (|I = 1, I_z = 1 > -| I = 1, I_z = 0>)$$

$$|np > = \frac{1}{\sqrt{2}} (|11 > -| 10>)$$

($\frac{1}{\sqrt{2}}$ is the normalization constant and the negative sign makes the wave function anti-symmetric as the np system is a system of fermions n and p).

Conclusion:

Since $| pp > = |11>, | nn > = |1 - 1>$ it follows that pp and nn systems have $I = 1$ but since $| np > -\frac{1}{\sqrt{2}} (|11 > -| 10>)$, it is clear that the np system has $I - 0$ part in addition to $I = 1$ part. So np interaction is slightly different compared to pp and nn interactions.

The difference can also be explained from the pattern or mode of the two interactions as described in Yukawa's meson theory of nuclear force (section 8.3).

The nn interaction and pp interaction are mediated by $\pi^0$ as follows

$$n \to n + \pi^0$$

$$n + \pi^0 \to n$$

$$p \to p + \pi^0$$

$$p + \pi^0 \to p$$

but np interaction is mediated by $\pi^\pm$ as

$$n \to p + \pi^-$$

$$n + \pi^+ \to p$$

$$p \to n + \pi^+$$

$$p + \pi^- \to n$$

Clearly there is a difference in the nature of interactions mediated by pions.

**Exercise 1.18** *Why is thin foil chosen in the Rutherford scattering experiment?*

Ans. In Rutherford's scattering experiment a thin metal foil (like gold foil) of thickness $\sim 10^{-7}$ m was chosen as scatterer because of the following reasons:

*(a)* The Rutherford scattering experiment involved study of atomic structure and hence the aim was to present a layer of atoms to the projectile ($\alpha$ particles). The thinner the layer, the better is the agreement.

*(b)* The thin layer would prevent $\alpha$ from being absorbed by the target gold foil. The emergent scattered $\alpha$ will carry information about the atomic structure.

*(c)* The thin layer would decrease ionization loss of energy of $\alpha$ and hence there would practically be no velocity decrease during passage through foil and so $\alpha$ will be able to emerge out of target.

*(d)* The thin layer would avoid scattering at multiple centres.

**Exercise 1.19** *Why was gold foil a better choice in the Rutherford $\alpha$ scattering experiment?*

Ans. Gold is malleable and ductile. So one can compress and stretch gold foil to a large extent and make it extremely thin ($\sim 2 \times 10^{-7}$ m) without causing cracks in it—atoms just spread out into a thin layer. This means a few atomic layers are presented for interaction with the projectile (figure 1.27). In other words, with gold foil an extremely thin target layer can be constructed and presented to the projectile $\alpha$ in the $\alpha$ scattering experiment.

**Exercise 1.20**

*(a)* Obtain the differential equation of the orbit followed by $\alpha$ in the Rutherford scattering from gold nucleus.

*(b)* Show that the path of $\alpha$ is a conic section.

*(c)* Find eccentricity $\varepsilon$ of the orbit and find the exact nature of orbit traced by $\alpha$.

*(d)* Establish the relation between impact parameter $R$ and angle of scattering $\theta$.

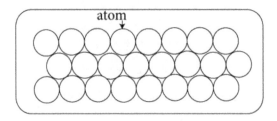

**Figure 1.27.** Thin layer of gold atoms presented to $\alpha$ projectile.

(e) Establish the relation between impact parameter $R$ and distance of closest approach $b$.

(f) Inter-relate the distance of closest approach $b$, impact parameter $R$ and angle of scattering $\theta$.

(g) Prove the Rutherford scattering formula (i.e. the differential scattering cross-section).

(h) Obtain the number of $\alpha$ particles falling upon the unit area of the detector used by Rutherford.

Ans. (a) Refer to figure 1.6 that schematically depicts Rutherford scattering of $\alpha$ by gold foil. The equation of motion (force = mass × acceleration) of $\alpha$ is

$$\vec{F} = F(r)\hat{r} = m_\alpha \vec{a} = m_\alpha[(\ddot{r} - r\dot{\phi}^2)\hat{r} + (r\ddot{\phi} + 2\dot{r}\dot{\phi})\hat{\phi}] \tag{1.184}$$

where the radial and transverse components of acceleration have been written explicitly.

$$[m_\alpha(\ddot{r} - r\dot{\phi}^2) - F(r)]\hat{r} + m_\alpha(r\ddot{\phi} + 2\dot{r}\dot{\phi})\hat{\phi} = 0$$

Setting the radial part equal to zero we have

$$m_\alpha(\ddot{r} - r\dot{\phi}^2) - F(r) = 0 \tag{1.185}$$

Setting the transverse part equal to zero we have

$$m_\alpha(r\ddot{\phi} + 2\dot{r}\dot{\phi}) = 0 \tag{1.186}$$

Multiplying by $r$ gives

$$m_\alpha(r^2\ddot{\phi} + 2r\dot{r}\dot{\phi}) = 0$$

$$m_\alpha r^2 \frac{d\dot{\phi}}{dt} + m_\alpha\left(\frac{d}{dt}r^2\right)\dot{\phi} = 0$$

$$\frac{d}{dt}(m_\alpha r^2 \dot{\phi}) = 0.$$

This means that $m_\alpha r^2 \dot{\phi}$ = conserved, and the quantity $m_\alpha r^2 \dot{\phi}$ can be identified as the angular momentum $L$. So the transverse equation leads to conservation of angular momentum

$$L = m_\alpha r^2 \dot{\phi} = \text{conserved} \tag{1.187}$$

The radial part equation (1.185) is

$$m_\alpha(\ddot{r} - r\dot{\phi}^2) = F(r) \tag{1.188}$$

Introduce a new variable

$$u = \frac{1}{r} \tag{1.189}$$

$$\frac{du}{d\phi} = -\frac{1}{r^2}\frac{dr}{d\phi} = -\frac{1}{r^2}\frac{dr}{dt}\frac{dt}{d\phi} = -\frac{1}{r^2}\frac{dr}{dt}\frac{1}{\frac{d\phi}{dt}} = -\frac{1}{r^2}\frac{\dot{r}}{\dot{\phi}}.$$

Using equation 1.187 viz. $L = m_\alpha r^2 \dot{\phi} \Rightarrow r^2 \dot{\phi} = \frac{L}{m_\alpha}$ we have

$$\frac{du}{d\phi} = -\frac{\dot{r}}{\frac{L}{m_\alpha}} = -\frac{m_\alpha \dot{r}}{L}. \tag{1.190}$$

Taking the derivative again we have

$$\frac{d^2 u}{d\phi^2} = \frac{d}{d\phi} \frac{du}{d\phi}$$

$$= \frac{d}{d\phi}\left(-\frac{m_\alpha \dot{r}}{L}\right) = \frac{d}{dt}\left(-\frac{m_\alpha \dot{r}}{L}\right)\frac{dt}{d\phi} = -\frac{m_\alpha}{L}\left(\frac{d}{dt}\dot{r}\right)\frac{1}{\frac{d\phi}{dt}} = -\frac{m_\alpha}{L}\ddot{r}\frac{1}{\dot{\phi}} = -\frac{m_\alpha \ddot{r}}{L\dot{\phi}}$$

Using equation 1.187 viz. $=m_\alpha r^2 \dot{\phi} \Rightarrow \dot{\phi} = \frac{L}{m_\alpha r^2}$, we have

$$\frac{d^2 u}{d\phi^2} = -\frac{m_\alpha \ddot{r}}{L\frac{L}{m_\alpha r^2}} = -\frac{m_\alpha^2 r^2 \ddot{r}}{L^2} \Rightarrow \ddot{r} = -\frac{L^2}{m_\alpha^2 r^2}\frac{d^2 u}{d\phi^2}. \tag{1.191}$$

Using equation (1.189) i.e. $r = \frac{1}{u}$ we have from equation (1.187) viz. $L = m_\alpha r^2 \dot{\phi}$

$$\dot{\phi} = \frac{L}{m_\alpha r^2} = \frac{Lu^2}{m_\alpha} \tag{1.192}$$

Also, using equation (1.189) $r = \frac{1}{u}$ in equation (1.191) we get

$$\ddot{r} = -\frac{L^2 u^2}{m_\alpha^2}\frac{d^2 u}{d\phi^2} \tag{1.193}$$

We are now in a position to recast the radial equation (1.188) viz. $m_\alpha(\ddot{r} - r\dot{\phi}^2) = F(r)$ as follows using equations (1.192) and (1.193).

$$m_\alpha\left(-\frac{L^2 u^2}{m_\alpha^2}\frac{d^2 u}{d\phi^2}\right) - m_\alpha\frac{1}{u}\left(\frac{Lu^2}{m_\alpha}\right)^2 = F\left(\frac{1}{u}\right)$$

$$-\frac{L^2 u^2}{m_\alpha}\frac{d^2 u}{d\phi^2} - \frac{L^2 u^3}{m_\alpha} = F\left(\frac{1}{u}\right)$$

This gives

$$\frac{d^2 u}{d\phi^2} + u = -\frac{m_\alpha}{L^2 u^2}F\left(\frac{1}{u}\right). \tag{1.194}$$

This is the differential equation of orbit followed by $\alpha$ under central force

$$F\left(\frac{1}{u}\right) = F(r) = \frac{2Z\,|\,e\,|^2}{4\pi\varepsilon_0 r^2} = \frac{|\,k\,|}{r^2} = |\,k\,|\,u^2 \text{ (equation 1.33)}$$

which is the Coulomb repulsive inverse square force.

*(b)* The force equation (1.194) is

$$\frac{d^2u}{d\phi^2} + u = -\frac{m_\alpha}{L^2 u^2} F\left(\frac{1}{u}\right)$$

With $F\left(\frac{1}{u}\right) = |k| u^2$ we have

$$\frac{d^2u}{d\phi^2} + u = -\frac{m_\alpha}{L^2 u^2} |k| u^2 \Rightarrow \frac{d^2u}{d\phi^2} + u = -\frac{m_\alpha |k|}{L^2}$$

$$\frac{d^2u}{d\phi^2} + \left(u + \frac{m_\alpha |k|}{L^2}\right) = 0 \tag{1.195}$$

Since $\frac{d^2}{d\phi^2}\left(\frac{m_\alpha |k|}{L^2}\right) = 0$ we can add $\frac{m_\alpha |k|}{L^2}$ in the first term and write the differential equation as

$$\frac{d^2}{d\phi^2}\left(u + \frac{m_\alpha |k|}{L^2}\right) + \left(u + \frac{m_\alpha |k|}{L^2}\right) = 0 \tag{1.196}$$

Defining

$$\xi - u + \frac{m_\alpha |k|}{L^2} \tag{1.197}$$

$$\frac{d^2\xi}{d\phi^2} + \xi = 0 \tag{1.198}$$

The solution of this equation (1.198) is

$$\xi = a \cos(\phi - \phi_0) \tag{1.199}$$

where $a$, $\phi_0$ are constants. This means using equation (1.197)

$$u + \frac{m_\alpha |k|}{L^2} = a \cos(\phi - \phi_0)$$

Using equation (1.189) we get

$$\frac{1}{r} = -\frac{m_\alpha |k|}{L^2} + a \cos(\phi - \phi_0).$$

Rearranging, we have

$$-\frac{L^2}{m_\alpha |k|}\frac{1}{r} = 1 - \frac{L^2 a}{m_\alpha |k|} \cos(\phi - \phi_0) \tag{1.200}$$

Putting

$$l = -\frac{L^2}{m_\alpha |k|}, \quad \varepsilon = -\frac{L^2 a}{m_\alpha |k|} \tag{1.201}$$

the solution of equation (1.200) becomes

$$\frac{l}{r} = 1 + \varepsilon \cos(\phi - \phi_0)$$

(1.202)

Again

$$\frac{\varepsilon}{l} = \left(-\frac{L^2 a}{m_\alpha \mid k \mid}\right)\left(-\frac{m_\alpha \mid k \mid}{L^2}\right) = a \text{ (one constant)}$$

(1.203)

By proper choice of axis we can make the other constant $\phi_0 = 0$ (or we can redefine $\phi - \phi_0 \rightarrow \phi$). This gives the solution of the radial equation (1.188) viz. $m_\alpha(\ddot{r} - r\dot{\phi}^2) = F(r)$ to be, from equation (1.202)

$$\frac{l}{r} = 1 + \varepsilon \cos \phi$$

(1.204)

which is the equation of a conic section in polar coordinates followed by $\alpha$ particle. We identify from equation (1.201) that

$$l = -\frac{L^2}{m_\alpha \mid k \mid} \text{is semi latus rectum}$$

(1.205)

$$\varepsilon = -\frac{L^2 a}{m_\alpha \mid k \mid} \text{ is eccentricity of orbit}$$

(1.206)

*(c)* To find the exact nature of conic section we have to determine the eccentricity of the orbit traced by the $\alpha$ particle. In equation (1.49) we showed that $v_a^2 = v^2 + \frac{2 \mid k \mid}{m_\alpha r}$ using energy conservation by equating energy at point $M$ and energy at point $P$ of the trajectory of the $\alpha$ particle (figure 1.6). The velocity of $\alpha$ at any point $P$, in terms of its radial and transverse components is

$$\vec{v} = \dot{r}\hat{r} + r\dot{\phi}\hat{\phi}$$

(1.207)

Squaring we have

$$v^2 = \dot{r}^2 + (r\dot{\phi})^2 = \left(\frac{dr}{dt}\right)^2 + r^2\dot{\phi}^2 = \left(\frac{dr}{d\phi}\frac{d\phi}{dt}\right)^2 + r^2\dot{\phi}^2$$

$$= \dot{\phi}^2[\left(\frac{dr}{d\phi}\right)^2 + r^2]$$

(1.208)

From the path equation (1.204) viz. $\frac{l}{r} = 1 + \varepsilon \cos \phi$ we can find $\frac{dr}{d\phi}$ through differentiation as follows. Consider the derivative

$$\frac{d}{d\phi}\frac{l}{r} = \frac{d}{d\phi}(1 + \varepsilon \cos \phi)$$

$$-\frac{l}{r^2}\frac{dr}{d\phi} = -\varepsilon \sin \phi$$

$$\frac{dr}{d\phi} = \frac{r^2}{l}\varepsilon \sin \phi.$$

Also, using equation 1.187 viz. $L = m_a r^2 \dot{\phi}^2$ i.e. $\dot{\phi}^2 = \frac{L}{m_a r^2}$ we have from equation (1.208)

$$v^2 = \left(\frac{L}{m_a r^2}\right)^2 \left[\left(\frac{r^2}{l}\varepsilon \sin \phi\right)^2 + r^2\right]$$

$$= \left(\frac{r^4 \varepsilon^2 \sin^2 \phi}{l^2} + r^2\right)\frac{L^2}{m_a^2 r^4} = \frac{L^2 \varepsilon^2}{m_a^2 l^2}\sin^2 \phi + \frac{L^2}{m_a^2 r^2}$$

Using from equation (1.204) $\frac{1}{r} = \frac{1 + \varepsilon \cos \phi}{l}$ we have

$$v^2 = \frac{L^2 \varepsilon^2}{m_a^2 l^2}\sin^2 \phi + \frac{L^2}{m_a^2}\left(\frac{1 + \varepsilon \cos \phi}{l}\right)^2$$

$$= \frac{L^2}{m_a^2 l^2}[\varepsilon^2 \sin^2 \phi + (1 + \varepsilon \cos \phi)^2]$$

$$= \frac{L^2}{m_a^2 l^2}[\varepsilon^2 \sin^2 \phi + 1 + \varepsilon^2 \cos^2 \phi + 2\varepsilon \cos \phi]$$

$$v^2 = \frac{L^2}{m_a^2 l^2}[\varepsilon^2 + 1 + 2\varepsilon \cos \phi] \qquad (1.209)$$

Putting this expression of $v^2$ in equation (1.49) viz. $v_\alpha^2 = v^2 + \frac{2|k|}{m_a r}$ and using equation (1.204) we get

$$v_\alpha^2 = \frac{L^2}{m_a^2 l^2}[\varepsilon^2 + 1 + 2\varepsilon \cos \phi] + \frac{2|k|}{m_a}\left(\frac{1 + \varepsilon \cos \phi}{l}\right)$$

Using equation (1.201) viz. $l = -\frac{L^2}{m_a |k|}$ we have

$$v_\alpha^2 = \frac{L^2}{m_a^2}\left(-\frac{m_a |k|}{L^2}\right)^2[\varepsilon^2 + 1 + 2\varepsilon \cos \phi] + \frac{2|k|}{m_a}(1 + \varepsilon \cos \phi)\left(-\frac{m_a |k|}{L^2}\right)$$

$$= \frac{|k|^2}{L^2}(\varepsilon^2 + 1 + 2\varepsilon \cos \phi) - \frac{2|k|^2}{L^2}(1 + \varepsilon \cos \phi)$$

$$= \frac{|k|^2}{L^2}[(\varepsilon^2 + 1 + 2\varepsilon \cos \phi) - 2(1 + \varepsilon \cos \phi)].$$

$$v_\alpha^2 = \frac{|k|^2}{L^2}(\varepsilon^2 - 1) \qquad (1.210)$$

$$\varepsilon^2 - 1 = \frac{L^2 v_\alpha^2}{|k|^2} \Rightarrow \varepsilon^2 = 1 + \frac{L^2 v_\alpha^2}{|k|^2}$$

$$\varepsilon = \sqrt{1 + \frac{L^2 v_\alpha^2}{|k|^2}} \tag{1.211}$$

Using $E_\alpha = \frac{1}{2} m_\alpha v_\alpha^2$ i.e. $v_\alpha^2 = \frac{2E_\alpha}{m_\alpha}$ we have

$$\varepsilon = \sqrt{1 + \frac{L^2}{|k|^2} \frac{2E_\alpha}{m_\alpha}}$$

$$\varepsilon = \sqrt{1 + \frac{2E_\alpha L^2}{m_\alpha |k|^2}} > 1 \tag{1.212}$$

Since $\frac{2E_\alpha L^2}{m_\alpha |k|^2} = +ve,$ $\varepsilon > 1$ means the conic section (equation (1.204)) $\frac{l}{r} = 1 + \varepsilon \cos \phi$ represents a hyperbolic orbit. Again at $r \to \infty$, $\phi \to \psi$ (figure 1.6) and so we have from equation (1.204)

$$\frac{l}{\infty} = 1 + \varepsilon \cos \psi \Rightarrow 0 = 1 + \varepsilon \cos \psi$$

$$\cos \psi = -\frac{1}{\varepsilon} \tag{1.213}$$

The negative sign indicates that $\alpha$ follows the negative branch of the hyperbola (figure 1.6).

(d) Relation between $R$ (impact parameter) and $\theta$( angle of scattering)

Let us write down the equation of the hyperbola in Cartesian coordinates viz.

$$\frac{x^2}{A^2} - \frac{y^2}{B^2} = 1 \tag{1.214}$$

and the equation of the asymptotes is

$$y = \pm \frac{B}{A} x \tag{1.215}$$

The slope of the asymptotes is (figure 1.28)

$$\pm \tan \psi = \pm \frac{B}{A} \tag{1.216}$$

Now $2\psi + \theta = \pi \Rightarrow \psi = \frac{\pi}{2} - \frac{\theta}{2}.$

$$\tan \psi = \tan \left( \frac{\pi}{2} - \frac{\theta}{2} \right) = \cot \frac{\theta}{2} \tag{1.217}$$

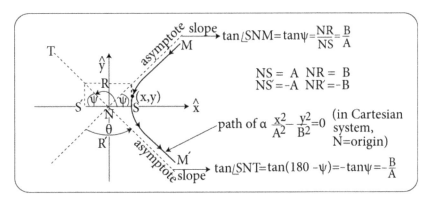

**Figure 1.28.** Slope of the asymptotes of hyperbolic trajectory.

For the hyperbola of equation (1.214) the eccentricity is given by

$$\varepsilon = \frac{\sqrt{A^2 + B^2}}{A} \Rightarrow \varepsilon^2 = \frac{A^2 + B^2}{A^2} = 1 + \frac{B^2}{A^2}$$

$$\varepsilon^2 - 1 = \frac{B^2}{A^2} \qquad (1.218)$$

From equation (1.216) $\tan^2 \psi = \frac{B^2}{A^2}$ and so equation (1.218) gives

$$\varepsilon^2 - 1 = \tan^2 \psi \qquad (1.219)$$

From equation (1.217) $\tan^2 \psi = \cot^2 \frac{\theta}{2}$ and so equation (1.219) gives

$$\varepsilon^2 - 1 = \cot^2 \frac{\theta}{2} \qquad (1.220)$$

Again from equation (1.212) viz. $\varepsilon = \sqrt{1 + \frac{2E_\alpha L^2}{m_\alpha \mid k \mid^2}}$

$$\varepsilon^2 - 1 = \frac{2E_\alpha L^2}{m_\alpha \mid k \mid^2}. \qquad (1.221)$$

Equating equations (1.220) and (1.221) we get

$$\varepsilon^2 - 1 = \cot^2 \frac{\theta}{2} = \frac{2E_\alpha L^2}{m_\alpha \mid k \mid^2} \qquad (1.222)$$

$$\cot \frac{\theta}{2} = \sqrt{\frac{2E_\alpha L^2}{m_\alpha \mid k \mid^2}} \qquad (1.223)$$

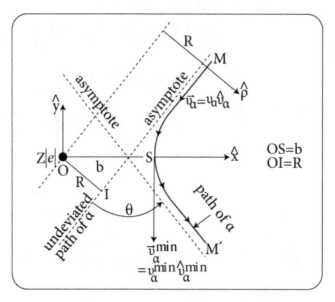

**Figure 1.29.** Evaluation of angular momenta at points M and S of $\alpha$ trajectory.

Under central force motion angular momentum

$$\vec{L} = \vec{r} \times \vec{p} = \vec{r} \times m_\alpha \vec{v}$$

is conserved. At the initial point $M$ the angular momentum is (figure 1.29)

$$\vec{L}\,\big|_M = \vec{r} \times m_\alpha \vec{v}\,\big|_M = R\hat{R} \times m_\alpha v_\alpha \hat{v}_\alpha = \text{constant}$$

This leads to

$$L\,\big|_M = L = R m_\alpha v_\alpha = \text{constant} \tag{1.224}$$

Also, using $|\,k\,| = \frac{2Z\,|\,e\,|^2}{4\pi\varepsilon_0}$, $E_\alpha = \frac{1}{2}m_\alpha v_\alpha^2$ we can write from equations (1.223) and (1.224)

$$\cot\frac{\theta}{2} = \sqrt{\frac{2\left(\frac{1}{2}m_\alpha v_\alpha^2\right)(R m_\alpha v_\alpha)^2}{m_\alpha\left(\frac{2Z\,|\,e\,|^2}{4\pi\varepsilon_0}\right)^2}} = \frac{4\pi\varepsilon_0 v_\alpha^2 R m_\alpha}{2Z\,|\,e\,|^2}$$

$$R = \frac{2Z\,|\,e\,|^2}{4\pi\varepsilon_0 v_\alpha^2 m_\alpha}\cot\frac{\theta}{2} = \frac{Z\,|\,e\,|^2}{4\pi\varepsilon_0\left(\frac{1}{2}m_\alpha v_\alpha^2\right)}\cot\frac{\theta}{2}$$

$$R = \frac{Z\,|\,e\,|^2}{4\pi\varepsilon_0 E_\alpha}\cot\frac{\theta}{2} \tag{1.225}$$

Using $|k| = \frac{2Z|e|^2}{4\pi\epsilon_0}$

$$R = \frac{|k|}{2E_\alpha} \cot \frac{\theta}{2} \tag{1.226}$$

This is the relation between impact parameter $R$ and angle of scattering $\theta$.

(e) Relation between $R$ (impact parameter $OI$) and $b$ (distance of closest approach $OS$)

Angular momentum $\vec{L} = \vec{r} \times \vec{p} = \vec{r} \times m_\alpha \vec{v}$ at the initial point $M$ is

$$\vec{L}\Big|_M = \vec{r} \times m_\alpha \vec{v}\Big|_M = R m_\alpha v_\alpha \text{ [equation (1.224)]}$$

Again at point S the angular momentum is (figure 1.29)

$$\vec{L}\Big|_S = \vec{r} \times m_\alpha \vec{v}\Big|_S = b\hat{x} \times m_\alpha v_\alpha^{\min} \hat{v}_\alpha^{\min}$$

$$L\Big|_S = b m_\alpha v_\alpha^{\min} \tag{1.227}$$

As angular momentum is conserved for central force motion, we can equate the two angular momenta $L\Big|_M = R m_\alpha v_\alpha$ and $L\Big|_S = b m_\alpha v_\alpha^{\min}$ to get

$R m_\alpha v_\alpha = b m_\alpha v_\alpha^{\min}$

$$\frac{v_\alpha^{\min}}{v_\alpha} = \frac{R}{b} \tag{1.228}$$

(f) Relation between distance of closest approach $b = OS$, impact parameter $R = OI$ and the angle of scattering $\theta$ (figure 1.29)

We showed in equation (1.49) that $v_\alpha^2 = v^2 + \frac{2|k|}{m_\alpha r}$. At $r = b$, $v = v_\alpha^{\min}$ we have

$v_\alpha^2 = (v^{\min})^2 + \frac{2|k|}{m_\alpha b}$.

For head-on collision, i.e. for perfect back scattering $\theta = \pi, v_\alpha^{\min} = 0$ we have

$$v_\alpha^2 = \frac{2|k|}{m_\alpha b} \Rightarrow b = \frac{2|k|}{m_\alpha v_\alpha^2} = \frac{|k|}{\frac{1}{2} m_\alpha v_\alpha^2}$$

$$b = \frac{|k|}{E_\alpha} \tag{1.229}$$

Now by equation (1.226) $= \frac{|k|}{2E_\alpha} \cot \frac{\theta}{2}$. This can be rewritten as

$$R = \frac{b}{2} \cot \frac{\theta}{2} \Rightarrow b = 2R \tan \frac{\theta}{2} \tag{1.230}$$

(g) Impact parameter is $R$. The $\alpha$ projected in between impact parameter $R$ and $R + dR$ (figure 1.30) sees a target area or cross-section of amount

$$d\sigma = 2\pi R dR \tag{1.231}$$

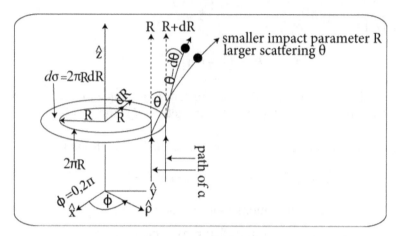

**Figure 1.30.** The smaller the impact parameter, the larger the scattering.

Clearly we got this by considering projectiles over all azimuthal angles $\phi$ ranging from $\phi = 0$ to $2\pi$ in the $x$–$y$ plane (azimuthal symmetry is there).

Now $d\sigma = \frac{d\sigma}{d\Omega}d\Omega$ where $d\Omega = \sin\theta d\theta d\phi$. As there is azimuthal symmetry we integrate over $\phi$ from 0 to $2\pi$. This gives $d\Omega \rightarrow \sin\theta d\theta \int_0^{2\pi} d\phi = 2\pi \sin\theta d\theta$ and so

$$d\sigma = \frac{d\sigma}{d\Omega}d\Omega = \frac{d\sigma}{d\Omega}2\pi \sin\theta d\theta \tag{1.232}$$

Equating with equation (1.231) viz. $d\sigma = 2\pi R dR$ we have

$$2\pi R dR = \frac{d\sigma}{d\Omega}2\pi \sin\theta d\theta \Rightarrow R dR = \frac{d\sigma}{d\Omega}\sin\theta d\theta$$

$$\frac{d\sigma}{d\Omega} = \frac{R}{\sin\theta}\frac{dR}{d\theta}$$

Using $R = \frac{|k|}{2E_\alpha}\cot\frac{\theta}{2}$ (equation (1.226)) we have

$$\frac{d\sigma}{d\Omega} = \frac{1}{\sin\theta}\frac{|k|}{2E_\alpha}\cot\frac{\theta}{2}\frac{d}{d\theta}\left(\frac{|k|}{2E_\alpha}\cot\frac{\theta}{2}\right) = \frac{1}{2\sin\frac{\theta}{2}\cos\frac{\theta}{2}}\left(\frac{|k|}{2E_\alpha}\right)^2\frac{\cos\frac{\theta}{2}}{\sin\frac{\theta}{2}}\left(-\frac{1}{2}\mathrm{cosec}^2\frac{\theta}{2}\right).$$

The negative sign signifies that if $R$ increases then $\theta$ decreases. We suppress the negative sign and write

$$\frac{d\sigma}{d\Omega} = \frac{1}{4}\left(\frac{|k|}{2E_\alpha}\right)^2\mathrm{cosec}^4\frac{\theta}{2}$$

Putting $|k| = \frac{2Z|e|^2}{4\pi\varepsilon_0}$ we get

$$\frac{d\sigma}{d\Omega} = \frac{1}{4}\left(\frac{1}{2E_\alpha}\right)^2\left(\frac{2Z|e|^2}{4\pi\varepsilon_0}\right)^2\mathrm{cosec}^4\frac{\theta}{2}$$

$$\frac{d\sigma}{d\Omega} = \left(\frac{Z \mid e \mid^2}{4\pi\varepsilon_0 2E_\alpha}\right)^2 \operatorname{cosec}^4\frac{\theta}{2}. \tag{1.233}$$

This is the differential scattering cross-section that gives the probability of scattering per unit solid angle.

*(h)* Let $n$ represent number of scattering nuclei per unit volume of foil and let $t$ be the foil thickness. Then the product $nt$ represents $\left(\frac{\text{number}}{\text{volume}}\right)$ (thickness) $= \frac{\text{number}}{\text{area}}$ of scattering nuclei of foil.

Let $\Delta N$ represent the number of $\alpha$ scattered along solid angle $d\Omega$ at angle $\theta$. This $\Delta N$ depends upon:

    (i) number of $\alpha$ incident on foil in a given time $N$;
    (ii) number of scattering nuclei per unit area of foil, i.e. $nt$;
    (iii) differential scattering cross-section $\frac{d\sigma}{d\Omega}$;
    (iv) solid angle $d\Omega$.

Hence

$$\Delta N = (N)(nt)\left(\frac{d\sigma}{d\Omega}\right)(d\Omega). \tag{1.234}$$

Consider a sphere drawn (figure 1.31) with centre at scattering nucleus (foil) and passing through the surface of the detector. So the radius of the sphere is the distance between nucleus (foil) and detector, say $r$. Area of the detector located in the direction $\theta$, $\phi$ is $r^2 d\Omega$. This area is responsible for detecting scattered particles from

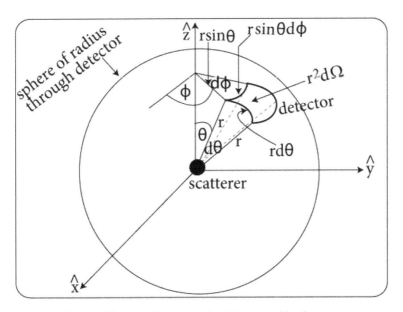

**Figure 1.31.** $\alpha$ particles scattered at $\theta$ intercepted by detector.

source nucleus/foil. Clearly the number of $\alpha$ particles along $\theta$ that are recorded by unit area of detector is given by

$$N_\theta = \frac{\Delta N}{r^2 d\Omega} = \frac{Nnt\frac{d\sigma}{d\Omega} d\Omega}{r^2 d\Omega} = \frac{Nnt}{r^2} \frac{d\sigma}{d\Omega}$$

$$N_\theta = \frac{Nnt}{r^2} \left( \frac{Z \mid e \mid^2}{4\pi\varepsilon_0 2E_\alpha} \right)^2 \text{cosec}^4 \frac{\theta}{2} \text{[using equation (1.233)].} \qquad (1.235)$$

This is the number of $\alpha$ recorded by the detector.

**Exercise 1.21** *Is nuclear boundary precisely defined?*

Ans. Let us note that the nucleus is not a solid sphere with abrupt boundary, which means that the radius is not a precisely well-defined quantity. Also, the radius that we measure depends on the kind of experiment performed—values are different in different experiments.

The nucleus is not a classical hard ball type system. The nucleus is a quantum system and so it does not have a sharp boundary but a diffused gradually fading boundary. The nuclear wave function gradually decreases in magnitude as one moves out from its interior.

**Exercise 1.22** *What should be the:*
 *(a) size of wavelength, and*
 *(b) momentum/energy*
  *of probe (say an electron) needed to see a nucleus?*

Ans. *(a)* A nucleus has size or extension $\sim 10^{-14}\,m = 10.10^{-15}\,m = 10\,fm$ (as per Rutherford's estimate). To see the details of nucleus, the probing agency should have smaller wavelength, i.e. $\lambda \leqslant 10^{-14}\,m = 10\,fm$, otherwise the effects of diffraction will partially or completely obscure the image.

 *(b)* According to the de Broglie hypothesis

$$\lambda = \frac{h}{p}$$

Multiplying the numerator and the denominator by $c$ (the speed of light in vacuum) we have $\lambda = \frac{hc}{pc}$

Using $hc = 1242\,MeV.\,fm$ *(exercise 1.2)*, we get

$$\lambda = \frac{hc}{pc} = \frac{1242\,MeV.\,fm}{pc}$$

So for $\lambda \sim 10\,fm$, $pc \sim \frac{1242\,MeV.\,fm}{10\,fm} = 124.2\,MeV$

$$p \sim 124.2 \frac{MeV}{c}.$$

So the required momentum is $p \geqslant 124.2 \frac{MeV}{c}$

For an ultra-relativistic particle $E \sim pc = 124.2\ MeV$. So the required energy is $E \geqslant 124.2\ MeV$

An accelerator can impart such energy to the electron.

**Exercise 1.23** *Can a nucleus be probed with electromagnetic radiation?*

A nucleus has size or extension $\sim 10^{-14}\ m$, while the most energetic electromagnetic radiation is the $\gamma$ ray that has $\lambda \sim 10^{-12}\ m$. So electromagnetic radiation (figure 1.32) will not be a good probe. Firing electrons of wavelength 10 *fm* having energy 124.2 *MeV* is a good choice (*exercise 1.22*).

**Exercise 1.24** *How did the number of $\alpha$ scattered in the Rutherford scattering experiment depend on the foil thickness and materials?*

Ans. Consult figure 1.3.

**Exercise 1.25** *The Rutherford scattering of $\alpha$ particles by gold nucleus corresponds to which value of total scattering cross-section?*

*(a)* $\infty$ *(b)* 0 *(c)* $\left( \frac{Z\mid e\mid^2}{4\pi\varepsilon_0\ m_a v^2} \right)^2 \mathrm{cosec}^4 \frac{\theta}{2}$ *(d)* $\frac{2Z\mid e\mid^2}{4\pi\varepsilon_0\ E_\alpha}$.

**Exercise 1.26** *In Rutherford scattering of $\alpha$ by gold foil, if the impact parameter reduces how do the scattering angle and number of scatterings change?*

*(a) both increase (b) both decrease (c) decrease, increase (d) increase, decrease.*

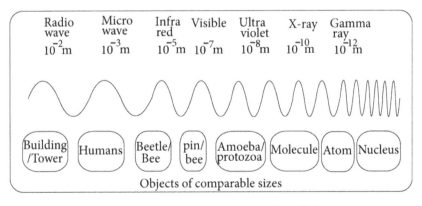

**Figure 1.32.** Wavelength of electromagnetic radiation.

**Exercise 1.27** *What is the evidence of the Rutherford model of a nuclear atom?*
*(a) Geiger Marsden experiment (b) np scattering experiment*
*(c) Bohr's theory* *(d) Franck–Hertz experiment.*

**Exercise 1.28** *In the Rutherford scattering experiment the number of $\alpha$ particles observed at $10°$ is $10^6$ per minute. What will be the number observed at an angle $90°$ and $180°$ ?*

Ans. Number of scattered $\alpha$ observed at angle $\theta$ is from equation (1.235)

$$N_\theta = \frac{Nnt}{r^2}\left(\frac{Z|e|^2}{4\pi\epsilon_0 2E_\alpha}\right)^2 \cosec^4\frac{\theta}{2} = k\cosec^4\frac{\theta}{2} \text{ where } k = \frac{Nnt}{r^2}\left(\frac{Z|e|^2}{4\pi\epsilon_0 2E_\alpha}\right)^2$$

For $\theta = 10°$, $N_{10} = 10^6 = k\cosec^4\frac{10}{2} = \frac{k}{\sin^4 5°} \Rightarrow k = 10^6 \sin^4 5° = 57.7$

For $\theta = 90°$, $N_{90°} = k\cosec^4\frac{90}{2} = \frac{57.7}{\sin^4 45°} = 230.8 \approx 230$

For $\theta = 180°$, $N_{180°} = k\cosec^4\frac{180}{2} = \frac{57.7}{\sin^4 90°} = 57.7 \approx 57$

**Exercise 1.29** *A fixed uranium nucleus scatters $\alpha$ of 5 MeV by $180°$. Find the value of the radius of nucleus.*

Ans. The distance of closest approach is using equation (1.52) (for $_{92}U$ as target)

$$b = \frac{2Z|e|^2}{4\pi\epsilon_0 E_\alpha} = \frac{2(92)(1.6 \times 10^{-19} \text{ C})^2}{4\pi\left(\frac{10^{-9}}{36\pi}\frac{F}{m}\right)(5 \times 10^6 \times 1.6 \times 10^{-19} \text{ J})} = 5.3 \times 10^{-14}\text{m} \Leftarrow \text{estimate for radius of nucleus}$$

**Exercise 1.30** *A 6 MeV$\alpha$ particle is back scattered by mercury nucleus $(Z = 80)$. Find:*
*(a) distance of closest approach, and*
*(b) the velocity of $\alpha$ at the point of closest approach.*

Ans. *(a)* The distance of closest approach is, using equation (1.52), (for $_{80}Hg$ as target)

$$b = \frac{2Z|e|^2}{4\pi\epsilon_0 E_\alpha} = \frac{2(80)(1.6 \times 10^{-19} \text{ C})^2}{4\pi\left(\frac{10^{-9}}{36\pi} Fm^{-1}\right)(6 \times 10^6 \times (1.6 \times 10^{-19} \text{ J})} = 3.84 \times 10^{-14} \text{ m}$$

*(b)* The velocity of $\alpha$ at the point of closest approach is

$$v_\alpha^{\min} = 0$$

at the point of closest approach where $\theta = 180°$ (back scattering).

**Exercise 1.31** *An $\alpha$ of energy $5 \times 10^{-13}$ J fired against uranium target is scattered at $90°$. Find the nearest distance of approach.*

Ans. Impact parameter is, using equation (1.225), for target $_{92}$U

$$R = \frac{Z|e|^2}{4\pi\varepsilon_0 E_\alpha} \cot \frac{\theta}{2} = \frac{(92)\left(1.6 \times 10^{-19}\, C\right)^2}{4\pi\left(\frac{10^{-9}}{36\pi}\, Fm^{-1}\right)(5 \times 10^{-13}\, J)} \cot \frac{90°}{2} = 4.2 \times 10^{-14}\, m$$

**Exercise 1.32** *Find the value of:*
*(a) impact parameter;*
*(b) cross-section;*
*(c) fraction of $\alpha$ scattered at angle $90°$ or more;*

*in the case of Rutherford scattering of $\alpha$ particles of energy $7.68$ MeV by thin gold foil of thickness $6 \times 10^{-7}$ m. (Given gold density $19.39$ g $cc^{-1}$, atomic weight of gold is $197.2$.)*

Ans. *(a)* The impact parameter is from equation (1.225) for target $_{79}$Au

$$R = \frac{Z|e|^2}{4\pi\varepsilon_0 E_\alpha} \cot \frac{\theta}{2} = \frac{79(1.6 \times 10^{-19}\, C)^2}{4\pi\left(\frac{10^{-9}}{36\pi}\, Fm^{-1}\right)(7.68 \times 10^6 \times 1.6 \times 10^{-19}\, J)} \cot \frac{90°}{2} = 1.48 \times 10^{-14}\, m$$

*(b)* Impact cross-section (equation (1.27)) for scattering at $90°$ or more is

$$\sigma = \pi R^2 = \pi (1.48 \times 10^{-14}\, m)^2 = 6.88 \times 10^{-28}\, m^2$$

The less the impact parameter, the more the scattering (figure 1.30).

*(c)* Let $n = \left(\begin{array}{c}\text{Number of atoms} \\ \text{per unit volume}\end{array}\right) = \frac{\text{density}}{\text{atomic weight}}\text{(Avogadro number)}$

$$= \frac{19.39\, g\, cc^{-1}}{197.2\, g}(6.023 \times 10^{23}) = \frac{19.39 \times 6.023 \times 10^{23}}{197.2 \times (10^{-2}\, m)^3} = 5.9 \times 10^{28}\, m^{-3}$$

$$\left(\begin{array}{c}\text{Number of atoms} \\ \text{in a thickness}\end{array}\right) = \left(\begin{array}{c}\text{Number of atoms} \\ \text{per unit volume}\end{array}\right)\text{(thickness)} = nt$$

$$nt = (5.9 \times 10^{28} m^{-3})(6 \times 10^{-7}\, m) = 3.54 \times 10^{22}\, m^{-2}$$

The fraction of $\alpha$ scattered at angle $90°$ or more is given by

$$\frac{N_{90°\text{or more}}}{N_i} = \sigma nt = (\pi R^2)nt = (6.89 \times 10^{-28}\, m^2)(3.54 \times 10^{22}\, m^{-2}) = 2.44 \times 10^{-5}$$

**Exercise 1.33** *Consider the experiment of scattering of $420$ MeV electrons by oxygen nucleus (figure 1.9).*
*(a) How do you know that the electrons recorded by the detector were coming from the oxygen nucleus and not from other electrons of the sample?*

*(b) In the experiment why don't we get a second minimum?*

*(c) If the experiment is done with a larger nucleus would you get a second minimum?*

[Ans.] *(a)* When the electrons strike the oxygen nucleus it is a collision between a light particle (i.e. electron) and a heavy nucleus (i.e. oxygen nucleus). So the highly energetic 420 *MeV* electrons will not lose much energy and are collected and identified as being scattered from the oxygen nucleus.

But when the electrons strike the outer shell electrons of the oxygen sample there will be large transfer of energy since it is then an elastic collision between particles of equal mass (electron–electron elastic collision). These electrons enter the detector with very small energy (due to such huge energy loss) and are easily segregated and rejected or ignored.

*(b)* The experiment of 420 *MeV* electron scattered by oxygen nucleus can be likened to electron wave of wavelength

$$\lambda = \frac{h}{p} = \frac{hc}{E} = \frac{1242 \ MeV. \ fm}{420 \ MeV} = 2.96 \ fm$$

being diffracted by a circular aperture leading to Airy diffraction pattern with first minimum (dark ring) at $\theta$ given by the relation

$$\sin\theta = \frac{1.22\lambda}{D} \Rightarrow D = \frac{1.22\lambda}{\sin\theta} = \frac{1.22(2.96 \ fm)}{\sin 43°} = 5.3 \ fm$$

and hence nuclear radius is

$$R = \frac{D}{2} = \frac{5.3 \ fm}{2} = 2.65 \ fm.$$

The second minimum (if we were to get it) occurs at $\theta$ given by the condition

$$\sin\theta = \frac{2.233\lambda}{D} = \frac{2.233(2.96 \ fm)}{5.3 \ fm} = 1.247 > 1.$$

Clearly the condition $\sin\theta > 1$ is not a possibility implying that such $\theta$ does not exist. Hence in the 420 *MeV* electron scattering by oxygen nucleus we do not get a second minimum.

*(c)* If the experiment is carried with a larger nucleus (*D* larger) we might get a second minimum as then $\frac{\lambda}{D}$ is smaller which might lead to $\sin\theta = \frac{2.233\lambda}{D} < 1$.

**Exercise 1.34** *Which of the following is not true in the description of various nuclei?*
*(a) Core size varies       (b) core is surrounded by shell of the same thickness*
*(c) nucleon density varies (d) proton density varies.*

**Exercise 1.35** *We can use the Woods–Saxon function $\rho = \dfrac{\rho_0}{1+e^{\frac{r-R}{a}}}$ to describe a nuclear structure. Which of the constants will not change for different nuclei?*
*(a) $\rho_0$ (b) R (c) a (d) all change.*

**Exercise 1.36** *Consider the nuclear charge distribution $\rho(r) = \dfrac{\rho_0}{1+e^{\frac{r-R}{a}}}$. Find the value of a if skin thickness t = 2.3 fm.*
Ans. From equations (1.71) and (1.72) we have

$$t = r_{02} - r_{01} = (R + 2.2a) - (R - 2.2a) = 4.4a$$

$$a = \frac{t}{4.4} = \frac{2.3\,fm}{4.4} = 0.5\,fm$$

**Exercise 1.37** *At what distance does the nucleon density fall to 90%,10% of the core value in an arbitrary nucleus?*
*(R= nuclear radius, a= constant of Woods–Saxon function $\rho = \dfrac{\rho_0}{1+e^{\frac{r-R}{a}}}$)*
*(a) R + 2.2a, R − 2.2a (b) R − 2.2a, R + 2.2a*
*(c) R + 4.4a, R − 2.2a (d) R − 4.4a, R + 4.4a.*

**Exercise 1.38** *The thickness of the shell lying outside the constant density nuclear core between which the core density falls from 90% to 10% of the core density is about*
*(a) 4.4 fm (b) 1 fm (c) 0.5 fm (d) 2.2 fm.*

**Exercise 1.39** *How do nuclear radius and nuclear volume vary with mass number?*
*(a) A, $A^{\frac{2}{3}}$ (b) $A^{\frac{1}{3}}$, A (c) $A^3$, $A^{\frac{1}{3}}$ (d) A, $A^{\frac{1}{3}}$*
Hint: $R = R_0 A^{\frac{1}{3}}$, $V = \frac{4}{3}\pi R^3 = \frac{4}{3}\pi\left(R_0 A^{\frac{1}{3}}\right)^3 = \frac{4}{3}\pi R_0^3 A$.

**Exercise 1.40** *Consider a uniformly charged sphere. Find its mean square charge radius.*
Ans. By definition the mean square charge radius is

$$<r^2> = \frac{\int_0^R r^2 dq}{Q}$$

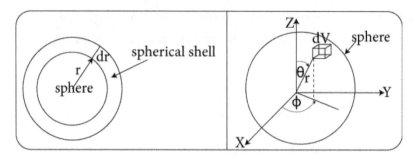

**Figure 1.33.** Showing spherical shell of area $4\pi r^2 dr$ and the volume element $dV$ of sphere

where $dq$ is the charge of an elemental spherical shell (figure 1.33) at distance $r$ from centre and so $dq = \rho 4\pi r^2 dr$ where $\rho$ is the volume charge density $=$ constant. $Q = \rho \frac{4}{3}\pi R^3$ represents the total charge of the uniformly charged sphere of radius $R$. Hence

$$<r^2> = \frac{\int_0^R r^2 \rho 4\pi r^2 dr}{\rho \frac{4}{3}\pi R^3} = \frac{3}{R^3} \int_0^R r^4 dr = \frac{3}{R^3} \frac{1}{5} R^5 = \frac{3}{5} R^2.$$

Alternatively:

$dq = \rho dV = \rho(r^2 dr \sin\theta d\theta d\phi)$ (figure 1.33)

$$<r^2> = \frac{\int_0^R r^2 dq}{Q} = \frac{\int_0^R r^2 \rho(r^2 dr \sin\theta d\theta d\phi)}{Q} = \frac{\rho \int_0^R r^4 dr \int_0^\pi \sin\theta d\theta \int_0^{2\pi} d\phi}{\rho \frac{4}{3}\pi R^3} = \frac{\frac{R^5}{5} \cdot 2.2\pi}{\frac{4}{3}\pi R^3} = \frac{3}{5} R^2$$

**Exercise 1.41** *Figure 1.34 shows multiple minima in the data corresponding to elastic scattering of electrons from $_{82}Pb^{208}$ that resembles a diffraction-like pattern with light incident on an opaque disc. Explain why these minima do not fall to zero.*

$\boxed{\text{Ans.}}$ The pattern lacks sharp minima because the nucleus does not have a sharp boundary.

**Exercise 1.42** *The radius of Ge nucleus is found to be twice that of $_4Be^9$. How many nucleons are expected to be found in it?*

$\boxed{\text{Ans.}}$ $\dfrac{R_{Ge}}{R_{Be}} = \dfrac{R_0 A_{Ge}^{1/3}}{R_0 A_{Be}^{1/3}} \Rightarrow 2 = \dfrac{A_{Ge}^{1/3}}{9^{1/3}}$

$$A_{Ge}^{1/3} = 2.9^{1/3}$$

$$A_{Ge} = (2.9^{1/3})^3 = 72. \text{ Symbol is } _{32}Ge_{40}^{72}.$$

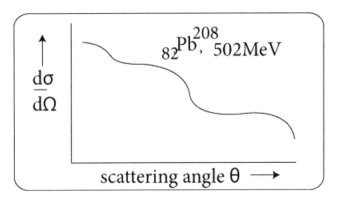

**Figure 1.34.** 502 MeV electron scattering by $_{82}Pb^{208}$ (*exercise 1.41*)

**Exercise 1.43** *How many times is the volume of $_8O^{16}$ nucleus compared to $_{54}Ge^{128}$?*
(a) 8  (b) 2  (c) 6.75  (d) 1.89.
Hint.

$$\frac{V_{Xe}}{V_O} = \frac{\frac{4}{3}\pi R_{Xe}^3}{\frac{4}{3}\pi R_O^3} = \frac{R_{Xe}^3}{R_O^3} = \frac{(R_0 A_{Xe}^{1/3})^3}{(R_0 A_{Xe}^{1/3})^3} = \frac{A_{Xe}}{A_O} = \frac{128}{16} = 8.$$

**Exercise 1.44** *Which one has a nuclear radius $\frac{1}{3}$ of $_8Os^{189}$ nucleus?*
(a) $Li^7$  (b) $O^{16}$  (c) $He^4$  (d) $N^{14}$.
Hint.

$$\frac{R}{R_{Os}} = \frac{R_0 A^{1/3}}{R_0 A_{Os}^{1/3}} = \frac{A^{1/3}}{189^{1/3}}$$

$$\frac{R_{Li}}{R_{Os}} = \frac{7^{1/3}}{189^{1/3}} = \frac{1}{3}, \quad \frac{R_O}{R_{Os}} = \frac{16^{1/3}}{189^{1/3}} = \frac{1}{2.277}, \quad \frac{R_{He}}{R_{Os}} = \frac{4^{1/3}}{189^{1/3}} = \frac{1}{3.615}, \quad \frac{R_N}{R_{Os}} = \frac{14^{1/3}}{189^{1/3}} = \frac{1}{2.38}.$$

**Exercise 1.45** (a) *The radius of $Ho^{165}$ is 7.731 fm. Find the radius of $He^4$.*
(b) *If radius of $_{29}X^{64}$ is $4.8 \times 10^{-15}$ m then find the radius of $_{12}Y^{27}$.*
[Ans.] (a) $\dfrac{R_{Ho}}{R_{He}} = \dfrac{R_0 A_{Ho}^{1/3}}{R_0 A_{He}^{1/3}} = \dfrac{165^{\frac{1}{3}}}{4^{1/3}} \Rightarrow \dfrac{7.731\,fm}{R_{He}} = \dfrac{165^{1/3}}{4^{1/3}}$

$R_{He} = 7.731 \dfrac{4^{1/3}}{165^{1/3}}\,fm = 2.238\,fm$

$$\frac{R_Y}{R_X} = \left(\frac{A_Y}{A_X}\right)^{1/3} \Rightarrow \frac{R_Y}{4.8 \times 10^{-15}\,cm} = \left(\frac{27}{64}\right)^{1/3}$$

$$R_Y = \left(\frac{27}{64}\right)^{1/3} (4.8 \times 10^{-15}\,m) = 3.6 \times 10^{-15}\,m$$

**Exercise 1.46** *What is the ratio of the sizes of $_{82}Pb^{208}$ and $_{12}Mg^{26}$ ?*

Ans. $\dfrac{R_{Pb}}{R_{Mg}} = \left(\dfrac{A_{Pb}}{A_{Mg}}\right)^{1/3} = \left(\dfrac{208}{26}\right)^{1/3} = 2$

**Exercise 1.47** *Is the nucleus visible with visible light? Explain.*

Ans. No. Because nuclear size $\sim fm = 10^{-15}\,m$, while wavelength of visible light is 400 *nm* to 700 *nm*, i.e. $4 \times 10^{-7}\,m$ to $7 \times 10^{-7}\,m$. As the dimensions are not comparable (since $\lambda_{\text{light}} \gg \lambda_{\text{nucleus}}$) it follows that visible light fails to probe nuclear size. A wave having wavelength $\sim fm$ can probe the nucleus, such as high energy electrons (accelerated by accelerators) which have wave property, and is associated with wavelength $\sim fm$.

**Exercise 1.48** *Find nuclear mass density of iron nucleus, given that mass of its nucleus is 55.85u and A = 56.*

Ans. Nuclear mass density

$$\rho_m = \frac{\text{Nuclear mass}}{\text{Nuclear volume}} = \frac{Am_p}{\frac{4}{3}\pi R^3} = \frac{m_{Fe}}{\frac{4}{3}\pi \left(R_0 A^{1/3}\right)^3}$$

$$= \frac{55.85u}{\frac{4}{3}\pi(1.2 \times 10^{-15}\,m)^3 A} = \frac{55.85(1.66 \times 10^{-27}\,kg)}{\frac{4}{3}\pi(1.2 \times 10^{-15}\,m)^3(56)} = 2.3 \times 10^{17}\,kg.\,m^{-3} \text{ (using equation (1.15))}$$

**Exercise 1.49** *(a) Calculate the electrostatic potential of an orbiting electron due to a uniformly charged spherical nucleus of radius R.*

*(b) Find the change in electrostatic potential and hence change in potential energy of an electron at a distance r from the origin if, it interacts with a point nucleus (at r = 0) and a uniformly charged spherical nucleus of radius R around r = 0.*

Ans. (a) Consider uniformly charged spherical nucleus of radius $R$ (figure 1.35). At outside point of sphere $r \geqslant R$, potential is

$$V(r \geqslant R) = \frac{Z\,|\,e\,|}{4\pi\varepsilon_0 r} \tag{1.236}$$

since w.r.t. an exterior point the spherical charge distribution can be considered to be concentrated at the centre, so the expression of potential has a point nucleus form.

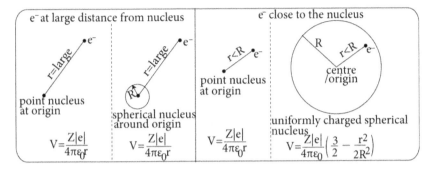

**Figure 1.35.** Evaluation of potential energy of an orbiting electron near the nucleus.

Now at inside point $r < R$ the field is given by

$$\vec{E}(\vec{r}) = \frac{Z \mid e \mid r}{4\pi\varepsilon_0 R^3} \hat{r}$$

and the potential difference between two points—interior point $r < R$ and point on surface $r = R$ is defined as

$$V(r) - V(R) = -\int_R^r \vec{E} \cdot \vec{dr} = -\int_R^r \frac{Z \mid e \mid r}{4\pi\varepsilon_0 R^3} \hat{r} \cdot \hat{r} dr = -\frac{Z \mid e \mid}{4\pi\varepsilon_0 R^3} \int_R^r r dr = -\frac{Z \mid e \mid}{4\pi\varepsilon_0 R^3} \left( \frac{r^2}{2} - \frac{R^2}{2} \right)$$

Putting $V(R)$ from equation (1.236) as

$$V(R) = \frac{Z \mid e \mid}{4\pi\varepsilon_0 R}$$

we get

$$V(r) - V(R) = V(r) - \frac{Z \mid e \mid}{4\pi\varepsilon_0 R} = -\frac{Z \mid e \mid}{4\pi\varepsilon_0 R^3} \left( \frac{r^2}{2} - \frac{R^2}{2} \right)$$

$$V(r) = \frac{Z \mid e \mid}{4\pi\varepsilon_0 R} - \frac{Z \mid e \mid}{4\pi\varepsilon_0 R^3} \left( \frac{r^2}{2} - \frac{R^2}{2} \right) = \frac{Z \mid e \mid}{4\pi\varepsilon_0 R} \left[ 1 - \left( \frac{r^2}{2R^2} - \frac{1}{2} \right) \right]$$

$$V = \frac{Z \mid e \mid}{4\pi\varepsilon_0 R} \left( \frac{3}{2} - \frac{r^2}{2R^2} \right) = V(r < R) \tag{1.237}$$

Note that potential at $r = R$ follows from equation (1.237) also since

$$V(r < R) \left.\right|_{r=R} = \frac{Z \mid e \mid}{4\pi\varepsilon_0 R} \left( \frac{3}{2} - \frac{R^2}{2R^2} \right) = \frac{Z \mid e \mid}{4\pi\varepsilon_0 R} = V(R).$$

So we write

$$V(r \leqslant R) = \frac{Z \mid e \mid}{4\pi\varepsilon_0 R} \left( \frac{3}{2} - \frac{r^2}{2R^2} \right) \tag{1.238}$$

This is the potential at an internal point.

Collecting the two results we write $V(r \geqslant R) = \frac{Z|e|}{4\pi\varepsilon_0 r}$, $V(r \leqslant R) = \frac{Z|e|}{4\pi\varepsilon_0 R}\left(\frac{3}{2} - \frac{r^2}{2R^2}\right)$.

(b) For $r \gg R$ region (i.e. at a large distance from nucleus) the electron considers the nucleus to be concentrated at a point (even if the nucleus has a spherical spread). The expression of electrostatic potential in both cases is

$$V(r = \text{large}) = \frac{Z|e|}{4\pi\varepsilon_0 r}.$$

So no change is there.

In the region $r < R$, the expressions are

$$V_1(r) = \frac{Z|e|}{4\pi\varepsilon_0 r}$$

for interaction between a point nucleus at $r = 0$ and an electron at distance $r$ and

$$V_2(r) = \frac{Z|e|}{4\pi\varepsilon_0 R}\left(\frac{3}{2} - \frac{r^2}{2R^2}\right)$$

for interaction between a uniformly charged spherical nucleus of radius $R$ and an electron at distance $r < R$.

The change in potential is thus the difference

$$\Delta V = V_2 - V_1 = \frac{Z|e|}{4\pi\varepsilon_0 R}\left(\frac{3}{2} - \frac{r^2}{2R^2}\right) - \frac{Z|e|}{4\pi\varepsilon_0 r}$$

$$\Delta V = \frac{Z|e|}{4\pi\varepsilon_0}\left(\frac{3}{2R} - \frac{r^2}{2R^3} - \frac{1}{r}\right)$$

and the change in potential energy is obtained by multiplying with $-|e|$ that gives

$$\Delta U = -|e|\left[\frac{Z|e|}{4\pi\varepsilon_0}\left(\frac{3}{2R} - \frac{r^2}{2R^3} - \frac{1}{r}\right)\right]$$

$$\Delta U = \frac{Z|e|^2}{4\pi\varepsilon_0}\left(\frac{1}{r} - \frac{3}{2R} + \frac{r^2}{2R^3}\right)$$

**Exercise 1.50** *The energy difference that a 1s electron would have if the nucleus has a spherical structure instead of being point-like is $\Delta E = \frac{Z^4|e|^2 R^2}{10\pi\varepsilon_0 a_0^3}$ (equation (1.90)). Can we use this formula to find nuclear radius R?*

Ans. From equation (1.91) it follows that

$$E_{1s} = E_{1s}^{\text{point nucleus}} + \frac{Z^4 |e|^2 R^2}{10\pi\varepsilon_0 a_0^3}$$

Real atoms have a spread-out nucleus (that we can assume to be spherical). If we had a supply of atoms with point-like nucleus then measuring $E_{1s}$ and $E_{1s}^{\text{point nucleus}}$ and taking their difference we could have found $R$. But since atoms having point nucleus are not available we cannot use this formula to find the nuclear radius.

**Exercise 1.51**

> (a) What is a negative muon?
>
> (b) What is a muonic atom?
>
> (c) How can it be used to know nuclear radius?

Ans. (a) A muon (symbol $\mu^-$) is a particle of the electron family (lepton), having

Mass $m_{\mu^-} = 207 m_e = 207 \times 0.511 \frac{MeV}{c^2} \approx 106 \frac{MeV}{c^2}$ (it is a heavy electron)

Charge $Q_{\mu^-} = Q_{e^-} = -|e|$

Spin $s_\mu = \frac{1}{2}$

It is not a stable particle. It is produced in the upper atmosphere from cosmic rays and is unstable. It decays as

$$\mu^- \xrightarrow{2.2\mu s} e^- + \bar{\nu}_e + \nu_\mu \text{ (weak decay)}$$

in rest frame

They can be produced in the laboratory and also in large accelerators. Nuclear reactions produce negative pion $\pi^-$ which decays as

$$\pi^- \xrightarrow{26ns} \mu^- + \bar{\nu}_\mu \text{ (weak decay)}.$$

The antiparticle of $\mu^-$ is positive muon $\mu^+$.

(b) A muonic atom can be formed (in the laboratory) when a proton ($p$) captures a negative muon ($\mu^-$) to form a bound system. It is thus similar to hydrogen atom with electron ($e^-$) replaced by a negative muon ($\mu^-$).

Radius of the first Bohr orbit of a muonic atom is

$$a_{\text{muonic atom}} = 4\pi\varepsilon_0 \frac{\hbar^2}{m_\mu |e|^2} = 4\pi\varepsilon_0 \frac{\hbar^2}{207 m_e |e|^2} = \frac{1}{207} a_0$$

where $a_0 = 4\pi\varepsilon_0 \frac{\hbar^2}{m_e |e|^2}$ is the radius of the first Bohr orbit in an electronic atom. Clearly thus, muonic orbits are 207 times closer to the nucleus than the electronic orbits. In fact, the muonic $1s$ orbit in a heavy nucleus like Pb has its mean radius inside the nuclear radius $R$ (figure 1.36) (radius of $_{82}Pb^{207}$ is $\sim 7\,fm$ and $a_{\text{muonic atom}} \sim fm$). Such is the degree of overlap between muonic wave function with the nuclear wave function that there is probability of the muon staying within the nucleus most of the time.

(c) The muon, except for its mass ($207 m_e$) is identical to the electron and so the muonic atom will have the same Coulomb interaction with nucleus and will have the same energy levels $1s$, $2p$,. etc. as in an electronic atom and hence we would have $K_\alpha$

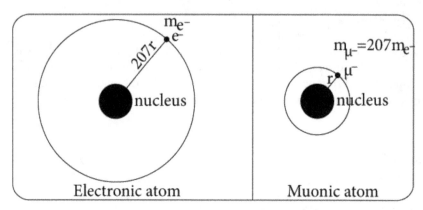

**Figure 1.36.** Muonic orbits are closer to the nucleus in a muonic atom than electronic orbits of an electronic atom. So interaction/overlap of the muon with the nucleus is much larger in a muonic atom.

x-rays for muonic transitions $2p \xrightarrow[K_\alpha x-\text{ray}]{\mu^- \text{ transition}} 1s$ (similar to what we described in electronic transitions $2p \xrightarrow[K_\alpha x-\text{ray}]{e^- \text{ transition}} 1s$) and the nuclear radius can be evaluated exactly as in the electronic atom case. Since for heavy nuclei the probability of a muon staying inside the nuclei is very large, the overlap interaction is large and so the shift or the change in energy due to finite size of nuclei in the muonic atom will be very large (as muon genuinely feels the spread or extension of the nucleus—it being inside nucleus). Muonic $K_\alpha$ x-rays have energies $\sim MeV$ (in contrast to the electronic $K_\alpha$ x-rays which have energy $\sim keV$). So observation and measurement is easier and hence evaluation of nuclear radius can be made with more ease.

**Exercise 1.52** *The proton radius and neutron radius are equal though heavy nuclei have excess neutrons. Explain.*

Ans. Heavy nuclei have excess neutrons and are expected to have larger neutron radius compared to proton radius. Actually proton–proton repulsion tends to push the protons outward. The neutron–proton force tends to pull the neutrons inward. Due to this outward push of protons and inward pull of neutrons there occurs a complete intermix of the nucleons and hence the proton radius and neutron radius become equal.

**Exercise 1.53** *The difference in Coulomb energy between the mirror nuclei $_{24}Cr^{49}_{25}$ and $_{25}Mn^{49}_{24}$ is 6 MeV. Find the radius of $_{25}Mn^{49}_{24}$ nucleus.*

Ans. The difference in electrostatic energies is given by equation (1.100)

$$\Delta E_C = \frac{1}{4\pi\varepsilon_0} \frac{3 \mid e \mid^2}{5 R_0 A^{1/3}} (A - 1)$$

with

$$\Delta E_C = 6 \, MeV = 6 \times 10^6 \times (1.6 \times 10^{-19} \, J), \; A = 49$$

Putting the values we get

$$6 \times 10^6 \times (1.6 \times 10^{-19}) J = \frac{1}{4\pi \left( \frac{10^{-9}}{36\pi} \frac{F}{m} \right)} \frac{3(1.6 \times 10^{-19} \, C)^2}{5R_0(49)^{1/3}} (49 - 1)$$

$$R_0 = 1.89 \times 10^{-15} m \Rightarrow R_0 = 1.89 \, fm$$

**Exercise 1.54** *Show that $\frac{B}{A}$ for $_2He^4_2$, $_{26}Fe^{56}_{30}$, $_{47}Ag^{107}_{60}$, $_1H^2_1$, $_3Li^7_4$, $_4Be^9_5$, $_4Be^8_4$, $_6C^{12}_6$, $_8O^{16}_8$ are respectively 7.07, 8.79, 8.56, 1.11, 5.74, 6.47, 6.8, 7.43, 7.73 $\frac{MeV}{nucleon}$ if nuclear masses are $_2He^4_2(4.001\ 506u)$, $_{26}Fe^{56}_{30}$ (55.920 69u), $_{47}Ag^{107}_{60}(106.879\ 34u)$, $_1H^2_1$ (2.013 553u), $_3Li^7_4$ (7.013 36u), $_4Be^9_5$ (9.009 99u), $_4Be^8_4$ (8.005 305 10u), $_6C^{12}_6$(12), $_8O^{16}_8$(15.994 9146u), n(1.008 665u), p(1.007 276u).*

**Ans.** From equations (1.106) and (1.107) we have

$$B = \left( Z m_p + N m_n \right) c^2 - M c^2 \Rightarrow \frac{B}{c^2} = Z m_p + N m_n - M$$

✓ For $_2He^4_2 = \alpha$

$$\frac{B}{c^2} = 2m_p + 2m_n - M_\alpha = 2(1.007\ 276u) + 2(1.008\ 665u) - 4.001\ 506u$$

$$= 0.030\ 376u = 0.0303\ 76 \times 931 \, MeV = 28.28 \, MeV$$

$$\frac{B}{A} = \frac{28.29 \, MeV}{4 \text{ nucleon}} = 7.07 \frac{MeV}{nucleon}.$$

✓ For $_{26}Fe^{56}_{30}$:

$$\frac{B}{c^2} = 26m_p + 30m_n - M_{Fe} = 26(1.007\ 276u) + 30(1.008\ 665u) - 55.920\ 69u$$

$$= 0.528\ 436u = 0.528\ 436 \times 931 \, MeV = 491.97 \, MeV$$

$$\frac{B}{A} = \frac{492.5 \, MeV}{56 \text{ nucleon}} = 8.79 \frac{MeV}{nucleon}.$$

✓ For $_{47}Ag^{107}_{60}$:

$$\frac{B}{c^2} = 47m_p + 60m_n - M_{Ag} = 47(1.007\ 276u) + 60(1.008\ 665u) - 106.879\ 34u$$

$$= 0.982\ 532u = 0.982\ 532 \times 931 \, MeV = 914.74 \, MeV$$

$$\frac{B}{A} = \frac{915.719 \, MeV}{107 \text{ nucleon}} = 8.55 \frac{MeV}{nucleon}.$$

✓ For $_1H_1^2 = d$:

$$\frac{B}{c^2} = m_p + m_n - M_d = 1.007\ 276u + 1.008\ 665u - 2.013\ 553u$$

$$= 0.002\ 388u = 0.002\ 388 \times 932\ MeV = 2.223\ MeV$$

$$\frac{B}{A} = \frac{2.225\ MeV}{2\ \text{nucleon}} = 1.11\frac{MeV}{\text{nucleon}}.$$

✓ For $_3Li_4^7$:

$$\frac{B}{c^2} = 3m_p + 4m_n - M_{Li} = 3(1.007\ 276u) + 4(1.008\ 665u) - 7.013\ 36u = 0.043\ 128u$$

$$= 0.043\ 128 \times 932\ MeV = 40.152\ MeV$$

$$\frac{B}{A} = \frac{40.195\ MeV}{7\ \text{nucleon}} = 5.74\frac{MeV}{\text{nucleon}}.$$

✓ For $_4Be_5^9$:

$$\frac{B}{c^2} = 4m_p + 5m_n - M_{Be} = 4(1.007\ 276u) + 5(1.008\ 665u) - 9.009\ 99u$$

$$= 0.062\ 439u = 0.062\ 439 \times 932\ MeV = 58.13\ MeV$$

$$\frac{B}{A} = \frac{58.193\ MeV}{9\ \text{nucleon}} = 6.46\frac{MeV}{\text{nucleon}}.$$

✓ For $_4Be_4^8$:

$$\frac{B}{c^2} = 4m_p + 4m_n - M_{Be} = 4(1.007\ 276u) + 4(1.008\ 665u) - 8.005\ 305\ 10u$$

$$= 0.058\ 4589u = 0.058\ 4589 \times 932\ MeV = 54.4252\ MeV$$

$$\frac{B}{A} = \frac{54.4837\ MeV}{8\ \text{nucleon}} = 6.8\frac{MeV}{\text{nucleon}}.$$

✓ For $_6C_6^{12}$:

$$\frac{B}{c^2} = 6m_p + 6m_n - M_C = 6(1.007\ 276u) + 6(1.008\ 66u) - 12u$$

$$= 0.095\ 646u = 0.095\ 646 \times 932\ MeV = 89.1421\ MeV$$

$$\frac{B}{A} = \frac{89.1421 \ MeV}{12 \ \text{nucleon}} = 7.42 \frac{MeV}{\text{nucleon}}.$$

✓ For $_8O_8^{16}$:

$$\frac{B}{c^2} = 8m_p + 8m_n - M_{Be} = 8(1.007\ 276u) + 8(1.008\ 665u) - 15.994\ 9146u$$

$$= 0.132\ 6134u = 0.132\ 6134 \times 932 \ MeV = 123.46 \ MeV$$

$$\frac{B}{A} = \frac{123.596 \ MeV}{16 \ \text{nucleon}} = 7.72 \frac{MeV}{\text{nucleon}}$$

**Exercise 1.55** *Why is $_4Be_4^8$ unstable though its $\frac{B}{A}$ is high?*

[Ans.] $_4Be_4^8 = 4p4n = 2p2n + 2p2n$, $(2p2n = {}_2^4He_2^4 = \alpha)$

Clearly $_4Be_4^8$ can be looked upon as a $2\alpha$ system. Again the total energy of $_4Be_4^8$ is nearly equal to that of two $\alpha$ particles. So the decay of $_4Be_4^8$ into two $\alpha$ particles is energetically favourable. The decay $_4Be_4^8 \rightarrow {}_2He_2^4 + {}_2He_2^4$ occurs with half life $8 \times 10^{-17} \ s$.

**Exercise 1.56** *(a) Write down separation energy of $\alpha$ from a nucleus. Obtain its form in terms of binding energies.*

*(b) Express separation energy of neutron and proton in terms of binding energy.*

[Ans.] *(a)* Separation energy of $\alpha \equiv {}_2H_2^4$ from a nucleus:

Separating $\alpha$ from nucleus $(Z, N, A)$ refers to the transformation

$$(Z, N, A) \rightarrow (Z - 2, N - 2, A - 4) + \alpha$$

And hence

$$S_\alpha = [M_{\text{final}} - M_{\text{initial}}]c^2$$

$$S_\alpha = [M(Z - 2, N - 2, A - 4) + m_\alpha - M(Z, N, A)]c^2$$

Let us add and subtract $(Zm_p + Nm_n)c^2$ to get

$$S_\alpha = \left[(Zm_p + Nm_n) - M(Z, N, A)\right]c^2 - \left[(Zm_p + Nm_n) - M(Z - 2, N - 2, A - 4)\right]c^2 + m_\alpha c^2$$

Using equation (1.106)

$$S_\alpha = B(Z, N, A) - \left[\{(Z - 2)m_p + (N - 2)m_n\} - M(Z - 2, N - 2, A - 4)\right]c^2 + (m_\alpha - 2m_p - 2m_n)c^2$$

$$= B(Z, N, A) - B(Z - 2, N - 2, A - 4) - [(2m_p + 2m_n) - m_\alpha]c^2$$

$$S_\alpha = B(Z, N, A) - B(Z - 2, N - 2, A - 4) - B_\alpha$$

(b) Neutron separation energy from equation (1.117) is

$$S_n = [M(Z, N - 1, A - 1) + m_n - M(Z, N, A)]c^2$$

Add and subtract $(Zm_p + Nm_n)c^2$ to get

$$S_n = \left[(Zm_p + Nm_n) - M(Z, N, A)\right]c^2 - \left[(Zm_p + Nm_n) - M(Z, N - 1, A - 1)\right]c^2 + m_n$$

Using equation (1.106)

$$S_n = B(Z, N, A) - \left[\left\{Zm_p + (N - 1)m_n\right\} - M(Z, N - 1, A - 1)\right]c^2$$

$$S_n = B(Z, N, A) - B(Z, N - 1, A - 1)$$

Proton separation energy from equation (1.120) is

$$S_p = \left[M(Z - 1, N, A - 1) + m_p - M(Z, N, A)\right]c^2$$

Add and subtract $(Zm_p + Nm_n)c^2$ to get

$$S_p = \left[(Zm_p + Nm_n) - M(Z, N, A)\right]c^2 - \left[(Zm_p + Nm_n) - M(Z - 1, N, A - 1)\right]c^2 + m_p$$

Using equation (1.106)

$$S_p = B(Z, N, A) - [\{(Z - 1)m_p + Nm_n\} - M(Z, N, A - 1)] c^2$$

$$S_p = B(Z, N, A) - B(Z - 1, N, A - 1)$$

**Exercise 1.57** *(a) Calculate the separation energy of neutron for $_{82}\text{Pb}^{209}_{127}$ if the mass of $_{82}\text{Pb}^{209}_{127}$ is 209.053 98u, $_{82}\text{Pb}^{208}_{126}$ is 208.047 54u and n is 1.008 665u*
*(b) Also find the same from binding energy expression.*
Ans. *(a)* Consider the interaction i.e.

$$_{82}\text{Pb}^{209}_{127} = {}_{82}\text{Pb}^{208}_{126} + {}_0n^1$$

$$M(Z, N, A) = M(Z, N - 1, A - 1) + m_n$$

Neutron separation energy from equation (1.117) is

$$\begin{aligned}
S_n &= [M(Z, N - 1, A - 1) + m_n - M(Z, N, A)]c^2 \\
&= [M({}_{82}\text{Pb}^{208}_{126}) + m_n - M({}_{82}\text{Pb}^{209}_{127})]c^2 \\
&= 208.047\ 54u + 1.008\ 665u - 209.053\ 98u \\
&= 0.002\ 225u = 0.002\ 225 \times 931\ MeV = 2.071\ 475\ MeV.
\end{aligned}$$

*(b)* From exercise 1.56 we can write

$$S_n = [M(Z, N - 1, A - 1) + m_n - M(Z, N, A)]c^2$$
$$= B(Z, N, A) - B(Z, N - 1, A - 1)$$
$$= 2.071\ 475\ MeV$$

**Exercise 1.58** *Explain how the following aspects regarding nucleus can be inter related:*

*binding energy per nucleon is constant (i.e. saturates) and nuclear force is short range.*

Ans. Suppose there are $A$ nucleons and let all nucleons interact with each other in the nucleus. So the amount of interaction or the binding energy $B$ will be proportional to the number of interacting pairs i.e. proportional to

$$^AC_2 = \frac{A!}{2!(A - 2)!} = \frac{A(A - 1)(A - 2)!}{2.\ (A - 2)!} = \frac{1}{2}A(A - 1).$$

Hence we can write

$$B \propto \frac{1}{2}A(A - 1) \approx \frac{1}{2}A^2$$

$$\frac{B}{A} \propto A$$

This means that the $\frac{B}{A}$ plot will be a straight line.

Figure 1.19 shows that $\frac{B}{A}$ versus $A$ plot is not a straight line and levels off at $8 \frac{MeV}{\text{nucleon}}$. So it follows that all nucleons do not interact with each other in the nucleus. In other words the interaction is short range (and not a long-range one). A nucleon interacts only with its immediate neighbours.

**Exercise 1.59** *If the binding energy per nucleon is large and we wish to strip off a nucleon from the nucleus, which of the following is true?*

*(a) Beta decay occurs; (b) easier to strip off;*
*(c) harder to strip off; (d) cannot be stripped off.*

**Exercise 1.60** *The atomic mass of $_{82}Pb_{125}^{207}$ is 206.975 897. What is the mass defect ?*

Ans. From equation (1.122) we have mass defect to be

$$\Delta m = Zm_p + Nm_n - M_{Pb}$$

$$\Delta m = 82m_p + 124m_n - M_{Pb}$$

$$= 82(1.007\ 276u) + 125(1.008\ 665u) - 206.975\ 897 = 1.703\ 86u$$

**Exercise 1.61** *Find the energy released in the beta decay*

$$_0n^1 \rightarrow\ _1p^1 +\ _{-1}e^0 + \bar{\nu}_e.$$

Ans. $\Delta m = m_n - m_p - m_e - m_{\bar{\nu}}$

$$= 1.008\ 665u - 1.007\ 276u - 0.000\ 547\ 21u - 0 = 0.000\ 841\ 79u.$$

$$B = \Delta mc^2 = 0.000\ 841\ 79 \times 931\ MeV = 0.7837\ MeV$$

**Exercise 1.62** *What is the energy released in the following reaction*
$_1H^2 +\ _1H^2 =\ _2He^4$
*where $_1H^2$ and $_2He^4$ possesses binding energy per nucleon as 1.125 MeV and 7.2 MeV respectively.*

Ans. $B(_1H^2) = A \times 1.125\ MeV = 2 \times 1.125\ MeV = 2.25\ MeV,$

$$B(_2He^4) = A \times 7.2\ MeV = 4 \times 7.2\ MeV = 28.8\ MeV$$

Energy released is

$$B(_2He^4) - 2B(_1H^2) = 28.8\ MeV - 2(2.25\ MeV) = 24.3\ MeV.$$

**Exercise 1.63** *An experimenter thinks that he has discovered a neutral particle with a mass of 2.027 33u and he assumes it to be two neutrons bound together (an nn system).*
*(a) Find the binding energy of the so called newly discovered particle?*
*(b) Comment on this proposition.*
Ans. *(a)* $n + n \rightarrow nn$ system

$$B = (m_n + m_n)c^2 - Mc^2 = 2m_nc^2 - Mc^2$$

$$B = 2(1.008\ 665u) - 2.02733u = -0.01u = -0.01 \times 931\ MeV = -9.31\ MeV$$

*(b)* Negative binding energy means an unbound system. This points to the fact that a di-neutron or nn bound system cannot exist.

**Exercise 1.64** *Show that binding energy can be written as*

$$B = [Zm_H + (A - Z)m_n - M_{atom}]c^2$$

*where $m_H$ is the mass of a hydrogen atom.*
Ans. By definition, from equation (1.106) we have

$$B = (Zm_p + Nm_n - M)c^2.$$

Adding and subtracting $Zm_e c^2$ we write

$$B = \left(Zm_p + Nm_n - M + Zm_e - Zm_e\right)c^2$$

$$= \left[Z(m_p + m_e) + Nm_n - (M + Zm_e)\right]c^2$$

We note that

$$m_p + m_e = m_H \text{ as } H = pe^-, \; N = A - Z \text{ and}$$

$$\text{Atom} = \text{Nucleus} + Z \text{ electrons, i.e. } m_{\text{atom}} = M + Zm_e$$

Hence

$$B = [Zm_H + (A - Z)m_n - M_{\text{atom}}]c^2$$

**Exercise 1.65** *Find the mass defect of* $_{10}\text{Ne}^{20}_{10}$, $_{20}\text{Ca}^{40}_{20}$, $_{82}\text{Pb}^{206}_{124}$, $_{92}\text{U}^{235}_{143}$, $_{92}\text{U}^{238}_{146}$.
*Hence find binding energy and binding energy per nucleon.*
*Given that* $M(_{10}\text{Ne}^{20}_{10}) = 19.992\,440u, M(_{20}\text{Ca}^{40}_{20}) = 39.962\,589u,$
$M(_{82}\text{Pb}^{206}_{124}) = 205.929\,52u,$ $M(_{92}\text{U}^{235}_{143}) = 235.113u,$ $M(_{92}\text{U}^{238}_{146}) = 238.0507u,$
$m_n = 1.008\,665u$, $m_p = 1.007276u$.

Ans. We shall use equation (1.122) for mass defect viz.

$$\Delta m = Zm_p + Nm_n - M$$

✓ For $_{10}\text{Ne}^{20}_{10}$

$$\Delta m = Zm_p + Nm_n - M_{\text{Ne}} = 10(1.007\,276u) + 10(1.008\,665u) - 19.992\,440u$$

$$= 0.166\,97u$$

$$B = \Delta mc^2 = 0.166\,97 \times 931 \; MeV = 155.449\,07 \; MeV$$

$$\frac{B}{A} = \frac{155.449\,07 \; MeV}{20} = 7.78\frac{MeV}{\text{nucleon}}.$$

✓ For $_{20}\text{Ca}^{40}_{20}$

$$\Delta m = 20m_p + 20m_n - M_{\text{Ca}} = 20(1.007\,276u) + 20(1.008\,665u) - 39.962\,589u$$

$$= 0.356\,231u$$

$$B = \Delta mc^2 = 0.356\,231 \times 931 \; MeV = 332 \; MeV$$

$$\frac{B}{A} = \frac{332 \; MeV}{40} = 8.3\frac{MeV}{\text{nucleon}}$$

✓ For $_{82}Pb_{124}^{206}$

$$\Delta m = 82m_p + 124m_n - M_{Pb} = 82(1.007\ 276u) + 124(1.008\ 665u) - 205.929\ 52u$$

$$= 1.741\ 572u$$

$$B = \Delta mc^2 = 1.741\ 572 \times 931\ MeV = 1621.403\ 532\ MeV$$

$$\frac{B}{A} = \frac{1621.403\ 532\ MeV}{206} = 7.87\frac{MeV}{nucleon}$$

✓ For $_{92}U_{143}^{235}$

$$\Delta m = 92m_p + 143m_n - M_U = 92(1.007\ 276u) + 143(1.008\ 665u) - 235.113\ 92u$$

$$= 1.794\ 567u$$

$$B = \Delta mc^2 = 1.794\ 567 \times 931\ MeV = 1670.741\ 877\ MeV$$

$$\frac{B}{A} = \frac{1670.741\ 877\ MeV}{235} = 7.1\frac{MeV}{nucleon}$$

✓ For $_{92}U_{146}^{238}$

$$\Delta m = 92m_p + 146m_n - M_U = 92(1.007\ 276u) + 146(1.008\ 665u) - 238.050\ 786u$$

$$= 1.883\ 696u$$

$$B = \Delta mc^2 = 1.883\ 696 \times 931\ MeV = 1753.720\ 976\ MeV$$

$$\frac{B}{A} = \frac{1753.720\ 976\ MeV}{238} = 7.4\frac{MeV}{nucleon}.$$

**Exercise 1.66** *A simultaneous flip of 2 coordinates x, y $\xrightarrow{\pi}$ x, −y corresponds to*
*(a) parity operation in 2D;  (b) 180° rotation;*
*(c) translation in 2D;      (d) reflection in 2D.*
Hint: 180° rotation as shown in figure 1.37

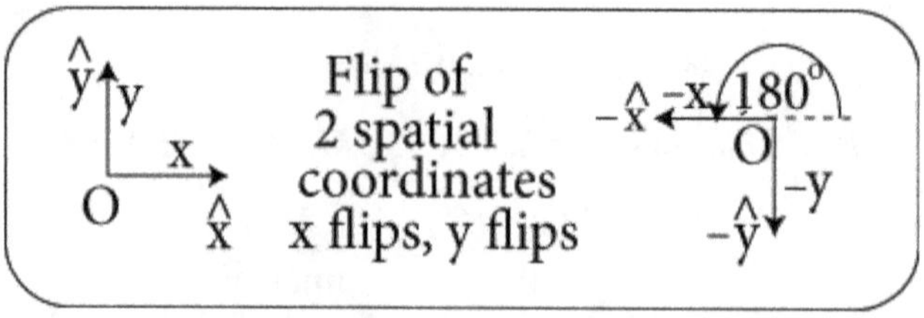

**Figure 1.37.** A simultaneous flip of two coordinates corresponds to 180° rotation.

**Exercise 1.67** *Find the eigenvalue of the parity operator.*

Ans. Eigen equation for the parity operator is

$$\hat{\pi}\psi(x) = \pi\psi(x) \tag{1.239}$$

where $\psi(x)$ is the parity eigen function.

Operating by $\hat{\pi}$ from the left gives

$$\hat{\pi}[\hat{\pi}\psi(x)] = \hat{\pi}[\pi\psi(x)]$$

$$\hat{\pi}^2\psi(x) = \pi[\hat{\pi}\psi(x)] = \pi[\pi\psi(x)]$$

$$\hat{\pi}^2\psi(x) = \pi^2\psi(x) \tag{1.240}$$

Again, parity operation $x \xrightarrow{\hat{\pi}} - x$ means

$$\hat{\pi}\psi(x) = \psi(-x).$$

Operating by $\hat{\pi}$ from the left we get

$$\hat{\pi}[\hat{\pi}\psi(x)] = \hat{\pi}[\psi(-x)] = \psi(x)$$

$$\hat{\pi}^2\psi(x) = \psi(x) \tag{1.241}$$

Comparison of equations (1.240) and (1.241) leads to

$$\hat{\pi}^2\psi(x) = \pi^2\psi(x) = \psi(x)$$

$$\pi^2 = 1 \Rightarrow \pi = \pm 1 \ (\Leftarrow \text{the eigenvalues}).$$

Hence the eigen equations of parity operator $\hat{\pi}$ can be written from equation (1.239) as

$$\hat{\pi}\psi(x) = \psi(x) \text{ for } \pi = +1 \text{ (even parity)} \tag{1.242}$$

$$\hat{\pi}\psi(x) = -\psi(x) \text{ for } \pi = -1 \text{ (odd parity)} \tag{1.243}$$

Writing the LHS as $\hat{\pi}\psi(x) = \psi(-x)$ gives from equations (1.242) and (1.243)

$$\psi(-x) = \psi(x) \text{ for } \pi = +1 \text{ (even parity)}$$

$$\psi(-x) = -\psi(x) \text{ for } \pi = -1 \text{ (odd parity)}$$

**Exercise 1.68** *Determine the parity of pion from a study of s wave capture of $\pi^-$ by deuteron that produces two neutrons.*

Ans. Consider the reaction

$$\pi^- + d \rightarrow n + n$$

Initial angular momentum is

$$\vec{J}_i = \vec{l}_i + \vec{s}_\pi + \vec{s}_d$$

Since it is an $s$ wave capture so $l_i = 0$. Also, pion spin $s_\pi = 0$, $s_d = 1$ (spin triplet).

$$\vec{J_i} = \vec{0} + \vec{0} + \vec{1} = \vec{1} \tag{1.244}$$

Parity of wave function of an $n$–$n$ system (final state) is

$$\psi_{nn} = (-1)^{l_f}(-1)^{1+S}(-1)^{1+I} \text{ [figures 8.5 and 8.8]} \tag{1.245}$$

where $S$ is spin and $I$ is isospin of the $n$–$n$ system (final state).

The $n$–$n$ system is a system of fermions and so system wave function will be antisymmetric. Hence parity of $\psi_{nn} \rightarrow (-1)$.

$I_z$ of the $n$–$n$ system is $-\frac{1}{2} - \frac{1}{2} = -1$ (since $|II_z > | = |\frac{1}{2} - \frac{1}{2}>$). This $I_z = -1$ can come from $I = 1$. Hence equation (1.245) can be rewritten as

$$(-1) = (-1)^{l_f}(-1)^{1+S}(-1)^{1+1} = (-1)^{l_f+S}(-1)^3$$

$$(-1)^{l_f+S} = 1$$

$$l_f + S = \text{even}$$

Final angular momentum is

$$\vec{J_f} = \vec{l_f} + \vec{s_n} + \vec{s_n} = \vec{l_f} + \vec{S}$$

$$\vec{S} = \vec{s_n} + \vec{s_n} = \frac{\vec{1}}{2} + \frac{\vec{1}}{2} = \begin{cases} 1 \\ 0 \end{cases}$$

$$\vec{J_f} = \vec{l_f} + \begin{cases} \vec{1} \\ \vec{0} \end{cases}$$

Angular momentum conservation gives

$$\vec{J_i} = \vec{J_f} = \vec{1} \text{ (equation 1.244)}$$

$$\vec{1} = \vec{l_f} + \begin{cases} \vec{1} \\ \vec{0} \end{cases}$$

$$\vec{l_f} = \vec{1} - \begin{cases} \vec{1} \rightarrow 2, 1, 0 \text{ for } S = 1 \\ \vec{0} \rightarrow 1 \quad\quad\text{ for } S = 0 \end{cases} \text{(from angular momentum addition rule)}$$

We choose that combination of $l_f$ and $S$ only, which gives $l_f + S =$ even. They are

| $l_f$ | $S$ | $l_f + S$ | odd/even | We choose |
|---|---|---|---|---|
| 2 | 1 | $2 + 1 = 3$ | odd | ✗ |
| 1 | 1 | $1 + 1 = 2$ | even | ✓ |
| 0 | 1 | $0 + 1 = 1$ | odd | ✗ |
| 1 | 0 | $1 + 0 = 1$ | odd | ✗ |

Clearly choice consistent with the conservation rules is

$$l_f = 1 \Rightarrow P \text{ state}, \ S = 1 \Rightarrow \text{Triplet},$$

Multiplicity $= 2S + 1 = 2.1 + 1 = 3$
Final $nn$ state is a $^3P_1$ state ($S = 1$, $2S + 1 = 3$, $l_f = 1$, $J_f = 1$)

$$|^3P_1 > = | \ l_f = 1, \ S = 1, \ J_f = 1>$$

The interaction is a strong interaction.
   Again parity of $n,d$ is positive $= (+1)$
   And parity is conserved so that we can equate parity of both sides of $\pi^- + d \rightarrow n + n$ to get
   $(-1)^{l_i}(\text{parity of } \pi^-)(+1) = (-1)^{l_f}(+1)(+1)$
   $(-1)^0(\text{parity of } \pi^-) = (-1)^1$
   Hence parity of $\pi^- = (-1)$. So pion parity is odd.

**Exercise 1.69** *The parity of the spin part of the wave function is called intrinsic parity. A deuteron d captures a $\pi^-$ in the $l = 1$ state and subsequently decays into a pair of neutrons through strong interaction. The spin wave function of the final state of the neutron will be which one of the following if it is known that the intrinsic parities of $\pi^-$, d, n, are $-1$, $+1$, $+1$ respectively:*
   *(a) linear combination of singlet and triplet; (b) singlet; (c) triplet; (d) doublet.*
   Hint: $\pi^- + d \rightarrow n + n$
   Parity is conserved in strong interaction. Hence we write taking into account that parities of $\pi^-$, $d$, $n$, are $-1$, $+1$, $+1$ respectively

$$(-1)^{l_i}(-1)(+1) = (-1)^{l_f}(+1)(+1).$$

With $l_i = 1$ it follows that

$$(-1)^{l_f} = 1.$$

So $l_f =$ even. Wave function of the $nn$ system is $\psi_{nn}$ and the parity equation is

$$\psi_{nn} = (-1)^{l_f}(-1)^{1+S}(-1)^{1+I} \quad \text{[figures 8.5 and 8.8]} \quad (1.246)$$

where $S$ is spin and $I$ is isospin of the $n$–$n$ system (final state).
   The $n$–$n$ system is a system of fermions and so system wave function will be anti-symmetric. This means $\psi_{nn} \rightarrow (-1)$.
   Again $I_z$ of $n$-$n$ system is $-\frac{1}{2} - \frac{1}{2} = -1$, since $|II_z > | = |\frac{1}{2} - \frac{1}{2}>_n$. This $I_z = -1$ can come from $I = 1$. We thus have from equation (1.246)
   $(-1) = (-1)^{\text{even}}(-1)^{1+S}(-1)^{1+1}$
   $(-1)^S = 1 \Rightarrow S = \text{even} = 0 \ (\text{say})$. So $nn$ is in a singlet state.

**Exercise 1.70** *What is the binding energy of a hydrogen atom sitting in ground state?* *(a)* 13.6 *eV* *(b)* −13.6 *MeV* *(c)*−2.225 *MeV* *(d)* +13.6 *eV*.
Hint: Hydrogen atom $\equiv pe^-$. Pumping +13.6 *eV* energy would disintegrate it.

Ans to Multiple Choice Questions

1.1*c*, 1.2*a,c*, 1.10*a*, 1.11*c,d*, 1.25*a*, 1.26*d*, 1.27*a*, 1.34*c*, 1.35*a,c*, 1.37*b*, 1.38*d*, 1.39*b*, 1.43*a*, 1.44*a*, 1.59*d*, 1.66*b*, 1.69*b*, 1.70*d*

## 1.37 Question bank

Q1.1   Mention areas where nuclear physics finds application.
Q1.2   Define *amu*. How is it related to mass of carbon 12, proton mass, neutron mass, electron mass? Express these masses in *MeV* also.
Q1.3   What are the two isotopes of hydrogen? Which one is a stable bound two-nucleon system and which one is radioactive?
Q1.4   There are 272 stable nuclei in nature. How do you classify them as even *A* and odd *A* nuclei? Give an example of even–even, even–odd, odd–even and odd–odd nuclei.
Q1.5   How many even–even and odd–odd nuclei are there? Which combination is preferred by nature?
Q1.6   What is a radio isotope? Give an example.
Q1.7   Distinguish between the terms isotope, isotone, isobar, isomer. Give an example of each.
Q1.8   How can isomeric states be differentiated experimentally?
Q1.9   How does an isomeric excited state differ from an ordinary excited state?
Q1.10  What are mirror nuclei? How is the charge symmetry hypothesis made from a study of mirror nuclei?
Q1.11  How can you determine nuclear radius from a study of mirror nuclei?
Q1.12  Justify why the isobars $_{20}Ca^{39}$ and $_{19}K^{39}$ are called mirror nuclei?
Q1.13  What symmetry is referred to in charge symmetry hypothesis?
Q1.14  What is charge independence hypothesis? What is its consequence?
Q1.15  What information about the nucleus do you get from the following hypotheses: charge symmetry hypothesis, charge independence hypothesis.
Q1.16  What is a nucleon? How do you define isospin quantum number?
Q1.17  Explain why neutron and proton are called iso-doublets?
Q1.18  Discuss why *nn*, *np* interactions are not exactly identical?
Q1.19  What is the cross-section of an interaction? What is its significance? What is a differential cross-section?
Q1.20  What is the SI unit of cross-section, differential cross-section? What is a barn? Why has it been defined?

Q1.21 Describe the Geiger, Marsden and Rutherford scattering experiment. How does it lead to a nuclear model of atom? How does the experiment depend upon foil thickness and materials?

Q1.22 Write down the important observations of Rutherford's gold foil experiment.

Q1.23 Most of $\alpha$ were forward scattered, very few $\alpha$ were back scattered in Rutherford's $\alpha$ scattering experiment. How does this lead to the concept of the nuclear model of an atom?

Q1.24 What are the arguments against the nuclear model of an atom of Rutherford?

Q1.25 Describe the nature of the path followed by an $\alpha$ due to scattering by gold foil in Rutherford's $\alpha$ scattering experiment. What is the equation of the path? Draw a schematic diagram.

Q1.26 Establish that the path followed by the $\alpha$ in Rutherford scattering experiment is a conic section. Find its eccentricity.

Q1.27 Write down the differential equation of the orbit followed by the $\alpha$ in Rutherford scattering from a gold nucleus.

Q1.28 Define the following in relation to Rutherford scattering experiment: impact parameter, distance of closest approach, differential scattering cross-section, total scattering cross-section. Write down their expressions.

Q1.29 Obtain an estimate of radius of nucleus as predicted by the Rutherford scattering experiment.

Q1.30 Show that the closer $\alpha$ is to the target, the greater is the deflection or scattering.

Q1.31 What are the various methods to obtain nuclear radius? Define the following: nuclear force radius, charge radius.

Q1.32 What information does departure from Rutherford's formula in his scattering experiment give?

Q1.33 Describe an experiment to measure charge radius of a nucleus.

Q1.34 What information is revealed regarding nucleon distribution within a nucleus from Woods–Saxon potential?

Q1.35 How is estimation of nuclear radius done from a study of atomic transition?

Q1.36 How is estimation of nuclear radius done from a study of mirror nuclei?

Q1.37 Calculate nuclear mass density, nucleon density.

Q1.38 What do you mean by binding energy of a nucleus? What is the significance of binding energy?

Q1.39 Show in a neat diagram the binding energy per nucleon against mass number of a nucleus. Identify region of constant $\frac{B}{A}$, region of fission, region of fusion. What is the value of the average binding energy per nucleon?

Q1.40 Define separation energy of a nucleon. Obtain its expression for proton and neutron.

Q1.41    What is mass defect? What is packing fraction? Plot packing fraction against mass number.

Q1.42    Give a schematic diagram of Segrè stability chart. Identify the belt of stability. How do the nuclei behave that do not lie on the stability curve?

Q1.43    Describe fine structure in the atomic case. What is meant by hyperfine structure? Show in a schematic diagram.

Q1.44    Define parity. Explain the following terms: space parity, intrinsic parity, total parity.

Q1.45    What is a Rabi magneton? Obtain its value.

Q1.46    Write down the expression of total magnetic moment of nucleus. How can you explain the magnetic moments $2.8\mu_N$ of a proton and $-1.9\mu_N$ of a neutron ?

Q1.47    What are the values of nuclear quadrupole moment of a nucleus having a shape that is identified to be spherical, prolate, oblate? Give examples of such nuclei.

# Further reading

[1] Krane S K 1988 *Introductory Nuclear Physics* (New York: Wiley)

[2] Tayal D C 2009 *Nuclear Physics* (Mumbai: Himalaya Publishing House)

[3] Prakash S 2005 *Nuclear Physics and Particle Physics* (New Delhi: Sultan Chand & Sons)

[4] Guha J 2019 *Quantum Mechanics: Theory, Problems and Solutions* 3rd edn (Kolkata: Books and Allied (P) Ltd))

[5] Lim Y-K 2002 *Problems and Solutions on Atomic, Nuclear and Particle Physics* (Singapore: World Scientific)

**IOP** Publishing

# Nuclear and Particle Physics with Cosmology, Volume 1
Nuclear physics
**Jyotirmoy Guha**

# Chapter 2

# Radioactivity

## 2.1 Introduction

Radioactivity was discovered by Becquerel in 1896.

Radioactivity refers to the emission of elementary particles $\alpha$, $\beta$, $\gamma$, by some unstable nuclei.

Stable nuclei are not radioactive—they do not emit particles.

| ☐ Stable nuclide | ☐ Unstable nuclide |
| --- | --- |
| Stable nuclide are stable against decay | Unstable nuclide are spontaneously radioactive |
| A stable nucleus cannot discover any avenue to shed off energy since its mass is already in the lowest state. | An unstable nucleus has an avenue to shed off its extra energy. In other words mass of unstable nucleus is larger than some other state to which it can transform. So it is in an excited state and needs to stabilize. |
| Binding energy of stable nucleus state is large. It is in ground state and holds no excess energy. | Binding energy of unstable nucleus is small. It is in excited state and has excess energy which it releases through emission of particles |

## 2.2 Process of shedding extra energy of a nucleus

Unstable nuclide can shed off their energy in five ways.

☐ $\alpha$ decay ☐ $\beta$ decay (three types) ☐ $\gamma$ decay.

Such transformation is represented as

$$\text{Parent} \xrightarrow{\ \alpha \text{ or } \beta \text{ or } \gamma \text{ decay}\ } \text{Daughter} \tag{2.1}$$

The parent is the nuclide before emission and the daughter is the nuclide after emission. The process of emission goes on till a stable nuclide is reached.

We give examples of the three types of decay.

- $\alpha$ decay

    Unstable nucleus emits $\alpha$ represented as

$$_Z X_N^A \rightarrow _{Z-2} Y_{N-2}^{A-4} + _2 He_2^4 \qquad (2.2)$$

as in

$$_{92} U_{146}^{238} \rightarrow _{90} Th_{144}^{234} + _2 He_2^4$$

Here $_Z X_N^A \equiv _{92} U_{146}^{238}$ is parent nucleus and $_{Z-2} Y_{N-2}^{A-4} \equiv _{90} Th_{144}^{234}$ is daughter nucleus.

Alpha emission occurs in a large parent nucleus and such decay reduces the size of the parent ($A \rightarrow A - 4$).

- $\beta$ decay (3 types)

☐ $\beta^-$ decay ☐ $\beta^+$ decay ☐ EC = Electron capture
☐ $\beta^-$ decay
Unstable nucleus emits $\beta^- \equiv e^- = _{-1} e^0$ represented as

$$_Z X_N^A \rightarrow _{Z+1} Y_{N-1}^A + _{-1} e^0 + \bar{\nu}_e \qquad (2.3)$$

as in

$$_6 C_8^{14} \rightarrow _7 N_7^{14} + _{-1} e^0 + \bar{\nu}_e$$

$\beta^-$ decay occurs because within nucleus neutron changes to a proton with the emission of electron $e^-$ and electron antineutrino $\bar{\nu}_e$ as

$$n \rightarrow p + e^- + \bar{\nu}_e$$

Hence neutron number decreases, proton number increases, i.e.

$$(Z, N, A) \xrightarrow{\beta^-} (Z + 1, N - 1, A) + e^- + \bar{\nu}_e$$

☐ $\beta^+$ decay
Unstable nucleus emits $\beta^+ \equiv e^+ = _{+1} e^0$ represented as

$$_Z X_N^A \rightarrow _{Z-1} Y_{N+1}^A + _{+1} e^0 + \nu_e \qquad (2.4)$$

as in

$$_{29} Cu_{35}^{64} \rightarrow _{28} Ni_{36}^{64} + e^+ + \nu_e$$

$\beta^+$ decay occurs because within the nucleus a proton changes to a neutron with the emission of positron $e^-$ and electron neutrino $\nu_e$ as

$$p \rightarrow n + e^+ + \nu_e$$

Hence neutron number increases, proton number decreases, i.e.

$$(Z, N, A) \xrightarrow{\beta^+} (Z - 1, N + 1, A) + e^+ + \nu_e$$

□ EC = Electron capture

An unstable nucleus captures an electron $e^-$. This is represented as

$$_Z X_N^A + _{-1}e^0 \rightarrow _{Z-1}Y_{N+1}^A + \nu_e \tag{2.5}$$

as in

$$_{29}Cu_{35}^{64} + _{-1}e^0 \rightarrow _{28}Ni_{36}^{64} + \nu_e$$

The nucleus has too many protons and one proton changes to a neutron through capture of an electron as

$$p + e^- \rightarrow n + \nu_e$$

Hence neutron number increases, proton number decreases, i.e.

$$(Z, N, A) + e^- \xrightarrow{EC} (Z - 1, N + 1, A) + \nu_e$$

✓ In $\beta^-$ decay, $\beta^+$ decay and $EC$ parent and daughter nuclei are isobars (same $A$).

• $\gamma$ decay

Unstable nucleus emits $\gamma$ represented as

$$(_Z X_N^A)^* \rightarrow _Z X_N^A + \gamma \tag{2.6}$$

as in

$$_{38}Sr_{49}^{87*} \rightarrow _{38}Sr_{49}^{87} + \gamma$$

The asterisk is used to denote explicitly that $_{38}Sr_{49}^{87}$ is in an excited state and it releases the extra energy directly through a radiative transition, i.e. by emission of a gamma.

✓ Comparison of penetrating powers of $\alpha, \beta, \gamma$.

Figure 2.1 shows the comparative ability of $\alpha, \beta$ and $\gamma$ to penetrate a medium.

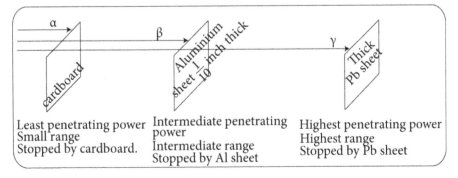

Figure 2.1. Penetrating power of $\alpha, \beta, \gamma$.

✓ Comparison of the ionising powers of $\alpha$, $\beta$, $\gamma$.

$\alpha$ particles are highly ionizing.

$$\text{Ionising power of } \alpha = 100 \text{ times more than the ionizing power of } \beta$$
$$= 10\,000 \text{ times more than the ionizing power of } \gamma \qquad (2.7)$$

On their passage through a medium (solid, liqiuid, gas) $\alpha$ particles ionize the medium through multiple collisions—in the process they gradually lose energy. When their energy falls below the ionization potential of the gas they stop ionizing and are converted to a neutral helium atom (through capture of two electrons) as

$$(_2\text{He}^4)^{++} + 2e^- \rightarrow {}_2\text{He}^4 \text{ atom}$$

## 2.3 Soddy Fajan's displacement law

In the periodic table

    ✓ During $\alpha$ decay daughter is displaced two places to the left as $Z$ decreases by 2 e.g.

$$_Z X \overset{\alpha}{\rightarrow} {}_{Z-2} Y \qquad (2.8)$$

as in

$$_{92}\text{U} \overset{\alpha}{\rightarrow} {}_{90}\text{Th}.$$

    ✓ During $\beta^-$ decay daughter is displaced one place to the right as $Z$ increases by unity e.g.

$$_Z X \overset{\beta^-}{\longrightarrow} {}_{Z+1} Y \qquad (2.9)$$

as in

$$_{90}\text{Th} \overset{\beta^-}{\longrightarrow} {}_{91}\text{Pa}.$$

    ✓ During $\beta^+$ decay daughter is displaced one place to the left as $Z$ decreases by unity e.g.

$$_Z X \overset{\beta^+}{\longrightarrow} {}_{Z-1} Y \qquad (2.10)$$

as in

$$_{29}\text{Cu} \overset{\beta^+}{\longrightarrow} {}_{28}\text{Ni}.$$

## 2.4 Radioactive series

There are four radioactive series or radioactive chains.

□ Thorium series (figure 2.2a)

First element is $_{90}\text{Th}^{232}_{142}$

Stable end product having $T_{1/2} = \infty$, $\lambda = 0$ is $_{82}\text{Pb}^{208}_{126}$

It is a $4n$ series starting at $A = 232 = 4(58)$ and ends with $A = 208 = 4(52)$.

□ Neptunium series (figure 2.2b)

First element is $_{93}\text{Np}^{237}_{144}$

Stable end product having $T_{1/2} = \infty$, $\lambda = 0$ is $_{83}\text{Bi}^{209}_{126}$

It is a $4n + 1$ series starting at $A = 237 = 4(59) + 1$ and ends with $A = 209 = 4(52) + 1$.

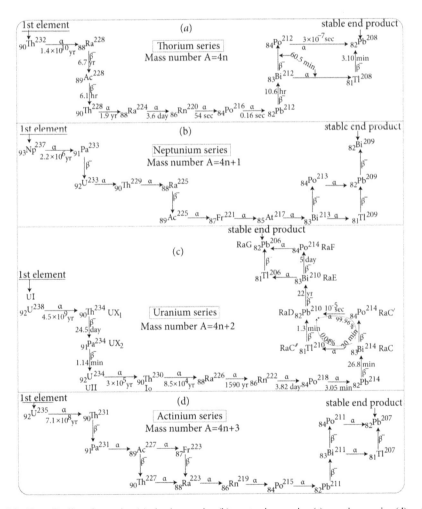

**Figure 2.2.** Four Radioactive series (a) thorium series (b) neptunium series (c) uranium series (d) actinium series.

☐ Uranium series (figure 2.2c)

First element is $_{92}U^{238}_{146}$

Stable end product having $T_{1/2} = \infty$, $\lambda = 0$ is $_{82}Pb^{206}_{124}$

It is a $4n + 2$ series starting at $A = 238 = 4(59) + 2$ and ends with $A = 206 = 4(51) + 2$.

☐ Actinium series (figure 2.2d)

First element is $_{92}U^{235}_{143}$

Stable end product having $T_{1/2} = \infty$, $\lambda = 0$ is $_{82}Pb^{207}_{125}$

It is a $4n + 3$ series starting at $A = 235 = 4(58) + 3$ and ends with $A = 207 = 4(51) + 3$.

## 2.5 Artificial radioactivity

Some nuclides can be artificially produced which are radioactive. Examples are as follows.

✓

$$_{2}He^4 + {_5}B^{10} \rightarrow {_7}N^{13} + {_0}n^1$$

$$_{7}N^{13} \rightarrow {_6}C^{13} + {_1}e^0 + \nu_e$$

$_{7}N^{13}$ is artificially produced unstable radioactive element. It decays through $e^+$ emission with $T_{1/2} = 14\,min$.

✓

$$_{0}n^1 + {_7}N^{14} \rightarrow {_6}C^{14} + p$$

$$_{6}C^{14} \rightarrow {_7}N^{14} + {_{-1}}e^0 + \bar{\nu}_e$$

$_{6}C^{14}$ is artificially produced unstable radioactive element. It decays through $e^-$ emission with $T_{1/2} = 5568$ year.

The artificially produced nuclides decay spontaneously according to the same laws that govern the disintegration of the naturally occurring radioactive elements viz.

$$N = N_0 e^{-\lambda t}, \ \lambda = \frac{0.693}{T_{1/2}}$$

that we discuss now.

- Above 1200 artificially radioactive nuclides are there.

## 2.6 Law of radioactivity or law of radioactive decay

### Survival equation

Consider a radioactive substance undergoing radioactive decay.

    ✓ Radioactive decay is a statistical process. Which nuclei would decay is subject to the law of chance.

    ✓ All nuclei of a radioactive sample have identical decay probability.

✓ Radioactive decay of a nucleus is independent of external excitation.
✓ The more time elapses, the more probable it is for a nucleus to exhibit decay. So probability $P$ of a nucleus to decay is proportional to time interval $dt$, i.e.

$$P \propto dt \tag{2.11}$$

Again the probability $P$ is, by definition

$$P = \frac{\text{number decaying}}{\text{total number}}. \tag{2.12}$$

Suppose number of radioactive nuclide present at time $t = 0$ (which is the point of time from which we start to observe) is $N_0$ and number present at time $t$ is $N(t)$. Obviously these nuclei have survived decay over time $t$ and are yet to decay. And in time $dt$ some more nuclei decay—the number of which is, say $-dN$ (minus signifying decay so that the number at time $t + dt$ is $N - dN$, i.e. number decreases).

Population of an unstable/radioactive sample diminishes in this way. So the probability of a nucleus to decay is

$$P = \frac{\text{number decaying}}{\text{total number}} = \frac{-dN}{N} \propto dt \tag{2.13}$$

$$\frac{-dN}{N} = \lambda dt \tag{2.14}$$

where we have introduced $\lambda$ as the proportionality constant called decay constant. $\lambda$ is a key factor in the survival equation that we derive in the following.

We rewrite equation (2.14) as

$$d\ln N = -\lambda dt \tag{2.15}$$

We integrate to get

$$\ln N = -\lambda t + C$$

where $C$ is a constant. At $t = 0$, $N = N_0$. So $C = \ln N_0$. With this we get

$$\ln N = -\lambda t + \ln N_0$$

$$\ln \frac{N}{N_0} = -\lambda t \tag{2.16}$$

This leads to

$$N = N_0 e^{-\lambda t} = N(t) \tag{2.17}$$

This is the number of nuclei surviving decay, i.e. the number of nuclei not decayed till time $t$. This equation is called the survival equation or law of radioactive disintegration or Rutherford–Soddy law.

It shows that the number of radioactive atoms/nuclei decreases exponentially with time. In other words with time the number of surviving nuclei $N(t)$ diminishes exponentionally due to radioactive decay.

We can derive the survival equation from the concept of probability also, as shown in *exercise 2.4*.

## 2.7 Half-life $T_{1/2}$

Half-life $T_{1/2}$ is the time interval in which the original number of nuclei is reduced to half. We give examples.

✓ Consider the radiactive decay

$$\text{Radium} \equiv {}_{88}\text{Ra}^{224}_{136} \xrightarrow{\alpha \text{ decay}} {}_{86}\text{Ra}^{220}_{134} \equiv \text{Radon}. \tag{2.18}$$

The half-life is $T_{1/2} = 3.6$ days. This means initial amount of radium always reduces to half in 3.6 days. If we start with an initial amount $N_0$ then the reduction of the amount of radium occurs as follows:

$$N_0 \xrightarrow{3.6 \text{ day}} \frac{N_0}{2} \equiv \frac{N_0}{2^1} \xrightarrow{3.6 \text{ day}} \frac{N_0/2}{2} \equiv \frac{N_0}{2^2} \xrightarrow{3.6 \text{ day}} \frac{N_0/2^2}{2} \equiv \frac{N_0}{2^3} \text{ and so on} \tag{2.19}$$

✓ Let us give another example

$$\text{Uranium} \equiv {}_{92}\text{U}^{238}_{146} \xrightarrow{\alpha \text{ decay}} {}_{90}\text{Th}^{234}_{144} \equiv \text{Thorium}.$$

The half-life is $T_{1/2} = 4.5 \times 10^9$ years.

This means that the amount of radium diminishes from amount $N_0$ to $\frac{N_0}{2}$ over a period of $4.5 \times 10^9$ years.

- In *exercise 2.1* we express the survival equation $N(t) = N_0 e^{-\lambda t}$ in terms of half-life $T_{1/2}$ as

$$N = N_0 2^{-t/T_{1/2}} \tag{2.20}$$

- Explicit expression of $\lambda$ in terms of half-life $T_{1/2}$

The Survival equation governing radioactive decay is, by equation (2.17) $N = N_0 e^{-\lambda t}$

$$\text{At } t = T_{1/2}, \ N = \frac{N_0}{2} \tag{2.21}$$

the survival equation gives

$$\frac{N_0}{2} = N_0 e^{-\lambda T_{1/2}} \ \Rightarrow \ \frac{1}{2} = e^{-\lambda T_{1/2}} \ \Rightarrow \ e^{\lambda T_{1/2}} = 2$$

$$\lambda T_{1/2} = ln2 \tag{2.22}$$

$$T_{1/2} = \frac{ln2}{\lambda} = \frac{0.693}{\lambda}$$

and so

$$\lambda = \frac{ln2}{T_{1/2}} = \frac{0.693}{T_{1/2}} \tag{2.23}$$

Clearly, $T_{1/2} \propto \frac{1}{\lambda}$ and $\lambda \propto \frac{1}{T_{1/2}}$

A graphical plot of the survival equation is shown in figure 2.3.

In *exercise 2.2* we show the plot of *lnN* versus *t*.

- Half-life of a radioactive substance can be experimentally measured.

- Significance of the decay constant $\lambda$
   1.      ✓

$$\lambda = \frac{-dN/N}{dt} \text{ (from equation (2.14))} \tag{2.24}$$

So decay constant $\lambda$ is the fraction of nuclei decaying per second. Clearly, $\lambda$ is the probability of decay per unit time of a particular radioactive nucleus.

✓ $\lambda$ is a measure of how much radioactively active an element is since $-\frac{dN}{dt} = \lambda N =$ rate of decay or activity = number of disintegrations per second.

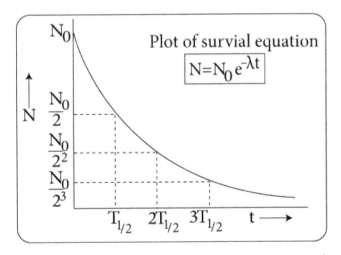

**Figure 2.3.** Plot of $N(t)$ versus $t$ as per survival equation $N(t) = N_0 e^{-\lambda t}$.

✓ Survival equation (2.17) is $N = N_0 e^{-\lambda t}$.

For $t = \frac{1}{\lambda}$, $N = N_0 e^{-\lambda \frac{1}{\lambda}} = N_0 e^{-1} \Rightarrow N = \frac{N_0}{e}$

$$\frac{N}{N_0} = \frac{1}{e} \tag{2.25}$$

So radioactive decay constant is the reciprocal of the time in which the number of atoms of a radioactive substance falls to $\frac{1}{e}$ times its initial value.

✓

$$\lambda = \frac{ln2}{T_{1/2}} = \frac{0.693}{T_{1/2}} \tag{2.26}$$

This means that the larger the decay constant $\lambda$ the smaller the half-life $T_{1/2}$ and the greater the chance of a given nucleus to decay in a certain time.

✓ Decay constant $\lambda$ is a measure of the speed of decay—slow or fast, i.e. controls the rate of decay. This is explained in figure 2.33 and in *exercise 2.3*.

✓ Inverse of decay constant $\frac{1}{\lambda}$ represents mean life $\tau$ of the radioactive sample, i.e.

$$\tau = 1/\lambda \tag{2.27}$$

We prove this relation now.

## 2.8 Mean life $\tau$

Define mean life as

$$\langle t \rangle \equiv \tau = \frac{\text{sum of lives of all radioactive nuclei}}{\text{total number of nuclei}} = \frac{\int_0^\infty t dN}{\int_0^\infty dN} \tag{2.28}$$

Using equation (2.17) viz. $N = N_0 e^{-\lambda t}$ we get

$$\tau = \frac{\int_0^\infty t d(N_0 e^{-\lambda t})}{N_0} = \int_0^\infty t de^{-\lambda t} \tag{2.29}$$

$$= \int_0^\infty t(-\lambda)e^{-\lambda t}dt = -\lambda \int_0^\infty te^{-\lambda t}$$

Let us emphasize that mean life $\tau$ is a positive definite quantity—it cannot be negative. The slope of the $N(t)$ versus $t$ plot (figure 2.3) is negative, signifying that as time $t$ increases the amount of radioactive substance diminishes. This negative slope is reflected in the negative sign before the decay constant $\lambda$ in the expression for mean life $\tau$ and hence should be ignored. We thus rewrite the expression for mean life

$$\tau = \lambda \int_0^\infty te^{-\lambda t}$$

We integrate by parts to get

$$\tau = \lambda \left[ t \int e^{-\lambda t} \, dt \; \Big|_0^\infty - \int_0^\infty \frac{d}{dt} t \left( \int e^{-\lambda t} \, dt \right) dt \right]$$

$$= \lambda \left[ t \frac{e^{-\lambda t}}{-\lambda} \Big|_0^\infty - \int_0^\infty \frac{e^{-\lambda t}}{-\lambda} dt \right] = \int_0^\infty e^{-\lambda t} \, dt$$

$$\tau = \frac{e^{-\lambda t}}{-\lambda} \; \Big|_0^\infty = \frac{e^{-\lambda \infty} - e^{\lambda 0}}{-\lambda} = \frac{0 - 1}{-\lambda} = \frac{1}{\lambda}$$

$$\tau = \frac{1}{\lambda} \tag{2.30}$$

So mean life is the inverse of decay constant.
- In *exercise 2.5* we evaluate $<t^2>$.

## 2.9 Activity in terms of Becquerel, Rutherford and Curie

Activity is defined as the rate of decay of a radioactive substance, i.e.

$$\text{Activity} = -\frac{dN}{dt} = \lambda N \tag{2.31}$$

The negative sign represents diminuation of nucleus population because of radioactive decay.
- The S.I unit of activity is the Becquerel, defined as

$$\text{Becquerel} \equiv 1\text{Bq} = 1 \frac{\text{decay}}{\text{second}}.$$

Let us mention the other units of radioactivity.
- ✓ 1 Rutherford $\equiv$ 1Rd represents that quantity of a radioactive substance which gives $10^6$ disintegrations per second.
- ✓ 1 Curie $\equiv$ 1 $Ci$ represents that quantity of a radioactive substance which gives $3.7 \times 10^{10}$ disintegrations per second.

- For 1 Curie of activity we require
  - small quantity of radioactive substance having small half life $T_{1/2}$ and
  - large quantity of radioactive substance having large half life $T_{1/2}$.

## 2.10 Law of successive disintegration

Suppose a parent nucleus is radioactive. It decays and we get a daughter nucleus. Suppose this daughter nucleus is also radioactive and it decays. The product nucleus —thus formed may also be radioactive and if so will decay. The process goes on till we get a stable nucleus. We thus have a radioactive series or decay chain.

When successive generations of nuclei are radioactive we describe the process as successive disintegration.

Let us now investigate what type of equilibrium is established in the decay chain, say

Radioactive parent $\xrightarrow[\text{occurs}]{\text{decay}}$ Radioactive daughter $\xrightarrow[\text{occurs}]{\text{decay}}$ stable nucleus (end product)

and let us rewrite it as

$$1 \xrightarrow{\lambda_1} 2 \xrightarrow{\lambda_2} 3 \tag{2.32}$$

where $\lambda_1$, $\lambda_2$ are the decay constants for the decays $1 \rightarrow 2$ and $2 \rightarrow 3$, respectively. Let at time $t$ the number of nuclei of type 1, 2, 3 be $N_1$, $N_2$, $N_3$, respectively.

Let at time $t = 0$ the number of type 1 nuclei be $N_{10}$.

The rate of disintegration or activity of type 1 nuclei (which is the parent) is given by

$$-\frac{dN_1}{dt} = \lambda_1 N_1 \tag{2.33}$$

$$\frac{dN_1}{N_1} = -\lambda_1 dt \Rightarrow dlnN_1 = -\lambda_1 dt.$$

Integrate to get

$$\int dlnN_1 = -\lambda_1 \int dt + C$$

$$lnN_1 = -\lambda_1 t + C$$

Let at $t = 0$, $N_1 = N_{10}$, $C = lnN_{10} = $ constant. This gives

$$ln\frac{N_1}{N_{10}} = -\lambda t$$

$$N_1 = N_{10}e^{-\lambda_1 t} \tag{2.34}$$

This is the number of type 1 nuclei at time $t$.

As parent (i.e. type 1 nuclei) is radioactive it decays and disappears and so type 2 nuclei (called daughter) appears.

Clearly the rate of appearance of 2 = rate of disappearance or activity of $1 = \lambda_1 N_1$.

Again, type 2 nuclei are also radioactive and decay to 3 which is the stable end product/ nuclei. The rate of disappearance or activity of 2 is $\lambda_2 N_2$. So net increase in 2 is governed by the equation

$$\frac{dN_2}{dt} = \text{creation of 2 minus disintegration of } 2 = \lambda_1 N_1 - \lambda_2 N_2.$$

We thus have

$$\frac{dN_2}{dt} = \lambda_1 N_1 - \lambda_2 N_2 \tag{2.35}$$

Using equation (2.34) we have

$$\frac{dN_2}{dt} = \lambda_1 N_{10} e^{-\lambda_1 t} - \lambda_2 N_2 \qquad (2.36)$$

$$\frac{dN_2}{dt} + \lambda_2 N_2 = \lambda_1 N_{10} e^{-\lambda_1 t}.$$

Multiply by integrating factor $e^{\lambda_2 t}$ to get

$$\frac{dN_2}{dt} e^{\lambda_2 t} + \lambda_2 N_2 e^{\lambda_2 t} = \lambda_1 N_{10} e^{-\lambda_1 t} e^{\lambda_2 t}$$

$$\frac{d}{dt}(N_2 e^{\lambda_2 t}) = \lambda_1 N_{10} e^{(\lambda_2 - \lambda_1)t}.$$

Integrating we get

$$N_2 e^{\lambda_2 t} = \lambda_1 N_{10} \int e^{(\lambda_2 - \lambda_1)t} \, dt + C$$

$$= \lambda_1 N_{10} \frac{e^{(\lambda_2 - \lambda_1)t}}{\lambda_2 - \lambda_1} + C$$

Let at $t = 0$, $N_2 = 0$, $C = -\frac{\lambda_1 N_{10}}{\lambda_2 - \lambda_1}$

$$N_2 e^{\lambda_2 t} = \lambda_1 N_{10} \frac{e^{(\lambda_2 - \lambda_1)t}}{\lambda_2 - \lambda_1} - \frac{\lambda_1 N_{10}}{\lambda_2 - \lambda_1}$$

$$= \frac{\lambda_1 N_{10}}{\lambda_2 - \lambda_1}(e^{(\lambda_2 - \lambda_1)t} - 1)$$

$$N_2 = \frac{\lambda_1 N_{10}}{\lambda_2 - \lambda_1}(e^{(\lambda_2 - \lambda_1)t} - 1)e^{-\lambda_2 t}$$

$$N_2 = \frac{\lambda_1 N_{10}}{\lambda_2 - \lambda_1}(e^{-\lambda_1 t} - e^{-\lambda_2 t}) \qquad (2.37)$$

This is the number of type 2 nuclei at time $t$.

As type 2 nuclei are radioactive they decay and disappear and so type 3 nuclei (stable end product) appear. Clearly

Rate of appearance of 3 = Rate of disappearance or activity of 2 = $\lambda_2 N_2$  (2.38)

So increase of 3 is governed by the equation

$$\frac{dN_3}{dt} = \text{creation of } 3 = \lambda_2 N_2. \qquad (2.39)$$

We thus have

$$\frac{dN_3}{dt} = \lambda_2 N_2$$

Putting $N_2$ from equation (2.37) we have

$$\frac{dN_3}{dt} = \lambda_2 \frac{\lambda_1 N_{10}}{\lambda_2 - \lambda_1}(e^{-\lambda_1 t} - e^{-\lambda_2 t}) \qquad (2.40)$$

We integrate to get

$$N_3 = \frac{\lambda_1 \lambda_2 N_{10}}{\lambda_2 - \lambda_1} \left( \int e^{-\lambda_1 t}\, dt - \int e^{-\lambda_2 t}\, dt \right) + C \qquad (2.41)$$

$$N_3 = \frac{\lambda_1 \lambda_2 N_{10}}{\lambda_2 - \lambda_1} \left( \frac{e^{-\lambda_1 t}}{-\lambda_1} - \frac{e^{-\lambda_2 t}}{-\lambda_2} \right) + C$$

$$N_3 = \frac{\lambda_1 \lambda_2 N_{10}}{\lambda_2 - \lambda_1} \left( \frac{e^{-\lambda_2 t}}{\lambda_2} - \frac{e^{-\lambda_1 t}}{\lambda_1} \right) + C$$

Let at $t = 0$, $N_3 = 0$, $C = \frac{\lambda_1 \lambda_2 N_{10}}{\lambda_2 - \lambda_1} \left( \frac{1}{\lambda_1} - \frac{1}{\lambda_2} \right)$.

$$N_3 = \frac{\lambda_1 \lambda_2 N_{10}}{\lambda_2 - \lambda_1} \left( \frac{e^{-\lambda_2 t}}{\lambda_2} - \frac{e^{-\lambda_1 t}}{\lambda_1} \right) + \frac{\lambda_1 \lambda_2 N_{10}}{\lambda_2 - \lambda_1} \left( \frac{1}{\lambda_1} - \frac{1}{\lambda_2} \right)$$

$$N_3 = \frac{\lambda_1 \lambda_2 N_{10}}{\lambda_2 - \lambda_1} \left( \frac{1 - e^{-\lambda_1 t}}{\lambda_1} - \frac{1 - e^{-\lambda_2 t}}{\lambda_2} \right) \qquad (2.42)$$

This is the number of type 3 nuclei at time $t$.
- Time $t_0$ in which type 2 nuclei reach a maximum in the radioactive chain $1 \to 2 \to 3$ where 1 and 2 are radioactive but 3 is the stable end product

In the radioactive chain $1 \to 2 \to 3$ we got in equation (2.37) $N_2 = \frac{\lambda_1 N_{10}}{\lambda_2 - \lambda_1} (e^{-\lambda_1 t} - e^{-\lambda_2 t})$. The time at which the amount of $N_2$ reaches maximum we should have

$$\frac{dN_2}{dt} = 0 \text{ (maximization of } N_2) \qquad (2.43)$$

This means

$$\frac{d}{dt} \frac{\lambda_1 N_{10}}{\lambda_2 - \lambda_1} (e^{-\lambda_1 t} - e^{-\lambda_2 t}) = 0 \Rightarrow \frac{d}{dt}(e^{-\lambda_1 t} - e^{-\lambda_2 t}) = 0$$

$$-\lambda_1 e^{-\lambda_1 t} - (-\lambda_2 e^{-\lambda_2 t}) = 0 \Rightarrow \lambda_1 e^{-\lambda_1 t} = \lambda_2 e^{-\lambda_2 t}$$

$$\frac{\lambda_2}{\lambda_1} = e^{(\lambda_2 - \lambda_1)t} \qquad (2.44)$$

Take *ln* to arrive at

$ln\frac{\lambda_2}{\lambda_1} = (\lambda_2 - \lambda_1)t$

$$t = \frac{1}{\lambda_2 - \lambda_1} ln\frac{\lambda_2}{\lambda_1} = t_0 \qquad (2.45)$$

This is the time $t_0$ at which type 2 nuclei will be maximum in number.
Using equation (2.30) we can write

$$\tau_1 = \frac{1}{\lambda_1}, \quad \tau_2 = \frac{1}{\lambda_2}$$

$$t_0 = \frac{\tau_1\tau_2}{\tau_1 - \tau_2} ln\frac{\tau_1}{\tau_2}$$

$$\text{Using } \tau = \frac{1}{\lambda} = \frac{T_{1/2}}{0.693}$$

$$t_0 = \tau_2\frac{(T_{1/2})_1}{(T_{1/2})_1 - (T_{1/2})_2} ln\frac{(T_{1/2})_1}{(T_{1/2})_2} \qquad (2.46)$$

Daughter activity is maximum at $t_0$ both at $(T_{1/2})_1 > (T_{1/2})_2$ and $(T_{1/2})_2 > (T_{1/2})_1$ (-there being no physical case for $(T_{1/2})_1 = (T_{1/2})_2$.

In *exercise 2.59* we show that $t_0 \cong \sqrt{\tau_1\tau_2}$

## 2.11 Ideal equilibrium

In the radioactive decay chain $1 \rightarrow 2 \rightarrow 3$ the concentration of 2 is maximum at time $t_0$. So $t = t_0$ corresponds to $\frac{dN_2}{dt} = 0$ or from equation (2.35)

$$\frac{dN_2}{dt} = \lambda_1 N_1 - \lambda_2 N_2 = 0$$

$$\lambda_1 N_1 = \lambda_2 N_2$$

which means

Parent activity = Daughter activity

i.e. there is a balance between production of 2 and disintegration of 2. This is called ideal equilibrium during which the slope of the $N_2$ versus $t$ plot (figure 2.4(a)) is zero.

✓

$$\text{At } t < t_0, \lambda_1 N_1 > \lambda_2 N_2, \frac{dN_2}{dt} > 0$$

as production of daughter dominates over decay of daughter. This is because the parent is more active. So $N_2$ grows with time.

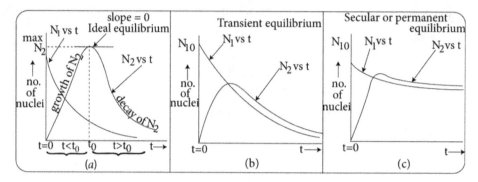

**Figure 2.4.** (a) Concentration of $N_2$ is maximum at time $t_0$ at which growth of $N_2$ and decay of $N_2$ are at equilibrium called ideal equilibrium. (b) Transient equilibrium. (c) Secular or permanent or long-term equilibrium.

Growth curve has positive slope.

✓

$$\text{At } t > t_0, \; \lambda_1 N_1 < \lambda_2 N_2, \; \frac{dN_2}{dt} < 0$$

as decay of daughter dominates over production of daughter. This is because the daughter is more active. So $N_2$ decays with time.

The decay curve has negative slope.

## 2.12 Short-lived parent

We investigate equilibrium if 1 is the short-lived compared to 2 in the radioactive chain $1 \to 2 \to 3$ (3 being the stable end product).

Consider the radioactive decay chain $1 \to 2 \to 3$ where 1 is the short-lived radioactive parent and the radioactive daughter 2 is longer lived while 3 is the stable end product.

The number of type 2 nuclei we obtained in equation (2.37) is $N_2 = \frac{\lambda_1 N_{10}}{\lambda_2 - \lambda_1} \left( e^{-\lambda_1 t} - e^{-\lambda_2 t} \right)$. Now

$$(T_{1/2})_1 \ll (T_{1/2})_2 \tag{2.47}$$

$$\lambda_1 \gg \lambda_2$$

$e^{-\lambda_1 t} \ll e^{-\lambda_2 t}$ for large $t$. This means we can neglect $e^{-\lambda_1 t}$ to get from equation (2.37)

$$N_2 \approx \frac{\lambda_1 N_{10}}{\lambda_2 - \lambda_1} \left( -e^{-\lambda_2 t} \right)$$

$$N_2 = \frac{\lambda_1 N_{10}}{\lambda_1 - \lambda_2} e^{-\lambda_2 t}$$

$$N_2 \cong N_{10} \, e^{-\lambda_2 t} \text{ as } \lambda_1 \gg \lambda_2 \text{ we took } \lambda_1 - \lambda_2 \cong \lambda_1 \tag{2.48}$$

So decay of 2 is determined by $\lambda_2$, i.e. by $(T_{1/2})_2 = \frac{0.693}{\lambda_2}$ that is, by its own half-life. So after a long time the activity of 2 becomes independent of the activity of 1. So there can be no balance—no equilibrium. This is because all parent nuclei get transformed into daughters.

## 2.13 Transient equilibrium. Relatively long-lived parent

Consider the radioactive decay chain $1 \rightarrow 2 \rightarrow 3$ where 1 is the long-lived radioactive parent and the radioactive daughter 2 is shorter lived while 3 is the stable end product. So

$$(T_{1/2})_1 > (T_{1/2})_2$$

$$\lambda_1 < \lambda_2 \Rightarrow \quad e^{-\lambda_1 t} > e^{-\lambda_2 t} \tag{2.49}$$

And for large $t$, $e^{-\lambda_1 t}$ is so large and $e^{-\lambda_2 t}$ is so small that we can neglet $e^{-\lambda_2 t}$. We derived the concentration of $N_1$ in equation (2.34) to be $N_1 = N_{10} e^{-\lambda_1 t}$. This means that decay of 1 is determined by $\lambda_1$, i.e.

$$(T_{1/2})_1 = \frac{0.693}{\lambda_1} \text{ — the half-life of 1}$$

By equation (2.37) viz. $N_2 = \frac{\lambda_1 N_{10}}{\lambda_2 - \lambda_1} \ (e^{-\lambda_1 t} - e^{-\lambda_2 t})$ we have on using equation (2.49) and neglecting $e^{-\lambda_2 t}$

$$N_2 \approx \frac{\lambda_1 N_{10}}{\lambda_2 - \lambda_1} \ e^{-\lambda_1 t}$$

Clearly the daughter decays with decay constant $\lambda_1$ and half-life $(T_{1/2})_1$ of the parent.

Using equation (2.34) we get

$$N_2 = \frac{\lambda_1 N_1}{\lambda_2 - \lambda_1} \tag{2.50}$$

This is the decay of 2 that is determined also by $\lambda_1$ after a long time.

Clearly the ratio of concentration of parent 1 to daughter 2 is after a long time

$$\frac{N_2}{N_1} = \frac{\lambda_1}{\lambda_2 - \lambda_1} = \text{constant} \tag{2.51}$$

Although $N_1$ and $N_2$ decrease with time their ratio is constant. This is reflective of the balance or equilibrium between parent 1 and daughter 2. This is referred to as transient equilibrium (figure 2.4(b)).

The ratio of activities of parent 1 and daughter 2 is after a long time

$$\frac{\lambda_2 N_2}{\lambda_1 N_1} = \frac{\lambda_2}{\lambda_1} \frac{\lambda_1}{\lambda_2 - \lambda_1} = \frac{\lambda_2}{\lambda_2 - \lambda_1} > 1 \tag{2.52}$$

So activity of the daughter is greater than the activity of the parent by a factor $\frac{\lambda_2}{\lambda_2 - \lambda_1}$.

Using $\lambda_1 = \frac{0.693}{(T_{1/2})_1}$, $\lambda_2 = \frac{0.693}{(T_{1/2})_2}$ we get from equation (2.52) after a long time

$$\frac{\lambda_2 N_2}{\lambda_1 N_1} = \frac{(T_{1/2})_1}{(T_{1/2})_1 - (T_{1/2})_2} > 1 \tag{2.53}$$

For example

$$_{88}\text{Ra}^{228} \xrightarrow[6.7\ \text{year}]{\beta^-} \left[ _{89}\text{Ac}^{228} \xrightarrow[6\ h]{\beta^-} \right] _{90}\text{Th}^{228} \xrightarrow[1.9\ h]{\alpha} {}_{88}\text{Ra}^{224}$$

As $_{89}\text{Ac}^{228}$ has negligible half-life we can ignore its presence and take the chain to be

$$_{88}\text{Ra}^{228} \xrightarrow[6.7\ \text{year}]{\beta^-} {}_{90}\text{Th}^{228} \xrightarrow[1.9\ h]{\alpha} {}_{88}\text{Ra}^{224}.$$

We note that

$$(T_{1/2})_{\text{Ra}} = 6.7\ \text{year} > (T_{1/2})_{\text{Th}} = 1.9\ h$$

So transient equilibrium is achieved.

## 2.14 Secular equilibrium. Extremely long-lived parent

Consider the radioactive decay chain $1 \to 2 \to 3$ where 1 is a very long-lived radioactive parent compared to radioactive daughter 2 while 3 is the stable end product. So

$$(T_{1/2})_1 > \gg (T_{1/2})_2 \Rightarrow \lambda_1 \ll < \lambda_2. \tag{2.54}$$

We can thus take
$(T_{1/2})_1 \approx \infty$, $\lambda_1 \approx 0$ and $(T_{1/2})_2 =$ very small $\to 0$, $\lambda_2 =$ very large $\to \infty$. So

$$e_1^{-\lambda_1 t} \approx e^{-(0)t} = 1, \quad e^{-\lambda_2 t} \approx e^{-(\infty)t} = 0. \tag{2.55}$$

We derived the concentrations to be

$$N_1 = N_{10} e^{-\lambda_1 t} \approx N_{10}$$

$$N_2 = \frac{\lambda_1 N_{10}}{\lambda_2 - \lambda_1} \left( e^{-\lambda_1 t} - e^{-\lambda_2 t} \right) \approx \frac{\lambda_1 N_{10}}{\lambda_2 - 0} (1 - 0) = \frac{\lambda_1 N_{10}}{\lambda_2} \equiv \frac{\lambda_1 N_1}{\lambda_2} = \text{constant.}$$

So the amount of daughter $N_2$ is practically constant. It follows that

$$\lambda_1 N_1 = \lambda_2 N_2 \tag{2.56}$$

i.e. activities of 1 and 2 are equal. This type of balance between parent and daughter is referred to as secular or permanent or long-term equilibrium and is shown in figure 2.4(c).

Example

$$_{88}Ra^{226} \xrightarrow[1590\ year]{\alpha} {}_{86}Rn^{222} \xrightarrow[3.82\ days]{\alpha} {}_{84}Po^{218}.$$

We note that

$$(T_{1/2})_{Ra} = 1590\ year \gg (T_{1/2})_{Rn} = 3.82\ day.$$

So secular equilibrium is achieved and

$$\lambda_{Ra}N_{Ra} = \lambda_{Rn}\ N_{Rn}. \tag{2.57}$$

- In case various members of the decay chain $1 \to 2 \to 3 \to 4 \to 5 \to \dots$ are in secular equilibrium we write

$$\lambda_1 N_1 = \lambda_2 N_2 = \lambda_3 N_3 = \lambda_4 N_4 = \lambda_5 N_5 = \dots$$

$$\frac{N_1}{(T_{1/2})_1} = \frac{N_2}{(T_{1/2})_2} = \frac{N_3}{(T_{1/2})_3} = \frac{N_4}{(T_{1/2})_4} = \frac{N_5}{(T_{1/2})_5} = \dots$$

In this sequence every member is present in a constant but different proportion. Abundance of different members is proportional to their half lives. Rate of decay and rate of production of each member are constant and activity is equal to the activity of the first member.

At secular equilibrium the rate of decay of any radioactive product is equal to its rate of production fom the previous member in the chain.

In *exercise 2.11* we discuss that due to secular equilibrium the percentage of radium contained in uranium is always experimentally found to be same viz. 1 gm of radium per 3.5 ton of pure uranium.

- In *exercise 2.13* we discuss radioactive dating (determination of age of earth $\sim 5.9 \times 10^9$ years.
- In *exercise 2.14* we discuss uranium dating (determination of age of minerals).
- In *exercises 2.15* and *2.16* we discuss radio carbon dating (determination of fossil or archeological sample).

## 2.15 Alpha decay

Consider $\alpha$ decay for nuclei having $A > 200$, $Z > 81$, $N > 120$ represented as in equation (2.2) viz.

$$_ZX^A \to {}_{Z-2}\ Y^{A-4} + {}_2He^4 \quad \text{for instance}$$

$$_{83}Bi^{212} \to {}_{81}Tl^{208} + {}_2He^4$$

Here $_{83}Bi^{212}$ is parent, $_{81}Tl^{208}$ is daughter. The $\alpha \equiv {}_2He^4$ that is emitted has certain discrete (fixed) values of energy (MeV). In other words $\alpha$ spectra is discrete, as shown in figure 2.5.

The implication of the existence of a discrete $\alpha$ spectrum (or fine structure in $\alpha$ energy spectrum) is that a nucleus exists in discrete energy levels (*exercise 2.19*).

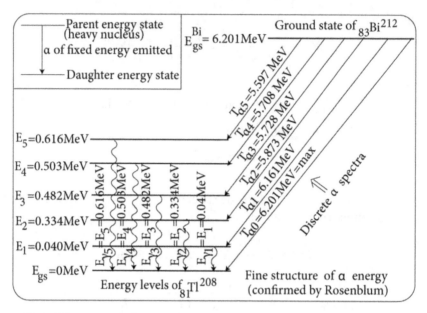

**Figure 2.5.** $\alpha$ spectra is discrete. $E$ is energy of level, $T_\alpha$ is kinetic energy of emitted $\alpha$.

Clearly $_{83}\mathrm{Bi}^{212}$ which is in the ground state emits $\alpha$ of energies $T_{\alpha 0}$, $T_{\alpha 1}$, $T_{\alpha 2}$, $T_{\alpha 3}$, $T_{\alpha 4}$, $T_{\alpha 5}$.

The $T_{\alpha 0} = 6.201$ $MeV$ is most energetic and obtained due to transformation from ground state $E_{gs}^{Bi} = 6.201$ $MeV$ of $_{83}\mathrm{Bi}^{212}$ to the ground state $E_{gs} = 0$ $MeV$ of $_{81}\mathrm{Tl}^{208}$ while $T_{\alpha 5} = 5.597$ $MeV$ is least energetic corresponding to transformation to the energy level $E_5 = 0.616$ $MeV$ of the daughter nucleus $_{81}\mathrm{Tl}^{208}$.

The other alphas emitted from ground state of $_{83}\mathrm{Bi}^{212}$ with various discrete energies 6.161, 5.873, 5.728, 5.708 $MeV$ give rise to daughter nucleus $_{81}\mathrm{Tl}^{208}$ in discrete states with energies $E_1 = 0.040$ $MeV$, $E_2 = 0.334$ $MeV$, $E_3 = 0.482$ $MeV$, $E_4 = 0.503$ $MeV$ respectively, which are different energy levels of the daughter nucleus $_{81}\mathrm{Tl}^{208}$.

The excited daughter nucleus $_{81}\mathrm{Tl}^{208}$ decays to its ground state $E_{gs} = 0$ $MeV$ by emission of $\gamma$ radiation $E_{\gamma 1}$, $E_{\gamma 2}$, $E_{\gamma 3}$, $E_{\gamma 4}$, $E_{\gamma 5}$ of energies 0.04, 0.334, 0.482, 0.503, 0.616 $MeV$ respectively as indicated in the figure 2.5.

## 2.16 Alpha disintegration energy $Q_\alpha$

Consider the $\alpha$ decay

$$_Z X^A \rightarrow {}_{Z-2} Y^{A-4} + {}_2\mathrm{He}^4 \qquad (2.58)$$

Assume that parent $_Z X^A$ is in ground state and suffers a spontaneous $\alpha$ decay. The daughter $_{Z-2} Y^{A-4}$ is also in the ground state. We rewrite the $\alpha$ decay process from the point of view of conservation of energy as

$$_Z X^A \rightarrow {}_{Z-2} Y^{A-4} + {}_2\mathrm{He}^4 + Q_\alpha \qquad (2.59)$$

If $Q_\alpha > 0$ this implies that rest mass energy of parent ${}_Z X^A$ is greater than the rest mass energy of the products (${}_{Z-2} Y^{A-4}$, ${}_2 He^4$) and the difference of rest mass energies is converted into kinetic energy of the products. The parent has thus an option to lower its rest mass energy and hence will be unstable and will suffer $\alpha$ decay.

Alpha disintegration energy $Q_\alpha$ is defined as the total energy released in the process of $\alpha$ decay — in fact it is the mass or energy difference on both sides called $Q$ value of $\alpha$ decay. So

$$Q_\alpha = (M_{\text{initial}} - M_{\text{final}})c^2 = (M_X - M_Y - M_\alpha)c^2 > 0 \tag{2.60}$$

Here $M$ is mass. By conservation of energy principle we write

$$E_X = E_Y + E_\alpha \tag{2.61}$$

where the formula is $E = T + Mc^2$, $T \to$ kinetic energy. So we have

$$T_X + M_X c^2 = T_Y + M_Y c^2 + T_\alpha + M_\alpha c^2 \tag{2.62}$$

There is no supply of energy to $X$ from outside and so $X$ is at rest in laboratory frame of reference. So we take
$T_X = 0$, $v_X = 0$. Also $T_Y = \frac{1}{2} M_Y v_Y^2$, $T_\alpha = \frac{1}{2} M_\alpha v_\alpha^2$. Putting these in the energy conservation equation we have

$$M_X c^2 = \frac{1}{2} M_Y v_Y^2 + M_Y c^2 + \frac{1}{2} M_\alpha v_\alpha^2 + M_\alpha c^2$$

$$(M_X - M_Y - M_\alpha)c^2 = \frac{1}{2} M_Y v_Y^2 + \frac{1}{2} M_\alpha v_\alpha^2 \tag{2.63}$$

Using equation (2.60) we identify LHS as $Q_\alpha$. Hence

$$Q_\alpha = \frac{1}{2} M_Y v_Y^2 + \frac{1}{2} M_\alpha v_\alpha^2 \tag{2.64}$$

Let us apply conservation of momentum

$$M_X v_X = M_Y v_Y + M_\alpha v_\alpha$$

With $v_X = 0$ we can write

$$M_\alpha v_\alpha = -M_Y v_Y$$

$$v_Y = -\frac{M_\alpha}{M_Y} v_\alpha \tag{2.65}$$

Daughter recoils with velocity $v_Y$. Put it in the $Q$ equation (2.64) to get

$$Q_\alpha = \frac{1}{2} M_Y \left( -\frac{M_\alpha}{M_Y} v_\alpha \right)^2 + \frac{1}{2} M_\alpha v_\alpha^2 = \frac{1}{2} \frac{M_\alpha^2 v_\alpha^2}{M_Y} + \frac{1}{2} M_\alpha v_\alpha^2 = \frac{1}{2} M_\alpha v_\alpha^2 \left( 1 + \frac{M_\alpha}{M_Y} \right)$$

$$Q_\alpha = T_\alpha \left( 1 + \frac{M_\alpha}{M_Y} \right) \tag{2.66}$$

$$T_\alpha = \frac{Q_\alpha}{1 + \frac{M_\alpha}{M_Y}} \tag{2.67}$$

This result holds if parent and daughter are both in the ground state. Let us replace the ratio of masses by the ratio of mass numbers, i.e.

$$\frac{M_\alpha}{M_Y} = \frac{4}{A - 4} \tag{2.68}$$

where $A$ is the mass number of parent. This gives

$$Q_\alpha = T_\alpha\left(1 + \frac{4}{A - 4}\right) \tag{2.69}$$

$$Q_\alpha = T_\alpha\frac{A}{A - 4} \Rightarrow T_\alpha = Q_\alpha\frac{A - 4}{A} \tag{2.70}$$

As $A$ is large for $\alpha$ decay we have $Q_\alpha \approx T_\alpha$. So the $\alpha$ particle carries away most of the disintegration energy.

If the daughter is in the excited state $E_n$ i.e. [instead of equation (2.58)] we have

$$_ZX^A \rightarrow (_{Z-2}Y^{A-4})^* + _2He^4 \tag{2.71}$$

then the conservation of energy equation has to include the excitation energy $E_n$ of the daughter. So instead of equation (2.61) we have

$$E_X = (E_Y + E_n) + E_\alpha \tag{2.72}$$

Hence instead of equation (2.63) we have

$$(M_X - M_Y - M_\alpha)c^2 = \frac{1}{2}M_Yv_Y^2 + \frac{1}{2}M_\alpha v_\alpha^2 + E_n \tag{2.73}$$

We note that the LHS of equation (2.73) is $Q_\alpha$ and using momentum conservation the first 2 terms of RHS give $T_\alpha(1 + \frac{M_\alpha}{M_Y})$ as the RHS of equation (2.66). Hence we have for an excited daughter the following relation instead of equation (2.66)

$$Q_\alpha = T_\alpha\left(1 + \frac{M_\alpha}{M_Y}\right) + E_n \tag{2.74}$$

✓ The kinetic energy of $\alpha$ in $\alpha$ decay, i.e the quantity $T_\alpha$ can be measured experimentally. The kinetic energy of recoiling nucleus $T_Y$ is small and difficult to detect directly as the recoiling nucleus is a part of the sample. If $A = 200$ we have from equation (2.70)

$$T_\alpha = Q_\alpha\frac{A - 4}{A} = Q_\alpha\frac{200 - 4}{200} = 0.98Q_\alpha \rightarrow 98\% \text{ of } Q_\alpha.$$

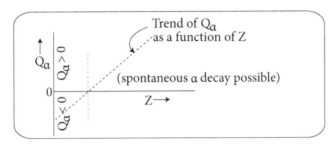

**Figure 2.6.** Trend of $Q_\alpha$ as a function of $Z$

Only $(100 - 98)\% = 2\%$ of $Q_\alpha$ will be shared by the daughter nucleus.

Knowing $T_\alpha$ we calculate the disintegration energy $Q_\alpha$ from equation (2.70) viz.

$Q_\alpha = T_\alpha \frac{A}{A-4}$. So we say that $Q_\alpha$ is a measurable quantity.

- No $Z$ is available for a high value of $A$ that corresponds to a stable nucleus in Segrè plot. Let us discuss why $\alpha$ emission is dominant for heavy nuclei.

Heavy nuclei having $A > 200$, $Z > 81$, $N > 120$ decays predominantly through $\alpha$ emission $_Z X^A \rightarrow \,_{Z-2}Y^{A-4} + \,_2He^4 + Q_\alpha$.

This is because the binding energy of four nucleons in $\alpha$ is large $\sim 28\ MeV$, i.e. $7\frac{MeV}{nucleon}$. The $\alpha$ or $_2He^4$ is a tightly bound system (doubly magic $Z = 2$, $N = 2$ nucleus) and so its rest mass energy is lower and this helps in making $Q_\alpha$ positive. In case other particles are emitted (instead of $\alpha$) the $Q_\alpha$ is not positive and so decay of heavy nucleus without $\alpha$ emission is not possible. For this reason heavy nuclei decays spontaneously through $\alpha$ emission. It follows from figure 2.6 that for heavy nuclei (say for $A > 150$) it is energetically favourable to spontaneously break via $\alpha$ emission since for them $Q_\alpha > 0$.

## 2.17 Nuclear potential barrier and alpha decay

Alpha decay occurs for nuclei with $A > 200$, $Z > 81$.

According to Frankel $\alpha$ particles do not pre-exist inside the nucleus, but just at the time of $\alpha$ decay, they are produced (*exercise 2.20*). We now discuss the process of $\alpha$ decay with the help of figure 2.7 and figure 2.8.

- ✓ When an $\alpha$ particle is inside the nucleus ($r \sim fm$, $r < R$) it is bound by a strong, short range nuclear attractive force that can be represented by a nuclear potential well $-V_0 \sim -40\ MeV$.
- ✓ After $\alpha$ decay $_Z X^A \rightarrow \,_{Z-2}Y^{A-4} + \,_2He^4$ the $\alpha$ is outside the nucleus and there is long range Coulomb repulsion between daughter nucleus $_{Z-2}Y^{A-4}$ and the emitted $\alpha$ at a separation $r$ from daughter at an instant, by a force

$$\vec{F} = \frac{1}{4\pi\varepsilon_0}\frac{(2\,|e|)(Z-2)|e|}{r^2} = \frac{1}{4\pi\varepsilon_0}\frac{2(Z-2)\,|e|^2}{r^2} = -\vec{\nabla}V = -\hat{r}\frac{dV}{dr} \qquad (2.75)$$

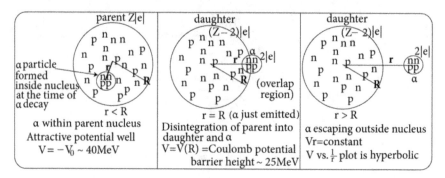

**Figure 2.7.** At the time of $\alpha$ decay $\alpha \equiv 2n2p$ particle is formed inside nucleus and emerges out. Here $R$ is nuclear radius and $r$ is the distance of $\alpha$ from centre of nucleus.

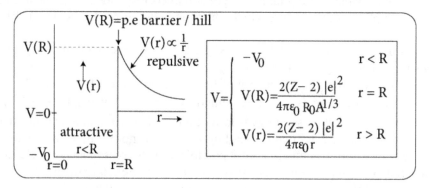

**Figure 2.8.** Potential as seen by $\alpha$ during its journey from inside nucleus to outside

where $\alpha = 2n2p$ system, $2|e| =$ charge of daughter nucleus, $Z =$ atomic number of parent nucleus, $\varepsilon_0 =$ permittivity of vacuum.

$$\frac{dV}{dr} = -\frac{1}{4\pi\varepsilon_0} \frac{2(Z-2)|e|^2}{r^2}$$

$$V = -\frac{2(Z-2)|e|^2}{4\pi\varepsilon_0} \int \frac{dr}{r^2} + C = \frac{2(Z-2)|e|^2}{4\pi\varepsilon_0 r} + C.$$

At $r \to \infty$, $V = 0$, $C = 0$. Thus

$$V = \frac{2(Z-2)|e|^2}{4\pi\varepsilon_0 r}\text{for } r > R. \tag{2.76}$$

The plot of $V$ versus $r$ is thus hyperbolic since $Vr =$ constant. This is the form of potential that $\alpha$ feels outside the nucleus. It is the Coulomb potential energy which is long range, repulsive (+ ve).

We can extrapolate this Coulomb repulsive energy backwards and find the value of potential hill or barrier at the surface of the parent nucleus of radius $R = R_0 A^{1/3}$ (with $R_0 = 1.4\,fm$). Substituting $r = R$ in the Coulomb repulsive energy expression of equation (2.76) we get

$$V(r = R) = \frac{2(Z - 2)\,|e|^2}{4\pi\varepsilon_0 R} = \frac{2(Z - 2)\,|e|^2}{4\pi\varepsilon_0 R_0 A^{1/3}} \qquad (2.77)$$

This is the barrier height that $\alpha$ had to surmount to emerge out of the nucleus (figure 2.8).

- Estimation of barrier height (*exercise 2.23*) for the $\alpha$ decay

$$_{84}Po^{210} \rightarrow {}_{82}Pb^{206} + {}_{2}He^4.$$

For $_{84}Po^{210}$ radius is

$$R = R_0 A^{1/3} = 1.4\,fm(210)^{1/3} = 8.32\,fm. \qquad (2.78)$$

We thus have

$$V(r = R) = \frac{1}{4\pi\varepsilon_0}\frac{2(Z - 2)\,|e|^2}{R} = \frac{1}{4\pi(\frac{10^{-9}}{36\pi}\frac{F}{m})}\frac{2(84 - 2)(1.6 \times 10^{-19}\,C)^2}{8.32 \times 10^{-15}\,m} = 4.5 \times 10^{-12}\,J$$

$$V(r = R) = \frac{4.5 \times 10^{-12}}{1.6 \times 10^{-19}}eV = 28\,MeV \qquad (2.79)$$

- The potential energy of interaction within nucleus $r < R = R_0 A^{1/3} = (1.4\,fm)A^{1/3}$ is strong, short range, attractive and hence negative. We denote this by $-V_0$ for $r < R$.
- We now discuss the mechanism of $\alpha$ emission from $r < R$ region (within the nucleus) to $r > R$ region (outside the nucleus). It requires effort to move from inside to the exterior and the energy needed for it is represented by a potential energy hill at $r = R$. During $\alpha$ decay the $\alpha$ emitted overcomes the potential energy hill $V(R) \sim 30\,MeV$ and emerges out of the influence of nuclear force (i.e. out of the parent nucleus). Once ejected it gets repelled electrostatically by the daughter nucleus. The experimentally obtained kinetic energy of $\alpha$ is between 4 and 9 $MeV$.
- Attempt to classically explain $\alpha$ decay.

Acording to the classical concept it is not possible for an $\alpha$ particle of energy $E \sim 4\,MeV$ to cross a barrier height of potential energy $V(R) \sim 30\,MeV$. It would bounce back at every attempt at crossing the barrier (figure 2.9). Put differently, $\alpha$ energy $E < V(R)$ and so it has to travel with kinetic energy $E - V(R) = -ve$ and imaginary velocity which is absurd. Classical concepts thus fail to explain emergence of an $\alpha$ particle from the nucleus.

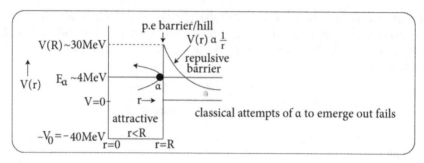

**Figure 2.9.** A classical $\alpha$ particle would bounce back from the wall when it tries to cross it.

## 2.18 Quantum mechanical explanation of $\alpha$ decay, Gammow, Gurney and Condon theory, quantum tunneling

Study of $\alpha$ decay suggests that there are three regions which the just formed $\alpha$ has to go through.

✓ The binding attractive force within the well $0 < r < R$ where $\alpha$ has energy $E > V$. $R$ is the nuclear radius and $V = -V_0$.

✓ As $\alpha$ of energy $E$ tries to emerge, it faces a p.e. hill $V(r)$ that starts at $r = R$, where $V = V(R)$—then the p.e. hill $V(r)$ diminishes and becomes $E = V(R_1)$ at $r = R_1$ as shown in figure 2.10. So $V(r) > E$ in the region $R < r < R_1$. In this region particle energy $E$ is less than $V(r)$ so that kinetic energy $E - V(r)$ is negative but the particle moves. This we cannot explain through classical concepts. In fact the maneuvers of the $\alpha$ particle to emerge out of the nucleus are of quantum nature and quantum calculations show that it emerges out of the well by a process called tunneling through the barrier. This phenomenon is called quantum tunneling.

✓ Beyond $R_1$, $\alpha$ feels the repulsive Coulomb barrier and escapes with +ve kinetic energy $E - V(r)$ since it has the requisite energy $E > V(r)$ in this region $r > R_1$.

It is a 3D potential and the motion of $\alpha$ (of mass $m$, energy $E$ which is $\sim Q_\alpha$) is governed by the 3D Schrödinger equation

$$\nabla^2 \psi(\vec{r}) + \frac{2m}{\hbar^2}[E - V(r)]\psi(\vec{r}) = 0 \qquad (2.80)$$

where as per the method of separation of variables we put

$$\psi(\vec{r}) = R(r)Y(\theta, \phi) \qquad (2.81)$$

and for spherically symmetric potential $Y(\theta, \phi) = Y_{lm_l}(\theta, \phi)$ which is the spherical harmonic. Hence

$$\psi(\vec{r}) = R(r)Y_{lm_l}(\theta, \phi). \qquad (2.82)$$

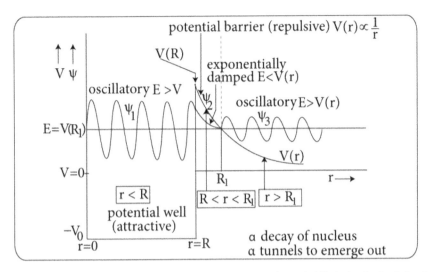

**Figure 2.10.** A quantum alpha particle would penetrate (with certain probability) the Coulomb barrier and emerge. There are three regions of interest in the analysis of $\alpha$ decay $0 < r < R$, $R < r < R_1$, $r > R_1$.

The radial part $R(r)$ satisfies the radial Schrödinger equation

$$\frac{d^2 R}{dr^2} + \frac{2}{r}\frac{dR}{dr} + \frac{2m}{\hbar^2}\left[E - V(r) - \frac{l(l+1)\hbar^2}{2mr^2}\right]R = 0 \tag{2.83}$$

With the standard substitution

$$R(r) = \frac{u(r)}{r} \tag{2.84}$$

the Schrödinger equation becomes

$$\frac{d^2 u}{dr^2} + \frac{2m}{\hbar^2}(E - V_{\text{eff}})R = 0 \tag{2.85}$$

where

$$V_{\text{eff}} = V(r) + \frac{l(l+1)\hbar^2}{2mr^2} \tag{2.86}$$

Equation (2.85) resembles a 1D equation. So the radial equation is similar to the 1D motion of a particle in a potential $V_{\text{eff}}$.

For simplicity we take $l = 0$ and work with the potential $V_{\text{eff}} = V(r) \equiv V$.

The motion of the $\alpha$ particle is governed by the Schrödinger equation

$$\frac{d^2 \psi}{dr^2} + \frac{2m}{\hbar^2}(E - V)\psi = 0 \tag{2.87}$$

In the analysis of $\alpha$ decay we have to consider three regions separately.

✓ $0 < r < R$ region (within the nucleus) in which $V = -V_0$, $E > V$. Here $\psi = \psi_1$ and the Schrödinger equation (2.87) is

$$\frac{d^2\psi_1}{dr^2} + \frac{2m}{\hbar^2}[E - (-V_0)]\psi_1 = 0 \Rightarrow \frac{d^2\psi_1}{dr^2} + \frac{2m}{\hbar^2}(E + V_0)\psi_1 = 0 \qquad (2.88)$$

Defining

$$k^2 = \frac{2m}{\hbar^2}(E + V_0) = +\text{ve} \Rightarrow k = \sqrt{\frac{2m}{\hbar^2}(E + V_0)} = \text{real} \qquad (2.89)$$

we have

$$\frac{d^2\psi_1}{dr^2} + k^2\psi_1 = 0 \qquad (2.90)$$

This is a simple harmonic equation and the solution $\psi_1$ is oscillatory (figure 2.10).
✓ $R < r < R_1$ region (just outside nucleus) where $R_1$ is defined as the distance where particle energy and barrier height are equal, i.e. where

$$E = V(R_1) \qquad (2.91)$$

In this classically forbidden region $E < V(r)$ and $V(r) = \frac{1}{4\pi\varepsilon_0}\frac{2(Z-2)|e|^2}{r}$, $\psi = \psi_2$ and the Schrödinger equation (2.87) is

$$\frac{d^2\psi_2}{dr^2} + \frac{2m}{\hbar^2}(E - V)\psi_2 = 0 \xrightarrow{\text{as } E < V(r)} \frac{d^2\psi_2}{dr^2} - \frac{2m}{\hbar^2}(V - E)\psi_2 = 0 \qquad (2.92)$$

Define

$$\beta^2 = \frac{2m}{\hbar^2}(V - E) = +\text{ve} \qquad (2.93)$$

we rewrite

$$\frac{d^2\psi_2}{dr^2} - \beta^2\psi_2 = 0 \qquad (2.94)$$

The solution $\psi_2$ is exponentially damped (figure 2.10).
✓ Region $r > R_1$ (outside the nuicleus) where $E > V(r)$ with $V(r) = \frac{1}{4\pi\varepsilon_0}\frac{2(Z-2)|e|^2}{r}$, $\psi = \psi_3$ the Schrödinger equation (2.87) is

$$\frac{d^2\psi_3}{dr^2} + \frac{2m}{\hbar^2}(E - V)\psi_3 = 0 \qquad (2.95)$$

Defining

$$k'^2 = \frac{2m}{\hbar^2}(E - V) = +\text{ve} \Rightarrow k' = \sqrt{\frac{2m}{\hbar^2}(E - V)} = \text{real}.$$

we have

$$\frac{d^2\psi_3}{dr^2} + k'^2\psi_3 = 0 \tag{2.96}$$

This is a simple harmonic equation the solution of which is oscillatory (figure 2.10).

Existence of solution outside the nucleus $r > R_1$ means $\alpha$ particle emerges out of the nucleus after it is created during $\alpha$ decay in spite of the existence of a p.e. hill ~30 $MeV$ at the nucleus surface and the fact that the $\alpha$ had energy ~4 $MeV$, not enough to climb atop the hill. Actually the nucleus and $\alpha$ are both quantum systems and the attempts that take it out of the periphery of the nucleus are referred to as quantum mechanical tunneling or leaking or bleeding—which is classically not visualisable.

Since the $\alpha$ has low energy than the potential energy hill, its emergence is not a free flow but a less probable phenomenon—the barrier suppresses $\alpha$ emission.

Let us estimate the transmission probability or the tunneling probability called the probability of escape. The relevant formula can be written down by comparing with the result of a rectangular barrier.

The results of a rectangular barrier suppressing particle flow (figure 2.11(a)) have been compared with the results of a nuclear barrier suppressing alpha emission (figure 2.11(b)). We would then generalize the tunneling probability from rectangular barrier and calculate the tunneling probability of $\alpha$ emission.

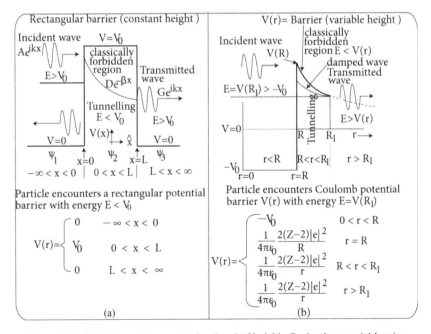

**Figure 2.11.** (a) Rectangular constant barrier. (b) Variable Coulomb potential barrier.

For a rectangular barrier suppressing particle flow the tunnelling probability is given by

$$T = \frac{\text{Transmitted flux}}{\text{Incident flux}} = e^{-2\beta L} \qquad (2.97)$$

where

$$\beta = \sqrt{\frac{2m}{\hbar^2}(V_0 - E)} \qquad (2.98)$$

and so

$$T = \exp\left[-2\sqrt{\frac{2m}{\hbar^2}(V_0 - E)}\,L\right] \Rightarrow T = \exp\left[-2\sqrt{\frac{2m}{\hbar^2}}(V_0 - E)^{1/2}\,L\right] \qquad (2.99)$$

This holds for rectangular barrier of constant height $V_0$ and width $L$.

Let us generalise this result for barrier of variable height (figure 2.12). Variable means we have to perform integration of $V(r)$ over $r$ from $R$ to $R_1$ and we have to replace $V_0$ by $V(r)$. So we have

$$T = \exp\left[-2\sqrt{\frac{2m}{\hbar^2}}\int_{r=R}^{R_1}[V(r) - E]^{1/2}\,dr\right] = e^{-G} \qquad (2.100)$$

We show in *exercise 2.22* that

$$G = \frac{2R_1\sqrt{2\,mE}}{\hbar}\left[\cos^{-1}\sqrt{\frac{R}{R_1}} - \sqrt{\frac{R}{R_1}\left(1 - \frac{R}{R_1}\right)}\right] \qquad (2.101)$$

This $T = e^{-G}$ is the tunneling probability of $\alpha$ from the nucleus.

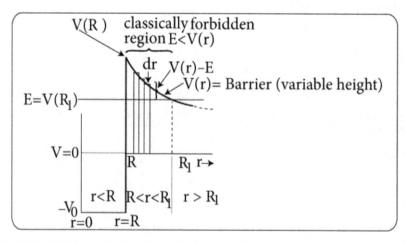

**Figure 2.12.** Divide the region $R < r < R_1$ into small parts and treat each of them as rectangular barriers of width $dr$ and height $V(r)$.

We can perform some calculations as in *exercise 2.22* and reduce it to a simpler form in terms of known parameters as

$T = e^{-G}$ where

$$G = A(Z - 2)E^{-1/2} - B(Z - 2)^{1/2}R^{1/2}; \qquad (2.102)$$

$$A = 1.587 \times 10^{-6}\, J^{1/2}, \ B = 94 \times 10^6\, m^{-1/2} \qquad (2.103)$$

Tunnelling probability thus depends upon
✓ energy $E$ of $\alpha$
✓ atomic number $Z$ of parent
✓ nuclear radius $R$.

Now barrier height at $R_1$ is obtained as follows

$$E = V(R_1) = \frac{2(Z - 2)\,|e|^2}{4\pi\varepsilon_0 R_1}$$

$$R_1 = \frac{2(Z - 2)\,|e|^2}{4\pi\varepsilon_0 E} \qquad (2.104)$$

And the width of potential barrier is

$$\text{Width} = R_1 - R \qquad (2.105)$$

- Height of potential barrier and width of potential barrier for $\alpha$ emitted from radon if $\alpha$ is emitted with kinetic energy 5.5 $MeV$.
  For radon $_{86}Rn^{222}$ the Coulomb barrier energy is

$$V(r = R) = \frac{1}{4\pi\varepsilon_0}\frac{2(Z - 2)\,|e|^2}{R_0 A^{1/3}} = \frac{1}{4\pi\left(\frac{10^{-9}}{36\pi}\frac{F}{m}\right)}\frac{2(86 - 2)(1.6 \times 10^{-19}\, C)^2}{(1.4 \times 10^{-15}\, m)(222)^{1/3}} = 4.57 \times 10^{-12}\, J$$

$$= \frac{4.57 \times 10^{-12}}{1.6 \times 10^{-19}}eV = 29\, MeV \qquad (2.106)$$

The barrier height at $r = R_1$ is

$$V(R_1) = \frac{1}{4\pi\varepsilon_0}\frac{2(Z - 2)\,|e|^2}{R_1} = E_\alpha \qquad (2.107)$$

$$R_1 = \frac{1}{4\pi\varepsilon_0}\frac{2(Z - 2)\,|e|^2}{E_\alpha} \qquad (2.108)$$

$$R_1 = \frac{1}{4\pi\left(\frac{10^{-9}}{36\pi}\frac{F}{m}\right)}\frac{2(86 - 2)(1.6 \times 10^{-19}\, C)^2}{5.5\, MeV} = \frac{1}{4\pi\left(\frac{10^{-9}}{36\pi}\frac{F}{m}\right)}\frac{2(86 - 2)(1.6 \times 10^{-19}\, C)^2}{5.5 \times 10^6 \times 1.6 \times 10^{-19}\, J}$$

$$= 43.985 \times 10^{-15} \, m = 43.985 \, fm \tag{2.109}$$

Now nuclear radius

$$R = R_0 A^{1/3} = (1.4 \, fm)(222)^{1/3} = 8.477 \, fm. \tag{2.110}$$

Thus width of potential barrier is

$$R_1 - R = 43.985 \, fm - 8.477 \, fm = 35.508 \, fm. \tag{2.111}$$

- We can interpret the transmission mechanism as follows and get a rough estimate of the half life of $\alpha$-radioactive parent.

  An $\alpha$ particle just after formation in the nucleus bounces back and forth knocking the potential energy barrier and after $e^{-G}$ hits the $\alpha$ emerges once out of the nucleus since $T = e^{-G}$. The velocity $v$ of $\alpha$ inside the nucleus can be taken to be the same as that just outside the nucleus (just after decay). So we can get an estimate of the velocity of $\alpha$ from the relation

$$E = \frac{1}{2}mv^2 \Rightarrow v = \sqrt{\frac{2E}{m}} \quad (m \text{ is mass of } \alpha) \tag{2.112}$$

The collision frequency $f$ is the number of times $\alpha$ hits the potential energy barrier per second. So

$$f = \frac{\text{velocity}}{\text{diameter of parent nucleus}} = \frac{v}{2R} \tag{2.113}$$

The probability of leaking out per second, i.e. number of decays per second is called decay constant. It is given by the following product:
$\lambda$ = (number of hits per second)(probability of leaking per hit)
= (collision frequency)(tunnelling probability)

$$\lambda = fT = \frac{v}{2R}e^{-G}. \tag{2.114}$$

Half-life of the $\alpha$-radioactive parent is

$$T_{1/2} = \frac{ln2}{\lambda} = \frac{0.693}{\lambda} \tag{2.115}$$

From analysis of $\alpha$ decay we are thus able to get a rough estimate of the half-life of $\alpha$-radioactive parent as in *exercise 2.24*.
- A rough estimate of half-life of $_{84}Po^{212}$.

  Consider the $\alpha$ decay

$$_{84}Po^{212} \rightarrow {}_{82}Pb^{208} + {}_2He^4$$

The experimental value of energy of $\alpha$ obtained is $8.95\,MeV$.

$$E = 8.95\,MeV = 8.95 \times 10^6 \times 1.6 \times 10^{-19}\,J = 1.432 \times 10^{-12}\,J.$$

Radius of parent nucleus is

$$R = R_0 A^{1/3} = (1.4\,fm)(212)^{1/3} = 8.34\,fm = 8.34 \times 10^{-15}\,m.$$

Atomic number of parent is $Z = 84$.

Using equations (2.102) and (2.103) we get the tunnelling probability to be $T = e^{-G}$ where

$$G = A(Z - 2)E^{-1/2} - B(Z - 2)^{1/2}R^{1/2}$$

$$G = (1.587 \times 10^{-6}\,J^{\frac{1}{2}})(84 - 2)(1.432 \times 10^{-12}\,J)^{-\frac{1}{2}} - (94 \times 10^6\,m^{-\frac{1}{2}})(84 - 2)^{\frac{1}{2}}(8.34 \times 10^{-15}\,m)^{\frac{1}{2}}$$

$$G = 108.75 - 77.74 = 33.56 \cong 31.$$

$$T = e^{-G} = e^{-31} = 3 \times 10^{-14} = \frac{3}{10^{14}}$$

Out of $10^{14}$ strikes $\alpha$ emerges three times only.

$$\text{Velocity } v = \sqrt{\frac{2E}{m}} = \sqrt{\frac{2(1.432 \times 10^{-12}\,J)}{4(1.67 \times 10^{-27}\,kg)}} = 2 \times 10^7\,ms^{-1}.$$

Collision frequency $f = \dfrac{v}{2R} = \dfrac{2 \times 10^7\,ms^{-1}}{2(8.34 \times 10^{-15}\,m)} = 1.2 \times 10^{21}\,s^{-1}.$

Decay constant $\lambda = fT = (1.2 \times 10^{21}\,s^{-1})(3 \times 10^{-14}) = 3.6 \times 10^6\,s^{-1}.$

$$\text{Half-life is } T_{1/2} = \frac{ln2}{\lambda} = \frac{0.693}{3.6 \times 10^6\,s^{-1}} = 0.19 \times 10^{-6}\,s.$$

The experimental value of half-life is $0.3 \times 10^{-6}\,s$. So we have obtained a rough estimate of half-life of $_{84}Po^{212}$.

- A rough estimate of collision frequency of $\alpha$ as it tries to emerge from the nucleus.

We use classical expression (figure 2.9)

$$\frac{1}{2}mv^2 = E - (-V_0) = E + V_0 \xrightarrow{E \sim Q_\alpha} Q_\alpha + V_0$$

$$v = \sqrt{\frac{2(Q_\alpha + V_0)}{m}} \tag{2.116}$$

So collision frequency is

$$f = \frac{v}{2R} = \frac{1}{2R}\sqrt{\frac{2(Q_\alpha + V_0)}{m}} = \sqrt{\frac{Q_\alpha + V_0}{2mR^2}} \tag{2.117}$$

- Expressions of (*i*) decay constant (*ii*) half-life.

✓

$$\lambda = fT = \frac{v}{2R}e^{-G} = \sqrt{\frac{Q_\alpha + V_0}{2mR^2}}\,e^{-G} \tag{2.118}$$

$$G = \frac{2R_1\sqrt{2\,mE}}{\hbar}\left[\cos^{-1}\sqrt{\frac{R}{R_1}} - \sqrt{\frac{R}{R_1}\left(1 - \frac{R}{R_1}\right)}\right] \tag{2.119}$$

✓

$$T_{1/2} = \frac{0.693}{\lambda} \tag{2.120}$$

## 2.19 Selection rule of $\alpha$ decay

Linear momentum, angular momentum and parity are conserved in $\alpha$ decay which is a strong interaction.

Consider even–even nucleus undergoing $\alpha$ decay $0^+ \rightarrow 0^+$ (figure 2.13(a)) and such decay from parent in state $J^\pi \equiv 0^+$ to daughter also in state $J^\pi \equiv 0^+$ can occur only through $\alpha$ having $l = 0$. Further $0^+ \rightarrow 0^+$ transition means no parity change. This is also ensured by the angular momentum $l = 0$ of $\alpha$ since its parity is $(-)^0 = +1$.

But transition involving parent in state $J^\pi \equiv 0^+$ to daughter in excited state say $2^+$ can occur through $\alpha$ having $l > 0$. If $l \neq 0$ then a potential barrier is raised by the centrifugal p.e. term $\frac{l(l+1)\hbar^2}{2mr^2}$ that reduces barrier tunneling probability and hence increases half-life. So the transition $0^+ \rightarrow 2^+$ can be effected by $\alpha$ having $l = 2$ and since it has parity $(-)^2 = +1$ it follows that parity is also conserved in the transition.

Let us discuss if the transition $0^+ \rightarrow 2^-$ is possible through $\alpha$ decay (figure 2.13(b)).

In the transition $0^+ \rightarrow 2^-$ conservation of angular momentum leads to $l = 2 + 0$ to $2 - 0$ in unit steps, i.e. $l = 2$. Again there is parity change from + to−. So the $\alpha$ effecting the transition (if possible) should have −ve parity. But parity of the $\alpha$ with $l = 2$ is $(-)^2 = +1$, i.e. + ve. This means the transition $0^+ \rightarrow 2^-$ cannot be effected – it is forbidden by $\alpha$ decay.

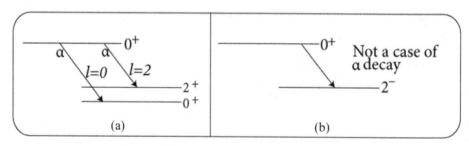

**Figure 2.13.** Transition using selection rule of $\alpha$ decay.

## 2.20 Range of $\alpha$

The distance which an $\alpha$ particle travels in a specified material before ceasing to ionize is called range of $\alpha$ in that material. Range is
   ✓ highest in gases $\sim$ few $cm$ (due to loose packing of molecules)
   ✓ less in liquids $\sim 10^{-3}\,mm$ (due to dense packing of molecules)
   ✓ least in solids $\lesssim 10^{-3}\,mm$ (due to most dense packing of molecules).

Range $R$ depends upon:
   (i) initial energy $E_\alpha$ (or velocity $v_\alpha$) of $\alpha$ particles, i.e.

$$R = R(E_\alpha), \;\; R = R(v_\alpha) \tag{2.121}$$

   (ii) ionisation potential of gas (medium)

   (iii) chance of collision between $\alpha$ and the medium (gas) particles.

Range is
   ✓ directly proportional to absolute temperature of gas
   ✓ inversely proportional to pressure of gas
   ✓ inversely proportional to density of gas

## 2.21 Geiger law

- We state Geiger law. For mono-energetic $\alpha$ particles of velocity $v_\alpha$ the range $R$ in standard air is

$$R \propto v_\alpha^3 \text{ i.e. } R = av_\alpha^3 (\Leftarrow\text{range–velocity relation}) \tag{2.122}$$

where

$$a = 9.42 \times 10^{-24}\,s^3\,m^{-2} = \text{constant}$$

Also $E_\alpha = \frac{1}{2}mv_\alpha^2$ and so $v_\alpha = \sqrt{\frac{2E_\alpha}{m}}$ ($m$ is mass of $\alpha$). So

$$R = av_\alpha^3 = a\left(\frac{2E_\alpha}{m}\right)^{3/2}$$

$$R = bE_\alpha^{3/2}(\Leftarrow\text{range energy relation}) \tag{2.123}$$

where $b = 3.15 \times 10^{-3}\,mMeV^{-3/2} = \text{constant}$.
- We now deduce Geiger law.

   $\alpha$ particles lose more energy per unit length near the end of their range, where they move more slowly and interact for a longer time with atoms. Clearly energy loss is more for larger path length and lower velocity, i.e.

$$-dE_\alpha \propto dx$$

$$-dE_\alpha \propto \frac{1}{v_\alpha}$$

Hence

$$-dE_\alpha = k\frac{dx}{v_\alpha}$$

$$-\frac{dE_\alpha}{dx} = \frac{k}{v_\alpha} \tag{2.124}$$

where $k$ = constant (depends on absorbing material). So

$$-\frac{d}{dx}\frac{1}{2}mv_\alpha^2 = \frac{k}{v_\alpha} \Rightarrow -\frac{1}{2}m2v_\alpha dv_\alpha = \frac{kdx}{v_\alpha} \Rightarrow -mv_\alpha^2 dv_\alpha = kdx$$

Assuming that at $x = 0$, velocity $= v_\alpha$ and at $x = R$ (Range), velocity $= 0$ we have

$$-m\int_{v_\alpha}^0 v_\alpha^2 dv_\alpha = k\int_0^R dx \Rightarrow -m\frac{v_\alpha^3}{3}\Big|_{v_\alpha}^0 = kR$$

$$R = \frac{mv_\alpha^3}{3k} = av_\alpha^3 \tag{2.125}$$

where $a = \frac{m}{3k}$.

- Geiger's law holds for any type of charged particle in any medium, provided velocity $\ll c$.

## 2.22 Geiger–Nuttall law

- We state Geiger–Nuttall law. The range $R$ of an $\alpha$ particle and the decay constant $\lambda$ of the emitting radioactive element are related as

$$ln\lambda = A + BlnR \tag{2.126}$$

where $A$, $B$ are constants having different values for different radioactive series.
  ✓ $lnR$ versus $ln\lambda$ plot is a straight line with slope $B = +$ ve and intercept $A$ from ordinate as shown in figure 2.14(a).
  ✓ Figure 2.14(b) shows $lnR$ versus $ln\lambda$ plot for three radioactive series. Slope $B$ is constant and so we have three parallel lines for the three series. The intercept $A$ is different for different series.
  ✓ The range energy relation given in equation (2.123) viz. $R = bE_\alpha^{3/2}$ can be used in equation (2.126) to get

$$ln\lambda = A + BlnbE_\alpha^{3/2} = C + DlnE_\alpha \tag{2.127}$$

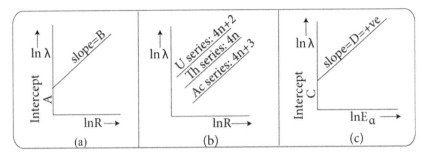

**Figure 2.14.** Plot of $lnR$ versus $ln\lambda$ and plot of $lnE_\alpha$ versus $ln\lambda$

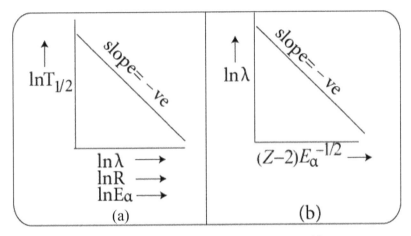

**Figure 2.15.** Plot of $lnR$ against $lnT_{1/2}$ and plot of $(Z-2)E_\alpha^{-1/2}$ against $ln\lambda$

where $C$, $D$ are constants. This is the relation between decay constant $\lambda$ and $\alpha$ energy $E_\alpha$. As shown in figure 2.14(c), the plot of $lnE_\alpha$ versus $ln\lambda$ is a straight line with positive slope ($D = +$ ve) and intercept $C$ from ordinate.

✓ Half-life is given by $T_{1/2} = \frac{0.693}{\lambda} = 0.693\lambda^{-1}$ and so taking $ln$ we have

$$lnT_{1/2} = E' - ln\lambda \ (E' = \text{constant}) \tag{2.128}$$

Using equation (2.126) viz. $ln\lambda = A + BlnR$ we get from equation (2.128)

$$lnT_{1/2} = E' - (A + BlnR) = F - BlnR \ (F = \text{constant}) \tag{2.129}$$

and using equation (2.127) viz. $ln\lambda = C + DlnE_\alpha$ we get from equation (2.128)

$$lnT_{1/2} = E' - (C + DlnE_\alpha) = H - DlnE_\alpha \ (H = \text{constant}) \tag{2.130}$$

Figure 2.15(a) shows the straight line nature of the plot of $ln\lambda$, $lnR$, $lnE_\alpha$ against $lnT_{1/2}$.

• We now deduce Geiger–Nuttall law.

Classical physics does not give a satisfactory explanation of Geiger–Nuttall law.

Geiger–Nuttall law can be arrived at from Gammow's theory of $\alpha$ decay which says that the $\alpha$ particle has to tunnel out of the parent nucleus after its formation. The expression of decay constant that we derived is

$$\lambda = \frac{v}{2R}e^{-G} \tag{2.131}$$

where $G = A(Z - 2)E^{-1/2} - B(Z - 2)^{1/2}R^{1/2}$

$$A = 1.587 \times 10^{-6}J^{1/2}, \; B = 94 \times 10^{6}m^{-1/2}.$$

Taking $ln$ of equation (2.131) we get

$$ln\lambda = ln\frac{v}{2R}e^{-G} = ln\frac{v}{2R} - G.$$

$$= ln\frac{v}{2R} - [A(Z - 2)E^{-1/2} - B(Z - 2)^{1/2}R^{1/2}]$$

$$ln\lambda = ln\frac{v}{2R} + B(Z - 2)^{1/2}R^{1/2} - A(Z - 2)E^{-1/2}$$

$$ln\lambda = P - Q(Z - 2)E_{\alpha}^{-1/2} \tag{2.132}$$

where $E \equiv E_{\alpha}$ is energy of $\alpha$ and $P$, $Q$ are constants.

Clearly the $ln\lambda$ versus $(Z - 2)E_{\alpha}^{-1/2}$ plot is a straight line with negative slope (figure 2.15(b)). This equation gives the relationship between decay constant $\lambda$ and the $\alpha$ energy $E_{\alpha}$. This is the Geiger–Nuttall law obtained from quantum mechanical analysis of $\alpha$ decay. This form is more accurate than the original form of Geiger–Nuttall law viz. $ln\lambda = C + DlnE_{\alpha}$ [equation (2.127)].

✓ We furnish two examples for comparison purpose.

| Element | Decay constant $\lambda$ | Half-life $T_{1/2}$ | $\alpha$ energy $E_{\alpha}$ |
|---|---|---|---|
| Decay of $_{90}Th^{232}$ | $\lambda = 5.3 \times 10^{-11}\,year^{-1}$ → small | $T_{1/2} = 1.3 \times 10^{10}$ year → large Slowest decay | $E_{\alpha} = 4.05\,MeV$ → small $R$ = small |
| Decay of $Po^{212}$ | $\lambda = 2.3 \times 10^{6}\,s^{-1}$ → large | $T_{1/2} = 3 \times 10^{-7}\,s$ → small Fastest decay | $E_{\alpha} = 8.95\,MeV$ → large $R$ = large |

- It is clear that if the $\alpha$ particles emitted from the parent have large $\lambda$, short $T_{1/2}$ then their range $R$ is large and vice versa.
- Geiger–Nuttall law is helpful in determining roughly the decay constants and hence also the half-lives of radioactive elements from knowledge of a range of $\alpha$ particles—which can be experimentally measured.

✓ Any theory describing $\alpha$ decay must be able to explain Geiger–Nuttall law for it to be accepted.

## 2.23 Beta decay

Beta decay is the process of spontaneous transformation of an unstable nucleus with atomic number $Z$, into an adjacent isobar (same $A$, i.e. same number of nucleons) with atomic number $Z \pm 1$ either through emission of electron (called $\beta^-$ decay) or through emission of positron (called $\beta^+$ decay) or through capture of an electron by the nucleus (called electron capture EC). The process of beta decay ($\beta^\pm$, EC) occurs through weak interaction—the life time being $10^{-13}$ $s$.

  ✓ Half-life of beta radioactive nuclei varies between $10^{-2}$ $s$ to $2 \times 10^{15}$ year.
  ✓ Energy of beta in beta decay lies between $18$ $keV$ ($_1\text{H}^3$) to $16$ $MeV$ ($_7\text{N}^{12}$)

- Characteristics of beta particles

  ✓ $\beta^- \equiv e^- \equiv {}_{-1}e^0$ is called electron, $\beta^+ \equiv e^+ \equiv {}_1e^0$ is called positron.
  ✓ All the beta particles emitted from a beta active parent do not have the same velocity but have a velocity spectrum between $0.3c$ to $0.99c$ (= speed of light in free space).
  ✓ $\beta$ has low ionizing power $\sim \frac{1}{100}$ times that of $\alpha$ particles.
  ✓ $\beta$ has high range, low mass and can travel fast.
  ✓ Penetrating power of $\beta$ is large $\sim 100$ times larger than $\alpha$.
  ✓ $\beta$ is charged particle and hence deflected by electric and magnetic field. Direction of deflection is opposite for $\beta^-$ and $\beta^+$.
  ✓ $\beta$ is spin $\frac{1}{2}$ particle. They are fermions.
  ✓ $\beta$ is a lepton, i.e. light particle, $m_\beta = m_e = 9.1 \times 10^{-31}$ $kg$.
  ✓ They do not take part in strong interaction.

- Illustration of $\beta^-$ decay
$$_Z X^A \rightarrow {}_{Z+1} Y^A + {}_{-1}e^0 + \bar{\nu}_e$$

$$_1\text{H}^3 \rightarrow {}_2\text{He}^3 + {}_{-1}e^0 + \bar{\nu}_e \xrightarrow{\text{we can rewrite this as}} (1p2n) + e^- \rightarrow (2p1n) + \bar{\nu}_e$$

$$_{16}\text{S}^{35} \rightarrow {}_{17}\text{Cl}^{35} + {}_{-1}e^0 + \bar{\nu}_e \xrightarrow{\text{we can rewrite this as}} (16p19n) + e^- \rightarrow (17p18n) + \bar{\nu}_e.$$

Clearly $Z$ increases by unity in the daughter as the parent sheds extra energy. Also evident is that in this interaction the neutron is converted to a proton as $n \rightarrow p + e^- + \bar{\nu}_e$

there being excess neutrons making the parent unstable. Emission of electron–antineutrino $\bar{\nu}_e$ in this interaction signifies that it is a weak interaction.

  ✓ Reason for emission of electron antineutrino in $\beta^-$ decay.

Electron lepton quantum number $L_e$ has to be conserved in all interactions. This is a fundamental requirement for any reaction to proceed. Now $L_e$ values are

$$L_e(e^-) = +1, \; L_e(\bar{\nu}_e) = -1, \; L_e(n) = 0, \; L_e(p) = 0.$$

So in the interaction $n \rightarrow p + e^- + \bar{\nu}_e$

$$L_e(\text{LHS}) = L_e(n) = 0 \text{ and}$$

$$L_e(\text{RHS}) = L_e(p) + L_e(e^-) + L_e(\bar{\nu}_e) = 0 + 1 - 1 = 0. \text{ Hence}$$

$$L_e(\text{LHS}) = L(\text{RHS}).$$

We note that $n$, $p$ are not leptons and we cannot define lepton quantum number for them.

The electron emerging from nuclide was generated during $\beta^-$ decay and has to come out since it cannot stay in the nucleus (*exercise 1.3*).

- Illustration of $\beta^+$ decay

$$_{Z}X^A \rightarrow \; _{Z-1}Y^A + \; _{+1}e^0 + \nu_e$$

$$_{9}F^{16} \rightarrow \; _{8}O^{16} + \; _{+1}e^0 + \nu_e \xrightarrow{\text{we can rewrite this as}} (9p7n) + e^+ \rightarrow (8p8n) + \nu_e$$

$$_{6}C^{11} \rightarrow \; _{5}B^{11} + \; _{+1}e^0 + \nu_e \xrightarrow{\text{we can rewrite this as}} (6p5n) + e^+ \rightarrow (5p6n) + \nu_e.$$

Clearly $Z$ decreases by unity in the daughter as the parent sheds extra energy.

Also evident is that in this interaction the proton is converted to a neutron as

$$p \rightarrow n + e^+ + \nu_e$$

there being excess protons making parent unstable. Emission of electron neutrino $\nu_e$ in this interaction, signifies that it is a weak interaction.

✓ Reason for emission of electron neutrino in $\beta^+$ decay

Conservation of electron lepton quantum number $L_e$ is a fundamental law of Nature. Electron lepton quantum number has to be conserved in all reactions. Now

$$L_e(e^+) = -1, \; L_e(\nu_e) = +1, \; L_e(n) = 0, \; L_e(p) = 0.$$

So in the interaction $p \rightarrow n + e^+ + \nu_e$

$$L_e(LHS) = L_e(p) = 0 \text{ and}$$

$$L_e(\text{RHS}) = L_e(n) + L_e(e^+) + L_e(\nu_e) = 0 - 1 + 1 = 0. \text{ Hence}$$

$$L_e(\text{LHS}) = L(\text{RHS}).$$

The positron emerging from the nuclide was generated during $\beta^+$ decay and has to come out since it cannot stay in the nucleus (*exercise 1.6*).

- Illustration of EC

$$_ZX^A + {}_{-1}e^0 \rightarrow {}_{Z-1}Y^A + \nu_e$$

$$_{30}Zn^{63} + {}_{-1}e^0 \rightarrow {}_{29}Cu^{63} + \nu_e \xrightarrow{\text{we can rewrite this as}} (30p33n) + e^- \rightarrow (29p34n) + \nu_e$$

$$_{13}Al^{24} + {}_{-1}e^0 \rightarrow {}_{12}Mg^{24} + \nu_e \xrightarrow{\text{we can rewrite this as}} (13p11n) + e^- \rightarrow (12p12n) + \nu_e$$

Clearly $Z$ decreases by unity in the daughter as the parent sheds extra energy.

The electron captured may be of the $K$ shell or $L$ shell. This process occurs in heavy unstable nuclei (that have a large number of protons and so a high Coulomb barrier) where the $K$ shell electron has a large probability ($\propto Z^4$) of being inside the nuclear volume and so the high Coulomb potential of parent nucleus can attract, capture and absorb an electron (say a $K$-shell electron) and decay to an isobaric daughter.

After electron capture (EC) a vacancy is created in the $K$ shell (if a $K$ electron is captured). So a higher shell electron will jump to fill the $K$-vacancy. As a result, a characteristic $X$-radiation, i.e. $X$-radiation of definite energy

$$E_{\text{shell}} - E_K = h\nu$$

and definite frequency

$$\frac{E_{\text{shell}} - E_K}{h} = \nu$$

will be released. In this interaction a proton is converted to a neutron

$$p + e^- \rightarrow n + \nu_e$$

Emission of an electron neutrino $\nu_e$ in this interaction signifies that it is a weak interaction.

✓ Reason for emission of an electron neutrino is the requirement of electron lepton number $L_e$ conservation in any reaction. Let us check the $L_e$ conservation using that

$L_e(e^-) = +1$, $L_e(\nu_e) = +1$, $L_e(n) = 0$, $L_e(p) = 0$. So in the interaction

$$p + e^- \rightarrow n + \nu_e$$

$$L_e(\text{LHS}) = L_e(p) + L_e(e^-) = 0 + 1 = 1 \text{ and}$$

$$L_e(\text{RHS}) = L_e(n) + L_e(\nu_e) = 0 + 1 = 1. \text{ Hence}$$

$$L_e(\text{LHS}) = L(\text{RHS}).$$

- Comparison of $\beta^+$ decay and $EC$

✓ $\beta^+$ decay and $EC$ are beta decays and are efforts to go to lower energy state. The processes are in competition with each other. That process occurs which takes a parent nucleus to a lower energy state which is then said to be an energetically favoured process.

✓ In both cases parent and daughter are same $_ZX^A$, $_{Z-1}Y^A$

✓ Both are weak interactions (as evident from electron neutrino emission)

✓ $\beta^+$ decay is $_ZX^A \rightarrow {}_{Z-1}Y^A + {}_{+1}e^0 + \nu_e$
   $EC$ is $_ZX^A + {}_{-1}e^0 \rightarrow {}_{Z-1}Y^A + \nu_e$

✓ The basic transformation in this process of $\beta^+$ decay is $p \rightarrow n + e^+ + \nu_e$

✓ The basic transformation in this process of $EC$ is $p + e^- \rightarrow n + \nu_e$

✓ In both the processes of $\beta^+$ decay and $EC$ the proton changes to a neutron.

✓ In $\beta^+$ decay a positron $e^+$ is emitted as a real particle.

In $EC$ a real particle electron $e^-$ is absorbed.

## 2.24 Energetics of $\beta$ decay

There are three processes of beta decay and generally they are in competition. Out of these three possibilities that beta decay process would occur, which would take the parent nucleus to the lowest energy state.

- Energy condition for the possibility of $\beta^-$ decay

$$\beta^- \text{ decay: } (Z, N) \rightarrow (Z + 1, N - 1) + e^- + \bar{\nu}_e \quad (\Leftarrow \text{Increase of } Z \text{ by unity}) \quad (2.133)$$

For $\beta^-$ decay to occur

$$M_{nu}(Z, N) > M_{nu}(Z + 1, N - 1) + m_e \quad (2.134)$$

where $M_{nu}$ is mass of nucleus.
   i.e. $M(\text{parent nucleus}) > M(\text{daughter nucleus}) + m_e$.
   Add $Zm_e$ to get

$$M_{nu}(Z, N) + Zm_e > M_{nu}(Z + 1, N - 1) + m_e + Zm_e$$

$$M_{atom}(Z, N) > M_{nu}(Z + 1, N - 1) + (Z + 1)m_e$$

where $M_{atom}$ is mass of atom

$$M_{atom}(Z, N) > M_{atom}(Z + 1, N - 1) \quad (2.135)$$

$$M(\text{parent atom}) > M(\text{daughter atom}).$$

- Energy condition for the possibility of $\beta^+$ decay

$$\beta^+ \text{ decay: } (Z, N) \rightarrow (Z - 1, N + 1) + e^+ + \nu_e \quad (\Leftarrow \text{Decrease of } Z \text{ by unity}) \quad (2.136)$$

For $\beta^+$ decay to occur

$$M_{\mathrm{nu}}(Z, N) > M_{\mathrm{nu}}(Z - 1, N + 1) + m_e \quad (M_{\mathrm{nu}} \text{ is mass of nucleus}) \qquad (2.137)$$

i.e. $M(\text{parent nucleus}) > M(\text{daughter nucleus}) + m_e$.

Add $Zm_e$ to get

$$M_{\mathrm{nu}}(Z, N) + Zm_e > M_{\mathrm{nu}}(Z - 1, N + 1) + m_e + Zm_e$$

$$M_{\mathrm{atom}}(Z, N) > M_{\mathrm{nu}}(Z - 1, N + 1) + (Z - 1)m_e + 2m_e (M_{\mathrm{atom}} \text{is mass of atom}) \quad (2.138)$$

$$M_{\mathrm{atom}}(Z, N) > M_{\mathrm{atom}}(Z - 1, N + 1) + 2m_e$$

$$M(\text{parent atom}) > M(\text{daughter atom}) + 2m_e \qquad (2.139)$$

- Energy condition for the possibility of EC

$$\text{EC} : (Z, N) + e^- \rightarrow (Z - 1, N + 1) + \nu_e \quad (\Leftarrow \text{Decrease of } Z \text{ by unity}) \quad (2.140)$$

For EC to occur

$$M_{\mathrm{nu}}(Z, N) + m_e > M_{\mathrm{nu}}(Z - 1, N + 1) \quad (M_{\mathrm{nu}} \text{ is mass of nucleus}) \qquad (2.141)$$

$M(\text{parent nucleus}) + m_e > M (\text{daughter nucleus})$
Add $(Z - 1)m_e$ to get
$M_{\mathrm{nu}}(Z, N) + m_e + (Z - 1)m_e > M_{\mathrm{nu}}(Z - 1, N + 1) + (Z - 1)m_e$

$$M_{\mathrm{nu}}(Z, N) + Zm_e > M_{\mathrm{nu}}(Z - 1, N + 1) + (Z - 1)m_e \qquad (2.142)$$

$$M_{\mathrm{atom}}(Z, N) > M_{\mathrm{atom}}(Z - 1, N + 1) \quad (M_{\mathrm{atom}} \text{ is mass of atom})$$

$$M(\text{parent atom}) > M(\text{daughter atom}).$$

✓ Energy liberated or Q value of $\beta$ decay can be expressed in terms of atomic masses that are tabulated and readily available.

- For $\beta^-$ decay $_ZX^A \rightarrow _{Z+1}Y^A + _{-1}e^0 + \bar{\nu}_e$ the Q value is

$$Q_{\beta^-} = [M_{\mathrm{nu}}(_ZX^A) - M_{\mathrm{nu}}(_{Z+1}Y^A) - m_e - m_{\bar{\nu}_e}]c^2 \qquad (2.143)$$

Taking $m_{\bar{\nu}_e} \sim \mathrm{eV} \rightarrow 0$

$$Q_{\beta^-} = [M_{\mathrm{nu}}(_ZX^A) - M_{\mathrm{nu}}(_{Z+1}Y^A) - m_e]c^2 \qquad (2.144)$$

In terms of atomic masses we write

$$Q_{\beta^-} = \left[ \left\{ M_{\mathrm{atom}}(_ZX^A) - Zm_e \right\} - \left\{ M_{\mathrm{atom}}(_{Z+1}Y^A) - (Z + 1)m_e \right\} - m_e \right]c^2$$

$$Q_{\beta^-} = [M_{\mathrm{atom}}(_ZX^A) - M_{\mathrm{atom}}(_{Z+1}Y^A)]c^2 \qquad (2.145)$$

$Q_{\beta-}$ is the difference of masses of parent and daughter.

• For $\beta^+$ decay $_ZX^A \rightarrow _{Z-1}Y^A + _{+1}e^0 + \nu_e$ the $Q$ value is

$$Q_{\beta^+} = [M_{nu}(_ZX^A) - M_{nu}(_{Z-1}Y^A) - m_e - m_\nu]c^2 \tag{2.146}$$

Taking $m_\nu \sim eV \rightarrow 0$

$$Q_{\beta^+} = [M_{nu}(_ZX^A) - M_{nu}(_{Z-1}Y^A) - m_e]c^2 \tag{2.147}$$

In terms of atomic masses we write

$$Q_{\beta^+} = \left[ \left\{ M_{atom}(_ZX^A) - Zm_e \right\} - \left\{ M_{atom}(_{Z-1}Y^A) - (Z-1)m_e \right\} - 2m_e \right]c^2$$

$$Q_{\beta^+} = [M_{atom}(_ZX^A) - M_{atom}(_{Z-1}Y^A) - 2m_e]c^2 \tag{2.148}$$

$Q_{\beta^+}$ is the difference of masses of parent and daughter minus twice the electron mass.

• For $EC$ $_ZX^A + _{-1}e^0 \rightarrow _{Z-1}Y^A + \nu_e$ the $Q$ value is

$$Q_{EC} = M_{nu}(_ZX^A) + m_e - \left\{ M_{nu}(_{Z-1}Y^A) + m_{\nu_e} \right\} \tag{2.149}$$

Taking $m_{\nu_e} \sim eV \rightarrow 0$

$$Q_{EC} = M_{nu}(_ZX^A) + m_e - M_{nu}(_{Z-1}Y^A) \tag{2.150}$$

In terms of atomic masses we write

$$Q_{EC} = \left[ \left\{ M_{atom}(_ZX^A) - Zm_e \right\} - \left\{ M_{atom}(_{Z-1}Y^A) - (Z-1)m_e \right\} \right]c^2$$

$$Q_{EC} = [M_{atom}(_ZX^A) - M_{atom}(_{Z-1}Y^A)]c^2 \tag{2.151}$$

$Q_{EC}$ is the difference of masses of parent and daughter.

Clearly it follows from equations (2.148) and (2.151) that

$$Q_{EC} > Q_{\beta^+}.$$

The same nucleus can decay through $\beta^+$ decay or $EC$. But as $Q_{EC}$ is larger more energy is available for decay through $EC$. So often $EC$ is energetically more probable than decay through $\beta^+$.

• Possibility of existence of two stable adjacent neighbouring isobars.

Let us consider isobars (same $A$) having atomic numbers differing by unity, i.e. we consider nuclides

(i) $Z - 1, N + 1, A$ (ii) $Z, N, A$ (iii) $Z + 1, N - 1, A$

The isobar with larger mass can decay into a neighbouring isobar with smaller mass through beta decay according to one of the following processes [equations (2.135), (2.139), and (2.142)]

(1) $\beta^-$ decay if $M_{atom}(Z, N, A) > M_{atom}(Z + 1, N - 1, A)$ (2.152)

(2) $\beta^+$ decay if $M_{atom}(Z, N, A) > M_{atom}(Z - 1, N + 1, A) + 2m_e$     (2.153)

(3) $EC$ if $M_{atom}(Z, N, A) > M_{atom}(Z - 1, N + 1, A)$     (2.154)

Clearly thus there cannot be two stable adjacent neighbouring isobars since the one with larger mass will decay into the one with smaller mass.
• If $\beta^+$ decay is energetically allowed then $EC$ is necessarily allowed.
If condition for $\beta^+$ decay [equation (2.153)] viz

$$M_{atom}(Z, N, A) > M_{atom}(Z - 1, N + 1, A) + 2m_e$$

is allowed then this automatically means that the condition for $EC$ [equation (2.154)] viz. $M_{atom}(Z, N, A) > M_{atom}(Z - 1, N + 1, A)$
is obeyed.
Example:
Consider beta decay of $_{25}Mn^{52}$, $_{30}Zn^{63}$, $_{35}Br^{80}$.
They undergo both $\beta^+$ and $EC$ because for them

$$M_{atom}(Z, N, A) > M_{atom}(Z - 1, N + 1, A) + 2m_e$$

is satisfied and hence condition for $EC$ is automatically satisfied.

$$_{25}Mn^{52} \rightarrow {}_{24}Cr^{52} + e^+ + \nu_e(35\%) \Leftarrow \beta^+ \text{ decay,}$$

$$_{25}Mn^{52} + e^- \rightarrow {}_{24}Cr^{52} + \nu_e(65\%) \Leftarrow EC$$

$$_{30}Zn^{63} \rightarrow {}_{29}Cr^{63} + e^+ + \nu_e \Leftarrow \beta^+ \text{ decay}$$

$$_{30}Zn^{63} + e^- \rightarrow {}_{29}Cr^{63} + \nu_e \Leftarrow EC$$

$$_{35}Br^{80} \rightarrow {}_{34}Se^{80} + e^+ + \nu_e \Leftarrow \beta^+ \text{ decay}$$

$$_{35}Br^{80} + e^- \rightarrow {}_{34}Se^{80} + \nu_e \Leftarrow EC$$

• If $EC$ is energetically allowed then $\beta^+$ decay is not necessarily allowed.
If the condition for $EC$ [equation (2.154)] viz. $M_{atom}(Z, N, A) > M_{atom}(Z - 1, N + 1, A)$ is allowed then this does not imply that the condition for $\beta^+$ decay [equation (2.153)] viz. $M_{atom}(Z, N, A) > M_{atom}(Z - 1, N + 1, A) + 2m_e$ is necessarily obeyed. We give an example.
Consider the beta decay for $_4Be^7$ (parent) that converts to $_3Li^7$ (daughter).

Atomic masses are $7.0169u$ for $_4Be^7$, $7.0160u$ for $_3Li^7$, $m_e = 0.000\,549u$.

□ Check the condition of $\beta^+$ decay
$_4Be^7 \rightarrow {}_3Li^7 + e^+ + \nu_e$ (if possible)
i.e we check whether $M_{atom}(Z, N, A) > M_{atom}(Z - 1, N + 1, A) + 2m_e$ [equation (2.153)] is satisfied.

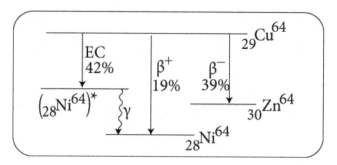

**Figure 2.16.** $_{29}Cu^{64}$ exhibits all types of beta decay.

$$LHS = M_{atom}(_4Be^7) = 7.0169u,$$
$$RHS = M_{atom}(_3Li^7) + 2m_e = 7.0160u + 2(0.000\ 549u)$$
$$= 7.017\ 098u$$

Clearly LHS < RHS $\Rightarrow \beta^+$ decay is not energetically favourable.
☐ Check the condition of EC
$_4Be^7 + e^- \rightarrow {}_3Li^7 + \nu_e$ (if possible)
i.e we check whether $M_{atom}(Z, N, A) > M_{atom}(Z - 1, N + 1, A)$ [equation (2.154)] is satisfied.

$$LHS = M_{atom}(_4Be^7) = 7.0169u$$

$$RHS = M_{atom}(_3Li^7) = 7.0160u$$

Clearly LHS > RHS $\Rightarrow$ EC is energetically favourable.
- Some nuclides exhibit all types of beta decay like $_{29}Cu^{64}$. For $_{29}Cu^{64}$ all the conditions viz. equations (2.152), (2.153), and (2.154) are satisfied simultaneously and so it exhibits all types of beta decay (figure 2.16).

## 2.25 Achievement of stability through beta decay

It follows from Bethe–Weizsacker's semi-empirical mass formula (we shall discuss it in chapter 3) that $M(Z, A)$ versus. $Z$ plot is parabolic in shape (figure 3.11).
✓ Case 1 : Odd $A$ = constant nuclei (figure 2.17)

All the Odd $A$ = constant nuclei fall upon a single parabola.
(a) The one isobar $M_0(Z_0, A)$ lying at the minimum of the mass parabola is least massive, most stable $(Z, A) > M(Z_0, A) \equiv M_0$.
(b) Isobars lying on $Z < Z_0$ side (left arm of parabola) decay successively by $\beta^-$ emission to reach the most stable isobar $M_0(Z_0, A)$.
(c) Isobars lying on $Z > Z_0$ side (right arm of parabola) decay successively by $\beta^+$ emission to reach the most stable isobar $M_0(Z_0, A)$.

We show mass parabola and beta decays corresponding to $A = 91, 135$ in figure 2.17 where the least massive and most stable isobars are, respectively, $_{40}Zr^{91}$ for $A = 91$ and $_{56}Ba^{135}$ for $A = 135$.

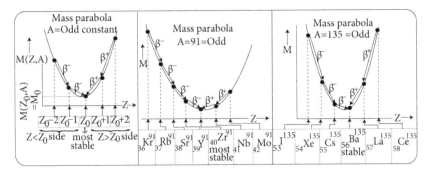

**Figure 2.17.** Mass parabola for odd $A$= constant nuclei.

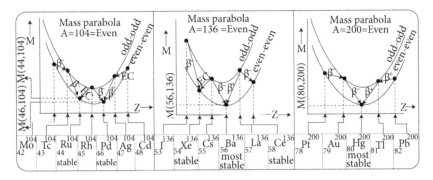

**Figure 2.18.** Mass parabola for even $A$= constant nuclei.

✓ Case 2 : Even $A$= constant nuclei (figure 2.18).

There are two types of even $A$ nuclei:

☐ All the odd–odd nuclei fall upon a parabola.

☐ All the even–even nuclei fall upon a parabola.

The even–even parabola is shifted slightly downwards w.r.t the odd–odd parabola.

- Beta decays go successively from one parabola to another until stable nuclei are reached. Accordingly, for even $A$ nuclides there can be one, two, three stable nuclides.

We show mass parabola and beta decays corresponding to $A = 104, 136, 200$ in figure 2.18.

✓ It is clear from figure 2.18 ($A = 104$) that there are two stable nuclei viz. $_{44}Ru^{104}$, $_{46}Pd^{104}$ as they cannot further decay since the adjacent nuclei are of higher mass.

✓ It follows from figure 2.18 ($A = 136$) that there are three stable nuclei viz. $_{54}Xe^{136}$, $_{56}Ba^{136}$, $_{58}Ce^{136}$ as they cannot further decay since the adjacent nuclei are of higher mass.

✓ In figure 2.18 ($A = 200$) we have one stable nuclide $_{80}Hg^{200}$ which is least massive and stable as it cannot decay to any other nuclei.

## 2.26 Double $\beta$ decay

Double beta decay is a nuclear transition in which a parent nucleus $(Z, A)$ decays to an isobaric daughter nucleus $(Z + 2, A)$ emitting 2 $\beta$s and 2 $\bar{\nu}_e$s, i.e. double beta decay $(2\nu\beta\beta)$ is represented by the following transition

$$(Z, A) \rightarrow (Z + 2, A) + 2e^- + 2\bar{\nu}_e \tag{2.155}$$

It occurs when energy conditions are (figure 2.19(a))

$$M(Z, A) < M(Z + 1, A)$$

$$M(Z, A) > M(Z + 2, A)$$

We can represent double beta decay through a Feynman diagram also (figure 2.19(b)).

This has also been indicated in the mass parabola of figure 2.19(c).

Double beta decay is a rare event manifested by very long half-lives ($\sim 10^{18}$ to $10^{24}$ year) and because of this rarity, measurement of double beta decay is extremely difficult. A large mass of nuclide is needed and highly precise instrumentation techniques are required.

- Example

$$_{48}Cd^{116} \rightarrow _{50}Sn^{116} + 2e^- + 2\bar{\nu}_e \text{ (half-life } 10^{19} \text{ year)}$$

The $_{48}Cd^{116}$ decays by double beta $(\beta\beta)$ directly to $_{50}Sn^{116}$. There is no transition of $_{48}Cd^{116}$ via single beta decay to $_{49}In^{116}$ (figure 2.19(d)) because

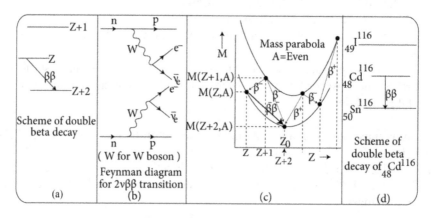

**Figure 2.19.** Double beta decay.

$M(_{48}Cd^{116}) < M(_{49}In^{116})$ and so the energy level of $_{49}In^{116}$ is higher than the energy level of $_{48}Cd^{116}$. So the transition $_{48}Cd^{116} \rightarrow _{49}In^{116}$ is forbidden.

Another example

$$_{34}Se^{82} \rightarrow _{36}Kr^{82} + 2e^- + 2\bar{\nu}_e \text{ (half-life } 10^{19} \text{ year)}$$

Double beta decay was predicted by Maria Goeppert Mayer and first observed by Steven Elliot.

## 2.27 Nature of beta spectra

Beta spectra are of two types

(1) Continuous $\beta$ spectra.

In continuous $\beta$ spectra $\beta$ particles are emitted with kinetic energy, that varies continuously from 0 to a maximum $Q_0$ called the end point energy or disintegration energy, as shown in figure 2.20(a).

Continuous $\beta$ spectra for $\beta^+$ decay are shown in figure 2.20(b).

$$_{29}Cu^{64} \rightarrow _{28}Ni^{64} + e^+ + \nu_e \qquad (2.156)$$

Continuous $\beta$ spectra for $\beta^-$ decay are shown in figure 2.20(c).

$$_{29}Cu^{64} \rightarrow _{30}Zn^{64} + e^- + \bar{\nu}_e. \qquad (2.157)$$

(2) Discrete $\beta$ spectra or sharp line $\beta$ spectra or characteristic $\beta$ spectra

In discrete $\beta$ spectra discrete lines are superposed on the continuous background —shown as spikes in figure 2.20(a) drawn over the continuous curve of $\beta$ spectra.

We shall discuss the origin of $\beta$ spectra first without considering neutrinos/antineutrinos. We shall then elaborate Pauli's attempts to explain $\beta$ spectra using his hypothesis of the existence of particles called neutrinos/antineutrinos.

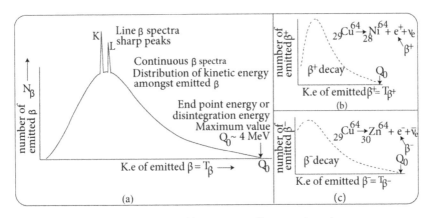

**Figure 2.20.** Two types of beta spectra—discrete and continuous.

## 2.28 Attempts to explain the origin of continuous beta spectra

Initially, the process of beta decay was thought to be emission of a beta particle $(\beta^+, \beta^-)$ with the emergence of a daughter nucleus, i.e. of the type

$$_zX^A \rightarrow _{z+1}Y^A + _{-1}e^0 \ (\beta^-\text{decay})$$

$$_zX^A \rightarrow _{z-1}Y^A + _{+1}e^0 \ (\beta^+ \text{ decay}).$$

There were several difficulties in explaning the above interactions. Let us mention some:

 ✓ Let us consider the energy distribution of $\beta$ particles, the reductioin in rest mass energies being $Q_0$ which is the available energy. The parent nucleus (unstable before decay) and the daughter nucleus (after decay) are in fixed energy states or levels $E_X$, $E_Y$.

Apply conservation of energy principle

$$E_X - E_Y = Q_0 \tag{2.158}$$

This energy $Q_0$ is released and may be assumed to be taken up or shared by the emerging $\beta$ and the daughter nucleus. Again energy of recoil of the daughter is very very small (as it is massive compared to $\beta$) and can be neglected (i.e. kept out of discussion). So the $\beta$ should be emitted with a fixed kinetic energy (figure 2.21(a))

$$T_\beta = Q_0 \tag{2.159}$$

i.e. monoenergetic spectrum is expected. But we observe that the $\beta$ energy $T_\beta$ is not discrete but assumes continuous values between 0 to $Q_0$ In fact, the experimentally obtained energy distribution of $\beta$ has been shown in figure 2.20 specifically for the decays

$$_{29}Cu^{64} \rightarrow _{28}Ni^{64} + e^+ \ \text{(figure 2.20(b))} \tag{2.160}$$

$$_{29}Cu^{64} \rightarrow _{30}Zn^{64} + e^- \ \text{(figure 2.20(c))} \tag{2.161}$$

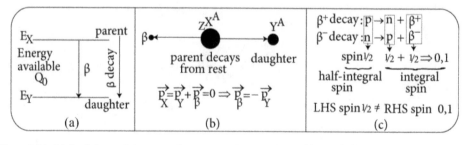

**Figure 2.21.** (a) In $\beta$ decay $\beta$ does not always carry the energy $Q_0$. (b) $\beta$ emission not always opposite to daughter nucleus as shown (c) LHS spin and RHS spins differ.

Clearly the energy $Q_0 - T_\beta$ appears to be missing. So mass energy conservation principle appears to be violated (figure 2.21(a)).

✓ Apply conservation of linear momentum principle. It is required that the emitted $\beta$ and the recoiling daughter nucleus $Y^A$ should be emitted in the opposite direction. But this seldom occurs. So linear momentum conservation principle appears to be violated (figure 2.21(b)).

✓ Apply conservation of angular momentum principle (figure 2.21c). The particles involved $n, p, \beta$ are all spin $\frac{1}{2}$ particles (fermions). With this the spins of both sides of the reaction appear to differ. So angular momentum conservation principle appears to be violated.

✓ Also then the statistics obeyed (i.e. the rules of distributing particles over energy levels) before and after $\beta$ decay appears to differ.

The above discussion hints at violation of energy, linear momentum, and angular momentum, which is not acceptable.

It follows that we are making some error or fallacy while trying to explain this interaction. This was a paradox or puzzle which was overcome by Pauli and we discuss how he did it.

## 2.29 Pauli's neutrino hypothesis

It occurred to Pauli that the residual kinetic energy and momentum must be taken away by a particle having certain properties that fits into the equation describing the interaction. Pauli assumed or predicted that in beta decay a neutral spin-half fermion called a neutrino $\nu_e$ or its antineutrino $\bar{\nu}_e$ is emitted along with beta.

Since these particles $\nu_e, \bar{\nu}_e$ are emitted along with an electron, positron they are called elctron neutrino $\nu_e$ and electron antineutrino $\bar{\nu}_e$. Hence the interactions of equations (2.160) and (2.161) would be

$$_{29}Cu^{64} \rightarrow {}_{28}Ni^{64} + e^+ + \nu_e \text{ for } \beta^+$$

decay [equation (2.156)] and

$$_{29}Cu^{64} \rightarrow {}_{30}Zn^{64} + e^- + \bar{\nu}_e \text{ for } \beta^-$$

decay [equation (2.157)]as depicted in figure 2.20(b),(c).

With this all the paradoxes could be explained as we describe now.

✓ Energy conservation.

The energy released in beta decay given by equation (2.158) viz. $E_X - E_Y = Q_0$ is shared between beta and neutrino. Hence

$$E_X - E_Y = Q_0 = T_\beta + T_\nu \tag{2.162}$$

So energy conservation holds. (In figure 2.22(a) the symbol for neutrino $\nu$ has been shown for representation purpose ($\nu \equiv \nu_e, \bar{\nu}_e$). Actually, electron neutrino $\nu_e$ and electron antineutrino $\bar{\nu}_e$ are emitted in $\beta^+, \beta^-$ decay, respectively.

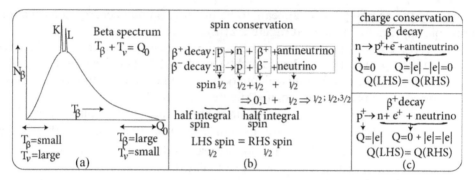

**Figure 2.22.** Pauli's explanation of beta decay using the neutrino $\nu_e$, antineutrino $\bar{\nu}_e$ concept.

✓ Linear momentum conservation.

$$\text{Since } \vec{P}_Y + \vec{P}_\beta + \vec{P}_{neutrino} = \vec{P}_X = 0 \qquad (2.163)$$

we should not expect that the emitted $\beta$ and the recoiling daughter nucleus $Y^A$ should be emitted in the opposite direction always. So the principle of linear momentum is obeyed.

✓ Angular momentum conservation

This is shown in figure 2.22(b). As the neutrino has spin half, angular momentum is conserved.

✓ Statistics conserved.

As both sides of the equation representing $\beta$ decay lead to spin half angular momentum (figure 2.22(b)), it follows that the same statistical laws operate on both sides, i.e. statistics is obeyed or conserved (i.e. the same rules of distributing particles over energy levels is operative on both sides). So statistics is conserved.

✓ The neutrino is electrically neutral and so charge conservation holds. (figure 2.22(c)).

✓ The neutrino being uncharged:

☐ it will not ionize matter

☐ it will not be deflected by electric and magnetic fields.

So it will be difficult to detect a neutrino as it does not participate in electromagnetic interaction.

✓ Mass of neutrino $m_\nu$ is very very small $\approx 0$. So neutrinos can be considered as practically zero rest mass particles and they can be assumed to travel with nearly the speed $c$ of light. The consequence of this is that neutrinos cannot transfer their energy to any other particle. So they are difficult to detect. However, from discovery of neutrino oscillations it follows that neutrinos have small finite mass $<0.5\frac{eV}{c^2}$.

✓ The antiparticle of a neutrino is the antineutrino. So $\nu_e$ and $\bar{\nu}_e$ are a particle–antiparticle pair.

✓ There are three types of neutrino or three flavours of neutrino:

☐ electron neutrino $\nu_e$ and its antiparticle electron antineutrino $\bar{\nu}_e$;
☐ Muon neutrino $\nu_\mu$ and its antiparticle muon antineutrino $\bar{\nu}_\mu$;
☐ tauon neutrino $\nu_\tau$ and its antiparticle tauon antineutrino $\bar{\nu}_\tau$.

✓ The phenomenon of interchange of flavor is referred to as neutrino oscillation, i.e. neutrino can oscillate from one specific flavor to another, i.e. a flavors oscillate. And this is possible if the neutrino has a small but finite mass. Neutrino oscillation is a quantum mechanical phenomenon (section 5.50).

✓ Neutrinos are leptons characterized by the following lepton quantum numbers:

Electron lepton quantum number

$$L_e(\nu_e) = +1 \tag{2.164}$$

$$L_e(\bar{\nu}_e) = -1 \tag{2.165}$$

As $L_e(e^-) = +1$ a $\bar{\nu}_e$ is emitted along with $e^-$ so that electron lepton quantum number is conserved as $L_e(e^-) + L_e(\bar{\nu}_e) = +1 - 1 = 0$.

Then $L_e^{\text{LHS}} = L_e^{\text{RHS}}$.

As $L_e(e^+) = -1$, a $\nu_e$ is always emitted along with $e^+$ so that electron lepton quantum number is conserved as $L_e(e^+) + L_e(\nu_e) = -1 + 1 = 0$.

Then $L_e^{\text{LHS}} = L_e^{\text{RHS}}$

We mention that lepton quantum number is not defined for neutron, proton. So they are not considered while studying lepton quantum number conservation.

✓ Neutrinos have practically no interaction with matter—so they are highly penetrative. A neutrino may pass through 1000 light years of solid iron without interaction. Sun emits a huge flux of neutrinos (called solar neutrinos) $\sim$ 300 million neutrinos cm$^{-2}$ of earth per second $10^{14}$ neutrinos pass through the human body per second.

✓ The only interaction which a neutrino can have with matter is the inverse beta decay viz.

$$p + \bar{\nu}_e \rightarrow n + e^+ \tag{2.166}$$

Based on this relation Reines and Cowans the detected the neutrino in 1956 ($\sim$ 26 years after the possibility of its existence was predicted by Pauli).

## 2.30 Fermi theory of beta decay

During $\beta$ decay the Coulomb barrier has no significant or decisive role and also there is no nuclear interaction since $\beta$ and $\nu/\bar{\nu}$ are leptons. Figure 2.22(a) shows the distribution of $\beta$ particles with energy they carry during $\beta$ decay.

The functional form of the spectra (i.e. the relative share of energy of $\beta$ decay between $\beta$ and $\nu/\bar{\nu}$) can be explained by Fermi theory of $\beta$ decay (1934) that is hinged on the time dependent perturbation theory of quantum mechanics. We state the Fermi golden rule in this context.

Suppose there is a system in initial state $|i>$ described by initial wave function $\psi_i$ and it makes a transition to a final state $|f>$ described by the wave function $\psi_f$. The probability of this transition is given by the transition rate equation

$$\lambda = \frac{2\pi}{\hbar} |H_{if}|^2 \frac{dn}{dE_f} \tag{2.167}$$

where $H_{if}$ is the interaction Hamiltonian that characterizes the transition and given by

$$H_{if} = <f|\hat{H}|i> = \int_{\tau} \psi_f^* \ \hat{H}\psi_i d\tau \tag{2.168}$$

$\tau$ is the volume of interaction, $\frac{dn}{dE_f}$ is called density of states at final energy $E_f$ that decides the shape of energy distribution.

The phenomenon of $\beta$ decay is due to weak interaction that is of very short range (~picometre) and is a consequence of $n \leftrightarrows p$ transition.

From the quark structure of neutron $n$ and proton $p$ viz.

$$n \equiv ddu, \ p \equiv uud \tag{2.169}$$

it follows from $n(ddu) \leftrightarrows p(uud)$ that there is conversion of quark from one flavor to another namely

$$d \leftrightarrows u \tag{2.170}$$

This is mediated by $W^{\pm}$ boson (having rest energy ~80 GeV). Interactions are

$$d \rightarrow u + W^-, \ W^- \rightarrow e^- + \bar{\nu} \text{ for } n \rightarrow p, \text{ i.e. } \beta^- \text{ decay and}$$

$$u \rightarrow d + W^+, \ W^+ \rightarrow e^+ + \nu \text{ for } p \rightarrow n, \text{ i.e. } \beta^+ \text{ decay.}$$

The order of range ~ picometre means it is a point interaction.

The charge of quarks are

$Q(d) = -\frac{1}{3}|e|, \ Q(u) = \frac{2}{3}|e|, Q(W^-)=-|e|, \ Q(W^+)=|e|$

Let $\psi_i \equiv \psi_P$ be the wave function of initial state, i.e. of the parent nucleus $\psi_f \equiv \psi_D\psi_\beta\psi_\nu$ be wave the function of the final state, i.e. of the daughter nucleus, beta particle (let us denote $\beta^{\pm}$ by $\beta$) and neutrino (let us denote $\nu_e$, $\bar{\nu}_e$ by $\nu$).

So $\psi_P$, $\psi_D$ are nuclear wave functions, $\psi_\beta$, $\psi_\nu$ are leptonic wave functions.

With this, from equation (2.168) we have

$$H_{if} = \int_{\tau} \psi_f^* \ \hat{H}\psi_i d\tau = \int_{\tau} (\psi_D\psi_\beta\psi_\nu)^* \ \hat{H}\psi_P d\tau = \int_{\tau} \psi_D^*\psi_\beta^*\psi_\nu^* \ \hat{H}\psi_P d\tau \tag{2.171}$$

We note that $\nu$ is very nearly free and $\beta$ feebly interacts with daughter nucleus. Treating $\beta$ and $\nu$ as free particles we write

$$\psi_\beta = \frac{1}{\sqrt{V}} e^{\frac{i}{\hbar} \, \vec{p}_\beta \cdot \vec{r}} \, , \, \psi_\nu = \frac{1}{\sqrt{V}} e^{\frac{i}{\hbar} \, \vec{p}_\nu \cdot \vec{r}} \tag{2.172}$$

where $V$ is nuclear volume.

✓ Let us make an estimate of the quantity $\frac{1}{\hbar} \, \vec{p}_\beta \cdot \vec{r}$ in the wave function $\psi_\beta = \frac{1}{\sqrt{V}} e^{\frac{i}{\hbar} \, \vec{p}_\beta \cdot \vec{r}}$.

Here $\vec{p}_\beta$ is linear momentum of $\beta$ that emerges out of the nucleus—say with kinetic energy

$$T \sim 1 \, MeV \tag{2.173}$$

We use the relativistic expression for total energy, namely

$$\sqrt{p_\beta^2 c^2 + m_e^2 c^4} = T + m_e c^2 \tag{2.174}$$

$$\sqrt{p_\beta^2 c^2 + (0.511 \, MeV)^2} = 1 \, MeV + 0.511 \, MeV$$

$$\sqrt{p_\beta^2 c^2 + (0.511 \, MeV)^2} = 1.511 \, MeV.$$

Squaring we have

$$p_\beta^2 c^2 + (0.511 \, MeV)^2 = (1.511 \, MeV)^2$$
$$p_\beta^2 c^2 = 2.022 \, MeV^2 \tag{2.175}$$
$$p_\beta c = 1.422 \, MeV$$

$$p_\beta = 1.422 \frac{MeV}{c}$$

Hence

$$\frac{1}{\hbar} \, \vec{p}_\beta \cdot \vec{r} \sim \frac{2\pi}{6.62 \times 10^{-34} \, J.s} 1.42 \frac{MeV}{c} \times 10^{-14} \, m = \frac{2\pi}{6.62 \times 10^{-34} \, J.s} 1.42 \frac{10^6 \times 1.6 \times 10^{-19} \, J}{3 \times 10^8 \, ms^{-1}} \times 10^{-14} \, m$$

$$\frac{1}{\hbar} \, \vec{p}_\beta \cdot \vec{r} \sim 0.072 = \text{very very small} \tag{2.176}$$

This means in the expansion of $\psi_\beta = \frac{1}{\sqrt{V}} e^{\frac{i}{\hbar} \, \vec{p}_\beta \cdot \vec{r}}$ and $\psi_\nu = \frac{1}{\sqrt{V}} e^{\frac{i}{\hbar} \, \vec{p}_\nu \cdot \vec{r}}$ we can make suitable approximations to retain only the first few terms. Let us approximate as

$$\psi_\beta = \frac{1}{\sqrt{V}} e^{\frac{i}{\hbar} \, \vec{p}_\beta \cdot \vec{r}} = \frac{1}{\sqrt{V}} \left( 1 + \frac{i}{\hbar} \, \vec{p}_\beta \cdot \vec{r} + \ldots \right) \cong \frac{1}{\sqrt{V}} \tag{2.177}$$

$$\psi_\nu = \frac{1}{\sqrt{V}} e^{\frac{i}{\hbar} \, \vec{p}_\nu \cdot \vec{r}} = \frac{1}{\sqrt{V}} \left( 1 + \frac{i}{\hbar} \, \vec{p}_\nu \cdot \vec{r} + \ldots \right) \cong \frac{1}{\sqrt{V}} \tag{2.178}$$

Hence from equation (2.168) we have

$$H_{if} = <f|\hat{H}|i> = \int_\tau \psi_D^* \psi_\beta^* \psi_\nu^* \; \hat{H} \psi_P d\tau = \int_\tau \psi_D^* \frac{1}{\sqrt{V}} \frac{1}{\sqrt{V}} \; \hat{H} \psi_P d\tau$$

$$H_{if} = \frac{1}{V} \int_\tau \psi_D^* \; \hat{H} \psi_P d\tau = \frac{M_{if}}{V} \tag{2.179}$$

where

$$M_{if} = \int_\tau \psi_D^* \; \hat{H} \psi_P d\tau \tag{2.180}$$

is the nuclear matrix element that is independent of lepton ($\beta$, $\nu$) momentum/energy and decided only by nuclear wave functions ($\psi_P$, $\psi_D$). Thus from equations (2.167) and (2.179) we get

$$\lambda = \frac{2\pi}{\hbar} |H_{if}|^2 \frac{dn}{dE_f} = \frac{2\pi}{\hbar} \left| \frac{M_{if}}{V} \right|^2 \frac{dn}{dE_f}$$

$$\lambda = \frac{2\pi}{\hbar V^2} |M_{if}|^2 \frac{dn}{dE_f}. \tag{2.181}$$

- State of a particle is characterized by its position $x, y, z$ and momenta $p_x, p_y, p_z$ and they are determined with an accuracy given by Heisenberg's uncertainty relations

$$\delta x \delta p_x \sim h, \;\; \delta y \delta p_y \sim h, \;\; \delta z \delta p_z \sim h \tag{2.182}$$

Cell volume in phase space is

$$h. \, h. \, h = h^3 \tag{2.183}$$

Configuration space volume is $V$.

Momentum space volume is the spherical shell volume $4\pi p^2 \, dp$ between $p$ and $p + dp$.

For a particle confined to a volume $V$ the number of quantum states within momenta range $p$ and $p + dp$ is given by

$$dn_0 = \frac{\text{phase space volume}}{\text{cell volume}} = \frac{4\pi p^2 \, dp \, V}{h^3}. \tag{2.184}$$

Clearly thus for $\beta$ we have $dn_\beta = \dfrac{4\pi p_\beta^2 \, dp_\beta \, V}{h^3}$ and for $\nu$ we have $n_\nu = \dfrac{4\pi p_\nu^2 \, dp_\nu \, V}{h^3}$

Hence for the leptons ($\beta$ and $\nu$) we have the number of quantum states given by

$$dn = dn_\beta dn_\nu = \frac{4\pi p_\beta^2 \, dp_\beta \, V}{h^3} \frac{4\pi p_\nu^2 \, dp_\nu \, V}{h^3} = \frac{(4\pi V)^2 p_\beta^2 p_\nu^2 \, dp_\beta}{h^6} dp_\nu \tag{2.185}$$

$$dn = \frac{(4\pi V)^2 p_\beta^2 p_\nu^2 \, dp_\beta}{h^6} \frac{dp_\nu}{dE_f} \, dE_f \qquad (2.186)$$

where $E_f$ is final energy.

$$\frac{dn}{dE_f} = \frac{(4\pi V)^2 p_\beta^2 p_\nu^2 \, dp_\beta}{h^6} \frac{dp_\nu}{dE_f} \qquad (2.187)$$

Now the final energy is

$$E_f = \text{ Energy of } \beta + \text{ Energy of } \nu = E_\beta + E_\nu \qquad (2.188)$$

$$E_f = (T_\beta + m_\beta c^2) + E_\nu \qquad (2.189)$$

where $m_\beta$ is $\beta$ rest mass. Now

$$E_\nu = \sqrt{p_\nu^2 c^2 + m_\nu^2 c^4} \qquad (2.190)$$

We neglect the rest mass of neutrino, i.e. take

$$m_\nu \cong 0 \qquad (2.191)$$

$$E_\nu = \sqrt{p_\nu^2 c^2 + m_\nu^2 c^4} \cong p_\nu c \qquad (2.192)$$

From equations (2.189) and (2.192)

$$E_f = T_\beta + m_\beta c^2 + p_\nu c \qquad (2.193)$$

This means for some $T_\beta$ the energy differential is

$$dE_f = dp_\nu c$$

$$\frac{dp_\nu}{dE_f} = \frac{1}{c} \qquad (2.194)$$

Hence we have from equations (2.187) and (2.194)

$$\frac{dn}{dE_f} = \frac{(4\pi V)^2 p_\beta^2 p_\nu^2 \, dp_\beta}{h^6} \frac{dp_\nu}{dE_f} = \frac{(4\pi V)^2 p_\beta^2 p_\nu^2 \, dp_\beta}{h^6} \frac{1}{c} = C_0 p_\beta^2 p_\nu^2 \, dp_\beta \propto p_\beta^2 p_\nu^2 \, dp_\beta \qquad (2.195)$$

where $C_0 = \frac{(4\pi V)^2}{ch^6} = \text{constant}$. So from equations (2.181) and (2.195)

$$\lambda = \frac{2\pi}{\hbar V^2} |M_{if}|^2 \frac{dn}{dE_f} = \frac{2\pi}{\hbar V^2} |M_{if}|^2 C_0 p_\beta^2 p_\nu^2 \, dp_\beta \propto p_\beta^2 p_\nu^2 \, dp_\beta \qquad (2.196)$$

And so the number of $\beta$ particles emerging at kinetic energy $T_\beta$ in the energy range $dT_\beta$ will also be $\propto p_\beta^2 p_\nu^2 \, dp_\beta$. Let us denote this by $N(T_\beta)dT_\beta$ and write

$$N(T_\beta)dT_\beta \propto p_\beta^2 p_\nu^2 \, dp_\beta \tag{2.197}$$

$$N(T_\beta)dT_\beta = C p_\beta^2 p_\nu^2 \, dp_\beta \tag{2.198}$$

where $C$ is a constant.

Kinetic energy of $\beta$ can vary from 0 to $Q_0$ (figure 2.20). Now

$$E_\beta = \sqrt{p_\beta^2 c^2 + m_\beta^2 c^4} = T_\beta + m_\beta c^2 \tag{2.199}$$

$$p_\beta^2 c^2 + m_\beta^2 c^4 = \left(T_\beta + m_\beta c^2\right)^2 = T_\beta^2 + 2 T_\beta m_\beta c^2 + m_\beta^2 c^4$$

$$p_\beta^2 c^2 = T_\beta^2 + 2 T_\beta m_\beta c^2$$

$$p_\beta^2 = \frac{1}{c^2} T_\beta \left(T_\beta + 2 m_\beta c^2\right) \text{ and so}$$

$$p_\beta = \frac{1}{c} \sqrt{T_\beta^2 + 2 T_\beta m_\beta c^2} \tag{2.200}$$

Hence taking differential

$$dp_\beta = \frac{1}{c} \frac{1}{2} \frac{2 T_\beta + 2 m_\beta c^2}{\sqrt{T_\beta \left(T_\beta + 2 m_\beta c^2\right)}} dT_\beta = \frac{1}{c} \frac{T_\beta + m_\beta c^2}{\sqrt{T_\beta \left(T_\beta + 2 m_\beta c^2\right)}} dT_\beta \tag{2.201}$$

As energy $Q_0$ is distributed between $\beta$ and $\nu$ (figure 2.20) we write

$$Q_0 = T_\beta + E_\nu \tag{2.202}$$

$$Q_0 = T_\beta + p_\nu c \tag{2.203}$$

$$p_\nu = \frac{1}{c}(Q_0 - T_\beta) \tag{2.204}$$

$$p_\nu^2 = \frac{1}{c^2}(Q_0 - T_\beta)^2 \tag{2.205}$$

Hence the number of $\beta$ particles emerging at kinetic energy $T_\beta$ in the energy range $dT_\beta$ is from equations (2.198), (2.200), (2.202), and (2.205)

$$N(T_\beta)dT_\beta = C p_\beta^2 p_\nu^2 \, dp_\beta$$

$$= C \frac{1}{c^2} T_\beta \left(T_\beta + 2 m_\beta c^2\right) \cdot \frac{1}{c^2} (Q_0 - T_\beta)^2 \cdot \frac{1}{c} \frac{T_\beta + m_\beta c^2}{\sqrt{T_\beta \left(T_\beta + 2 m_\beta c^2\right)}} dT_\beta$$

$$N(T_\beta)dT_\beta = C'\sqrt{T_\beta(T_\beta + 2m_\beta c^2)}(Q_0 - T_\beta)^2(T_\beta + m_\beta c^2)dT_\beta \qquad (2.206)$$

with $C'$ = constant. So

$$N(T_\beta) = C'\sqrt{T_\beta(T_\beta + 2m_\beta c^2)}(Q_0 - T_\beta)^2(T_\beta + m_\beta c^2). \qquad (2.207)$$

This is the number of $\beta$ particles emerging with kinetic energy $T_\beta$.

A plot of $N(T_\beta)$ against $T_\beta$ from equation (2.207) gives the shape of the energy distribution curve (figure 2.20(a)) provided we take into account the Coulomb interaction between daughter nucleus and $\beta^-$ (in the case of $\beta^-$ decay) and between daughter nucleus and $\beta^+$ (in the case of $\beta^+$ decay).

- For $\beta$ decay of $_{29}Cu^{64}$ let us use the following data in equation (2.207).

$Q_0 = 0.6\ MeV$, $T_\beta = 0$, 0.05, 0.1, 0.15, 0.2, 0.25, 0.3, 0.35, 0.4, 0.45, 0.5, 0.55, 0.6 $MeV$.

The calculation details have been done in figure 2.23 taking $m_\beta c^2 = 0.511\ MeV$. And we get the plot of $\dfrac{N(T_\beta)}{C'}$ [from equation (2.207)] against $T_\beta$ as shown in figure 2.23.

Clearly experimental curves of $\beta^-$ decay, $\beta^+$ decay of figure 2.20(b),(c) are not reproduced. We thus conclude that equation (2.207) obtained from time dependent perturbation theory cannot describe $\beta$ decay properly.

$\beta^-$ and $\beta^+$ weak decays differ in the Coulomb interaction between daughter nucleus and the emitted $\beta$. This we have to incorporate into Fermi theory and so the wave function $\psi_\beta = \dfrac{1}{\sqrt{V}}e^{\frac{i}{\hbar}\vec{p}_\beta \cdot \vec{r}}$ needs modification. Quantum mechanical calculation shows that the necessary modification can be done by multiplying results by a factor $F(Z_D, T_\beta)$ that is dependent upon atomic number of daughter nucleus $Z_D$ and kinetic energy of $\beta$, i.e. on $T_\beta$ and is called Fermi factor or Fermi function. It takes care of

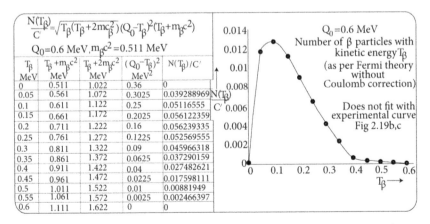

**Figure 2.23.** Plot of $N(T_\beta)$ versus $T_\beta$ as per Fermi theory without Coulomb correction.

the interaction between daughter nucleus and the emerging $\beta$. Non-relativistic calculation gives

$$F(Z_D, T_\beta) = \frac{2\pi\eta}{1 - e^{-2\pi\eta}} \tag{2.208}$$

where

$$\eta = \pm\frac{Z_D |e|^2}{4\pi\varepsilon_0 \hbar v} \tag{2.209}$$

Using $\frac{|e|^2}{4\pi\varepsilon_0} = 1.44 \ MeV.fm$ (*exercise 1.1*) we have from equation (2.209)

$$\eta = \pm\frac{Z_D}{\hbar v}(1.44 \ MeV.fm)$$

Using $v = \sqrt{2T_\beta/m_\beta}$ (velocity of $\alpha$)

$$\eta = \pm\frac{Z_D}{\hbar\sqrt{2T_\beta/m_\beta}}(1.44 \ MeV.fm) \tag{2.210}$$

$$\eta = \pm\frac{Z_D(1.44 \ MeV.fm)}{\hbar c}\sqrt{\frac{m_\beta c^2}{2T_\beta}} = \pm\frac{Z_D(1.44 \ MeV.fm)}{197 \ MeV.fm}\sqrt{\frac{0.511 \ MeV}{2T_\beta}}$$

$$\eta = \pm0.003\ 6948\frac{Z_D}{\sqrt{T_\beta}} \tag{2.211}$$

where we used $\hbar c = 197 \ MeV.fm$ (*exercise 1.2*) and $T_\beta$ is in $MeV$. Also $\eta$ is positive for $\beta^-$ decay and negative for $\beta^+$ decay. Incorporating the Fermi function in equation (2.207) we can write down the number of $\beta$ particles emerging with kinetic energy $T_\beta$

$$N(T_\beta)F(Z_D, T_\beta) = C'\sqrt{T_\beta(T_\beta + 2m_\beta c^2)}(Q_0 - T_\beta)^2(T_\beta + m_\beta c^2)F(Z_D, T_\beta) \equiv NF \tag{2.212}$$

Though energy associated with $\beta, v$ is $\sim MeV$ we have incorporated a non-relativistic expression for $\eta$. This is because Coulombic interaction significantly affects the low energy part of $\beta$ spectrum. In other words if $\beta$ emerges with low velocity then Coulomb correction is more needed.

Let us use the following data in equation (2.212).

$Q_0 = 0.6 \ MeV$, $T_\beta = 0.02.0.04, \ldots\ldots0.6 \ MeV$. The calculation details have been done in figure 2.24 taking $m_\beta c^2 = 0.511 \ MeV$ and using equation (2.211). And we get the plot of $\frac{NF}{C'}$ (from equation (2.212) as per Fermi's theory of $\beta$ decay) against $T_\beta$ as shown in figure 2.24 ($\beta^\pm$ decay of $_{29}Cu^{64}$). The plot agrees with experimental curves of figure 2.20(b),(c).

| $T_\beta$ MeV | $T_\beta+m_\beta c^2$ MeV | $T_\beta+2m_\beta c^2$ MeV | $(Q_0-T_\beta)^2$ MeV² | $N/C'$ | For $\beta^+$ $\eta=-0.0036948 Z_D\sqrt{}/T_\beta$ $Z_D=28$ $\eta$ | $F=\dfrac{2\pi\eta}{1-e^{-2\pi\eta}}$ | $\dfrac{NF}{C'}$ | For $\beta^-$ $\eta=+0.0036948 Z_D\sqrt{}/T_\beta$ $Z_D=30$ $\eta$ | $F=\dfrac{2\pi\eta}{1-e^{-2\pi\eta}}$ | $\dfrac{NF}{C'}$ |
|---|---|---|---|---|---|---|---|---|---|---|
| 0.02 | 0.531 | 1.042 | 0.3364 | 0.025786914 | −0.73153 | 0.046843 | 0.00121 | 0.783785 | 4.960709 | 0.12792 |
| 0.04 | 0.551 | 1.062 | 0.3136 | 0.035613931 | −0.51727 | 0.131088 | 0.00467 | 0.55422 | 3.592698 | 0.12795 |
| 0.06 | 0.571 | 1.082 | 0.2916 | 0.042424124 | −0.42235 | 0.200938 | 0.00852 | 0.452519 | 3.019077 | 0.12808 |
| 0.08 | 0.591 | 1.102 | 0.2704 | 0.047449316 | −0.36577 | 0.256607 | 0.01218 | 0.391893 | 2.69177 | 0.12772 |
| 0.1 | 0.611 | 1.122 | 0.25 | 0.05116555 | −0.32715 | 0.301792 | 0.01544 | 0.35052 | 2.476085 | 0.12669 |
| 0.12 | 0.631 | 1.142 | 0.2304 | 0.053818986 | −0.29865 | 0.339303 | 0.01826 | 0.319979 | 2.321374 | 0.12493 |
| 0.14 | 0.651 | 1.162 | 0.2116 | 0.055560191 | −0.27649 | 0.37107 | 0.02062 | 0.296243 | 2.203988 | 0.12245 |
| 0.16 | 0.671 | 1.182 | 0.1936 | 0.056493256 | −0.25864 | 0.398424 | 0.02251 | 0.27711 | 2.111288 | 0.11927 |
| 0.18 | 0.691 | 1.202 | 0.1764 | 0.056697657 | −0.24384 | 0.422307 | 0.02394 | 0.261262 | 2.035858 | 0.11543 |
| 0.2 | 0.711 | 1.222 | 0.16 | 0.056239336 | −0.23133 | 0.443403 | 0.02494 | 0.247855 | 1.973037 | 0.11096 |
| 0.22 | 0.731 | 1.242 | 0.1444 | 0.055176825 | −0.22057 | 0.462223 | 0.0255 | 0.23632 | 1.919736 | 0.10592 |
| 0.24 | 0.751 | 1.262 | 0.1296 | 0.053564892 | −0.21118 | 0.479153 | 0.02567 | 0.226259 | 1.873821 | 0.10037 |
| 0.26 | 0.771 | 1.282 | 0.1156 | 0.051456815 | −0.20289 | 0.494494 | 0.02545 | 0.217383 | 1.833764 | 0.09436 |
| 0.28 | 0.791 | 1.302 | 0.1024 | 0.048905865 | −0.19551 | 0.508486 | 0.02487 | 0.209475 | 1.798443 | 0.08795 |
| 0.3 | 0.811 | 1.322 | 0.09 | 0.045966318 | −0.18888 | 0.521316 | 0.02396 | 0.202373 | 1.767011 | 0.08122 |
| 0.32 | 0.831 | 1.342 | 0.0784 | 0.042694146 | −0.18288 | 0.533141 | 0.02276 | 0.195946 | 1.738818 | 0.07424 |
| 0.34 | 0.851 | 1.362 | 0.0676 | 0.039147521 | −0.17742 | 0.544088 | 0.0213 | 0.190096 | 1.713354 | 0.06707 |
| 0.36 | 0.871 | 1.382 | 0.0576 | 0.035387176 | −0.17242 | 0.554261 | 0.01961 | 0.18474 | 1.690214 | 0.05981 |
| 0.38 | 0.891 | 1.402 | 0.0484 | 0.031476677 | −0.16783 | 0.563749 | 0.01774 | 0.179813 | 1.669072 | 0.05254 |
| 0.4 | 0.911 | 1.422 | 0.04 | 0.027482622 | −0.16358 | 0.572627 | 0.01574 | 0.17526 | 1.649661 | 0.04534 |
| 0.42 | 0.931 | 1.442 | 0.0324 | 0.023474804 | −0.15963 | 0.580959 | 0.01364 | 0.171036 | 1.631762 | 0.03831 |
| 0.44 | 0.951 | 1.462 | 0.0256 | 0.019526325 | −0.15596 | 0.588799 | 0.0115 | 0.167104 | 1.615192 | 0.03154 |
| 0.46 | 0.971 | 1.482 | 0.0196 | 0.015713695 | −0.15254 | 0.596195 | 0.00937 | 0.163431 | 1.599797 | 0.02514 |
| 0.48 | 0.991 | 1.502 | 0.0144 | 0.012116906 | −0.14932 | 0.603188 | 0.00731 | 0.15999 | 1.585447 | 0.01921 |
| 0.5 | 1.011 | 1.522 | 0.01 | 0.00881949 | −0.14631 | 0.609814 | 0.00538 | 0.156757 | 1.572031 | 0.01386 |
| 0.52 | 1.031 | 1.542 | 0.0064 | 0.005908572 | −0.14347 | 0.616104 | 0.00364 | 0.153713 | 1.559453 | 0.00921 |
| 0.54 | 1.051 | 1.562 | 0.0036 | 0.003474902 | −0.14078 | 0.622086 | 0.00216 | 0.15084 | 1.547632 | 0.00538 |
| 0.56 | 1.071 | 1.582 | 0.0016 | 0.001612897 | −0.13825 | 0.627785 | 0.00101 | 0.148122 | 1.536496 | 0.00248 |
| 0.58 | 1.091 | 1.602 | 0.0004 | 0.000420659 | −0.13584 | 0.633223 | 0.00027 | 0.145545 | 1.525982 | 0.00064 |
| 0.6 | 1.111 | 1.622 | 0 | 0 | −0.13356 | 0.638419 | 0 | 0.143099 | 1.516036 | 0 |

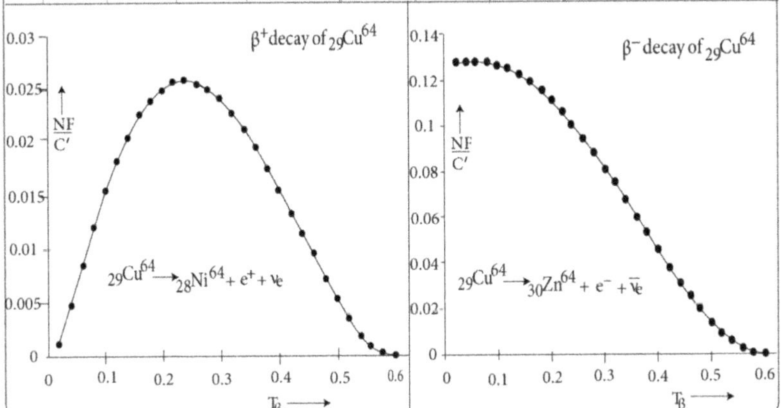

**Figure 2.24.** Plot of $N(T_\beta)F(Z_D, T_\beta)$ against $T_\beta$ as per Fermi's theory of $\beta$ decay.

Incorporating the Fermi function $F(Z_D, T_\beta)$ of equation (2.208) into equation (2.198) we have for the number of $\beta$ within momenta range $p_\beta$ and $p_\beta + dp_\beta$ to be

$$N(p_\beta)dp_\beta = C'' p_\beta^2 p_\nu^2 \, F(Z_D, T_\beta)dp_\beta$$

$$N(p_\beta) = C'' p_\beta^2 p_\nu^2 \, F(Z_D, T_\beta) \qquad (2.213)$$

where $C''$ is a constant. This gives the momentum distribution of $\beta$.

## 2.31 Fermi–Kurie plot

Let us try to obtain a linear plot from this momentum distribution as then any deviation from linearity is easily identifiable and can be analysed. Now

$$\frac{N(p_\beta)}{p_\beta^2 F(Z_D, T_\beta)} \propto p_\nu^2 \Rightarrow \sqrt{\frac{N(p_\beta)}{p_\beta^2 F(Z_D, T_\beta)}} \propto p_\nu \qquad (2.214)$$

And this relation has been derived based upon the assumption that the rest mass of a neutrino is zero viz. equation (2.191). We use equations (2.192) and (2.204) to write

$$E_\nu = p_\nu c = Q_0 - T_\beta \qquad (2.215)$$

And so

$$\sqrt{\frac{N(p_\beta)}{p_\beta^2 F(Z_D, T_\beta)}} \propto p_\nu \propto Q_0 - T_\beta \qquad (2.216)$$

$$\sqrt{\frac{N}{p_\beta^2 F}} = C_1(Q_0 - T_\beta) \qquad (2.217)$$

where $C_1$ is a constant. So the plot of $\sqrt{\frac{N}{p_\beta^2 F}}$ against $T_\beta$ will lead to a straight line with negative slope. Such a plot is referred to as a Fermi–Kurie plot.

✓ It is a standard way to check if $\beta$ decay can be described by Fermi theory.
✓ Any deviation from linearity would mean failure of the theory to describe the $\beta$ decay under consideration
✓ Any deviation from linearity would mean that $\nu$ possesses rest mass.
✓ The plot of $\sqrt{\frac{N}{p^2 F}}$ versus $T_\beta + m_\beta c^2$ for the $\beta^+$ decay of

$$_{31}Ga^{66} \rightarrow _{30}Zn^{66} + e^+ + \nu_e$$

has been depicted in figure 2.25(a) There is deviation at low energy due to scattering of $\beta$ inside radioactive nucleus at low energy. At larger energy,

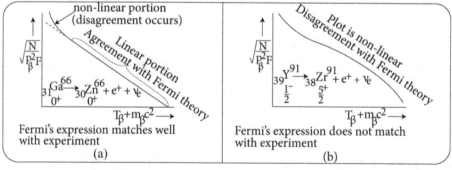

**Figure 2.25.** Plot of $\sqrt{\frac{N}{p^2 F}}$ versus $T_\beta + m_\beta c^2$ for (a) $\beta^+$ decay of $_{31}Ga^{66}$ (b) $\beta^+$ decay of $_{39}Y^{91}$.

scattering inside the source material is not very effective and we get a linear plot that agrees with Fermi's theory of $\beta$ decay.

✓ The plot of $\sqrt{\dfrac{N}{p^2 F}}$ versus $T_\beta + m_\beta c^2$ for the $\beta^+$ decay of

$$_{39}Y^{91} \rightarrow {_{38}}Zr^{91} + e^+ + \nu_e$$

has been depicted in figure 2.25(b). The plot is nonlinear and is not as per Fermi's theory.

Clearly thus there are a number of $\beta$ decays that follow Fermi's theory, but there are also a number of $\beta$ decays that do not follow Fermi's theory, as is clear from the Fermi–Kurie plot of figure 2.25. This agreement/disagreement can be explained studying the angular momentum and parity of parent and daughter nuclei that we shall discuss now.

## 2.32 Selection rules of beta decay

Consider the $\beta$ decay

$$\text{Parent } P \rightarrow \text{ daughter } D + \text{ lepton pair, i.e. } X \rightarrow Y + \text{lepton pair} \qquad (2.218)$$

Use angular momentum conservation relation
$\vec{J}_P = \vec{J}_D + \vec{J}_{\text{lepton pair}} \ (\vec{J} \rightarrow$ angular momentum vector)
$\vec{J}_P - \vec{J}_D = \vec{J}_{\text{lepton pair}} = $ total angular momentum carried away by lepton pair

$$= \vec{l} + \vec{s}_e + \vec{s}_\nu \qquad (2.219)$$

where $\vec{l} \rightarrow$ orbital angular momentum part and
$\vec{s}_e + \vec{s}_\nu \rightarrow$ spin angular momentum of leptons.

The motion of the lepton pair (having reduced mass $\mu$) in a potential well $V$ is governed by the Schrödinger equation

$$\frac{d^2 u}{dr^2} + \frac{2\mu}{\hbar^2}\left[E - \left(V + \frac{l(l+1)\hbar^2}{2\mu r^2}\right)\right]u = 0 \qquad (2.220)$$

with $\psi = \dfrac{u}{r} = $ particle wave function
where $V = $ attractive well ($V = -$ve)

$$\frac{l(l+1)\hbar^2}{2\mu r^2} = \text{ repulsive centrifugal potential energy } (+\text{ve}).$$

The lepton pair generated during $\beta$ decay has to overcome this effective potential barrier or $p. \, e.$ hill viz.

$$V_{\text{eff}} = V + \frac{l(l+1)\hbar^2}{2\mu r^2} \qquad (2.221)$$

in order to emerge (figure 2.26).

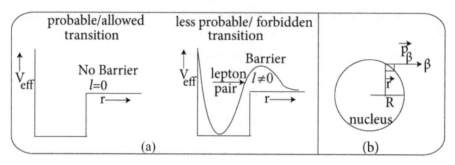

**Figure 2.26.** (a) Barrier faced by lepton pair during $\beta$ decay. (b) $\beta$ emerging out of a nucleus with momentum $\vec{P}_\beta$.

It is clear that the height of barrier depends on $l$ value. If $l$ is low the barrier height is small and if $l$ is high the barrier height is large. So low $l$ transitions are favourable, i.e. more probable, and high $l$ transitions are forbidden, i.e. suppressed or less probable. We thus can classify transitions according to their probability of occurrence in terms of $l$ value.

Parity of the transition is governed by $l$. $\Delta\pi = 0$ means no parity change; $\Delta\pi \neq 0$ means parity changes.

- We now define what we mean by allowed transition ($l = 0$), first forbidden transition ($l = 1$), second forbidden transition ($l = 2$) along with corresponding parity values.

| Angular momentum | Barrier height faced by lepton pair | Probability of transition | Name of transition | Does parity change? $\Delta\pi = 0$ for $l$ = even, $\Delta\pi \neq 0$ for $l$ = odd |
|---|---|---|---|---|
| $l = 0$ | Zero | Maximum | Allowed | No, $\Delta\pi = 0$ |
| $l = 1$ | High | Small | 1st forbidden | Yes, $\Delta\pi \neq 0$ |
| $l = 2$ | Higher | Smaller | 2nd forbidden | No, $\Delta\pi = 0$ |

Allowed transition is nearly 100 times more probable than first forbidden transition.

First forbidden transition is nearly 100 times more probable than second forbidden transition.

- Let us next define what we mean by Fermi transition and Gammow–Teller transition.

Angular momentum conservation relation applied to the $\beta$ decay

$P \rightarrow D + \beta + \nu$ ($P$ is parent, $D$ is daughter, $\beta$ is $e^-$ or $e^+$, $\nu$ is neutrino or antineutrino) is given by

$$\vec{J}_P = \vec{J}_D + \vec{l} + \vec{s}_e + \vec{s}_\nu$$

where $\vec{l}$ is orbital angular momentum, $s_e = \frac{1}{2}$, $s_\nu = \frac{1}{2}$ are spins of electron, neutrino, respectively. Now

$$\vec{J}_P - \vec{J}_D = \vec{l} + \vec{s}_e + \vec{s}_\nu = \vec{l} + \frac{\overline{1}}{2} + \frac{\overline{1}}{2} \tag{2.222}$$

$$\vec{J}_P - \vec{J}_D = \begin{cases} \vec{l} + \vec{0} & \text{(antiparallel spins } \Leftarrow \text{ singlet : Fermi transition)} \\ \vec{l} + \vec{1} & \text{(parallel spins } \Leftarrow \text{ triplet : Gammow–Tellar transition)} \end{cases} \tag{2.223}$$

Fermi transition corresponds to antiparallel spin combination of $\beta$ and $\nu$ that leads to $S = 0$, multiplicity $2S + 1 = 2.0 + 1 = 1$, i.e. singlet state.

Gammow–Teller transition corresponds to parallel spin combination of $\beta$ and $\nu$ that leads to $S = 1$, multiplicity $2S + 1 = 2.1 + 1 = 3$, i.e. triplet state.

✓ It also follows from equation (2.223) that for Fermi transition maximum value of $\Delta J$ is $l$ and so

$$\Delta J \leqslant l.$$

In Gammow–Teller transition maximum value of $\Delta J$ is $l + 1$ and so

$$\Delta J \leqslant l + 1.$$

Allowed transition means

$$l = 0, \quad \Delta \pi = 0 \tag{2.224}$$

$$\text{First forbidden means } l = 1, \Delta \pi \neq 0 \tag{2.225}$$

$$\text{Second forbidden means } l = 2, \quad \Delta \pi = 0. \tag{2.226}$$

Fermi transition means

$$\vec{s}_e + \vec{s}_\nu \rightarrow 0 \text{(antiparallel spins)} \tag{2.227}$$

Gammow Teller transition means

$$\vec{s}_e + \vec{s}_\nu \rightarrow 1 \text{(parallel spins)} \tag{2.228}$$

- We thus have the following selection rules for beta decay. The selection rules follow from analysis of the matrix element $M_{if}$ [equation (2.180)] used in Fermi's theory of beta decay.

☐ Allowed Fermi transition
$\vec{J}_P - \vec{J}_D = \vec{l} + \vec{s}_e + \vec{s}_\nu = \vec{0} + \vec{0} = \vec{0}$ (as allowed means $l = 0$, Fermi means $\vec{s}_e + \vec{s}_\nu \rightarrow 0$)

$$\vec{J}_P = \vec{J}_D, J_P - J_D = \Delta J = 0$$

So the selection rule for allowed Fermi transition is

$$l = 0, \ \Delta\pi = 0, \ \Delta J = 0 (\text{antiparallel spins} \ \ \vec{s}_e + \vec{s}_\nu \to 0) \tag{2.229}$$

Example

$_8O^{14} \to (_7N^{14})^* + e^+ + \nu_e$ (ground state of $_7N^{14}$ has spin 1 but 2.31 $MeV$ excited state $_7N^{14}$ has spin 0)

$$_{13}Al^{26} \to {}_{12}Mg^{26} + e^- + \bar{\nu}_e$$

$$_{17}Cl^{34} \to {}_{16}S^{34} + e^- + \bar{\nu}_e$$

These are $0^+ \to 0^+$ $\beta$ transitions (*exercise 2.53*) which are allowed Fermi transitions.

☐ Allowed Gammow–Teller transition

$$\vec{J}_P - \vec{J}_D = \vec{l} + \vec{s}_e + \vec{s}_\nu = \vec{0} + \vec{1} = \vec{1}$$

(as allowed means $l = 0$, Gammo–Teller means $\vec{s}_e + \vec{s}_\nu \to 1$)

$$\vec{J}_P = \vec{J}_D + \vec{1} \Rightarrow J_P = J_D + 1, \ J_D, \ | \ J_D - 1 \ |,$$

i.e. $J_P - J_D = \Delta J = 0, \pm 1$. Also, $J_P = 0 \to J_D = 0$ is a forbidden transition

So the selection rule for allowed Gammow–Teller transition is

$$l = 0, \ \Delta\pi = 0, \ \Delta J = 0, \pm 1 \ (\text{parallel spins} \ \vec{s}_e + \vec{s}_\nu \to 1) \ 0 \leftrightarrow 0 \ \text{forbidden} \tag{2.230}$$

Example

$$_2He^6 \to {}_3Li^6 + e^- + \bar{\nu}_e$$

$$_6C^{14} \to {}_7N^{14} + e^- + \bar{\nu}_e$$

These are $0^+ \to 1^+$ $\beta$ transition (*exercise 2.50)* which are allowed Gammow–Teller transitions.

$$_{32}Ge^{75} \to {}_{33}As^{75} + e^- + \bar{\nu}_e$$

This is a $\frac{1}{2}^- \to \frac{3}{2}^-$ transition (*exercise 2.54*) which is an allowed Gammow–Teller transition.

✓ Other examples

$$_0n^1 \to p + e^- + \bar{\nu}_e$$

$$_1H^3 \to {}_2He^3 + e^- + \bar{\nu}_e$$

These are $\frac{1}{2}^+ \to \frac{1}{2}^+$ transitions (*exercise 2.49*) which are both allowed Fermi and allowed Gammow–Teller transitions.

$$_{16}S^{35} \to {}_{17}Cl^{35} + e^- + \bar{\nu}_e$$

This is $\frac{3}{2}^+ \to \frac{3}{2}^+$ transition (*exercise 2.55*) which are both allowed Fermi and allowed Gammow–Teller transition.

□ First forbidden Fermi transitions

$$\vec{J}_P - \vec{J}_D = \vec{l} + \vec{s}_e + \vec{s}_\nu = \vec{1} + \vec{0} = \vec{1}$$

(as first forbidden means $l = 1$, Fermi means $\vec{s}_e + \vec{s}_\nu \to 0$)

$$\vec{J}_P = \vec{J}_D + \vec{1}$$

$$J_P = J_D + 1, J_D, |J_D - 1| \text{ i.e. } J_P - J_D = \Delta J = 0, \pm 1$$

$J_P = 0 \leftrightarrow J_D = 0$ forbidden. So the selection rule for first forbidden Fermi transition is

$$l = 1, \Delta \pi \neq 0, \Delta J = 0, \pm 1, 0 \leftrightarrow 0 \text{ forbidden} \tag{2.231}$$

(antiparallel spins $\vec{s}_e + \vec{s}_\nu \to 0$).
For example

$$_6C^{14} \to {}_7N^{14} + e^- + \bar{\nu}_e$$

This is $0^+ \to 1^-$ (*exercise 2.56*) which is the first forbidden Fermi transition.
□ First forbidden Gammow–Teller transition

$$\vec{J}_P - \vec{J}_D = \vec{l} + \vec{s}_e + \vec{s}_\nu = \vec{1} + \vec{1} = 2, 1, 0 \text{ i.e. } \vec{J}_P - \vec{J}_D = \vec{2}, \vec{1}, \vec{0}$$

(as first forbidden means $l = 1$, Gammow–Teller means $\vec{s}_e + \vec{s}_\nu \to 1$)

$\vec{J}_P = \vec{J}_D + \vec{0}$ leads to $\Delta J = 0$, $\vec{J}_P = \vec{J}_D + \vec{1}$ leads to $\Delta J = 0, \pm 1$, $\vec{J}_P = \vec{J}_D + \vec{2}$ leads to

$$J_P = J_D + 2, J_D + 1, J_D, |J_D - 1|, |J_D - 2| \text{ i.e. } J_P - J_D = \Delta J = 0, \pm 1, \pm 2.$$

$0 \leftrightarrow 0, \frac{1}{2} \leftrightarrow \frac{1}{2}, 0 \leftrightarrow 1$ transitions are forbidden.
So the selection rule for first forbidden Gammow–Teller transition is

$$l = 1, \Delta \pi \neq 0, \Delta J = 0, \pm 1, \pm 2 \tag{2.232}$$

(parallel spins $\vec{s}_e + \vec{s}_\nu \to 1$) $0 \leftrightarrow 0, \frac{1}{2} \leftrightarrow \frac{1}{2}, 0 \leftrightarrow 1$ forbidden
For example

$$_{18}Ar^{39} \to {}_{19}K^{39} + e^- + \bar{\nu}_e$$

This is $\frac{7}{2}^- \to \frac{3}{2}^+$ $\beta$ transition (*exercise 2.48*) which is first forbidden Gammo–Teller transition.
✓ Another example

$$_{17}Cl^{38} \to {}_{18}Ar^{38} + e^- + \bar{\nu}_e$$

This is a $2^- \to 2^+$ $\beta$ transition (*exercise 2.57*) which are both first forbidden Fermi as well as first forbidden Gammo–Teller transitions.

- For $l$th forbidden transition the selection rule is as follows.

$$\Delta J = \pm l, \pm(l - 1) \text{ is the Fermi selection rule} \tag{2.233}$$

$\Delta J = \pm l, \pm(l + 1)$ is the Gammow–Teller selection rule $\qquad$ (2.234)

Parity does not change $\Delta \pi = 0$ for even $l$
Parity changes $\Delta \pi \neq 0$ for odd $l$
Accordingly let us write down some other selection rules.
☐ Second forbidden ($l = 2$) Fermi transition
Selection rule is

$$l = 2, \ \Delta \pi = 0, \ \Delta J = \pm 1, \pm 2, \ 0 \rightarrow 1, \ 1 \rightarrow 0 \text{ forbidden} \qquad (2.235)$$

☐ Second forbidden ($l = 2$) Gammow–Teller transition
Selection rule is

$$l = 1, \ \Delta \pi \neq 0 = \Delta J = \pm 2, \pm 3, \ 0 \rightarrow 0, \ 0 \rightarrow 2, \ 2 \rightarrow 0 \text{ forbidden} \qquad (2.236)$$

For example

$$_{11}\text{Na}^{22} \rightarrow {}_{10}\text{Ne}^{22} + e^- + \bar{\nu}_e$$

This is $3^+ \rightarrow 0^+$ $\beta$ transition (*exercise 2.58*), the second forbidden Gammow–Teller transition.
- Other rules
  $J_P \neq J_D \neq k$ is allowed by Gammow–Teller transition
  $J_P = J_D = 0$ is allowed by Fermi transition
  $J_P = J_D = k \neq 0$ is allowed by both Fermi and Gammow–Teller transition.
- In Fermi's theory of $\beta$ decay we made the approximation that

$$\psi_\beta = \frac{1}{\sqrt{V}} e^{\frac{i}{\hbar} \vec{p}_\beta \cdot \vec{r}} = \frac{1}{\sqrt{V}} \left( 1 + \frac{i}{\hbar} \vec{p}_\beta \cdot \vec{r} + \ldots \right) \cong \frac{1}{\sqrt{V}} \qquad (2.237)$$

$$\psi_\nu = \frac{1}{\sqrt{V}} e^{\frac{i}{\hbar} \vec{p}_\nu \cdot \vec{r}} = \frac{1}{\sqrt{V}} \left( 1 + \frac{i}{\hbar} \vec{p}_\nu \cdot \vec{r} + \ldots \right) \cong \frac{1}{\sqrt{V}} \qquad (2.238)$$

These can be looked upon as

$$\psi_\beta = \frac{1}{\sqrt{V}} e^{\frac{i}{\hbar} \vec{p}_\beta \cdot \vec{r}} \xrightarrow{r = 0} \frac{1}{\sqrt{V}} \qquad \psi_\nu = \frac{1}{\sqrt{V}} e^{\frac{i}{\hbar} \vec{p}_\nu \cdot \vec{r}} \xrightarrow{r = 0} \frac{1}{\sqrt{V}} \qquad (2.239)$$

which means that $\psi_\beta, \psi_\nu$ represent wave function at origin implying that the $\beta$ and $\nu$ are created at the origin $r = 0$ (centre of nucleus) and emerge out from the center of the nucleus. So w.r.t the daughter nucleus the orbital angular momentum of the lepton pair $\beta$, $\nu$ is $\vec{l} = 0$, i.e. $l = 0$. Clearly Fermi's theory of $\beta$ decay corresponds to decays involving $l = 0$, i.e. for allowed transitions (as shown in figure 2.26(a)).

$\beta$ decays with $l \neq 0$ show deviations from Fermi's theory of $\beta$ decay.
- We can get a rough estimate of the probability that a lepton pair emerges with non-zero angular momentum.

Consider figure 2.26(b) showing emergence of a $\beta$ from a spherical nucleus of radius $R$. The $\beta$ is created within the nucleus and emerges with angular momentum $\vec{l}$ in some direction. Now

$$\vec{l} = \vec{r} \times \vec{p}_\beta \rightarrow r p_\beta \qquad (2.240)$$

Maximum angular momentum occurs for $r = R$ and hence

$$l = R p_\beta \qquad (2.241)$$

With kinetic energy $T_\beta = 1 \, MeV$ [equation (2.173)] and using

$$\sqrt{p_\beta^2 c^2 + m_\beta^2 c^4} = T_\beta + m_\beta c^2 \qquad (2.242)$$

we have shown that the magnitude of linear momentum is $p_\beta = 1.422 \, \frac{MeV}{c}$ [equation (2.175)]. Hence taking $R \sim 5 \, fm$ we have

$$l = R p_\beta \sim 5 \, fm. \, 1.422 \frac{MeV}{c}. \, = 7 \frac{MeV. \, fm}{c}. \qquad (2.243)$$

Replacing by eigenvalue we get

$$\vec{l} \xrightarrow{\text{eigenvalue}} \sqrt{l(l+1)} \, \hbar = 7 \frac{MeV. \, fm}{c}$$

$$\sqrt{l(l+1)} = \frac{7}{\hbar c} MeV. \, fm = \frac{7 \, MeV. \, fm}{197 \, MeV. \, fm} = 0.0355 \, (\text{exercise 1.2}).$$

So

$$l(l+1) \sim 0.0355^2 = 0.001\,26$$

$$l \ll 1 \qquad (2.244)$$

So it is quite unlikely, i.e. there is very low probability for $l \neq 0$. In other words $\beta$, $\nu$ are less likely to carry off orbital angular momenta $l > 0$. So $l = 0$ is the most prominent mode of transition.

However, some $\beta$ decays are energetically permitted that occur with high $l$ value. For them the approximations equations (2.177) and (2.178) viz.

$$\psi_\beta = \frac{1}{\sqrt{V}} e^{\frac{i}{\hbar} \vec{p}_\beta \cdot \vec{r}} \xrightarrow{r=0} \frac{1}{\sqrt{V}}, \, \psi_\nu = \frac{1}{\sqrt{V}} e^{\frac{i}{\hbar} \vec{p}_\nu \cdot \vec{r}} \xrightarrow{r=0} \frac{1}{\sqrt{V}}$$

do not hold and contribution from higher order terms cannot be neglected.

## 2.33 Effect of neutrino mass

In Fermi's theory of $\beta$ decay we approximated the mass of $\nu$ to be [equation (2.191)]

$$m_\nu = 0$$

and took the final energy of $\nu$ to be [equation (2.192)]

$$E_\nu = \sqrt{p_\nu^2 c^2 + m_\nu^2 c^4} \xrightarrow{m_\nu = 0} p_\nu c \tag{2.245}$$

And this led to [equation (2.194)]

$$\frac{dp_\nu}{dE_f} = \frac{1}{c} \tag{2.246}$$

But this will not be valid if $\nu$ rest energy is not zero. Then we cannot use the formula $E_\nu = p_\nu c = \frac{h}{\lambda}c = h\nu$

If $m_\nu \neq 0$ then we can use the formula

$$E_\nu = T_\nu + m_\nu c^2 \tag{2.247}$$

In the $N(T_\beta) \equiv N_\beta$ versus $T_\beta$ plot (figure 2.20) we see that at high kinetic energy

$$T_\beta \cong Q_0 \tag{2.248}$$

And this means that the $\nu$ energy is close to zero. Since kinetic energy is small, let us write

$$T_\nu = \frac{p_\nu^2}{2m_\nu} \tag{2.249}$$

And so from equation (2.247)

$$E_\nu = \frac{p_\nu^2}{2m_\nu} + m_\nu c^2 \tag{2.250}$$

Using $E_\beta = T_\beta + m_\beta c^2$ the available final energy is from equation (2.188)

$$E_f = E_\beta + E_\nu = T_\beta + m_\beta c^2 + \frac{p_\nu^2}{2m_\nu} + m_\nu c^2 \tag{2.251}$$

At a fixed kinetic energy of $\beta$ we thus get

$$dE_f = d\left(T_\beta + m_\beta c^2 + \frac{p_\nu^2}{2m_\nu} + m_\nu c^2\right) = d\frac{p_\nu^2}{2m_\nu} = \frac{p_\nu dp_\nu}{m_\nu}$$

$$\frac{dp_\nu}{dE_f} = \frac{m_\nu}{p_\nu} \tag{2.252}$$

(instead of equation (2.246), i.e. $\frac{dp_\nu}{dE_f} = \frac{1}{c}$). This leads to [instead of equation (2.195)].

$$\frac{dn}{dE_f} = \frac{(4\pi V)^2 p_\beta^2 p_\nu^2 \, dp_\beta}{h^6} \frac{dp_\nu}{dE_f} = \frac{(4\pi V)^2 p_\beta^2 p_\nu^2 \, dp_\beta}{h^6} \frac{m_\nu}{p_\nu} = C_0' p_\nu p_\beta^2 \, dp_\beta \propto p_\nu p_\beta^2 \, dp_\beta \tag{2.253}$$

where $C_0' = \frac{(4\pi V)^2 m_\nu}{h^6} =$ constant. Hence

$$N(T_\beta)dT_\beta \propto p_\nu p_\beta^2 \, dp_\beta$$

$$N(T_\beta)dT_\beta = C' p_\nu p_\beta^2 \, dp_\beta \text{ [instead of equation (2.198)]} \qquad (2.254)$$

where $C'$ is a constant.

We obtained in equations (2.200) and (2.201)

$$p_\beta = \frac{1}{c}\sqrt{T_\beta^2 + 2T_\beta m_\beta c^2}, \; dp_\beta = \frac{1}{c}\frac{T_\beta + m_\beta c^2}{\sqrt{T_\beta(T_\beta + 2m_\beta c^2)}}dT_\beta.$$

Again from equation (2.202)

$$Q_0 = T_\beta + T_\nu$$

Using $T_\nu = \frac{p_\nu^2}{2m_\nu}$

$$Q_0 = T_\beta + \frac{p_\nu^2}{2m_\nu} \Rightarrow \frac{p_\nu^2}{2m_\nu} = Q_0 - T_\beta \Rightarrow p_\nu = \sqrt{2m_\nu(Q_0 - T_\beta)} \qquad (2.255)$$

Hence from equations (2.254), (2.255), (2.200), and (2.201) we have

$$N(T_\beta)dT_\beta = C' . \sqrt{2m_\nu(Q_0 - T_\beta)} . \frac{1}{c^2}T_\beta(T_\beta + 2m_\beta c^2) . \frac{1}{c}\frac{T_\beta + m_\beta c^2}{\sqrt{T_\beta(T_\beta + 2m_\beta c^2)}}dT_\beta$$

$$N(T_\beta)dT_\beta = C'' \sqrt{T_\beta(T_\beta + 2m_\beta c^2)} \sqrt{Q_0 - T_\beta}(T_\beta + m_\beta c^2) \, dT_\beta \text{ (for } m_\nu \neq 0) \text{ (2.256)}$$

in contrast to equation (2.206) viz.

$$N(T_\beta)dT_\beta = C' \sqrt{T_\beta(T_\beta + 2m_\beta c^2)}(Q_0 - T_\beta)^2(T_\beta + m_\beta c^2)dT_\beta \text{ (for } m_\nu = 0) \quad (2.257)$$

The difference of these two expressions, namely

$$N(T_\beta) \propto \sqrt{Q_0 - T_\beta} \text{ for } m_\nu \neq 0 \qquad (2.258)$$

$$N(T_\beta) \propto (Q_0 - T_\beta)^2 \text{ for } m_\nu = 0 \qquad (2.259)$$

is manifested in the tail region of the $N_\beta$ versus $T_\beta$ plot of figure 2.27. The deviation is exhibited at a point decided by the neutrino mass.

It is easier to discuss in terms of slope of $N(T_\beta)$ versus $T_\beta$ plot, i.e. in terms of $\frac{dN(T_\beta)}{dT_\beta}$ especially finding the slope in the tail region specifically at $T_\beta = Q_0$.

Writing $\sqrt{T_\beta(T_\beta + 2m_\beta c^2)} \; (T_\beta + m_\beta c^2) = S$ we rewrite from equations (2.256) and (2.257)

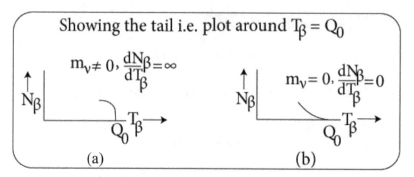

**Figure 2.27.** The shape of tail of $N_\beta$ versus $T_\beta$ plot for $m_\nu \neq 0$ and for $m_\nu = 0$

$$N(T_\beta) = C'' S \sqrt{Q_0 - T_\beta} \text{ (for } m_\nu \neq 0) \tag{2.260}$$

$$N(T_\beta) = C'S(Q_0 - T_\beta)^2 \text{ (for } m_\nu = 0) \tag{2.261}$$

Hence for $m_\nu \neq 0$ we have from equation (2.260)

$$\frac{dN(T_\beta)}{dT_\beta} = C'' S \frac{-1}{2\sqrt{Q_0 - T_\beta}} + C'' \sqrt{Q_0 - T_\beta} \frac{dS}{dT_\beta} \xrightarrow{T_\beta = Q_o} \infty \tag{2.262}$$

and for $m_\nu = 0$ we have from equation (2.261)

$$\frac{dN(T_\beta)}{dT_\beta} = C'S2(Q_0 - T_\beta)(-1) + C'(Q_o - T_\beta)^2 \frac{dS}{dT_\beta} \xrightarrow{T_\beta = Q_o} 0 \tag{2.263}$$

Clearly for $m_\nu \neq 0$ since $\frac{dN(T_\beta)}{dT_\beta} = \infty$ at $T_\beta = Q_o$ the $N(T_\beta)$ versus $T_\beta$ plot will have a tail that meets $T_\beta$ axis perpendicularly (figure 2.27(a)). And for $m_\nu = 0$ since $\frac{dN(T_\beta)}{dT_\beta} = 0$ at $T_\beta = Q_o$ the $N(T_\beta)$ versus $T_\beta$ plot will have a tail that meets $T_\beta$ axis tangentially (figure 2.27(b)). So study of the tail of $N(T_\beta)$ versus $T_\beta$ plot gives the neutrino rest mass $m_\nu$.

One rough estimate of neutrino rest mass is $0.05\,eV$, i.e. $\frac{0.05\,eV}{c^2} = 0.05 \times \frac{1.6 \times 10^{-19}\,J}{(3 \times 10^8\,ms^{-1})^2} \sim 10^{-38}\,kg$.

## 2.34 Origin of line beta spectra: internal conversion

Suppose a nucleus is in an unstable excited state. It can go over to lower state by releasing the excitation energy, i.e. by de-excitation. This excitation energy may not be carried away by a gamma photon—it may be a non-radiative process. In other words the energy difference does not come out as electromagnetic radiation. It can be directly transferred to an orbital electron (either $K$ or $L$)—which is thus emitted. Such emission of an electron during nuclear transition of states is called internal conversion.

In fact, wave functions of nucleus and the *s* electron overlap, i.e. they interact, making it possible for the *s* electron to directly receive energy of the nuclear transition.

It is a direct or single-step process. The transition of nucleus between two states leading to the release of energy and the emission of electron occurs simultaneously—in a single step. So to say the energy involved in the transition is converted into an electron —hence the name internal conversion and the electron is called a converted electron.

The orbital electron (called the converted electron) is emitted with a kinetic energy *T* given by

*T* = (Excitation energy of nucleus *E*) – (Binding energy of electron in the orbit $E_K$)

(assuming that its binding energy $\sim keV$ is less than the received energy $\sim MeV$ of nuclear transition).

$$T = E - E_K \qquad (2.264)$$

if the *K* shell electron is emitted. Since $E$, $E_K$ have discrete or fixed energies $E - E_K$ = fixed and so we would get a sharp discrete line. This is called *K* line or *K* peak.

If the electron emitted comes from (i.e. if the energy is converted to) an *L* electron we would get a line spectrum. This is called *L* line or *L* peak.

This sharp $\beta$ lines (*K* line, *L* line,....) constitute the beta line spectrum and is superimposed upon the continuous $\beta$ spectrum.

We discuss in *exercise 2.34* how Auger effect differs from internal conversion.

## 2.35 Gamma decay

When a nucleus decays by $\alpha$ emission or by $\beta$ emission it is left in an excited state. If it is energetically impossible for the nucleus to emit another particle, say $\alpha$ or $\beta$, then the nucleus sheds off its extra energy directly through emission of electromagnetic radiation (called radiative transition), i.e. by the emission of a $\gamma$ photon. This is shown in figure 2.28.

The order of the energy of $\gamma$ during nuclear transition

$$(_zX^A)^* \rightarrow {_z}X^A + \gamma$$

is $E_\gamma = h\nu \sim MeV$. So $\nu \sim$ large, $\lambda \sim$ small $\approx$ picometer to fermi.

- For atomic transition the energy of transition $\sim keV$. As $\gamma$ decay or transition occurs between two nuclear energy states $\gamma$ spectrum is discrete (not continuous) and occurs in a very small time $\sim$ nanosecond ($10^{-9}$ s), pico-second ($10^{-12}$s).
- In $\alpha$ decay, $\beta$ decay parent nucleus and daughter nucleus are different, but in $\gamma$ decay the nucleus remains the same—parent and daughter nuclei are structurally the same but exist in different energy levels.
- Another competiting process through which such extra nuclear energy can be shed off is internal conversion. The one that is energetically favourable will occur.
- Difference of $\gamma$-ray and x-ray

$\gamma$-ray and x-ray both are electromagnetic radiation.

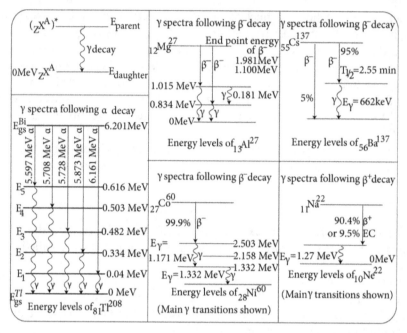

**Figure 2.28.** $\gamma$ decay: Nucleus may find itself in excited state after undergoing $\alpha$ decay or $\beta$ decay. Such a nucleus, if unstable may de-excite from excited state to ground state or to a lower excited state and emits a photon.

The origins of $\gamma$-ray and x-ray are different. X-ray is due to electronic transition from inner shells while $\gamma$-ray is due to de-excitation of the nucleus.

$\gamma$-ray has higher energy, higher frequency. X-ray has lower energy, lower frequency.

## 2.36 Selection rules for $\gamma$ decay

Consider the $\gamma$ decay

$$(_ZX^A)^* \rightarrow \ _ZX^A + \gamma \text{ (star is for excited)} \tag{2.265}$$

We write this as

$$P \rightarrow D + \gamma(P \rightarrow \text{ excited parent } (_ZX^A)^*, \ D \rightarrow \text{ daughter } _ZX^A) \tag{2.266}$$

Energy conservation equation gives
$$E_{\text{initial}} = E_{\text{final}} \ \Rightarrow \ E_P = E_D + E_\gamma$$

$$m_Pc^2 = m_Dc^2 + T_D + h\nu \tag{2.267}$$

where $m_P$, $m_D$ are masses of parent and daughter nuclei, $T_D$ is the recoil kinetic energy of daughter. This is very small ($T_D \sim eV$) and can be neglected.

Momentum conservation equation is

$$\vec{p}_{\text{initial}} = \vec{p}_{\text{final}} \text{ where } \vec{p}_i = \vec{p}_P, \vec{p}_f = \vec{p}_\gamma + \vec{p}_D. \tag{2.268}$$

In the centre of mass frame

$$\vec{p}_i = \vec{p}_P = 0 \text{ and } \vec{p}_f = \vec{p}_\gamma + \vec{p}_D = 0 \tag{2.269}$$

Here $\vec{p}_D$ is momentum of the recoil nucleus (daughter). The $\gamma$ momentum is

$$p_\gamma = \frac{E_\gamma}{c}. \tag{2.270}$$

This $\gamma$ quantum carries away the overwhelming part of the nuclear excitation energy. Also, $\vec{p}_D = -\vec{p}_\gamma$

$$p_D^2 = p_\gamma^2 = \left(\frac{E_\gamma}{c}\right)^2 \tag{2.271}$$

Now

$$T_D = \frac{p_D^2}{2M} = \frac{p_\gamma^2}{2M} = \frac{(E_\gamma/c)^2}{2M} = \frac{E_\gamma^2}{2Mc^2} \tag{2.272}$$

where $M$ is the mass of daughter nucleus.

Angular momentum conservation gives

$$\vec{J}_P = \vec{J}_D + \vec{J}_\gamma \Rightarrow \vec{J}_P - \vec{J}_D = \vec{J}_\gamma \tag{2.273}$$

It is customary to denote $J_\gamma$ by $l$, i.e. we write

$$J_\gamma \equiv l \tag{2.274}$$

Hence we rewrite

$$\vec{J}_P - \vec{J}_D = \vec{l} \tag{2.275}$$

By angular momentum selection rule the quantum number $l$ will be

$$l = (J_P + J_D), (J_P + J_D - 1), \ldots\ldots, |J_P - J_D|$$

- Let us discuss the possible values of $l$.

$$\vec{J}_\gamma \equiv \vec{l} = \text{total angular momentum carried away by the } \gamma \text{ photon} \tag{2.276}$$

$$= \text{orbital angular momentum} + \text{spin angular momentum}$$

As photon spin is 1 we have

$$\vec{J}_\gamma \equiv \vec{l} = \text{orbital angular momentum} + 1 \tag{2.277}$$

It clearly follows that a $\gamma$ photon must carry away at least one unit of angular momentum. So $l = 0$ is not possible for a $\gamma$ transition photon. $\gamma$ transitions should correspond to $l \neq 0$.

This $l \neq 0$ again means that a $0 \rightarrow 0$ transition is not possible by a real $\gamma$ photon. In other words, a radiative process cannot connect two states having $J_i = 0$ and $J_f = 0$.

- Note that a $0 \rightarrow 0$ transition is possible through a virtual photon during internal conversion process and this is a non-radiative process.
- Gamma transition (or radiative transfer of energy) is possible through ($l \neq 0$).

$l = 1$ is called dipole transition
$l = 2$ is called quadrupole transition
$l = 3$ is called octupole transition.

- Two types of $\gamma$ transitions are there: electric multipole type and magnetic multipole type. We discuss this now.

Electromagnetic radiation comes from oscillating charges and currents, which are changing with time.

✓ A charge distribution is a superposition of multipoles of various orders: monopole ($l = 0$), dipole ($l = 1$), quadrupole ($l = 2$) and so on.

If monopole moment ($\sim Q-$ the total charge) is zero then dipole moment $\int \rho(\vec{r})\ \vec{r}\, d\tau$ ($\rho(\vec{r})$ = charge density, $\vec{r}$ = field coordinate) dominates.
If dipole moment = 0 then quadrupole moment ($\Leftarrow$ rank 2 tensor) dominates.

If charges oscillate they create radiation at a large distance except the monopole. (Even if the monopole oscillates the field at large distance is the same as that due to total charge placed at the centre of charge distribution—and so it does not give off radiation.) We do not have electric monopole radiation $E0$.

An oscillating charge distribution that has dipole moment will give off radiation at large distance called electric dipole radiation (denoted by $El \equiv E1$).

An oscillating charge distribution that has quadrupole moment will give off radiation at large distance called electric quadrupole radiation (denoted by $El \equiv E2$).

An oscillating charge distribution that has octupole moment will give off radiation at large distance called electric octupole radiation (denoted by $El \equiv E3$) and so on.

✓ A current distribution can be looked upon as a superposition of magnetic multipoles, the lowest order multipole being a magnetic dipole—there being no magnetic monopole. When these magnetic moments ($\sim$ the magnetic dipole moment, magnetic quadrupole moment, magnetic octupole moment etc) oscillate, they give off radiation at large distance that are called magnetic dipole radiation, magnetic quadrupole radiation, magnetic octupole radiation etc, denoted by $M1$, $M2$, $M3$, ... , respectively.

So the $\gamma$ photon that comes out in nuclear $\gamma$ decay carries angular momentum characterized by the quantum number $l$ and can be of two types:

$El$ (electric multipole radiation)

$Ml$ (magnetic multipole radiation)

Initial and final states decide what angular momentum $l$ will the photon carry (i.e. if it is a dipole radiation or a quadrupole radiation or an octupole radiation).

We compare $El$ transition and $Ml$ transition.

| ☐ $El$ transition | ☐ $Ml$ transition |
|---|---|
| If $\gamma$ is emitted due to a redistribution of electric charges in the nucleus, i.e. due to a periodic vibration or oscillation of charge density of nuclear fluid we call it an electric multipole transition $El$. | If $\gamma$ is emitted due to a redistribution of spin and orbital magnetic moment, i.e. due to a periodic vibration or oscillation of current density of nuclear fluid we call it magnetic multipole transition $Ml$. |
| $E1$ is called electric dipole transition. $E2$ is called electric quadrupole transition. | $M1$ is called magnetic dipole transition. $M2$ is called magnetic quadrupole transition |

- Parity selection rule

Parity selection rule is shown in figure 2.29.

Parity of photon coming out depends on angular momentum which it carries, i.e. on $l$ and whether it is electric or magnetic transition.

Parity of electric multipole transition is decided by the factor $(-1)^l$ and parity of magnetic multipole transition is decided by the factor $(-1)^{l+1}$.

- As $l \neq 0$ it follows that $E0$, $M0$ transitions are not possible. In other words, $0 \to 0$ transition is forbidden by $\gamma$ transition and can proceed through internal conversion.
- $l = 0$ to $l = 0$ occurs only through internal conversion as a $\gamma$ cannot have $l = 0$. (figure 2.30). In other transitions internal conversion and $\gamma$ decay are in competition. Which would occur is decided by a conversion coefficient

$$\eta = \frac{\text{probability of internal conversion}}{\text{probability of } \gamma \text{ decay}}.$$

| Parity of electric multipole transition | | Parity of magnetic multipole transition | |
|---|---|---|---|
| $(-)^l$ | | $(-)^{l+1}$ | |
| $l$ = even =2,4,6,... | $l$ = odd =1,3,5,... | $l$ = even =2,4,6,... | $l$ = odd =1,3,5,... |
| No parity change | Parity changes | Parity changes | No parity change |

**Figure 2.29.** Parity selection rule for electric and magnetic multipole transition.

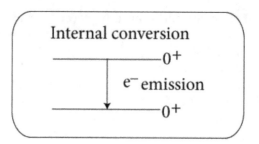

**Figure 2.30.** $0 \to 0$ is internal conversion.

There is some $\gamma$ decay and some internal conversion.

- For the same $l$ electric transitions are more probable, i.e. high $l$ magnetic transitions are highly suppressed. In fact, for the same $l$

$$\frac{\text{probability of } El}{\text{probability of } Ml} \sim 10^2 \tag{2.278}$$

  i.e. $El$ is more probable compared to $Ml$.

- For different multipolarities, i.e. for different $\gamma$, decay through lower multipolarity is favoured. In fact, for the same type of transition (either electric or magnetic)

$$\frac{\text{probability of } l+1 \text{ transition}}{\text{probability of } l \text{ transition}} \sim 10^{-5}. \tag{2.279}$$

  So higher $l$ transitions are suppressed, i.e. $l$ transition is more probable than $l+1$ transition.

- Let us analyse whether the shown transitions of figure 2.31 are $\gamma$ transitions and if so we shall indicate their nature.

(a) Transition $1^+ \to 1^-$

    As per the angular momentum addition rule the transition $1^+ \to 1^-$ means $l \to (1+1)$ to $|1-1|$ in unit steps, i.e. $l = 2, 1, 0$. But $l = 0$ is forbidden. So $l = 2, 1$. This corresponds to the multipole transitions $E2$, $E1(\Leftarrow$ electric multipole transitons) $M2$, $M1(\Leftarrow$ magnetic multipole transitions).

    We now employ parity selection rule. The transition $1^+ \to 1^-$ means parity changes. From figure 2.29 $l =$ odd for electric multipole transition $El$ and $l =$ even for magnetic multipole transition $Ml$. This means the allowed $\gamma$ transitions are $E1$ (electric dipole transition), $M2$ (magnetic quadrupole transition). The electric dipole transition $E1$ is most probable.

(b) Transition $2^+ \to 0^+$

    We employ angular momentum selection rule. The transition $2^+ \to 0^+$ means

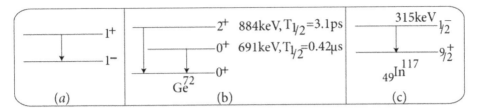

**Figure 2.31.** Some possible $\gamma$ transitions.

$l \to (2 + 0)$ to $|2 - 0|$ in unit steps, i.e. $l = 2$. This corresponds to quadrupole transition $E2$, $M2$.

We now employ parity selection rule. The transition $2^+ \to 0^+$ means parity does not change. From figure 2.29 $l =$ even for electric multipole transition $El$ and $l =$ odd for magnetic multipole transition $Ml$. This means the allowed $\gamma$ transition is $E2$ electric quadrupole transition.

Transition $0^+ \to 0^+$

This transition corresponds to internal conversion. It cannot be a $\gamma$ transition.

(c) Transition $\frac{1}{2}^- \to \frac{9}{2}^+$

We employ angular momentum selection rule. The transition $\frac{1}{2}^- \to \frac{9}{2}^+$ means $l \to (\frac{1}{2} + \frac{9}{2})$ to $|\frac{1}{2} - \frac{9}{2}|$ in unit steps, i.e. $l = 5, 4$. This corresponds to multipole transitions $E5$, $M5$, $E4$, $M4$.

We now employ parity selection rule. The transition $\frac{1}{2}^- \to \frac{9}{2}^+$ means parity changes. From figure 2.29 $l =$ odd for electric multipole transition $El$ and $l =$ even for magnetic multipole transition $Ml$. This means the allowed $\gamma$ transitions are $E5$, $M4$. The $l$ value (of 4 or 5) is very large and so probability is very small. This means the lifetime is large ($\sim$ minutes) and these are isomeric states. Again out of $E5$, $M4$ the dominant (i.e. more probable) transition is for $M4$ (i.e. magnetic multipole transition with $l = 4$).

## 2.37 Mossbauer effect

Consider a nucleus in an excited state (figure 2.32(a)). It emits $\gamma$ and such de-excitation brings it to the ground state. Energy available in the process is the difference in the energy levels $E_0$ and by conservation of energy we have

$$E_0 = E_\gamma + E_R \tag{2.280}$$

where $E_\gamma = h\nu$ is the energy of $\gamma$ and $E_R$ is the kinetic energy of recoiling nucleus given by

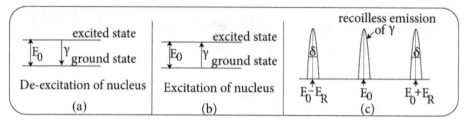

**Figure 2.32.** Nuclear transitions and Mossbauer effect

$$E_R = \frac{p^2}{2M} \tag{2.281}$$

assuming that the nucleus is free to recoil with momentum $\vec{p}$ in the backward direction. Also, as the recoil velocity is very very small ($\sim eV$) we can use a non-relativistic formula. So

$$E_0 = E_\gamma + \frac{p^2}{2M} \tag{2.282}$$

From the momentum conservation in centre of mass frame we write

$$\vec{P}_{\text{initial}} = \vec{P}_{\text{final}} \implies 0 = \vec{P}_\gamma + \vec{p}$$

where

$$P_\gamma = \frac{E_\gamma}{c}$$

Hence

$$\vec{p} = -\vec{P}_\gamma \implies p^2 = P_\gamma^2 = \left(\frac{E_\gamma}{c}\right)^2 \tag{2.283}$$

We thus have from equations (2.282) and (2.283)

$$E_0 = E_\gamma + \frac{E_\gamma^2}{2Mc^2}$$

$$E_\gamma = E_0 - \frac{E_\gamma^2}{2Mc^2} = E_0 - E_R \tag{2.284}$$

Clearly $\gamma$ energy is less than energy level difference by an amount

$$E_R = \frac{E_\gamma^2}{2Mc^2} \sim eV \tag{2.285}$$

And this is plotted in figure 2.32(c). There is broadening around the energy $E_0 - E_R$ due to Heisenberg uncertainty principle and it can be large due to thermal energy and Doppler shift. Let us now discuss if this $\gamma$ can be absorbed by another nucleus lying in the ground state (figure 2.32(b)). To take a nucleus from ground state (where it is at rest $\vec{p_i} = 0$) to the excited state energy needed is $E_0$ and this energy should be delivered by a photon (having momentum $p_\gamma = \frac{E_\gamma}{c}$) and absorbed by the nucleus which in the process would recoil in the forward direction with energy $E_R = \frac{p^2}{2M}$. By energy conservation

$$E_\gamma = E_0 + E_R = E_0 + \frac{p^2}{2M} \qquad (2.286)$$

and by momentum conservation

$$\vec{p}_{\text{initial}} = \vec{p}_{\text{final}} \Rightarrow \vec{p_\gamma} + \vec{p_i} = \vec{p_f} \qquad (2.287)$$

As $\vec{p_i} = 0$, $\vec{p_f} \equiv \vec{p}$ we have

$$\vec{p_\gamma} = \vec{p} \Rightarrow p = \frac{E_\gamma}{c} \qquad (2.288)$$

From equations (2.286) and (2.288)

$$E_\gamma = E_0 + E_R = E_0 + \frac{E_\gamma^2}{2Mc^2} \qquad (2.289)$$

And this is plotted in figure 2.32(c). There is broadening around the energy $E_0 + E_R$ due to Heisenberg uncertainty principle and it can be large due to thermal energy and Doppler shift.

Clearly energy available for absorption is $E_0 - E_R$ and energy needed is $E_0 + E_R$.

If $E_R \leqslant \delta$ where $\delta$ is line width and there is overlap of the wave profiles at $E \pm E_R$ then absorption takes place. And we call it resonant absorption. In the discussion we took the nucleus to be free to recoil.

If nuclear energy levels are very sharply defined resonant absorption occurs only if an emitted photon carries exact transition energy causing the nucleus to transit from ground state to excited state. For a free nucleus or atom, resonant absorption cannot occur since the emitted energy is less than the resonance energy by an amount equal to recoil kinetic energy of source nucleus viz. $E_R = \frac{E_\gamma^2}{2Mc^2}$.

Clearly recoil energy $E_R$ is inversely proportional to the nuclear mass $M$. So if $M$ is increased the recoil energy diminishes and if $M \to \infty$ is accomplished by embedding the nucleus in a crystalline lattice then $E_R \to 0$ and then resonant absorption can occur—which is called Mössbauer effect.

• Suppose the nucleus is embedded in a crystal lattice. Then the energy $E_0 - E_\gamma = E_R$ is to be taken up by the entire lattice. Energy of lattice vibration can increase only in quantum, i.e. in discrete steps—the minimum energy being the phonon energy. If the energy $E_R$ is less than this minimum the lattice will not accept it and then $E_0 = E_\gamma$, i.e. $\gamma$ will come out with the full energy $E_0$— there being no recoil. This is referred to as recoilless emission of $\gamma$—indicated by the spread around

$E_0$ in figure 2.32(c). Similarly, recoilless absorption can occur ($E_0 = E_\gamma$) when this $\gamma$ photon, carrying the full nuclear transition energy strikes another similar nucleus also embedded in a tight crystalline lattice. The recoil free emission and absorption of $\gamma$-ray by nuclei or atoms held tightly in a crystalline lattice is called Mössbauer effect. It is a tool of experimental physics in Mössbauer spectroscopy.

## 2.38 Pair production

A high energy radiation having photon energy $h\nu \geqslant 1.02 \ MeV$ may interact with a nucleus $N$ and get converted into electron $e^-$ and positron $e^+$ pair. Such materialization of energy according to the equation $E = mc^2$ is called pair production.

$$h\nu + N \rightarrow e^- + e^+ + N$$

The nucleus $N$ serves to conserve momentum. Rest energy of $e^\pm$ is $m_0 c^2 = 0.51 \ MeV$.

- Possibility of pair production occurring in vacuum

Consider the decay

$$\gamma \rightarrow e^- + e^+ \tag{2.290}$$

in vacuum (if possible). Let us study the energy and momentum conservations and investigate if such decay can occur. The energy conservation relation is

$$E_\gamma = E_{e^+} + E_{e^-}$$

$$= (T_{e^+} + m_0 c^2) + (T_{e^-} + m_0 c^2) = T_{e^+} + T_{e^-} + 2m_0 c^2$$

$$E_\gamma = T + 2m_0 c^2 \ (\text{where } T = T_{e^+} + T_{e^-} \rightarrow \text{ kinetic energy of } e^\pm)$$

$$E_\gamma = T + 2(0.51 \ MeV)$$

$$E_\gamma = T + 1.02 \ MeV$$

For $T = 0$, i.e. if $e^\pm$ are emitted with zero kinetic energy
$E_\gamma = 1.02 \ MeV =$ Threshold energy for $e^+ e^-$ pair production with zero kinetic energy.
Clearly for pair production the energy condition is

$$E_\gamma \geqslant 1.02 \ MeV \tag{2.291}$$

The momentum conservation principle is

$$\vec{P}_{\text{initial}} = \vec{P}_{\text{final}}$$

$$\vec{P}_\gamma = \vec{P}_{e^+} + \vec{P}_{e^-} \tag{2.292}$$

Let us view the interaction in centre of mass frame where c.m is at rest (always), i.e. the centre of mass momentum is zero (always).

$$\vec{P}_\gamma^{cm} = \vec{P}_{e^+}^{cm} + \vec{P}_{e^-}^{cm} = 0 \Leftarrow \text{ all quantities are w.r.t. c.m.} \qquad (2.293)$$

But there is no reference frame in which $\gamma$ photon can be brought to rest. So $\vec{P}_\gamma^{cm} = 0$ is not possible—this is forbidden by Einstein's second postulate of relativity.

So the decay $\gamma \to e^- + e^+$ cannot take place in vacuum. We can come to the same conclusion starting from four-momentum conservation also and we show it in *exercise 2.32*.

Also according to Einstein's second postulate of relativity, light speed is unchangeable. Again, electromagnetic radiation $\gamma$ travels with speed of light $c$. So a $\gamma$ cannot split up on its own and convert to matter.

Clearly $\gamma \to e^- + e^+$ is forbidden by momentum conservation and also by Einstein's second postulate of special theory of relativity.

## 2.39 Exercises

**Exercise 2.1** *Express the survival equation in terms of half-life.*
$\boxed{Ans}$ Survival equation is $N = N_0 e^{-\lambda t}$.
Use

$$\lambda = \frac{ln2}{T_{1/2}}$$

$$N = N_0 \exp\left(-\frac{ln2}{T_{1/2}} t\right).$$

Let

$$n = \frac{t}{T_{1/2}} = \text{number of half-lives within the time } t. \text{ So}$$

$$N = N_0 e^{-n ln2} \Rightarrow N = N_0 e^{ln2^{-n}}$$

$$N = N_0 2^{-n}$$

$$N = N_0 2^{-t/T_{1/2}}$$

This is the survival equation in terms of half-life.

**Exercise 2.2** *Show that the survival equation leads to a straight line of slope $\lambda$.*
$\boxed{Ans}$ The survival equation is $N = N_0 e^{-\lambda t}$. Taking $ln$ we get

$$lnN = lnN_0 e^{-\lambda t}.$$

$$lnN = lnN_0 - \lambda t \Rightarrow lnN = -\lambda t + lnN_0.$$

So plot of $lnN$ versus $t$ will be a straight line of slope $\lambda$ as shown in figure 2.33.

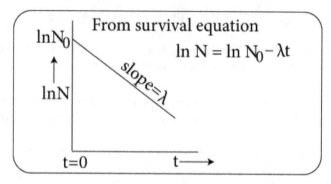

**Figure 2.33.** Plot of $lnN$ versus $t$ from survival equation $lnN = lnN_0 - \lambda t$.

**Exercise 2.3** *Show through diagram that decay constant $\lambda$ indicates the rapidiy of decay of a radioactive substance.*

*Ans* Figure 2.34 shows explicitly that decay constant $\lambda$ indicates the rapidity of decay of a radioactive substance—value of $\lambda$ is a measure of how fast or how slow decay of radioactive substance occurs.

**Exercise 2.4** *Find the survival probability of one nucleus in a radioactive sample. Hence deduce the radioactive law of disintegration.*

*Ans* Consider decay of a radioactive sample in time $t$.

We divide $t$ into $n$ intervals each of duration $\Delta t$ say. So

$$t = n\Delta t \Rightarrow \Delta t = \frac{t}{n}$$

Probability $P$ of a nucleus to decay in a time interval $\Delta t$ is proportional to time interval $\Delta t$

$$P \propto \Delta t$$

$$P = \lambda \Delta t$$

where $\lambda$ is a constant of proportionality.

Probability that the given nucleus survives the first interval $\Delta t$ is

$$1 - P = 1 - \lambda \Delta t = 1 - \lambda \frac{t}{n}.$$

Probability that the given nucleus survives the second interval $2\Delta t$ is

$$(1 - P)(1 - P) = (1 - P)^2 = \left(1 - \lambda \frac{t}{n}\right)^2$$

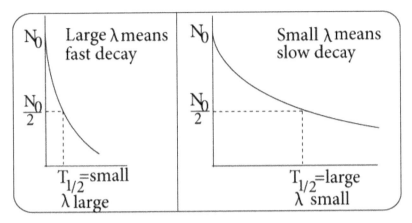

**Figure 2.34.** For fast decay $\lambda$ is large and for slow decay $\lambda$ is small.

since survival in the first interval and survival in the second interval are independent events.

Probability for a nucleus to survive the $n$th interval is

$$(1 - P)^n = \left(1 - \lambda\frac{t}{n}\right)^n \xrightarrow[\Delta t \to 0]{n \to \infty} e^{-\lambda t}$$

This is survival probability of a nucleus from decay.

Again by definition probability $= \dfrac{\text{number of nuclei at time } t}{\text{number of nuclei at time } t = 0} = \dfrac{N(t)}{N_0}$.

Hence $e^{-\lambda t} = \frac{N(t)}{N_0}$

$$N = N_0 e^{-\lambda t}$$

This number of nuclei survives decay in time $t$ and it is called the survival equation.

**Exercise 2.5** *For radioactive substance mean life is $<t> = \frac{1}{\lambda}$. Find $<t^2>$.*
[Ans] Mean square life is

$$<t^2> = \left| \frac{\int_0^\infty t^2\, dN}{\int_0^\infty dN} \right|$$

Using $N = N_0 e^{-\lambda t}$

$$<t^2> = \left| \frac{\int_0^\infty t^2\, d(N_0 e^{-\lambda t})}{N_0} \right| = \left| \int_0^\infty t^2\, de^{-\lambda t} \right| = \left| \int_0^\infty t^2 (-\lambda) e^{-\lambda t} dt \right|.$$

$$<t^2> = \left| -\lambda \int_0^\infty t^2 e^{-\lambda t}\, dt \right|.$$

Integrate by parts to get

$$<t^2> = \left| -\lambda \left[ t^2 \int e^{-\lambda t}\, dt \Big|_0^\infty - \int_0^\infty \frac{d}{dt} t^2 \left( \int e^{-\lambda t}\, dt \right) dt \right] \right|$$

$$= \left| -\lambda \left[ t^2 \frac{e^{-\lambda t}}{-\lambda} \Big|_0^\infty - 2 \int_0^\infty t\, \frac{e^{-\lambda t}}{-\lambda} dt \right] \right|$$

$$<t^2> = \left| -2 \int_0^\infty t\, e^{-\lambda t}\, dt \right| = \left| -2 \left[ t \frac{e^{-\lambda t}}{-\lambda} \Big|_0^\infty - \int_0^\infty \frac{d}{dt} t \left( \int e^{-\lambda t}\, dt \right) dt \right] \right|$$

$$= \left| 2 \int_0^\infty \frac{e^{-\lambda t}}{-\lambda} dt \right| = \left| -\frac{2}{\lambda} \int_0^\infty e^{-\lambda t}\, dt \right|$$

$$= \left| -\frac{2}{\lambda} \frac{e^{-\lambda t}}{-\lambda} \Big|_0^\infty \right| = \left| \frac{2}{\lambda^2} (e^{-\lambda\infty} - e^{\lambda 0}) \right| = \left| -\frac{2}{\lambda^2} \right|.$$

So $<t^2> = \frac{2}{\lambda^2}$

Using $\tau = \frac{1}{\lambda}$ ($\tau$ = mean life)

$$<t^2> = 2\tau^2$$

**Exercise 2.6** *What is single decay mode? What is radioactive branching? Elucidate. How is the decay constant and the mean life of the branches related to the branching ratio?*
    [Ans] Single decay mode
    In some radioactive transformation either an $\alpha$ or a $\beta$ is emitted—never both or more than one particle of each kind is emitted. This is called decay by single mode or no branching mode (figure 2.35(a)).

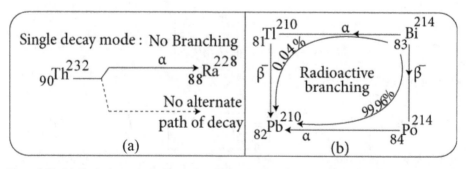

**Figure 2.35.** (a) Single decay mode has no branching. (b) Radioactive branching with two decay paths.

- Radioactive branching

There are a few radioactive substances which decay by two different modes or alternate paths—called branches and they give rise to the same end product.

One path corresponds to $\alpha$ emission followed by $\beta$ emission while the other path corresponds to $\beta$ emission followed by $\alpha$ emission.

This is shown in figure 2.35(b).

The decay constants are $\lambda_\alpha$, $\lambda_\beta$

mean lives are $\tau_\alpha = \frac{1}{\lambda_\alpha}$, $\tau_\beta = \frac{1}{\lambda_\beta}$ and

activities are $\lambda_\alpha N$, $\lambda_\beta N$.

The relative probability of decay by two alternative decay paths is called branching ratio which is the ratio of the decay constants $\frac{\lambda_\alpha}{\lambda_\beta}$.

The total activity is

$$-\frac{dN}{dt} = \lambda_\alpha N + \lambda_\beta N = (\lambda_\alpha + \lambda_\beta)N = \lambda N$$

where

$$\lambda = \lambda_\alpha + \lambda_\beta$$

is the net decay constant. The mean life is

$$\tau = \frac{1}{\lambda} = \frac{1}{\lambda_\alpha + \lambda_\beta} = \frac{1}{\frac{1}{\tau_\alpha} + \frac{1}{\tau_\beta}}$$

$$\frac{1}{\tau} = \frac{1}{\tau_\alpha} + \frac{1}{\tau_\beta}$$

**Exercise 2.7** *Find the time at which $_{84}Po^{210}$ reaches maximum concentration in the reaction $_{83}Bi^{210} \xrightarrow[\text{5 day}]{\beta^-} {}_{84}Po^{210} \xrightarrow[\text{138 day}]{\alpha} {}_{82}Pb^{206}$.*

$\boxed{Ans}$ In the radioactive chain $1 \to 2 \to 3$ the time at which 2 has maximum concentation is

$t = \frac{1}{\lambda_2 - \lambda_1} ln\frac{\lambda_2}{\lambda_1} = t_0$ [equation (2.45)]. So time at which the concentration of $_{84}Po^{210}$ reaches maximum in the above radiactive chain will be

$$t_0 = \frac{1}{\lambda_{Po} - \lambda_{Bi}} ln\frac{\lambda_{Po}}{\lambda_{Bi}}$$

$$\lambda_{Po} = \frac{0.693}{(T_{1/2})_{Po}} = \frac{0.693}{138 \text{ day}} = 5 \times 10^{-3} \text{ day}^{-1}$$

$$\lambda_{Bi} = = \frac{0.693}{(T_{1/2})_{Bi}} = \frac{0.693}{5 \text{ day}} = 0.1386 \text{ day}^{-1}$$

Hence

$$t_0 = \frac{1}{\lambda_{Po} - \lambda_{Bi}} ln\frac{\lambda_{Po}}{\lambda_{Bi}} = \frac{1}{(5 \times 10^{-3} - 0.1386)day} ln\frac{5 \times 10^{-3}}{0.1386} = 25 \text{ day}$$

**Exercise 2.8** *Is equilibrium possible in the radioactive decay chain* $1 \to 2 \to 3$. *(3 being the stable end product) if*
   *(a)* 1 *is short lived*
   *(b)* 1 *is long lived* [vide sections 2.10 to 2.14].

**Exercise 2.9** *Starting from* 10 *gm radon how much radon will be there after* 16 *days? Half-life of radon is* 4 *days.*
   Ans $\lambda = \frac{0.693}{T_{1/2}} = \frac{0.693}{4 \text{ day}} = 0.173 \ 25 \text{ day}^{-1}$
   We get from survival equation

$$N = N_0 e^{-\lambda t} = (10 \ gm)e^{-(0.173 \ 25 \ day^{-1})(16 \ day)} = 0.625 \ gm$$

**Exercise 2.10** *In* 1920 *the American people gave Madam Curie* 1 *gm of* $Ra^{226}$ *(half-life* 1620 *yrs). How much of it will be left in* 2040 *?*
   Ans $\lambda = \frac{0.693}{T_{1/2}} = \frac{0.693}{1620 \ yr} = 4.28 \times 10^{-4} \ yr^{-1}$.
   We get from survival equation

$$N = N_0 e^{-\lambda t} = (1 \ gm)e^{-(4.28 \times 10^{-4} yr^{-1})(2040 - 1920)} = 0.949 \ gm.$$

**Exercise 2.11** *Why is the percentage of radium contained in uranium always experimentally found to be same viz.* 1 *gm of radium per* 3.5 *ton of pure uranium?*
   Ans Consider the following part of uranium series (figure 2.2)

$$_{92}U^{238} \xrightarrow[4.5 \times 10^9 \ yr]{\alpha} {}_{90}Th^{234} \xrightarrow{\beta^-} {}_{91}Pa^{234} \xrightarrow{\beta^-} {}_{92}U^{234} \xrightarrow{\alpha} {}_{90}Th^{230} \xrightarrow{\alpha} {}_{88}Ra^{226} \xrightarrow[1590 \ yr]{\alpha} {}_{88}Rn^{222}$$

In the uranium series, uranium has the longest half-life. Hence all the products which are successively formed, like radium (except end product $_{82}Pb^{206}$) are in secular equilibrium with uranium, e.g.

$$\lambda_U N_U = \lambda_{Ra} N_{Ra}$$

and so their percentage content in a given sample of uranium is constant and is equal to the ratio of their half-lives. So

$$\frac{N_{Ra}}{N_U} = \frac{\lambda_U}{\lambda_{Ra}} = \frac{(T_{1/2})_{Ra}}{(T_{1/2})_U} = \frac{1590 \ yr}{4.5 \times 10^9 \ yr} = 3.5 \times 10^{-7}$$

$$N_{Ra} = 3.5 \times 10^{-7} N_U.$$

For $N_U = 3.5$ ton $= 3.5 \times 10^6 \ gm$ we have

$$N_{Ra} = 3.5 \times 10^{-7}(3.5 \times 10^6 \ gm) = 1.225 \ gm \approx 1 \ gm.$$

**Exercise 2.12** *Calculate the amount of* $Ra^{226}$ *in secular equilibrium with 1 kg of pure* $U^{238}$.

*Given* $T_{1/2}$ *for* $Ra^{226}$ *and* $U^{238}$ *are 1620 yr and* $4.5 \times 0^9$ *yr respectively.*

[Ans] Condition of secular equilibrium

$$\lambda_U N_U = \lambda_{Ra} N_{Ra}$$

$$\frac{N_{Ra}}{N_U} = \frac{\lambda_U}{\lambda_{Ra}} = \frac{(T_{1/2})_{Ra}}{(T_{1/2})_U}$$

$$N_{Ra} = \frac{(T_{1/2})_{Ra}}{(T_{1/2})_U} N_U = \frac{1620 \ yr}{4.5 \times 10^9 \ yr} 1 \ kg = 3.6 \times 10^{-7} \ kg$$

**Exercise 2.13** *What is radioactive dating?*

[Ans] Decay of radioactive elements is independent of the physical and chemical conditions imposed on them. Age of earth is estimated from the relative abundance of the two isotopes of Uranium $_{92}U^{238}$, $_{92}U^{235}$. This method of dating, i.e. obtaining the age of a sample (say earth) is referred to as radioactive dating.

$$(T_{1/2})_{U-238} = 4.5 \times 10^9 \ yr = \frac{0.693}{\lambda_{238}} \Rightarrow \lambda_{238} = \frac{0.693}{4.5 \times 10^9 \ yr}$$

$$(T_{1/2})_{U-235} = 7 \times 10^8 \ yr = \frac{0.693}{\lambda_{235}} \Rightarrow \lambda_{235} = \frac{0.693}{7 \times 10^8 \ yr}$$

Let us assume that earth was formed at an instant denoted by $t = 0$. So $N(U^{238})_{t=0} = N(U^{235})_{t=0} = N_0 \Leftarrow$ same amount existed at the initial point of time. At present time $t$ the natural uranium on earth has $N(U^{238})_t \equiv N^{238} = 99.3\%$, $N(U^{235})_t \equiv N^{235} = 0.7\%$. Hence

$$\frac{N^{238}}{N^{235}} = \frac{N_0 e^{-\lambda_{238} t}}{N_0 e^{-\lambda_{235} t}} = e^{(\lambda_{235} - \lambda_{238}) t}$$

$$\frac{99.3}{0.7} = e^{(\lambda_{235} - \lambda_{238}) t} \Rightarrow ln \frac{99.3}{0.7} = (\lambda_{235} - \lambda_{238}) t. \text{ So}$$

$$t = \frac{1}{\lambda_{235} - \lambda_{238}} ln \frac{99.3}{0.7}$$

Using $\lambda_{235} - \lambda_{238} = \frac{0.693}{7 \times 10^8 \ yr} - \frac{0.693}{4.5 \times 10^9 \ yr} = 8.36 \times 10^{-10} \ yr^{-1}$ we have

$t = \frac{1}{8.36 \times 10^{-10} \ yr^{-1}} ln \frac{99.3}{0.7} = 5.9 \times 10^9 \ yr = \text{Age of earth.}$

**Exercise 2.14** *What is uranium dating?*

[Ans] If a rock, formed as a result of pre-historic volcanic eruption contained a small amount of uranium it would decay to lead. By measuring the ratio of uranium to lead in a rock sample an exact estimate of the age of minerals can be obtained. This method of dating, i.e. obtaining the age of a sample (say a rock or mineral) is referred to as uranium lead dating.

We can liken the conversion $_{92}U^{238} \rightarrow {}_{82}Pb^{206}$ to the transformation $1 \rightarrow 2$. We derived in equation (2.37)

$$N_2 = \frac{\lambda_1 N_{10}}{\lambda_2 - \lambda_1} (e^{-\lambda_1 t} - e^{-\lambda_2 t}).$$

Taking $1 \equiv {}_{92}U^{238} =$ radioactive, $2 \equiv {}_{82}Pb^{206} =$ stable end product we can write
$N_{Pb} = \frac{\lambda_U N_{U0}}{\lambda_{Pb} - \lambda_U} (e^{-\lambda_U t} - e^{-\lambda_{Pb} t}).$

Using $(T_{1/2})_U = 4.5 \times 10^9 \ yr = \frac{0.693}{\lambda_U}$

$$(T_{1/2})_{Pb} = \infty = \frac{0.693}{\lambda_{Pb}} \Rightarrow \lambda_{Pb} \approx 0 \text{ we have}$$

$$N_{Pb} = \frac{\lambda_U N_{U0}}{-\lambda_U} (e^{-\lambda_U t} - e^{-(0) t})$$

$$N_{Pb} = N_{U0} (1 - e^{-\lambda_U t}) \tag{2.294}$$

$$\begin{array}{c} \text{Number} \\ \text{of atoms} \end{array} = \begin{array}{c} \text{Present number} \\ \text{of Pb atoms} \end{array} + \begin{array}{c} \text{Present number} \\ \text{of U atoms} \end{array}$$

$N_{U0} = N_{Pb} + N_U$
Using this in equation (2.294) we get

$$N_{Pb} = (N_{Pb} + N_U)(1 - e^{-\lambda_U t}) = N_{Pb} + N_U - (N_{Pb} + N_U) e^{-\lambda_U t}.$$

Hence

$$N_U = (N_{Pb} + N_U) e^{-\lambda_U t}$$

$$e^{\lambda_U t} = \frac{N_{Pb} + N_U}{N_U} \Rightarrow \lambda_U t = ln\frac{N_{Pb} + N_U}{N_U}$$

$$t = \frac{1}{\lambda_U} ln\frac{N_{Pb} + N_U}{N_U}$$

This equation enables us to find the age of minerals.

**Exercise 2.15** *What is radio carbon dating?*

*Ans* Determination of archeological and geological times (not more than 30 000 yrs old) can be done using radio isotopes (particularly radio carbon) and is called radio carbon dating.

There are two isotopes of carbon

☐$_6C^{12}$ ☐$_6C^{14}$.

This $_6C^{14}$ is radioactive and so called radio carbon. It decays as

$$_6C^{14} \rightarrow {_7N^{14}} + e^- + \bar{\nu}_e \text{ with } (T_{1/2})_{C-14} = 5700 \; yr.$$

In the process exponential decay law is obeyed viz.

$N = N_0 e^{-\lambda t}$, $\lambda$ = decay constant of radio carbon, $N_0$ = Number of radio carbons initially present at $t = 0$, $N$ = Number of radio carbons surviving decay at time $t$.

Production of $_6C^{14}$ and its subsequent decay are in equilibrium in the Earth's atmosphere. Meanwhile, the radio carbon atoms get incorporated into atmospheric carbon dioxide and when plants use $CO_2$ for their growth, radio carbon is incorporated in a plant's body and hence in the animal's body that feeds on these plants. (So plants are slightly radioactive.)

Carbon content of a living plant has a definite ratio of $_6C^{14}$ and $_6C^{12}$ atoms. When a plant dies, no additional $_6C^{14}$ is taken in and that within the plant body begins to decay without being replaced. So by measuring the ratio of the amounts of $_6C^{14}$ and $_6C^{12}$ present in an organic archeological sample the age of fossil (sample) is obtained.

Radio carbon dating method is not good for materials more than 30 000 years old. This is because the amount of $_6C^{14}$ becomes immeasurably small in 30 000 years.

**Exercise 2.16** *An archeologist finds a piece of wood in an excavated house. It weighs 50 gm and shows C-14 activity of 320 disintegrations per minute. Establish the length of time which has elapsed since the wood was part of a living tree assuming living plants show C-14 activity of 12 disintegrations per minute per gm ($T_{1/2}$ for C-14 is 5730 years.)*

*Ans* C-14 activity in living plants = 12 disintegrations/min/gm = $N_0$.

C-14 activity in excavated house = 320 disintegrations/min/50 gm

$= \frac{320}{50}$ disintegration/min/gm = 6.4 disintegration/min/gm = $N$.

Now for $C$ -14

$$\lambda = \frac{0.693}{T_{1/2}} = \frac{0.693}{5730 \ yr} = 1.2 \times 10^{-4} \ yr^{-1}.$$

Decay is governed by the survival equation $N = N_0 e^{-\lambda t}$. Let $t$ = age of specimen (in years). Hence

$$6.4 = 12 e^{-(1.2 \times 10^{-4})t}$$

$$\ln\frac{12}{6.4} = 1.2 \times 10^{-4} t$$

$$t = \frac{1}{1.2 \times 10^{-4}} \ln\frac{12}{6.4} \ yr = 5238.4 \ yr.$$

**Exercise 2.17** *Consider a radioactive series of three elements. The first two members have half lives of 5 and 12 hours respectively while the third member is stable. Initially there are $10^6$ atoms of the first member and none of the second and third. After what period of time will the number of atoms of the second element reach their maximum value? Find the number of atoms.*

    Ans Radioactive decay chain is $1 \rightarrow 2 \rightarrow 3$.

$$(T_{1/2})_1 = 5 \ h \Rightarrow \lambda_1 = \frac{0.693}{(T_{1/2})_1} = \frac{0.693}{5 \ h} = 0.1386 \ h^{-1}$$

$$(T_{1/2})_2 = 12 \ h \Rightarrow \lambda_2 = \frac{0.693}{(T_{1/2})_2} = \frac{0.693}{12 \ h} = 0.057 \ 75 \ h^{-1}$$

For the decay the time at which number of type 2 nuclei will peak is by equation (2.45)

$$t_0 = \frac{1}{\lambda_2 - \lambda_1} \ln\frac{\lambda_2}{\lambda_1} = \frac{1}{0.057 \ 75 \ h^{-1} - 0.1386 \ h^{-1}} \ln\frac{0.057 \ 75 \ h^{-1}}{0.1386 \ h^{-1}} = 10.8 \ h.$$

Number of atoms of type 2 is given by equation (2.37)

$$N_2 = \frac{\lambda_1 N_{10}}{\lambda_2 - \lambda_1} (e^{-\lambda_1 t} - e^{-\lambda_2 t})$$

$$= \frac{(0.1386 \ h^{-1}) 10^6}{0.057 \ 75 \ h^{-1} - 0.1386 \ h^{-1}} (e^{-(0.1386 \ h^{-1})(10.8 \ h)} - e^{-(0.057 \ 75 \ h^{-1})(10.8 \ h)})$$

$$N_2 = \frac{0.1386 \times 10^6}{0.057 \ 75 - 0.1386} (e^{-0.1386 \times 10.8} - e^{-0.057 \ 75 \times 10.8})$$

$$N_2 = 5.4 \times 10^5 \ atoms.$$

**Exercise 2.18** *The half-life of radium (atomic mass 226 ) is 1600 years and that for radon (atomic mass 222 ) is 3.8 days. Calculate the volume of radon gas at N. T. P that would be in equilibrium with 1 g of radium.*

$\boxed{Ans}$ $_{88}Ra^{226}\xrightarrow[1600\,yr]{\alpha}$ $_{86}Rn^{222}\xrightarrow[3.82\,day]{\alpha}$

$$\text{Clearly } (T_{1/2})_{Ra}\gg(T_{1/2})_{Rn}$$

i.e. parent $_{88}Ra^{226}$ is very long-lived compared to daughter $_{86}Rn^{222}$ that is short-lived. This situation leads to secular or permanent equilibrium, i.e.

$$\lambda_{Ra}N_{Ra} = \lambda_{Rn}N_{Rn} \Rightarrow \frac{N_{Ra}}{(T_{1/2})_{Ra}} = \frac{N_{Rn}}{(T_{1/2})_{Rn}}$$

$$N_{Rn} = N_{Ra}\frac{(T_{1/2})_{Rn}}{(T_{1/2})_{Ra}} = N_{Ra}\frac{3.82\,day}{1600\,yr} = N_{Ra}\frac{3.82\,day}{1600(365\,day)} = 6.54 \times 10^{-6}N_{Ra}$$

226g Ra has $N_A$ atoms ($N_A$ = Avogadro number).

So 1 *g* Ra has $\frac{N_A}{226} = N_{Ra}$ atoms. This means

$$N_{Rn} = 6.54 \times 10^{-6}N_{Ra} = 6.54 \times 10^{-6}\frac{N_A}{226} = 2.9 \times 10^{-8}N_A$$

Again $N_A$ atoms exist in 222 *gm* of Rn.

So $2.9 \times 10^{-8}N_A$ atoms would exist in $\frac{222}{N_A}(2.9 \times 10^{-8}N_A)$ *gm* = $6.4 \times 10^{-6}$ *gm* of Rn.

Since 222 *gm* corresponds to 22.4 *lt*, it follows that

$$6.4 \times 10^{-6}\,gm \text{ corresponds to } \frac{22.4\,lt}{222\,gm}(6.4 \times 10^{-6}\,gm)$$

$$= 6.46 \times 10^{-7}\,lt = 6.46 \times 10^{-7} \times 1000\,cc = 6.46 \times 10^{-4}\,cc.$$

**Exercise 2.19** *Consider the decay $_{83}Bi^{212} \rightarrow _{81}Tl^{208} + _2He^4$ . Determine the energy levels of $_{81}Tl^{208}$ from the energy spectra of $\alpha$ . Given that the disintegration energy of $\alpha$ is 6.320 25 MeV . Also find the energy of the gamma transitions. (Energies of $\alpha$ are $T_\alpha$ : 6.201, 6.161, 5.873, 5.728, 5.708, 5.597 MeV.)*

$\boxed{Ans}$ Consider the decay

$$_{83}Bi^{212} \rightarrow _{81}Tl^{208} + _2He^4$$

The energies of $\alpha$ are $T_\alpha$ s and they have been shown in figure 2.36. Given

$$Q_\alpha = 6.320\,25\,MeV$$

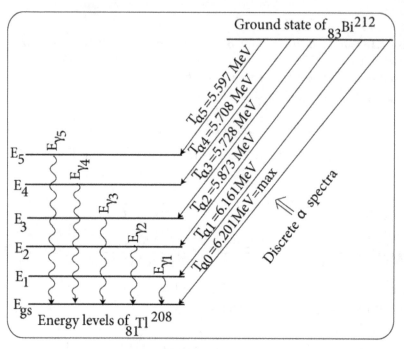

**Figure 2.36.** Determination of energy levels of $_{81}Tl^{208}$ from $\alpha$ spectra.

We are to find the energy levels of $_{81}Tl^{208}$. The disintegration energy for $\alpha$ decay

$_ZX^A \rightarrow {}_{Z-2}Y^{A-4} + {}_2He^4$ is given by equation (2.74)

$Q_\alpha = T_\alpha\left(1 + \frac{M_\alpha}{M_Y}\right) + E_n$ where $E_n$ is excitation energy of the daughter.

$$E_n = Q_\alpha - T_\alpha\left(1 + \frac{M_\alpha}{M_Y}\right)$$

With $M_\alpha = 4$, $M_Y = 208$

$$E_n = 6.320\,25\;MeV - T_\alpha\left(1 + \frac{4}{208}\right)$$

$$E_n = 6.320\,25\;MeV - \frac{212}{208}T_\alpha \quad (T_\alpha \text{ in } MeV).$$

Let us find the values of $E_n$ knowing $T_\alpha$.

| $T_\alpha$ values | $E_n$ = energy levels of daughter $_{81}Tl^{208}$ |
|---|---|
| 6.201 $MeV$ | $6.320\,25 - (212/208)6.201\ MeV = 0 \equiv E_{gs}$ |
| 6.161 $MeV$ | $6.320\,25 - (212/208)6.161\ MeV = 0.04\ MeV \equiv E_1$ |
| 5.873 $MeV$ | $6.320\,25 - (212/208)5.873\ MeV = 0.334\ MeV \equiv E_2$ |
| 5.728 $MeV$ | $6.320\,25 - (212/208)5.728\ MeV = 0.482\ MeV \equiv E_3$ |
| 5.708 $MeV$ | $6.320\,25 - (212/208)5.708\ MeV = 0.503\ MeV \equiv E_4$ |
| 5.597 $MeV$ | $6.320\,25 - (212/208)5.597\ MeV = 0.616\ MeV \equiv E_5$ |

Energy of the gamma transitions are

$$E_1 - E_{gs} = E_{\gamma 1} = 0.04\ MeV - 0 = 0.04\ MeV$$

$$E_2 - E_{gs} = E_{\gamma 2} = 0.334\ MeV - 0 = 0.334\ MeV$$

$$E_3 - E_{gs} = E_{\gamma 3} = 0.482\ MeV - 0 = 0.482\ MeV$$

$$E_4 - E_{gs} = E_{\gamma 4} = 0.503\ MeV - 0 = 0.503\ MeV$$

$$E_5 - E_{gs} = E_{\gamma 5} = 0.616\ MeV - 0 = 0.616\ MeV.$$

The discrete $\alpha$ spectrum or fine structure in $\alpha$ energy spectrum gives conclusive evidence that a nucleus exists in discrete energy levels.

**Exercise 2.20** *Justify if the alpha particle is formed in the nucleus during $\alpha$ decay.*

[Ans] Size of nucleus is $\sim 10^{-14}\ m$. If $\alpha$ stays within the nucleus it can exist anywhere within it. So the uncertainty in position of $\alpha$ is $\Delta x \sim 10^{-14}\ m$. It is connected to the momentum uncertainty of $\alpha$ viz. $\Delta p$ as

$$\Delta x\,\Delta p \sim \hbar$$

$$\Delta p \sim \frac{\hbar}{\Delta x} = \frac{h}{2\pi\Delta x} = \frac{6.626 \times 10^{-34}\ J.\ s}{2\pi(10^{-14}\ m)} = 1.055 \times 10^{-20}\ kg.\ m.\ s^{-1}$$

Thus momentum of $\alpha$ is

$$p + \Delta p \xrightarrow{\text{take } p = 0} \Delta p = M_\alpha v_\alpha$$

$$v_\alpha = \frac{\Delta p}{M_\alpha} = \frac{\Delta p}{4m_p} \text{ as}$$

$$M_\alpha = 2m_p + 2m_n \approx 4m_p = 4 \times 1.67 \times 10^{-27} \, kg. \text{ So}$$

$$v_\alpha = \frac{\Delta p}{4m_p} = \frac{1.055 \times 10^{-20} \, kg. \, m. \, s^{-1}}{4 \times 1.67 \times 10^{-27} \, kg} = 1.58 \times 10^6 \, m \, s^{-1} \ll c$$

where $c = 3 \times 10^8 \, m \, s^{-1}$.

This means motion of $\alpha$ can be treated as non relativistic. Hence kinetic energy that an $\alpha$ needs to stay within the nucleus, as estimated from Heisenberg uncertainty principle is

$$T_\alpha = \frac{(\Delta p)^2}{2M_\alpha} = \frac{(\Delta p)^2}{2.4m_p} = \frac{(1.055 \times 10^{-20} \, kg \, m \, s^{-1})^2}{8 \times 1.67 \times 10^{-27} \, kg}$$

$$= 8.33 \times 10^{-15}J = \frac{8.33 \times 10^{-15}}{1.6 \times 10^{-19}} eV = 52 \times 10^3 \, eV = 52 \, keV.$$

Again in $\alpha$ decay say $_{92}U^{238} \rightarrow _{90}Th^{234} + _2He^4$ the $\alpha$ emitted has energy $\sim 4 \, MeV > 52 \, keV$. It follows therefore that $\alpha$ can be formed inside the nucleus just prior to $\alpha$ decay.

   ✓ When $\alpha$ is formed from a neutron and a proton within the nucleus at the time of $\alpha$ decay energy is released which is taken up by the $\alpha$ as its kinetic energy to escape out of nucleus. So $\alpha$ makes use of the energy of its own formation.

For a proton or a neutron to escape out of the nucleus kinetic energy is needed also. But it has to be supplied from outside which may not be available. So they cannot escape.

**Exercise 2.21** *How do you know that the alpha particle is a helium nucleus?*
   [Ans] Rutherford and Royd performed an experiment that showed that alpha particles are helium nuclei.
   Radon gas was put in a thin walled glass tube $A$ as shown in figure 2.37. Radon decays to polonium emitting $\alpha$ as

$$_{86}Rn^{222} \rightarrow _{84}Po^{218} + _2He^4 \text{ with } T_{1/2} = 3.82 \text{ day.}$$

The emitted $\alpha$ passes through the thin wall of $A$ and gets collected in glass tube $B$. After about 6 days enough $\alpha$ gas is collected in tube $B$. By introducing Hg in tube $B$, $\alpha$ gas is forced into capillary tube $C$.

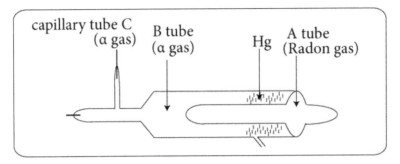

**Figure 2.37.** Rutherford and Royd's experimental demonstration that $\alpha$ and He are same.

High voltage is applied across the electrodes of tube $C$ to produce electric discharge through $\alpha$ gas. The light coming from the tube $C$ was examined with a spectroscope which clearly showed the spectral lines of helium.

This spectroscopic evidence conclusively proves that $\alpha$ particles are helium nuclei. So

$$q_\alpha \equiv q_{\text{He nucleus}} \equiv 2\,|e| = 2(1.6 \times 10^{-19}\ C) = 3.2 \times 10^{-19}\ C.\ \text{And}$$

$$M_\alpha = M_{\text{He nucleus}} \cong 4m_p = 4 \times 1.67 \times 10^{-27}\ kg = 6.68 \times 10^{-27}\ kg.$$

**Exercise 2.22** *The tunneling probability of $\alpha$ from the nucleus is given by equation (2.100)*

$$T = \exp[-2\sqrt{\frac{2m}{\hbar^2}} \int_R^{R_1} [V(r) - E]^{1/2}\ dr].$$

*Express this result in terms of the known parameters viz ✓ energy $E \equiv E_\alpha$ of $\alpha$ ✓ atomic number Z of parent ✓ nuclear radius R.*

$\boxed{Ans}$ $T = \exp[-2\sqrt{\frac{2m}{\hbar^2}} \int_{r=R}^{R_1} [V(r) - E]^{\frac{1}{2}}\ dr]$

Using equations (2.76), (2.91), and (2.107) we construct

$$\frac{V(r)}{E} = \frac{V(r)}{V(R_1)} = \frac{2(Z-2)\,|e|^2}{4\pi\varepsilon_0 r} \cdot \frac{4\pi\varepsilon_0 R_1}{2(Z-2)\,|e|^2} = \frac{R_1}{r}.\ \text{So}$$

$$T = \exp[-2\sqrt{\frac{2\,mE}{\hbar^2}} \int_{r=R}^{R_1} [\frac{V(r)}{E} - 1]^{1/2}\ dr]$$

$$T = \exp[-2\sqrt{\frac{2\,mE}{\hbar^2}} \int_{r=R}^{R_1} \left(\frac{R_1}{r} - 1\right)^{\frac{1}{2}}\ dr] = e^{-G}$$

where

$$G = 2\sqrt{\frac{2\,mE}{\hbar^2}} \int_{r=R}^{R_1} \left(\frac{R_1}{r} - 1\right)^{\frac{1}{2}} dr.$$

Put $r = R_1 \cos^2 \theta \Rightarrow dr = -2R_1 \cos \theta \sin \theta d\theta$. Hence

$$G = 2\sqrt{\frac{2\,mE}{\hbar^2}} \int_{r=R}^{R_1} \left(\frac{R_1}{R_1 \cos^2 \theta} - 1\right)^{\frac{1}{2}} (-2R_1 \cos \theta \ \sin \theta d\theta)$$

$$G = -\frac{4R_1\sqrt{2\,mE}}{\hbar} \int_{r=R}^{R_1} \left(\frac{1 - \cos^2 \theta}{\cos^2 \theta}\right)^{\frac{1}{2}} (\cos \theta \ \sin \theta d\theta) d\theta = -\frac{4R_1\sqrt{2\,mE}}{\hbar} \int_{r=R}^{R_1} \sin^2 \theta d\theta$$

$$G = -\frac{2R_1\sqrt{2\,mE}}{\hbar} \int_{r=R}^{R_1} (1 - \cos 2\theta) d\theta = -\frac{2R_1\sqrt{2\,mE}}{\hbar} (\theta - \frac{\sin 2\theta}{2})\Big|_{r=R}^{R_1}$$

$$G = -\frac{2R_1\sqrt{2\,mE}}{\hbar} (\theta - \sin \theta \cos \theta)\Big|_{r=R}^{R_1}.$$

Use $\gamma = R_1 \cos^2 \theta \Rightarrow \cos \theta = \sqrt{\frac{r}{R_1}}$, $\sin \theta = \sqrt{1 - \cos^2 \theta} = \sqrt{1 - \frac{r}{R_1}}$

$$G = -\frac{2R_1\sqrt{2\,mE}}{\hbar} (\cos^{-1}\sqrt{\frac{r}{R_1}} - \sqrt{1 - \frac{r}{R_1}} \sqrt{\frac{r}{R_1}})\Big|_{r=R}^{R_1}.$$

Now put the limits to get

$$G = -\frac{2R_1\sqrt{2\,mE}}{\hbar} \left[\left(\cos^{-1} 1 - 0\right) - \cos^{-1}\sqrt{\frac{R}{R_1}} + \sqrt{1 - \frac{R}{R_1}} \sqrt{\frac{R}{R_1}}\right]$$

$$G = \frac{2R_1\sqrt{2\,mE}}{\hbar} \left[\cos^{-1}\sqrt{\frac{R}{R_1}} - \sqrt{\frac{R}{R_1}\left(1 - \frac{R}{R_1}\right)}\right]$$

Define $x = \sqrt{\frac{R}{R_1}}$, $x^2 = \frac{R}{R_1}$

$$G = \frac{2R_1\sqrt{2\,mE}}{\hbar} \left[\cos^{-1} x - x\sqrt{1 - x^2}\right]$$

As $R \ll R_1 \Rightarrow x \to$ small. We thus approximate

$$sinx \cong x. \quad \text{Again}$$

$$sinx = \cos \left(\frac{\pi}{2} - x\right). \quad \text{So}$$

$$x = \cos \left(\frac{\pi}{2} - x\right) \Rightarrow \cos^{-1} x = \frac{\pi}{2} - x.$$

Also $x^2 \to$ smaller. Hence $1 - x^2 \cong 1$. With this we write

$$G = \frac{2R_1\sqrt{2\,mE}}{\hbar}\left[\left(\frac{\pi}{2} - x\right) - x\right]$$

$$G = \frac{2R_1\sqrt{2\,mE}}{\hbar}\left(\frac{\pi}{2} - 2x\right)$$

$$G = \frac{2R_1\sqrt{2\,mE}}{\hbar}\left(\frac{\pi}{2} - 2\sqrt{\frac{R}{R_1}}\right) \tag{2.295}$$

$$G = \frac{2R_1\sqrt{2\,mE}}{\hbar}\frac{\pi}{2} - \frac{2R_1\sqrt{2\,mE}}{\hbar}2\sqrt{\frac{R}{R_1}} = \frac{\pi R_1\sqrt{2\,mE}}{\hbar} - \frac{4\sqrt{2\,mERR_1}}{\hbar}.$$

Using equation (2.208) viz $R_1 = \frac{2(Z-2)|e|^2}{4\pi\varepsilon_0 E}$ we have

$$G = \frac{\pi\sqrt{2\,mE}}{\hbar} \cdot \frac{2(Z-2)|e|^2}{4\pi\varepsilon_0 E} - \frac{4\sqrt{2\,mER}}{\hbar} \cdot \sqrt{\frac{2(Z-2)|e|^2}{4\pi\varepsilon_0 E}}$$

$$G = \frac{|e|^2}{\hbar\varepsilon_0}\sqrt{\frac{m}{2}}(Z-2)E^{-\frac{1}{2}} - \frac{4|e|}{\hbar}\sqrt{\frac{m}{\pi\varepsilon_0}}(Z-2)^{\frac{1}{2}}R^{\frac{1}{2}}$$

Let us rewrite this as

$$G = A(Z-2)E^{-\frac{1}{2}} - B(Z-2)^{\frac{1}{2}}R^{\frac{1}{2}}$$

where

$$A = \frac{|e|^2}{\hbar\varepsilon_0}\sqrt{\frac{m}{2}}, \quad B = \frac{4|e|}{\hbar}\sqrt{\frac{m}{\pi\varepsilon_0}}.$$

We can put the values of the physical constants to get

$$A = \frac{|e|^2}{\hbar\varepsilon_0}\sqrt{\frac{m}{2}} = \frac{(1.6 \times 10^{-19}\ C)^2}{(\frac{6.626 \times 10^{-34}\ J.s}{2\pi})(\frac{10^{-9}}{36\pi}\ Fm^{-1})}\sqrt{\frac{4 \times 1.67 \times 10^{-27}\ kg}{2}} = 1.587 \times 10^{-6}\ J^{1/2} \text{ and}$$

$$B = \frac{4|e|}{\hbar}\sqrt{\frac{m}{\pi\varepsilon_0}} = \frac{4(1.6 \times 10^{-19}\ C)}{\frac{6.626 \times 10^{-34}\ J.s}{2\pi}}\sqrt{\frac{4 \times 1.67 \times 10^{-27}\ kg}{\pi(\frac{10^{-9}}{36\pi}\ Fm^{-1})}} = 94 \times 10^6\ m^{-1/2}.$$

**Exercise 2.23** *Estimate barrier height and barrier width for the $\alpha$ decay*
$_{92}U^{238} \to {}_{90}Th^{234} + {}_2He^4$.
$\boxed{Ans}$ From equation (2.77) we have

$$V(r = R) = \frac{1}{4\pi\varepsilon_0}\frac{2(Z-2)|e|^2}{R_0 A^{1/3}} = \frac{1}{4\pi(\frac{10^{-9}}{36\pi}\ \frac{F}{m})}\frac{2(92-2)(1.6 \times 10^{-19}\ C)^2}{(1.4 \times 10^{-15}\ m)(210)^{1/3}}$$

$$V(r = R) = 4.98 \times 10^{-12} \, J = \frac{4.98 \times 10^{-12}}{1.6 \times 10^{-19}} eV = 31 \, MeV \Leftarrow \text{ barrier height.}$$

The barrier height at $r = R_1$ is

$$V(r = R_1) = \frac{1}{4\pi\varepsilon_0} \frac{2(Z - 2) \, |e|^2}{R_1} = E_\alpha$$

$$R_1 = \frac{1}{4\pi\varepsilon_0} \frac{2(Z - 2) \, |e|^2}{E_\alpha} \quad \text{[equation (2.104)]}$$

$$R_1 = \frac{1}{4\pi(\frac{10^{-9}}{36\pi} \frac{F}{m})} \frac{2(92 - 2)(1.6 \times 10^{-19} \, C)^2}{5 \, MeV} = \frac{1}{4\pi(\frac{10^{-9}}{36\pi} \frac{F}{m})} \frac{2(92 - 2)(1.6 \times 10^{-19} \, C)^2}{5 \times 10^6 \times 1.6 \times 10^{-19} \, J} = 51.84 \, fm.$$

Now nuclear radius is
$R_0 A^{1/3} = (1.4 \, fm)(238)^{1/3} = 8.68 \, fm.$
Thus width of potential barrier is by equation (2.105)
$R_1 - R = 51.84 \, fm - 8.68 \, fm = 43.16 \, fm.$

**Exercise 2.24** *In the $\alpha$ decay $_{92}U^{238} \rightarrow _{90}Th^{234} + _2He^4$ $\alpha$ has energy $E = 4.4 \, MeV$. Obtain a rough estimate of half life of parent?*
$\boxed{Ans}$ Consider the $\alpha$ decay

$$_{92}U^{238} \rightarrow _{90}Th^{234} + _2He^4.$$

$$E = 4.4 \, MeV = 4.4 \times 10^6 \times 1.6 \times 10^{-19} \, J = 7 \times 10^{-13} \, J.$$

Radius of parent $R = R_0 A^{1/3} = (1.4 \, fm)(238)^{1/3} = 8.7 \, fm = 8.7 \times 10^{-15} \, m.$
Atomic number of parent is $Z = 92$. Tunnelling probability is (*exercise 2.22*)
$T = e^{-G}$ where $G = A(Z - 2)E^{-1/2} - B(Z - 2)^{1/2}R^{1/2}$

$$G = (1.587 \times 10^{-6} \, J^{\frac{1}{2}})(92 - 2)(7 \times 10^{-13} \, J)^{-\frac{1}{2}} - (94 \times 10^6 \, m^{-\frac{1}{2}})(92 - 2)^{\frac{1}{2}}(8.7 \times 10^{-15} \, m)^{\frac{1}{2}}$$

$$G = 171 - 83 = 88.$$

$$T = e^{-G} = e^{-91} = 6 \times 10^{-39} = \frac{6}{10^{39}}$$

Out of $10^{39}$ strikes $\alpha$ emerges six times only. Velocity is

$$v = \sqrt{\frac{2E}{m}} = \sqrt{\frac{2(6.72 \times 10^{-13} \, J)}{4(1.67 \times 10^{-27} \, kg)}} = 1.42 \times 10^7 \, m \, s^{-1}.$$

Collision frequency

$$f = \frac{v}{2R} = \frac{1.42 \times 10^7 \, m \, s^{-1}}{2(8.7 \times 10^{-15} \, m)} = 8.2 \times 10^{20} \, s^{-1}.$$

Decay constant

$$\lambda = fT = (8.2 \times 10^{20} \ s^{-1})(6 \times 10^{-39}) = 4.92 \times 10^{-18} \ s^{-1}.$$

Half-life is

$$T_{1/2} = \frac{ln2}{\lambda} = \frac{0.693}{4.92 \times 10^{-18} \ s^{-1}} = 1.4 \times 10^{17} \ s = \frac{1.4 \times 10^{17}}{365 \times 24 \times 60 \times 60} yr = 4.4 \times 10^{9} \ yr.$$

The experimental value of half-life is $4.5 \times 10^{9} \ yr$. So we have obtained a rough estimate of half life of $_{92}U^{238}$.

**Exercise 2.25** *Show that in $\alpha$ decay the tunneling probability can be expressed as* $T = e^{-G}$ , *where* $G = \frac{(Z-2)|e|^2}{\hbar v_\alpha \varepsilon_0}$ *under certain approximation.*

$\boxed{Ans}$ In *exercise 2.22 we got equation (2.295) namely* $G = \frac{2R_1\sqrt{2\,mE}}{\hbar}\left(\frac{\pi}{2} - 2\sqrt{\frac{R}{R_1}}\right).$ Now by equations (2.77) and (2.107) we have

$$\frac{V(R_1)}{V(R)} = \frac{E}{V(R)} = \frac{4\pi\varepsilon_0 R}{2(Z-2)|e|^2} \cdot \frac{2(Z-2)|e|^2}{4\pi\varepsilon_0 R_1} = \frac{R}{R_1}.$$

Again using equation (2.104)

$$G = \frac{2}{\hbar}\sqrt{2m}\sqrt{R_1^2 E}\left(\frac{\pi}{2} - 2\sqrt{\frac{E}{V(R)}}\right) = \frac{2}{\hbar}\sqrt{2m}\sqrt{\left(\frac{2(Z-2)|e|^2}{4\pi\varepsilon_0 E}\right)^2 E}\left(\frac{\pi}{2} - 2\sqrt{\frac{E}{V(R)}}\right)$$

$$G = \frac{2}{\hbar}\sqrt{\frac{2m}{E}}\frac{2(Z-2)|e|^2}{4\pi\varepsilon_0}\left(\frac{\pi}{2} - 2\sqrt{\frac{E}{V(R)}}\right) \qquad (2.296)$$

Since $Q_\alpha$ is the energy available to $\alpha$ we can write

$$\frac{E}{V(R)} \xrightarrow{E \sim Q_\alpha} \frac{Q_\alpha}{V(R)} \text{ and so}$$

$$G = \frac{2}{\hbar}\sqrt{\frac{2m}{Q_\alpha}}\frac{2(Z-2)|e|^2}{4\pi\varepsilon_0}\left(\frac{\pi}{2} - 2\sqrt{\frac{Q_\alpha}{V(R)}}\right) \qquad (2.297)$$

Now $\frac{\pi}{2} \cong 1.57$, $2\sqrt{\frac{Q_\alpha}{V(R)}} \cong 2\sqrt{\frac{4\ MeV}{23\ MeV}} = 0.8$. Hence we approximate

$$G = \frac{2}{\hbar}\sqrt{\frac{2m}{Q_\alpha}}\frac{2(Z-2)|e|^2}{4\pi\varepsilon_0}\frac{\pi}{2}.$$

Use $E \sim Q_\alpha = \frac{1}{2}mv_\alpha^2$. Hence

$$G = \frac{2}{\hbar}\sqrt{\frac{2m}{\frac{1}{2}mv_\alpha^2}}\frac{2(Z-2)|e|^2}{4\pi\varepsilon_0}\frac{\pi}{2} = \frac{(Z-2)|e|^2}{\hbar v_\alpha \varepsilon_0}. \text{ Hence}$$

$$T = e^{-G} = e^{-\frac{(Z-2)|e|^2}{\hbar v_\alpha \varepsilon_0}}.$$

This is tunneling probability in terms of velocity $v_\alpha$ of $\alpha$ particle under a certain approximation that we made.

**Exercise 2.26** *Calculate half-life $T_{1/2}$ of $\alpha$ radioactive sample (typical Coulomb barrier ~25 MeV, $Z \sim 94$) for Q value $Q_\alpha \sim 4, 5, 6, 7$ MeV and comment on the variation of $T_{1/2}$ with $Q_\alpha$.*

$\boxed{\text{Ans}}$ We write the results for $\alpha$ decay equations (2.297), (2.118), and (2.120)

$$G = \frac{2}{\hbar}\sqrt{\frac{2m}{Q_\alpha}}\frac{2(Z-2)|e|^2}{4\pi\varepsilon_0}\left(\frac{\pi}{2} - 2\sqrt{\frac{Q_\alpha}{V(R)}}\right), \quad T = e^{-G}, \quad \lambda = \sqrt{\frac{Q_\alpha + V_0}{2mR^2}}e^{-G}, \quad T_{1/2} = \frac{0.693}{\lambda}$$

With $m = m_{4p} = 4 \times 1.67 \times 10^{-27}$ kg, $Z = 94$, $V(R) = 25$ MeV, $V_0 = 40$ MeV, $R = 1\,fm = 10^{-15}$ m we have

$$G = \frac{2}{\frac{6.626 \times 10^{-34}}{2\pi}}\sqrt{\frac{2 \times 4 \times 1.67 \times 10^{-27}\ kg}{Q_\alpha \times 10^6 \times 1.6 \times 10^{-19}\ J}}\frac{2(94-2)(1.6 \times 10^{-19}\ C)^2}{4\pi \times \frac{10^{-9}}{36\pi}\frac{F}{m}}\left(\frac{\pi}{2} - 2\sqrt{\frac{Q_\alpha}{25\ MeV}}\right),$$

$$G = \frac{232.33}{\sqrt{Q_\alpha}}\left(1.57 - 0.4\sqrt{Q_\alpha}\right) \tag{2.298}$$

$$\lambda = \sqrt{\frac{(Q_\alpha + 40) \times 10^6 \times 1.6 \times 10^{-19}\ J}{2 \times 4 \times 1.67 \times 10^{-27}kg \times (10^{-15}\ m)^2}}e^{-G}$$

$$\lambda = 3.46 \times 10^{21}\sqrt{Q_\alpha + 40}\,e^{-G} \tag{2.299}$$

We calculate $G, T^{-G}, \lambda, T_{1/2}$ for various values of $Q_\alpha$ as follows,

| $Q_\alpha$ (in MeV) | G | $T = e^{-G}$ | $\lambda$ | $T_{1/2} = \frac{0.693}{\lambda}$ |
|---|---|---|---|---|
| 4 MeV | 89.447 | $1.424 \times 10^{-39}$ | $3.27 \times 10^{-17}\ s^{-1}$ | $2.1 \times 10^{16}\ s$ |
| 5 MeV | 70.19 | $3.29 \times 10^{-31}$ | $7.64 \times 10^{-9}\ s^{-1}$ | $90.7 \times 10^6\ s$ |
| 6 MeV | 55.98 | $4.88 \times 10^{-25}$ | $0.011\,45\ s^{-1}$ | $60.52s$ |
| 7 MeV | 44.93 | $3.1 \times 10^{-20}$ | $735.34\ s^{-1}$ | $9.4 \times 10^{-4}\ s$ |

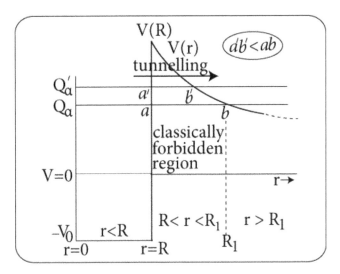

**Figure 2.38.** Slight increase in $Q_\alpha$ reduces barrier width and vastly enhances tunneling probability and reduces half-life.

- There are nuclei with mass number $A > 150$ for which it is energetically favourable to suffer $\alpha$ decay since for them the sum of the rest energies of daughter nucleus and helium nuclei is less than the original parent nuclei. $\alpha$ is pre-formed inside the parent nucleus. As it tries to emerge out of the parent nucleus it encounters Coulomb potential barrier (figure 2.38) of height ~25 *MeV*. For small changes in $Q_\alpha$ (say 4 *MeV* → 7 *MeV*) there is huge change in $T_{1/2}$ ($10^{16}$ *s* → $10^{-4}$ *s*) and this could be explained through Gammow's theory of $\alpha$ decay which gives the barrier tunneling probability — the phenomenon being a quantum mechanical effect. A slight increase of $Q_\alpha$ will lead to reduction in the width of Coulomb barrier (from $ab$ → $a'b'$) to be crossed and this will vastly enhance the probability of tunneling, reducing half-life significantly.

**Exercise 2.27** *Show in a plot the variation of intensity versus range for $\alpha$ particle. What is straggling effect?*

[Ans] Plot of intensity versus distance travelled by $\alpha$ is shown in figure 2.39. As $\alpha$ beam moves through a medium it is observed that the number of ions per unit length, i.e the intensity is initially constant (as represented by portion $OA$).

At the end of the path of $\alpha$ called range, the $\alpha$ particles have a low velocity and so greater probability of colliding with the gas atoms. Hence ionization increases fast (along curve $AB$) to reach a maximum at $B$— called Bragg hump. But soon the $\alpha$ energy falls below the ionization potential of gas. This is indicated by a steep fall of the curve $BL$.

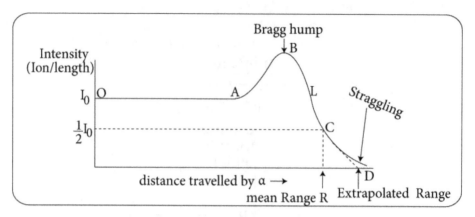

**Figure 2.39.** Intensity versus distance travelled by $\alpha$.

As the collisions between $\alpha$ and gas molecules is a random statistical process the fall in intensity is not abrupt but occurs with a finite slope as represented by the portion $LC$.

At the end of the path of $\alpha$ the curve tails off (portion $CD$). Actually all the $\alpha$ particles in the $\alpha$ beam were not emitted with the same initial velocity (at a given point of time) and hence all of them do not reach the end of the range simultaneously. They lose their energy after travelling various distances within the portion $CD$. So the curve shows a tail $CD$. This phenomenon of extension of range is called straggling effect or straggling of range.

**Exercise 2.28** *Calculate the energy released in the decay of* $U^{238}$ *from the following data:*

| Nucleus | Mass–excess ($MeV$) |
|---|---|
| $U^{238}$, $Th^{234}$, $He^4$ | 47.3, 40.6, 2.4 |

*Also evaluate the height of the Coulomb barrier between the daughter nucleus and the $\alpha$ particle taking the radius parameter to be 1.2 fm.*

$\boxed{Ans}$ $_{92}U^{238} \rightarrow {}_{90}Th^{234} + {}_2He^4$

$$47.3 \ MeV = 40.6 \ MeV + 2.4 \ MeV + E_\alpha$$

$$E_\alpha = (47.3 - 40.6 - 2.4) \ MeV = 4.3 \ MeV.$$

$$V = \frac{2(Z-2)\,|e|^2}{4\pi\varepsilon_0 R_0 A^{1/3}} = \frac{2(92-2)\,|1.6 \times 10^{-19} \ C|^2}{4\pi\left(\frac{10^{-9}}{36\pi}\frac{F}{m}\right)(1.2 \times 10^{-15} \ m)(238)^{1/3}}$$

$$= 5.58 \times 10^{-12} \, J = \frac{5.58 \times 10^{-12}}{1.6 \times 10^{-19}} eV = 34.875 \, MeV$$

Comment: $E_\alpha = 4.3 \, MeV$ $\alpha$ surmounts $34.875 \, MeV$ potential barrier through tunneling.

**Exercise 2.29** *Potassium is present in the human body to the extent of about* $0.35\%$ *of the body weight. Calculate the total activity due to* $K^{40}$ *in a man weighing* $70 \, kg$, *assuming relative abundance of* $K^{40}$ *in potassium to be* $0.012\%$ *(Half-life of* $K^{40}$ *is* $1.3 \times 10^9 \, yr$).

[Ans] Weight of man $= 70 \, kg$. And $0.35\%$ of body weight is potassium.
Amount of potassium in the body of the man is
$0.35\%$ of $70 \, kg = \frac{0.35}{100} \times 70 \, kg = 0.245 \, kg$

Relative abundance of $K^{40}$ in potassium $= 0.012\%$
Amount of $K^{40}$ in the body of man is $0.012\%$ of $0.245 \, kg$

$$= \frac{0.012}{100} \times 0.245 \, kg = 2.94 \times 10^{-5} \, kg = 2.94 \times 10^{-2} \, g.$$

Number of atoms $N_0$ in $2.94 \times 10^{-2} \, g$ of $K^{40}$ is

$$N_0 = \frac{6.023 \times 10^{23}}{40} 2.94 \times 10^{-2} \, g = 4.4 \times 10^{20}$$

$$\text{Total activity } \lambda N_0 = \frac{0.693}{T_{1/2}} N_0 = \frac{0.693}{1.3 \times 10^9 \, yr} 4.4 \times 10^{20}$$

$$\lambda N_0 = \frac{0.693}{1.3 \times 10^9 \times 365 \times 24 \times 60 \times 60} 4.4 \times 10^{20} \frac{\text{disintegration}}{s} = 7437.65 \frac{\text{disintegration}}{s}$$

$$= \frac{7437.65}{3.7 \times 10^{10}} \, Curie = 2 \times 10^{-7} \, Curie$$

**Exercise 2.30** *In a sample of archeological wood, the ratio of* $C^{14}$ *to* $C^{12}$ *is only* $\frac{1}{8}$ *of what is found in a sample of present-day wood. The half-life of* $C^{14}$ *is* $5570 \, yrs$. *What is the age of the archeological wood?*

[Ans] $C^{14}$ decays $T_{1/2} = 5570 \, yr$.
According to survival equation $N = N_0 e^{-\lambda t} \Rightarrow \frac{N}{N_0} = e^{-\lambda t}$

$$\frac{1}{8} = e^{-\lambda t} \Rightarrow e^{\lambda t} = 8 \Rightarrow \lambda t = \ln 8.$$

$$\lambda = \frac{0.693}{T_{1/2}} = \frac{0.693}{5570 \ yr}. \ \text{So}$$

$$\frac{0.693}{5570 \ yr} t = ln8$$

$$t = \frac{5570 \ yr}{0.693} ln8 = 16\,714 \ yr.$$

**Exercise 2.31** *A nucleus decays by emission of a gamma from*
*(i) an excited state 2$^+$ to the ground state 0$^+$*
*(ii) an excited state 3$^+$ to the state 1$^-$*
*Identify the transition type:*
*(a) electric dipole, (b) magnetic dipole, (c) electric quadrupole, (d) magnetic quadrupole.*

[Ans] (i) We employ the angular momentum selection rule.
The transition $2^+ \rightarrow 0^+$ means
$l \rightarrow (2 + 0)$ to $| \ 2 - 0 \ |$ in unit steps, i.e. $l = 2$.
This corresponds to the multipole transitions
$E2$( electric quadrupole transiton), $M2$( magnetic quadrupole transition).
We now employ parity selection rule.
The transition $2^+ \rightarrow 0^+$ means parity does not change. So $l =$ even for electric multipole transition $El$ and $l =$ odd for magnetic multipole transition $Ml$.
This means the allowed $\gamma$ transition is $E2$ (electric quadrupole transition).
(ii) We employ angular momentum selection rule.
The transition $3^+ \rightarrow 1^-$ means
$l \rightarrow (3 + 1)$ to $| \ 3 - 1 \ |$ in unit steps, i.e. $l = 4, 3, 2$.
This corresponds to the multipole transitions
$E4, \ E3, \ E2$ ( electric multipole transitons) $M4, \ M3, \ M2$( magnetic multipole transitions).
We now employ parity selection rule.
The transition $3^+ \rightarrow 1^-$ means parity changes. So $l =$ odd for electric multipole transition $El$ and $l =$ even for magnetic multipole transition $Ml$. This means the allowed $\gamma$ transitions are $E3, M4, M2$. The magnetic quadrupole transition $M2$ is most probable.

**Exercise 2.32** *Show that the decay $\gamma \rightarrow e^+ + e^-$ in vacuum violates 4-momentum conservation and hence cannot proceed (in vacuum).*

[Ans] Consider 4-momentum conservation in vacuum for the decay

$$\gamma \rightarrow e^+ + e^- \ \text{(if possible)}$$

$$(p_\gamma)_\mu = (p_{e^+})_\mu + (p_{e^-})_\mu$$

where $\mu \to 1, 2, 3, 4 \equiv i, 4$ in the $diag(1, 1, 1, 1)$ representation of Poincaré

$$p_\mu = (p_i, p_4) = \left(\vec{p}, \frac{iE}{c}\right) = \text{4-momentum vector (time-like)}.$$

Squaring we get

$$(p_\gamma)_\mu(p_\gamma)_\mu = \left[(p_{e^+})_\mu + (p_{e^-})_\mu\right]\left[(p_{e^+})_\mu + (p_{e^-})_\mu\right].$$

$$(p_\gamma)_\mu(p_\gamma)_\mu = (p_{e^+})_\mu(p_{e^+})_\mu + (p_{e^-})_\mu(p_{e^-})_\mu + 2(p_{e^+})_\mu(p_{e^-})_\mu.$$

Use $p_\mu p_\mu = -m_0^2 c^2$ (time like), $m_0 = $ rest mass of particle. So

$$-(m_\gamma)_0^2 c^2 = -(m_{e^+})_0^2 c^2 - (m_{e^-})_0^2 c^2 + 2[(p_{e^+})_i(p_{e^-})_i + (p_{e^+})_4(p_{e^-})_4].$$

Using $(m_\gamma)_0 = 0$ (as photon rest mass is zero) and $(m_{e^+})_0 = (m_{e^-})_0 = (m_e)_0$ we have

$$0 = -(m_e)_0^2 c^2 - (m_e)_0^2 c^2 + 2[(p_{e^+})_i(p_{e^-})_i + (p_{e^+})_4(p_{e^-})_4]$$

$$0 = -(m_e)_0^2 c^2 - (m_e)_0^2 c^2 + 2(\vec{p}_{e^+} \cdot \vec{p}_{e^-} + \frac{iE_{e^+}}{c}\frac{iE_{e^-}}{c})$$

$$2(m_e)_0^2 c^2 = 2(p_{e^+}p_{e^-}\cos\theta - \frac{E_{e^+}E_{e^-}}{c^2})$$

$$(m_e)_0^2 c^2 = p_{e^+}p_{e^-}\cos\theta - \frac{E_{e^+}E_{e^-}}{c^2}. \text{ So}$$

$$\cos\theta = \frac{1}{p_{e^+}p_{e^-}}[\frac{E_{e^+}E_{e^-}}{c^2} + (m_e)_0^2 c^2]$$

$$\cos\theta = \frac{E_{e^+}E_{e^-} + (m_e)_0^2 c^4}{p_{e^+}p_{e^-}c^2}.$$

Let us use the energy formula

$$E = \sqrt{p^2 c^2 + m_0^2 c^4} \text{ to get}$$

$$\cos\theta = \frac{\sqrt{p_{e^+}^2 c^2 + (m_e)_0^2 c^4}\sqrt{p_{e^-}^2 c^2 + (m_e)_0^2 c^4} + (m_e)_0^2 c^4}{p_{e^+}p_{e^-}c^2} \geqslant 1$$

This is impossible.

So the decay $\gamma \to e^+ + e^-$ cannot conserve 4-momentum and hence cannot occur in vacuum. Pair production of $e^+$, $e^-$ can take place in the vicinity of a nucleus as

$$\gamma + N \rightarrow e^+ + e^- + N \ (N \rightarrow \text{nucleus}).$$

From momentum conservation we can write (in the centre of mass frame)

$$\vec{p}_\gamma^{\,cm} + \vec{p}_N^{\,cm} = \vec{p}_{e^+}^{\,cm} + \vec{p}_{e^-}^{\,cm} + \vec{p}_N^{\,'cm} = 0$$

This is possible because nucleus $N$ absorbs the momentum change and the reaction proceeds.

**Exercise 2.33** *A 40 keV electron collides with a positron of same energy. What are the products and their energy?*
$\boxed{Ans}$ The reaction is

$$e^+ + e^- \rightarrow \gamma + \gamma.$$

Energy conservation

$$(m_{e^+}c^2 + T_{e^+}) + (m_{e^-}c^2 + T_{e^-}) = E_\gamma + E_\gamma$$

$$(0.51 \ MeV + 40 \ keV) + (0.51 \ MeV + 40 \ keV) = E_\gamma + E_\gamma$$

$$2E_\gamma = 1.02 \ MeV + 80 \ keV = 1.02 \times 10^6 \ eV + 80 \times 10^3 \ eV$$

$$= 1.1 \times 10^6 \ eV = 1.1 \ MeV$$

$$E_\gamma = 0.55 \ MeV.$$

So the products are two photons of energy $0.55 \ MeV$ each.

**Exercise 2.34** *What is Auger effect?*
$\boxed{Ans}$ Suppose $EC$ occurs. A $K$ electron gets captured by nucleus. A vacancy is produced in the $K$-shell. Suppose a higher level electron ($i$ state) makes a downward transition $i \rightarrow K$ and so energy released is $E_i - E_K$ that falls in the $X$-region. Two possibilities are there.
- A photon carries away this energy $h\nu = E_i - E_K$. This is called $X$-fluorescence. It is a line spectrum ($\nu =$ constant).
- ✓ The energy released $E_i - E_K$ is converted to an electron, i.e. the energy released $E_i - E_K$ is directly associated with the release of an electron in the $f$ orbit (say). This $f$ electron is thus ejected with kinetic energy

$$T = E_i - E_K - (B.\ E)_f.$$

This is called Auger effect. The electron ejected is called Auger electron (figure 2.40). Auger effect resembles internal conversion of nucleus.

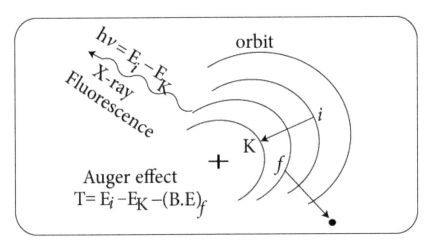

**Figure 2.40.** Auger effect.

**Exercise 2.35** *Show that in $\beta^-$ transformation*

$$_ZX^A \rightarrow _{Z+1}Y^A + \beta^- + \bar{\nu}_e$$

*the kinetic energy of recoil of Y is given by*

$$E_Y = \frac{Q + 2m_0c^2}{2M_Yc^2}T_{\max}$$

*where $Q$ is the $\beta$ disintegration energy, $T_{\max}$ is the maximum kinetic energy of the $\beta$ particle. Assume the recoil to be non-relativistic.*

[Ans] The $\beta$ disintegration energy is shared between the recoil nucleus $Y$, the $\beta^-$ particle and the electron anti-neutrino. In the case where kinetic energy of $\bar{\nu}_e$ is zero ($T_{\bar{\nu}_e} \cong 0$) the kinetic energy of the $\beta^-$ particle is maximum—say $T_{\max}$. Now
$Q = T_Y + T_{\max} \Rightarrow T_{\max} = Q - T_Y$ ($T_Y$ = kinetic energy of recoil nucleus).
Conservation of linear momentum gives with $p_{\bar{\nu}_e} \cong 0$

$$\vec{p}_Y + \vec{p}_\beta = 0 \Rightarrow \vec{p}_Y = -\vec{p}_\beta \Rightarrow p_Y^2 = p_\beta^2.$$

Again $p_Y = \sqrt{2M_YT_Y}$ (We use classical relation as the daughter mass $M_Y$ = large). Total energy of $\beta$ particle (rest mass $m_0$) is

$$T_{\max} + m_0c^2 = \sqrt{p_\beta^2c^2 + m_0^2c^4}.$$

Squaring

$$(T_{\max} + m_0c^2)^2 = p_\beta^2c^2 + m_0^2c^4.$$

This leads to

$$T_{max}^2 + 2T_{max}m_0c^2 + m_0^2c^4 = p_\beta^2 c^2 + m_0^2 c^4$$

$$p_\beta^2 = \frac{2T_{max}m_0c^2 + T_{max}^2}{c^2} = \frac{T_{max}(T_{max} + 2m_0c^2)}{c^2}$$

Hence

$$p_Y^2 = p_\beta^2 = 2M_Y T_Y = \frac{T_{max}(T_{max} + 2m_0c^2)}{c^2} = \frac{T_{max}(Q - T_Y + 2m_0c^2)}{c^2}.$$

As $T_Y$ is small compared to $Q$ and $2m_0c^2$ we neglect it. So

$$2M_Y T_Y \approx \frac{T_{max}(Q + 2m_0c^2)}{c^2}$$

$$T_Y = \frac{T_{max}(Q_Y + 2m_0c^2)}{2M_Yc^2}.$$

**Exercise 2.36** *Calculate the maximum energy of antineutrinos emitted in the beta decay of $_5B^{12}$. The atomic masses are* 12.014 44u *for* $_5B^{12}$ *and* 12.0000u *for* $_6C^{12}$.
[Ans] Beta decay of $_5B^{12}$ is

$$_5B^{12} = {}_6C^{12} + {}_{-1}e^0 + \bar{\nu}_e.$$

Energy released is

$$T_0 = (\text{atomic mass of } _5B^{12}) - (\text{atomic mass of } _6C^{12})$$

$$= 12.014\,44u - 12.000u = 0.014\,44 \times 931 \; MeV = 13.443\,64 \; MeV$$

This energy $T_0$ is shared by the lepton pair $e^-$, $\bar{\nu}_e$. Again when $e^-$ carries no kinetic energy this energy $T_0$ is carried off by $\bar{\nu}_e$.

So the maximum energy of antineutrino is 13.443 64 $MeV$.

**Exercise 2.37** *The specific charge of a beta particle is*
*(a)* $1.76 \times 10^{11}C.kg^{-1}$ *(b)* $1.6 \times 10^{-19}C$ *(c)* $1.6 \times 10^{-11}C$ *(d)* $1.76 \times 10^{-11}C.kg^{-1}$
Hint. Specific charge of electron is
$$\frac{|e|}{m_e} = \frac{1.6 \times 10^{-19}\,C}{9.1 \times 10^{-31}\,kg} = 1.76 \times 10^{11} \; C.kg^{-1}$$

**Exercise 2.38** *Does the specific charge of electron change with electronic movement?*
*(a) No change, (b) decreases, (c) increases, (d) decreases only for speed* $\sim c$.

Hint. Mass changes with speed as $=\dfrac{m_0}{\sqrt{1-\frac{v^2}{c^2}}}$. So as speed increases, mass increases

so specific charge $\dfrac{|e|}{m}$ decreases.

**Exercise 2.39** *The beta line spectrum is due to the emission of*
(a) *electron,* (b) *positron,* (c) *both (beta particles),* (d) *internal electron.*

**Exercise 2.40** *Identify the correct choice regarding internal conversion.*
(a) *radiative process,*
(b) *non-radiative process,*
(c) *involves real photons,*
(d) *involves virtual photons.*

**Exercise 2.41** *The parity selection rules for $\gamma$ decay are $\pi^f = \pi^i(-)^l, \pi^f = \pi^i(-)^{l+1}$ where $\pi \to$ parity eigenvalue, $l \to$ angular momentum quantum number, $i \to$ initial state, $f \to$ final state. Identify the nature of $\gamma$ transition.*
(a) *El, Ml;* (b) *Ml, El,* (c) *El + 1, Ml;* (d) *Ml + 1, El.*

**Exercise 2.42** *Suppose a radioactive nucleus is active (i.e. escaped decay) at $t = 0$. What is the probabilty that it would decay in the next time interval dt ?*
Ans Probability $= \lambda dt$ since $\lambda =$ probability of radioactive decay per unit time.

**Exercise 2.43** *Show the relation between $Q_\alpha$ and $T_{1/2}$ as per the Geiger–Nuttall relation.*
Ans $Q_\alpha$ for $\alpha$ decay (that varies from 4 to 10 *MeV*) and half-life $T_{1/2}$ (that varies from $10^{-6}$ to $10^{17}$ *s*) are related as shown in figure 2.41 and is referred to as the

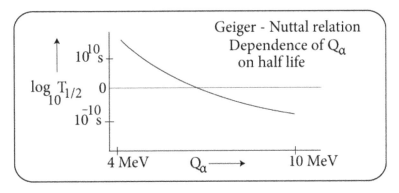

**Figure 2.41.** Dependence of $Q_\alpha$ on half-life $T_{1/2}$.

Geiger–Nuttall relation. The relation between them is of inverse nature. It follows that if $Q_\alpha$ is varied by a little amount $T_{1/2}$ changes by a huge amount.

**Exercise 2.44** *Which electron 1s, 2s, 2p is more likely to be captured by nucleus in EC?*

[Ans] As 1s electron is closer to the nucleus it is most likely to be absorbed by the nucleus. Among 2s and 2p electrons we note that the wave function corresponding to a 2s electron is non-zero at origin while the wave function for a 2p electron is zero at origin. In fact, the wave function of s electrons overlaps with the wave function of the nucleus, i.e. spends more time close to the nucleus and so the 2s electron is more likely to be captured by the nucleus. On the other hand, the 2p electron wave function is zero at the origin. So the 2p electron has less likelihood of being captured.

**Exercise 2.45** *Most penetrative is*
    *(a) α, (b) β, (c) γ, (d) ₁H².*

**Exercise 2.46** *Most ionizing is*
    *(a) α, (b) β, (c) γ, (d) ₁H².*

**Exercise 2.47** *In β decay the transition $2^+ \rightarrow 3^+$ is*
    *(a) allowed both by Fermi and Gammow–Teller selection rule,*
    *(b) allowed by Fermi selection rule and not allowed by Gammow–Teller selection rule,*
    *(c) not allowed by Fermi selection rule but allowed by Gammow–Teller selection rule,*
    *(d) not allowed both by Fermi and Gammow–Teller selection rule.*

Hint. Transition $2^+ \rightarrow 3^+$ means parity does not change. $\Delta\pi = 0$. So it is an allowed transition.

Again angular momentum selection rules gives $\Delta J = 1, 2, 3, 4, 5$.

The selection rule for allowed Fermi transition is $\Delta J = 0$ and for allowed Gammow–Teller transition is $\Delta J = 0, \pm 1$.

Occurrence of $\Delta J = 1$ means that the transition $2^+ \rightarrow 3^+$ is allowed Gammow–Teller transition.

✓ We can make a physical verification also. The transition is

$$\vec{J_P} \rightarrow \vec{J_D} + \vec{l} + \vec{s_e} + \vec{s_\nu} \Rightarrow 2^+ \rightarrow 3^+ + \vec{l} + \vec{s_e} + \vec{s_\nu} \tag{2.300}$$

For allowed Fermi transition $l = 0$, $\vec{s}_e + \vec{s}_\nu = \vec{0}$ and so
RHS of equation (2.300) is $3 + 0 + 0 \rightarrow 3$. LHS $= 2$
As LHS $\neq$ RHS it is not an allowed Fermi transition.
For an allowed Gammow–Teller transition $l = 0$, $\vec{s}_e + \vec{s}_\nu = \vec{1}$ and so
RHS of equation (2.300) is $3 + 0 + 1 \rightarrow 2, 3, 4$, i.e. 2 can be generated.
As LHS $= 2 =$ RHS it is an allowed Gammow–Teller transition.

**Exercise 2.48** *The β decay transition*

$$_{18}\text{Ar}^{39} \rightarrow {}_{19}\text{K}^{39} + e^- + \bar{\nu}_e$$

*is a transition* $\frac{7}{2}^- \rightarrow \frac{3}{2}^+$.
*Then it is*
 *(a) allowed both by Fermi and Gammow–Teller selection rule,*
 *(b) allowed by Fermi selection rule,*
 *(c) the first forbidden Fermi selection rule,*
 *(d) the first forbidden Gammow–Teller selection rule.*

Hint. Here $J_P = \frac{7}{2}$, $J_D = \frac{3}{2}$. Transition $\frac{7}{2}^- \rightarrow \frac{3}{2}^+$ means parity changes. $\Delta\pi \neq 0$. So
it is not an allowed transition. We can take it to be the first forbidden transition.
Again, addition of angular momentum gives $| \vec{J}_P - \vec{J}_D | = \Delta J = 2, 3, 4, 5$.
The selection rule for the first forbidden Fermi transition is $\Delta J = 0, \pm 1$ and for
the first forbidden Gammow–Teller transition is $\Delta J = 0, \pm 1, \pm 2$.
So the transition is the first forbidden Gammow–Teller transition.
✓ We can make a physical verification also. The transition is

$$\vec{J}_P \rightarrow \vec{J}_D + \vec{l} + \vec{s}_e + \vec{s}_\nu \Rightarrow \frac{7}{2} \rightarrow \frac{3}{2} + \vec{l} + \vec{s}_e + \vec{s}_\nu \qquad (2.301)$$

For the first forbidden Fermi transition $l = 1$, $\vec{s}_e + \vec{s}_\nu = \vec{0}$ and so
RHS of equation (2.301) is $\frac{3}{2} + \vec{1} + \vec{0} \rightarrow \frac{5}{2}, \frac{3}{2}, \frac{1}{2}$. LHS $= \frac{7}{2}$.
As LHS $\neq$ RHS it is not the first forbidden Fermi transition.
For the first forbidden Gammow–Teller transition $l = 1$, $\vec{s}_e + \vec{s}_\nu = \vec{1}$ and so
RHS of equation (2.301) is $\frac{3}{2} + \vec{1} + \vec{1} \rightarrow \frac{3}{2} + 2, 1, 0 \rightarrow \frac{7}{2}$ etc, i.e. $\frac{7}{2}$ can be generated.
As LHS $= \frac{7}{2} =$ RHS it is the first forbidden Gammow–Teller transition.

**Exercise 2.49** *The β decay* $_1\text{H}^3 \rightarrow {}_2\text{He}^3 + e^- + \bar{\nu}_e$ *is a* $\frac{1}{2}^+ \rightarrow \frac{1}{2}^+$ *transition. It is*
 *(a) allowed both by Fermi and Gammow–Teller selection rule,*
 *(b) allowed by Fermi selection rule and not allowed by Gammow–Teller selection rule,*

*not allowed by Fermi selection rule but allowed by Gammow–Teller selection rule, not allowed both by Fermi and Gammow–Teller selection rule.*

Hint. Here $J_P = \frac{1}{2}$, $J_D = \frac{1}{2}$. Transition $\frac{1}{2}^+ \to \frac{1}{2}^+$ means parity does not change. $\Delta \pi = 0$. So it is an allowed transition.

Again, addition of angular momentum gives $| \vec{J}_P - \vec{J}_D | = \Delta J = 0$.

The selection rule for the allowed Fermi transition is $\Delta J = 0$ and for the allowed Gammow–Teller transition is $\Delta J = 0, \pm 1, 0 \to 0$ forbidden.

Occurrence of $\Delta J = 0$ means that the transition $\frac{1}{2}^+ \to \frac{1}{2}^+$ is allowed both by Fermi and Gammow–Teller selection rule.

✓ We can make a physical verification also. The transition is

$$\vec{J}_P \to \vec{J}_D + \vec{l} + \vec{s}_e + \vec{s}_\nu \Rightarrow \frac{\vec{1}}{2} \to \frac{\vec{1}}{2} + \vec{l} + \vec{s}_e + \vec{s}_\nu \qquad (2.302)$$

For allowed Fermi transition $l = 0$, $\vec{s}_e + \vec{s}_\nu = \vec{0}$ and so

RHS of equation (2.302) is $\frac{\vec{1}}{2} + \vec{0} + \vec{0} \to \frac{1}{2}$.

As LHS $= \frac{1}{2}$ =RHS it is allowed by Fermi transition.

For allowed Gammow–Teller transition $l = 0$, $\vec{s}_e + \vec{s}_\nu = \vec{1}$ and so

RHS of equation (2.302) is $\frac{\vec{1}}{2} + \vec{0} + \vec{1} = \frac{1}{2}, \frac{3}{2}$, i.e. $\frac{1}{2}$ can be generated.

As LHS $= \frac{1}{2}$ = RHS, hence it is an allowed Gammow–Teller transition also.

**Exercise 2.50** *The $\beta$ decay $0^+ \to 1^+ + e^- + \bar{\nu}_e$ is*

(a) *allowed both by Fermi and Gammow–Teller selection rule,*

(b) *allowed by Fermi selection rule and not allowed by Gammow–Teller selection rule,*

(c) *not allowed by Fermi selection rule but allowed by Gammow–Teller selection rule,*

(d) *not allowed both by Fermi and Gammow–Teller selection rule.*

Hint. Here $J_P = 0$, $J_D = 1$. Transition $0^+ \to 1^+$ means parity does not change. $\Delta \pi = 0$. So it is an allowed transition.

The angular momentum addition rule gives $| \vec{J}_P - \vec{J}_D | = \Delta J = 1$.

The selection rule for an allowed Fermi transition is $\Delta J = 0$ and for an allowed Gammow–Teller transition is $\Delta J = 0, \pm 1, 0 \to 0$ forbidden.

Occurrence of $\Delta J = 1$ means that the transition $0^+ \to 1^+$ is not allowed by Fermi but allowed by Gammow–Teller selection rule.

✓ We can make a physical verification also. The transition is

$$\vec{J}_P \rightarrow \vec{J}_D + \vec{l} + \vec{s}_e + \vec{s}_\nu \Rightarrow \vec{0} \rightarrow \vec{1} + \vec{l} + \vec{s}_e + \vec{s}_\nu \tag{2.303}$$

For allowed Fermi transition $l = 0$, $\vec{s}_e + \vec{s}_\nu = \vec{0}$ and so
RHS of equation (2.303) is $\vec{1} + \vec{0} + \vec{0} \rightarrow 1$. LHS = 0
As LHS $\neq$ RHS it is not allowed by Fermi transition.
For an allowed Gammow–Teller transition $l = 0$, $\vec{s}_e + \vec{s}_\nu = \vec{1}$ and so
RHS of equation (2.303) is $\vec{1} + \vec{0} + \vec{1} \rightarrow 2, 1, 0$, i.e. 0 can be generated.
As LHS $= 0 =$ RHS it is allowed by Gammow–Teller selection rule.

**Exercise 2.51** *What are the possible spin parity values of a $\frac{3}{2}^+$ state that decays by the first forbidden $\beta$ transition.?*

Ans. Let the spin parity of the final state be $J^\pi$. Then the decay is

$$\frac{3^+}{2} \rightarrow J^\pi + e^- + \bar{\nu}_e$$

First forbidden transition means $l = 1$ which means parity changes. So $J^\pi = J^-$
The transition can be written as

$$\vec{J}_P \rightarrow \vec{J}_D + \vec{l} + \vec{s}_e + \vec{s}_\nu \Rightarrow \frac{\vec{3}}{2} \rightarrow J^\pi + \vec{l} + \vec{s}_e + \vec{s}_\nu \tag{2.304}$$

For the first forbidden Fermi transition $l = 1$, $\vec{s}_e + \vec{s}_\nu = \vec{0}$ and so from equation (2.304)

$$\frac{\vec{3}}{2} \rightarrow J^\pi + \vec{1} + \vec{0} \Rightarrow J^\pi = \frac{\vec{3}}{2} - \vec{1} - \vec{0} = \frac{1}{2}, \frac{3}{2}, \frac{5}{2}$$

For the first forbidden Gammow–Teller transition $l = 1$, $\vec{s}_e + \vec{s}_\nu = \vec{1}$ and so from equation (2.304)

$$\frac{\vec{3}}{2} \rightarrow J^\pi + \vec{1} + \vec{1} \Rightarrow J^\pi = \frac{\vec{3}}{2} - \vec{1} - \vec{1} \rightarrow \frac{\vec{3}}{2} - 2, 1, 0, \text{ i.e.}$$

$$J^\pi = \frac{\vec{3}}{2} - \vec{2}, \frac{\vec{3}}{2} - \vec{1}, \frac{\vec{3}}{2} - \vec{0} \Rightarrow J^\pi = \rightarrow \frac{1}{2}, \frac{3}{2}, \frac{5}{2}, \frac{7}{2}; \frac{1}{2}, \frac{3}{2}, \frac{5}{2}; \frac{3}{2}$$

Hence $J^-$ values are $\frac{1}{2}^-, \frac{3}{2}^-, \frac{5}{2}^-, \frac{7}{2}^-$ (considering all the possibilities).

**Exercise 2.52** *Find the nature of $\beta$ decay $_{16}S^{35} \rightarrow _{17}Cl^{35} + e^- + \bar{\nu}_e$.*
Ans. We first find the spin parity $J^\pi$ of parent and daughter.
Let us find the spin parity of $_{16}S^{35}_{19}$

Shell structure of $16p$ (section 3.20)

$$(1s_{1/2})^2(1p_{3/2})^4(1p_{1/2})^2(1d_{5/2})^6(2s_{1/2})^2. \text{ This is } 0^+.$$

Shell structure of $19n$

$$(1s_{1/2})^2(1p_{3/2})^4(1p_{1/2})^2(1d_{5/2})^6(2s_{1/2})^2(1d_{3/2})^3. \text{ This leads to } J^\pi = \frac{3^+}{2}$$

Let us find the spin parity of $_{17}Cl_{18}^{35}$
Shell structure of $18n$

$$(1s_{1/2})^2(1p_{3/2})^4(1p_{1/2})^2(1d_{5/2})^6(2s_{1/2})^2(1d_{3/2})^4. \text{ This is } 0^+$$

Shell structure of $17p$

$$(1s_{1/2})^2(1p_{3/2})^4(1p_{1/2})^2(1d_{5/2})^6(2s_{1/2})^2(1d_{3/2})^1. \text{ This leads to } J^\pi = \frac{3^+}{2}$$

With this let us rewrite the equation $_{16}S^{35} \rightarrow {}_{17}Cl^{35} + e^- + \bar{\nu}_e$ as

$$\frac{3^+}{2} \rightarrow \frac{3^+}{2} + e^- + \bar{\nu}_e.$$

This $\frac{3^+}{2} \rightarrow \frac{3^+}{2}$ $\beta$ transition is both an allowed Fermi as well as an allowed Gammow–Teller transition as shown in *Exercise 2.55*.


**Exercise 2.53** *Ground state of $_7N^{14}$ has spin 1 but excited state of $_7N^{14}$ has spin 0 . Consider the $\beta$ decay $_8O^{14} \rightarrow (_7N^{14})* + e^+ + \nu_e$ which is a $0^+ \rightarrow 0^+$ transition. It is*
   *(a) allowed both by Fermi and Gammow–Teller selection rule,*
   *(b) allowed by Fermi selection rule only,*
   *(c) allowed by Gammow–Teller selection rule,*
   *(d) not allowed both by Fermi and Gammow–Teller selection rule.*

Hint. $J_P = 0$, $J_D = 0$. The transition $0^+ \rightarrow 0^+$ means parity does not change. $\Delta\pi = 0$. So it is an allowed transition.
The angular momentum addition rule gives $| \vec{J}_P - \vec{J}_D |= \Delta J = 0$.
The selection rule for allowed Fermi transition is $\Delta J = 0$ and for allowed Gammow–Teller transition is $\Delta J = 0, \pm 1$, $0 \rightarrow 0$ forbidden.
So the transition $0^+ \rightarrow 0^+$ is allowed by Fermi selection rule.

✓ We can make a physical verification also. The transition is

$$\vec{J}_P \rightarrow \vec{J}_D + \vec{l} + \vec{s}_e + \vec{s}_\nu \Rightarrow \vec{0} \rightarrow \vec{0} + \vec{l} + \vec{s}_e + \vec{s}_\nu \tag{2.305}$$

For allowed Fermi transition $l = 0$, $\vec{s}_e + \vec{s}_\nu = \vec{0}$ and so
RHS of equation (2.305) is $\vec{0} + \vec{0} + \vec{0} \rightarrow 0$.
As LHS $= 0 =$ RHS it is an allowed Fermi transition.

For allowed Gammow–Teller transition $l = 0$, $\vec{s}_e + \vec{s}_\nu = \vec{1}$ and so
RHS of equation (2.305) is $\vec{0} + \vec{0} + \vec{1} \rightarrow 1$. LHS $= 0$
As LHS $\neq$ RHS it is not an allowed Gammow–Teller transition.

**Exercise 2.54** *Consider the* transition $_{32}Ge^{75} \rightarrow {}_{33}As^{75} + e^- + \bar{\nu}_e$ which is a $\frac{1}{2}^- \rightarrow \frac{3}{2}^-$ transition. It is allowed by *(a) Fermi, (b) Gammow–Teller, (c) both, (d) none.*

Hint. Here $J_P = \frac{1}{2}$, $J_D = \frac{3}{2}$. The transition $\frac{1}{2}^- \rightarrow \frac{3}{2}^-$ means parity does not change. $\Delta\pi = 0$. So it is an allowed transition.

Again, addition of angular momentum gives
$| \vec{J}_P - \vec{J}_D | = \Delta J = \frac{3}{2} - \frac{1}{2} = 1$, $\frac{1}{2} + \frac{3}{2} = 2$, i.e. 1, 2.

The selection rule for allowed Fermi transition is $\Delta J = 0$ and for allowed Gammow–Teller transition is $\Delta J = 0, \pm 1, 0 \rightarrow 0$ forbidden.

Occurrence of $\Delta J = 1$ means it is an allowed Gammow–Teller transition.

✓ We can make a physical verification also. The transition is

$$\vec{J}_P \rightarrow \vec{J}_D + \vec{l} + \vec{s}_e + \vec{s}_\nu \Rightarrow \frac{\vec{1}}{2} \rightarrow \frac{\vec{3}}{2} + \vec{l} + \vec{s}_e + \vec{s}_\nu \qquad (2.306)$$

For allowed Fermi transition $l = 0$, $\vec{s}_e + \vec{s}_\nu = \vec{0}$ and so
RHS of equation (2.306) is $\frac{\vec{3}}{2} + \vec{0} + \vec{0} \rightarrow \frac{3}{2}$. LHS $= \frac{1}{2}$.
As LHS $\neq$ RHS it is not an allowed Fermi transition.
For allowed Gammow–Teller transition $l = 0$, $\vec{s}_e + \vec{s}_\nu = \vec{1}$ and so
RHS of equation (2.306) is $\frac{\vec{3}}{2} + \vec{0} + \vec{1} \rightarrow \frac{1}{2}, \frac{3}{2}$, i.e. $\frac{1}{2}$ can be generated.
As LHS $= \frac{1}{2} =$ RHS it is an allowed Gammow–Teller transition.

**Exercise 2.55** *Consider the* transition $_{16}S^{35} \rightarrow {}_{17}Cl^{35} + e^- + \bar{\nu}_e$ which is a $\frac{3}{2}^+ \rightarrow \frac{3}{2}^+$ transition. It is allowed by *(a) Fermi, (b) Gammow–Teller, (c) both, (d) none.*

Hint. Here $J_P = \frac{3}{2}$, $J_D = \frac{3}{2}$. Transition $\frac{3}{2}^+ \rightarrow \frac{3}{2}^+$ means parity does not change. $\Delta\pi = 0$. So it is an allowed transition.

Again angular momentum addition rule gives $| \vec{J}_P - \vec{J}_D | = \Delta J = 0, 1, 2, 3$.

The selection rule for allowed Fermi transition is $\Delta J = 0$ and for allowed Gammow–Teller transition is $\Delta J = 0, \pm 1, 0 \rightarrow 0$ forbidden.

Occurrence of $\Delta J = 0$ means it can both be an allowed Fermi transition as well as an allowed Gammow–Teller transition.

✓ We can make a physical verification also. The transition is

$$\vec{J}_P \rightarrow \vec{J}_D + \vec{l} + \vec{s}_e + \vec{s}_\nu \Rightarrow \frac{\vec{3}}{2} \rightarrow \frac{\vec{3}}{2} + \vec{l} + \vec{s}_e + \vec{s}_\nu \qquad (2.307)$$

For allowed Fermi transition $l = 0$, $\vec{s}_e + \vec{s}_\nu = \vec{0}$ and so
RHS of equation (2.307) is $\frac{3}{2} + \vec{0} + \vec{0} \to \frac{3}{2}$.
As LHS $= \frac{3}{2} =$ RHS it is an allowed Fermi transition.
For allowed Gammow–Teller transition $l = 0$, $\vec{s}_e + \vec{s}_\nu = \vec{1}$ and so
RHS of equation (2.307) is $\frac{3}{2} + \vec{0} + \vec{1} \to \frac{1}{2}, \frac{3}{2}, \frac{5}{2}$, i.e. $\frac{3}{2}$ can be generated.
As LHS $= \frac{3}{2} =$ RHS it is an allowed Gammow–Teller transition also.

**Exercise 2.56** *Consider the* transition $_6C^{14} \to {}_7N^{14} + e^- + \bar{\nu}_e$ described by $0^+ \to 1^-$. It is

   *(a) First forbidden Fermi, (b) First forbidden Gammow–Teller, (c) both, (d) none.*
Hint. Here $J_P = 0$, $J_D = 1$.
Transition $0^+ \to 1^-$ means parity changes. $\Delta\pi \neq 0$. So it is the first forbidden transition $l = 1$.
Again angular momentum addition rule gives $| \vec{J}_P - \vec{J}_D | = \Delta J = 1$.
The selection rule for the first forbidden Fermi transition is $\Delta J = 0, \pm 1$ and for the first forbidden Gammow–Teller transition is $\Delta J = 0, \pm 1, \pm 2, 0 \leftrightarrow 1$ forbidden.
So it is a first forbidden Fermi transition.
✓ We can make a physical verification also. The transition is

$$\vec{J}_P \to \vec{J}_D + \vec{l} + \vec{s}_e + \vec{s}_\nu \Rightarrow \vec{0} \to \vec{1} + \vec{l} + \vec{s}_e + \vec{s}_\nu \tag{2.308}$$

For the first forbidden Fermi transition $l = 1$, $\vec{s}_e + \vec{s}_\nu = \vec{0}$ and so
RHS of equation (2.308) is $\vec{1} + \vec{1} + \vec{0} \to 2, 1, 0$, i.e. 0 can be generated.
As LHS $= 0 =$ RHS it is a first forbidden Fermi transition.
And $0 \leftrightarrow 1$ is not a permitted first forbidden Gammow–Teller transition.

**Exercise 2.57** *Consider the* transition $_{17}Cl^{38} \to {}_{18}Ar^{38} + e^- + \bar{\nu}_e$ described by $2^- \to 2^+$. It is

   *(a) allowed Fermi, (b) first forbidden Fermi,*
   *(c) first forbidden Gammow–Teller, (d) first forbidden Fermi, Gammow–Teller.*
Hint. Here $J_P = 2$, $J_D = 2$.
Transition $2^- \to 2^+$ means parity changes. $\Delta\pi \neq 0$. So it is first forbidden transition $l = 1$.
Again, angular momentum addition rule gives $| \vec{J}_P - \vec{J}_D | = \Delta J = 0$.
The selection rule for the first forbidden Fermi transition is $\Delta J = 0, \pm 1$ and for the first forbidden Gammow–Teller transition is $\Delta J = 0, \pm 1, \pm 2, 0 \leftrightarrow 1$ forbidden.
So the transition is both first forbidden Fermi transition as well as first forbidden Gammow–Teller transition.
✓ We can make a physical verification also. The transition is

$$\vec{J}_P \to \vec{J}_D + \vec{l} + \vec{s}_e + \vec{s}_\nu \Rightarrow \vec{2} \to 2 + \vec{l} + \vec{s}_e + \vec{s}_\nu \tag{2.309}$$

For first forbidden Fermi transition $l = 1$, $\vec{s}_e + \vec{s}_\nu = \vec{0}$ and so
RHS of equation (2.309) is $\vec{2} + \vec{1} + \vec{0} \to 3, 2, 1$, i.e. 2 is generated.

As LHS = 2 = RHS it is a first forbidden Fermi transition.

For first forbidden Gammow–Teller transition $l = 1$, $\vec{s}_e + \vec{s}_\nu = \vec{1}$ and so RHS of equation (2.309) is $\vec{2} + \vec{1} + \vec{1} \to \vec{2} + 2, 1, 0 \to 2$ can be generated.

As LHS = 2 = RHS it is the first forbidden Gammow–Teller transition also.

**Exercise 2.58** *Consider the* transition $_{11}\mathrm{Na}^{22} \to {}_{10}\mathrm{Ne}^{22} + e^- + \bar{\nu}_e$ *described by* $3^+ \to 0^+$. *It is*

*(a) allowed Fermi, (b) first forbidden Fermi,*

*(c) first forbidden Gammow–Teller, (d) second forbidden Gammow–Teller.*

Hint. Here $J_P = 3$, $J_D = 0$. Transition $3^+ \to 0^+$ means parity does not change. $\Delta\pi = 0$. So it can be second forbidden transition $l = 2$.

Again, angular momentum addition rule gives $|\vec{J}_P - \vec{J}_D| = \Delta J = 3$.

The selection rule for the second forbidden Fermi transition is $\Delta J = \pm l, \pm(l - 1) = \pm 2, 1$ and for the allowed Gammow–Teller transition it is $\Delta J = \pm l, \pm(l + 1) = \pm 2, \pm 3$ [equations (2.233) and (2.234)].

So it is the second forbidden Gammow–Teller transition.

✓ We can make a physical verification also. The transition is

$$\vec{J}_P \to \vec{J}_D + \vec{l} + \vec{s}_e + \vec{s}_\nu \Rightarrow \vec{3} \to \vec{0} + \vec{l} + \vec{s}_e + \vec{s}_\nu \tag{2.310}$$

For the second forbidden Fermi transition $l = 2$, $\vec{s}_e + \vec{s}_\nu = \vec{0}$ and so RHS of equation (2.310) is $\vec{0} + \vec{2} + \vec{0} \to 2$. LHS = 3

As LHS $\neq$ RHS it is not a second forbidden Fermi transition.

For the second forbidden Gammow–Teller transition $l = 2$, $\vec{s}_e + \vec{s}_\nu = \vec{1}$ and so RHS of equation (2.310) is $\vec{0} + \vec{2} + \vec{1} \to 3, 2, 1$, i.e. 3 can be generated.

As LHS = 3 = RHS it is a second forbidden Gammow–Teller transition.

**Exercise 2.59** *Show that the time $t_0$ in which type 2 nuclei reach a maximum in the radioactive chain $1 \to 2 \to 3$ where 1 and 2 are radioactive but 3 is stable end product is $t_0 \cong \sqrt{\tau_1 \tau_2}$ if the half-lives are nearly equal.*

Ans. For the radioactive chain $1 \to 2 \to 3$ where 1 and 2 are radioactive but 3 is stable end product we got equation (2.46) viz. $t_0 = \tau_2 \dfrac{(T_{1/2})_1}{(T_{1/2})_1 - (T_{1/2})_2} ln \dfrac{(T_{1/2})_1}{(T_{1/2})_2}$

Define $\dfrac{(T_{1/2})_1}{(T_{1/2})_2} = 1 + \delta$ where $\delta \ll 1$ as the half-lives are nearly equal.

$$(T_{1/2})_1 = (T_{1/2})_2(1 + \delta)$$

$$t_0 = \tau_2 \frac{(T_{1/2})_2(1 + \delta)}{(T_{1/2})_2 \delta} ln(1 + \delta) = \tau_2 \frac{1 + \delta}{\delta} ln(1 + \delta)$$

$$= \tau_2 \frac{1 + \delta}{\delta}\left(\delta - \frac{\delta^2}{2} + \frac{\delta^3}{3} - \ldots\right) = = \tau_2(1 + \delta)\left(1 - \frac{\delta}{2} + \frac{\delta^2}{3} - \ldots\right)$$

$$= \tau_2\left(1 - \frac{\delta}{2} + \delta - \ldots\right) = \tau_2(1 + \delta)^{1/2} = \tau_2\sqrt{\frac{(T_{1/2})_1}{(T_{1/2})_2}}$$

$$\text{Using } \tau = \frac{1}{\lambda} = \frac{T_{1/2}}{0.693}$$

$$t_0 = \tau_2\sqrt{\frac{\tau_1}{\tau_2}} = \sqrt{\tau_1\tau_2}$$

### Ans to Multiple Choice Questions

2.31 $c,d$, 2.37$a$, 2.38$b$, 2.39$a$, 2.40$b,d$, 2.41$a$, 2.45$c$, 2.46$a$, 2.47$c$, 2.48$d$, 2.49$a$, 2.50$c$, 2.53$b$, 2.54$b$, 2.55$c$, 2.56$a$, 2.57$d$, 2.58$d$.

## 2.40 Question bank

Q2.1 What are the processes by which an unstable excited nucleus goes to ground state? What is Soddy–Fajan's displacement law?

Q2.2 What are the three types of $\beta$ decay.?

Q2.3 Sketch the four radioactive series?

Q2.4 What is artificial radioactivity? Give an example.

Q2.5 What does survival equation indicate in the case of radioactive decay?

Q2.6 What happens in one half-life to a radioactive element?

Q2.7 What does decay constant of a radioactive substance indicate?

Q2.8 Define what is meant by mean life of a radioactive element. Obtain its value.

Q2.9 Define the term activity. Express activity in terms of Becquerel, Rutherford and Curie.

Q2.10 Obtain the amount of radioactive subtances 1, 2, 3 left in time $t$ in the decay

$$1 \rightarrow 2 \rightarrow 3$$

where elements 1, 2 are radioactive and 3 is stable.

Q2.11 Define the terms ideal equilibrium, transient equilibrium, secular equilibrium.

Q2.12 Discuss the characteristic features in the radioactive series

$$1 \rightarrow 2 \rightarrow 3$$

where elements 1, 2 are radioactive and 3 is stable if the parent 1 is (a) shortlived, (b) relatively long-lived, (c) extremely long-lived.

Q2.13 Discuss the differences between the terms radioactive dating, uranium dating, carbon dating.

Q2.14 Find alpha disintegration energy.

Q2.15 Discuss the mechanism of alpha decay. What prevents a classical explanation of $\alpha$ decay?

Q2.16 Give a quantum mechanical explanation of $\alpha$ decay.

Q2.17 What is quantum tunneling?

Q2.18 Obtain an estimate of collision frequency of $\alpha$ as it tries to emerge out of the nucleus.

Q2.19 Obtain an estimate of half-life of $_{84}Po^{212}$ from Gammow's theory of $\alpha$ decay.

Q2.20 Which of $\alpha$, $\beta$, $\gamma$ has the longest range?

Q2.21 Explain Geiger law, Geiger–Nuttall law.

Q2.22 Compare $\beta^{+}$ decay and EC.

Q2.23 How is it ensured which $\beta$ decay ($\beta^{\pm}$, EC) would occur? Obtain the relevant conditions considering the energetics.

Q2.24 Obtain the amount of energy liberated in the process of $\beta$ decay ($\beta^{\pm}$, EC).

Q2.25 How is it possible to achieve stability through beta decay of a nucleus with odd $A$, even $A$? Sketch relevent mass parabolas in this context.

Q2.26 Is double beta decay possible? Give an example.

Q2.27 How is it established that $\alpha$ spectra are discrete?

Q2.28 What is the nature of $\beta$ spectra? Discrete/continuous.

Q2.29 Explain the origin of discrete/continuous $\beta$ spectra.

Q2.30 How did Pauli neutrino hypothesis help explain $\beta$ spectra?

Q2.31 What is a neutrino? What role did it have in the explanation of continuous $\beta$ spectra?

Q2.32 Outline briefly Fermi's theory of beta decay.

Q2.33 What does a Fermi–Curie plot represent? Is it that all $\beta$ decays follow Fermi's theory? Comment.

Q2.34 Which selection rules govern $\beta$ decay?

Q2.35 Explain what is meant by Fermi transition, Gammow–Teller transition in $\beta$ decay.

Q2.36 Explain what we mean by allowed and forbidden $\beta$ transition?

Q2.37 How can one obtain neutrino mass from a study of $\beta$ spectra?

Q2.38 What is internal conversion? Is it a radiative/non-radiative process?

Q2.39 When does nuclear transition occur? When does $\alpha$, $\beta$, $\gamma$ decay occur?

Q2.40 What is the nature of the following spectra: $\alpha$ spectra, $\beta$ spectra, $\gamma$ spectra?

Q2.41 What are the selection rules of $\gamma$ decay?

Q2.42 What is meant by multipolarity of $\gamma$ transition?

Q2.43 Mention the parities of the following gamma transition $El$, $Ml$? Which transition will be more favoured?

Q2.44   Argue whether $0 \rightarrow 0$ transition can occur through gamma decay/internal conversion?

Q2.45   Discuss the nature (multipolarity) of the following $\gamma$ transitions: $1^- \rightarrow 1^+, \frac{1}{2}^+ \rightarrow \frac{9}{2}^+$.

Q2.46   Describe in brief what is Mössbauer effect?

Q2.47   What is meant by the phenomenon called pair production? Can it occur in vacuum?

## Further reading

[1]  Krane S K 1988 *Introductory Nuclear Physics* (New York: Wiley)

[2]  Tayal D C 2009 *Nuclear Physics* (Mumbai: Himalaya Publishing House)

[3]  Satya P 2005 *Nuclear Physics and Particle Physics* (New Delhi: Sultan Chand & Sons)

[4]  Guha J 2019 *Quantum Mechanics: Theory, Problems and Solutions* 3rd edn (Kolkata: Books and Allied (P) Ltd))

[5]  Lim Y-K 2002 *Problems and Solutions on Atomic, Nuclear and Particle Physics* (Singapore: World Scientific)

# Chapter 3

# Nuclear models

## 3.1 Introduction

The exact nature of nuclear force prevailing in the nucleus is not known with certainty. Nuclear force that binds nucleons in a nucleus is very complicated. It is vastly different from the electromagnetic or gravitational force which are inverse square in nature. In fact, there is no specific neat relation between the nuclear force and distance between two nucleons. The nucleus in general contains a large number of nucleons and hence it is a many-body problem. It is very difficult to develop a concrete theory of the nucleus based on rigorous treatment.

To explain the different properties of a nucleus, several models have been proposed. But none of the models explains all the properties satisfactorily. These models are different from each other in their structure and purpose and do not agree in all aspects with one another. The important thing is that each of the nuclear models is limited in scope and explains satisfactorily some extraordinary features of the nucleus but cannot explain the vast variety of features or too many properties of the nucleus.

In this chapter we deal with a few nuclear models like the Fermi gas model (section 3.2), liquid drop model (section 3.3), shell model (section 3.20), collective model (section 3.31)

## 3.2 Fermi gas model

The Fermi gas model was proposed by Enrico Fermi. This model resembles the free-electron gas model for conduction electrons in a metal. We give an outline of the model.

✓ Nucleons do not interact with one another and move independently in the nucleus. Individual wave functions are taken to be plane waves.
✓ Nucleons being fermions obey Pauli exclusion principle.

✓ The nucleus is taken to be composed of a degenerate Fermi gas of neutrons and protons—degenerate because the particles crowd into the lowest possible or allowable levels as per Pauli exclusion principle.

✓ The nucleons move freely within a spherical potential well.

✓ In the ground state all the lower energy levels are filled up to a certain level called Fermi energy level. All the states above the Fermi energy level are empty.

✓ The potential energy well is filled separately with nucleons of each type. We thus have two separate Fermi gases—Fermi gas of neutron and Fermi gas of proton.

Consider the phase space built of cells of volume $h^3$.

Configuration space volume is $V = \frac{4}{3}\pi R^3$

As $R = R_0 A^{1/3}$ is nuclear radius we can write

$$V = \frac{4}{3}\pi \left(R_0 A^{1/3}\right)^3 = \frac{4}{3}\pi R_0^3 A$$

which is the nuclear volume.

The momentum space volume is the volume of the annular region of radii lying between momenta $p$ and $p + dp$ given by $4\pi p^2\, dp$. So the phase space volume is

Configuration space volume × momentum space volume $= V.\, 4\pi p^2\, dp$

and number of cells in phase space is

$$\frac{\text{phase space volume}}{\text{volume per cell}} = \frac{V.\, 4\pi p^2\, dp}{h^3} \qquad (3.1)$$

Since nucleons (proton and neutron) are fermions with spin $s = \frac{1}{2}$ (and hence $m_s = \pm\frac{1}{2}$, multiplicity $= 2s + 1 = 2\frac{1}{2} + 1 = 2$) they can exist either in spin up state or in spin down state each corresponding to number of states given by equation (3.1). So total number of states in phase space is given by

$$dN = 2\frac{V.\, 4\pi p^2\, dp}{h^3}$$

Let us consider neutron states. The number of neutrons having momentum between $p$ and $p_{\max} = \sqrt{2m_n E_F^n}$ where $E_F^n$ is Fermi energy for neutrons, is given by

$$N = \int_0^{p\max} dN = \int_0^{p\max} 2.\frac{V.\, 4\pi p^2\, dp}{h^3}$$

$$N = \frac{8\pi V}{h^3} \int_0^{p\max} p^2\, dp = \frac{8\pi V p_{\max}^3}{3h^3} = \frac{V p_{\max}^3}{3\pi^2 \hbar^3}\left(\text{using } \hbar = \frac{h}{2\pi}\right) \qquad (3.2)$$

$$p_{\max} = (3\pi^2)^{1/3}\hbar(N/V)^{1/3} \tag{3.3}$$

Fermi energy for neutron gas is

$$E_F^n = \frac{p_{\max}^2}{2m_n} = \frac{1}{2m_n}\left[(3\pi^2)^{1/3}\hbar\left(\frac{N}{V}\right)^{1/3}\right]^2 = \frac{\hbar^2}{2m_n}(3\pi^2)^{2/3}\left(\frac{N}{V}\right)^{2/3}$$

$$E_F^n = \frac{\hbar^2}{2m_n}(3\pi^2)^{2/3}\left(\frac{N}{\frac{4}{3}\pi R_0^3 A}\right)^{2/3}$$

$$= \frac{\hbar^2}{2m_n R_0^2}\left(\frac{9\pi}{4}\right)^{2/3}\left(\frac{N}{A}\right)^{2/3} \tag{3.4}$$

Similarly for proton gas the Fermi energy is

$$E_F^p = \frac{p_{\max}^2}{2m_p} = \frac{\hbar^2}{2m_p R_0^2}\left(\frac{9\pi}{4}\right)^{2/3}\left(\frac{Z}{A}\right)^{2/3} \tag{3.5}$$

To get a rough estimate let us take $N = Z = \frac{A}{2}$, $R_0 = 1.2\,fm$

$$E_F^n = \frac{\hbar^2}{2m_n R_0^2}\left(\frac{9\pi}{4}\right)^{2/3}\left(\frac{N}{A}\right)^{2/3} = \frac{(\hbar c)^2}{2(m_n c^2)R_0^2}\left(\frac{9\pi}{4}\right)^{2/3}\left(\frac{1}{2}\right)^{2/3}$$

$$= \frac{(197\ MeV.fm)^2}{2(939\ MeV)(1.2\ fm)^2}\left(\frac{9\pi}{8}\right)^{\frac{2}{3}} = 33\ MeV = E_F^p \quad \text{(using exercise 1.2)}$$

However, in the case of protons, electrostatic repulsion occurs and so symmetry is destroyed. The model thus involves two separate gases contained in different wells and having different energy levels.

Taking rough average values $\frac{N}{A} = \frac{1}{1.8}$, $\frac{Z}{A} = \frac{1}{2.2}$, we have

$$E_F^n = \frac{\hbar^2}{2m_n R_0^2}\left(\frac{9\pi}{4}\right)^{2/3}\left(\frac{N}{A}\right)^{2/3} = \frac{(\hbar c)^2}{2(m_n c^2)R_0^2}\left(\frac{9\pi}{4}\right)^{2/3}\left(\frac{1}{1.8}\right)^{2/3} = \frac{(197\ MeV.fm)^2}{2(939\ MeV)(1.2\ fm)^2}\left(\frac{9\pi}{7.2}\right)^{2/3} = 36\ MeV$$

$$E_F^p = \frac{\hbar^2}{2m_p R_0^2}\left(\frac{9\pi}{4}\right)^{2/3}\left(\frac{N}{A}\right)^{2/3} = \frac{(\hbar c)^2}{2(m_p c^2)R_0^2}\left(\frac{9\pi}{4}\right)^{2/3}\left(\frac{1}{2.2}\right)^{2/3} = \frac{(197\ MeV.fm)^2}{2(938\ MeV)(1.2\ fm)^2}\left(\frac{9\pi}{8.8}\right)^{2/3} = 31\ MeV$$

where *exercise 1.2* viz. $\hbar c = 197\ MeV.fm$ has been used.

Using the mean binding energy per nucleon $8\frac{MeV}{nucleon}$ the Fermi potential wells are of depth $36 + 8 = 44\ MeV$ for neutrons and $31 + 8 = 39\ MeV$ for protons as shown in figure 3.1.

We note that a neutron is neutral but a proton is positively charged. The protons prevent entry of additional protons from outside and this is attributed to the existence of an additional Coulomb energy barrier $E_C$ in the proton well.

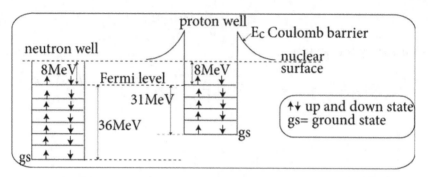

**Figure 3.1.** Square well potentials for neutron and proton in Fermi gas model.

Neutrons feel only attractive nuclear force. But protons feel in addition repulsive Coulomb force. For stable nuclei there will thus be an excess of neutrons. For this reason a neutron well is of greater depth than the proton well.

The Fermi energies for both neutrons and protons are represented by the same horizontal line. If the Fermi energy of the neutron well and proton well are at different depths below the top of the well then the nucleons of one type from higher Fermi level would make spontaneous transition to lower Fermi level for the other type of nucleons by $\beta$ decay so as to equalize the Fermi levels.

The Fermi gas model is not useful since it fails to predict the energy levels of excited states of nuclei accurately.

## 3.3 Liquid drop model

The liquid drop model was proposed by Bohr and Wheeler. Here we will see that some special combinations of the number of protons and number of neutrons will lead to a stable bound nucleus.

We shall work based upon certain assumptions. They are as follows.
   ✓ Nucleons (neutrons and protons) within the nucleus interact strongly. So the nuclear model we are going to discuss is a strong interaction model.
   ✓ The nucleus is analogous to a liquid drop.
   ✓ Nucleons are analogous to molecules of liquid.

In *exercise 3.4* we discuss the basis of these assumptions.

## 3.4 Bethe–Weizsacker's semi-empirical mass formula—based on liquid drop model

The relation between mass of a nucleus having $Z$ protons and $N$ neutrons and its binding energy $B(Z, A)$ is given by ($Z$ = atomic number, $A$ = mass number = $Z + N$)

$$M(Z, A) = Zm_p + Nm_n - \frac{B(Z, A)}{c^2} \tag{3.6}$$

where

$$B(Z, A) = a_V A - a_S A^{2/3} - a_C \frac{Z(Z-1)}{A^{1/3}} \tag{3.7}$$

with

$$a_V = 15.75 \, MeV, \, a_S = 17.8 \, MeV, \, a_C = 0.71 \, MeV \tag{3.8}$$

Including contributions from Pauli exclusion principle and quantum mechanical effects, two more terms are to be added to equation (3.7). Then we get

$$B(Z, A) = a_V A - a_S A^{2/3} - a_C \frac{Z(Z-1)}{A^{1/3}} - a_{\text{asym}} \frac{(A-2Z)^2}{A} + \frac{\delta}{A^{3/4}} \tag{3.9}$$

where

$$a_{\text{asym}} = 23.7 \, MeV, \, |\delta| = 34 \, MeV, \, \delta = \begin{cases} |\delta| & \text{for even} - \text{even nuclei} \\ 0 & \text{for odd A nuclei} \\ -|\delta| & \text{for odd} - \text{odd nuclei} \end{cases} \tag{3.10}$$

- Clearly $B$ is energy in $MeV$, $A$ is dimensionless. So $a$ has unit of energy, i.e. $MeV$.

## 3.5 Identification of various terms and explanation of Bethe–Weizsacker's semi-empirical mass formula

Bethe–Weizsacker's formula contains many terms and our task now is to identify and interpret each term. This will increase our understanding of the formula.

Figure 3.2 shows the experimentally obtained plot of binding energy per nucleon $\frac{B}{A}$ against mass number $A$.

We shall try to theoretically reproduce this curve through Bethe–Weizsacker's semi-empirical mass formula based on the liquid drop model.

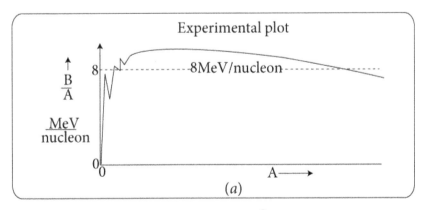

**Figure 3.2.** Experimental plot of $\frac{B}{A}$ against $A$.

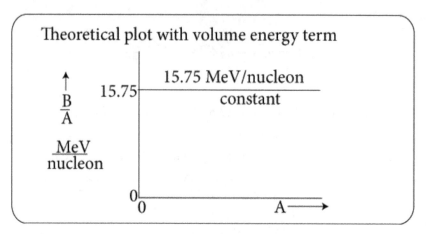

**Figure 3.3.** Theoretical plot of $\frac{B}{A}$ against $A$ considering volume energy term only.

## 3.6 Volume energy term

The first term in equations (3.7) and (3.9), i.e. $a_V A$ is called volume energy term.

The larger the size of nucleus, i.e. the larger the mass number $A$ of nucleus, the more difficult it will be to remove a nucleon (i.e. an individual proton or neutron) from the nucleus. Thus the binding energy is directly proportional to the total number of nucleons $A$. So we can write

$$B \propto A \;\Rightarrow\; B = a_V A \qquad (3.11)$$

where $a_V$ is a constant, its empirical value being $a_V = 15.75\ MeV$. This means that

$$\frac{B}{A} = \text{constant} = a_V = 15.75\ MeV \qquad (3.12)$$

This is the theoretical prediction—the plot has been shown in figure 3.3. We can compare this theoretical plot figure 3.2 with the experimental plot of figure 3.3.

We note that the actual experimental plot is different. $\frac{B}{A}$ is actually not a constant — $\frac{B}{A}$ varies with $A$. For $A \geqslant 16$ the average value of $\frac{B}{A}$ is approximately constant at $8\frac{MeV}{\text{nucleon}}$.

Volume energy term is the largest term in the binding energy expression.

To explain the actually observed $\frac{B}{A}$ versus $A$ plot we have to make some correction over this volume energy term.

## 3.7 Surface energy correction term

A nucleon at the interior of a nucleus is completely surrounded by other neighbouring nucleons—and pulled in all directions. This is pictorially shown for the nucleon marked 1 in figure 3.4(a). So it is difficult for the interior nucleon to escape.

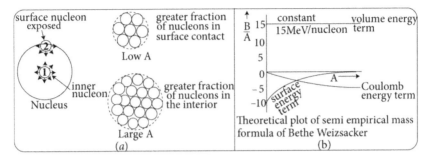

**Figure 3.4.** (a) Explaining surface effect. (b) Plot of volume energy term, surface energy correction term and Coulomb energy correction term in the $\frac{B}{A}$ versus $A$ plot.

However a surface nucleon is not completely surrounded by nucleons on all sides, as shown for the nucleon marked 2 in figure 3.4(a). A part of it is exposed and so the surface nucleon has a tendency to escape. This is called surface effect. This escaping tendency reduces the binding energy $B$. This surface energy term is to be subtracted from the volume energy term of equation (3.11). So we can attach a correction term to the RHS of equation (3.11) as

$$B = a_V A - \text{(surface effect)} \qquad (3.13)$$

Number of surface nucleons depends on surface area of the nucleus which is given by

$$4\pi R^2 = 4\pi (R_0 A^{1/3})^2 = 4\pi R_0^2 A^{2/3} \propto A^{2/3}. \qquad (3.14)$$

Hence the surface energy term is $a_S A^{2/3}$ where $a_S$ is a proportionality constant. Hence the modified expression of binding energy after surface energy correction is

$$B = a_V A - a_S A^{2/3} \qquad (3.15)$$

where $a_S = 17.8 \, MeV$ (empirical value). The surface energy correction term is plotted in figure 3.4(b).

*Exercise 3.5* discusses why surface effect is dominant for small $A$ nuclei.

## 3.8 Coulomb energy correction term

A nucleus contains multiple protons. They repel each other electrostatically as per Coulomb's law and try to destabilize the nucleus. This repulsive Coulomb interaction between protons tends to unbind them and hence decreases the binding energy.

The Coulomb interaction energy is proportional to

✓ number of interacting pairs of protons which is ${}^Z C_2 = \frac{1}{2} Z(Z-1)$

✓ $\dfrac{1}{\text{pp separation}} \cong \dfrac{1}{R} = \dfrac{1}{R_0 A^{1/3}}$

Thus the Coulomb interaction energy $\propto \dfrac{1}{2} Z(Z-1) . \dfrac{1}{R_0 A^{1/3}} = \text{constant} \dfrac{Z(Z-1)}{A^{1/3}}$ (3.16)

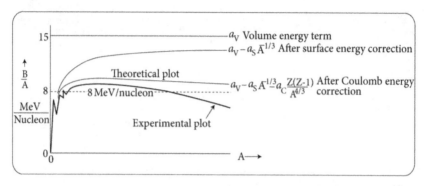

**Figure 3.5.** Relative contributions of various correction terms to the volume energy term in Bethe–Weizsacker's semi-empirical mass formula based on classical liquid drop model leading to theoretical plot of $\frac{B}{A}$ versus $A$.

Binding energy is weakened due to this term and so we have to subtract this term from the RHS of equation (3.15) to get

$$B \equiv B(Z, A) = a_V A - a_S A^{2/3} - a_C \frac{Z(Z - 1)}{A^{1/3}} \tag{3.17}$$

where $a_C = 0.71 \ MeV$ (empirical value). The Coulomb energy correction term is plotted in figure 3.4(b).

Equation (3.17) gives binding energy as predicted by Bethe–Weizsacker's semi-empirical mass formula based on the classical liquid drop model.

The theoretical plot of Bethe–Weizsacker's semi-empirical mass formula based on the classical liquid drop model that incorporates classical effects of volume energy along with modifications incorporating surface energy correction and Coulomb energy correction is depicted in figure 3.5. There are disagreements with the experimental plot and we have to improve upon the model further.

We introduce some quantum effects that are inescapable when we deal with a nuclear system.

## 3.9 Quantum mechanical corrections on Bethe–Weizsacker's semi-empirical mass formula

Bethe–Weizsacker's semi-empirical mass formula based on the classical liquid drop model cannot exactly reproduce the experimental plot of $\frac{B}{A}$ versus $A$ as shown in figure 3.5. This is because the nucleus is a quantum mechanical system and so we cannot leave out or ignore certain quantum mechanical principles that the nucleus obeys. Accordingly, to generate the experimentally observed $\frac{B}{A}$ versus $A$ plot we have to use the following quantum mechanical corrections.

  ✓ Pauli exclusion principle.
    The nucleus consists of protons and neutrons which are fermions. They obey Pauli exclusion principle which we have to consider.
  ✓ Spin dependence of nuclear force.

Nature allows only one bound two-nucleon system of deuteron in spin triplet state with a non-central potential. This means nuclear force depends on spin state which we have to consider.

## 3.10 Asymmetry energy correction term

The asymmetry energy correction term arises due to the asymmetry in the number of protons $Z$ and the number of neutrons $N$ in a nucleus.

Among the stable light nuclei there is a clear tendency for the number of protons $Z$ and number of neutrons $N$ to be equal.

Example

$$_6C^{12}(Z = 6, A = 12, N = 12 - 6 = 6)$$

$$_7N^{14}(Z = 7, A = 14, N = 14 - 7 = 7)$$

$$_8O^{16}(Z = 8, A = 16, N = 16 - 8 = 8)$$

For lighter nuclei the stability curve is the $45°$ line which is the $N = Z$ line. (figure 3.6(a)).

Let us investigate what happens to the stability aspect if we try to build a heavy nucleus, i.e. nucleus with large $A$ by populating it with an equal number of protons and neutrons, i.e. we investigate whether $N = Z$ is maintainable for large $A$.

If a large $A$ nucleus has an equal number of protons and neutrons then the number $N = Z = \frac{A}{2}$ is also large. This large number of protons will mutually repel and due to this large Coulomb repulsive $pp$ interaction the nucleus will become unstable. In fact, repulsive $pp$ Coulomb force will dominate over the attractive

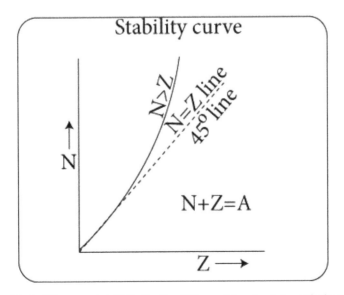

**Figure 3.6.** Stability curve is initially the $N = Z$ line and then bends towards the $N$ axis.

strong interactions between nucleons, i.e. over $pp$, $np$ and $nn$ strong interactions. This means the $N = Z$ relation cannot be maintained in a large $A$ nucleus for it to be stable. In other words a large $A$ nucleus cannot be symmetric in $N$ and $Z$.

We now try to build a stable nucleus with large $A$.

To build a stable nucleus with $A=$ large it is thus necessary to reduce the number of protons and also shield them from each other so that the repulsive Coulomb interaction becomes less. This is accomplished by taking off protons and introducing neutrons without affecting the total number $A$ of nucleons. This means a stable high $A$ nucleus requires asymmetry, i.e. neutron excess.

Now the neutrons are fermions. So these excess neutrons should be introduced into the system obeying Pauli exclusion principle. As lower energy levels are filled with neutrons these excess neutrons have to be placed in higher energy states.

Clearly the requirement of giving the excess neutrons higher energy means introducing instability into the system. Increase of instability (also called neutron instability) reduces binding energy. In a nutshell, we can say that:

    ✓ A stable high $A$ nucleus requires neutron–proton asymmetry $N > Z$. The $N$ versus $Z$ stability curve bends towards the $N$-axis (figure 3.6).

    ✓ Neutron–proton asymmetry ($N > Z$), i.e. putting excess neutrons to high energy states as per Pauli exclusion principle reduces binding energy of the nucleus. This correction term follows from Pauli exclusion principle and so it is a quantum mechanical effect.

We now illustrate the neutron–proton asymmetry through an example. We shall then deduce the amount of reduction of binding energy due to $np$ asymmetry. We first explain through an example and then address the problem in a more general manner.

    • We explain through a hypothetical example referring to figure 3.7.

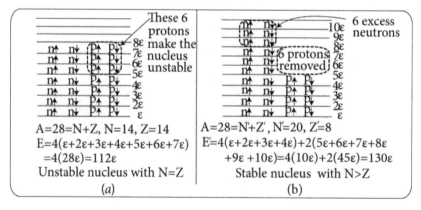

Figure 3.7. (a) Symmetry in $N$ and $Z$ in a high $A$ nucleus leads to instability. (b) Asymmetry in $N$ and $Z$ in a high $A$ nucleus leads to stability at the cost of reducing binding energy by a term called asymmetry energy.

Let us consider an $A = 28$ nucleus and assume that it is a very large number. For a nucleus, symmetric in $n$ and $p$ we should have $Z = \frac{A}{2} = \frac{28}{2} = 14$ protons and $N = \frac{A}{2} = \frac{28}{2} = 14$ neutrons. They populate the 7 lowest energy levels $\varepsilon$, $2\varepsilon$, $3\varepsilon$, $4\varepsilon$, $5\varepsilon$, $6\varepsilon$, $7\varepsilon$ according to Pauli exclusion principle and energy of system is $E = 112\varepsilon$ where $\varepsilon$ is energy spacing. This is shown in figure 3.7(a).

We assume in this example that too many protons are involved and the nucleus with $A = 28$, $Z = 14$, $N = 14$ is unstable (since we assume this number to be very large). We further assume that 6 protons make the nucleus $(A, Z, N) = (28, 14, 14)$ unstable, inviting their removal.

To keep $A = 28$ and to restore stability we have to remove the 6 protons and introduce 6 neutrons instead. This is shown in figure 3.7(b) and so the new stable nucleus will have $A = 28$, $Z' = Z - 6 = 14 - 6 = 8$, $N' = N + 6 = 14 + 6 = 20$. This makes the nucleus $(A, Z', N') = (28, 8, 20)$ stable as per our assumption. So neutron excess is 6 (in this example).

We define neutron excess as

$$\frac{N' - Z'}{2} = \frac{20 - 8}{2} = \frac{12}{2} = 6 \text{ (in this problem)}. \tag{3.18}$$

This number of protons should be taken off and replaced by neutrons to make the system stable.

These excess neutrons $\frac{N' - Z'}{2} = 6$ have to be accommodated in the system obeying Pauli exclusion principle. So they cannot sit in the $5\varepsilon$, $6\varepsilon$, $7\varepsilon$ levels (as these levels are already populated by neutrons). Clearly these excess 6 neutrons have to be accommodated in the higher levels viz. $8\varepsilon$, $9\varepsilon$, $10\varepsilon$. Then the energy of the system becomes $E' = 130\varepsilon$ (which is greater than $E = 112\varepsilon$). The asymmetry needed to stabilize high $A$ nucleus thus necessitates that some higher levels are populated by excess neutrons. Again populating higher levels of a system means greater excitation energy which lessens the binding, i.e. it reduces binding energy.

We note that the asymmetry energy is the increase in excitation energy by an amount given by

$$E' - E = 130\varepsilon - 112\varepsilon = 18\varepsilon \tag{3.19}$$

We are thus in a position to build a formula considering that neutron excess is $\frac{N' - Z'}{2} = 6$ and energy spacing is $\varepsilon$.

$$\frac{\text{Increase in excitation energy}}{\text{Neutron excess}} = \frac{18\varepsilon}{6} = 6\frac{1}{2}\varepsilon = (\text{neutron excess})\frac{1}{2}(\text{energy spacing})$$

Increase in excitation energy $= (\text{neutron excess})^2\frac{1}{2}(\text{energy spacing})$

$$= \left(\frac{N' - Z'}{2}\right)^2 \frac{1}{2}\varepsilon \tag{3.20}$$

Since

$$N' - Z' = (A - Z') - Z' = A - 2Z'$$

we can write from equation (3.20)

$$\text{Increase in excitation energy } = \frac{1}{8}(A - 2Z')^2\varepsilon \xrightarrow[Z'\to Z]{\text{redefine}} \frac{1}{8}(A - 2Z)^2\varepsilon \quad (3.21)$$

Again, the greater the number of nucleons $A$ in a nucleus, the smaller is the energy level spacing $\varepsilon$, i.e. $\varepsilon \propto \frac{1}{A}$. Taking this fact into account in equation (3.21) we have

$$\text{Asymmetry energy term } \propto \frac{(A - 2Z)^2}{A}. \quad (3.22)$$

Introducing the proportionality constant $a_{\text{asym}}$ we write

$$\text{Asymmetry energy correction term } = a_{\text{asym}}\frac{(A - 2Z)^2}{A} \quad (3.23)$$

- Let us now proceed in a general manner to find the asymmetry energy term.

Let energy level separation be $\varepsilon$ (assuming equispaced energy level structure). Pauli exclusion principle is to be followed.

Consider a large $A$ nucleus with $N = \frac{A}{2}, Z = \frac{A}{2}$, i.e. $N = Z$ (symmetric) as shown in figure 3.8(a). Let us discuss w.r.t. a base level or reference level marked $XX$. Both above and below the level $XX$ there are nucleons.

In the levels below the base level $XX$ suppose $\frac{A}{2} - \nu$ neutrons and $\frac{A}{2} - \nu$ protons are accommodated. The remaining $\frac{A}{2} - \left(\frac{A}{2} - \nu\right) = \nu$ neutrons and $\frac{A}{2} - \left(\frac{A}{2} - \nu\right) = \nu$ protons, i.e. $2\nu$ nucleons populate the upper levels. In each level

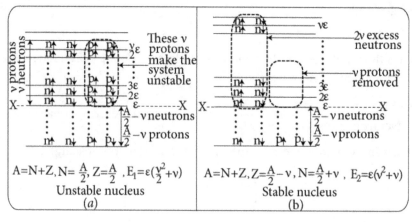

**Figure 3.8.** (a) High $A$ nucleus having symmetry in $N$ and $Z$ is unstable. (b) High $A$ nucleus having asymmetry in $N$ and $Z$ is stable.

we put 4 nucleons (2 protons—up and down spin oriented and 2 neutrons—up and down spin oriented). So there are $\frac{2\nu}{4} = \frac{\nu}{2}$ levels above $XX$. Suppose their energies are $\varepsilon, 2\varepsilon, 3\varepsilon, \ldots$

Total energy in this case is

$$E_1 = 4\varepsilon + 4.2\varepsilon + 4.3\varepsilon + \ldots \frac{\nu}{2} \text{ terms} = 4\varepsilon\left(1 + 2 + 3 + \ldots \frac{\nu}{2} \text{terms}\right)$$

$$E_1 = 4\varepsilon \frac{\frac{\nu}{2}\left(\frac{\nu}{2} + 1\right)}{2} = \varepsilon\left(\frac{\nu^2}{2} + \nu\right) \tag{3.24}$$

Suppose the $\nu$ protons above $XX$ make the system unstable and have to be removed and $\nu$ neutrons are to be inserted instead as per Pauli exclusion principle.

Consider now a nucleus with $N > Z$ (asymmetric) $N = \frac{A}{2} + \nu$, $Z = \frac{A}{2} - \nu$ and so

$$N - Z = \left(\frac{A}{2} + \nu\right) - \left(\frac{A}{2} - \nu\right) = 2\nu = \text{ neutron excess} \tag{3.25}$$

as shown in figure 3.8(b). So now we would have $2\nu$ excess neutrons that make the nucleus stable. Clearly there are a greater number of neutrons compared to proton population.

$$N = \frac{A}{2} + \nu = \left(\frac{A}{2} - \nu\right) + 2\nu \tag{3.26}$$

Now $\frac{A}{2} - \nu$ neutrons and $\frac{A}{2} - \nu$ protons are accommodated in levels below the base level $XX$. The remaining $2\nu$ excess neutrons populate the upper levels. In each level we put 2 neutrons and there are $\frac{2\nu}{2} = \nu$ levels. Total energy in this case is

$$E_2 = 2\varepsilon + 2.2\varepsilon + 2.3\varepsilon + \ldots \nu \text{ terms} = 2\varepsilon(1 + 2 + 3 + \ldots \nu \text{ terms})$$

$$E_2 = 2\varepsilon . \frac{\nu(\nu + 1)}{2} = \varepsilon(\nu^2 + \nu) \tag{3.27}$$

Increase in energy due to asymmetry in number of neutrons and protons is

$$E_2 - E_1 = \varepsilon(\nu^2 + \nu) - \varepsilon\left(\frac{\nu^2}{2} + \nu\right) = \varepsilon\frac{\nu^2}{2}. \tag{3.28}$$

This means binding energy is reduced by this corrretion term. With equation (3.25), i.e. $N - Z = 2\nu$ we write

$$\varepsilon\frac{\nu^2}{2} = \varepsilon\frac{(N - Z)^2}{8} \xrightarrow{N = A - Z} \varepsilon\frac{(A - 2Z)^2}{8} \tag{3.29}$$

Again, the greater the number of nucleons $A$ in a nucleus, the smaller is the energy level spacing $\varepsilon$, i.e. $\varepsilon \propto \frac{1}{A}$. So the asymmetry energy term is $\propto \frac{(A - 2Z)^2}{A}$. Introducing

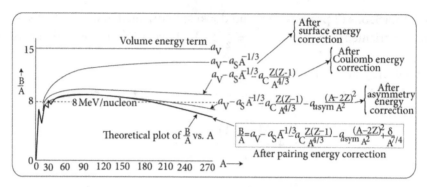

**Figure 3.9.** Relative contributions of various correction terms to the volume energy term in Bethe–Weizsacker's semi-empirical mass formula leading to theoretical plot of $\frac{B}{A}$ versus $A$.

proportionality constant $a_{asym}$ the asymmetry energy correction term becomes $a_{asym}\frac{(A-2Z)^2}{A}$.

Binding energy is weakened due to this term $a_{asym}\frac{(A-2Z)^2}{A}$ and so we have to subtract this term from equation (3.17) to get

$$B \equiv B(Z, A) = a_V A - a_S A^{2/3} - a_C \frac{Z(Z-1)}{A^{1/3}} - a_{asym}\frac{(A-2Z)^2}{A} \qquad (3.30)$$

$a_{asym} = 23.7 \ MeV$ (empirical value).

Figure 3.9 shows the $\frac{B}{A}$ versus $A$ plot after taking into account the asymmetry energy term.

- *Exercise* 3.8 shows that for lighter nuclei $A = 2Z$ holds.

## 3.11 Pairing energy correction term

It is found that all nuclei are not similarly stable. For instance even $Z$-even $N$ nuclei are most stable, odd $A$ nuclei (i.e. odd $Z$-even $N$ or even $Z$-odd $N$) are of intermediate stability, while odd $Z$-odd $N$ nuclei are least stable. This can be explained through spin dependence of nuclear force. In other words, the different stabilities of various nuclei are explained with the help of the pairing energy term.

We mention that study of data relating to neutron separation energy $S_n$ and proton separation energy $S_p$ for various nuclei reveals the following.

☐ For even $N$, $S_n=$ large. This implies that neutrons are paired and it is difficult to separate a neutron.

☐ For odd $N$, $S_n=$ low. This implies that there exists an unpaired neutron that can be separated easily.

☐ For even $Z$, $S_p=$ large. This implies that protons are paired and it is difficult to separate a proton.

☐ For odd $Z$, $S_p=$ low. This implies that there exists an unpaired proton that can be separated easily.

- *Exercise* 3.9 explains what we mean by pairing of nucleons.

Let us explain stability of various nuclei using the concept of spin dependence of nuclear force.

- ✓ Even–even nuclei. These are even $A$ nuclei

  All even numbers of neutrons pair up $n \uparrow n\downarrow$. All even numbers of protons pair up $p \uparrow p\downarrow$. Clearly such pairing leads to lowering of energy and strong binding. Binding energy is thus large and the nucleus has a stable configuration. So even–even nuclei like

  $_2\text{He}^4$ ($Z = 2$, $N = 2$), $_6\text{C}^{12}$ ($Z = 6$, $N = 6$), $_8\text{O}^{16}$ ($Z = 8$, $N = 8$)

  appear as peaks on the $\frac{B}{A}$ versus $A$ curve. This explains the kinks or the initial zigzag portion of the $\frac{B}{A}$ versus $A$ plot of figure 3.2.

- ✓ Odd $A$ nuclei

  - ☐ Odd $Z$-even $N$ nuclei. All even numbers of neutrons pair up. All protons except one pair up. One proton is left alone. Due to the existence of one unpaired proton the nucleus becomes less stable.
  - ☐ Even $Z$-odd $N$ nuclei. All even numbers of protons pair up. All neutrons except one pair up. One neutron is left alone. Due to the existence of one unpaired neutron the nucleus becomes less stable.

- ✓ Odd–odd nuclei. These are even $A$ nuclei.

All protons except one pair up. One proton is left alone. All neutrons except one pair up. One neutron is left alone. Due to the existence of two unpaired nucleons (one unpaired proton and one unpaired neutron) the nucleus becomes much less stable.

Mathematical analysis leads to a pairing energy term $\frac{\delta}{A^{3/4}}$

where $\delta$ is given by equation (3.10).

Incorporating this in equation (3.30) the binding energy becomes

$$B(Z, A) = a_V A - a_S A^{2/3} - a_C \frac{Z(Z - 1)}{A^{1/3}} - a_{\text{asym}} \frac{(A - 2Z)^2}{A} + \frac{\delta}{A^{3/4}} \quad (3.31)$$

with

$a_V = 15.75\ MeV$, $a_S = 17.8\ MeV$, $a_C = 0.71\ MeV$, $a_{\text{asym}} = 23.7\ MeV$, $|\delta| = 34\ MeV$

- Figure 3.9 shows relative contributions of volume energy term, surface energy correction term, Coulomb energy correction term, asymmetry energy correction term and pairing energy correction term.
- Considering all the correction terms (surface energy term + Coulomb energy term + asymmetry energy term + pairing energy term) we get the experimentally obtained $\frac{B}{A}$ versus $A$ plot.
- Atomic masses are known accurately and therefrom binding energies are calculated. Therefrom the constants $a_V$, $a_S$, $a_C$, $a_{\text{asym}}$, $|\delta|$ are adjusted so as to fit the binding energy data. In other words the constants appearing in Bethe–Weizsacker's mass formula are empirically determined.

- In *exercise* 3.10 we address why this formula of equation (3.31) is called the semi-empirical mass formula.

## 3.12 Merits and demerits of Bethe–Weizsacker's semi-empirical mass formula

Merits of Bethe–Weizsacker's semi-empirical mass formula involve successful explanation of the following.

✓ $\frac{B}{A}$ fails to grow with nuclear size, i.e. approximate constancy of binding energy per nucleon.
✓ Approximate constancy of density of nuclear matter.
✓ Nuclear fission (symmetric case).
✓ Stability against $\alpha$, $\beta$ decay.
✓ Mass parabola.
✓ Neutron drip line, proton drip line.
✓ Existence of neutron star.

Demerits of Bethe–Weizsacker's semi-empirical mass formula involve failure to explain the following.

☐ It cannot predict the binding energies of magic nuclei.

| Magic nuclei | Predicted B.E | Experimental B.E |
|---|---|---|
| $_{28}Ni^{56}$ ($Z = 28$, $N = 28$) | 477.7 *MeV* | 484 *MeV* |
| $_{50}Sn^{132}$ ($Z = 50$, $N = 82$) | 1084 *MeV* | 1110 *MeV* |

The departure of predictions of semi-empirical mass formula from experiment can be measured by studying the plot of $\Delta$ versus $A$ where

$$\Delta = \frac{B.\,E}{A}\bigg|_{\substack{\text{measured} \\ \text{experimentally}}} - \frac{B.\,E}{A}\bigg|_{\substack{\text{semi-empirical mass} \\ \text{formula prediction}}} \tag{3.32}$$

The deviations are prominent at certain regions where shell effects are important, for instance at regions where the shell closes—where binding energy is large and the nucleus becomes extra stable and inactive. This occurs at magic numbers. This is shown in a plot in figure 3.10.

☐ It fails to explain spin of nuclei/nucleons.
☐ It fails to explain magnetic moment of nuclei.
☐ It fails to explain asymmetric nuclear fission.

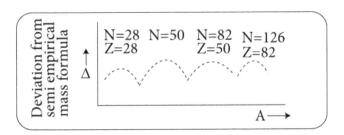

**Figure 3.10.** Deviations from Bethe–Weizsacker's semi-empirical mass formula are peaked at magic numbers.

## 3.13 Mass parabola

Let us find the combination of $Z$ and $A$ that gives a stable bound nucleus for a fixed value of $A$. Bethe–Weizsacker's semi-empirical mass formula is given by equation (3.31). Putting $B$ in the expression of $M$ as written in equation (3.6) viz. $M(Z, A) = Zm_p + Nm_n - \frac{B}{c^2}$ we have

$$M(Z, A) = Zm_p + Nm_n - \frac{1}{c^2}(a_V A - a_S A^{2/3} - a_C \frac{Z(Z-1)}{A^{1/3}} - a_{asym}\frac{(A-2Z)^2}{A} + \frac{\delta}{A^{3/4}}) \quad (3.33)$$

Put $N = A - Z$

$$M(Z, A) = Zm_p + (A - Z)m_n - \frac{a_V}{c^2}A + \frac{a_S}{c^2}A^{2/3} + a_C\frac{Z^2 - Z}{c^2 A^{1/3}}$$

$$+ \frac{a_{asym}}{c^2 A}(A^2 - 4AZ + 4Z^2) - \frac{\delta}{c^2 A^{3/4}}$$

$$M(Z, A) = A\left[m_n - (\frac{a_V}{c^2} - \frac{a_S}{c^2 A^{1/3}} - \frac{a_{asym}}{c^2})\right] + Z\left[(m_p - m_n) - \frac{a_C}{c^2 A^{1/3}} - 4\frac{a_{asym}}{c^2}\right]$$

$$+ Z^2\left(\frac{a_C}{c^2 A^{1/3}} + 4\frac{a_{asym}}{c^2 A}\right) - \frac{\delta}{c^2 A^{3/4}} \quad (3.34)$$

$$M(Z, A) = \alpha A + \beta Z + \gamma Z^2 - \frac{\delta}{c^2 A^{3/4}} \quad (3.35)$$

where

$$\alpha = m_n - \left(\frac{a_V}{c^2} - \frac{a_S}{c^2 A^{1/3}} - \frac{a_{asym}}{c^2}\right), \quad \beta = (m_p - m_n) - \frac{a_C}{c^2 A^{1/3}} - 4\frac{a_{asym}}{c^2}, \quad \gamma = \frac{a_C}{c^2 A^{1/3}} + 4\frac{a_{asym}}{c^2 A} \quad (3.36)$$

Let us consider isobaric nuclei. So for $A = $ constant, $\alpha = $ constant, $\beta = $ constant, $\gamma = $ constant. It follows from equation (3.35) that $M$ versus $Z$ plot for a given $A$ is parabolic in shape and is referred to as mass parabola, as shown in figure 3.11(a).

The mass parabola has a minimum corresponding to that value of $Z = Z_0$ which gives the minimum $M$ and corresponds to the most stable isobar in an isobaric family of $A = $ constant. We can mathematically evaluate this minimum $Z = Z_0$ by minimizing $M(Z, A)$, i.e. by setting

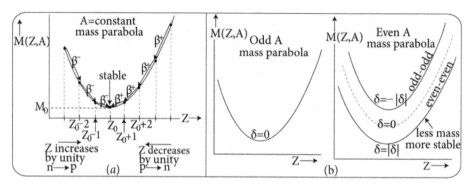

**Figure 3.11.** (a) Plot of $M(Z, A)$ versus $Z$ for $A$ = constant gives mass parabola. (b) One mass parabola for odd $A$ = constant nuclei and two mass parabola for even $A$ = constant nuclei.

$$\frac{d}{dZ}M(Z, A) = 0 \qquad (3.37)$$

In *exercise 3.12* we show that

$$Z_0 = \frac{A/2}{1 + 0.0075A^{2/3}} \qquad (3.38)$$

and for light nuclei $A \leqslant 20$, $Z_0 \cong \frac{A}{2}$ but for heavier nuclei $Z_0 < \frac{A}{2}$.

In figure 3.11(a) the points in the plot correspond to integral values of $Z$. For a particular $A$ = constant we have a series of nuclei with different $Z$ and $N$ but the same $A$, called isobars ($A$ = fixed, $Z$, $N \to$ different). The nucleus sitting at (or near) the minimum having $Z = Z_0$ is stable and the nuclei on the left portion of parabola suffer $\beta^-$ decay and those on the right portion of parabola suffer $\beta^+$ decay in their effort to gain stability by shedding off extra energy and approach a stable state ($M_0$, $Z_0$).

In $\beta^-$ decay we move towards right along the $Z$-axis because the corresponding $n \to p$ transformation involves increase of $Z$ by unity.

In $\beta^+$ decay we move towards left along the $Z$-axis because the corresponding $p \to n$ transformation involves decrease of $Z$ by unity.

The number of parabolas predicted by equation (3.35) has been shown in figure 3.11(b).

✓ For odd $A$= constant nuclei $\delta = 0$. Equation (3.35) becomes

$$M(Z, A) = \alpha A + \beta Z + \gamma Z^2 \qquad (3.39)$$

It represents one parabola corresponding to odd $A$, as shown in figure 3.11(b).

✓ For even $A$= constant nuclei

☐ Even–even nuclei has $\delta = |\delta|$. Equation (3.35) becomes

$$M(Z, A) = \alpha A + \beta Z + \gamma Z^2 - \frac{|\delta|}{c^2 A^{3/4}} \qquad (3.40)$$

□ Odd–odd nuclei has $\delta = -|\delta|$. Equation (3.35) becomes

$$M(Z, A) = \alpha A + \beta Z + \gamma Z^2 + \frac{|\delta|}{c^2 A^{3/4}} \tag{3.41}$$

Clearly we have two parabolas for even $A$ nuclei. Equation (3.40) represents one parabola—the lower parabola in figure 3.11(b) for even–even nuclei and equation (3.41) represents another parabola—the upper parabola in figure 3.11(b) for odd–odd nuclei.

- In figure 3.11(b) we note that the upper even $A$ odd–odd parabola is displaced upwards w.r.t. $\delta = 0$ by an amount $\frac{|\delta|}{c^2 A^{3/4}}$ while the lower even $A$ even–even parabola is displaced downwards w.r.t. $\delta = 0$ by an amount $\frac{|\delta|}{c^2 A^{3/4}}$. So the two even $A$ parabolas are displaced w.r.t. each other by an amount

$$\frac{|\delta|}{c^2 A^{3/4}} + \frac{|\delta|}{c^2 A^{3/4}} = \frac{2|\delta|}{c^2 A^{3/4}} \tag{3.42}$$

The even–even parabola is displaced downwards as even–even nuclei have lower mass—they are more bound and are more stable w.r.t. odd–odd parabola.

The odd–odd parabola is displaced upwards as each element of it has higher mass—they are less bound and are less stable by an amount $\frac{2|\delta|}{c^2 A^{3/4}}$ w.r.t. even–even parabola.

- It is clear that isobars ($A$= constant) reside on mass parabolas.
- For small $A$ parabolas are steeper.
- The isobar with lowest energy will be stable and others will be radioactive ($\beta$ active).
- For odd $A$ mass parabola we can find one stable isotope. But for even $A$ mass parabola we can have one, two or three stable isotopes (figures 2.17, 2.18).

## 3.14 Neutron drip line, proton drip line

Consider the $N$ versus $Z$ plot of figure 3.12 that shows the stability curve. Stable nuclei are located on it or around it.

We here discuss whether we can go on putting neutrons arbitrarily and form a stable nucleus with a fixed number of protons.

Let us consider a vertical line at any $Z$ value, say $Z = Z_1$, as shown in figure 3.12(a). Fixed $Z = Z_1$ means we are keeping proton number fixed at $Z_1$. As we move up the $Z = Z_1$ line neutron number $N$ increases. As the $Z = Z_1$ line crosses the stability curve we find nuclei lying to the left of the stability curve that are neutron rich, unstable and prone to $n \rightarrow p$ transformation, i.e. they are $\beta^-$ active. As we move up further along the $Z = Z_1$ vertical line nuclei become increasingly richer in neutrons and at point $l_1$ corresponding to $Z = Z_1$, $N = N_1$ we reach a limit at which neutron emission becomes energetically favourable instead of $n \rightarrow p$ transformation (i.e. instead of $\beta^-$ emission). So now at point $l_1$ neutron number has to reduce for attainment of stability—the nucleus will not accept any more neutrons. In

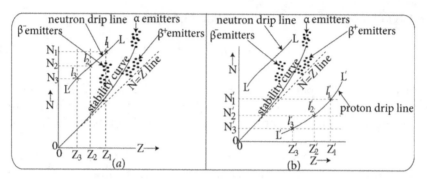

**Figure 3.12.** (a) Neutron drip line $LL$. (b) Proton drip line $L'L'$.

other words, the process of acceptance of neutron $_ZX_N^A + n \rightarrow \,_ZX_{N+1}^{A+1}$ is energetically not favoured since

$$M(_ZX_N^A) + m_n < M(_Z X_{N+1}^{A+1}).$$

So the neutron can no longer be introduced into the nucleus (neutron separation energy $S_n$ is negative). It is clear that with a particular value of $Z$ we cannot keep on arbitrarily increasing neutrons and get a bound nucleus—there is a limit to putting neutrons in a nucleus with some fixed $Z$.

In *exercise 3.19* we show from the semi-empirical mass formula that for a fixed value of $Z$ as we go on increasing neutron number in a nucleus a limit is reached beyond which neutron emission becomes energetically favourable.

It is thus evident that we cannot form a nucleus with an arbitrary number of neutrons for a fixed number of protons. There is a limit to the number of neutrons that can be pumped into a nucleus for a fixed proton number. As shown in figure 3.12(a) for $Z = Z_1, Z_2, Z_3...$ there are no nuclei with neutron number $N \geqslant N_1, N_2, N_3...$, respectively. So $N_1, N_2, N_3...$ represent the limit of the number of neutrons that are possible in a nucleus and the corresponding limit points being $l_1, l_2, l_3...$. Joining these limit points $l_1, l_2, l_3...$ gives a line called neutron drip line $LL$. Above and to the left of the neutron drip line $LL$ we cannot form any nucleus—stable or unstable—because if we try to push neutrons into the nucleus the pushed neutrons will simply drip out or fall off. We cannot go beyond the limiting capacity of a nucleus regarding its neutron number.

From a similar discussion it follows that we cannot go on putting protons arbitrarily and form a stable nucleus with a fixed number of neutrons—in fact, there is a limit to the number of protons that can be pumped into a nucleus for a fixed neutron number.

This is depicted through a similar construction in figure 3.12(b). For $N = N_1', N_2', N_3', ...$ there are no nuclei with proton number $Z \geqslant Z_1', Z_2', Z_3'...$ respectively. So $Z_1', Z_2', Z_3'...$ represent the limit of number of protons that are possible in a nucleus and the corresponding limit points being $l_1', l_2', l_3'...$. Joining these limit points $l_1', l_2', l_3', ...$ gives a line called proton drip line $L'L'$. Below and to the right of the proton drip line $L'L'$ we cannot form any nucleus—stable or unstable—because protons will simply drip out or fall off. We cannot go beyond the limiting capacity of a nucleus regarding its proton number.

## 3.15 Prediction of mass of neutron star from semi-empirical mass formula

From the semi-empirical mass formula we can set a limit on the number of neutrons in a nucleus for a fixed proton number, as shown in figure 3.12(a). Let us extrapolate our result into a very big nucleus.

Stars are big celestial bodies that are formed when hydrogen, H, and helium, He, combine in space and shrink or collapse due to gravitational attraction so as to increase the density of the core, and temperature shoots up consequently (as gravitational potential energy is converted into thermal energy). The density and temperature may become so high that nuclear fusion initiates and a star is born. The process of fusion can involve conversions from

H → He → C → O → Ne → Mg → ...→Fe.

Nuclear fusion cannot proceed further after formation of Fe. The nuclear conversions can stop at any stage depending on the mass of the star. For instance for sun the process is H → He → C but for stars of mass $M > M_{sun}$ the process can proceed further and in the end nuclear fusion at the core stops.

When the process of nuclear fusion occurs, the energy that is radiated outward balances the inward gravitational attraction and there is a balance, i.e. equilibrium is maintained. But when nuclear fusion stops gravitational attraction dominates and gravitational collapse occurs fast—and a hard dense core results. Outer layers of the star collide with the hard dense core—then rebound—and these outer layers get thrown off into outer space—there is a supernova explosion.

The dense core that is left keeps collapsing further because of gravitational attraction. The core density becomes enormous, pressure builds up and a stage comes when the reaction

$$p + e^- \rightarrow n + \nu_e \tag{3.43}$$

starts, i.e. proton $p$ and electron $e^-$ combine to generate neutron $n$ and electron neutrino $\nu_e$. The $\nu_e$ escape out of the star quickly because of their high speed $\sim c$.

The star now consists predominantly of neutrons barring a few surface layers and this star is called a neutron star.

Let us make a rough estimate of the number of neutrons in a neutron star whose radius is around 15 $km$.

Taking mass of neutron star $\sim$ solar mass $= M_{sun} \cong 2 \times 10^{30}$ $kg$ and neutron mass $m_n = 1.675 \times 10^{-27}$ $kg$ the number of neutrons in the neutron star can be calculated to be

$$\frac{M_{sun}}{m_n} = \frac{2 \times 10^{30} \ kg}{1.675 \times 10^{-27} \ kg} = 10^{57} \tag{3.44}$$

So $10^{57}$ neutrons are packed in a region of radius 15 $km$ in the neutron star. We can think of the neutron star as a big sized stable nucleus built with neutrons only with a density of

$$\frac{10^{57}}{\frac{4}{3}\pi(15\,km)^3} = \frac{3 \times 10^{57}}{4\pi(15 \times 10^3\,m)^3} \sim 10^{43} \tag{3.45}$$

Let us compare this result with a real nucleus on earth.

✓ A stable nucleus is formed with protons and neutrons and for a particular $Z$ there is a limit on the number of neutrons beyond which neutron drips off. Nucleon density $\sim 10^{17}\,m^{-3}$ which is constant for most nuclei.

✓ A nucleus has diameter $\sim fm$ in which nuclear force $\gg$ gravitational force and hence gravitational potential energy is not considered in the calculation of binding energy of the nucleus.

We note that extraordinary density in a neutron star is due to gravitational attraction and so we have to include the gravitational potential energy term in the expression of binding energy.

• Let us extend the semi-empirical mass formula (applicable for a nucleus of size $\sim fm$) to a neutron star, likened to a big sized nucleus incorporating gravitational potential energy.

  We shall investigate whether observed properties of a neutron star can be reproduced from the semi-empirical mass formula.

• Derivation of gravitational potential energy.

Gravitational potential energy is the work done to bring the constituent parts of the nucleus from infinitely diluted state to its present location.

Consider a spherical mass $M$ of radius $R$, volume $\frac{4}{3}\pi R^3$ and constant mass density $\rho = \frac{M}{\frac{4}{3}\pi R^3}$.

We can construct the sphere by bringing the layers of mass from infinity to its present location. Work done to add mass element $dm$ to form a sphere of radius $r$ of mass $m$, as shown in figure 3.13, is

$$dW = dm[V(r) - V(\infty)] \tag{3.45}$$

where

$V(r) = -\frac{Gm}{r}$ = gravitational potential at $r$ and

$V(\infty) = 0$ = gravitational potential at $r = \infty$.

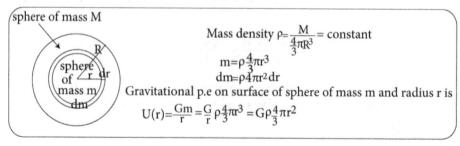

Figure 3.13. Spherical mass of radius $R$, volume $\frac{4}{3}\pi R^3$ and mass $M$.

$$dW = dm\left(-\frac{Gm}{r}\right) = -\frac{1}{r}Gmdm \tag{3.46}$$

For spherical mass $m$ of radius $r$ we have

$$m = \rho\,\frac{4}{3}\pi\,r^3 \tag{3.47}$$

$$dm = d\left(\rho\frac{4}{3}\pi\,r^3\right) = \rho 4\pi r^2 dr \tag{3.48}$$

$$dW = -\frac{1}{r}G.\,\rho\frac{4}{3}\pi\,r^3.\,\rho 4\pi r^2 dr = -\,G\frac{(4\pi\rho)^2}{3}r^4 dr$$

$$= -G\frac{(4\pi)^2}{3}\left(\frac{M}{\frac{4}{3}\pi\,R^3}\right)^2 r^4 dr = -\frac{3GM^2}{R^6}r^4 dr \tag{3.49}$$

Total work done to build up the sphere is

$$W = \int_0^R dW = -\int_0^R \frac{3GM^2}{R^6}r^4 dr = -\frac{3GM^2}{R^6}\frac{r^5}{5}\Big|_0^R$$

$$W = -\frac{3GM^2}{5R} \tag{3.50}$$

This work done to assemble the elemental masses is stored up as its gravitational self-potential energy

$$U_{self} = -\frac{3GM^2}{5R} \tag{3.51}$$

If we supply to it energy of amount $\frac{3GM^2}{5R}$ it disintegrates (as then its energy will be zero). Clearly this energy $\frac{3GM^2}{5R}$ binds the system and is called the binding energy due to gravitational attraction $B_g$, i.e.

$$B_g = \frac{3GM^2}{5R} \tag{3.52}$$

The semi-empirical mass formula [equation (3.31)] along with the gravitational binding energy (equation (3.52)) as extended to the neutron star is given by

$$B_{star}(Z, A) = a_V A - a_S A^{2/3} - a_C\frac{Z(Z-1)}{A^{1/3}} - a_{asym}\frac{(A-2Z)^2}{A} + \frac{\delta}{A^{3/4}} + \frac{3GM^2}{5R} \tag{3.53}$$

✓ As no protons are present in the core of neutron star $Z = 0$ and hence the Coulomb energy term

$$a_C \frac{Z(Z-1)}{A^{1/3}} = 0 \qquad (3.54)$$

✓ A neutron star has no protons. So $Z = 0$, $A - 2Z = A$. Hence the asymmetry term is

$$a_{\text{asym}} \frac{(A - 2Z)^2}{A} = a_{\text{asym}} A \qquad (3.55)$$

✓ Consider the surface energy term. The surface of a spherical system having surface area $\sim 4\pi R^2 \propto R^2$ is significant when size of the system is small. But for a large system surface contribution is less important since the volume $\sim \frac{4}{3}\pi R^3 \propto R^3$ is more important now (*exercise 3.5*, figure 3.4(a)). So for a neutron star the surface energy term can be dropped to achieve simplification.

✓ Consider the pairing energy term $\sim \frac{\delta}{A^{3/4}}$. It is an even–odd effect and shows up when we add one nucleon to the system. Now in an assembly of a huge number of neutrons, adding or subtracting a single neutron will not change the energy of system. So we ignore this pairing term.

✓ Mass of a neutron star is $M = A m_n$ as there are no protons in a neutron star. Taking the radius to be $R = R_0 A^{1/3}$, $R_0 = 1.2\,fm$ we write for the gravitational binding energy

$$\frac{3GM^2}{5R} = \frac{3G(A m_n)^2}{5R_0 A^{1/3}} = \frac{3G A^{5/3} m_n^2}{5R_0}$$

With these modifications we rewrite equation (3.53) as [using equations (3.54) and (3.55)]

$$B(Z, A) = a_V A - a_{\text{asym}} A + \frac{3G A^{5/3} m_n^2}{5R_0} \qquad (3.56)$$

with $a_V = 15.75\,MeV$, $a_{\text{asym}} = 23.7\,MeV$

For a neutron star to be a bound system we should have a positive binding energy. Thus

$$B(Z, A) \geqslant 0 \qquad (3.57)$$

$$(a_V - a_{\text{asym}})A + \frac{3G A^{5/3} m_n^2}{5R_0} > 0$$

$$(15.75 - 23.7)\,MeV + \frac{3G A^{2/3} m_n^2}{5R_0} \geqslant 0$$

$$\frac{3GA^{2/3}m_n^2}{5R_0} \geqslant 7.95 \ MeV$$

$$A^{2/3} \geqslant 7.95 \ MeV \frac{5R_0}{3Gm_n^2} = \frac{7.95 \times 10^6 \times 1.6 \times 10^{-19} \ J \times 5 \times 1.2 \times 10^{-15} \ m}{3 \times 6.67 \times 10^{-11} \frac{Nm^2}{kg^2} \times (1.675 \times 10^{-27} \ kg)^2} = 1.36 \times 10^{37}$$

$$A \geqslant (1.36 \times 10^{37})^{3/2} = 5 \times 10^{55}.$$

$$A \geqslant 10^{55}. \tag{3.58}$$

This result nearly matches our earlier estimate in equation (3.44).
Again, the mass of a neutron star with $10^{55}$ neutrons is

$$M = Am_n = 10^{55}m_n = 10^{55} \times 1.675 \times 10^{-27} \ kg$$

$$M = 1.675 \times 10^{28} \ kg \sim 10^{28} \ kg \tag{3.59}$$

Neutron stars are observed with mass somewhat larger than solar mass $M_{sun} \sim 10^{30} \ kg$. So nearly the correct order of magnitude for the mass of a neutron star has been predicted.

We see that the semi-empirical mass formula that holds for a tiny nucleus can be extended to astronomical objects involving large mass.

## 3.16 Concepts of shells in atomic structure

In an atom there is a positively charged nucleus and around the nucleus electrons revolve in different orbits due to electrostatic attraction of the nucleus. The structure of a hydrogen atom is that it consists of one electron and one proton (which is the nucleus), as shown in figure 3.14(a). The Hamiltonian of the hydrogen atom system is

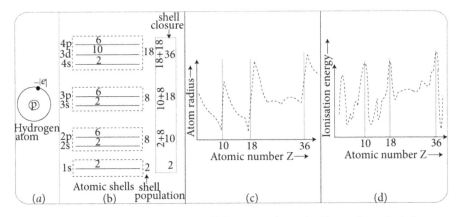

**Figure 3.14.** (a) H atom. (b) Atomic shells. (c),(d) Some experimental evidence of atomic shell structure.

$$H = T_1 + V_1 = -\frac{\hbar^2}{2m}\nabla_1^2 - \frac{Z\,|\,e\,|^2}{4\pi\varepsilon_0 r_1} \tag{3.60}$$

where $r_1$ is electron coordinate, $m$ is electron mass.

For a multi-electron system we have to write the kinetic energies of each electron and their potential energies. There will be an interaction energy term present in the Hamiltonian

$$H = \sum_i \left[ -\frac{\hbar^2}{2m}\nabla_i^2 - \frac{Z\,|\,e\,|^2}{4\pi\varepsilon_0 r_i} \right] + \sum_i \sum_{j\neq i} \frac{|\,e\,|^2}{4\pi\varepsilon_0\,|\,\vec{r}_i - \vec{r}_j\,|} \tag{3.61}$$

It is evident from the structure of this Hamiltonian that it is difficult to handle and so we have to look for manageable ways to get energy eigenvalue of a typical electron of the system. The approach one follows is to focus attention on an electron and assume it to move in a potential created by the entire atom, i.e. by the nucleus plus the remaining electrons, and the requirement is that the potential should be effective, central, average, single-particle potential $V(r)$ that is non-Coulombic, where $r$ is the coordinate of the representative electron. Various potentials have been suggested and tested. We need not consider the coordinates of all the electrons which would be an impossible task. With this central single-particle potential $V(r)$ the atomic Hamiltonian is

$$H = T + V = -\frac{\hbar^2}{2m}\nabla^2 + V(r) \tag{3.62}$$

The time-independent Schrödinger equation is

$$H\psi(\vec{r}) = E\psi(\vec{r}) \tag{3.63}$$

We can solve the Schrödinger equation and get the energy eigenvalues corresponding to any electron of the atomic system. The energy eigenvalues depend on two quantum numbers viz.

✓ $n$ = principal quantum number
✓ $l$ = orbital angular momentum quantum number.

So we denote the energy eigenvalue as $E = E_{nl}$.

The various energy levels appear with varying energy gaps or energy separations. If energy gaps of two levels are very small then we say that they belong to the same shell, while if gaps are very large we say that the energy levels belong to different shells. Accordingly, atomic shell structure arises as shown in the diagram of figure 3.14(b).

Different energy levels arise for different combinations of $n$ and $l$. They are populated by electrons of the atom as per Pauli exclusion principle since electrons are fermions. According to the Pauli exclusion principle, only one fermion can populate a quantum state. Electron capacity is 2 for $ns$, 6 for $np$, 10 for $nd$ etc. Shell population is 2, 8, 8, 18... and shell closure occurs at 2, 10, 18, 36, ...

The central potential $V(r)$ has no spin considerations. Hence the levels designated by the quantum numbers $n$ and $l$ are in fact degenerate in other quantum numbers like $m_l$, $m_s$. For instance, levels with $| l = 1 >$ are a composite of three levels, viz.

$| l = 1, m_l = 1 >$ , $| l = 1, m_l = 0 >$, $| l = 1, m_l = -1 >$

Further considering electron spin $s = \frac{1}{2}$ the state $|s = \frac{1}{2}>$ is two-fold degenerate, i.e. a mixture of two states viz.

$$|s = \frac{1}{2}, m_s = \frac{1}{2} > = \chi_{\frac{1}{2}, \frac{1}{2}} \text{ and } |s = \frac{1}{2}, m_s = -\frac{1}{2} > = \chi_{\frac{1}{2}, -\frac{1}{2}} \qquad (3.64)$$

Figure 3.14(c),(d) shows some experimental evidence of atomic shell structure.

✓ Changes in atomic radii of atoms occur abruptly around $Z = 10, 18, 36, ...$ indicating change in atomic behavior at shell closures. Abrupt changes suggest that we are very close to one shell and far away from the other (figure 3.14(c)).

✓ Ionization energy is the minimum energy required to remove an electron from atom. Ionization energy shows a remarkable fall after $Z = 10, 18, 36, 54$ indicating existence of stable shell structure in atoms (figure 3.14(d)). When a shell closes, it is stable and its ionization energy is high. But when one electron is added it is loosely bound and so it is easy to dislodge that electron indicating very low values of ionization energy just after a peak. So when a new shell starts, ionization energy drops abruptly and sharply.

This indicates existence of shell structure in an atom.


## 3.17 Concepts of shells in nucleus

Electrons are arranged in the extra nuclear space as per the Bohr model. The extra nuclear electrons are arranged in different energy levels in accordance with the Pauli exclusion principle. Some energy levels are not filled to their capacity while some are completely filled (e.g. the inert gas atom). This gives rise to atomic shells.

The nuclear shell model also predicts a similar shell structure for the proton and neutron within the nucleus.

The protons and neutrons (i.e. nucleons) are fermions and according to Pauli exclusion principle only one Fermion can populate one quantum state. So only one proton can stay in one quantum state. Similarly, only one neutron can populate one quantum state.

The neutrons and protons constituting the nucleus are supposed to be arranged in energy levels which give rise to subshells and shells. These shells get closed with a suitable number of protons and neutrons.

So the nucleus has a shell structure just like the extra nuclear electrons.

## 3.18 Definition of magic number, magic nuclei, doubly magic nuclei, semi-magic nuclei

• Experimentally it has been observed that nuclei possessing a certain number of protons and /or certain number of neutrons exhibit more than average stability. The numbers are

$$2, 8, 20, 50, 82, 126 \tag{3.65}$$

These numbers are referred to as magic numbers.

• A nucleus having a magic number of neutrons or magic number of protons is called a magic nucleus.

Examples of magic nuclei are

$$_2\text{He}_2^4, \ _8\text{O}_8^{16}, \ _{20}\text{Ca}_{20}^{40}, \ _{28}\text{Ni}_{50}^{78}, \ _{50}\text{Sn}_{69}^{119}, \ _{82}\text{Pb}_{126}^{208}$$

• Magic nuclei have markedly different properties from nuclei with neighbouring values of $N$ or $Z$. The magic nuclei are like inert gas atoms that have closed electronic shells.

Nuclei containing both proton number and neutron number equal to magic number exhibit extraordinary stability and are called doubly magic.

Examples of doubly magic nuclei are

$$_2\text{He}_2^4, \ _8\text{O}_8^{16}, \ _{20}\text{Ca}_{20}^{40}, \ _{82}\text{Pb}_{126}^{208}.$$

• Nuclei having nucleon number 28 or 40 show stability to some extent. So they are called semi-magic numbers.

In the nuclear shell model the significance of magic numbers is that shell closure occurs at the magic numbers. This also explains the high degree of stability of the magic nuclei.

## 3.19 Evidence of shell structure in nucleus

✓ The simplest evidence for the nuclear shell model comes from the discovery of magic numbers 2, 8, 20, 28, 50, 82, 126.
✓ The binding energies of magic nuclei are much larger than the neighbouring nuclei.
✓ A larger energy is required to separate a nucleon from magic nuclei.
✓ The lower excited states of magic nuclei are separated from their ground state by a large amount of energy.
✓ The number of stable nuclei with a given value of $Z$ and $N$ equal to the magic number is much more than the number of stable nuclei with neighbouring values of $Z$ and $N$. This is shown in figure 3.15.

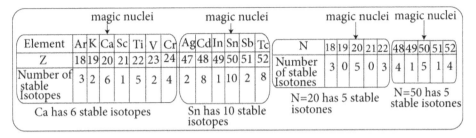

**Figure 3.15.** Magic nuclei with $Z = 20, 50$ have more isotopes than their neighbours. Magic nuclei with $N = 20, 50$ have more isotones than their neighbours.

For instance:

☐ Among the set of nuclei between $Z = 18$ to $Z = 24$ magic nuclei Ca with $Z = 20$ have a maximum number of 6 stable isotopes.

☐ Among the set of nuclei between $Z = 47$ to $Z = 52$ magic nuclei Sn with $Z = 50$ have a maximum number of 10 stable isotopes.

☐ Among the set of nuclei between $N = 18$ to $N = 22$ magic nuclei with $N = 20$ have a maximum number of 5 stable isotones.

☐ Among the set of nuclei between $N = 48$ to $N = 52$ magic nuclei with $N = 50$ havc a maximum number of 5 stable isotones.

✓ Radioactive elements emit $\alpha$ which is a bound system of $2n - 2p$, i.e. a $_2\text{He}_2^4$ nucleus. This suggests that $_2\text{He}_2^4$ is an especially stable state. Also, we note that $Z = 2$, $N = 2$ are magic numbers implying thereby that $_2\text{He}_2^4$ is doubly magic which indicates extra stability.

✓ All three radioactive series ends with lead, i.e. Pb is the stable end product.

☐ Uranium series $_{92}\text{U}^{238} \rightarrow {}_{82}\text{Pb}^{206}$
☐ Actinium series $_{92}\text{U}^{235} \rightarrow {}_{82}\text{Pb}^{207}$
☐ Thorium series $_{90}\text{Th}^{232} \rightarrow {}_{82}\text{Pb}^{208}$.

Pb has $Z = 82$ which is a magic number. Further, thorium series has the stable end product $_{82}\text{Pb}^{208}$ which has $Z = 82$, $N = 208 - 82 = 126$. As the numbers 82, 126 both are magic numbers this implies that $_{82}\text{Pb}^{208}$ is doubly magic and hence indicates extra stability.

✓ Some nuclei have higher abundance in nature. For instance, it has been observed that
$_{50}\text{Sn}$ has 10 stable isotopes with $A = 112, 114, 115, 116, 117, 118, 119, 120, 122, 124$ and
$_{20}\text{Ca}$ has 6 stable isotopes with $A = 40, 42, 43, 44, 46, 48$.
So the elements with $Z = 50$ and $Z = 20$ are more than usually stable.

✓ Out of the 6 stable isotopes of $_{20}\text{Ca}$ the isotope $_{20}\text{Ca}^{40}$ has an abundance of 97% in the natural mixture of calcium isotopes. We note that $Z = 20$, $N = 40 - 20 = 20$. As 20 is a magic number $_{20}\text{Ca}^{40}$ is doubly magic and hence indicates extra stability.

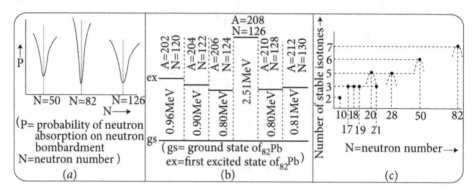

**Figure 3.16.** (a),(b),(c) Evidence of shell model.

✓ It is observed that plots of many nuclear properties against $Z$ or $N$ show characteristic peaks or point of inflexion at magic numbers. This shows that magic numbers are especially favoured w.r.t. certain nuclear properties.

✓ Most abundant nuclei are spherical in shape as evident from the zero value of their quadrupole moment. These elements have nuclei that contain a magic number of protons and/or magic number of neutrons—hence they are called magic nuclei.

✓ It is experimentally observed that the neutron absorption cross-section of nuclei dips sharply when nuclei possess a magic number of neutrons, say $N = 50, 82, 126$ signaling that neutron shells suffer closure and become inactive at magic numbers (figure 3.16(a)).

✓ All lead isotopes have $Z = 82$ which is a magic number. But the isotope $_{82}Pb^{208}$ shows extra stability since it has $N = A - Z = 208 - 82 = 126$ neutrons. This is clear from the fact that it has its first excited state much separated from ground state as compared to other isotopes. Obviously possessing 2 magic numbers $Z = 82$ and $N = 126$ makes $_{82}Pb^{208}$ extra stable (figure 3.16(b)).

✓ The plot of the number of stable isotones for a particular neutron number $N$ shows clear peaks at $N = 20, 28, 50, 82$, i.e. at the magic numbers (figure 3.16(c)).

## 3.20 Nuclear shell model by Mayer and Jensen, assumptions of shell model

✓ The nucleus is a quantum system. Nucleons have energy level structure (or shell structure) that resembles the atomic electron energy level structure. Shells or the energy levels are filled by nucleons (neutrons and protons) as per Pauli exclusion principle (as nucleons are fermions).

Consequence:

As nucleons move in orbits, the probability of collision between nucleons is very small, i.e. nucleons have a long mean free path (greater than nuclear radius). So nucleons interact weakly. Hence this model is called the weak interaction model. In other words, the shell model assumes nucleons to be non-colliding (or weakly interacting) as they occupy non-intersecting orbits.

✓ Each nucleon moves independently of all the other nucleons inside the nucleus in a fixed orbit about the centre of mass of the nucleus under the influence of a spherically symmetric (i.e. central) self-consistent average nuclear potential, say $V(r)$, produced by the average interaction between all the remaining $A - 1$ nucleons in it, where $A$ is mass number. In other words, $V(r)$ is a single-nucleon potential. For this assumption the nuclear shell model is also known as the independent particle model or IPM.

✓ The nucleus is a many-body problem having in general a large number of interacting nucleons. Clearly setting up of the Schrödinger equation for such a system is tough and obtaining a solution is virtually impossible. In practice one proceeds by choosing a potential that would suit the problem. Two common choices are made.

☐ We can choose the average potential to be the 3D harmonic oscillator potential, i.e.

$$V(r) = \frac{1}{2}mw^2r^2 \tag{3.66}$$

It is a central potential where $r$ is the coordinate of nucleon, $m$ is nucleon mass, $w$ is angular frequency of rotation of a nucleon in the orbit of the shell (the shell is a group of levels).

☐ We can choose the average potential to be infinite square well potential as in *exercise 3.28.*

Success of choice of a potential would depend upon its ability to generate experimentally observed magic numbers.

## 3.21 Generation of magic numbers with harmonic oscillator potential

Let us work with the 3D harmonic oscillator potential $V(r) = \frac{1}{2}mw^2r^2$. The nuclear Hamiltonian is

$$H_0 = T + V(r) = -\frac{\hbar^2}{2m}\nabla^2 + V(r)$$

$$H_0 = -\frac{\hbar^2}{2m}\nabla^2 + \frac{1}{2}mw^2r^2 \tag{3.67}$$

where $T$ is kinetic energy of a nucleon. The corresponding time-independent Schrödinger equation is

$$H_0\psi = E\psi \tag{3.68}$$

$$-\frac{\hbar^2}{2m}\nabla^2\psi + V(r)\psi = E\psi \tag{3.69}$$

where $\psi$ is the nuclear wave function, $E$ is energy eigenvalue. This can be rewritten as

$$\nabla^2 \psi + \frac{2m}{\hbar^2}[E - V(r)]\psi = 0 \tag{3.70}$$

Form of $\nabla^2$ in spherical polar coordinates is

$$\nabla^2 = \frac{1}{r^2}\frac{\partial}{\partial r}\left(r^2 \frac{\partial}{\partial r}\right) + \frac{1}{r^2}\left[\frac{1}{\sin\theta}\frac{\partial}{\partial\theta}\left(\sin\theta\frac{\partial}{\partial\theta}\right) + \frac{1}{\sin^2\theta}\frac{\partial^2}{\partial\phi^2}\right] \tag{3.71}$$

Again, using the square of angular momentum operator

$$L^2 = -\hbar^2\left[\frac{1}{\sin\theta}\frac{\partial}{\partial\theta}\left(\sin\theta\frac{\partial}{\partial\theta}\right) + \frac{1}{\sin^2\theta}\frac{\partial^2}{\partial\phi^2}\right] \tag{3.72}$$

we get

$$\nabla^2 = \frac{1}{r^2}\frac{\partial}{\partial r}\left(r^2 \frac{\partial}{\partial r}\right) - \frac{1}{r^2}\frac{L^2}{\hbar^2}$$

Hence equation (3.70) can be written as

$$\left[\frac{1}{r^2}\frac{\partial}{\partial r}\left(r^2 \frac{\partial}{\partial r}\right) - \frac{1}{r^2}\frac{L^2}{\hbar^2}\right]\psi + \frac{2m}{\hbar^2}[E - V(r)]\psi = 0 \tag{3.73}$$

Use method of separation of variables assuming

$$\psi(\vec{r}) = R(r)Y(\theta, \phi) \tag{3.74}$$

This gives

$$\left[\frac{1}{r^2}\frac{\partial}{\partial r}\left(r^2 \frac{\partial}{\partial r}\right) - \frac{1}{r^2}\frac{L^2}{\hbar^2}\right]RY + \frac{2m}{\hbar^2}[E - V(r)]RY = 0$$

$$Y\frac{1}{r^2}\frac{d}{dr}\left(r^2 \frac{d}{dr}\right)R - \frac{R}{r^2}\frac{L^2 Y}{\hbar^2} + \frac{2m}{\hbar^2}\left[E - \frac{1}{2}mw^2r^2\right]RY = 0$$

Dividing by $RY$ gives

$$\frac{1}{R}\frac{1}{r^2}\frac{d}{dr}\left(r^2 \frac{d}{dr}\right)R - \frac{1}{Y}\frac{1}{r^2}\frac{L^2 Y}{\hbar^2} + \frac{2m}{\hbar^2}\left(E - \frac{1}{2}mw^2r^2\right) = 0$$

$$\frac{1}{R}\frac{d}{dr}\left(r^2 \frac{d}{dr}\right)R + \frac{2mr^2}{\hbar^2}\left(E - \frac{1}{2}mw^2r^2\right) = \frac{1}{Y}\frac{L^2 Y}{\hbar^2} \tag{3.75}$$

The LHS is a function of $r$ only and represents the radial part while the RHS is a function of angles and is the angular part. As $V = V(r)$ is spherically symmetric potential solution to the angular part will be $Y = Y_{lm_l}(\theta, \phi)$ called spherical harmonic. Also, the operator $L^2$ has eigenvalues $l(l + 1)\hbar^2$ with $l = 0, 1, 2, 3, ... =$ orbital angular momentum quantum number and $Y_{lm_l}(\theta, \phi)$ as its eigen function, i.e.

$$L^2 Y_{lm_l}(\theta, \phi) = l(l + 1)\hbar^2 Y_{lm_l}(\theta, \phi) \tag{3.76}$$

From equation (3.75) using equation (3.76) we have

$$\frac{1}{R}\frac{d}{dr}\left(r^2\frac{d}{dr}\right)R + \frac{2mr^2}{\hbar^2}\left(E - \frac{1}{2}mw^2r^2\right) = \frac{1}{Y}\frac{l(l + 1)\hbar^2 Y}{\hbar^2}$$

$$= l(l + 1) \tag{3.77}$$

This is the radial equation satisfied by $R(r)$.

$$\frac{d}{dr}\left(r^2\frac{d}{dr}\right)R + \frac{2mr^2}{\hbar^2}\left(E - \frac{1}{2}mw^2r^2 - \frac{l(l+1)\hbar^2}{2mr^2}\right)R = 0$$

$$r^2\frac{d^2R}{dr^2} + 2r\frac{dR}{dr} + \frac{2mr^2}{\hbar^2}\left(E - \frac{1}{2}mw^2r^2 - \frac{l(l+1)\hbar^2}{2mr^2}\right)R = 0$$

$$\frac{d^2}{dr^2}R(r) + \frac{2}{r}\frac{d}{dr}R(r) + \frac{2m}{\hbar^2}\left[E - \frac{1}{2}mw^2r^2 - \frac{l(l + 1)\hbar^2}{2mr^2}\right]R(r) = 0. \tag{3.78}$$

Solving, we get energy eigenvalue, i.e. discrete energy levels of the 3D harmonic oscillator as

$$E_{ml} = \left(2m + l + \frac{3}{2}\right)\hbar w \text{ where } m = 0, 1, 2, ... \qquad l = 0, 1, 2, .... \tag{3.79}$$

- In *exercise 3.26* we have employed Cartesian coordinates to solve the Schrödinger equation (3.70) using harmonic oscillator potential as average potential.

The energy eigenvalue turns out to be

$$E_N = \hbar w\left(N + \frac{3}{2}\right) \text{ where } N = 0, 1, 2, 3, ... \text{is the total quantum number.} \tag{3.80}$$

Comparing equations (3.79) and (3.80) we have

$$N = 2m + l \tag{3.81}$$

Define for convenience the quantum number $n = m + 1$. Thus $m = n - 1$. So from equation (3.81) $N = 2(n - 1) + l$. So

$$N = 2n + l - 2 \text{ where } n = 1, 2, 3, ...., \quad l = 0, 1, 2, ....N = 0, 1, 2, ..... \tag{3.82}$$

✓ It is observed that the eigenvalues in spherical polar coordinates depend on two quantum numbers $n$ and $l$.

✓ Each $l$ is degenerate and corresponds to $2l + 1$ values of $m_l = -l, -l + 1, ....0, .......l - 1, l$. $m_l$ is magnetic quantum number.

✓ Again, protons and neutrons are fermions with spin $s = \frac{1}{2}$. So there are $2s + 1 = 2\frac{1}{2} + 1 = 2$ possible spin states corresponding to $m_s = \pm\frac{1}{2}$. This means we have to double each of the $2l + 1$ states, i.e. there will be $2(2l + 1)$ states. As spin states are doubly degenerate, the degree of degeneracy or subshell occupancy will be

$$\sum_l 2(2l + 1) \qquad (3.83)$$

Our aim is to reproduce the magic numbers. Applying Pauli exclusion principle we obtain the number of proton or neutron populating subshells and shells as shown in figure 3.17.

| $N=2n+l-2$ | $n,l$ | State symbol $n$ ($l$ state) | Individual subshell occupancy $\sum_l 2(2l+1)$ | Shell closure occurs at nucleon number |
|---|---|---|---|---|
| $0 = 2.1 + 0-2$ | $1,0$ | $1s$ | $2(2.0+1)=2$ | ②= magic number |
| $1 = 2.1 + 1-2$ | $1,1$ | $1p$ | $2(2.1+1)=6$ | $2+6=$⑧$=$ magic number |
| $2 = 2.1 + 2-2$ | $1,2$ | $1d$ | $2(2.2+1)$ | |
| $2 = 2.2 + 0-2$ | $2,0$ | $2s$ | $+2(2.0+1)=10+2=12$ | $8+12=$⑳$=$ magic number |
| $3 = 2.1 + 3-2$ | $1,3$ | $1f$ | $2(2.3+1)$ | |
| $3 = 2.2 + 1-2$ | $2,1$ | $2p$ | $+2(2.1+1)=14+6=20$ | $20+20=$㊵$=$ magic number |
| $4 = 2.1 + 4-2$ | $1,4$ | $1g$ | $2(2.4+1)$ | |
| $4 = 2.2 + 2-2$ | $2,2$ | $2d$ | $+2(2.2+1)$ | $40+30=$⑦⓪ |
| $4 = 2.3 + 0-2$ | $3,0$ | $3s$ | $+2(2.0+1)$ $=18 + 10 +2 =30$ | Not a magic number |

**Figure 3.17.** Reproducing magic numbers 2, 8, 20, 40 using 3D harmonic oscillator potential $V(r) = \frac{1}{2}mw^2r^2$ using the solution of the Schrödinger equation in spherical polar coordinates.

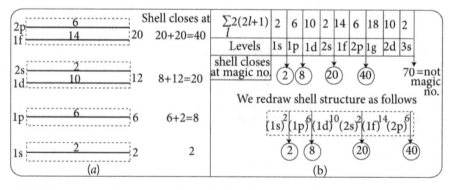

**Figure 3.18.** Shell structure showing generation of the first few magic numbers based on harmonic oscillator potential. Shell closes at magic numbers 2, 8, 20, 40. After that no magic number is produced with a pure harmonic oscillator potential.

Magic numbers 2, 8, 20, 40 are reproduced. The level structure is shown in figure 3.18(a). We can represent the scheme of things in a simpler way, as shown in figure 3.18(b). Other magic numbers are not reproduced with this Hamiltonian $H_0$ of equation (3.67).

The observed magic numbers are 2, 8, 20, (28), (40), 50, 82, 126.

The theory based on harmonic oscillator potential $V(r) = \frac{1}{2}mw^2r^2$ could reproduce only the first few magic numbers viz. 2, 8, 20, 40. Other magic numbers are not generated.

This means we have to improve upon our theory by introducing correction to the Hamiltonian $H_0$ of equation (3.67). The potential $V(r) = \frac{1}{2}mw^2r^2$ is only partially successful.

## 3.22 Generation of magic numbers by introducing spin–orbit correction to harmonic oscillator potential

In our effort to generate the magic numbers we used 3D harmonic oscillator potential $\frac{1}{2}mw^2r^2$ as the spherically symmetric (i.e. central) self-consistent average single-particle nuclear potential. Let us add to this central potential a non-central spin–orbit interaction term

$$V_{so} = -\xi(r)\vec{l} \cdot \vec{s} \tag{3.84}$$

where $\xi(r)$ is some function of $r$, $\vec{l}$ is orbital angular momentum of a nucleon, $\vec{s}$ is intrinsic spin angular momentum of a nucleon. This spin–orbit potential is felt by each nucleon of the nucleus. The nuclear Hamiltonian thus becomes

$$H = H_0 + V_{so} = T + V(r) - \xi(r)\vec{l} \cdot \vec{s} \tag{3.85}$$

$$= -\frac{\hbar^2}{2m}\nabla^2 + \frac{1}{2}mw^2r^2 - \xi(r)\vec{l} \cdot \vec{s} \tag{3.86}$$

$$= H_0 - \xi(r)\vec{l} \cdot \vec{s} \tag{3.87}$$

$H_0 = -\frac{\hbar^2}{2m}\nabla^2 + \frac{1}{2}mw^2r^2$ commutes with $l^2$, $l_z$, $s^2$, $s_z$. So we have definite values of energy eigen function along with definite values of $l^2$, $m_l$, $s^2$, $m_s$.

But since $\vec{l} = \left(l_x, l_y, l_z\right)$, $\vec{s} = (s_x, s_y, s_z)$ it follows that $\vec{l} \cdot \vec{s}$ and hence $H$ will not commute with $l_z$, $s_z$ though $H$ commutes with $l^2$, $s^2$. This means that though we have definite values of $l^2$, $s^2$ we do not have definite values of $m_l$, $m_s$. In fact, $m_l$, $m_s$ get mixed up.

However, $\vec{l} \cdot \vec{s}$ and hence $H$ commutes with $j^2$, $j_z$ where

$$\vec{j} = \vec{l} + \vec{s} \tag{3.88}$$

This is shown in *exercise 3.27*.

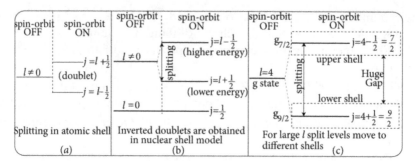

**Figure 3.19.** (a) Atomic doublet. (b) Inverted nuclear doublet. (c) Splitting of levels with large $l$.

So we can define state in terms of quantum numbers $j$, $j_z$ corresponding to a single nucleon. Eigenvalue of $j^2$ is $j(j + 1)\hbar^2$ and that of $j_z$ is $m_j\hbar$ and $m_j$ has $2j + 1$ definite values $-j$ to $+j$ in unit steps. We note that $m_l$ and $m_s$ are not definite and get mixed up to give definite values of $m_j$ according to the relation

$$m_l + m_s = m_j \tag{3.89}$$

- Since spin–orbit coupling has been introduced let us mention the rule for addition of angular momentum of $\vec{l}$ and $\vec{s}$ as depicted in figure 3.19(b) with $s = \frac{1}{2}$ and compare with atomic case shown in figure 3.19(a).

$$j = \begin{cases} l \pm s = l \pm \dfrac{1}{2} & \text{for } l \neq 0 \\[2mm] s = \dfrac{1}{2} & \text{for } l = 0 \end{cases} \tag{3.90}$$

This is referred to as fine structure of $l \neq 0$ energy level.
- State corresponding to $H = H_0 - \xi(r)\vec{l} \cdot \vec{s}$ is thus $| n\,l\,s\,j\,m_j \rangle$.

We calculate the average of $\vec{l} \cdot \vec{s}$ in the state $|n\,l\,s\,j\,m_j\rangle$

$$\langle \vec{l} \cdot \vec{s} \rangle = \langle n\,l\,s\,j\,m_j \,|\, \vec{l} \cdot \vec{s} \,|\, n\,l\,s\,j\,m_j \rangle \tag{3.91}$$

Using the result of *exercise 3.27* viz. $\vec{l} \cdot \vec{s} = \frac{1}{2}(\vec{j}^2 - \vec{l}^2 - \vec{s}^2)$ we can write

$$\langle \vec{l} \cdot \vec{s} \rangle = \langle n\,l\,s\,j\,m_j|\frac{1}{2}(\vec{j}^2 - \vec{l}^2 - \vec{s}^2)\,|\,n\,l\,s\,j\,m_j\rangle$$

$$= \langle \frac{1}{2}(\vec{j}^2 - \vec{l}^2 - \vec{s}^2)\rangle \tag{3.92}$$

Replacing $j^2$, $l^2$, $s^2$ by respective eigenvalues, as shown in *exercise 3.27*, we have $\langle \vec{l} . \vec{s} \rangle \equiv \vec{l} . \vec{s}$

$$\vec{l} . \vec{s} = \begin{cases} \dfrac{\hbar^2}{2} l & \text{for } j = l + \dfrac{1}{2} \\ -\dfrac{\hbar^2}{2}(l+1) & \text{for } j = l - \dfrac{1}{2} \end{cases} \qquad (3.93)$$

We note that the unit of $\vec{l} . \vec{s}$ is angular momentum square. We get energy when we multiply $\vec{l} . \vec{s}$ by a suitable factor $\xi(r)$. Actually the spin–orbit potential is defined as in equation (3.84) and we show in *exercise 3.27* that

$$V_{so} = -\xi(r)\vec{l} . \vec{s} = \begin{cases} -\text{ve} & \text{for } j = l + \dfrac{1}{2} \text{ (shifts down)} \\ +\text{ve} & \text{for } j = l - \dfrac{1}{2} \text{ (shifts up)} \end{cases} \qquad (3.94)$$

It follows from here that level with higher $j = l + \frac{1}{2}$ has lower energy and level with lower $j = l - \frac{1}{2}$ has higher energy.

The Schrödinger equation is

$$\hat{H}\psi = E\psi \qquad (3.95)$$

The solution leads to energy eigenvalues having the following characteristics and leads to a shell structure that we describe in the following.

✓ Each $l \neq 0$ level is split up into a doublet denoted by $j = l \pm \frac{1}{2}$ (with $j = l - \frac{1}{2}$ having higher energy). This is called fine structure, as shown in figure 3.19(b). This is in contrast to the atomic case where $j = l + \frac{1}{2}$ has higher energy, as shown in figure 3.19(a). For this reason the nuclear doublets for $l \neq 0$ are called inverted doublets.

✓ The $l = 0$ (*s* state) is not split and is designated by $j = \frac{1}{2}$. This is also shown pictorially in figure 3.19(b).

✓ Amount of splitting is small for small $l$ but it increases with increase of $l$. For large $l$ splitting is so large and levels are so spread-out that the split levels move apart from each other and become members of different shells. This implies cross-over of energy levels to a different shell. This is depicted in figure 3.19(c).

✓ A group of energy levels (i.e. a bunch of energy levels) is called a shell. Two adjacent shells are separated by a large energy gap.

✓ In the ground state the nucleons fill up the available energy level from the bottom upwards according to the Pauli exclusion principle (as nucleons are fermions).

| $N=2n+l-2$ | $n, l, j$ | State symbol $n(l)_j$ | Arranging shells in ascending order of energy | Individual shell occupancy $\sum_j(2j+1)$ | Shell closure occurs at magic number of nucleon |
|---|---|---|---|---|---|
| $0 = 2.1 + 0-2$ | $1,0,\tfrac12$ | $1s_{1/2}$ | $1s_{1/2}$ | 2 | ② |
| $1 = 2.1 + 1-2$ | $1,1,\tfrac12,\tfrac32$ | $1p_{1/2},1p_{3/2}$ | $1p_{3/2},1p_{1/2}$ | $4+2=6$ | $2+6=⑧$ |
| $2 = 2.1 + 2-2$ / $2 = 2.2 + 0-2$ | $1,2,\tfrac32,\tfrac52$ / $2,0,\tfrac12$ | $1d_{3/2},1d_{5/2}$ / $2s_{1/2}$ | $1d_{5/2},2s_{1/2},1d_{3/2}$ | $6+2+4=12$ | $8+12=⑳$ |
| $3 = 2.1 + 3-2$ / $3 = 2.2 + 1-2$ | $1,3,\tfrac52,\tfrac72$ / $2,1,\tfrac12,\tfrac32$ | $1f_{5/2},1f_{7/2}$ / $2p_{1/2},2p_{3/2}$ | $1f_{7/2}$ | 8 | $20+8=㉘$ |
|  |  |  | $2p_{3/2},1f_{5/2},2p_{1/2}$ | $4+6+2=12$ | $12+28=㊵$ |
| $4= 2.1 + 4-2$ / $4= 2.2 + 2-2$ / $4= 2.3 + 0-2$ | $1,4,\tfrac72,\tfrac92$ / $2,2,\tfrac32,\tfrac52$ / $3,0,\tfrac12$ | $1g_{7/2},1g_{9/2}$ / $2d_{3/2},2d_{5/2}$ / $3s_{1/2}$ | $1g_{9/2}$ | 10 | $10+40=㊿$ |
|  |  |  | $1g_{7/2},2d_{5/2},2d_{3/2},3s_{1/2},1h_{11/2}$ | $8+6+4+2+12=32$ | $50+32=㊿②$ |
| $5= 2.1 + 5-2$ / $5= 2.2 + 3-2$ / $5= 2.3 + 1-2$ | $1,5,\tfrac92,\tfrac{11}{2}$ / $2,3,\tfrac52,\tfrac72$ / $3,1,\tfrac12,\tfrac32$ | $1h_{9/2},1h_{11/2}$ / $2f_{7/2},2f_{5/2}$ / $3p_{3/2},3p_{1/2}$ | $1h_{9/2},2f_{7/2},2f_{5/2},3p_{3/2},3p_{1/2},1i_{13/2}$ | $10+8+6+4+2+14=44$ | $82+44=⑫⑥$ |
| $6= 2.1 + 6-2$ | $1,6,\tfrac{11}{2},\tfrac{13}{2}$ | $1i_{11/2},1i_{13/2}$ |  |  |  |

**Figure 3.20.** Generating all the magic numbers by shell model with harmonic oscillator potential (central potential) corrected by non-central spin–orbit potential.

- We are thus in a position to develop the nuclear shell structure based on harmonic oscillator potential (central potential) corrected by non-central spin–orbit potential. The sequence of the resulting nuclear shell structure is developed in figure 3.20 and the nuclear shell structure is shown in figure 3.21. A very handy way of writing the energy levels along with their occupancy is depicted in figure 3.22.
- The arrangement predicted by the nuclear shell model holds for $A \leqslant 50$. The ordering of levels is not very strictly followed by nuclei beyond $A = 50$. For instance, the ordering of levels $1d_{5/2}2s_{1/2}1d_{3/2}$ changes to $1d_{5/2}1d_{3/2}2s_{1/2}$ etc. However, such a shift of levels is small and so the levels do not leave the shell and so magic number is reproduced.
- The average separation between energy levels decreases as mass number $A$ increases.
- Let $L$ = total orbital angular momentum of nucleus, $S$ = total intrinsic spin angular momentum of nucleus, $J$ = total angular momentum of nucleus.
- This $J$ is also called spin in literature.
- Parity is denoted by

$$\pi = (-1)^L \tag{3.96}$$

If $L$ is even, parity is positive and if $L$ is odd, parity is negative.

Spin parity is denoted by $J^\pi$.

- The nuclear shell model is successful in determining spin parity $J^\pi$ of nuclei in ground state. And this is discussed by employing the concept of pairing that has been explained in *exercise 3.9*. Pairing occurs when two nucleons couple

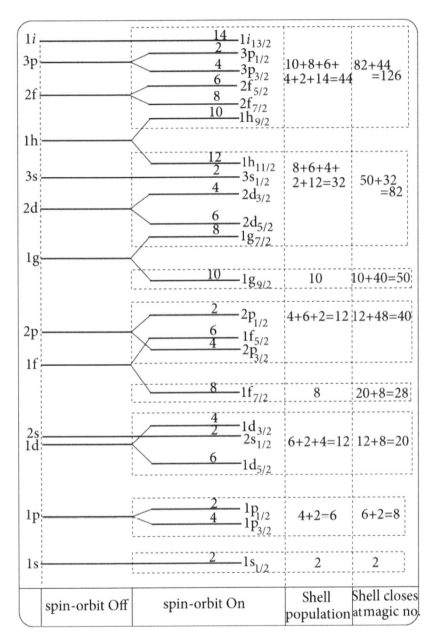

**Figure 3.21.** Shell structure of nucleus. All magic numbers are reproduced using harmonic oscillator potential (central potential) corrected by non-central spin–orbit potential.

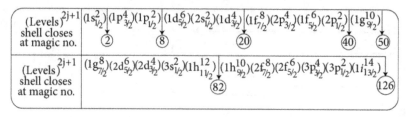

**Figure 3.22.** A handy way of writing the energy levels along with their occupancy depicting nuclear shell structure producing magic numbers.

or combine to produce zero angular momentum leading to lowering of energy.

- We would discuss the extreme single-particle model in which we consider the extreme situation where there is the possibility of pairing between two protons or pairing between two neutrons. Since it requires two particles to pair up, it follows that an even number of protons will pair up, and an even number number of neutrons will also pair up. Only when there is an odd number of protons will one proton be left unpaired. Similarly, only when there is an odd number of neutrons will one neutron be left unpaired.

## 3.23 Spin parity of even–even nuclear ground state from shell structure (even $Z$, even $N$)

Consider an even–even nucleus.

As $Z$ is even, all protons pair up.

As $N$ is even, all neutrons pair up.

Hence $L = 0$, $S = 0$, $J = L + S = 0 + 0 = 0$.

Parity $\pi = (-)^L = (-)^0 = +\text{ve}$.

Ground state spin parity of an even–even nucleus is

$J^\pi = 0^+$

We give some examples.

✓ $_{14}\text{Si}^{30}$ is an even–even nucleus. Arrangement of protons and neutrons in different levels as per the nuclear shell model is shown in figure 3.23(a).

As $Z = 14$, $N = 16$, $A = 30$ there are 14 protons and 16 neutrons.

14 protons are arranged as

$$(1s_{1/2})^2(1p_{3/2})^4(1p_{1/2})^2(1d_{5/2})^6.$$

So this portion of the nuclear shell structure is closed through pairing.

16 neutrons are arranged as

$$(1s_{1/2})^2(1p_{3/2})^4(1p_{1/2})^2(1d_{5/2})^6(2s_{1/2})^2.$$

So this portion of the nuclear shell structure is closed through pairing.

Due to complete pairing, nucleons becomes satisfied with themselves—i.e. they are inactive and hence stable and do not look for any vacant states. Hence

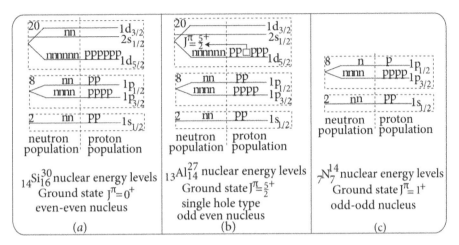

**Figure 3.23.** Arrangement of protons and neutrons in different levels as per nuclear shell model of some nuclei, e.g. (a)$_{14}$Si$_{16}^{30}$ (b)$_{13}$Al$_{14}^{27}$ (c) $_7$N$_7^{14}$.

$$L = 0, \ S = 0, \ J = L + S = 0 + 0 = 0.$$

Parity $\pi = (-)^L = (-)^0 = +\text{ve}$.

So ground state spin parity of $_{14}$Si$^{30}$ nucleus is $J^\pi = 0^+$.

✓ Similarly $_2$He$^4$, $_8$O$^{16}$, $_{38}$Sr$^{88}$, $_{76}$Os$^{192}$, $_{82}$Pb$^{208}$, $_{92}$U$^{238}$ have spin parity value $J^\pi = 0^+$. They are like atoms of inert gas which have closed shells. They are inactive and hence most stable.

## 3.24 Spin parity of odd $A$ nuclear ground state from shell structure (odd $Z$, even $N$ or even $Z$, odd $N$)

Consider odd $A$ nucleus.

We can think of two cases. ☐ Single-particle nucleus type ☐ Single-hole nucleus type.

✓ Single-particle nucleus type

All nucleons in the nucleus, except the last odd nucleon are bound in pairs forming the inert core.

The total angular momentum (called spin) of nucleus is determined by the total angular momentum (or spin) $j$ of the last unpaired nucleon.

Parity of nucleus is determined by the $l$ value of the last unpaired nucleon. Parity is given by $(-1)^l$.

We give an example.

Consider the nucleus $_8$O$^{17}$($Z = 8$, $N = 9$, $A = 17$). Arrangement of protons and neutrons in different levels as per nuclear shell model is shown in figure 3.36(b).

8 protons are arranged as

$$(1s_{1/2})^2(1p_{3/2})^4(1p_{1/2})^2.$$

So this portion of the nuclear shell structure is closed through pairing.

9 neutrons are arranged as

$$(1s_{1/2})^2(1p_{3/2})^4(1p_{1/2})^2(1d_{5/2})^1.$$

The portion of the nuclear shell structure $(1s_{1/2})^2(1p_{3/2})^4(1p_{1/2})^2$ is closed through pairing.

The unpaired and hence active part is $(1d_{5/2})^1$. The electron in state $1d_{5/2}$ is not paired and always looks to obtain a partner. So it is active. As the remaining portion is inactive and $(1d_{5/2})^1$ is active this neutron will determine the spin parity of the entire nucleus.

Now the angular momentum or spin of the state $1d_{5/2}$ is $J = \frac{5}{2}$. Also, $d$ state means $l = 2$. So parity will be $= (-1)^l = (-1)^2 = +$ve. Hence

$$J^\pi \text{of nucleus} = J^\pi \text{ of } 1d_{5/2} = \frac{5^+}{2}.$$

✓ Single-hole nucleus type

All nucleons in the nucleus, except the last topmost shell are bound in pairs forming the inert core. The topmost shell is one nucleon short of its full quota, i.e. there is a single vacancy or hole.

The total angular momentum (called spin) of the nucleus is determined by the total angular momentum (or spin) $j$ of the hole, i.e. $j$ of the state where the hole is situated.

Parity of nucleus is determined by the $l$ value of the hole, i.e. $l$ value of the state where the hole is situated. Parity is given by $(-1)^l$.

We give an example.

Consider the nucleus $_{13}Al^{27}$ $Z = 13$, $N = 14$, $A = 27$. Arrangement of protons and neutrons in different levels as per the nuclear shell model is shown in figure 3.23(b).

14 neutrons are arranged as

$$(1s_{1/2})^2(1p_{3/2})^4(1p_{1/2})^2(1d_{5/2})^6.$$

So this portion of the nuclear shell structure is closed through pairing.

13 protons are arranged as

$$(1s_{1/2})^2(1p_{3/2})^4(1p_{1/2})^2(1d_{5/2})^5.$$

The portion of the nuclear shell structure $(1s_{1/2})^2(1p_{3/2})^4(1p_{1/2})^2$ is closed through pairing.

The remaining part, i.e. the topmost shell $1d_{5/2}$ has a full quota of 6 but 5 slots are filled and so we write $(1d_{5/2})^5$. Clearly there is one proton deficiency. A proton could make it full, closed, inert and stable. So it always looks for a proton to fill its vacancy. As the remaining portion is inactive and portion $(1d_{5/2})^5$ is active, the hole in $1d_{5/2}$ will determine the spin parity of the entire nucleus.

Now the total angular momentum or spin of the state $1d_{5/2}$ is $J = \frac{5}{2}$. Also, $d$ state means $l = 2$. So parity will be $=(-1)^l = (-1)^2 = +$ve. Hence

$$J^\pi \text{of nucleus} = J^\pi \text{ of } 1d_{5/2} = \frac{5^+}{2}.$$

In *exercise 3.31* we have explained how the ground state spin parity is $\frac{3}{2}^-$ for nickel 61.

## 3.25 Spin parity of odd–odd nuclear ground state from shell structure (odd $Z$, odd $N$)

The total angular momentum of an odd–odd nucleus is determined by Nordheim's empirical rule.

Let the last odd proton have orbital angular momentum $\vec{l_1}$, total angular momentum $\vec{j_1}$.

Let the last odd neutron have orbital angular momentum $\vec{l_2}$, total angular momentum $\vec{j_2}$.

The rule to determine spin parity is decided by a number called Nordheim's number, defined as

$$N = j_1 + j_2 + l_1 + l_2 \tag{3.97}$$

- Two rules are there depending on $N$.
- ✓ Strong rule.

$$J = |j_1 - j_2| \text{ if } N = \text{even number.}$$

- ✓ Weak rule.

$J$ lies between $|j_1 - j_2|$ and $j_1 + j_2$ in unit steps but approaches $j_1 + j_2$ if $N =$ odd number.
- Parity rule

Parity is $\pi = \pi_p \pi_n = (-1)^{l_1}(-1)^{l_2} = (-)^{l_1+l_2}$
We give two examples.
□ Consider the nucleus $_7N^{14}(Z = 7, A = 14)$ which is an odd–odd nucleus $Z = 7$, $N = 7$. Arrangement of protons and neutrons in different levels as per the nuclear shell model is shown in figure 3.23(c). Apply Nordheim's rule.

7 protons are arranged as $(1s_{1/2})^2(1p_{3/2})^4(1p_{1/2})^1$.

7 neutrons are arranged as $(1s_{1/2})^2(1p_{3/2})^4(1p_{1/2})^1$

The portion of nuclear shell structure $(1s_{1/2})^2(1p_{3/2})^4$ is closed through pairing which is inactive and hence is stable. The last odd unpaired nucleons are

one proton in state $1p_{1/2} \Rightarrow j_1 = \frac{1}{2}$, $l_1 = 1$ and

one neutron in state $1p_{1/2} \Rightarrow j_2 = \frac{1}{2}$, $l_2 = 1$ (as $p$ state is $l = 1$).
Now Nordheim's number is given by (equation (3.97))
$N = j_1 + j_2 + l_1 + l_2 = \frac{1}{2} + \frac{1}{2} + 1 + 1 = 3 = $ odd. So the weak rule is applicable.
$J = |\frac{1}{2} - \frac{1}{2}|$ to $\frac{1}{2} + \frac{1}{2} = 0$, 1 and parity is given by
$\pi = (-)^{l_1 + l_2} = (-)^{1+1} = +ve = $ even parity. Hence

$$J^\pi = 0^+, \ 1^+.$$

The more likely value is $J^\pi = 1^+$.

☐ Consider the nucleus $_5B^{10}(Z = 5, A = 10)$ which is an odd–odd nucleus. Apply Nordheim's rule.

5 protons are arranged as $(1s_{1/2})^2(1p_{3/2})^3$.

5 neutrons are arranged as $(1s_{1/2})^2(1p_{3/2})^3$

The portion of the nuclear shell structure $(1s_{1/2})^2(1p_{3/2})^2$ is closed through pairing, i.e. inactive and hence stable. The last odd unpaired nucleons are

one proton in state $1p_{3/2} \Rightarrow j_1 = \frac{3}{2}$, $l_1 = 1$ and

one neutron in state $1p_{3/2} \Rightarrow j_2 = \frac{3}{2}$, $l_2 = 1$. (as $p$ state is $l = 1$)

Now Nordheim's number is given by

$N = j_1 + j_2 + l_1 + l_2 = \frac{3}{2} + \frac{3}{2} + 1 + 1 = 5 = $ odd. So the weak rule is applicable.

$J = |\frac{3}{2} - \frac{3}{2}|$ to $\frac{3}{2} + \frac{3}{2} = 0$, 3 and $\pi = (-)^{l_1 + l_2} = (-)^{1+1} = +ve \Leftarrow$ even parity.

Hence $J^\pi = 0^+$, $3^+$. The more likely value is $J^\pi = 3^+$.

## 3.26 Explanation of excited states by shell model

We can explain some excited states by nuclear shell model also. This is illustrated in *exercise 3.32* in which we explain spin parity of first and second excited states of $_8O_9^{17}$ (figure 3.36).

## 3.27 Magnetic moment of nucleus from nuclear shell model— Schmidt limits

✓ Magnetic moment of a proton is

$$\mu_p = 2.79\mu_N \tag{3.98}$$

✓ Magnetic moment of a neutron is

$$\mu_n = -1.91\mu_N \tag{3.99}$$

where

$$\mu_N = \frac{|e|\hbar}{2m_p} \tag{3.100}$$

is the nuclear magneton.

- Expression of spin magnetic moment of a nucleon is

$$\vec{\mu}_s = g_s \mu_N \frac{\vec{s}}{\hbar} \tag{3.101}$$

and the $z$ component of spin magnetic moment of a nucleon is (taking dot product with $\hat{k}$)

$$\vec{\mu}_s \cdot \hat{k} = g_s \mu_N \frac{\vec{s} \cdot \hat{k}}{\hbar}$$

$$\mu_{sz} = g_s \mu_N \frac{s_z}{\hbar} \tag{3.102}$$

Replace by eigenvalues to get

$$\mu_{sz} = g_s \mu_N \frac{m_s \hbar}{\hbar}$$

$$\mu_{sz} = g_s \mu_N m_s \tag{3.103}$$

where $g_s$ is gyromagnetic ratio.

For $m_s = \frac{1}{2}$ we have from equation (3.103)

$$\mu_{sz} = \frac{1}{2} g_s \mu_N \tag{3.104}$$

Let us find the gyromagnetic ratios for proton and neutron.

✓ For a proton we have from equations (3.104) and (3.98)

$$\mu_p = \frac{1}{2} g_p \mu_N$$

$$2.79 \mu_N = \frac{1}{2} g_p \mu_N \Rightarrow g_p = 2 \times 2.79$$

$$g_p = 5.58 \mu_N (\text{actual value is } g_p = 5.5857) \tag{3.105}$$

✓ For a neutron we have from equations (3.104) and (3.99)

$$\mu_n = \frac{1}{2} g_n \mu_N$$

$$-1.91 \mu_N = \frac{1}{2} g_n \mu_N \Rightarrow g_n = 2 \times (-1.91)$$

$$g_n = -3.82 (\text{actual value is } g_n = -3.8260) \tag{3.106}$$

For a nucleus the total magnetic moment is obtained by adding orbital magnetic moment $\vec{\mu}_l$ and the spin magnetic moment $\vec{\mu}_s$ i.e.

$$\vec{\mu} = \vec{\mu}_l + \vec{\mu}_s = g_l \mu_N \frac{\vec{l}}{\hbar} + g_s \mu_N \frac{\vec{s}}{\hbar} \tag{3.107}$$

We note that coupling occurs between $\vec{l}$ and $\vec{s}$ and not between $\vec{\mu}_l$ and $\vec{\mu}_s$. We cannot treat $\vec{l}$ and $\vec{s}$ independently because we are considering $l$ -$s$ coupling. Define

$$\vec{\mu}_j = g_j \mu_N \frac{\vec{j}}{\hbar} = (\vec{\mu} \cdot \hat{j})\hat{j} = \mu_j \hat{j} \tag{3.108}$$

Consider the component of $\vec{\mu}$ along $\vec{j}$ i.e. take dot product of $\vec{\mu}$ with $\hat{j}$ to get

$$\mu_j = \vec{\mu} \cdot \hat{j} = \left( g_l \mu_N \frac{\vec{l}}{\hbar} + g_s \mu_N \frac{\vec{s}}{\hbar} \right) \cdot \hat{j} = \frac{g_l \mu_N}{\hbar} \vec{l} \cdot \hat{j} + \frac{g_s \mu_N}{\hbar} \vec{s} \cdot \hat{j} \tag{3.109}$$

$$\mu_j = \frac{g_l \mu_N}{\hbar} l \cos \angle \vec{l}, j + \frac{g_s \mu_N}{\hbar} s \cos \angle \vec{s}, j \tag{3.110}$$

Now

$$\vec{j} = \vec{l} + \vec{s} \Rightarrow \vec{s} = \vec{j} - \vec{l} \tag{3.111}$$

Squaring

$$s^2 = j^2 + l^2 - 2jl \cos \angle \vec{l}, \hat{j}$$

$$\cos \angle \vec{l}, \hat{j} = \frac{j^2 + l^2 - s^2}{2jl} \tag{3.112}$$

Again,

$$\vec{j} = \vec{l} + \vec{s} \Rightarrow \vec{l} = \vec{j} - \vec{s} \tag{3.113}$$

Squaring,

$$l^2 = j^2 + s^2 - 2js \cos \angle \vec{s}, \vec{j}$$

$$\cos \angle \vec{s}, \vec{j} = \frac{j^2 + s^2 - l^2}{2js} \tag{3.114}$$

From equations (3.110), (3.113), and (3.114) we have

$$\mu_j = \frac{g_l \mu_N}{\hbar} l \frac{j^2 + l^2 - s^2}{2jl} + \frac{g_s \mu_N}{\hbar} s \frac{j^2 + s^2 - l^2}{2js}$$

$$= g_l\mu_N \frac{j}{\hbar} \cdot \frac{j^2 + l^2 - s^2}{2j^2} + g_s\mu_N \frac{j}{\hbar} \cdot \frac{j^2 + s^2 - l^2}{2j^2} \qquad (3.115)$$

Replace by eigenvalues

$$\mu_j = g_l\mu_N \sqrt{j(j+1)} \cdot \frac{j(j+1) + l(l+1) - s(s+1)}{2j(j+1)}$$
$$+ g_s\mu_N \sqrt{j(j+1)} \cdot \frac{j(j+1) + s(s+1) - l(l+1)}{2j(j+1)}$$

$$= \left[ g_l \frac{j(j+1) + l(l+1) - s(s+1)}{2j(j+1)} + g_s \frac{j(j+1) + s(s+1) - l(l+1)}{2j(j+1)} \right] \mu_N \sqrt{j(j+1)}$$

$$\mu_j = g_j\mu_N \sqrt{j(j+1)} \qquad (3.116)$$

where

$$g_j = g_l \frac{j(j+1) + l(l+1) - s(s+1)}{2j(j+1)} + g_s \frac{j(j+1) + s(s+1) - l(l+1)}{2j(j+1)} \qquad (3.117)$$

$$g_j = g_l \left( \frac{1}{2} + \frac{l(l+1) - s(s+1)}{2j(j+1)} \right) + g_s \left( \frac{1}{2} - \frac{l(l+1) - s(s+1)}{2j(j+1)} \right)$$

$$g_j = \frac{1}{2}(g_l + g_s) + \frac{g_l - g_s}{2j(j+1)} [l(l+1) - s(s+1)] \qquad (3.118)$$

For $s = \frac{1}{2}$

$$g_j = \frac{1}{2}(g_l + g_s) + \frac{g_l - g_s}{2j(j+1)} \left[ l(l+1) - \frac{3}{4} \right] \qquad (3.119)$$

For $l = 0, j = s = \frac{1}{2}$

$$g_j = \frac{1}{2}(g_l + g_s) - \frac{1}{2}(g_l - g_s) = g_s \qquad (3.120)$$

The magnetic moment of the nucleus can be predicted from a study of the nuclear shell model, particularly the pairing term.

- Consider even $Z$ even $N$ nucleus.

All protons pair up and for each pair the total angular momentum is zero. So the magnetic moment of protons is also zero.

Similarly, all neutrons pair up and the total angular momentum and magnetic moment of neutrons are zero.

So according to the nuclear shell model the total magnetic moment of the even–even nucleus in ground state is zero, i.e.

$$<\mu_z> = 0 \qquad (3.121)$$

This result does not hold in the excited state since in the excited state pairs can be broken and a particle can move to a higher level.

• For odd $A$ nuclei

All but one nucleon get paired up. The paired up part has zero angular momentum and so no magnetic moment. Only the unpaired nucleon contributes to magnetic moment.

In experiment we measure the $z$ component of magnetic moment.

Considering the component of $\vec{\mu}$ (equation (3.107)) along $\hat{z}$ i.e. taking dot product of $\vec{\mu}$ with $\hat{z}$

$$\vec{\mu} \cdot \hat{z} = g_l \mu_N \frac{\vec{l} \cdot \hat{z}}{\hbar} + g_s \mu_N \frac{\vec{s} \cdot \hat{z}}{\hbar}$$

$$\mu_z = g_l \mu_N \frac{l_z}{\hbar} + g_s \mu_N \frac{s_z}{\hbar} = (g_l l_z + g_s s_z)\frac{\mu_N}{\hbar} \tag{3.122}$$

where $g_l, g_s$ are gyromagnetic ratio corresponding to orbital and intrinsic spin part, respectively.

✓ For proton

$$g_l = 1,\ g_S = 5.5857 \text{ [equation (3.105)]} \tag{3.123}$$

✓ For neutron

$$g_l = 0,\ g_S = -3.8260 \text{ [equation (3.106)]} \tag{3.124}$$

The last odd nucleon is characterized by $\vec{l} + \vec{s} = \vec{j}$ due to spin–orbit interaction $\vec{l} \cdot \vec{s}$ and the state is defined as

$$| l s j m_j > \equiv | j m_j > \tag{3.125}$$

where

$$m_l + m_s = m_j \tag{3.126}$$

In experiment what is measured is the average value of magnetic moment which is expressed as

$$\langle \mu_z \rangle = \langle j m_j | \mu_z | j m_j \rangle \tag{3.127}$$

Since measurement is along the $z$-axis we have to pick up $m_j = j$ value.

Contribution to magnetic moment comes from $l \neq 0$ for which $j = l \pm \frac{1}{2}$. So we have to consider four cases. We outline them now.

$j = l \pm \frac{1}{2}$ for unpaired proton or unpaired neutron.

• Consider the case of $j = l + \frac{1}{2}$ (called stretch case)

Now $m_j = m_l + m_s$ and we consider $m_j = j$. So

$$m_j = j = m_l + m_s = l + \frac{1}{2} \tag{3.128}$$

$m_l$ takes values $-l$ to $+l$ in unit steps and $m_s = \pm\frac{1}{2}$. Hence the only possible combination of $m_l$ and $m_s$ that leads to equation (3.128) is $m_l = l$ and $m_s = \frac{1}{2}$.

Using equation (3.122) in equation (3.127) we have

$$<\mu_z> = <jm_j \,|\, \mu_z \,|\, jm_j> \; = \; <jm_j \,|\, (g_l l_z + g_s s_z)\frac{\mu_N}{\hbar} \,|\, jm_j>$$

$$= g_l < jm_j \,|\, l_z \,|\, jm_j > \frac{\mu_N}{\hbar} + g_s < jm_j \,|\, s_z \,|\, jm_j > \frac{\mu_N}{\hbar}$$

$$= g_l < l_z > \frac{\mu_N}{\hbar} + g_s < s_z > \frac{\mu_N}{\hbar} \tag{3.129}$$

As $m_l = l$, $<l_z> = m_l \hbar = l\hbar$ and as $m_s = \frac{1}{2}$, $<s_z> - \frac{\hbar}{2}$. So from equation (3.129)

$$<\mu_z> = g_l l\hbar \frac{\mu_N}{\hbar} + g_s \frac{\hbar}{2} \frac{\mu_N}{\hbar} = \left( g_l l + \frac{1}{2} g_s \right) \mu_N \tag{3.130}$$

✓ If the last nucleon is a proton then equation (3.130) becomes with equation (3.123)

$$<\mu_z> = \left( g_l l + \frac{1}{2} g_s \right) \mu_N = \left( 1. \, l + \frac{1}{2} 5.5857 \right) \mu_N$$

Using $j = l + \frac{1}{2} \Rightarrow l = j - \frac{1}{2}$

$$<\mu_z> = \left( j - \frac{1}{2} + \frac{1}{2} 5.5857 \right) \mu_N$$

$$<\mu_z> = (j + 2.292\,85)\, \mu_N \tag{3.131}$$

✓ If the last nucleon is a neutron then equation (3.130) becomes with equation (3.124)

$$<\mu_z> = \left( g_l l + \frac{1}{2} g_s \right) \mu_N = \left( 0. \, l - \frac{1}{2} 3.8260 \right) \mu_N = -1.913\, \mu_N \tag{3.132}$$

- Consider the case of $j = l - \frac{1}{2}$ (called the jack-knife case)

Now $m_j = m_l + m_s$ and we consider $m_j = j$. So

$$m_j = j = m_l + m_s = l - \frac{1}{2} \tag{3.133}$$

$m_l$ takes values $-l$ to $+l$ in unit steps and $m_s = \pm\frac{1}{2}$. Hence the only possible combinations of $m_l$ and $m_s$ that lead to equation (3.133) are

✓ $m_l = l$ and $m_s = -\frac{1}{2}$ corresponding state is $| \psi_1 > = | m_l = l \; m_s = -\frac{1}{2} >$ (3.134)

✓ $m_l = l - 1$ and $m_s = \frac{1}{2}$ corresponding state is $| \psi_2 > = | m_l = l - 1 \; m_s = \frac{1}{2} >$ (3.135)

Clearly thus the state $| j \; m_j = j>$ is a linear combination or a mixture of these states written as

$$| j \; m_j = j > = a \, | \psi_1 > + b \, | \psi_2> \tag{3.136}$$

$$\left| j \; m_j = j > = a \right| m_l = l \; m_s = -\frac{1}{2} > +b \, | m_l = l - 1 \; m_s = \frac{1}{2}> \tag{3.137}$$

Hence equation (3.127) gives upon using equation (3.136)

$$\left\langle \mu_z \right\rangle = \left\langle jm_j \left| \mu_z \right| jm_j \right\rangle = \left\langle a\psi_1 + b\psi_2 \left| \mu_z \right| a\psi_1 + b\psi_2 \right\rangle$$

$$= \left\langle a\psi_1 \left| \mu_z \right| a\psi_1 \right\rangle + \left\langle b\psi_2 \left| \mu_z \right| b\psi_2 \right\rangle + \left\langle a\psi_1 \left| \mu_z \right| b\psi_2 \right\rangle + \left\langle b\psi_2 \left| \mu_z \right| a\psi_1 \right\rangle$$

The last two terms on the RHS are zero due to orthogonality of $| \psi_1>$ and $| \psi_2>$ since

$$<a\psi_1 \left| \mu_z \right| b\psi_2 > \; \sim \; < \psi_1 | \psi_2 > = 0$$

$$<b\psi_2 \left| \mu_z \right| a\psi_1 > \; \sim \; < \psi_2 | \psi_1 > = 0$$

So we are left with

$$<\mu_z > = | a |^2 < \psi_1 \left| \mu_z \right| \psi_1 > + | b |^2 < \psi_2 \left| \mu_z \right| \psi_2> \tag{3.138}$$

Now using equation (3.122) we have

$$<\psi_1 \left| \mu_z \right| \psi_1 > \; = \; <m_l = l \; m_s = -\frac{1}{2} | \, (g_l l_z + g_s s_z)\frac{\mu_N}{\hbar} \, | \, m_l = l \; m_s = -\frac{1}{2}>$$

$$= \; <m_l = l \; m_s = -\frac{1}{2} | \, g_l l \hbar \frac{\mu_N}{\hbar} \, | \, m_l = l \; m_s = -\frac{1}{2}>$$

$$+ <m_l = l \ m_s = -\frac{1}{2} \mid g_s\left(-\frac{\hbar}{2}\right)\frac{\mu_N}{\hbar} \mid m_l = l \ m_s = -\frac{1}{2}>$$

$$<\psi_1 \mid \mu_z \mid \psi_1> = \left(g_l l - \frac{g_S}{2}\right)\mu_N \qquad (3.139)$$

$$<\psi_2 \mid \mu_z \mid \psi_2> = <m_l = l - 1 \ m_s = \frac{1}{2} \mid (g_l l_z + g_s s_z)\frac{\mu_N}{\hbar} \mid m_l = l - 1 \ m_s = \frac{1}{2}>$$

$$<\psi_2 \mid \mu_z \mid \psi_2> = \left[g_l(l - 1) + \frac{g_S}{2}\right]\mu_N \qquad (3.140)$$

Recall the raising ladder operator

$$j_+ = j_x + ij_y \qquad (3.141)$$

$$j_+\mid j \ m_j > = \sqrt{(j - m_j)(j + m_j + 1)}\mid j \ m_j + 1> \qquad (3.142)$$

Now apply $j_+$ on both sides of equation (3.137)

$$j_+ \mid j \ m_j = j> = aj_+ \mid m_l = l \ m_s = -\frac{1}{2} > +bj_+ \mid m_l = l - 1 \ m_s = \frac{1}{2}> \qquad (3.143)$$

As $m_j = j$ the operator $j_+$ cannot raise $\mid j \ m_j = j >$ and so the LHS of equation (3.143) is

$$j_+\mid j \ m_j = j> = 0 \qquad (3.144)$$

Consider the first term on the RHS of equation (3.143) and use $j_+ = l_+ + s_+$ to get

$$j_+ \mid m_l = l \ m_s = -\frac{1}{2} > = (l_+ + s_+)\mid m_l = l \ m_s = -\frac{1}{2}>$$

$$= l_+ \mid m_l = l \ m_s = -\frac{1}{2} > +s_+ \mid m_l = l \ m_s = -\frac{1}{2}> \qquad (3.145)$$

As $m_l = l$ the operator $l_+$ cannot raise $\mid m_l = l \ m_s = -\frac{1}{2}>$ and so

$$l_+ \mid m_l = l \ m_s = -\frac{1}{2} > = 0 \qquad (3.146)$$

Now $s_+\mid sm_s > = \sqrt{(s - m_s)(s + m_s + 1)}\mid sm_s + 1>$ [like equation (3.142)]. Hence

$$s_+ \mid s = \frac{1}{2}m_s = -\frac{1}{2} > = \sqrt{\left(\frac{1}{2} - \left(-\frac{1}{2}\right)\right)\left(\frac{1}{2} - \frac{1}{2} + 1\right)}\mid s = \frac{1}{2}m_s = -\frac{1}{2} + 1>$$

$$= \mid s = \frac{1}{2}m_s = \frac{1}{2}> \qquad (3.147)$$

Thus

$$s_+ \mid m_l = l \; m_s = -\frac{1}{2} > \; = \mid m_l = l \; m_s = \frac{1}{2} > \tag{3.148}$$

From equations (3.145), (3.146), and (3.148) we get

$$j_+ \mid m_l = l \; m_s = -\frac{1}{2} > \; = \mid m_l = l \; m_s = \frac{1}{2} > \tag{3.149}$$

Consider the second term on the RHS of equation (3.143)

$$j_+ \mid m_l = l - 1 \; m_s = \frac{1}{2} > \; = (l_+ + s_+) \mid m_l = l - 1 \; m_s = \frac{1}{2} >$$

$$= l_+ \mid m_l = l - 1 \; m_s = \frac{1}{2} > \; + s_+ \mid m_l = l - 1 \; m_s = \frac{1}{2} > \tag{3.150}$$

Using an equation similar to equation (3.142)

$$l_+ \mid l \; m_l = l - 1 > \; = \sqrt{(l - (l-1))(l + (l-1) + 1)} \mid l \; m_l = l - 1 + 1 >$$

$$l_+ \mid l \; m_l = l - 1 > \; = \sqrt{2l} \mid l \; m_l = l >$$

Hence

$$l_+ \mid m_l = l - 1 \; m_s = \frac{1}{2} > \; = \sqrt{2l} \mid m_l = l \; m_s = \frac{1}{2} > \tag{3.151}$$

$$\text{Also } s_+ \mid m_l = l - 1 \; m_s = \frac{1}{2} > \; = 0 \tag{3.152}$$

since $m_s$ is already at the highest value of $\frac{l}{2}$ and it cannot be raised further.
From equations (3.150), (3.151), and (3.152) we get

$$j_+ \mid m_l = l - 1 \; m_s = \frac{1}{2} > \; = \sqrt{2l} \mid m_l = l \; m_s = \frac{1}{2} > \tag{3.153}$$

From equations (3.143), (3.144), (3.149), and (3.153) we have

$$0 = a \mid m_l = l \; m_s = \frac{1}{2} > \; + b\sqrt{2l} \mid m_l = l \; m_s = \frac{1}{2} > \tag{3.154}$$

$$(a + b\sqrt{2l}) \mid m_l = l \; m_s = \frac{1}{2} > \; = 0$$

$$a + b\sqrt{2l} = 0$$

$$a = -b\sqrt{2l} \tag{3.155}$$

Again, normalization restriction is

$$| a |^2 + | b |^2 = 1 \tag{3.156}$$

Using equation (3.155) we get

$$|-b\sqrt{2l}|^2 + | b |^2 = 1 \Rightarrow | b |^2 (2l + 1) = 1$$

$$| b |^2 = \frac{1}{2l + 1} \tag{3.157}$$

From equation (3.155)

$$| a |^2 = | -b\sqrt{2l} |^2 = 2l | b |^2$$

$$| a |^2 = \frac{2l}{2l + 1} \tag{3.158}$$

The expectation value of $<\mu_z>$ is from equations (3.138), (3.139), (3.140), (3.157), (3.158) for the case $j = l - \frac{1}{2}$ is

$$<\mu_z> = | a |^2 <\psi_1 | \mu_z | \psi_1> + | b |^2 <\psi_2 | \mu_z | \psi_2>$$

$$= \frac{2l}{2l + 1}(g_l l - \frac{g_S}{2})\mu_N + \frac{1}{2l + 1}\left[ g_l(l - 1) + \frac{g_S}{2} \right] \mu_N$$

$$<\mu_z> = \frac{1}{2l + 1}\left( g_l 2l^2 - l g_S + g_l l - g_l + \frac{g_S}{2} \right) \mu_N$$

$$<\mu_z> = \frac{1}{2l + 1}\left[ g_l(2l^2 + l - 1) + g_S\left(-l + \frac{1}{2}\right) \right]\mu_N \tag{3.159}$$

As $j = l - \frac{1}{2}$ we put $l = j + \frac{1}{2}$ to get

$$<\mu_z> = \frac{1}{2(j + \frac{1}{2}) + 1}\left[ g_l\left( 2\left(j + \frac{1}{2}\right)^2 + j + \frac{1}{2} - 1 \right) + g_S\left[ -\left(j + \frac{1}{2}\right) + \frac{1}{2} \right] \right]\mu_N$$

$$= \frac{1}{2j + 2}[g_l\left( 2\left(j^2 + j + \frac{1}{4}\right) + j - \frac{1}{2} \right) - jg_S]\mu_N$$

$$= \frac{1}{2j + 2}\left[ g_l\left( 2j^2 + 2j + \frac{1}{2} + j - \frac{1}{2} \right) - jg_S \right] = \frac{1}{2j + 2}[g_l(2j^2 + 3j) - jg_S)]\mu_N$$

$$<\mu_z> = \frac{j}{2j + 2}[g_l(2j + 3) - g_S)]\mu_N \tag{3.160}$$

✓ If the last nucleon is a proton then equation (3.160) becomes using equation (3.123)

$$<\mu_z> = \frac{j}{2j+2}[1.(2j+3) - 5.5857)]\mu_N = \frac{j}{2j+2}(2j - 2.5857)\mu_N$$

$$<\mu_z> = \frac{1}{j+1}(j - 1.292\ 85)\mu_N \tag{3.161}$$

✓ If the last nucleon is a neutron then equation (3.160) becomes using equation (3.124)

$$<\mu_z> = \frac{j}{2j+2}[0.(2j+3) - (3.8260)]\mu_N = \frac{j}{j+1}1.9130\mu_N \tag{3.162}$$

☐ Let us make a plot of the magnetic moments $<\mu_z>$ of all odd $A$ nuclei having a neutron as the last nucleon (i.e. for $Z$ = even, $N$ = odd) against $j$ and we get the plot as shown in figure 3.24(a).

For one single nucleon $j$= half integral as nucleons are fermions having $s = \frac{1}{2}$. The $j$ values are $j = \frac{1}{2}, \frac{3}{2}, \frac{5}{2}, \frac{7}{2}, \frac{9}{2}, \frac{11}{2}\cdots$

✓ For the case $j = l + \frac{1}{2}$ according to equation (3.132) $<\mu_z> = -1.913\mu_N$ which is independent of $j$. This case corresponds to

$$p_{3/2}\left(l = 1, j = 1 + \frac{1}{2} = \frac{3}{2}\right), d_{5/2}\left(l = 2, j = 2 + \frac{1}{2} = \frac{5}{2}\right) \text{etc.}$$

And this gives a straight line $ab$ at $<\mu_z> = -1.913\mu_N$.

✓ For the case $j = l - \frac{1}{2}$ according to equation (3.162) $<\mu_z> = \frac{j}{j+1}1.9130\mu_N$ which depends on $j$.

So for $p_{1/2}(l = 1, j = \frac{1}{2}), <\mu_z> = \frac{1/2}{\frac{1}{2}+1}1.9130\mu_N = \frac{1}{3}(1.9130\mu_N) = 0.64\mu_N$.

So for $d_{3/2}(l = 2, j = \frac{3}{2}), <\mu_z> = \frac{3/2}{\frac{3}{2}+1}1.9130\mu_N = \frac{3}{5}(1.9130\mu_N) = 1.15\mu_N$.

For $j = \frac{5}{2}, <\mu_z> = \frac{5}{7}(1.9130\mu_N) = 1.37\mu_N$,

for $j = \frac{7}{2}, <\mu_z> = \frac{7}{9}(1.9130\mu_N) = 1.49\mu_N$,

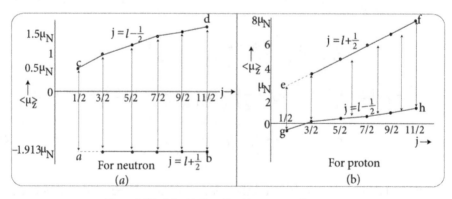

**Figure 3.24.** Schmidt lines for (a) neutrons (b) protons.

for $j = \frac{9}{2}$, $<\mu_z> = \frac{9}{11}(1.9130\mu_N) = 1.57\mu_N$,

for $j = \frac{11}{2}$, $<\mu_z> = \frac{11}{13}(1.9130\mu_N) = 1.62\mu_N$.

This leads to the line $cd$ of figure 3.24(a).

Lines $ab$ and $cd$ are called Schmidt lines and the experimental values of magnetic moments lie in between these lines which thus serve as limits. Experimental values of the magnetic moments fall on the double-headed vertical lines as shown in the figure 3.24(a). No value of magnetic moment lies beyond the Schmidt limits $ab$ and $cd$.

☐ Let us next make a plot of the magnetic moments $<\mu_z>$ of all odd $A$ nuclei having a proton as the last nucleon (i.e. for $N=$ even, $Z=$ odd) against $j$ and we get the plot shown in figure 3.24(b).

✓ For the case $j = l + \frac{1}{2}$ according to equation (3.131) $<\mu_z> = (j + 2.2928)\mu_N$ which is a linear function of $j$.

At $j = l + \frac{1}{2} = 1 + \frac{1}{2} = \frac{3}{2}$, $<\mu_z> = \left(\frac{3}{2} + 2.2928\right)\mu_N = 3.7928\mu_N$. For $j = 2 + \frac{1}{2} = \frac{5}{2}$, $<\mu_z> = \left(\frac{5}{2} + 2.2928\right)\mu_N = 4.7928\mu_N$. Similarly for $j = \frac{7}{2}, \frac{9}{2}, \frac{11}{2}$ the values of $<\mu_z>$ are, respectively, $5.7928\mu_N$, $6.7928\mu_N$, $7.7928\mu_N$ and this gives a straight line $ef$.

✓ For the case $j = l - \frac{1}{2}$ according to equation (3.161) $<\mu_z> = \frac{1}{j+1}(j - 1.2928)\mu_N$.

For $j = l - \frac{1}{2} = 1 - \frac{1}{2} = \frac{1}{2}$

$$<\mu_z> = \frac{1}{j+1}(j - 1.2928)\mu_N = \frac{1}{\frac{1}{2}+1}\left(\frac{1}{2} - 1.2928\right)\mu_N = -0.53\mu_N.$$

For $j = 2 - \frac{1}{2} = \frac{3}{2}$

$$<\mu_z> = \frac{1}{\frac{3}{2}+1}\left(\frac{3}{2} - 1.2928\right)\mu_N = 0.08\mu_N.$$

Similarly, for $j = \frac{5}{2}, \frac{7}{2}, \frac{9}{2}, \frac{11}{2}$ $<\mu_z> = 0.35\mu_N, 0.49\mu_N, 0.58\mu_N, 0.65\mu_N$ respectively leading to the line $gh$.

Lines $ef$ and $gh$ are called Schmidt lines and the experimental values of magnetic moments lie in between these lines which thus serve as limits. Experimental values of the magnetic moments fall on the double-headed vertical lines in figure 3.24(b). No value lies beyond the Schmidt limits $ef$ and $gh$.

## 3.28 Success of the nuclear shell model

✓ It explains the existence of magic nuclei.

✓ It predicts spin parity of ground state nucleus with remarkable success.

✓ It explains greater stability of even–even nuclei and lower stability of odd–odd nuclei.

✓ It explains the zero quadrupole moment of magic nuclei.
✓ The nuclear shell model is partly successful to predict the observed values of magnetic moments.
✓ The nuclear model is well justified for light nuclei in the ground state.

## 3.29  Failures of the nuclear shell model

✓ It fails to explain the vibrational and rotational levels.
✓ It gives drastically reduced value of quadrupole moment for nuclei other than magic nuclei.
✓ It fails to explain excited states of nuclei.
✓ The extreme single-particle model is itself not the right description. There can be more coupling between the so-called paired nucleons.
✓ The value of $g_s$ for neutron and proton may be different when they are free and when they are part of a nucleus. We have used the values in the free state. And if we use values that are reduced from the free state values, the Schmidt lines come closer and agreement with experiment is better.

## 3.30  Comparison of liquid drop model and nuclear shell model

We present a comparison as follows.

| ☐ Liquid drop model | ☐ Nuclear shell model |
| --- | --- |
| It is a strong interaction model. | It is a weak interaction model. |
| Nucleus is likened to a drop of liquid—we speak of nuclear fluid—and no shell structure is thought of. | Nucleons stay in shells and hence do not interact (though the interacting tendency or property is there) as per Pauli exclusion principle. |
| The first three terms of Bethe–Weizsacker's semi-empirical mass formula are derived based on classical ideas and liquid drop model. The asymmetry energy term and pairing energy term follow from quantum ideas and the Pauli exclusion principle. | Nuclear shell model is based on quantum ideas and concepts. |
| Explains binding energy, nuclear fission. | Explains magic number. Predicts spin parity of ground state of nuclei. |

## 3.31  Collective model

In even–even nuclei the ground state is always $0^+$. All protons and neutrons are paired up.

If a pair gets broken, a nucleon of the broken pair can go to an excited state. Energy needed to break the pair and then take it to the excited state is quite large ∼ 2 or 3 *MeV*. But in many even–even nuclei, the excited state occurs at a much lower energy. These excited states are not due to excitation of nucleons in shell model orbitals but involve other types of motion.

In a collective model all the level structures of the shell model are accepted.

We here consider collective motion of the whole nucleus.

Motion of nucleus is considered to be of two types.

✓ Shape vibrations

Equilibrium shape of most nuclei is spherical. However, the shape of the surface of the nucleus may undergo vibration leading to deviation from spherical shape. It might get squeezed or it may bulge out—leading to oscillation of the shape and the surface. Energy will be increased due to such shape vibrations.

The vibrational motion involves vibrational energy state $E_v$ added to nucleonic energy state $E_n$.

As nuclear size is of the order of Fermi we expect that the vibrations are quantized. Different modes of vibrations occur in different quantized energy. If this energy happens to be less than the energy when one nucleon goes from one sub-level or shell to another sub-level or shell then extra energy levels will be seen in between.

✓ Rotation of nucleus

Rotation of a spherical nucleus about a diameter will not change anything as relative configurations are not altered.

But if a non-spherical nucleus is rotated about an axis which is not the symmetry axis then this leads to emergence of energy levels called rotational energy levels.

Rainwater pointed out that a nucleus having large quadrupole moment, e.g. $_{71}Lu^{175}$ is non-spherical, i.e. permanently deformed by nucleons in the outermost shell. The shell model fails to describe such large nuclear deformation.

The rotational motion involves rotational energy states $E_r$ added to nucleonic energy state $E_n$.

Hence total energy will be

$$E = E_n + E_v + E_r \qquad (3.163)$$

Such vibrational and rotational motion of nuclei involves displacements of nucleons and these are referred to as nuclear collective motion. This was discussed by Aage, Bohr and Mottelson and is known as the collective model.

As collective motion of all nucleons is considered, this model amounts to a combination of the concepts of liquid drop model and shell model. For this reason, the collective model is able to account for a large number of experimental observations.

• Deformation of nucleus leads to two types of collective effects or collective modes of excitation—one is collective oscillation and the other is collective rotation.

For a closed-shell configuration the deformimg effects of nucleons on the average cancel out since orbitals are randomly oriented.

Pairing forces couple two similar nucleons to make their net angular momentum zero—and have the tendency to establish a state of spherical symmetry. But the presence of nucleons outside the closed-shell core try to deform the core—resulting in non-zero quadrupole moment of nucleus due to deviation from spherical shape. The deformation tendency increases as more and more nucleons populate levels outside the closed-shell core and eventually the nucleus gets permanently deformed.

We now discuss vibrational and rotational motion separately.

## 3.32 Vibrational energy levels in the collective model

The shell model is based on a nuclear potential that is assumed to be static, spherically symmetric, average potential—it can be taken to be harmonic oscillator potential. Such average field or potential depend on motion of nucleons in the nucleus that introduces time-dependent fluctuations in the average equilibrium potential. The nucleus as a whole takes part in generating such surface vibrations. The situation can be compared to surface waves in a liquid drop.

If a spherical nucleus vibrates, i.e. undergoes surface oscillations then that means its shape is changing from spherical to non-spherical as a function of time. In fact, a nucleus can be thought to be like a continuous liquid drop—which when excited gets deformed from spherical shape executing surface vibrations.

For a spherical surface the distance to any surface point from the centre along any direction $\theta$, $\phi$ is $R$ and

$$R \neq R(\theta, \phi) \tag{3.164}$$

But in the case of surface oscillations the shape of the nucleus gets deformed and the non-spherical surface can be described in terms of distance $R$ of a point on surface from the centre of the nucleus. Now we expect

$$R = R(\theta, \phi) \tag{3.165}$$

The spherical harmonics $Y_{lm_l}(\theta, \phi)$ form the basis in $\theta$, $\phi$ space. Any function of $\theta$, $\phi$ can be expanded in terms of spherical harmonics $Y_{lm_l}(\theta, \phi)$ where
$l = 0, 1, 2, \ldots\infty$ and $m_l = -l$ to $+l$ in unit steps.
where the symbols $l$, $m_l$ are orbital angular momentum quantum number and the magnetic quantum number, respectively, corresponding to nucleons of nucleus.

Since we are now considering vibrations over and above the shell model results (that use the symbols $l$, $m_l$) we cannot use the symbols $l$, $m_l$. We have to use different symbols which we take to be $\lambda$, $\mu$, i.e. spherical harmonics are represented as

$Y_{\lambda\mu}(\theta, \phi)$ where $\lambda = 0, 1, 2, \ldots\infty$ and $\mu = -l$ to $+l$ in unit steps. (3.166)

Let us expand $R(\theta, \phi)$ in terms of $Y_{\lambda\mu}(\theta, \phi)$ as follows

$$R(\theta, \phi) = R_0 \left[ 1 + \sum_{\lambda=0}^{\infty} \sum_{\mu=-\lambda}^{\lambda} a_{\lambda\mu} Y_{\lambda\mu}(\theta, \phi) \right] \tag{3.167}$$

where $R_0$ is the radius of spherical nucleus, $a_{\mu\nu}$ are expansion coefficients with the help of which a general function $R(\theta, \phi)$ can be expanded in terms of the complete set of spherical harmonics $Y_{\lambda\mu}(\theta, \phi)$.

The expansion coefficient $a_{\mu\nu}$ takes care of the oscillation in the shape of nucleus —i.e. of the vibrations induced around the spherical shape. They are also called the deformation parameters. Since shape changes as a function of time the expansion coefficients $a_{\lambda\mu}$ are time dependent, i.e.

$$a_{\mu\nu} = a_{\mu\nu}(t) \tag{3.168}$$

and represent instantaneous orientation of the nuclear surface.

$$\text{If } \lambda = 0, \mu = 0, \quad Y_{00} = \frac{1}{\sqrt{4\pi}} = \text{constant.} \tag{3.169}$$

So we identify $Y_{00}$ as average radius of undeformed spherical nucleus around which vibrations are induced (figure 3.25(a)).

$$\text{If } \lambda = 1, \mu = 0, \quad Y_{10}(\theta, \phi) = \sqrt{\frac{3}{4\pi}} \cos\theta \tag{3.170}$$

This is identified as representing shift in the centre of mass of the nucleus. So this is not shape oscillation and represents motion of the nucleus as a whole. This term will thus not represent or lead to excited state. As we are concerned with internal motion leading to shape oscillations we ignore the $\lambda = 1$ term.

The term $Y_{2\mu}$ corresponding to $\lambda = 2$ is called quadrupole mode of vibration and describes shape oscillations, as shown in figure 3.25(b).

The term corresponding to $\lambda = 3$ is called octupole mode of vibration and describes bulging of surface, as shown in figure 3.25(c).

The nucleus is a quantum system whose energy cannot change in a continuous fashion. The vibrations are quantized in units of $\hbar w$ where vibrational frequency is $w$. Vibrational levels are evenly spaced.

Shape oscillation involves motion of nucleons inside the nucleus and so nucleon orbital angular momentum also corresponds to shape oscillations.

It is clear thus that the energy and parity predicted by the shell model will get modified due to shape oscillation.

☐ A popular way to describe vibration of the nucleus is in terms of a phonon.

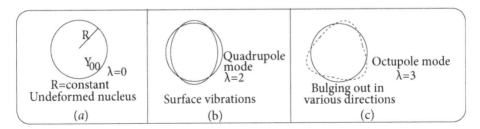

**Figure 3.25.** Vibratory motion leads to deformation of nucleus.

Setting a nucleus into vibration means we have created a phonon of energy $\hbar w$. In other words, vibration of the nucleus with energy $\hbar w$ is described as creation of a phonon of energy $\hbar w$.

- Quadrupole vibration

Quadrupole vibration corresponds to $\lambda = 2$ type vibration. It involves vibrations that can be described through different phonon vibrations.

✓ The lowest energy vibration mode is a one-phonon vibration of $\lambda = 2$ type having energy $\hbar w_2$. In other words, in this vibration mode one phonon is created. The angular momentum is $\vec{2}$. The parity is $(-1)^{\lambda=2} = +$ve.

So spin parity of one-phonon vibration of $\lambda = 2$ type is $2^+$ of energy $\hbar w_2$.

✓ The next higher energy vibration mode is a two-phonon vibration of $\lambda = 2$ type having energy $2\hbar w_2$. In other words, in this vibration mode two phonons are created. The angular momentum is obtained by addition of the angular momenta of the two phonons each with angular momentum $\vec{2}$. Hence addition gives $\vec{2} + \vec{2} \rightarrow 0, 1, 2, 3, 4$.

Since the combining phonons are identical particles, odd numbers are to be ignored as described in *exercise 3.37* where we discuss the addition of angular momenta of two identical particles. So $\vec{2} + \vec{2} \rightarrow 0, 2, 4$ and parity is $+$ve since each phonon has $+$ve parity.

So spin parity of two-phonon vibration of $\lambda = 2$ type is $0^+$, $2^+$, $4^+$ of energy $2\hbar w_2$.

✓ The next higher energy vibration mode is a three-phonon vibration of $\lambda = 2$ type having energy $3\hbar w_2$. In other words, in this vibration mode three phonons are created. The angular momentum is obtained by addition of the angular momenta of the three phonons each with angular momentum $\vec{2}$. Addition $\vec{2} + \vec{2} + \vec{2}$ can be performed and the spin parity of three-phonon vibration of $\lambda = 2$ type turns out to be (can be shown) $0^+$, $2^+$, $3^+$, $4^+$, $6^+$ of energy $3\hbar w_2$.

- Octupole vibration

Octupole vibration corresponds to $\lambda = 3$ type vibration. It involves vibrations that can be described through different phonon vibrations.

✓ The lowest energy vibration mode is a one-phonon vibration of $\lambda = 3$ type having energy $\hbar w_3$. In other words, in this vibration mode one phonon is created. Angular momentum is $\vec{3}$. Parity is $(-1)^{\lambda=3} = -$ve.

So spin parity of one-phonon vibration of $\lambda = 3$ type is $3^-$ of energy $\hbar w_3$.

✓ The next higher energy vibration mode is a two-phonon vibration of $\lambda = 3$ type having energy $2\hbar w_3$. In other words, in this vibration mode two phonons are created.

We can proceed similarly.

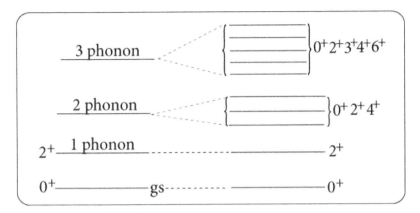

**Figure 3.26.** Splitting of vibrational levels.

For even–even nuclei, ground state as predicted by the shell model is $0^+$. The low-lying excited states can be explained by considering

$\lambda = 2$ one-quadrupole phonon vibration ($2^+$, $\hbar w_2$)

$\lambda = 2$ two-quadrupole phonon vibration ($0^+$, $2^+$, $4^+$, $2\hbar w_2$): triplet of states

$\lambda = 2$ three-quadrupole phonon vibration ($0^+$, $2^+$, $3^+$, $4^+$, $6^+$, $3\hbar w_2$): quintet of states

$\lambda = 3$ one-octupole phonon vibration ($3^-$, $\hbar w_3$).

No nucleus is like a pure harmonic oscillator. And in a pure spherical harmonic oscillator the energy levels are generated by one-phonon vibration of energy $\hbar w$, two-phonon vibration of energy $2\hbar w$, three-phonon vibration of energy $3\hbar w$ and so on. The spin parities of energy levels of even–even nucleus are $2^+$ corresponding to one-quadrupole phonon; $0^+$, $2^+$, $4^+$ degenerate triplet corresponding to two-quadrupole phonon; $0^+$, $2^+$, $3^+$, $4^+$, $6^+$ degenerate quintet corresponding to three-quadrupole phonon.

Real nuclei involve anharmonic vibrational terms. The anharmonic terms remove the degeneracies (figure 3.26) of the two-quadrupole phonon triplet, three-quadrupole phonon quintet. The magnitude of the splitting is a measure of the deviation of a nucleus from a pure harmonic oscillator.

With this we can explain some of the excited energy levels of nuclei like $_{18}Ar^{38}$, $_{52}Te^{120}$, $_{44}Ru^{102}$ as shown in figure 3.27.

So from a study of oscillation of nuclear shape we can understand the low-lying energy levels.

## 3.33 Rotational energy levels in the collective model

Consider the energy levels shown for $_{74}W^{174}$ in figure 3.28(a). Energy value is small and the spacing between energy levels goes on increasing. Energy difference between $0^+$ and $2^+$ is 0.1 $MeV$. The nucleus $_{74}W^{174}$ has a large value of quadrupole moment meaning that its equilibrium shape is not spherical. As more and more nucleons are added outside the closed shells deformation of the nucleus increases and it gets permanently deformed. So the nuclei are heavily deformed—it is non-spherical in shape. Any rotation is then discernible—rotation changes the configuration and corresponds to energy changes.

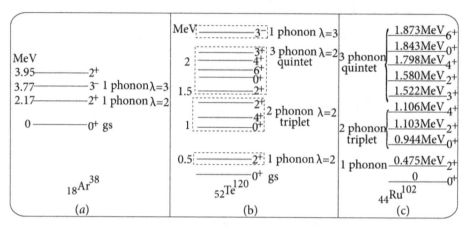

**Figure 3.27.** Explanation using collective model the vibrational energy levels of nuclei above the $0^+$ ground state as predicted by shell model for (a)$_{18}Ar^{38}$ (b) $_{52}Te^{120}$ (c) $_{44}Ru^{102}$.

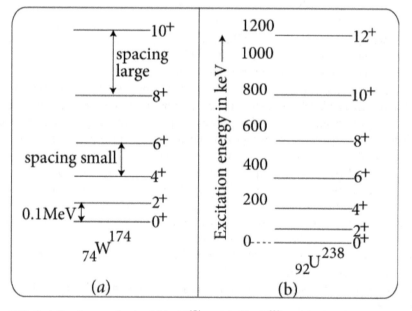

**Figure 3.28.** Rotational energy levels of (a) $_{74}W^{174}$ nuclei. (b) $_{92}U^{238}$ nuclei using the collective model.

Such a nucleus in space can be described by a set of angles. Again, azimuthal angle $\phi$ and the corresponding angular momentum $l_z$ are related by Heisenberg's uncertainty relation

$$\Delta\phi\Delta l_z \geqslant \hbar$$

implying that angular momentum cannot be restricted to one value—a number of angular momentum states exist—called rotational states. Rotational energy levels for $_{92}U^{238}$ are shown in figure 3.28(b).

Rotational spectrum is characterized by the following:
- ✓ All rotational levels have the same parity.
- ✓ Angular momentum increases by $2\hbar$.
- ✓ Spacing between adjacent levels increases with increasing spin.
- ✓ Rotation is quantized in quantum mechanics.

We have shown in *exercise* 3.46 that there can be no collective rotation about symmetry axis.

However, collective rotation of the deformed nucleus about an axis perpendicular to symmetry axis is possible. Let symmetry axis be $\hat{z}$ and rotation axis be perpendicular to it, say $\hat{x}$, as shown in figure 3.29. Intrinsic spin angular momentum of the nucleus is zero. So $\vec{l}$ is the total angular momentum.

Energy of rotation is given by

$$E = \frac{1}{2}Iw^2 \qquad (3.171)$$

where $I$ is the moment of inertia or rotational inertia about the axis of rotation and $w$ is the angular frequency of rotation. In terms of angular momentum

$$l = Iw \Rightarrow w = \frac{l}{I} \qquad (3.172)$$

So we have

$$E = \frac{1}{2}Iw^2 = \frac{1}{2}I\left(\frac{l}{I}\right)^2 = \frac{l^2}{2I} \qquad (3.173)$$

Rotational Hamiltonian is

$$H = \frac{l^2}{2I} \qquad (3.174)$$

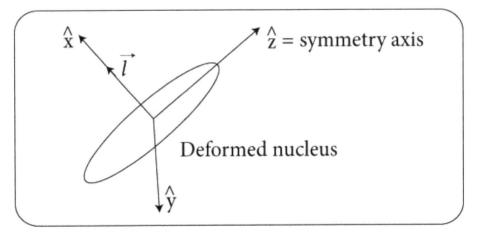

**Figure 3.29.** Permanently deformed axially symmetric nucleus. $\vec{l}$ is not along $\hat{z}$ but perpendicular to it.

where $l^2$ is square of angular momentum operator.

The Schrödinger equation is

$$H\psi = E\psi \Rightarrow \frac{l^2}{2I}\psi = E\psi \qquad (3.175)$$

Obviously $\psi$ is the eigen function of $l^2$ and is spherical harmonics, i.e. $\psi = Y_{lm_l}(\theta, \phi)$.

Putting the eigenvalue of $l^2$ viz. $(l + 1)\hbar^2$, $l = 0, 1, 2, \ldots$ we get

$$E = \frac{\hbar^2}{2I}l(l + 1) \equiv E_l, \, l = 0, 1, 2, \ldots \qquad (3.176)$$

Clearly angular momentum is quantized, rotational energy levels are also quantized.

Parity is $(-1)^l$.

For $l=$ odd, parity $=(-1)^{odd} = -ve$ and for $l=$ even, parity $=(-1)^{even} = +ve$.

Again, from figure 3.29 it follows that the situation is symmetrical about the $XY$ plane, i.e. the nucleus has mirror symmetry and is of definite parity. The wave function should not change sign under reflection from the $XY$ plane. So odd values of $l$ giving odd parity are not acceptable. Allowed values of $l$ are

$$l = \text{even i. e.} \, l = 0, 2, 4, 6, 8, 10, \ldots \qquad (3.177)$$

Hence all states are of even parity.

Energy eigenvalues are

$$E = \frac{\hbar^2}{2I}l(l + 1) \equiv E_l, \, l = 0, 2, 4, 6, 8, 10, \ldots \qquad (3.178)$$

We have to incorporate these results on the shell model ground state $0^+$ of even–even nucleus.

For $l = 0$, $E = 0$ is the ground state $0^+$ and corresponds to no rotation.

For $l = 2$, $E = \frac{\hbar^2}{2I}2(2 + 1) = 6\frac{\hbar^2}{2I}$ and parity $(-1)^2 = +ve$. So spin parity is $2^+$

For $l = 4$, $E = \frac{\hbar^2}{2I}4(4 + 1) = 20\frac{\hbar^2}{2I}$ and parity $(-1)^2 = +ve$. So spin parity is $4^+$

For $l = 6$, $E = \frac{\hbar^2}{2I}6(6 + 1) = 42\frac{\hbar^2}{2I}$ and parity $(-1)^2 = +ve$. So spin parity is $6^+$

Similarly, spin parities of other rotational levels are $8^+$, $10^+$ and so on.

It is clear that all rotational levels are of even parity:$0^+$, $2^+$, $4^+$, $6^+$, $8^+$, $10^+$ ....; angular momentum increases by $2\hbar$ since $l = 0\hbar, 2\hbar, 4\hbar, \ldots$ The ratio of energies eigenvalues is

$$0: 6\frac{\hbar^2}{2I}: 20\frac{\hbar^2}{2I}: 42\frac{\hbar^2}{2I}: \ldots = 0: 6: 20: 42: \ldots\ldots \qquad (3.179)$$

The ratio of energy level spacing is

$$(6 - 0)\frac{\hbar^2}{2I}: (20 - 6)\frac{\hbar^2}{2I}: (42 - 20)\frac{\hbar^2}{2I}: \ldots\ldots = 6: 14: 22: \ldots \qquad (3.180)$$

which means the energy level spacing increases as we move up to higher quantum numbers. These are rotational energy levels for deformed nuclei executing rotation and appear as a rotational band. These energies are quite low—much below the vibrational excitation. So the energy levels shown in figure 3.28 can be explained.

A spherical nucleus being undeformed, any rotation will not change its configuration and hence rotation will not be discernible. So no rotational band will appear for a spherical nucleus. Rotational energy levels do not involve breaking of pairs of a paired set of nucleons but arise due to rotation of a deformed non-spherical nucleus about an axis which is not the symmetry axis—say rotation about an axis perpendicular to symmetry axis.

Rotational spectra predicted by equation (3.178) are in general agreement with experimental observation. However, the computed values are somewhat higher than experimental spectra. This can be explained taking into account centrifugal stretching of the nucleus.

## 3.34 Exercises

**Exercise 3.1** *Why do we use the term nuclear fluid to describe a nucleus?*
[Ans] To describe a nucleus we use the term nuclear fluid because a nucleus can be compared to a liquid drop. Nucleus has mass density $\rho_m \sim 10^{17}\ kg\ m^{-3} = $ constant and nucleon density $\rho \sim 10^{44}$ nucleon $m^{-3}$ implying that the nucleus is incompressible.

It seems that there is virtually no place within the nucleus, which is empty, i.e. where mass is not present. Mass is essentially continuously distributed throughout the nucleus—like a fluid (liquid). So we can look upon an assembly of nucleons in a nucleus as a nuclear fluid.

**Exercise 3.2** *Identify the strong interaction model from the following options.*
*(a)* Shell model, *(b)* liquid drop model, *(c)* both *(d)* none.

**Exercise 3.3** *Identify the weak interaction model from the following options.*
*(a)* Shell model, *(b)* liquid drop model, *(c)* both, *(d)* None.

**Exercise 3.4** *What is the basis of the assumption of the liquid drop model?*
[Ans] In the liquid drop model a nucleus is likened to a liquid drop and nucleons are considered analogous to molecules of liquid. The basis of this assumption is that the two systems have marked similarities, as mentioned in the following table.

| ☐ Nucleus | ☐ Liquid drop |
|---|---|
| Nucleus is assumed to be spherical in shape in the ground state. | A small liquid drop is spherical in shape due to surface tension forces that dominate over the pull of gravity. |
| Nucleons cannot generally escape from the nucleus due to existence of a potential barrier at the surface. | Molecules cannot generally escape from the liquid drop due to existence of surface tension prevailing at the liquid skin. |
| Nuclear fluid is incompressible. Density $\rho_m \sim 10^{17}\ kg\ m^{-3}$ | Liquid is incompressible. Density is constant and independent of the volume and size of the liquid drop. |
| Density is independent of the mass number $A$ (i.e. independent of the size of nucleus). | |
| Nuclear force is of short-range type. | Inter-molecular forces in a liquid are short-range Van der Waals force. |
| Nucleons interact only with their immediate neighbours. This leads to saturation of nuclear force and a constant binding energy per nucleon. | Molecules in a liquid drop interact only with their immediate neighbours. |
| Nucleus has constant binding energy per nucleon. $\frac{B}{A} = 8\frac{MeV}{\text{nucleon}}$. This is independent of size of nuclei i.e. independent of mass number $A$. | Liquid has a fixed latent heat of vapourization. This is independent of size of liquid drop. |
| Supply of binding energy dismantles the nucleus and the nucleons are freed. | Supply of this latent heat converts liquid to gas at that temperature. The molecules bound in liquid are freed. |
| A radioactive nuclide disintegrates by emission of particles—this is called radioactivity. | Liquid molecules evaporate from a liquid drop. |
| Two nuclei can interact to form a compound nucleus—which may subsequently decay. | Liquid drops can condense into a larger drop. |
| When $_{92}U^{235}$ captures a thermal neutron it gets excitation energy that is sufficient to cause oscillations within it so that it suffers fission and disintegrates into smaller fragments of nearly equal size. | When a small liquid drop is allowed to oscillate it breaks up into smaller drops of nearly equal size. |

We now mention some dissimilarities between nuclear fluid and liquid drop.

| ☐ Nucleus | ☐ Liquid drop |
|---|---|
| Nuclear force operative between nucleons in a nucleus is strong force. | Van der Waals force operative between molecules in liquid drop is a weak force. |
| Average kinetic energy of nucleons in nuclei ~10 $MeV$ | Average kinetic energy of molecules of liquid ~0.1 $eV$ |
| Corresponding deBroglie wavelength $\lambda_{\text{nucleon}} \sim 10^{-15}\,m$ | Corresponding deBroglie wavelength $\lambda_{\text{molecule}} \sim 10^{-11}\,m$ |
| Inter-nucleon distance $d \sim 10^{-15}\,m \equiv fm$ | Inter-molecular distance $d \sim 10^{-10}\,m \equiv \text{Å}$ |
| Since $d \sim \lambda_{\text{nucleon}}$ motion of nucleons in the nucleus has to be treated quantum mechanically. | Since $d > \lambda_{\text{molecule}}$ motion of molecules in the liquid is treated classically. |

**Exercise 3.5** *For which nuclei, small A or large A is the surface effect dominant?*
[Ans] The fraction of nucleons which are in contact with the surface is given by the

ratio $\dfrac{\text{surface area of nucleons}}{\text{nuclear volume}} = \dfrac{4\pi R^2}{\frac{4}{3}\pi R^3} = \dfrac{3}{R} = \dfrac{3}{R_0 A^{1/3}} \propto \dfrac{1}{A^{1/3}}$

| ☐ For light nuclei | ☐ For heavy nuclei |
|---|---|
| As $A$= small, the fraction of nucleons in contact with the exterior surface is large. The surface is more populated as shown in figure 3.4(a). As these surface nucleons are more exposed to the external world the tendency to escape is more for a low $A$ surface nucleon. This destabilizing effect reduces the binding energy. So the surface energy correction term is more significant for lighter nuclei. In other words a low $A$ nucleus is less able to hold on to surface nucleons. | As $A$ = large, the fraction of nucleons staying on the surface is small. The interior is more populated as shown in figure 3.4(a). Since the relative number of surface nucleons compared to its internal population is small, it follows that the tendency shown by surface nucleons to escape has less impact on the binding energy. Binding energy is thus not significantly decreased due to surface effect. So the surface energy correction term is less significant for heavy nuclei. |

**Exercise 3.6** *The ratio of surface energy per nucleon in a liquid drop model of $_{13}Al^{27}$ and $_{30}Zn^{64}$ will be which of the following?*
  *(a)* 4: 3,   *(b)* 5 : 3,   *(c)* 3 : 2,   *(d)* 2 : 3.
  Hint: Surface energy term is $a_S A^{2/3}$.

Surface energy per nucleon is obtained by dividing by $A$, i.e.

$\frac{a_S A^{2/3}}{A} = a_S A^{-1/3}$.

And so the ratio of surface energy per nucleon of $_{13}\text{Al}^{27}$ and $_{30}\text{Zn}^{64}$ will be

$\frac{a_S A_{Al}^{-1/3}}{a_S A_{Zn}^{-1/3}} = \frac{27^{-1/3}}{64^{-1/3}} = \left(\frac{64}{27}\right)^{1/3} = \left(\frac{4^3}{3^3}\right)^{1/3} = \frac{4}{3}$

**Exercise 3.7** *Establish that knowing atomic masses of elements is enough to get a fairly accurate estimate of binding energy of nucleus.*

$\boxed{Ans}$ Relation between mass $M_{nu}$ and binding energy $B$ of a nucleus $_Z X_N^A \equiv (Z, N, A) \equiv (Z, N)$ is

$$M_{nu}(Z, N)c^2 = Zm_p c^2 + Nm_n c^2 - B \tag{3.181}$$

$$B = Zm_p c^2 + Nm_n c^2 - M_{nu}(Z, N)c^2 \tag{3.182}$$

Again, atomic masses are more easily and accurately measured experimentally and tabulated compared to the mass of nuclei. In view of this, let us express the formula in terms of atomic masses. Add mass of $Z$ electrons, i.e. $Zm_e$ on both sides of equation (3.181) to get

$$M_{nu}(Z, N)c^2 + Zm_e c^2 = Zm_p c^2 + Zm_e c^2 + Nm_n c^2 - B$$

$$= Z(m_p c^2 + m_e c^2) + Nm_n c^2 - B \tag{3.183}$$

The relation between mass $M_{atom}$ and binding energy $B_{atom}$ of an atom having $Z$ protons, $N$ neutrons, $Z$ electrons is

$$M_{atom}(Z, N)c^2 = M_{nu}(Z, N)c^2 + Zm_e c^2 - B_{atom}$$

$$M_{atom}(Z, N)c^2 + B_{atom} = M_{nu}(Z, N)c^2 + Zm_e c^2 \tag{3.184}$$

Relation between mass $M_H$ and binding energy $B_H$ of a hydrogen atom (system of $p, e^-$) is

$$M_H c^2 = m_p c^2 + m_e c^2 - B_H$$

$$m_p c^2 + m_e c^2 = M_H c^2 + B_H \tag{3.185}$$

From equations (3.184) and (3.183)

$$M_{atom}(Z, N)c^2 + B_{atom} = Z(m_p c^2 + m_e c^2) + Nm_n c^2 - B$$

Using equation (3.185) we have

$$M_{atom}(Z, N)c^2 + B_{atom} = Z(M_H c^2 + B_H) + Nm_n c^2 - B$$

$$M_{atom}(Z, N)c^2 = ZM_H c^2 + Nm_n c^2 - B + ZB_H - B_{atom}$$

Binding energy of atom $\sim eV$ and further the terms $ZB_H$ and $B_{atom}$ are of comparable order and so we can neglect $ZB_H - B_{atom}$. Thus

$$M_{atom}(Z, N)c^2 = ZM_Hc^2 + Nm_nc^2 - B$$

$$B = ZM_Hc^2 + Nm_nc^2 - M_{atom}(Z, N)c^2 \qquad (3.186)$$

This equation (3.186) resembles equation (3.182). We have $M_{atom}(Z, N)$ instead of $M_{nu}(Z, N)$ and $M_H$ instead of $m_p$. Knowing the masses of corresponding neutral atom $M_{atom}(Z, N)$ and hydrogen $M_H$ atom we can obtain binding energy $B$ of the nucleus.

**Exercise 3.8** *Show that for lighter nuclei $A = 2Z$ holds.*

[Ans] The asymmetry energy correction term that tries to destabilize a nuclei due to disparity in the number of protons and neutrons is given by

$$B_{asym} = a_{asym}\frac{(A - 2Z)^2}{A} \text{ (magnitude)}$$

where $a_{asym} = 23.7\ MeV$.

For stability $B_{asym}$ should be minimum. To minimize $B_{asym}$ we set

$$\frac{d}{dZ}B_{asym} = 0 \ \Rightarrow \ \frac{d}{dZ}a_{asym}\frac{(A - 2Z)^2}{A} = 0$$

$$\frac{a_{asym}}{A}2(A - 2Z)(-2) = 0$$

$$A - 2Z = 0$$

$$Z = \frac{A}{2}$$

Also let us check the second derivative at $Z = \frac{A}{2}$

$$\frac{d^2}{dZ^2}B_{asym} = \frac{d}{dZ}\frac{d}{dZ}B_{asym} = \frac{d}{dZ}\frac{a_{asym}}{A}2(A - 2Z)(-2).$$

$$= \frac{a_{asym}}{A}2(-2)(-2) = \frac{8a_{asym}}{A} = +ve.$$

So $B_{asym}$ will be minimum and hence binding energy will be large for $Z = \frac{A}{2}$ leading to a stable nuclei. Hence for lighter nuclei $Z = \frac{A}{2}$ makes it a stable nucleus.

**Exercise 3.9** *What is meant by pairing of nucleons?*

[*Ans*] If two protons having equal and opposite angular momenta combine to form a system with zero angular momentum then we say that we have a paired set of protons. If angular momenta of the two protons are $\vec{j_1}$, $\vec{j_2}$ then the *pp* system has angular momentum $\vec{j} = \vec{j_1} + \vec{j_2}$. If the coupling is such that $\vec{j} = 0$ then that corresponds to lowest energy and the two protons are said to be paired.

Similarly for neutrons.

If two neutrons having equal and opposite angular momenta combine to form a system with zero angular momentum then we say that we have a paired set of neutrons. Such coupling or combination leads to lowering of energy, i.e. helps attain energy that is lower than any other type of combination. In other words pairing lowers energy.

**Exercise 3.10** *Why is the formula of equation (3.31) called the semi-empirical mass formula?*

[*Ans*] This formula of equation (3.31) is called the semi-empirical mass formula since the formula is based upon some qualitative (though not rigorous) theoretical arguments and some calculations, as well as involving fitting of data. The term empirical means only data fitting and no theoretical background.

**Exercise 3.11** *Binding energy of a nucleus is given by which of the following options?*
(a) $B = ZM_{\mathrm{H}}c^2 + Nm_nc^2 - M_{\mathrm{atom}}(Z, N)c^2$
(b) $B = Zm_pc^2 + Nm_nc^2 + M_{nu}(Z, N)c^2$
(c) $B = -ZM_{\mathrm{H}}c^2 - Nm_nc^2 + M_{\mathrm{atom}}(Z, N)c^2$
(d) $B = Zm_pc^2 + Nm_nc^2 - M_{nu}(Z, N)c^2$

**Exercise 3.12** *Show that the minimum of mass parabola occurs at* $Z_0 = \dfrac{A/2}{1 + 0.0075A^{2/3}}$ *and for light nuclei* $A \leqslant 20$, $Z_0 \cong \frac{A}{2}$ *but for heavier nuclei* $Z_0 < \frac{A}{2}$. *Also, find* $Z_0$ *for* $A = 6, 140$.

[*Ans*] The relation between $M(Z, A)$, $Z$ and $A$ is given by equation (3.35)

$$M(Z, A) = \alpha A + \beta Z + \gamma Z^2 - \frac{\delta}{c^2 A^{\frac{3}{4}}}$$

$$\alpha = m_n - \left(\frac{a_V}{c^2} - \frac{a_{\mathrm{asym}}}{c^2} - \frac{a_S}{c^2 A^{1/3}}\right), \beta = (m_p - m_n) - 4\frac{a_{\mathrm{asym}}}{c^2} - \frac{a_C}{c^2 A^{1/3}}, \gamma = \frac{a_C}{c^2 A^{1/3}} + 4\frac{a_{\mathrm{asym}}}{c^2 A}$$

The value of $Z = Z_0$ at which the equation (3.35) leads to a minimum, for some fixed value of $A$ is obtained by setting $\frac{d}{dZ} M(Z, A) = 0$ (equation (3.37)). This leads to

$$\frac{d}{dZ}\left[\alpha A + \beta Z + \gamma Z^2 - \frac{\delta}{c^2 A^{\frac{3}{4}}}\right] = 0$$

$$\beta + 2\gamma Z = 0$$

$$Z = -\frac{\beta}{2\gamma} = Z_0$$

Putting the values of $\beta$, $\gamma$ we get

$$Z_0 = -\frac{(m_p - m_n) - 4\frac{a_{\text{asym}}}{c^2} - \frac{a_C}{c^2 A^{1/3}}}{2\left(\frac{a_C}{c^2 A^{1/3}} + 4\frac{a_{\text{asym}}}{c^2 A}\right)} = \frac{(m_n - m_p)c^2 + 4a_{\text{asym}} + \frac{a_C}{A^{1/3}}}{\frac{2a_C}{A^{1/3}} + \frac{8a_{\text{asym}}}{A}} \qquad (3.187)$$

Now $a_{\text{asym}} = 23.7\ MeV$, $a_C = 0.71\ MeV$

$$\frac{a_C}{A^{1/3}} = \frac{0.71}{A^{1/3}}\ MeV \leqslant 0.71\ MeV.$$

Also,

$$(m_n - m_p)c^2 \leqslant 1\ MeV.$$

This means we can approximately write for the numerator of equation (3.187) $(m_n - m_p)c^2 + 4a_{\text{asym}} + \frac{a_C}{A^{1/3}} \cong 4a_{\text{asym}}$. Thus from equation (3.187) we have

$$Z_0 = \frac{4a_{\text{asym}}}{\frac{2a_C}{A^{1/3}} + \frac{8a_{\text{asym}}}{A}}.$$

Multiplying numerator and denominator by $A$

$$Z_0 = \frac{4Aa_{\text{asym}}}{2A^{2/3}a_C + 8a_{\text{asym}}}.$$

Dividing by $a_{\text{asym}}$ we have

$$Z_0 = \frac{4A}{2A^{2/3}\frac{a_C}{a_{\text{asym}}} + 8}.$$

Taking 8 common in the denominator and putting $a_{\text{asym}} = 23.7\ MeV$, $a_C = 0.71\ MeV$

$$Z_0 = \frac{A}{2}\frac{1}{1 + \frac{1}{4}\frac{a_C}{a_{\text{asym}}}A^{2/3}} = \frac{A/2}{1 + \frac{1}{4}\frac{0.71\ MeV}{23.7\ MeV}A^{2/3}}$$

$$Z_0 = \frac{A/2}{1 + 0.0075A^{2/3}} \qquad (3.188)$$

✓ For light nuclei $A \leqslant 20$,

$$Z_0 = \frac{A/2}{1 + 0.0075A^{2/3}} \cong \frac{A}{2} \tag{3.189}$$

So for light nuclei the nucleon population in nucleus is symmetric w.r.t. neutron and proton and that gives stable nuclei.

✓ For heavier nuclei the factor $1 + 0.0075A^{2/3}$ contributes and so

$$Z_0 < \frac{A}{2} \tag{3.190}$$

So for heavier nuclei neutron population is larger compared to proton population for stability.

☐ For $A = 6$, equation (3.188) gives

$$Z_0 = \frac{6/2}{1 + 0.0075 6^{2/3}} \cong 3.$$

So the most stable nucleus for $A = 6$ is ${}_3\text{Li}^6$. (Note that here $Z_0 = \frac{A}{2} = \frac{6}{2} = 3$)

☐ For $A = 140$, equation (3.188) gives

$$Z_0 = \frac{140/2}{1 + 0.0075 140^{2/3}} \cong 58.$$

So the most stable nucleus for $A = 140$ is ${}_{58}\text{Li}^{140}$. (Note that here $Z_0 < \frac{A}{2} = \frac{140}{2} = 70$)

**Exercise 3.13** *Calculate the binding energy of α particle from the following data.*

$$M_{\text{atom}}({}_2\text{He}^4) = 4.002\ 60u,\ m_n = 1.008\ 665u,\ M_{\text{atom}}({}_1\text{H}^1) = 1.007\ 825u.$$

Ans $B = ZM_{\text{H}}c^2 + Nm_n c^2 - M_{\text{atom}}(Z, N)c^2$ (Equation (3.186))

$$= 2(1.007\ 825u)c^2 + 2(1.008\ 665u)c^2 - 4.002\ 60uc^2$$

$$= 0.030\ 38uc^2$$

$$\xrightarrow{1uc^2 = 931\ MeV} 0.030\ 38 \times 931\ MeV = 28.3\ MeV.$$

**Exercise 3.14** *Binding energy is a consequence of attractive nuclear force between nucleons within the nucleus. Argue if all nucleons in a nucleus contribute to its binding energy.*

*Ans* Consider a nucleus with $A$ nucleons having $^{A}C_2 = \frac{A!}{2!(A-2)!} = \frac{A(A-1)(A-2)!}{2(A-2)!} = \frac{1}{2}A(A-1)$ pairs. While Coulomb interaction occurs only between the $pp$ pair, the nuclear interaction can occur between all $pp$, $nn$, $np$ pairs.

If each pair of the $\frac{1}{2}A(A-1)$ pairs contributes to nuclear binding energy then the binding energy should be proportional to the number of pairs, i.e.

$$B \propto \frac{1}{2}A(A-1), \text{ i.e. } B \propto A^2$$

$$\frac{B}{A} \propto A$$

So the $\frac{B}{A}$ versus $A$ variation is expected to be linear if a nucleon interacts with all other nucleons (since we have considered all pairs). This again implies long-range interaction.

However, the experimental plot of $\frac{B}{A}$ versus $A$ is constant at $8\frac{MeV}{\text{nucleon}}$ (figure 3.2).

This means all pairs of nucleons will not interact to contribute to the binding energy. In other words nuclear interaction is not of long-range but of short-range nature. A nucleon will interact with another nucleon situated close to it in its immediate surrounding.

Suppose each nucleon intercts with $n$ nucleons in its immediate neighbourhood. So $A$ nucleons will interact with $An$ nucleons. But each pair is counted twice. So the number of pairs would be $\frac{1}{2}An$. Clearly then the binding energy will be

$$B \propto \frac{1}{2}An \Rightarrow \frac{B}{A} \propto n$$

$$\frac{B}{A} \propto A^0 \Rightarrow \frac{B}{A} = \text{constant}$$

This is what is observed experimentally demonstrating that nuclear force is of short-range nature.

**Exercise 3.15** *The Bethe–Weizsacker's formula for binding energy of a nucleus is*

$$B(A, Z) = a_V A - a_S A^{\frac{2}{3}} - a_C \frac{Z^2}{A^{1/3}} - a_{\text{asym}} \frac{(A-2Z)^2}{A} + \frac{\delta}{A^{1/2}} \text{ where}$$

$a_V = 15.835 \ MeV$, $a_S = 18.33 \ MeV$, $a_C = 0.714 \ MeV$, $a_{\text{asym}} = 23.20 \ MeV$,

$\delta = 11.2 \ MeV$ *(even–even)*, $\delta = 0$ *(even–odd)*, $\delta = -11.2 \ MeV$ *(odd–odd)*.
  *(a) Evaluate $a_C$ assuming the nucleus to be a uniformly charged sphere of charge $Z|e|$ and radius $R$*

(b) *From the above formula show that for N = Z nuclei, the average binding energy per nucleon is a maximum around Z = 26 . Ignore the pairing energy term in the binding energy formula.*

Ans (a) Assume that the nucleus is a uniformly charged sphere of charge $Z|e|$ and radius $R = R_0 A^{1/3}$, $R_0 = 1.2\,fm$. Electrostatic self-energy is

$$W_{\text{self}} = \frac{1}{4\pi\varepsilon_0}\frac{3}{5}\frac{(Z\mid e\mid)^2}{R} = \frac{1}{4\pi\varepsilon_0}\frac{3\mid e\mid^2 Z^2}{5R_0 A^{1/3}} = \left(\frac{1}{4\pi\varepsilon_0}\frac{3\mid e\mid^2}{5R_0}\right)\frac{Z^2}{A^{1/3}}$$

Equating $W_{\text{self}}$ with the Coulomb energy term $a_C\frac{Z^2}{A^{1/3}}$ we have

$$W_{\text{self}} = \left(\frac{1}{4\pi\varepsilon_0}\frac{3\mid e\mid^2}{5R_0}\right)\frac{Z^2}{A^{1/3}} = a_C\frac{Z^2}{A^{1/3}}$$

$$a_C = \frac{1}{4\pi\varepsilon_0}\frac{3\mid e\mid^2}{5R_0} = \frac{3}{5}\left(\frac{1}{4\pi\varepsilon_0}\frac{\mid e\mid^2}{\hbar c}\right)\frac{\hbar c}{R_0} = \frac{3}{5}\alpha\frac{\hbar c}{R_0}$$

where $\alpha = \frac{1}{4\pi\varepsilon_0}\frac{\mid e\mid^2}{\hbar c} = \frac{1}{137} =$ fine structure constant.

Now $a_C = \frac{3}{5}\alpha\frac{\hbar c}{R_0} = \frac{3}{5}\frac{1}{137}\frac{197\ MeV.fm}{1.2\ fm}$ using $\hbar c = 197\ MeV.fm$ (exercise 1.2)

$$a_C = 0.7\ MeV$$

(b) The Bethe–Weizsacker's semi-empirical mass formula for binding energy of a nucleus is

$$B(A, Z) = a_V A - a_S A^{2/3} - a_C\frac{Z^2}{A^{1/3}} - a_{\text{asym}}\frac{(A - 2Z)^2}{A} + \frac{\delta}{A^{1/2}}$$

Neglecting the pairing energy term and assuming the nucleus to have no asymmetry, i.e. $N = Z$, $A - 2Z = 0 \Rightarrow A = 2Z$. Hence

$$B(A, Z) = a_V A - a_S A^{2/3} - a_C\frac{Z^2}{A^{1/3}}$$

$$\frac{B}{A} = a_V - a_S A^{\frac{2}{3}-1} - a_C Z^2 A^{-\frac{1}{3}-1} = a_V - a_S A^{-1/3} - a_C Z^2 A^{-4/3}$$

Use $A = 2Z$ to get

$$\frac{B}{A} = a_V - a_S(2Z)^{-1/3} - a_C Z^2(2Z)^{-4/3}$$

$$\frac{B}{A} = a_V - 2^{-1/3}a_S Z^{-1/3} - 2^{-4/3}a_C Z^{2/3}$$

Maximize $\frac{B}{A}$ by setting $\frac{d}{dZ}\frac{B}{A} = 0$

$$\frac{d}{dZ}(a_V - 2^{-1/3}a_S Z^{-1/3} - 2^{-4/3}a_C Z^{2/3}) = 0$$

$$0 - 2^{-\frac{1}{3}}a_S\left(-\frac{1}{3}\right)Z^{-\frac{1}{3}-1} - 2^{-\frac{4}{3}}a_C\frac{2}{3}Z^{\frac{2}{3}-1} = 0$$

$$a_S Z^{-4/3} = a_C Z^{-1/3}$$

$$Z = \frac{a_S}{a_C} = \frac{18.33\ MeV}{0.714\ MeV} \cong 26.$$

So average binding energy per nucleon is maximum around 26 for $N = Z$ nuclei.

**Exercise 3.16** *The binding energy of a light nuclei* $(Z, A)$ *in MeV is given approximately by*

$$B(A, Z) \cong 16A - 20A^{2/3} - \frac{3}{4}Z^2 A^{-1/3} + 30\frac{(N-Z)^2}{A}.$$

For a given $A$, which of the following options give the atomic number of the most stable isobar

(a) $\frac{A}{2}\left(1 - \frac{A^{2/3}}{160}\right)^{-1}$  (b) $\frac{A}{2}$  (c) $\frac{A}{2}\left(1 - \frac{A^{2/3}}{120}\right)^{-1}$  (d) $\frac{A}{2}\left(1 + \frac{A^{4/3}}{64}\right)^{-1}$

Hint. For a given $A$ the most stable isobar is given by
$\frac{dB}{dZ} = 0$.

With $N = A - Z \Rightarrow N - Z = A - 2Z$ we get

$$\frac{d}{dZ}\left[16A - 20A^{2/3} - \frac{3}{4}Z^2 A^{-1/3} + 30\frac{(A - 2Z)^2}{A}\right] = 0$$

$$-\frac{3}{4}2Z A^{-1/3} + 30\frac{2(A - 2Z)(-2)}{A} = 0$$

$$ZA^{-1/3} + 80\frac{(A - 2Z)}{A} = 0 \Rightarrow ZA^{1-\frac{1}{3}} + 80(A - 2Z) = 0$$

$$ZA^{2/3} + 80A - 160Z = 0 \Rightarrow \frac{ZA^{2/3}}{160} + \frac{80}{160}A - Z = 0$$

$$Z - \frac{ZA^{\frac{2}{3}}}{160} = \frac{1}{2}A \Rightarrow Z\left(1 - \frac{A^{2/3}}{160}\right) = \frac{1}{2}A$$

$$Z = \frac{A/2}{\left(1 - \frac{A^{2/3}}{160}\right)} \Rightarrow Z = \frac{A}{2}\left(1 - \frac{A^{2/3}}{160}\right)^{-1}$$

**Exercise 3.17** *If binding energy of a light nucleus (Z, A) in MeV is given by*

$$B = a_V A - a_S A^{2/3} - a_C Z^2 A^{-1/3} - a_{asym} \frac{(2Z - A)^2}{A}$$

where $a_V = 16$ *MeV*, $a_S = 16$ *MeV*, $a_C = 0.75$ *MeV*, $a_{asym} = 24$ *MeV* then the most stable isobar for $A = 216$ has atomic number (identify the correct choice)
 (a) 68, (b) 77, (c) 84, (d) 92.
 Repeat for $A = 77$ and justify why $_{33}As^{77}, _{35}Br^{77}$ are unstable but $_{34}Se^{77}$ is stable.
 Hint. ✓ For a given $A$ the most stable isobar is given by

$$\frac{dB}{dZ} = 0$$

$$\frac{d}{dZ}\left[ a_V A - a_S A^{2/3} - a_C Z^2 A^{-1/3} - a_{asym}\frac{(2Z - A)^2}{A} \right] = 0$$

$$-a_C 2ZA^{-1/3} - a_{asym}\frac{2(2Z - A)2}{A} = 0 \Rightarrow a_C ZA^{-1/3} + a_{asym}\frac{4Z - 2A}{A} = 0$$

$$\frac{4Z}{A}a_{asym} - 2a_{asym} + a_C ZA^{-1/3} = 0 \Rightarrow \left(\frac{4}{A}a_{asym} + a_C A^{-1/3}\right)Z = 2a_{asym}$$

For $A = 216$

$$\left[\frac{4}{216}24 \; MeV + 0.75 \; MeV(216)^{-\frac{1}{3}}\right]Z = 2(24 \; MeV)$$

$$Z = \frac{2(24)}{\left[\frac{4}{216}24 + 0.75(216)^{-1/3}\right]} = 84.$$

✓ For $A = 77$

$$Z = \frac{2(24)}{\left[\frac{4}{77}24 + 0.75(77)^{-1/3}\right]} = 34.$$

So $_{34}Se^{77}$ is stable but the adjacent isobars $_{33}As^{77}, _{35}Br^{77}$ are unstable.


**Exercise 3.18** *Make a plot of M versus Z where M is given by Bethe–Weizsacker's formula and it is given that*
 $m_p = 1.007\,825\ u$, $m_n = 1.008\,665\ u$, $a_V = 15.75$ *MeV*, $a_S = 17.8$ *MeV*,
 $a_C = 0.71$MeV, $a_{asym} = 23.7$ *MeV*. $1uc^2 = 931.5$ *MeV*.
*Consider $A = 27$, $Z = 11, 12, 13, 14, 15, 16$.*
 [Ans] Bethe–Weizsacker's formula is

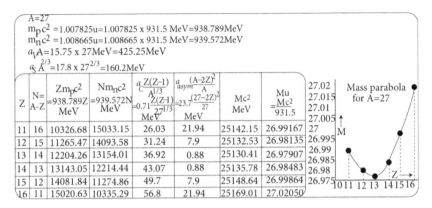

**Figure 3.30.** Plot of $M(Z, A)$ versus $Z$ for $A = 27$.

$$M(Z, A)c^2 = Zm_pc^2 + Nm_nc^2 - a_V A + a_S A^{\frac{2}{3}} + a_C \frac{Z(Z-1)}{A^{1/3}} + a_{asym}\frac{(A-2Z)^2}{A} - \frac{\delta}{A^{3/4}}$$

For odd $A = 27$, $\delta = 0$.

$$M(Z, A)c^2 = Zm_pc^2 + Nm_nc^2 - a_V A + a_S A^{2/3} + a_C \frac{Z(Z-1)}{A^{1/3}} + a_{asym}\frac{(A-2Z)^2}{A}$$

Using the supplied data we calculate each term of the RHS of the Bethe–Weizsacker's semi-empirical mass formula and make a plot of $M(Z, A)$ against $Z$. Plot of $M(Z, A) \equiv M(Z, 27)$ versus $Z$ for $Z = 11$ to $16$ is shown in figure 3.30. Minimum occurs at $Z_0 = 13$ that corresponds to stable nuclei.

**Exercise 3.19** *For a fixed value of $Z = 18$ investigate from the semi-empirical mass formula whether neutron emission is favoured for $A = 40, 45, 50, 60$.*

*Ans* Using semi-empirical mass formula

$$M(Z, A)c^2 = Zm_pc^2 + Nm_nc^2 - a_V A + a_S A^{\frac{2}{3}} + a_C \frac{Z(Z-1)}{A^{1/3}} + a_{asym}\frac{(A-2Z)^2}{A}$$

we calculate $M(Z, N)c^2$ for nucleus with $Z = 18$ having $A = 40, 45, 50, 60$ where

$$m_p = 1.007\,825u, \; m_n = 1.008\,665u, \; a_v = 15.75\;MeV, \; a_S = 17.8\;MeV,$$

$$a_C = 0.71\;MeV, \; a_{asym} = 23.7\;MeV.1uc^2 = 931.5\;MeV.$$

We then calculate $M(Z, N - 1)c^2 + m_nc^2$. If

$M(Z, N)c^2 < M(Z, N - 1)c^2 + m_nc^2$ neutron emission is not favoured.

$M(Z, N)c^2 > M(Z, N - 1)c^2 + m_nc^2$ neutron emission is favoured.

Figure 3.31 shows the calculations and we can conclude as follows. Final results are in $u$.

Z=18
$m_pc^2$ =1.0078u=1.0078 x 931.5 MeV=938.7657MeV
$Zm_pc^2$=18(938.7657)MeV=16897.8MeV
$a_VA$=15.75 x 27MeV=425.25MeV
$m_n$=1.008665u
$m_nc^2$ =1.008665u=1.008665 x 931.5 MeV=939.572MeV

$a_C\frac{Z(Z-1)}{A^{1/3}} = \frac{0.71\text{x}18(18-1)}{A^{1/3}}$

$a_{asym}\frac{(A-2Z)^2}{A} = \frac{23.7(A-2\text{x}18)^2}{A}$

| A | N= A−Z | $Nm_nc^2$ =939.572N MeV | $a_VA$ =15.75A MeV | $a_S A^{2/3}$ =17.8$A^{2/3}$ MeV | $\frac{217.26}{A^{1/3}}$ MeV | MeV | $Mc^2$ MeV | Mu $=\frac{Mc^2}{931.5}$ | $M+m_n$ u | $M_N - M_{N-1}$ |
|---|---|---|---|---|---|---|---|---|---|---|
| 40 | 22 | 20670.58 | 630 | 208.19 | 63.53 | 9.48 | 37219.58 | 39.95661 | | −0.009 |
| 39 | 21 | 19731.01 | 614.25 | 204.71 | 64.07 | 5.47 | 36288.8 | 38.95738 | 39.96605 | |
| 45 | 27 | 25368.44 | 708.75 | 225.19 | 61.08 | 42.66 | 41886.43 | 44.96665 | | −0.005 |
| 44 | 26 | 24428.87 | 693 | 221.85 | 61.54 | 34.47 | 40951.53 | 43.96299 | 44.97166 | |
| 50 | 32 | 30066.3 | 787.5 | 241.58 | 58,97 | 92.90 | 46570.06 | 49.99470 | | −0.002 |
| 49 | 31 | 29126.73 | 771.75 | 238.35 | 59.37 | 81.74 | 45632.25 | 48.98792 | 49.99658 | |
| 60 | 42 | 39462.02 | 945 | 272.81 | 55.5 | 227.52 | 55970.65 | 60.08658 | | 0.002 |
| 59 | 41 | 38522.45 | 929.25 | 269.77 | 55.81 | 212.49 | 55029.07 | 59.07576 | 60.08443 | |

**Figure 3.31.** Investigating the limit beyond which neutron emission is energetically favourable for a fixed proton number as discussed in *exercise 3.19*.

$M({}_{18}X_{22}^{40}) = 39.956\,61u < M({}_{18}X_{21}^{40}) + m_n = 39.966\,05u$. So ${}_{18}X_{22}^{40} \rightarrow {}_{18}X_{21}^{40} + n$ does not occur.

$M({}_{18}X_{27}^{45}) = 44.966\,65u < M({}_{18}X_{26}^{44}) + m_n = 44.971\,66u$. So ${}_{18}X_{27}^{45} \rightarrow {}_{18}X_{26}^{44} + n$ does not occur.

$M({}_{18}X_{32}^{50}) = 49.994\,70u < M({}_{18}X_{31}^{49}) + m_n = 49.996\,58u$. So ${}_{18}X_{32}^{50} \rightarrow {}_{18}X_{31}^{49} + n$ does not occur.

$M({}_{18}X_{42}^{60}) = 60.086\,58u > M({}_{18}X_{41}^{59}) + m_n = 60.084\,43u$. So ${}_{18}X_{42}^{60} \rightarrow {}_{18}X_{41}^{59} + n$ can occur.

So if we go on increasing neutron number in a nucleus, keeping proton number $Z$ = constant, then after a limit neutron emission becomes energetically favourable.

**Exercise 3.20** *There are many heavy nuclei with* $A > 120$ *that are* $\alpha$ *radioactive. However, many heavy nuclei with* $A > 120$ *are also stable. Explain.*

Ans Consider the $\alpha$ decay process of a heavy nucleus with $A > 120$

$$ {}_ZX^A \rightarrow {}_{Z-2}Y^{A-4} + {}_2\text{He}^4 $$

Suppose mass of parent ${}_ZX^A$ is $M_1$ and combined mass of daughter ${}_{Z-2}Y^{A-4}$ plus $\alpha$ is $M_2$. If $M_1 > M_2$ then $\alpha$ decay is energetically favourable and it occurs for many heavy nuclei with $A > 120$.

We note that transition from

$$| \text{ initial state} > \equiv | _ZX^A > \text{ to } |\text{final state} > \equiv | _{Z-2} Y^{A-4} + {}_2He^4 >$$

involves intermediate deformation of the parent nucleus $_ZX^A$. As $\alpha$ tries to escape there might be bulging of a portion of the parent nucleus $_ZX^A$. We can write it as

$$| _ZX^A > \rightarrow |\text{intermediate state} > \rightarrow | _{Z-2} Y^{A-4} + {}_2He^4 >$$

Let mass of this intermediate stage be $M_3$. And so the transformation is

$$| M_1 > \rightarrow | M_3 > \rightarrow | M_2>$$

It may so happen that $M_1 < M_3$ and so parent nucleus $_ZX^A$ cannot go over to intermediate state. This means $| _ZX^A > \rightarrow | _{Z-2} Y^{A-4}>$ transition (though energetically favourable) is prevented and the heavy nucleus is stable. In other words the intermediate state so to say creates a barrier that the $\alpha$ cannot overcome to emerge out.

**Exercise 3.21** *What is the difference between $\alpha$ decay and nuclear fission?*

[*Ans*] In $\alpha$ decay $\alpha$ comes out of the parent nucleus and the daughter is created almost of the same size as (slightly less than) the parent.

In nuclear fission the parent nucleus disintegrates into two nearly equal sized daughter nuclei.

**Exercise 3.22** *What is weak interaction paradox in case of nuclear shell model? How is the paradox resolved?*

[*Ans*] Several experiments (like scattering experiments etc) suggest that nucleons are strongly interacting. But the shell model is a weak interaction model where nucleons are assumed to occupy non-intersecting orbits. So the shell model assumes that nucleons are weakly interacting (non-colliding). This apparent contradiction is called the weak interaction paradox.

This paradox can be resolved by Pauli exclusion principle. For a nucleus in the ground state (or having very low excitation energy) the nucleons fill up all the lower levels according to Pauli exclusion principle and there is no vacancy in the lower most levels. The nucleons do interact strongly but as there is no vacancy in the lower shell structure nucleons cannot fall down through loss of energy (i.e. via inter-nucleon collision). Clearly thus the strong interaction between nucleons is very much there but fails to manifest through a shift of nucleon states due to the Pauli exclusion principle.

**Exercise 3.23** *Why is the nuclear shell model also called the independent particle model?*

[*Ans*] Each nucleon moves independently of all the other nucleons in a spherically symmetric (i.e. central) self-consistent average nuclear potential (field), say $V(r)$, produced by the action of all the nucleons. Hence the nuclear shell model is also called independent particle model, or IPM. It is the potential that any nucleon (proton / neutron) experiences in the nucleus. In other words $V(r)$ is the single-nucleon potential.

**Exercise 3.24** *Which of the following sets consists of a number that is not a magic number?*
*(a)* 50, 82, 126, *(b)* 2, 8, 20, *(c)* 20, 40, 128, *(d)* 8, 82, 126.

**Exercise 3.25** *Nucleons stay in the nucleus and interact with each other as suggested by scattering data from the nucleus. According to the shell model, nucleons stay in non-intersecting shells and hence do not interact. Which is true regarding this apparent contradiction?*
*(a)* *This is a drawback of the shell model.*
*(b)* *This is an assumption in the shell model.*
*(c)* *If we use Pauli exclusion principle there is no contradiction.*
*(d)* *The shell model was later improved in the form of the collective model where this contradiction was overcome.*

**Exercise 3.26** *Reproduce the magic numbers choosing the average potential to be harmonic oscillator potential and solving the Schrödinger equation in Cartesian coordinates.*
‎ ‎ ‎ ‎ $\boxed{Ans}$ Let us work with isotropic 3D harmonic oscillator potential

$$V(r) = \frac{1}{2}mw^2r^2 \tag{3.191}$$

The nuclear Hamiltonian is

$$H_0 = T + V(r) = -\frac{\hbar^2}{2m}\nabla^2 + \frac{1}{2}mw^2r^2$$

where $T$ is kinetic energy of a nucleon. The time-independent Schrödinger equation (3.69) is

$$H_0\psi = E\psi$$

$$\left[-\frac{\hbar^2}{2m}\nabla^2 + V(r)\right]\psi(\vec{r}) = E\psi(\vec{r})$$

where $\psi$ is nuclear wave function, $E$ is energy eigenvalue. In Cartesian coordinates we have

$$\left[ -\frac{\hbar^2}{2m}\left( \frac{\partial^2}{\partial x^2} + \frac{\partial^2}{\partial y^2} + \frac{\partial^2}{\partial z} \right) + \frac{1}{2}mw^2(x^2 + y^2 + z^2) \right]\psi(\vec{r}) = E\psi(\vec{r})$$

In method of separation of variables we put

$$\psi(\vec{r}) = \psi(x, y, z) = \psi_1(x)\psi_2(y)\psi_3(z)$$

$$\left[ -\frac{\hbar^2}{2m}\left( \frac{\partial^2}{\partial x^2} + \frac{\partial^2}{\partial y^2} + \frac{\partial^2}{\partial z} \right) + \frac{1}{2}mw^2(x^2 + y^2 + z^2) \right]\psi_1(x)\psi_2(y)\psi_3(z) = E\psi_1(x)\psi_2(y)\psi_3(z)$$

$$-\frac{\hbar^2}{2m}\left[ \psi_2\psi_3\frac{d^2\psi_1}{dx^2} + \psi_1\psi_3\frac{d^2\psi_2}{dx^2} + \psi_1\psi_2\frac{d^2\psi_3}{dx^2} \right] + \frac{1}{2}mw^2(x^2 + y^2 + z^2)\psi_1\psi_2\psi_3 = E\psi_1\psi_2\psi_3$$

Divide by $\psi_1\psi_2\psi_3$ and rearrange to get

$$\frac{1}{\psi_1}\left( -\frac{\hbar^2}{2m}\frac{d^2}{dx^2} + \frac{1}{2}mw^2x^2 \right)\psi_1 + \frac{1}{\psi_2}\left( -\frac{\hbar^2}{2m}\frac{d^2}{dy^2} + \frac{1}{2}mw^2y^2 \right)\psi_2 + \frac{1}{\psi_3}\left( -\frac{\hbar^2}{2m}\frac{d^2}{dz^2} + \frac{1}{2}mw^2z^2 \right)\psi_3 = E$$

The bracketed terms are functions of $x,y,z$, respectively, and are equal to constants $E_1$, $E_2$, $E_3$, respectively. Hence

$$\frac{1}{\psi_1}\left( -\frac{\hbar^2}{2m}\frac{d^2}{dx^2} + \frac{1}{2}mw^2x^2 \right)\psi_1 = E_1 \tag{3.192}$$

$$\frac{1}{\psi_2}\left( -\frac{\hbar^2}{2m}\frac{d^2}{dy^2} + \frac{1}{2}mw^2y^2 \right)\psi_2 = E_2 \tag{3.193}$$

$$\frac{1}{\psi_3}\left( -\frac{\hbar^2}{2m}\frac{d^2}{dz^2} + \frac{1}{2}mw^2z^2 \right)\psi_3 = E_3 \tag{3.194}$$

with

$$E_1 + E_2 + E_3 = E \tag{3.195}$$

We get from equations (3.192), (3.193), and (3.194)

$$\frac{d^2\psi_1}{dx^2} + \frac{2m}{\hbar^2}\left( E_1 - \frac{1}{2}mw^2x^2 \right)\psi_1 = 0$$

$$\frac{d^2\psi_2}{dy^2} + \frac{2m}{\hbar^2}\left( E_2 - \frac{1}{2}mw^2y^2 \right)\psi_2 = 0,$$

$$\frac{d^2\psi_3}{dz^2} + \frac{2m}{\hbar^2}\left( E_3 - \frac{1}{2}mw^2z^2 \right)\psi_3 = 0$$

These are three 1D harmonic oscillator equations. So the eigenvalues of the 3D harmonic oscillator will be the sum of the three 1D oscillator eigen energies. We thus get the discrete energy levels of 3D oscillator as follows

$$E_{n_1 n_2 n_3} = \hbar w\left(n_1 + \frac{1}{2}\right) + \hbar w\left(n_2 + \frac{1}{2}\right) + \hbar w\left(n_3 + \frac{1}{2}\right)$$

$$= \hbar w\left(n_1 + n_2 + n_3 + \frac{3}{2}\right) \tag{3.196}$$

where

$$n_1, n_2, n_3 = 0, 1, 2, 3, \ldots$$

Defining

$$N = n_1 + n_2 + n_3 \tag{3.197}$$

we write equation (3.196) as

$$E = \hbar w\left(N + \frac{3}{2}\right) \tag{3.198}$$

$$N = 0, 1, 2, 3, \ldots \text{is the total quantum number} \tag{3.199}$$

It is clear that $E$ depends on $N$ and not explicitly on $n_1$, $n_2$, $n_3$ implying that there is degeneracy of energy levels. The $N$th level is $\frac{1}{2}(N + 1)(N + 2)$-fold degenerate.

Our aim is to reproduce the magic numbers. Again, protons and neutrons are fermions having spin $s = \frac{1}{2}$. So there are $2s + 1 = 2\frac{1}{2} + 1 = 2$ possible spin states corresponding to $m_s = \pm\frac{1}{2}$. Applying Pauli exclusion principle we obtain the number of protons or the number of neutrons populating subshells and shells, as shown in figure 3.32. As spin states are doubly degenerate the degree of degeneracy or subshell occupancy will be

| N | $n_1$ $n_2$ $n_3$ | Subshell occupancy $(N+1)(N+2)$ | Shell closure occurs at nucleon number |
|---|---|---|---|
| 0 | 0 0 0 | $(0+1)(0+2)=2$ | ②= magic number |
| 1 | 1 0 0<br>0 1 0<br>0 0 1 | $(1+1)(1+2)=6$ | 2+6=⑧<br>= magic number |
| 2 | 1 1 0   2 0 0<br>1 0 1   0 2 0<br>0 1 1   0 0 2 | $(2+1)(2+2)=12$ | 8+12=⑳<br>= magic number |
| 3 | 1 1 1   2 0 1<br>2 1 0   1 0 2<br>1 2 0   3 0 0<br>0 2 1   0 3 0<br>0 1 2   0 0 3 | $(3+1)(3+2)=20$ | 20+20=㊵<br>= magic number |
| 4 | 2 2 0   3 0 1   1 2 1<br>0 2 2   1 0 3   2 1 1<br>2 0 2   0 1 3   4 0 0<br>3 1 0   0 3 1   0 4 0<br>1 3 0   1 1 2   0 0 4 | $(4+1)(4+2)=30$ | 40+30=⑦⓪<br>= Not a magic number |

**Figure 3.32.** Reproducing magic numbers 2, 8, 20, 40 using 3D harmonic oscillator potential using the solution of the Schrödinger equation in Cartesian coordinates

$$2. \frac{1}{2}(N + 1)(N + 2) = (N + 1)(N + 2) \qquad (3.200)$$

So magic numbers 2, 8, 20, 40 are reproduced. Other magic numbers are not reproduced in this model.

**Exercise 3.27** *Show that spin–orbit interaction commutes with $\vec{j}^{2}, j_{z}$ . Hence show that introduction of spin–orbit coupling removes degeneracy w.r.t. spin. Also, show that spacing between the split levels is proportional to $l$ +1 . Also, show that higher $j$ value corresponds to smaller energy of level.*

[Ans] ✓ The spin–orbit interaction or coupling term (as in equation (3.84)) is proportional to $\vec{l} \cdot \vec{s}$.

Now, the relation between total angular momentum $\vec{j}$ , orbital angular momentum $\vec{l}$, and spin angular momentum $\vec{s}$ is given by

$$\vec{j} = \vec{l} + \vec{s}$$

Now consider

$$\vec{j} \cdot \vec{j} = (\vec{l} + \vec{s}) \cdot (\vec{l} + \vec{s})$$

$$\vec{j}^2 = \vec{l}^2 + \vec{s}^2 + \vec{l} \cdot \vec{s} + \vec{s} \cdot \vec{l} = \vec{l}^2 + \vec{s}^2 + 2\vec{l} \cdot \vec{s} \text{ (as } \vec{l}, \vec{s} \text{ commutes)}$$

$$2\vec{l} \cdot \vec{s} = \vec{j}^2 - \vec{l}^2 - \vec{s}^2$$

$$\vec{l} \cdot \vec{s} = \frac{1}{2}(\vec{j}^2 - \vec{l}^2 - \vec{s}^2) \qquad (3.201)$$

It is clear that

$$[\vec{l} \cdot \vec{s}, \vec{j}^2] = \left[\frac{1}{2}(\vec{j}^2 - \vec{l}^2 - \vec{s}^2), \vec{j}^2\right] = \frac{1}{2}[\vec{j}^2, \vec{j}^2] - \frac{1}{2}[\vec{l}^2, \vec{j}^2] - \frac{1}{2}[\vec{s}^2, \vec{j}^2] = 0$$

$$[\vec{l} \cdot \vec{s}, j_z] = \left[\frac{1}{2}(\vec{j}^2 - \vec{l}^2 - \vec{s}^2), j_z\right] = \frac{1}{2}[\vec{j}^2, j_z] - \frac{1}{2}[\vec{l}^2, j_z] - \frac{1}{2}[\vec{s}^2, j_z] = 0$$

i.e. $\vec{l} \cdot \vec{s}$ commutes with $\vec{j}^{2}, j_z$.

✓ We can calculate the average of $\vec{l} \cdot \vec{s}$ in the state $| n \, l \, s \, j \, m_j>$ by replacing $\vec{j}^2, \vec{l}^2, \vec{s}^2$ by corresponding eigenvalues. This gives from equation (3.201)

$$\vec{l} \cdot \vec{s} = \frac{1}{2}\left(\vec{j}^2 - \vec{l}^2 - \vec{s}^2\right) \rightarrow \frac{1}{2}[j(j + 1)\hbar^2 - l(l + 1)\hbar^2 - s(s + 1)\hbar^2] \qquad (3.202)$$

Again, from the angular momentum addition rule given by equation (3.90) (figure 3.19) we have

$$j = \begin{cases} l \pm s = l \pm \frac{1}{2} & \text{for } l \neq 0 \\ s = \frac{1}{2} & \text{for } l = 0 \end{cases}$$

Putting $s = \frac{1}{2}$, $j = l + \frac{1}{2}$ in equation (3.202) we have

$$\vec{l} \cdot \vec{s} = \frac{1}{2}\left[\left(l + \frac{1}{2}\right)\left(l + \frac{1}{2} + 1\right) - l(l+1) - \frac{1}{2}\left(\frac{1}{2} + 1\right)\right]\hbar^2$$

$$= \frac{1}{2}\left[\left(l + \frac{1}{2}\right)\left(l + \frac{3}{2}\right) - l(l+1) - \frac{3}{4}\right]\hbar^2 = \frac{1}{2}\left[l^2 + 2l + \frac{3}{4} - l^2 - l - \frac{3}{4}\right]\hbar^2 = \frac{1}{2}l\hbar^2$$

Putting $s = \frac{1}{2}$, $j = l - \frac{1}{2}$ in equation (3.202) we have

$$\vec{l} \cdot \vec{s} = \frac{1}{2}\left[\left(l - \frac{1}{2}\right)\left(l - \frac{1}{2} + 1\right) - l(l+1) - \frac{1}{2}\left(\frac{1}{2} + 1\right)\right]\hbar^2$$

$$= \frac{1}{2}\left[\left(l - \frac{1}{2}\right)\left(l + \frac{1}{2}\right) - l(l+1) - \frac{3}{4}\right]\hbar^2 = \frac{1}{2}\left[l^2 - \frac{1}{4} - l^2 - l - \frac{3}{4}\right]\hbar^2 = -\frac{1}{2}(l+1)\hbar^2$$

Clearly the result

$$\vec{l} \cdot \vec{s} = \begin{cases} \dfrac{\hbar^2}{2}l & \text{for } j = l + \dfrac{1}{2} \\[2mm] -\dfrac{\hbar^2}{2}(l+1) & \text{for } j = l - \dfrac{1}{2} \end{cases} \tag{3.203}$$

shows that the spin–orbit force removes degeneracy of levels since the values of $\vec{l} \cdot \vec{s}$ are different for $j = l \pm \frac{1}{2}$.

✓ Let us find the spacing or splitting between $j = l \pm \frac{1}{2}$ levels. It follows from equation (3.203) that

$$\vec{l} \cdot \vec{s}\,|_{l+\frac{1}{2}} - \vec{l} \cdot \vec{s}\,|_{l-\frac{1}{2}} = \frac{\hbar^2}{2}l - \left[-\frac{\hbar^2}{2}(l+1)\right] = \frac{\hbar^2}{2}(2l+1) \propto (2l+1)$$

Clearly the spacing between the split levels is proportional to $2l + 1$.

It shows that the larger the $l$, the larger is the splitting. So splitting is smaller for low $l$ (say for $l = 1, 2, ..$) but larger for large $l$ (say $l = 3, 4, ..$).

✓ The spin–orbit interaction potential is by equation (3.84) taken as (with $\xi(r) = $ positive)

$$V_{so} = -\xi(r)\vec{l} \cdot \vec{s} = \begin{cases} -\xi(r)\dfrac{\hbar^2}{2}l & = -\text{ve} \quad \text{for } j = l + \dfrac{1}{2} \quad \text{(shifts down)} \\[3mm] \xi(r)\dfrac{\hbar^2}{2}(l+1) & = +\text{ve} \quad \text{for } j = l - \dfrac{1}{2} \quad \text{(shifts up)} \end{cases} \tag{3.204}$$

It follows from equation (3.204) that a level with higher $j = l + \frac{1}{2}$ has lower energy and a level with lower $j = l - \frac{1}{2}$ has larger energy.

**Exercise 3.28** *(a) Explain the nuclear shell structure with infinite square well potential.*

*(b) Why do you think that infinite square well potential is not a good description of nuclear potential?*

*(c) Suggest suitable potentials to describe a nucleus. Comment on how nuclear shell structure is obtained.*

$\boxed{Ans}$ (a) Infinite square well potential (figure 3.33(a)) is defined by

$$V(r) = \begin{cases} 0 & r < r_0 \\ \infty & r \geq r_0 \end{cases}.$$

It is a central potential—does not depend on angular coordinates $\theta$, $\phi$ (so there is no directional dependence). Let $r$ be the coordinate of nucleon and $m$ is nucleon mass.

The nuclear Hamiltonian is

$$H = T + V(r)$$

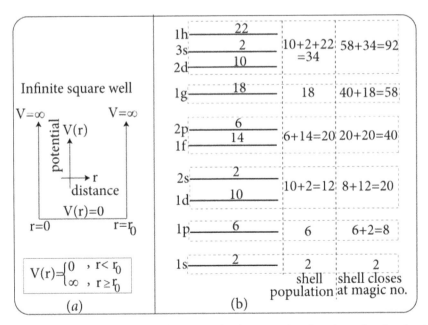

**Figure 3.33.** (a) Infinite square well. (b) Shell structure showing generation of magic numbers based on infinite square well potential.

where $T$ is kinetic energy of a nucleon. The corresponding time-independent Schrödinger equation is

$$H\psi = E\psi$$

where $\psi$ is the nuclear wave function, $E$ is energy eigenvalue. Written explicitly

$$\nabla^2\psi + \frac{2m}{\hbar^2}[E - V(r)]\psi = 0$$

Using the method of separation of variables

$$\psi = \psi(r, \theta, \phi) = R(r)Y(\theta, \phi)$$

As $V = V(r)$ is the spherically symmetric potential solution to the angular part is spherical harmonic $Y_{lm_l}(\theta, \phi)$ and so

$$\psi = R(r)Y_{lm_l}(\theta, \phi)$$

Putting $R(r) = \frac{u(r)}{r}$ the radial part $u(r)$ satisfies the radial equation (*exercises 6.4, 6.5*)

$$\frac{d^2}{dr^2}u(r) + \frac{2m}{\hbar^2}\left[E - V(r) - \frac{l(l+1)\hbar^2}{2mr^2}\right]u(r) = 0$$

At $r > r_0$, $V(r) = \infty$ and so $u = 0$.
At $r = r_0$, $u = 0$ from continuity requirement.
For $r < r_0$ the solution can be shown to be the spherical Bessel function viz.
$u(r) = rj_l(\text{kr})$
where $k = \sqrt{\frac{2mE}{\hbar^2}}$.
At $r = r_0$, $u = 0 \Rightarrow j_l(kr_0) = 0$.
From the expression of Bessel function we can get the value of $k$ and hence evaluate the energy eigenvalue (i.e. energy levels of nucleon) viz. $E$, which are a set of levels (characterized by quantum number)
$n = 1, 2, 3, \ldots$ for each value of $l = 0, 1, 2, \ldots$.
Unlike the atomic case, here all combinations of $n, l$ are possible. This leads to the following shell structure which generates the first 4 magic numbers only.

$$n = 1, l = 0 \Rightarrow 1s$$

$$n = 1, l = 1 \Rightarrow 1p$$

$$n = 1, l = 2 \Rightarrow 1d$$

$$n = 1, l = 3 \Rightarrow 1f$$

$$n = 1, l = 4 \Rightarrow 1g \text{ etc,}$$

are allowed but in the atomic case only $n = 1, l = 0 \Rightarrow 1s$ is allowed.

Each state can accommodate two protons and also two neutrons—one with spin up and another with spin down due to Pauli exclusion principle. So 1*s* means two quantum states and so on. Figure 3.33(b) shows the shell structure as obtained based on the infinite square well potential. As magic numbers are not reproduced we conclude that it is not the right kind of potential to choose for describing a nucleus.

(b) The infinite potential well corresponds to the fact that if a particle is trapped within, it cannot come out as the edges are of infinite height, i.e. $V(r)$ is infinity at $r_0$. The infinitely high fence cannot be overcome by the particle. But in a real nucleus, under certain situations of $\alpha$ decay or $\beta$ decay nucleons can come out. So an infinite well fails to describe a nucleus.

(c) To describe the real nucleus we can think of potentials that are:

- ✓ finite potential so that nucleons are able to emerge out under special circumstances, say during $\alpha$ decay or $\beta$ decay;
- ✓ the edges should be rounded to take care of the fact that the nucleus does not possess a sharp boundary.

Examples of such potentials are:
(a) harmonic oscillator potential well (figure 3.34(a));
(b) Woods–Saxon potential well (figure 3.34(b)).

However, none of these potentials could generate all the observed magic numbers. Magic numbers were generated by the nuclear shell model when one takes the spin–orbit interaction term $V_{so} = -\xi(r)\,\vec{l}.\,\vec{s}$ into consideration as a correction to either the harmonic oscillator potential or the Woods–Saxon potential. We have discussed this in figure 3.18.

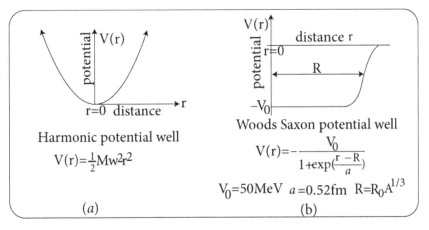

Figure 3.34. (a) Harrmonic potential well. (b) Woods–Saxon potential well.

**Exercise 3.29** *Can you describe neutron–proton asymmetry through Woods–Saxon single-particle potential?*

[Ans] The Woods–Saxon potential well is depicted in figure 3.34(b).

The neutron does not experience any Coulomb potential and interacts with the rest of the nucleons through nuclear potential.

The proton interacts with the rest of the nucleons through nuclear potential over and above the interaction with the rest of the protons through Coulomb potential.

Clearly potential is different for neutrons and protons.

The difference is not marked and can be ignored for low $A$, i.e. for lighter nuclei, and hence for them energy levels are similar.

But for large $A$ nuclei Coulomb interaction is significant. For a spherically symmetric charge distribution $Z \mid e \mid$ of radius $R$ Coulomb potential energy is given by

$$V = \frac{Z \mid e \mid^2}{4\pi\varepsilon_0 R}\left[\frac{3}{2} - \frac{1}{2}\left(\frac{r}{R}\right)^2\right] = w\left[\frac{3}{2} - \frac{1}{2}\left(\frac{r}{R}\right)^2\right] = +\text{ve for } r < R$$

$$V \propto \frac{1}{r} \text{ for } r > R$$

where $w = \frac{Z \mid e \mid^2}{4\pi\varepsilon_0 R}$, $r$ is radial distance. Clearly

$$\text{at } r = 0, \ V = \frac{Z \mid e \mid^2}{4\pi\varepsilon_0 R}\frac{3}{2} = \frac{3}{2}w \text{ and}$$

$$\text{at } r = R, \ V = \frac{Z \mid e \mid^2}{4\pi\varepsilon_0 R} = w$$

In other words, the potential for the proton is shifted up at $r = 0$ as well as at $r = R$ and falls as $\frac{1}{r}$ for $r > R$, as shown in figure 3.35. In other words, potential is modified for the proton compared to the neutron potential.

Energy levels of the portion $ab$ are missing for proton potential.

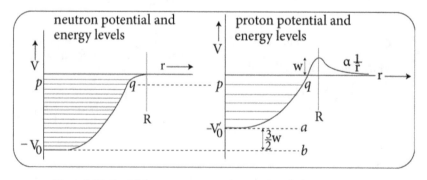

**Figure 3.35.** Explaining *np* asymmetry through Woods–Saxon potential.

Because of lowering of depth, spacing will be larger for proton potential.

If nucleons fill up up to a certain energy level, say $pq$, then there will be a greater number of energy levels in the case of neutron potential than proton potential. So the number of neutrons will be larger than the number of protons, i.e. $N > Z$, i.e. asymmetry in $n$ and $p$ occurs.

**Exercise 3.30** *Find the ground state spin parity of carbon 6 and silicon 27.*

$\boxed{Ans}$ ✓ $_6C_7^{13}$ is a single-particle nucleus type ($Z = 6$, $N = 7$, $A = 13$). The arrangement of neutrons and protons is shown in figure 3.36(a).

6 protons are arranged as $(1s_{1/2})^2 (1p_{3/2})^4$. This part is closed through pairing.

7 neutrons are arranged as $(1s_{1/2})^2 (1p_{3/2})^4(1p_{1/2})^1$. The part $(1s_{1/2})^2 (1p_{3/2})^4$ is closed through pairing. The unpaired and hence active part is $(1p_{1/2})^1$. This electron in state $1p_{1/2}$ is not paired and always looks to obtain a partner. So it is active. As the remaining portion is inactive and $(1p_{1/2})^1$ is active, this neutron will determine the spin parity of the entire nucleus. Now $1p_{1/2}$ state has $j = \frac{1}{2}$ and $l = 1$. So parity is $\pi = (-1)^l = (-1)^1 = -\text{ve}$. Hence

$$J^\pi \text{ of nucleus } = J^\pi \text{of } 1p_{1/2} = \frac{1^-}{2}.$$

✓ $_{14}Si_{13}^{27}$ is a single-hole nucleus type ($Z = 14$, $N = 13$, $A = 27$). The arrangement of neutrons and protons is shown in figure 3.36(b).

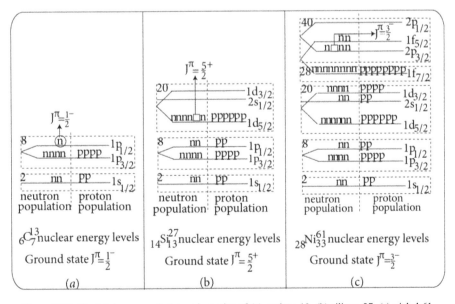

**Figure 3.36.** Explaining ground state spin parity of (a) carbon 13, (b) silicon 27, (c) nickel 61.

14 protons are arranged as

$$(1s_{1/2})^2 \, (1p_{3/2})^4 (1p_{1/2})^2 (1d_{5/2})^6.$$

This portion is closed through pairing.
13 neutrons are arranged as

$$(1s_{1/2})^2 \, (1p_{3/2})^4 (1p_{1/2})^2 (1d_{5/2})^5.$$

The part $(1s_{1/2})^2 \, (1p_{3/2})^4 (1p_{1/2})^2$ is closed through pairing. The remaining part, i.e. the topmost shell $(1d_{5/2})^5$ has a full quota of six but five slots are filled and there is one proton deficiency. A proton could make it full and closed, inert and stable. So it always looks for a proton to fill its vacancy. As the remaining portion is inactive and portion $(1d_{5/2})^5$ is active, the hole in $(1d_{5/2})^5$ will determine the spin parity of the entire nucleus. Now $1d_{5/2}$ state has $j = \frac{5}{2}$ and $l = 2$. So parity is $\pi = (-1)^l = (-1)^2 = +$ve. Hence

$$J^\pi \text{ of nucleus} = J^\pi \text{ of } 1d_{5/2} = \frac{5}{2}^+$$

**Exercise 3.31** *Explain the ground state spin parity* $\frac{3}{2}^-$ *of nickel* 61.

$\boxed{Ans}$ The energy levels of $_{28}\text{Ni}^{61}_{33}$ are shown in figure 3.36(c). The arrangement of neutrons and protons is shown.

28 protons are paired up and arranged as follows

$$(1s_{1/2})^2 \, (1p_{3/2})^4 (1p_{1/2})^2 (1d_{5/2})^6 (2s_{1/2})^2 (1d_{3/2})^4 (1f_{7/2})^8.$$

This part is closed through pairing.
33 neutrons are arranged as follows

$$(1s_{1/2})^2 \, (1p_{3/2})^4 (1p_{1/2})^2 (1d_{5/2})^6 (2s_{1/2})^2 (1d_{3/2})^4 (1f_{7/2})^8 \, (2p_{3/2})^3 (1f_{5/2})^2.$$

This part except $(2p_{3/2})^3$ is closed through pairing. We note that the higher state $1f_{5/2}$ contains two paired neutrons but the lower $2p_{3/2}$ state has a hole. Actually, $_{28}\text{Ni}^{61}_{33}$ attains lower energy in this configuration in which pairing occurs in $1f_{5/2}$ state keeping a hole in $2p_{3/2}$. Hence spin parity $J^\pi$ of ground state is the spin parity of the hole in $2p_{3/2}$.

The competing configuration of

$$(\text{closed part})(2p_{3/2})^4(1f_{5/2})^1$$

in which pairing occurs in $2p_{3/2}$ state, keeping an unpaired neutron in $1f_{5/2}$ is not favoured as it leads to higher energy.

Now the $2p_{3/2}$ state has $j = \frac{3}{2}$ and has $l = 1$ leading to parity $(-1)^l = -$ve. Hence $J^\pi$ of ground state of nickel 61 is $\frac{3}{2}^-$.

**Exercise 3.32** *Find the spin parity of ground state, first excited state and second excited state of $_8O_9^{17}$ from the nuclear shell model.*

[*Ans*] ✓ The energy levels of $_8O_9^{17}$ are shown in figure 3.37(a).

✓ Spin parity of ground state of $_8O_9^{17}$ is $\frac{5}{2}^+$ is shown in figure 3.37(b).

✓ Due to excitation the $1d_{5/2}$ neutron can jump to $2s_{1/2}$ state as shown in figure 3.33(c) and spin parity is then $J^\pi = \frac{1}{2}^+$. This is the first excited state.

✓ If the *nn* pair in $1p_{1/2}$ state (figure 3.37(b)) is broken and one neutron jumps from $1p_{1/2}$ state to $1d_{5/2}$ state to get paired with the existing neutron there then the energy of $_8O_9^{17}$ is lowered. This is the second excited state of $_8O_9^{17}$ and due to hole in $1p_{1/2}$ the spin parity is $\frac{1}{2}^-$. This is depicted in figure 3.37(d).

**Exercise 3.33** *Explain the spin parity $\frac{5}{2}^-$, $\frac{3}{2}^-$, $\frac{3}{2}^+$ of the higher excited states of $_8O_9^{17}$ from the nuclear shell model.*

[*Ans*] The energy levels of $_8O_9^{17}$ are shown in figure 3.37(a). And figure 3.37(b) shows the ground state.

✓ Refer to figure 3.38(a). Suppose the neutron in $1d_{5/2}$ (figure 3.37(b)) goes to $2s_{1/2}$. Pairing in $1p_{1/2}$ is broken and one neutron is lifted to $1d_{5/2}$. So now we have three unpaired neutrons in the states $1p_{1/2}$, $1d_{5/2}$ and $2s_{1/2}$. The corresponding three angular momenta $j_1^- = \frac{1}{2}^-$, $j_2^+ = \frac{5}{2}^+$, $j_3^+ = \frac{1}{2}^+$ combine to give $(-1)(+1)(+1) = -ve$, i.e. negative parity. Also angular momenta $\frac{1}{2}$ and $\frac{1}{2}$ give 1, 0 and angular momenta of 0 and $\frac{5}{2}$ give $\frac{5}{2}$. Hence resultant spin parity is $\frac{5}{2}^-$.

✓ Refer to figure 3.38(b). Suppose the pair in $1p_{1/2}$ (figure 3.37(b)) is broken and one neutron travels to $1d_{5/2}$ and pairs up with a neutron there. Again, one

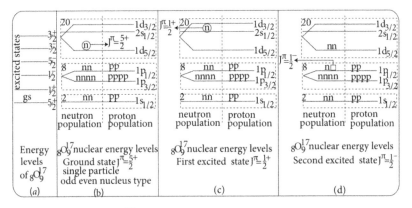

**Figure 3.37.** (a) Energy levels of $_8O_9^{17}$. (b) Spin parity of ground state : single-particle odd–even nucleus type. (c) Spin parity of first excited state. (d) Spin parity of second excited state of $_8O_9^{17}$ from the nuclear shell model.

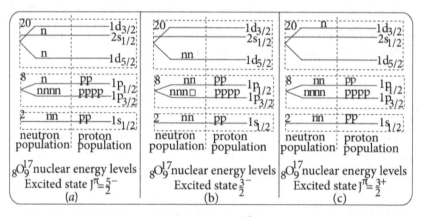

**Figure 3.38.** Spin parity of excited states of $_8O_9^{17}$ from the nuclear shell model.

pair in $1p_{3/2}$ gets broken and one neutron moves to $1p_{1/2}$ and pairs up with a neutron there. A hole is created in the state $1p_{3/2}$. So the spin parity of this excited state is the spin parity of the hole in $1p_{3/2}$. Spin is $\frac{3}{2}$ and $p$ state has $l = 1$ leading to parity $(-1)^1 = -$ve. Hence spin parity is $\frac{3}{2}^-$.

✓ Refer to figure 3.38(c). Suppose neutron in $1d_{5/2}$ (figure 3.37(b)) gets excited to the state $d_{3/2}$. The spin parity of this excited state will be $\frac{3}{2}^+$ since $d$ state has $l = 2$ and parity is $(-1)^2 = +$ve.

**Exercise 3.34** *According to the shell model the spin parity of $_9F^{19}$, $_{11}Na^{23}$ and $_{25}Mn^{55}$ should have been $\frac{5}{2}^+, \frac{5}{2}^+, \frac{7}{2}^+$ but the observed spin parities are $\frac{1}{2}^+, \frac{3}{2}^+, \frac{5}{2}^+$. Explain.*

$\boxed{Ans}$ Figure 3.39 shows the shell structure of $_9F_{10}^{19}$, $_{11}Na_{12}^{23}$, $_{25}Mn_{30}^{55}$.

✓ Figure 3.39(a): The last odd proton of $_9F_{10}^{19}$ instead of going to $1d_{5/2}$ level goes to $2s_{1/2}$ level and so $J^\pi = \frac{1}{2}^+$.

✓ Figure 3.39(b): The last 3 odd protons of $_{11}Na^{23}$ populate $1d_{5/2}$ level. If two protons are completely paired and one remains unpaired then we would get $J = \frac{5}{2}$. But complete pairing does not occur—there is some residual interaction and the three protons interact in such a way as to produce angular momentum $\frac{3}{2}$. $\left( \frac{5}{2} + \frac{5}{2} + \frac{5}{2} \rightarrow 5, 4, 3, 2, 1, 0 + \frac{5}{2} \rightarrow \frac{5}{2} - 1, \ldots\ldots = \frac{3}{2} \right)$

Again $d$ state has $l = 2$. So parity is $(-1)^2 = +$ve. So $J^\pi = \frac{3}{2}^+$.

✓ Figure 3.39(c): The last 5 odd protons of $_{25}Mn_{30}^{55}$ populate $1f_{7/2}$ level. If 4 protons are completely paired and one remains unpaired then we would get $J = \frac{7}{2}$. But complete pairing does not occur—there are some residual interactions and the 5 protons interact in such a way as to produce angular momentum $\frac{5}{2}$. Again, $f$ state has $l = 3$. So parity is $(-1)^3 = -$ve. So $J^\pi = \frac{5}{2}^-$.

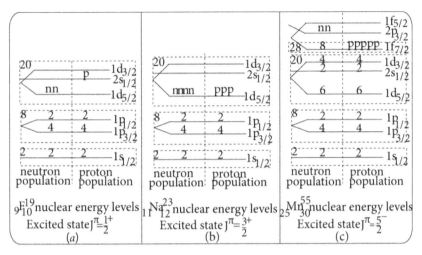

**Figure 3.39.** Nuclear energy shells of $_9F^{19}$, $_{11}Na^{23}$, $_{25}Mn^{55}$

**Exercise 3.35** *Find the magnetic moment of* $_{13}Al^{27}$, $_7N^{15}$, $_{15}P^{31}$ *according to the shell model.*

$\boxed{Ans}$ ✓ Spin parity of $_{13}Al^{27}$ is $\frac{5}{2}^+$ as shown in figure 3.23(b). It is determined by proton deficiency in $1d_{5/2}$ state. Clearly for the odd proton in $d_{5/2}$ state

$$j = \frac{5}{2} = 2 + \frac{1}{2} = l + \frac{1}{2}.$$

The magnetic moment as per equation (3.131) is

$$<\mu_z> = (j + 2.292\,85)\mu_N = \left(\frac{5}{2} + 2.292\,85\right)\mu_N = 4.792\,85\mu_N.$$

✓ Spin parity of $_7N^{15}$ is $\frac{1}{2}^-$ as shown in *exercise 3.41*. Clearly for the odd proton in $p_{1/2}$ state

$$j = \frac{1}{2} = 1 - \frac{1}{2} = l - \frac{1}{2}.$$

The magnetic moment as per equation (3.161) is

$$<\mu_z> = \frac{j}{j+1}(j - 1.292\,85)\mu_N = \frac{\frac{1}{2}}{\frac{1}{2}+1}\left(\frac{1}{2} - 1.292\,85\right)\mu_N = -0.264\,28\mu_N.$$

✓ Spin parity of $_{15}P^{31}$ is $\frac{1}{2}^+$ as shown in *exercise 3.41*. The odd proton is in $2s_{1/2}$ state. The magnetic moment as per equation (3.124) is

$$<\mu_z> = g_n\mu_N = -3.8260\mu_N$$

**Exercise 3.36** *Find the energy difference between the $1d_{3/2}$ and $1f_{7/2}$ neutron shells if it is known that the masses of $_{20}Ca^{40}$, $_{20}Ca^{41}$ and $_{20}Ca^{39}$ are 39.962 589u, 40.962 275u and 38.970 691u respectively. Mass of the neutron is 1.008 665u.*

$\boxed{Ans}$ As shown in figure 3.40, $_{20}Ca^{39}$ has a hole in $1d_{3/2}$ (figure 3.40(a) $_{20}Ca^{40}$ is doubly magic having a closed $1d_{3/2}$ shell (figure 3.40(b)) and $_{20}Ca^{41}$ has one neutron in $1f_{7/2}$ shell (figure 3.40(c)).

The binding energy of a $1d_{3/2}$ neutron in $_{20}Ca^{40}$ is

$$B.\ E]_d = M(\ _{20}Ca^{39}) + m_n - M(\ _{20}Ca^{40})$$

$$= 38.970\ 691u + 1.008\ 665u - 39.962\ 589u$$

$$= 0.016\ 767 \times 931\ MeV = 15.61\ MeV$$

The binding energy of a $1f_{7/2}$ neutron in $_{20}Ca^{41}$ is

$$B.\ E]_f = M(\ _{20}Ca^{40}) + m_n - M(\ _{20}Ca^{41})$$

$$= 39.962\ 589u + 1.008\ 665u - 40.962\ 275u$$

$$= 0.008\ 979 \times 931\ MeV = 8.36\ MeV$$

The energy difference between the $1d_{3/2}$ and $1f_{7/2}$ shells is

$$\Delta = B.\ E]_d - B.\ E]_f = 15.61\ MeV - 8.36\ MeV = 7.25\ MeV$$

**Exercise 3.37** *Show that for identical particles $j_1 = 2$, $j_2 = 2$ the resultant is $j = 0, 2, 4$.*

$\boxed{Ans}$ $j_1 = 2 \Rightarrow m_j = 2, 1, 0, -1, -2$ and

$$j_2 = 2 \Rightarrow m_j = 2, 1, 0, -1, -2$$

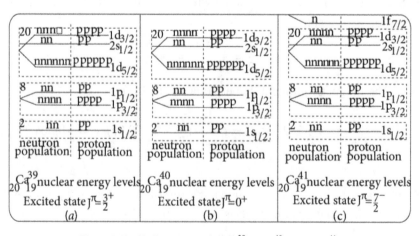

**Figure 3.40.** Shell structure of $_{20}Ca^{39}$, $_{20}Ca^{40}$ and $_{20}Ca^{41}$.

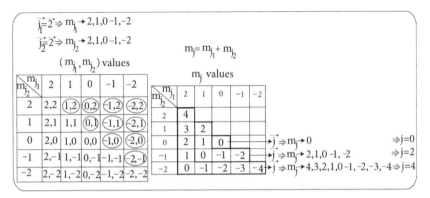

**Figure 3.41.** Angular momentum addition for two identical particles with $\vec{j}_1 = \vec{2}$, $\vec{j}_2 = \vec{2}$ leads to $\vec{j} = \vec{j}_1 + \vec{j}_2$ where $j = 0, 2, 4$.

Now $\vec{j} = \vec{j}_1 + \vec{j}_2$ and $m_j = m_{j_1} + m_{j_2}$

The number of cases will become limited for identical particles.

In finding $m_j$ the combination $(m_{j_1}, m_{j_2})$ should not repeat as we are considering identical particles. For instance, we have to consider the options $(m_{j_1}, m_{j_2})$ →(1, 2) and (2, 1) as one possibility since we cannot determine which phonon corresponds to $m_{j_1} = 1$ and $m_{j_2} = 2$ and vice versa. All the options have been shown in figure 3.41. The combinations that repeat have been encircled and discarded. We have shown the accepted values of $m_j = m_{j_1} + m_{j_2}$ that give the $j$ value to be

$j = 0, 2, 4$

The odd ones $j = 1, 3$ are discarded.

**Exercise 3.38** *The shell model explains spin parity of all nuclei in which state?*
*(a) Ground state, (b) first excited state, (c) even–even nuclei, (d) odd A nuclei.*

**Exercise 3.39** *Compare the shells of atomic and nuclear models.*
Ans Let us present a comparison as follows.

| Atomic shells | Nuclear shells |
| --- | --- |
| Electrons move nearly independently in a Coulomb potential produced by the nucleus and other electrons. | Nucleons move independently of all the other nucleons in a spherically symmetric (central) self-consistent average nuclear potential $V(r)$ produced by the action of all the other nucleons. |

*(Continued)*

(*Continued*)

| Atomic shells | Nuclear shells |
|---|---|
| Nucleus acts as a force centre—so the system is spherically symmetric and force is central. | There is no common centre of force. For multi-nucleon nuclei, potential is not spherically symmetric. Force is non-central. |
| Atom-to-atom variation in binding energy is large. | Nuclei-to-nuclei variation in binding energy is small. |
| Electron spin–orbit coupling is weak. | Nucleon spin–orbit coupling is strong and is of opposite sign. |
| Higher $j$ levels have higher energy. This is shown in figure 3.19(a). | Higher $j$ levels have lower energy. Doublets are inverted w.r.t. their atomic counterparts. This is shown in figure 3.19(b). |
| Stable atoms with closed shells occur for $Z = 2, 10, 18, 36, 54, 86$. | Magic nuclei with closed-shell structure having high degree of stabiltiy occur for magic numbers 2, 8, 50, 82, 126. |
| Electromagnetic interactions are of long range. | Nuclear interactions are of short range. |

**Exercise 3.40** *Find the spin parity of the following nuclei as per the nuclear shell model in the ground state. (a)* $_{22}\text{Ti}^{47}$ *(titanium), (b)* $_{3}\text{Li}^{6}$ *(lithum), (c)* $_{1}\text{H}^{2}$ *(deuteron).*
[Ans] *(a)* $_{22}\text{Ti}^{47}$ $(Z = 22, N = 25, A = 47)$ is odd $A$ nucleus. The arrangement of levels is as follows.

22 protons are arranged as

$$(1s_{1/2})^2 \, (1p_{3/2})^4(1p_{1/2})^2(1d_{5/2})^6(2s_{1/2})^2(1d_{3/2})^4(1f_{7/2})^2$$

This part is closed through pairing.
25 neutrons are arranged as

$$(1s_{1/2})^2 \, (1p_{3/2})^4(1p_{1/2})^2(1d_{5/2})^6(2s_{1/2})^2(1d_{3/2})^4(1f_{7/2})^5$$

The part except $(1f_{7/2})^5$ is closed through pairing. The remaining part, i.e. the topmost shell $(1f_{7/2})^5$ has 5 neutrons out of which 4 are paired and one unpaired. The unpaired neutron makes the portion $(1f_{7/2})^5$ active and this state determines the spin parity of the entire nucleus. The $1f_{7/2}$ state has $=\frac{7}{2}$. Also, the $f$ state has $l = 3$ and so parity is $\pi = (-1)^l = (-1)^3 = -\text{ve}$, i.e. odd parity. Hence $J^\pi$ of nucleus $= J^\pi$ of $1f_{7/2}$ $= \frac{7}{2}^-$.

*(b)* $_{3}\text{Li}^{6}$ $(Z = 3, N = 3, A = 6)$ is odd–odd nucleus. Apply Nordheim's rule.

3 protons are arranged as $(1s_{1/2})^2 (1p_{3/2})^1$ while 3 neutrons are arranged as $(1s_{1/2})^2 (1p_{3/2})^1$. The portion $(1s_{1/2})^2$ is closed through pairing, i.e. inactive and hence stable. The last odd, i.e. unpaired, nucleons are

one proton in $1p_{3/2}$ state which has $j_1 = \frac{3}{2}$, $l_1 = 1$ and

one neutron in state $1p_{3/2}$ which has $j_2 = \frac{3}{2}$, $l_2 = 1$.

Now Nordheim's number is

$N = j_1 + j_2 + l_1 + l_2 = \frac{3}{2} + \frac{3}{2} + 1 + 1 = 5 =$ odd. So weak rule is applicable.

$J = |\frac{3}{2} - \frac{3}{2}|$ to $\frac{3}{2} + \frac{3}{2} = 0, 3$ and $\pi = (-)^{l_1+l_2} = (-)^{1+1} = (-)^2 = +ve$, i.e. even parity. Hence $J^\pi = 0^+, 3^+$. The more likely value is $J^\pi = 3^+$.

(c) $_1\text{H}^2$ ($Z = 1$, $N = 1$, $A = 2$) is odd–odd nucleus. Apply Nordheim's rule.

One proton state is $(1s_{1/2})^1$ and one neutron state is $(1s_{1/2})^1$. The odd unpaired nucleons are

one proton in $1s_{1/2}$ state which has $j_1 = \frac{1}{2}$, $l_1 = 0$ and

one neutron in $1s_{1/2}$ state which has $j_1 = \frac{1}{2}$, $l_1 = 0$.

Now Nordheim's number is

$N = j_1 + j_2 + l_1 + l_2 = \frac{1}{2} + \frac{1}{2} + 0 + 0 = 1 =$ odd. So weak rule is applicable.

$J = |\frac{1}{2} - \frac{1}{2}|$ to $\frac{1}{2} + \frac{1}{2} = 0, 1$ and $\pi = (-)^{l_1+l_2} = (-)^{0+0} = (-)^0 = +ve$, i.e. even parity. Hence $J^\pi = 0^+, 1^+$. The more likely value is $J^\pi = 1^+$.

And experimentally obtained spin parity of deuteron is $J^\pi = 1^+$.

**Exercise 3.41** *According to nuclear shell model the spin parity of $_7\text{N}^{15}$, $_6\text{C}^{12}$, $_{15}\text{P}^{31}$ are*

*(a) $\frac{1}{2}^+$, $0^+$, $\frac{1}{2}^-$ (b) $\frac{1}{2}^-$, $0^+$, $\frac{1}{2}^+$ (c) $\frac{1}{2}^+$, $0^-$, $\frac{1}{2}^+$ (d) $\frac{1}{2}^-$, $0^-$, $\frac{1}{2}^-$*

Hint. ✓ $_7\text{N}^{15}$ has 8 neutrons having shell structure $(1s_{1/2})^2 (1p_{3/2})^4 (1p_{1/2})^2$ which is closed through pairing.

7 protons have the shell structure $(1s_{1/2})^2 (1p_{3/2})^4 (1p_{1/2})^1$. The portion $(1s_{1/2})^2 (1p_{3/2})^4$ is closed through pairing. The unpaired and active part is $(1p_{1/2})^1$. Also, $p$ state means $l = 1$. Parity is $(-)^l = -ve$.

$$J^\pi \text{ of } _7\text{N}^{15} \text{ nucleus} = J^\pi \text{ of } 1p_{1/2} \text{ state i.e. } \frac{1}{2}^- .$$

✓ $_6\text{C}^{12}$ is even–even nucleus. So $J^\pi = 0^+$. Shell structure of 6 protons and 6 neutrons are $(1s_{1/2})^2 (1p_{3/2})^4$.

✓ $_{15}\text{P}^{31}$ has 16 neutrons having shell structure

$$(1s_{1/2})^2 (1p_{3/2})^4 (1p_{1/2})^2 (1d_{5/2})^6 (2s_{1/2})^2$$

which is closed through pairing.

15 protons have the shell structure $(1s_{1/2})^2 (1p_{3/2})^4 (1p_{1/2})^2 (1d_{5/2})^6 (2s_{1/2})^1$. The portion except $(2s_{1/2})^1$ is closed through pairing. The unpaired and active part is $(2s_{1/2})^1$. Also, $s$ state means $l = 0$. Parity is $(-1)^0 = +ve$.

$J^\pi$ of $_{15}P^{31}$ nucleus $= J^\pi$ of $2s_{1/2}$ state, i.e. $\dfrac{1}{2}^+$.

**Exercise 3.42** *Which of the following statements is consistent with the shell model.*
  *(a) Nucleons have continuous energy states.*
  *(b) Nucleons move independently in a common static spherical potential.*
  *(c) Nucleons move independently in a distorted spherical potential*
  *(d) Nucleons move under symmetric potential.*

**Exercise 3.43** *According to the nuclear shell model the spin parities of $_{16}S^{33}$, $_{18}Ar^{41}$ are*

  *(a)* $\dfrac{3}{2}^-, \dfrac{7}{2}^+$ *(b)* $\dfrac{3}{2}^+, \dfrac{7}{2}^+$ *(c)* $\dfrac{3}{2}^+, \dfrac{7}{2}^-$ *. (d)* $\dfrac{3}{2}^-, \dfrac{7}{2}^-$

Hint. ✓ $_{16}S^{33}$ has 16 protons having shell structure

$$(1s_{1/2})^2 (1p_{3/2})^4 (1p_{1/2})^2 (1d_{5/2})^6 (2s_{1/2})^2$$

which is closed through pairing.

17 neutrons have shell structure

$$(1s_{1/2})^2 (1p_{3/2})^4 (1p_{1/2})^2 (1d_{5/2})^6 (2s_{1/2})^2 (1d_{3/2})^1.$$

The part $(1s_{1/2})^2 (1p_{3/2})^4 (1p_{1/2})^2 (1d_{5/2})^6 (2s_{1/2})^2$ is closed through pairing. The unpaired and active part is $(1d_{3/2})^1$. Also, $d$ state means $l = 2$. Parity is $(-)^2 = +ve$.

$$J^\pi \text{ of } _{16}S^{33} \text{ nucleus } = J^\pi \text{ of } 1d_{3/2} \text{ state i.e. } \dfrac{3}{2}^+.$$

✓ $_{18}Ar^{41}$ has 18 protons having shell structure

$$(1s_{1/2})^2 (1p_{3/2})^4 (1p_{1/2})^2 (1d_{5/2})^6 (2s_{1/2})^2 (1d_{3/2})^2$$

which is closed through pairing.

23 neutrons have shell structure

$$(1s_{1/2})^2 (1p_{3/2})^4 (1p_{1/2})^2 (1d_{5/2})^6 (2s_{1/2})^2 (1d_{3/2})^4 (1f_{7/2})^3.$$

The part $(1s_{1/2})^2 \, (1p_{3/2})^4 (1p_{1/2})^2 (1d_{5/2})^6 (2s_{1/2})^2 (1d_{3/2})^4$ is closed through pairing.

The part $(1f_{7/2})^3$ contains 3 neutrons of which one is unpaired and hence active. Also, $f$ state means $l = 3$. Parity is $(-)^3 = -ve$.

$J^\pi$ of $_{18}Ar^{41}$ nucleus $= J^\pi$ of $1f_{7/2}$ state i.e. $\dfrac{7^-}{2}$.

**Exercise 3.44** *Show that energy eigenvalues describing rotational spectrum corresponding to collective nuclear motion is $E_J = \frac{1}{6}J(J + 1)E_2$, the first excited state being $E_2$.*

[Ans] Rotational spectrum corresponding to collective rotational motion of nucleus is described by the energy eigenvalue given by equation (3.176) viz. $E_l = \frac{\hbar^2}{2I}l(l + 1)$ with $l = 0, 2, 4, 6, 8, 10, \ldots$ Here $\vec{l}$ is the orbital angular momentum and since we considered nucleus with zero intrinsic spin, $\vec{l}$ represents the total angular momentum. So $\vec{l} = \vec{J}$ and so we can rewrite the energy eigenvalue as

$$E_J = \frac{\hbar^2}{2I}J(J + 1), \; J = 0, 2, 4, 6, 8, 10, \ldots$$

Clearly for $J = 2$, $E_2 = \frac{\hbar^2}{2I}2(2 + 1) = 6\frac{\hbar^2}{2I}$.

Hence $\frac{\hbar^2}{2I} = \frac{1}{6}E_2$. We thus have

$$E_J = \frac{\hbar^2}{2I}J(J + 1) = \frac{1}{6}J(J + 1)E_2$$

**Exercise 3.45** *Justify from collective model whether $_{82}Pb^{208}$ will show a rotational spectrum.*

[Ans] $_{82}Pb^{208}$ is an even–even nucleus with $Z = 82$, $N = 126$. It consists of 2 magic numbers and is a magic nuclei. It is a spherical nucleus.

The energy eigenvalue describing the rotational spectrum corresponding to collective rotational motion of the nucleus is given by equation (3.176) viz. $E_l = \frac{\hbar^2}{2I}l(l + 1)$ with $l = 0, 2, 4, 6, 8, 10, \ldots$ Now $I$ is rotational inertia associated with nuclear deformation. It is zero for even–even nuclei which are spherical nuclei. Obviously $I = 0$ for $_{82}Pb^{208}$ and so it is expected not to show any rotational spectrum as nuclear matter does not rotate about the symmetry axis of $_{82}Pb^{208}$.

**Exercise 3.46** *Show that there can be no collective rotation of the nucleus about symmetry axis.*

$\boxed{Ans}$ Refer to figure 3.29. Let $\hat{z}$ be the symmetry axis of the nucleus. Angle of rotation about the symmetry axis is $\phi$. For an axially symmetric nucleus wave function $\psi$ is independent of $\phi$ i.e. $\frac{\partial \psi}{\partial z} = 0$.

Now the angular momentum operator along $\hat{z}$ is $l_z = -i\hbar \frac{\partial}{\partial \phi}$. Thus

$$\hat{l}_z \psi = -i\hbar \frac{\partial \psi}{\partial \phi} = 0$$

Thus axial symmetry implies that the component of angular momentum along the symmetry axis is zero. So there is no collective rotation about the symmetry axis.

**Exercise 3.47** *Which nucleus is more stable $_4Be^9$ or $_4Be^{10}$, $_3Li^7$ or $_3Li^8$?*

$\boxed{Ans}$ The most stable nucleus for a given mass number $A$ has the atomic number given by

$Z_0 = \frac{A/2}{1 + 0.0075A^{2/3}}$ [equation (3.188)]

For $A = 9$, $Z_0 = \frac{9/2}{1 + 0.0075 \times 9^{2/3}} = 4.36$

For $A = 10$, $Z_0 = \frac{10/2}{1 + 0.0075 \times 10^{2/3}} = 4.83$

As 4.36 is nearer to 4, $_4Be^9$ is more stable.

For $A = 7$, $Z_0 = \frac{7/2}{1 + 0.0075 \times 7^{2/3}} = 3.4$

For $A = 8$, $Z_0 = \frac{8/2}{1 + 0.0075 \times 8^{2/3}} = 3.88$

As 3.4 is nearer to 3, $_3Li^7$ is more stable.

**Exercise 3.48** *Which nucleus is more stable $_2He^6$, $_2Be^6$ or $_3Li^6$?*

$\boxed{Ans}$ The most stable nucleus for a given mass number $A$ has the atomic number given by

$Z_0 = \frac{A/2}{1 + 0.0075A^{2/3}} \xrightarrow[\text{light nuclei}]{A \cong \text{small}} \frac{A}{2}$ [equation (3.188)]

For $A = $ small, i.e. for light nuclei $Z_0 = \frac{A/2}{1 + 0.0075A^{2/3}} \cong \frac{A}{2}$

For $_3Li^6$ has $Z = 3 = \frac{6}{2} = \frac{A}{2} = Z_0$. So it is most stable.

**Exercise 3.49** *Study of rotational energy level shows that the first excited state of $_{74}W^{182}$ is $2^+$ and is 0.1 MeV above the ground state. Estimate the energies of the lowest lying $4^+$ and $6^+$ states of $_{74}W^{182}$.*

$\boxed{Ans}$ Energy values of rotational states are

$E_l = \frac{\hbar^2}{2I} l(l + 1)$, $l = 0, 2, 4, \ldots$ [equation (3.176)]

$$E_2 = \frac{\hbar^2}{2I}2(2 + 1) = 6\frac{\hbar^2}{2I}.$$

Given $E_2 = 0.1\ MeV$. So

$$6\frac{\hbar^2}{2I} = 0.1\ MeV$$

$$\frac{\hbar^2}{2I} = \frac{0.1}{6}\ MeV$$

$$E_4 = \frac{\hbar^2}{2I}4(4 + 1) = 20\frac{\hbar^2}{2I} = 20 \times \frac{0.1}{6}\ MeV = \frac{1}{3}\ MeV =\ \text{energy of } 4^+$$

$$E_6 = \frac{\hbar^2}{2I}6(6 + 1) = 42\frac{\hbar^2}{2I} = 42 \times \frac{0.1}{6}\ MeV = 0.7\ MeV =\ \text{energy of } 6^+$$

### Ans to Multiple Choice Questions

3.2 *b*, 3.3 *a*, 3.6 *a*, 3.11 *a*, *d*, 3.16 *a*, 3.17 *c*, 3.24 *c*, 3.25*c*, 3.38 *a*, 3.41 *b*, 3.42 *b*, 3.43 *c*.

## 3.35 Question bank

Q3.1   What do nuclear models wish to describe?

Q3.2   Describe the Fermi gas model.

Q3.3   Show in a schematic diagram the square well potentials for neutron and proton in Fermi gas model. Why is it that the Fermi energy of the neutron and proton wells occurs at the same level though their depths are different?

Q3.4   Obtain an estimate of Fermi energy for proton gas and neutron gas in the Fermi gas model.

Q3.5   What are the assumptions of the liquid drop model?

Q3.6   Write down Bethe–Weizsacker's semi-empirical mass formula. Explain each term. Why is the mass formula called semi-empirical?

Q3.7   Show in a diagram the contributions of various terms of Bethe–Weizsacker's semi-empirical mass formula and compare with the experimentally obtained $\frac{B}{A}$ versus $A$ curve.

Q3.8   What do we mean by asymmetry energy correction in Bethe–Weizsacker's semi-empirical mass formula? Discuss why neutron excess is needed for a stable nucleus. Explain with an example how neutron excess reduces the amount of binding energy of a nucleus.

Q3.9   What is the role of the pairing energy correction term in Bethe–Weizsacker's semi-empirical mass formula?

Q3.10   Write down Bethe–Weizsacker's semi-empirical mass formula based on the liquid drop model. How was it corrected introducing quantum mechanical considerations?

Q3.11   Mention the merits and demerits of Bethe–Weizsacker's semi-empirical mass formula.

Q3.12   Can magic numbers be explained by Bethe–Weizsacker's semi-empirical mass formula?

Q3.13   What do we mean by mass parabola? What is its significance?

Q3.14   Show how one can derive a parabola starting from Bethe–Weizsacker's semi-empirical mass formula.

Q3.15   Establish that there is one parabola for odd $A$ isobars and two parabolas for even $A$ isobars. Which of the latter two parabolas would be shifted up?

Q3.16   Discuss the terms proton drip line, neutron drip line by means of a suitable diagram.

Q3.17   How can one predict a stable bound nucleus from among a set of isobaric nuclei?

Q3.18   Starting from Bethe–Weizsacker's semi-empirical mass formula how can one predict the mass of a neutron star?

Q3.19   Why would we think of shells in a nucleus? What is the evidence of shell structure in a nucleus?

Q3.20   Define the following terms. Magic nuclei, semi-magic nuclei, doubly magic nuclei. Give examples.

Q3.21   What are the assumptions of the shell model? Justify them.

Q3.22   What is the importance of magic numbers?

Q3.23   Can you generate magic numbers with a harmonic oscillator potential? Justify.

Q3.24   Can you generate all the magic numbers with a harmonic oscillator potential modified with spin–orbit correction?Justify.

Q3.25   Show in a neat diagram the shell structure of the nucleus. Indicate the magic numbers in the diagram.

Q3.26   Obtain the spin parity of even–even nucleus according to the shell model.

Q3.27   Obtain the spin parity of an odd $A$ nucleus according to the shell model.

Q3.28   Obtain the spin parity of an odd–odd nucleus according to the shell model.

Q3.29   Find spin parity of ground state from the shell structures of $_{76}Os^{192}$, $_8O^{17}$, $_{13}Al^{27}$, $_5B^{10}$.

Q3.30   What is Nordheim's rule?

Q3.31   What are Schmidt limits? Sketch a schematic diagram and explain. How does it tally with experimental results of magnetic moments of the nucleus?

Q3.32   How successful is the shell model? What are its failures?

Q3.33  Make a short comparison of the liquid drop model and the nuclear shell model.

Q3.34  Why is the collective model needed?

Q3.35  What effect on nuclear energy levels or shells do rotation and vibration of nucleus have?

## Further reading

[1]  Krane S K 1988 *Introductory Nuclear Physics* (New York: Wiley)

[2]  Tayal D C 2009 *Nuclear Physics* (Mumbai: Himalaya Publishing House)

[3]  Satya P 2005 *Nuclear Physics and Particle Physics* (New Delhi: Sultan Chand & Sons)

[4]  Guha J 2019 *Quantum Mechanics: Theory, Problems and Solutions* 3rd edn (Kolkata: Books and Allied (P) Ltd))

[5]  Lim Y-K 2002 *Problems and Solutions on Atomic, Nuclear and Particle Physics* (Singapore: World Scientific)

**IOP** Publishing

# Nuclear and Particle Physics with Cosmology, Volume 1
Nuclear physics
**Jyotirmoy Guha**

# Chapter 4

# Nuclear reaction

## 4.1 Introduction

In chemical reactions constituent molecules interact and there is redistribution or rearrangement of atoms or ions. In a nuclear reaction, constituent atoms interact and there is redistribution or rearrangement of nucleons

We mention some nuclear reactions that have made an important impact in the study of nuclear physics.

- $\alpha + {}_7N^{14} \rightarrow {}_8O^{17} + p$

  This was the first artificial nuclear reaction performed by Rutherford in 1919 in which 7.68 $MeV$ $\alpha$ fuses with ${}_7N^{14}$ and generates ${}_8O^{17}$ and a proton. This reaction showed how one element can be converted to create another element and also led to the discovery of the proton.

- $\alpha + {}_4Be^9 \rightarrow {}_6C^{12} + n$

  This nuclear reaction was studied by Chadwick and led to the discovery of the neutron in 1932.

- $p + {}_3Li^7 \rightarrow \alpha + \alpha$

  Proton was accelerated to 0.5 $MeV$ energy with the help of an accelerator invented by Cockcroft and Walton in 1932. These protons bombarded ${}_3Li^7$ as a result of which two $\alpha$ particles were produced having kinetic energy of 8.9 $MeV$. The remarkable feature was that this reaction showed that using a low energy projectile ($\sim$0.5 $MeV$) it is possible, through nuclear reaction, to obtain higher output energy ($\sim$17.8 $MeV$).

- $\alpha + {}_{13}Al^{27} \rightarrow {}_{15}P^{30} + n$

  $${}_{15}P^{30} \rightarrow {}_{14}Si^{30} + e^+ + \nu_e$$

Frideric and Irene Joliot Curie produced the first radioactive nuclei ${}_{15}P^{30}$ by bombarding ${}_{13}Al^{27}$ with $\alpha$ in 1934.

doi:10.1088/978-0-7503-5027-3ch4  4-1

## 4.2 Definition of incident and exit channels

A nuclear reaction is represented as

$$a + X \rightarrow Y + b \quad X(a, b) Y \ (a, b) \tag{4.1}$$

- ✓ Here $a$ represents a projectile which is the bombarding particle, say a nucleon, i.e. a proton or a neutron, or it may be a multi-nucleon system, say a deuteron, an $\alpha$, $_3Li$ or a heavy ion.

  It is of small mass—much much lighter compared to the target $X$, i.e.

$$m_a \ll m_X.$$

- ✓ $X$ is target nucleus. It is generally a heavy nucleus—it is part of a material in the form of a plate, block or film kept fixed in the laboratory—it may be in an evacuated chamber.

  Example of target is, say Au nucleus of gold foil.
- ✓ $Y$ is product nucleus.

  Target nucleus $X$ suffers a change and transforms into $Y$ and this $Y$ either remains in the target material or may emerge out.
- ✓ $b$ is the product particle that emerges out of the reaction zone, i.e. gets scattered in various directions $\theta$ and can be captured by a detector placed at a long distance along the $\theta$ direction.

The detector performs the following functions.
- ☐ It counts the number of product particles.
- ☐ It determines the kinetic energy of the product particles.

So we can get the number of particles as a function of kinetic energy.

The quantities $a$, $X$ are the pre-reaction entities. They occur in the LHS of the reaction and we say that they constitute the incident channel, as shown in figure 4.1(a).

On the other hand, $Y$, $b$ are the post reaction quantities. They occur in the RHS of the reaction and we say that they constitute the exit channel, as shown in figure 4.1(b).

The projectile beam emerging from accelerator or source that is focused on the target has a very small cross-section $\sim mm^2$ and since the detector is far away from

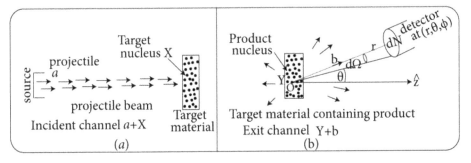

**Figure 4.1.** (a) Incident channel $a + X$. (b) Exit channel $Y + b$ of a nuclear reaction.

the interaction zone $\sim 100\ mm$ we can consider the interaction site to be a point $O$. Angular position of the detector is $(\theta, \phi)$. The detector window makes a solid angle $d\Omega = \sin\theta d\theta d\phi$ at point $O$ of the target.

$dN$ is the number of product particles $b$ detected per unit of time.

## 4.3 Nuclear reaction cross-section

The projectile particle passes through the interaction zone and is supposed to strike the target. But all the projectile particles that are incident on the target do not react with the target—a fraction interact. The probability of reaction is measured by a quantity called reaction cross-section. It is the effective area that the target presents to the projectile for reaction. In other words, the reaction cross-section $\sigma$ is the area of cross-section of an imaginary disc associated with each target nucleus such that if the incident projectile passes through it reaction occurs.

Reaction cross-section depends upon:
- ✓ energy of the projectile;
- ✓ nature of the projectile;
- ✓ nature of the target.

Consider the reaction $a + X \rightarrow Y + b$
- Let a beam of the projectile be incident on a sheet of the target. Suppose each projectile or particle interacts only once. As the beam passes through a distance $dx$ suppose $dN$ number of projectile particles react. Hence we attach $-ve$ sign to it and write as $-dN$.

  Let $N$ be the number of projectile particles and target area be $A$.

  Clearly the greater the number of projectile particles incident per unit area, the greater is $\frac{N}{A}$—the greater the reaction and hence

$$-dN \propto \frac{N}{A} \tag{4.2}$$

  Let number density of target nuclei be denoted by $n$.

  Clearly the greater the number of target nuclei the greater the reaction. Now the beam passes through a distance $dx$ through a volume $Adx$. So the number of target nuclei in this volume is

$$\left(\frac{\text{number of target nuclei}}{\text{volume}}\right)(Adx) = nAdx.$$

Clearly then

$$-dN \propto nAdx \tag{4.3}$$

Combining the relations of equations (4.2) and (4.3) we can write

$$-dN \propto \frac{N}{A}nAdx$$

$$-dN \propto Nndx$$

Attaching the proportionality constant $\sigma$ which is defined as the nuclear reaction cross-section we have

$$-dN = \sigma N n dx$$

$$-\frac{dN}{N} = \sigma n dx \tag{4.4}$$

If at $x = 0$, $N = N_0$ and if at $x = x$, $N = N_i$ then upon integration we get

$$\int_{N_0}^{N_i} \frac{dN}{N} = -n\sigma \int_0^x dx$$

$$\ln N_i - \ln N_0 = -n\sigma(x - 0)$$

$$\ln \frac{N_i}{N_0} = -n\sigma x$$

$$N_i = N_0 e^{-n\sigma x} \tag{4.5}$$

Here $N_0$, $N_i$ represent the number of projectile particles hitting the target at $x = 0$, $x$, respectively. The expression of number density of target nuclei $n$ is given by

$$n = \frac{\rho N_A}{M} \tag{4.6}$$

where $\rho =$ density of target, $N_A =$ Avogadro's number, $M =$ molecular weight of target.

Let us check the unit of $n$.

$$n = \frac{\rho N_A}{M} \rightarrow \frac{kg}{m^3}\left(\frac{\text{number}}{\text{mole}}\right)\frac{1}{kg} \sim \frac{1}{m^3}$$

We can recast equation (4.5) in terms of intensity also as

$$I = I_0 e^{-n\sigma x} \tag{4.7}$$

where $I$, $I_0$ represent intensity of incident projectile at $x = 0$, $x$, respectively. Intensity means number per unit area per second, i.e. $I \rightarrow \frac{\text{number}}{m^2 \, s}$.

- Let number $N_i$ of projectile particles $a$ striking per unit area of target per second, i.e. intensity of projectile be $I_i$.

Let number $N_b$ of product particles $b$ emerging out per unit area per second be $I_b$.

If $n$ is number of target nuclei per unit volume and $x$ is thickness then $I_i$ and $I_b$ are related as

$$I_b = I_i \sigma n x \tag{4.8}$$

Let us check validity of this formula from a dimensional point of view by finding the unit of reaction cross-section $\sigma$ substituting the units of other quantities viz. $I_0$, $I_i$, $n$, $x$.

$$\frac{\text{number}}{m^2 \, s} = \frac{\text{number}}{m^2 \, s} \sigma \frac{1}{m^3} m$$

$$\sigma \sim m^2$$

We can rewrite this formula also by replacing Intensity $I(\sim \frac{\text{number}}{m^2 \, s})$ by number $N$.

$$N_b = N_i \sigma n x \qquad (4.9)$$

where $N_b$ = number of product particles $b$ emerging out, $N_i$ = number of projectile particles striking the target.

## 4.4 Partial cross-sections

Consider a beam of projectiles incident on a target. The beam may interact with the target in a variety of ways. Some particles might get scatterered—elastically and/or inelastically—the cross-section of which is $\sigma_{sc}$. Some particles might react—the cross-section of which is $\sigma_r$. We refer to the cross-section of the different processes as partial cross-sections.

If we consider all such processes, we speak of total cross-section $\sigma_t$.

The total cross-section $\sigma_t$ and the partial cross-sections can be related as

$$\sigma_t = \sigma_{sc} + \sigma_r \qquad (4.10)$$

In other words the total cross-section is the sum over all the partial cross-sections.

We define the partial and total cross-sections as follows.

$$\sigma_{sc} = \frac{\text{number of particles scattered per second}}{\text{number of particles incident on target per unit area per second}} \text{(unit } m^2)$$

$$\sigma_r = \frac{\text{number of particles undergoing reaction per second}}{\text{number of particles incident on target per unit area per second}} \text{ (unit } m^2)$$

Assuming that scattering and reaction are the only processes that occur we define

$$\sigma_t = \frac{\text{number of particles suffering scattering and reaction per second}}{\text{number of particles incident on target per unit area per second}} \text{ (unit } m^2)$$

## 4.5 Lab frame of reference or L-frame

Consider the interaction $a + X \rightarrow X + a$ which is a perfectly elastic collision.

We can study this interaction $a + X \rightarrow X + a$ from a frame of reference fixed to the laboratory. This is called laboratory frame or lab frame or L-frame. Generally the target is at rest in L-frame. A pictorial representation of interaction in L-frame is shown in figure 4.2.

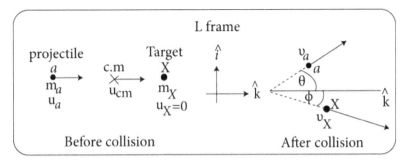

**Figure 4.2.** L-frame of reference.

Let the incident direction of projectile $a$ be $\hat{k}$. The initial velocity of projectile $a$ is $u_a$. The initial velocity of target is $u_X = 0$ since the target is at rest in L-frame.

The final velocities are $v_a$, $v_X$, respectively, for $a$ and $X$.

The angle $\theta$ by which projectile $a$ gets deflected is called scattering angle of projectile $a$. Due to the interaction the target recoils—the angle of recoil of target $X$ is $\phi$ in L-frame.

## 4.6 Centre-of-mass frame of reference or C-frame

Consider again the same interaction $a + X \to X + a$ which is a perfectly elastic collision.

We can study the same interaction $a + X \to X + a$ from a frame of reference in which the centre of mass of projectile $a$ and target $X$ is at rest. This is called C-frame or centre-of-mass frame of reference. In other words, the origin of C-frame is located at the centre of mass of the projectile–target system always (both before and after collision). A pictorial representation of interaction in C-frame is shown in figure 4.3.

We note that the interaction is viewed from centre of mass and since it has no velocity w.r.t. itself, velocity of centre of mass is zero, i.e.

$$U_{CM} = 0$$

Also thus linear momentum of the centre of mass is zero.

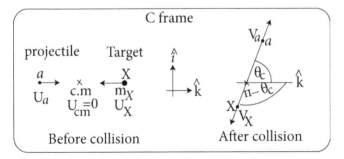

**Figure 4.3.** C-frame of reference.

It thus follows from the conservation of linear momentum that linear momentum before and after collision will be zero.

Hence to make total momentum zero in the C-frame, the projectile and target will move in opposite directions, i.e. will have equal and opposite momenta. This is evident from the following discussion.

$\hat{k}$ is inicident direction.

$\theta_c$ is angle of scattering of the projectile in C-frame. It is the polar angle ranging from 0 to $\pi$.

The initial velocities of projectile $a$ and target $X$ are $U_a$, $U_X$, respectively.

The final velocities of projectile $a$ and target $X$, respectively, are $V_a$, $V_X$.

The centre of mass of projectile $a$ and target $X$ obviously moves w.r.t. the L-frame with velocity, say $u_{cm}$ (figure 4.2)—but does not move w.r.t. itself, i.e. w.r.t. C-frame and so $U_{cm} = 0$.

- In C-frame the centre of mass of the projectile–target system is at rest—both before and after collision. Clearly the position vector of the centre of mass w.r.t. itself (C-frame, i.e. the centre of mass) is zero, i.e.

$$\vec{R}_{cm} = 0$$

the centre-of-mass velocity is

$$\vec{U}_{cm} = \frac{d}{dt}\vec{R}_{cm} = 0.$$

The centre-of-mass momentum in C-frame is thus

$$\vec{P}_{cm} = 0 \qquad (4.11)$$

Let $\vec{p}_a$, $\vec{p}_X$ represent pre-collision momenta of $a$ and $X$ in C-frame.

By conservation of linear momentum

$$\vec{p}_a + \vec{p}_X = 0 \ \text{(before collision)}$$

$$\vec{p}_a = -\vec{p}_X \qquad (4.12)$$

Let $\vec{p}'_a$, $\vec{p}'_X$ represent post-collision momenta of $a$ and $X$ in C-frame.

By conservation of linear momentum

$$\vec{p}'_a + \vec{p}'_X = 0 \ \text{(after collision)}$$

$$\vec{p}'_a = -\vec{p}'_X \qquad (4.13)$$

It follows clearly from equations (4.12) and (4.13) that in C-frame projectile $a$ and target $X$ will move in opposite directions, i.e. will have equal and opposite momenta both before and after collision.

Since $\vec{R}_{cm} = 0$, $\vec{U}_{cm} = 0$, $\vec{P}_{cm} = 0$ C-frame is called zero reference frame.

Since $\frac{d}{dt}\vec{U}_{cm} = 0$ C-frame is unaccelearted, i.e. an inertial frame of reference.

## 4.7 Justification of having two frames L-frame and C-frame

☐ Actual laboratory experiment is done in L-frame in which the target is fixed and only the projectile moves. But analysis of collision and calculations can be done more easily in C-frame. Equations describing nuclear reactions are easy to derive and work with in C-frame.

We can inter-relate the parameters in the two frames.

☐ In L-frame we have to handle 6 coordinates. 3 coordinates are needed to describe center of mass while 3 more coordinates are needed to describe relative motion in L-frame.

But centre of mass is at rest in C-frame. So its coordinates get fixed at (0,0,0). We therefore need to handle only 3 coordinates to describe motion relative to centre of mass. This leads to great simplification.

In C-frame $\vec{p}_{cm} = 0$. This leads to a huge simplification.

In *exercise 4.13* we show that L-frame involves 4 unknowns and there are 3 equations. So only 3 unknowns can be found.

In *exercise 4.21* we show that the speeds of particles are not affected due to collision in C-frame. Thus another great simplification is achieved. $(U_a = V_a, U_X = V_X)$.

Calculation is simpler in C-frame and so it is popular.

- We have made a list of symbols used in figure 4.4 for easy reference. We have used small letters for L-frame and capital letters for C-frame.

## 4.8 Relation between kinetic energy in L-frame and C-frame

We consider the collision $a + X \rightarrow X + a$.

Total initial kinetic energy in L-frame

$$\frac{1}{2}m_a u_a^2 + \frac{1}{2}m_X u_X^2 \xrightarrow{u_X=0} \frac{1}{2}m_a u_a^2 = T_{lab} \qquad (4.14)$$

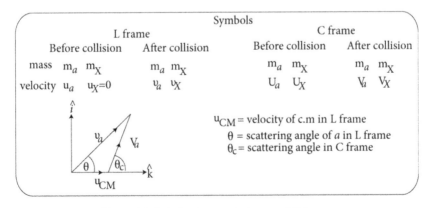

**Figure 4.4.** Relation between L-frame and C-frame.

This is the energy given to the projectile. Most nuclear reactions involve energy $\sim MeV$. However, let us use the non-relativistic formula.

Velocity of centre of mass of projectile $a$ and target $X$ in L-frame is

$$u_{CM} = \frac{m_a u_a + m_X u_X}{m_a + m_X} \xrightarrow{u_X = 0} \frac{m_a u_a}{m_a + m_X} \tag{4.15}$$

But in C-frame the centre of mass is at rest which means we have to add $-u_{CM}$ to the velocities in L-frame.

So the velocity of the projectile in C-frame is

$$U_a = u_a - u_{CM} = u_a - \frac{m_a u_a}{m_a + m_X} = \frac{m_X u_a}{m_a + m_X} \tag{4.16}$$

And the velocity of the target $X$ in C-frame is

$$U_X = u_X - u_{CM} = 0 - \frac{m_a u_a}{m_a + m_X} = -\frac{m_a u_a}{m_a + m_X} \tag{4.17}$$

Total kinetic energy in C-frame is

$$T_{CM} = \frac{1}{2} m_a U_a^2 + \frac{1}{2} m_X U_X^2 = \frac{1}{2} m_a \left( \frac{m_X u_a}{m_a + m_X} \right)^2 + \frac{1}{2} m_X \left( -\frac{m_a u_a}{m_a + m_X} \right)^2$$

$$= \frac{1}{2} \frac{m_a m_X u_a^2}{(m_a + m_X)^2} (m_a + m_X)$$

$$= \frac{1}{2} m_a u_a^2 \frac{m_X}{m_a + m_X}$$

Using equation (4.14)

$$T_{CM} = T_{\text{lab}} \frac{m_X}{m_a + m_X} \tag{4.18}$$

$$\frac{T_{CM}}{T_{\text{lab}}} = \frac{m_X}{m_a + m_X} < 1 \tag{4.19}$$

$$T_{CM} < T_{\text{lab}} \tag{4.20}$$

So total initial kinetic energy in C-frame is less than the total initial kinetic energy in L-frame.

## 4.9 Relation between scattering angle in L-frame and C-frame

Figure 4.4 shows the relation between scattering angle $\theta$ in L-frame and scattering angle $\theta_c$ in C-frame. It follows from figure 4.2 that the final velocity of projectile $a$ in L-frame $v_a$ makes angle $\theta$ with $\hat{k}$ (the direction of incidence of the projectile) while $u_{CM}$ is along $\hat{k}$. Again, it follows from figure 4.3 that the final velocity of the projectile $a$ in C-frame $V_a$ makes angle $\theta_c$ with $\hat{k}$. Clearly we can construct a diagram as in figure 4.4 showing $v_a$, $V_a$, $\theta$, $\theta_c$ and $u_{CM}$. Consider

$$\tan \theta = \frac{\sin \theta}{\cos \theta} = \frac{v_a \sin \theta}{v_a \cos \theta} \text{ (multiplying numerator and denominator by } v_a) \quad (4.21)$$

Resolving $v_a$ along $\hat{k}$ we have

$$v_a \cos \theta = u_{Cm} + V_a \cos \theta_c \quad (4.22)$$

Resolving $v_a$ along $\hat{i}$ we have

$$v_a \sin \theta = V_a \sin \theta_c. \quad (4.23)$$

From equations (4.21), (4.22), (4.23) we have

$$\tan \theta = \frac{v_a \sin \theta}{v_a \cos \theta} = \frac{V_a \sin \theta_c}{u_{Cm} + V_a \cos \theta_c}$$

$$\tan \theta = \frac{\sin \theta_c}{\frac{u_{Cm}}{V_a} + \cos \theta_c} \quad (4.24)$$

Using equation (4.15) viz. $u_{CM} = \frac{m_a u_a}{m_a + m_X}$ and recalling from equation (4.16) that

$$u_a = u_{CM} + U_a \quad (4.25)$$

we write

$$u_{CM} = \frac{m_a}{m_a + m_X} u_a = \frac{m_a}{m_a + m_X}(u_{CM} + U_a)$$

$$u_{CM}\left(1 - \frac{m_a}{m_a + m_X}\right) = \frac{m_a}{m_a + m_X} U_a$$

$$u_{CM}\frac{m_X}{m_a + m_X} = \frac{m_a}{m_a + m_X} U_a$$

$$u_{CM} m_X = m_a U_a \quad (4.26)$$

$$\frac{u_{CM}}{U_a} = \frac{m_a}{m_X} \quad (4.27)$$

Again, in *exercise 4.21* we show that the speeds of particles are not affected due to collision in C-frame. Hence

$$U_a = V_a \quad (4.28)$$

Hence equation (4.27) can be rewritten as

$$\frac{u_{CM}}{V_a} = \frac{m_a}{m_X} \quad (4.29)$$

Using equation (4.29) in equation (4.24) we have

$$\tan \theta = \frac{\sin \theta_c}{\frac{m_a}{m_X} + \cos \theta_c} \tag{4.30}$$

## 4.10 Charged particle approaching target

Consider a charged projectile approaching a similarly charged target. There is Coulomb repulsion before nuclear interaction starts. So there is a Coulomb barrier through which the projectile has to tunnel to interact strongly with the nucleus (figure 4.5).

If the projectile carries small energy it cannot tunnel through. It gets scattered elastically (as in Rutherford scattering) and so nuclear interaction does not occur.

For nuclear interaction, the projectile has to carry large energy so as to get closer to the nucleus.

- Consider single-channel interaction

$$a + X \rightarrow X + a$$

This interaction can occur in two ways.

We discuss perfectly elastic collision or scattering in section 4.11 and inelastic collision or scattering in section 4.13. We also compare the two processes.

## 4.11 Elastic collision or scattering

Consider a perfectly elastic collision

$$a + X \rightarrow X + a \tag{4.31}$$

It has the following features.

There is no change in channel. It is a one-channel process.

Incident and exit particles are identical in internal structure and composition.

Total energy, linear momentum, angular momentum are conserved.

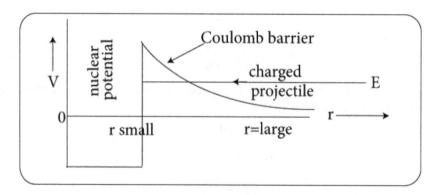

**Figure 4.5.** A charged projectile has to tunnel through Coulomb barrier to interact with nucleus.

Kinetic energy is conserved, i.e.

$$T_{LHS} = T_{RHS} \qquad (4.32)$$

$$\frac{1}{2}m_a u_a^2 + \frac{1}{2}m_X u_X^2 = \frac{1}{2}m_a v_a^2 + \frac{1}{2}m_X v_X^2 \qquad (4.33)$$

$$\frac{1}{2}m_a u_a^2 = \frac{1}{2}m_a v_a^2 + \frac{1}{2}m_X v_X^2 \qquad (4.34)$$

where $m_a, m_X$ are masses of $a$, $X$, respectively, and $u_a$, $u_X = 0$ are pre-collision velocities, while $v_a$, $v_X$ are post-collision velocities of $a$, $X$, respectively, in L-frame. Figure 4.4 defines all symbols.

The target remains in the same nuclear state—does not absorb energy and so cannot go to an excited state—there is no excitation, as shown in figure 4.6(a).

Energy brought by projectile $a$ is shared or redistributed between $a$ and $X$. There is no energy transfer.

Figure 4.6(a) shows a light target nucleus $X$ that has excited levels $E_1$, $E_2$ much away from its ground state, i.e. $E_1 - E_{gs}$= large. So energy $\Delta E$ taken up by $X$ from projectile cannot carry it to an excited state $E_1$ since $E_1 > \Delta E$.

So $X$ spends this energy by gaining a momentum, i.e. momentum is transferred from projectile to target. Also, the target recoils with the energy that it gets. And in the process kinetic energy is conserved. The kinetic energy brought about by the projectile is not absorbed by the target which continues to stay in the ground state after scattering. This is elastic scattering.

Example of elastic scattering:
  ✓ collision of gas molecules with the wall of container;
  ✓ Rayleigh scattering—scattering of $X$ ray photon by a bound electron or atom;
  ✓ Compton scattering—scattering of $X$ ray photon by a free electron.

Example of an elastic channel
  ☐ $\alpha + B^{10} \rightarrow B^{10} + \alpha$
    After scattering $\alpha$ follows a hyperbolic trajectory which means the direction of $\alpha$ particle changes after scattering.

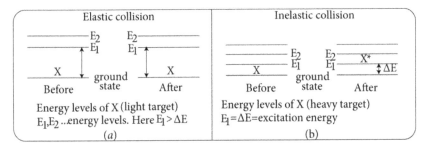

**Figure 4.6.** (a) Elastic scattering. (b) Inelastic scattering.

☐ $\alpha + {}_{79}Au^{197} \to {}_{79}Au^{197} + \alpha$ (Rutherford scattering)

☐ ${}_{0}n^{1} + {}_{82}Pb^{208} \to {}_{82}Pb^{208} + {}_{0}n^{1}$

☐ Fast neutrons are robbed of their energy through elastic collisions as they pass through paraffin (light nuclei).

## 4.12 Inelastic collision or scattering

Consider inelastic collision

$$a + X \to X* + a \tag{4.35}$$

It has the following features.

There is no change in channel. It is a one-channel process.

Incident and exit particles are identical in internal structure and composition.

Total energy, linear momentum, angular momentum are conserved.

Kinetic energy is not conserved.

$$T_{LHS} > T_{RHS}, \; T_{LHS} = T_{RHS} + \Delta E \tag{4.36}$$

$$\frac{1}{2}m_a u_a^2 + \frac{1}{2}m_X u_X^2 = \frac{1}{2}m_a v_a^2 + \frac{1}{2}m_X v_X^2 + \Delta E \tag{4.37}$$

where $\Delta E$ is the excitation energy absorbed by the target.

If $\Delta E = 0$ it is an elastic collision.

The projectile loses energy and the target absorbs this energy and goes to an excited state. This higher quantum state, i.e. the excited state, is not stable and immediately the excited target de-excites and emits a $\gamma$ ray.

Energy brought by projectile $a$ is shared or redistributed between $a$ and $X$ and a portion $\Delta E$ is absorbed by $X$.

Figure 4.6(b) shows a heavy target nucleus $X$ that has excited levels $E_1, E_2$ close to its ground state, i.e. $E_1 - E_{gs}$= small and $E_1 = \Delta E$. So energy $\Delta E$ absorbed by $X$ from a projectile can take it to an excited state $E_1$. And if $\Delta E = E_2$ then $X$ goes to excited state $E_2$.

So $X$ absorbs energy $\Delta E$ and gets excited. Clearly kinetic energy is not conserved. This is inelastic scattering.

Examples of inelastic scattering:

✓ Raman scattering (scattering of photon by a molecule);

✓ photoelectric effect (absorption of photon by electron).

Examples of inelastic channel (asterisk * means excited state)

☐ $\alpha + B^{10} \to (B^{10})* + \alpha$
$(B^{10})* \to B^{10} + \gamma$

☐ ${}_{0}n^{1} + {}_{82}Pb^{208} \to ({}_{82}Pb^{208})* + {}_{0}n^{1}$
$({}_{82}Pb^{208})* \to {}_{82}Pb^{208} + \gamma$

☐ ${}_{1}p^{1} + {}_{3}Li^{7} \to ({}_{3}Li^{7})* + {}_{1}p^{1}$
$({}_{3}Li^{7})* \to {}_{3}Li^{7} + \gamma$

☐ Fast $n$ while passing through an assembly of heavy nuclei transfer their energy through a few inelastic collisions (followed by elastic collisions).

After a few successive inelastic collisions the projectile is robbed of most energy and now elastic collision occurs.

- Ineleastic collision is of two types.

✓ Partly inelastic collision. It has the following features.

Particles do not stick after collision. So masses of colliding particles $m_a$, $m_X$ remain the same before and after collision.

Kinetic energy is not conserved. Loss of kinetic energy occurs.

✓ Perfectly inelastic or plastic collision. It has the following features.

Particles stick after collision. So pre-collision masses $m_a$, $m_X$ combine and the post-collision mass is $m_a + m_X$.

Kinetic energy is not conserved. Loss of kinetic energy is maximum.

## 4.13 Reaction channel

When a projectile nucleus bombards a target nucleus as a result of which its mass number, atomic number changes we say a nuclear reaction occurs. This process is written as

$$a + X \rightarrow Y + b \quad \text{or} \quad X(a, b)Y \tag{4.38}$$

called an $(a, b)$ reaction.

In a nuclear reaction, incident and exit channels are different.

Examples of reaction channels are as follows:

One incident channel may give rise to multiple exit channels.

$$\alpha + B^{10} \rightarrow \begin{cases} C^{12} + d \\ C^{13} + p \\ N^{13} + n \end{cases}$$

Multiple incident channels may lead to a single exit channel

$$\left. \begin{array}{l} \alpha + B^{10} \\ \alpha + C^{12} \\ p + C^{13} \end{array} \right\} \rightarrow C^{12} + d$$

☐ The process $a + X \rightarrow Y + b$ is a nuclear reaction if $a \neq b$, $X \neq Y$. Then there is change in channel.

☐ Clearly $a + X \rightarrow Y + b$ is a scattering process if

$$a = b, X = Y$$

as there is no change in channel in this case.

Some other example of reaction are

$$_2He^4 + {_7}N^{14} \rightarrow {_8}O^{17} + {_1}H^1$$

$$_0n^1 + {}_{82}Pb^{208} \rightarrow ({}_{81}Tl^{1208})^* + {}_1p^1$$

$$({}_{81}Tl^{1208})^* \rightarrow {}_{81}Tl^{1208} + \gamma$$

## 4.14 Radiative capture

A particle may combine with (i.e. get absorbed into) a nucleus to generate a compound nucleus in the excited state. The excess energy is emitted in the form of $\gamma$ ray photon. This is called radiative capture.

Example

$$p + {}_{12}Mg^{26} \rightarrow ({}_{13}Al^{27})^* \rightarrow {}_{13}Al^{27} + \gamma$$

## 4.15 Photodisintegration

Here projectile is the $\gamma$ ray that is absorbed by the target. The target thus goes to an excited state. If $\gamma$ energy $h\nu$ is sufficiently high the target may disintegrate. A good example is photodisintegration of deuteron nucleus.

$$\gamma + d \rightarrow n + p \tag{4.39}$$

## 4.16 Transmutation

Transmutation refers to conversion of one element (atom/nucleus) into another. Let us furnish examples.

✓ Natural radioactivity

$$_{92}U^{238} \xrightarrow{\alpha} {}_{90}Th^{234} \xrightarrow{\beta} \ldots$$

✓ Artificial radioactivity

$$\alpha + {}_5B^{10} \rightarrow {}_7N^{13} + {}_0n^1, \ {}_7N^{13} \rightarrow {}_6C^{13} + {}_1e^0 + \nu_e$$

✓ Nuclear reaction $\alpha + {}_7N^{14} \rightarrow {}_8O^{17} + p$

## 4.17 Conserved quantities in nuclear reaction

☐ The following quantities are conserved in a nuclear reaction.

✓ Conservation of electric charge $Q$, atomic number, proton number.

Total electric charge of $LHS$ = Total electric charge of $RHS$ i.e.

$$Q(LHS) = Q(RHS) \tag{4.40}$$

Consider the nuclear reaction

$$_2He^4 + {}_5B^{10} \rightarrow ({}_7N^{14})^* \rightarrow {}_6C^{13} + {}_1H^1.$$

$$Q(\text{LHS}) = 7 \mid e \mid (7 \text{ protons are there})$$

$$Q \text{ (compound nucleus)} = 7 \mid e \mid (7 \text{ protons are there})$$

$$Q(\text{RHS}) = 7 \mid e \mid (7 \text{ protons are there}).$$

So $Q$ = conserved or $\sum Z$ = conserved. $\tag{4.41}$

✓ Conservation of mass number $A$, i.e.

$$A(LHS) = A(RHS) \text{ or in other words } \sum A = \text{conserved} \qquad (4.42)$$

Consider the nuclear reaction $_2\text{He}^4 + _5\text{B}^{10} \rightarrow (_7\text{N}^{14})^* \rightarrow _6\text{C}^{13} + _1\text{H}^1$

$$A(LHS) = 4 + 10 = 14 \text{ nucleons}$$

$$A \text{ (compound nucleus)} = 14 \text{ nucleons}$$

$$A(RHS) = 13 + 1 = 14 \text{ nucleons.}$$

$$\text{Also } \sum Z = \text{conserved.} \qquad (4.43)$$

In this example

$$Z(LHS) = 7$$

$$Z \text{ (compound nucleus)} = 7$$

$$Z(RHS) = 7$$

$$\text{Also } \sum N = \text{conserved.} \qquad (4.44)$$

In this example

$$N(LHS) = 7$$

$$N \text{ (compound nucleus)} = 7$$

$$N(RHS) = 7.$$

✓ Conservation of baryon number $B$.

$$B(LHS) = B(RHS) \qquad (4.45)$$

In the nuclear reaction $_2\text{He}^4 + _5\text{B}^{10} \rightarrow (_7\text{N}^{14})^* \rightarrow _6\text{C}^{13} + _1\text{H}^1$

$$B(LHS) = 14$$

$$B(RHS) = 14$$

since there are 14 nucleons and nucleons are baryons.
✓ Conservation of lepton number.
   Lepton number is also conserved.
✓ Conservation of energy.
   In the nuclear reaction $a + X \rightarrow Y + b$ the energy conservation relation is

$$(m_a c^2 + T_a) + (m_X c^2 + T_X) = (m_Y c^2 + T_Y) + (m_b c^2 + T_b) \qquad (4.46)$$

✓ Conservation of linear momentum $\vec{p}$

$$\vec{p}_{LHS} = \vec{p}_{RHS}. \tag{4.47}$$

In centre-of-mass frame of reference total linear momentum is zero.

✓ Conservation of angular momentum $\vec{J}$

$$\vec{J}_{LHS} = \vec{J}_{RHS} \tag{4.48}$$

Consider the nuclear reaction

$$\alpha + B^{10} \rightarrow (N^{14})^* \rightarrow C^{13} + p$$

Let us find the angular momentum of the exit channel using angular momentum conservation relation.

$$\vec{J}_{initial} = \vec{l}_i + \vec{S}_{B^{10}} + \vec{S}_\alpha$$

$$\vec{J}_{final} = \vec{l}_f + \vec{S}_{C^{13}} + \vec{S}_p$$

Let us consider capture of incident particle $\alpha$ for $E_\alpha \leqslant 10\ MeV$. For such low energy the capture is assumed to be $S$ wave capture, i.e.

$$l_i = 0.$$

Also, it is known that $S_\alpha = 0$, $S_{B^{10}} = 3$, $S_{C^{13}} = \frac{1}{2}$, $S_p = \frac{1}{2}$

$$\vec{J}_{initial} = \vec{l}_i + \vec{S}_{B^{10}} + \vec{S}_\alpha = \vec{0} + \vec{3} + \vec{0} = \vec{3} \tag{4.49}$$

$$\vec{J}_{final} = \vec{l}_f + \vec{S}_{C^{13}} + \vec{S}_p = \vec{l}_f + \frac{\vec{1}}{2} + \frac{\vec{1}}{2} \tag{4.50}$$

From equations (4.48), (4.49), (4.50) we have

$$\vec{3} = \vec{l}_f + \frac{\vec{1}}{2} + \frac{\vec{1}}{2}$$

$$\vec{l}_f = \vec{3} - \left( \frac{\vec{1}}{2} + \frac{\vec{1}}{2} \right)$$

$$\vec{l}_f = \vec{3} - \begin{cases} \vec{0} \\ \vec{1} \end{cases} = \begin{cases} \vec{3} - \vec{0} \\ \vec{3} - \vec{1} \end{cases} \rightarrow \begin{cases} 3 \\ 4,3,2 \end{cases} \text{ (by angular momentum addition rule)} \tag{4.51}$$

Clearly the exit channel angular momentum $\vec{l}_f$ is restricted by the angular momentum conservation rule since it can assume certain fixed values

$$l_f = 2,3,4 \tag{4.52}$$

and not any arbitrary value.

✓ Conservation of parity

Nuclear reactions occurring via strong interactions conserve parity. Let $\pi$ denote intrinsic parity. Then

$$\pi(\text{initial state}) = \pi(\text{final state}) \tag{4.53}$$

Let us give an example.

Consider the nuclear reaction $_2\text{He}^4 + _5\text{B}^{10} \rightarrow (_7\text{N}^{14})^* \rightarrow _6\text{C}^{13} + _1\text{H}^1$.

$l_i$, $l_f$ represent relative orbital angular momenta of incident and exit channels. Let us find the value of $l_f$ that conservation of parity rule allows.

Parity of initial state $= \pi(_2\text{He}^4)\pi(_5\text{B}^{10})(-)^{l_i}$. (parity is multiplicative)

Parity of final state $= \pi(_6\text{C}^{13})\pi(_1\text{H}^1)(-)^{l_f}$.

From the shell model parity of ground state of $_2\text{He}^4$, $_5\text{B}^{10}$ and $_1\text{H}^1$are

$\pi(_2\text{He}^4) = +ve$, $\pi(_5\text{B}^{10}) = +ve$, $\pi(_6\text{C}^{13}) = -ve$ and $\pi(_1\text{H}^1) = +ve$.

For low energy of incident $_2\text{He}^4$ i.e. for $E_a \leqslant 10 \; MeV$, $S$ wave capture occurs and $l_i = 0$

So input parity $= \pi(_2\text{He}^4)\pi(_5\text{B}^{10})(-)^{l_i} = (+)(+)(-)^0 = +ve$

and final parity $= \pi(_6\text{C}^{13})\pi(_1\text{H}^1)(-)^{l_f} = (-)(+)(-)^{l_f}$

From parity conservation rule of equation (4.53) it follows that

$$( + )=(-)(+)(-)^{l_f}$$

$$l_f = \text{odd} \tag{4.54}$$

Choosing from equation (4.52) we have $l_f = 3$.

Hence we get $l_f = 3$ as suggested by angular momentum conservation rule and parity selection rule.

- In weak interaction parity conservation is violated.

✓ Conservation of isotopic spin

In nuclear reaction involving strong reaction isotopic spin is conserved.

✓ Conservation of statistics or the spin character.

Overall statistics followed on both sides of a nuclear reaction is same. This means overall spin is same before and after reaction.

☐ The folllowing quantities are not conserved in a nuclear reaction.

✓ Magnetic dipole moment of reacting nuclei

✓ Electric quadrupole moment of reacting nuclei.

These moments depend upon the internal distribution of mass, charge and current within nuclei and are not subject to conservation laws.

## 4.18 $Q$ value of nuclear reaction

Consider a nuclear reaction

$$a + X \rightarrow Y + b \quad \text{i.e.} \quad X(a, b)Y \tag{4.55}$$

Let us view the reaction in the laboratory frame of reference in which the target $X$ is at rest.

So kinetic energy of $X$, i.e. $T_X = 0$.

Let us write down the conservation of energy relation equation (4.46) where $mc^2$ is rest energy of rest mass $m$, $T$ is kinetic energy of mass

$$(m_a c^2 + T_a) + (m_X c^2 + T_X) = (m_Y c^2 + T_Y) + (m_b c^2 + T_b) \tag{4.56}$$

$$(m_a + m_X)c^2 - (m_Y + m_b)c^2 = (T_b + T_Y) - (T_a + T_X) \tag{4.57}$$

With $T_X = 0$ in L-frame we recast the relation of equation (4.57) as

$$(m_a + m_X)c^2 - (m_Y + m_b)c^2 = T_b + T_Y - T_a \tag{4.58}$$

$$T_{\text{final}} - T_{\text{initial}} = (m_{\text{initial}} - m_{\text{final}})c^2 \tag{4.59}$$

where

$$T_{\text{initial}} = T_a + T_X = T_a \text{ as } T_X = 0$$

$$T_{\text{final}} = T_b + T_Y$$

$$m_{\text{initial}} = m_a + m_X$$

$$m_{\text{final}} = m_b + m_Y$$

✓ $Q$ value of reaction is defined as the difference between the final and initial kinetic energies (i.e. difference between kinetic energies of products and reactants) [equation (4.59)]

✓ $Q$ value of reaction is defined as the difference between the initial and final rest mass energies (i.e. difference between masses of reactants and products) [equation (4.59)].

Hence

$$Q = T_{\text{final}} - T_{\text{initial}} = (m_{\text{initial}} - m_{\text{final}})c^2 \tag{4.60}$$

$$Q = T_b + T_Y - T_a = (m_a + m_X)c^2 - (m_Y + m_b)c^2 \tag{4.61}$$

☐ Significance of $Q$ value is that it indicates energy balance in an interaction.
✓ $Q = 0$

$Q = 0$ corresponds to perfectly elastic collision where kinetic energy is conserved.

$$m_{\text{initial}} = m_{\text{final}} \tag{4.62}$$

$$T_{\text{initial}} = T_{\text{final}} \tag{4.63}$$

So there is no rest mass deficit, no loss of kinetic energy but there is only a change in the direction of motion of the particles.

An example is:

$\alpha + {}_{79}\text{Au}^{197}_{118} \rightarrow {}_{79}\text{Au}^{197}_{118} + \alpha$. This is Rutherford $\alpha$ scattering.

- For $Q \neq 0$ kinetic energy is not conserved. Two cases arise.

✓ $Q > 0$ i.e. $Q = |Q|$

$Q > 0$ i.e. $Q = |Q|$ corresponds to exoergic or exothermic reaction. From equations (4.60) and (4.61) it follows that

$$m_{\text{initial}} > m_{\text{final}} \text{ i.e. } m_a + m_X > m_b + m_Y \tag{4.64}$$

$$T_{\text{final}} > T_{\text{initial}} \text{ i.e. } T_b + T_Y > T_a \tag{4.65}$$

Clearly there is excess of energy

$$(m_{\text{initial}} - m_{\text{final}})c^2 \tag{4.66}$$

before the start of the reaction. So energy release occurs due to nuclear interaction.

We can rewrite the energy relation from equations (4.60) and (4.61) as

$$m_{\text{initial}}c^2 = m_{\text{final}}c^2 + |Q| \tag{4.67}$$

$$(m_a + m_X)c^2 = (m_Y + m_b)c^2 + |Q| \tag{4.68}$$

$$T_{\text{final}} = T_{\text{initial}} + |Q| \tag{4.69}$$

$$T_b + T_Y = T_a + |Q| \tag{4.70}$$

It is clear that an exoergic reaction can take place spontaneously—even when projectile kinetic energy $T_a = 0$, i.e. with no external input of energy since it follows from equation (4.70) that

$$T_b + T_Y = |Q| \text{ for } T_a = 0. \tag{4.71}$$

So interaction can be spontaneous, i.e. can occur on contact—pumping energy from outside is not needed.

Examples of energy releasing processes are:

Cellular respiration: Glucose + oxygen→$CO_2$ + water + energy (used for cell activities).

Other examples are:

Rusting of iron, combustion reactions, sweating, freezing or solidification etc.

The nuclear reaction

$${}_6\text{C}^{12}(n, \alpha){}_4\text{Be}^9$$

Fission of ${}_{92}\text{U}^{235}_{143}$ corresponds to $Q > 0$.

✓ $Q < 0$ i.e. $Q = -|Q|$

$Q < 0$ i.e. $Q = -|Q|$ corresponds to an endoergic or endothermic reaction.

From equations (4.60) and (4.61) it follows that

$$m_{initial} < m_{final} \qquad (4.72)$$

$$m_a + m_X < m_b + m_Y \qquad (4.73)$$

$$T_{final} < T_{initial} \qquad (4.74)$$

$$T_b + T_Y < T_a \qquad (4.75)$$

Clearly there is deficiency of energy before the start of the reaction, which means the reaction cannot start unless and until energy is supplied from outside.

For $Y$ and $b$ to emerge they have to emit with positive kinetic energy, i.e.

$$T_Y + T_b \geqslant 0 \qquad (4.76)$$

This is the condition for the reaction to occur. Again from equations (4.60) and (4.61) with $Q = -|Q|$ we have

$$-|Q| = T_b + T_Y - T_a \qquad (4.77)$$

$$-|Q| + T_a = T_b + T_Y \geqslant 0 \qquad (4.78)$$

This is a necessary condition for the reaction to occur.

Occurrence of $-ve$ sign in equations (4.77) and (4.78) reminds us that we have to supply energy externally because there is shortage of energy to initiate the reaction.

So to initiate the reaction, energy has to be supplied, i.e. energy is absorbed during the nuclear reaction and we have from equations (4.60) and (4.61)

$$m_{initial}c^2 + |Q| = m_{final}c^2 \qquad (4.79)$$

$$(m_a + m_X)c^2 + |Q| = (m_Y + m_b)c^2 \qquad (4.80)$$

Interaction is not spontaneous—it does not occur on its own. We have to pump energy from outside, i.e. throw the projectile with kinetic energy—the environment cools down.

Examples of energy consuming processes are:

Photosynthesis, melting or liquefaction, cooking egg etc.

The nuclear reaction

$_6C^{13}(p, \alpha)_5B^{10}$ has $Q = -4 \, MeV$.

In the formula of equation (4.61) viz. $Q = T_b + T_Y - T_a$ the quantity $T_Y$ is the recoil kinetic energy of the product nucleus. Normally nucleus is heavy and hence $T_Y$ is very small and so hard to measure. So using this formula we cannot determine the $Q$ value of a nuclear reaction. We should eliminate $T_Y$ and replace it with measurable quantities to get $Q$ value.

We shall evaluate the explicit expression of $Q$ value in this chapter and also evaluate the minimum energy required to initiate an endoergic reaction.

In figure 4.7 we have shown various $Q$ values and the interaction they represent.

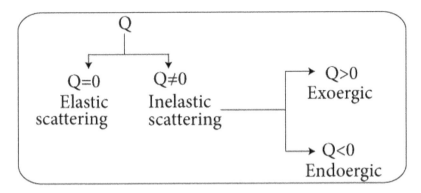

**Figure 4.7.** Significance of $Q$ value.

## 4.19 Threshold energy for an endoergic reaction

For an endoergic reaction it follows from equations (4.79) and (4.80) that

$$m_{\text{initial}} < m_{\text{final}} \tag{4.81}$$

$$m_a + m_X < m_Y + m_b \tag{4.82}$$

So the reaction cannot proceed spontaneously. To initiate the reaction the projectile has to be projected with some kinetic energy—the minimum value of which is written as$(T_a)_{\text{min}}$.

The minimum energy required to initiate an endoergic (energy absorbing) reaction is called threshold energy and we denote it by

$$E_{th} = (T_a)_{\text{min}}.$$

In other words, at threshold, the endoergic reaction just commences.

- The reaction can occur directly in which interacting entities (projectile and target) react to form product nucleus and product particle.
- The projectile might get absorbed by the target and an intermediate compound nucleus generated with enough excitation energy for it to disintegrate into product nucleus and product particle. In that case the energy supplied to initiate the endoergic reaction should be sufficient so that it is possible for these processes to happen. In other words, the threshold energy $E_{th}$ is used up in two ways

✓ To supply momentum, i.e. kinetic energy to intermediate compound nucleus $\frac{1}{2}m_C v_C^2$. This part is not available for a nuclear reaction.

✓ To give excitation energy to the compound nucleus causing it to break up. This part is available as $Q$ value for nuclear reaction. In fact, this is the $Q = |Q|$ part. Hence

$$E_{th} = (T_a)_{\text{min}} = \frac{1}{2}m_C v_C^2 + |Q| \tag{4.83}$$

## 4.20 Types of nuclear reaction

The advent of high energy accelerators (like Van de graff generator, Cockroft–Walton generator, cyclotron, betatron, synchrocyclotron etc) opened the window for studying various nuclear reactions. With their help, high-energy bombarding particles like proton, neutron, deuteron, alpha partcle, electron were used to initiate nuclear reactions employing various elements of the periodic table as targets.

An understanding of nucleus and nuclear force is facilitated from the study of cross-section and angular distribution of the products of nuclear reactions.

We can think of two types of nuclear reactions.
- ✓ Compound nuclear reaction.
- ✓ Direct nuclear reaction.

  Direct nuclear reactions are classified into two more types of reactions. They are
- Transfer reaction
- Knoc kout reaction.

Transfer reactions can be classified into two types of reactions. We mention them in the following.
- ☐ Stripping reaction
- ☐ Pick up reaction.

This classification of various types of nuclear reactions is presented in figure 4.8.

In a compound nuclear reaction, interaction occurs between the projectile and the entire target nucleus, while in a direct nuclear reaction, interaction occurs between the projectile and a few nucleons of the target nucleus.
- Let us address what decides the type of nuclear reaction—i.e. discuss whether the nuclear reaction would be a compound nuclear reaction or direct nuclear reaction.

Whether a compound nuclear reaction or direct nuclear reaction would occur is decided by the relative size of de Broglie wavelength of the projectile and the size of the target nucleus.

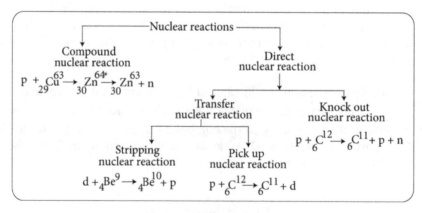

**Figure 4.8.** Classification of nuclear reactions.

To get an estimate consider proton $p$ as the projectile having $m_p c^2 = 938\ MeV$. The de Broglie wavelength of the projectile proton is given by

$$\lambda = \frac{h}{p} = \frac{2\pi\hbar}{\sqrt{2mT}} = \frac{2\pi\hbar c}{\sqrt{2mc^2 T}}$$

$$= \frac{2\pi(197\ MeV.fm)}{\sqrt{2(938\ MeV)T}} \text{ (taking } \hbar c \sim 197\ MeV.fm, \text{ exercise 1.2)}$$

✓ For $T \sim 10\ MeV$

$$\lambda = \frac{2\pi(197\ MeV.fm)}{\sqrt{2(938\ MeV)(10\ MeV)}} = 9\ fm$$

✓ For $T \sim 20\ MeV$

$$\lambda = \frac{2\pi(197\ MeV.fm)}{\sqrt{2(938\ MeV)(20\ MeV)}} = 6\ fm$$

This $\lambda$ is of the order of the target nucleus.

So a projectile carrying low energy $T \leqslant 20\ MeV$ would interact with the entire target by getting absorbed. It will share its energy with nucleons of the nucleus, losing its identity. It would excite the target to form a compound nucleus

$$a + X \rightarrow C^*$$

The compound nucleus in excited state would then decay

$$C^* \rightarrow Y + b$$

This is a slow process because of the intermediate step of compound nucleus formation. It occurs in a time $\sim 10^{-13}\ s$.

✓ For projectile carrying high energy $T \geqslant 20\ MeV$ or more, de Broglie wavelength $\lambda$ goes down and becomes comparable to the size of a few nucleons. So now the projectile will directly interact with a few surface nucleons and emerge out with large energy in the forward direction. Obviously it occurs directly and very fast because no intermediate formation of a compound nucleus is there. It occurs in a time $\sim 10^{-22}\ s$ [equation (4.109)].

## 4.21 Compound nuclear reaction

Consider the nuclear reaction which involves one incident channel leading to various exit channels, i.e. showing various decay patterns

$$a + X \rightarrow C^\star \rightarrow \begin{cases} Y + b \\ Y' + b' \\ Y'' + b'' \\ Y''' + b''' \end{cases} \tag{4.84}$$

For instance

$$p + {}_{29}Cu^{63} \rightarrow ({}_{30}Zn^{64})^* \rightarrow \begin{cases} {}_{30}Zn^{63} + n \\ {}_{30}Zn^{62} + n + n \\ {}_{29}Cu^{62} + p + n \end{cases} \tag{4.85}$$

And various incident channels might lead to the same exit channel as follows

$$\left.\begin{array}{c} a + X \\ a' + X' \\ a'' + X'' \\ a''' + X''' \end{array}\right\} \rightarrow C^* \rightarrow Y + b \tag{4.86}$$

For instance

$$\left.\begin{array}{c} p + {}_{29}Cu^{63} \\ \alpha + {}_{28}Ni^{60} \end{array}\right\} \rightarrow ({}_{30}Zn^{64})^* \rightarrow {}_{30}Zn^{63} + n \tag{4.87}$$

Here $({}_{30}Zn^{64})^*$ refers to a compound nucleus in excited state.

Such reactions exhibit narrow resonances (section 4.35). These reactions were studied by Bohr through his model called the Bohr compound nuclear model. It is based on Bohr's independence hypothesis.

## 4.22 Bohr's independence hypothesis

The nuclear reaction

$$a + X \rightarrow C^* \rightarrow Y + b \tag{4.88}$$

is a two-step process—which are independent of each other but one follows the other.

☐ The first step is formation of the compound nucleus

For projectile energy $\leqslant 10 \ MeV$ the projectile $a$ is absorbed totally by the target nucleus $X$. An intermediate compound nucleus $C$ in a highly excited virtual metastable (quasi-stationary) state is formed. As the projectile enters the nucleus it is subjected to strong force and it shares the energy it carries with nucleons of the nucleus through a large number of collisions and energy exchanges. Obviously there is a through mixing of energy and this means mean free path for collisions is very small (much much less than compound nuclear radius $R$). The identity of the projectile particle is lost completely. As a result, all information regarding the particular channel which served as the entrance channel is forgotten completely—memory is lost. The constants of motion (e.g. total energy, linear momentum, angular momentum etc) are the only memory the compound nucleus retains.

The same excited compound nucleus $C^*$ can be created starting from various channels $a + X$, $a' + X'$, $a'' + X''$, $a''' + X'''$ etc as indicated in equation (4.86). We note that to create $C^*$ in the same excited state projectile $a$ (for the channel $a + X$), projectile $a'$ (for the channel $a' + X'$)... require different excitation energies.

☐ The second step is disintegration of compound nucleus

The process of thorough mixing and creation of an excited compound nucleus $C^*$ takes a long time $\sim 10^{-13}\,s$ (long compared to the time scale of direct reaction which is $10^{-22}\,s$). Ultimately enough energy is concentrated on one or a group of particles (nucleons) which by statistical fluctuation finds itself very close to or on the surface of the nucleus possessing enough energy to escape. As the excited compound nucleus $C^*$ has no memory of how it received the excitation energy, it is free to decay through any arbitrary channel, for instance $C*$ can decay into $Y + b$, $Y' + b'$, $Y'' + b''$, $Y''' + b'''$ etc as indicated in equation (4.84). Decay through any channel is possible with equal probability. The decay conserves total energy, linear and angular momentum etc.

Clearly formation of compound nucleus $C^\star$ is possible through any incident channel and decay of the compound nucleus $C^\star$ is possible through any exit channel. This is shown in equation (4.89)

$$\left.\begin{array}{c} a + X \\ a' + X' \\ a'' + X'' \\ a''' + X''' \end{array}\right\} \rightarrow C^\star \rightarrow \left\{\begin{array}{l} Y + b \\ Y' + b' \\ Y'' + b'' \\ Y''' + b''' \end{array}\right. \tag{4.89}$$

And the processes are independent—the only condition being that decay should follow formation, i.e. chronological order should be maintained for it to be a meaningful interaction and total energy, linear and angular momentum should be conserved.

This is Bohr's independence hypothesis.

## 4.23 Analysis of compound nuclear reaction using Bohr's independence hypothesis

Let us study the following reactions

$$p + {}_{29}\text{Cu}^{63} \rightarrow ({}_{30}\text{Zn}^{64})^* \rightarrow \left\{\begin{array}{l} {}_{30}\text{Zn}^{63} + n \\ {}_{30}\text{Zn}^{62} + n + n \\ {}_{29}\text{Cu}^{62} + p + n \end{array}\right. \tag{4.90}$$

$$\alpha + {}_{28}\text{Ni}^{60} \rightarrow ({}_{30}\text{Zn}^{64})^* \rightarrow \left\{\begin{array}{l} {}_{30}\text{Zn}^{63} + n \\ {}_{30}\text{Zn}^{62} + n + n \\ {}_{29}\text{Cu}^{62} + p + n \end{array}\right. \tag{4.91}$$

Let us find the kinetic energies $T_p$, $T_\alpha$ of the incident projectiles $p$, $\alpha$ that lead to the same excited state of the compound nucleus $({}_{30}\text{Zn}^{64})^*$.

We use energy conservation relation in centre of mass frame in which the total momentum before and after the nuclear reaction is zero.

Consider the nuclear reaction of equation (4.90) in C-frame.

$$\vec{P}_p + \vec{P}_{Cu} = 0 \text{ (total initial momentum before reaction)}$$

$$\vec{P}_p = -\vec{P}_{Cu}$$

$$P_p = P_{Cu} \equiv p = \text{momentum magnitude of } p, \text{ Cu.} \tag{4.92}$$

Also

$$P_{Zn} = 0 \text{ (due to momentum conservation)} \tag{4.93}$$

Hence for $p + {}_{29}Cu^{63} \rightarrow ({}_{30}Zn^{64})^*$ [equation (4.90)] the energy conservation relation becomes

$$(m_p c^2 + T_p) + (m_{Cu} c^2 + T_{Cu}) = (m_{Zn} c^2 + T_{Zn}) + \Delta E \tag{4.94}$$

where $\Delta E$ is excitation energy. Putting

$$T_p = \frac{p_p^2}{2m_p} \equiv \frac{p^2}{2m_p}, \quad T_{Cu} = \frac{p_{Cu}^2}{2m_{Cu}} \equiv \frac{p^2}{2m_{Cu}} \quad \text{and} \quad T_{Zn} = 0 \tag{4.95}$$

we get from equation (4.94)

$$\left(m_p c^2 + \frac{p^2}{2m_p}\right) + \left(m_{Cu} c^2 + \frac{p^2}{2m_{Cu}}\right) = m_{Zn} c^2 + \Delta E. \tag{4.96}$$

$$(m_p + m_{Cu} - m_{Zn})c^2 + \frac{p^2}{2m_p}\left(1 + \frac{m_p}{m_{Cu}}\right) = \Delta E$$

Using $T_p = \frac{p^2}{2m_p}$ [equation (4.95)] we have

$$T_p\left(1 + \frac{m_p}{m_{Cu}}\right) = \Delta E - (m_p + m_{Cu} - m_{Zn})c^2$$

$$T_p = \frac{\Delta E - (m_p + m_{Cu} - m_{Zn})c^2}{1 + \frac{m_p}{m_{Cu}}} \tag{4.97}$$

Similarly

$$T_\alpha = \frac{\Delta E - (m_\alpha + m_{Cu} - m_{Zn})c^2}{1 + \frac{m_\alpha}{m_{Cu}}}. \tag{4.98}$$

Putting the values of $m_p$, $m_{Cu}$, $m_{Zn}$, $m_\alpha$ we can find $T_p$, $T_\alpha$ for a fixed $\Delta E$ from equations (4.97) and (4.98).

## 4.24 Experimental verification by Ghoshal

For the interaction

$$a + X \rightarrow C^* \rightarrow Y + b \tag{4.99}$$

the compound nuclear cross-section is defined through the relation

$$\sigma(a, b) = \sigma_C(a)G_C(b) \tag{4.100}$$

where

$\sigma(a, b)$ is the cross-section for the reaction $X(a, b)Y$ and represents the probability of occurrence of the nuclear reaction.

$\sigma_C(a)$ is the cross-section for the formation of compound nucleus $C^*$ by particle $a$ absorbed by $X$ and

$G_C(b)$ is the relative probability of decaying of $C^*$ by emission of product particle $b$ along with product nucleus $Y$.

This is how equation (4.100) describes the nuclear reaction of equation (4.99) since $\sigma_C(a)$ clearly represents that the incident channel is $a + X \rightarrow C^*$ and $G_C(b)$ represents that the exit channel is $C^* \rightarrow Y + b$.

Accordingly we have for

$$p + {}_{29}\text{Cu}^{63} \rightarrow ({}_{30}\text{Zn}^{64})^* \rightarrow {}_{30}\text{Zn}^{63} + n \text{ the relation } \sigma(p, n) = \sigma_{Zn}(p)G_{Zn}(n)$$

$$p + {}_{29}\text{Cu}^{63} \rightarrow ({}_{30}\text{Zn}^{64})^* \rightarrow {}_{30}\text{Zn}^{62} + n + n \text{ the relation } \sigma(p, nn) - \sigma_{Zn}(p)G_{Zn}(nn)$$

$$p + {}_{29}\text{Cu}^{63} \rightarrow ({}_{30}\text{Zn}^{64})^* \rightarrow {}_{29}\text{Cu}^{62} + p + n \text{ the relation } \sigma(p, pn) = \sigma_{Zn}(p)G_{Zn}(pn)$$

And also we have for

$$\alpha + {}_{28}\text{Ni}^{60} \rightarrow ({}_{30}\text{Zn}^{64})^* \rightarrow {}_{30}\text{Zn}^{63} + n \text{ the relation } \sigma(\alpha, n) = \sigma_{Zn}(\alpha)G_{Zn}(n)$$

$$\alpha + {}_{28}\text{Ni}^{60} \rightarrow ({}_{30}\text{Zn}^{64})^* \rightarrow {}_{30}\text{Zn}^{62} + n + n \text{ the relation } \sigma(\alpha, nn) = \sigma_{Zn}(\alpha)G_{Zn}(nn)$$

$$\alpha + {}_{28}\text{Ni}^{60} \rightarrow ({}_{30}\text{Zn}^{64})^* \rightarrow {}_{29}\text{Cu}^{62} + p + n \text{ the relation } \sigma(\alpha, pn) = \sigma_{Zn}(\alpha)G_{Zn}(pn)$$

Now consider the ratio

$$\frac{\sigma(p, n)}{\sigma(p, nn)} = \frac{\sigma_{Zn}(p)G_{Zn}(n)}{\sigma_{Zn}(p)G_{Zn}(nn)} = \frac{G_{Zn}(n)}{G_{Zn}(nn)}$$

Multiplying the numerator and denominator by $\sigma_{Zn}(\alpha)$ we have

$$\frac{\sigma(p, n)}{\sigma(p, nn)} = \frac{G_{Zn}(n)}{G_{Zn}(nn)} = \frac{\sigma_{Zn}(\alpha)G_{Zn}(n)}{\sigma_{Zn}(\alpha)G_{Zn}(nn)}$$

$$\frac{\sigma(p, n)}{\sigma(p, nn)} = \frac{\sigma(\alpha, n)}{\sigma(\alpha, nn)} \tag{4.101}$$

$$\sigma(p, n): \sigma(p, nn) = \sigma(\alpha, n): \sigma(\alpha, nn) \tag{4.102}$$

Again, consider the ratio

$$\frac{\sigma(p, nn)}{\sigma(p, pn)} = \frac{\sigma_{Zn}(p)G_{Zn}(nn)}{\sigma_{Zn}(p)G_{Zn}(pn)} = \frac{G_{Zn}(nn)}{G_{Zn}(pn)}$$

Multiplying the numerator and denominator by $\sigma_{Zn}(\alpha)$ we have

$$\frac{\sigma(p,\,nn)}{\sigma(p,\,pn)} = \frac{G_{Zn}(nn)}{G_{Zn}(pn)} \, \frac{\sigma_{Zn}(\alpha)G_{Zn}(nn)}{\sigma_{Zn}(\alpha)G_{Zn}(pn)}$$

$$\frac{\sigma(p,\,nn)}{\sigma(p,\,pn)} = \frac{\sigma(\alpha,\,nn)}{\sigma(\alpha,\,pn)} \tag{4.103}$$

$$\sigma(p,\,nn):\sigma(p,\,pn) = \sigma(\alpha,\,nn):\sigma(\alpha,\,pn) \tag{4.104}$$

The results we got in equations (4.102) and (4.104) can be combined and written as

$$\sigma(p,\,n):\sigma(p,\,nn):\sigma(p,\,pn) = \sigma(\alpha,\,n):\sigma(\alpha,\,nn):\sigma(\alpha,\,pn) \tag{4.105}$$

This is the mathematical statement of Bohr independence hypothesis. It follows that probability of formation and probability of decay are independent of each other.

This was experimentally verified in Ghoshal's experiment.

Figure 4.9 shows two plots.

One is a plot of cross-section $\sigma$ against kinetic energy of $p$, i.e. $T_p$ [equation (4.97)] and the other plot is of cross-section $\sigma$ against kinetic energy of $\alpha$, i.e. $T_\alpha$ [equation (4.98)].

Each point in the plot of $\sigma$ versus $T_p$, $T_\alpha$ corresponds to a particular excitation energy. The plot shows how cross-section or reaction probability changes with energy.

We note that various channels $p + {}_{29}Cu^{63}$, $\alpha + {}_{28}Ni^{60}$ may lead to the same excited compound nucleus $({}_{30}Zn^{64})^*$. To create $C^*$ in the same excited state projectile $p$ for the channel

$$p + {}_{29}Cu^{63} \rightarrow ({}_{30}Zn^{64})^*$$

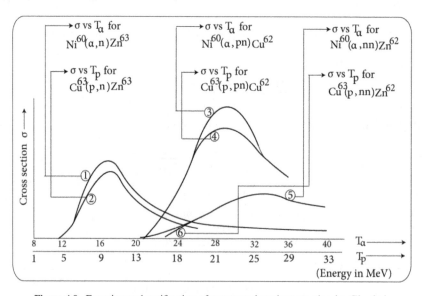

**Figure 4.9.** Experimental verification of compound nuclear reaction by Ghoshal.

and projectile $\alpha$ for the channel

$$\alpha + {}_{28}Ni^{60} \rightarrow ({}_{30}Zn^{64})^*$$

require different excitation energies.

For this reason in figure 4.9 we have used two different energy scales in the abscissa, one for energy of $\alpha$ (8 to 40 $MeV$) and another for energy of $p$ (1 to 33 $MeV$).

The two horizontal energy scales are shifted relative to one another to match the energy available for excitation so that the excited compound nuclear state $({}_{30}Zn^{64})^*$ is formed in the same excited state. This enables us to compare the two incident channels.

We are comparing the cross-section for the same final product [equations (4.90) and (4.91)].

The curve ① showing $\sigma$ versus $T_\alpha$ and the curve ② showing $\sigma$ versus $T_p$ correspond to the same final products ${}_{30}Zn^{63} + n$ obtained from two different incident channels viz. $\alpha + {}_{28}Ni^{60}$ and $p + {}_{29}Cu^{63}$. We note that the patterns are very similar. So the exit channel ${}_{30}Zn^{63} + n$ is independent of the incident channels.

Similarly, the curve ③ showing $\sigma$ versus $T_\alpha$ and the curve ④ showing $\sigma$ versus $T_p$ correspond to the same final products ${}_{29}Cu^{62} + p + n$ obtained from two different incident channels viz. $\alpha + {}_{28}Ni^{60}$ and $p + {}_{29}Cu^{63}$. We note that the patterns are very similar. So the exit channel ${}_{29}Cu^{62} + p + n$ is independent of the incident channels.

Similarly, the curve ⑤ showing $\sigma$ versus $T_\alpha$ and the curve ⑥ showing $\sigma$ versus $T_p$ correspond to the same final products ${}_{30}Zn^{62} + n + n$ obtained from two different incident channels viz. $\alpha + {}_{28}Ni^{60}$ and $p + {}_{29}Cu^{63}$. We note that the patterns are very similar. So the exit channel ${}_{30}Zn^{62} + n + n$ is independent of the incident channels.

Ghoshal's experiment thus proves that the decay of a compound nucleus is independent of how the compound nucleus was formed. Exit channel and incident channel are independent of one another as assumed by Bohr.

☐ A similar conclusion follows from the following example also.

$$\left.\begin{array}{c} \alpha + B^{10} \\ d + C^{12} \\ p + C^{13} \end{array}\right\} \rightarrow (N^{14})^* \rightarrow \left\{\begin{array}{c} B^{10} + \alpha \\ C^{12} + d \\ C^{13} + p \\ N^{13} + n \end{array}\right. \tag{4.106}$$

and we end up with

$$\sigma(\alpha\alpha): \sigma(dd): \sigma(\alpha p): \sigma(\alpha n) = \sigma(d\alpha): \sigma(dd): \sigma(dp): \sigma(dn)$$

$$= \sigma(p\alpha): \sigma(pd): \sigma(pp): \sigma(pn) \tag{4.107}$$

## 4.25 Direct reaction

Consider the nuclear reaction $a + X \rightarrow Y + b$.

For large projectile energy $T_a \geqslant 20\ MeV$ the reaction does not proceed through compound nucleus formation. This is because the projectile is so energetic that it cannot wait so as to share its energy with all nucleons of the nucleus. It interacts with the nucleus for a time during which it crosses the nucleus.

Let us get an estimate of the time of direct interaction of the projectile with the nucleus.

Direct reaction is a one-step process—it provides a passage from initial to final state directly—there being no intermediate state.

Suppose the projectile is a proton $p$ with kinetic energy $T_p \sim 20 \ MeV$. Using non-relativistic formula (as we are interested to get a rough estimate of the order of magnitude of lifetime or duration of reaction)

$$\frac{1}{2} m_p v^2 = 20 \ MeV$$

$$v = \sqrt{\frac{2(20 \ MeV)}{m_p}} \tag{4.108}$$

$$v = \sqrt{\frac{2(20 \ MeV)}{m_p c^2}} \, c \cong \sqrt{\frac{2(20 \ MeV)}{938 \ MeV}} \, c = 0.2c \ \ m_p c^2 \cong 938 \ MeV)$$

Time taken to cross the nucleus say of size $R \sim 10 \ fm$ is

$$\frac{R}{v} = \frac{10 \ fm}{0.2c} = \frac{10 \times 10^{-15} \ m}{0.2 \times 3 \times 10^8 \ m \ s^{-1}} \sim 10^{-22} \ s. \tag{4.109}$$

This is the lifetime of direct interaction.

Clearly as no compound nucleus is formed and the reaction takes place directly, it is a fast process.

- Direct reactions can be of two types.

Transfer reaction and knock out reaction. We discuss them now.

## 4.26 Transfer reaction

Transfer reaction is a direct reaction in which a nucleon or a group of nucleons gets transferred or shifted between the projectile and the target.

Transfer reaction is of two types.

One is stripping reaction and the other is pick up reaction. We discuss them now.

## 4.27 Stripping reaction

Stripping reaction is a direct transfer reaction in which a projectile is stripped of $x$ nucleons (say) so that the target nucleus $X^A$ has $x$ nucleons added to it. The residual nucleus is $Y^{A+x}$.

A stripping reaction can be written as

$$\underset{b+x}{a} + \underset{A}{X} \rightarrow \underset{A+x}{Y} + b \ \text{ i.e } X(a, b)Y \tag{4.110}$$

where $A$ is mass number of target $X$.

We give an example of stripping reaction.

✓ $(d, p)$ reaction: $d + {}_4\text{Be}^9 \rightarrow {}_4\text{Be}^{10} + p$.

In this reaction $d \equiv np$ is stripped of the constituent neutron $n$. So proton $p$ is emitted in the exit channel. The target ${}_4\text{Be}^9$ absorbs the neutron $n$ so that the daughter nucleus becomes ${}_4\text{Be}^{10}$ which is a constituent of the exit channel.

We furnish a few more examples:

✓ $(d, n)$ reaction : $d + {}_4\text{Be}^9 \rightarrow {}_5\text{B}^{10} + n$

✓ $(\alpha, d)$ reaction : $\alpha + {}_6\text{C}^{12} \rightarrow {}_7\text{N}^{14} + d$, $\alpha + {}_6\text{C}^{12} \rightarrow {}_6\text{C}^{13} + \text{He}^3$

✓ $(d, p)$ reaction : $d + {}_{29}\text{Cu}^{63} \rightarrow {}_{29}\text{Cu}^{64} + p$

## 4.28 Pick up reaction

Pick up reaction is a direct transfer reaction in which a projectile pulls out or picks up $x$ nucleons (say) from the target so that the target nucleus $X^A$ transforms to $Y^{A-x}$.

A pick up reaction can be written as

$$a + \underset{A}{X} \rightarrow \underset{A-x}{Y} + \underset{a+x}{b} \quad \text{i.e } X(a, b)Y \tag{4.111}$$

where $A$ is mass number of target $X$.

We give an example of a pick up reaction.

✓ $(p, d)$ reaction : $p + {}_6\text{C}^{12} \rightarrow {}_6\text{C}^{11} + d$.

Here the projectile proton $p$ picks up a neutron $n$ from target and hence a deuteron $d \equiv np$ gets emitted in the exit channel while the target ${}_6\text{C}^{12}$ converts to ${}_6\text{C}^{11}$ which is a constituent of the exit channel.

Other examples are

$$p + {}_3\text{Li}^7 \rightarrow {}_3\text{Li}^6 + d$$

$$p + {}_4\text{Be}^9 \rightarrow {}_4\text{Be}^8 + d.$$

## 4.29 Knock out reaction

Knock out reaction is a direct reaction in which a projectile knocks a nucleon or a group of nucleons out of the nucleus causing multi-particle emission in general. A knock out reaction can be written as

$$a + \underset{A}{X} \rightarrow \underset{A-x}{Y} + a + x \text{ nucleons} \tag{4.112}$$

where $A$ is the mass number of the target $X$ and the projectile $a$ knocks out $x$ nucleons from the target—so that target mass number becomes $A - x$ and there is multi-particle emission in the exit channel.

We give an example of knock out reaction.

✓ $(p, pp)$ reaction. $p + {}_6\text{C}^{12} \rightarrow {}_5\text{B}^{11} + p + p$.

The projectile proton $p$ knocks out one proton from the target ${}_6\text{C}^{12}$ and so two protons $pp$ are emitted in the exit channel. The target upon losing one proton becomes ${}_5\text{B}^{11}$ as the daughter nucleus—a constituent of the exit channel.

Other examples are
✓ $(p, pn)$ reaction. $p+ {}_6C^{12} \rightarrow {}_6C^{11} + p + n$
✓ $(\alpha, 2\alpha)$ ✓ $(p, 2n)$ ✓ $(p, 3p)$ etc.

## 4.30 Comparison of reactions

- Knock out and pick up reactions

✓ ${}_6C^{12}(p, 2p){}_5B^{11}$, ${}_6C^{12}(p, pn){}_6C^{11}$ are knock out reactions while ${}_6C^{12}(p, d){}_6C^{11}$ is a pick up reaction.
✓ In both cases particles are removed from the target nucleus.
✓ Energy spectra are similar.

So these reactions give rather similar information about the effect of making holes in the target when the target is in ground state.
- Stripping reaction and pick up reactions
  ✓ ${}_6C^{12}(d, p){}_6C^{13}$ is a stripping reaction and ${}_6C^{12}(p, d){}_6C^{11}$ is a pick up reaction.

  These are inverse processes of each other.

  Similarly $(d, n)$ is a stripping reaction and $(n, d)$ is a pick up reaction and they are the inverse of each other.
  ✓ Stripping reaction gives information about the effect of adding particles to target. This is because the projectile gets stripped, i.e. a portion of it is removed and the removed part is added to the target. Pick up reaction gives information about the effect of making holes in the target. This is because the projectile picks up a portion of the target and combines with it drilling holes in the target.
- Compound nuclear reaction and direct nuclear reaction.

| ☐ Compound nuclear reaction | ☐ Direct nuclear reaction |
|---|---|
| Intermediate compound nucleus is formed. | Does not proceed via compound nucleus formation. |
| Projectile $a$ is absorbed by the target nucleus $X$ to form a compound nucleus in highly excited state $C^* \Leftarrow$ first step. After a time delay the compound nucleus de-excites by the emission of particles $\Leftarrow$ second step. Compound nuclear reaction is a two-step process (formation of compound nucleus and decay of compound nucleus) | Passage from initial to final state occurs directly. It is a one-step process. (Lifetime of compound nucleus is zero.) Direct reaction is a single-step process. |
| More likely to occur at lower projectile energy $T_a \leqslant 20\ MeV$. | More likely to occur at higher projectile energy $T_a \geqslant 20\ MeV$. |

(*Continued*)

(*Continued*)

| ☐ Compound nuclear reaction | ☐ Direct nuclear reaction |
|---|---|
| At this energy de Broglie wavelength of projectile is comparable to the size of nucleus. | At this energy de Broglie wavelength of projectile is comparable to the size of the nucleon. |
| Slow process. | Fast process. |
| Lifetime $10^{-13}\,s$. | Lifetime $10^{-22}\,s$. This is the nuclear transit time. |
| Compound nucleus is formed which is in zero momentum state in C-frame. After thorough mixing of energy there is statistical equilibrium and there is equal probability for the compound nucleus to decay in any direction. This is also because it cannot retain any memory regarding how it was formed. So it shows no preference to decay along any particular direction. | In C-frame projectile and target approach each other and collide along a line. So direction is important here. |
| Angular distribution is almost isotropic as shown in figure 4.10. | Projectile directly hits target with huge energy and moves forward (in L-frame). Angular distribution is forward peaked as shown in figure 4.10. |
| Example $p + {}_{29}Cu^{63} \rightarrow ({}_{30}Zn^{64})^{*} \rightarrow {}_{30}Zn^{63} + n$ $\alpha + {}_{28}Ni^{60} \rightarrow ({}_{30}Zn^{64})^{*} \rightarrow {}_{30}Zn^{63} + n$ | Example Knock out: $(p, 2p)$, $(p, pn)$ Pick up:　　$(p, d)$, $(n, d)$ Stripping:　$(d, p)(d, n)$ |

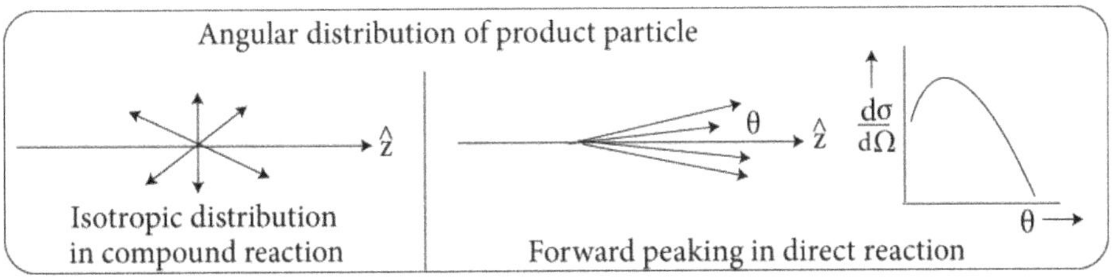

Figure 4.10. Angular distribution of product particle in compound nuclear reaction and direct reaction.

## 4.31 Two-body non-relativistic $Q$ value equation

Nuclear reactions are governed by conservation laws. We will now discuss how conservation of energy and conservation of momentum impose restrictions on nuclear reactions—these restrictions are referred to as kinematic restrictions.

Consider the nuclear reaction $a + X \rightarrow Y + b$ i.e. $X(a, b)Y$ in lab frame as shown in figure 4.11. Use conservation of linear momentum

$$\vec{P}_a = \vec{P}_Y + \vec{P}_b. \tag{4.113}$$

Resolving the vectors of equation (4.113) along the $x$-axis we have

$$P_a = p_b \cos\theta + p_Y \cos\phi$$

$$p_Y \cos\phi = P_a - p_b \cos\theta \tag{4.114}$$

Resolving the vectors of equation (4.113) along the $y$-axis we have

$$0 = p_b \sin\theta - p_Y \sin\phi$$

$$p_Y \sin\phi = p_b \sin\theta \tag{4.115}$$

Squaring and adding equations (4.114) and (4.115) we have

$$p_Y^2 = \left(P_a - p_b \cos\theta\right)^2 + \left(p_b \sin\theta\right)^2$$

$$p_Y^2 = p_a^2 + p_b^2 - 2p_a p_b \cos\theta \tag{4.116}$$

With

$$p_Y = \sqrt{2m_Y T_Y}, \; p_a = \sqrt{2m_a T_a}, \; p_b = \sqrt{2m_b T_b} \tag{4.117}$$

we rewrite equation (4.116) as

$$2m_Y T_Y = 2m_a T_a + 2m_b T_b - 2\sqrt{2m_a T_a}\sqrt{2m_b T_b} \cos\theta$$

$$T_Y = T_a \frac{m_a}{m_Y} + T_b \frac{m_b}{m_Y} - \frac{2}{m_Y}\sqrt{m_a m_b T_a T_b} \cos\theta \tag{4.118}$$

The energy conservation relation gives (with $T_X = \frac{p_X^2}{2m_X} = 0$)

$$(m_a c^2 + T_a) + m_X c^2 = (m_Y c^2 + T_Y) + (m_b c^2 + T_b) \tag{4.119}$$

$$(m_a + m_X)c^2 - (m_Y + m_b)c^2 = T_b + T_Y - T_a = Q \; [\text{by equation (4.61)}] \tag{4.120}$$

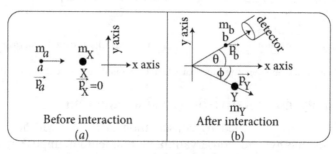

**Figure 4.11.** Projectile target interaction in L-frame.

where $Q$ is called $Q$ value of the nuclear reaction $X(a, b)Y$.

✓ Note that if we assume that daughter nucleus $Y$ goes to excited state with an excitation energy $\Delta E$ then the energy conservation relation becomes (with $T_X = 0$)

$$(m_a c^2 + T_a) + m_X c^2 = (m_Y c^2 + T_Y + \Delta E) + (m_b c^2 + T_b) \qquad (4.121)$$

[instead of equation (4.119)]

$$(m_a + m_X)c^2 - (m_Y + m_b)c^2 = T_b + T_Y - T_a + \Delta E = Q \text{ (redefine)} \qquad (4.122)$$

If $\Delta E = 0$ i.e. if daughter nucleus $Y$ is in the ground state then equation (4.122) reduces to equation (4.120). In view of this we can say that $Q - \Delta E$ is the $Q$ value of the ground state.

Let us use equation (4.120) to further our analysis.

From equation (4.120) we write

$$Q = T_b + T_Y - T_a = T_b - T_a + T_Y$$

Putting $T_Y$ from equation (4.118) we have

$$Q = T_b - T_a + T_a \frac{m_a}{m_Y} + T_b \frac{m_b}{m_Y} - \frac{2}{m_Y} \sqrt{m_a m_b T_a T_b} \cos \theta$$

$$Q = T_b \left(1 + \frac{m_b}{m_Y}\right) - T_a \left(1 - \frac{m_a}{m_Y}\right) - \frac{2}{m_Y} \sqrt{m_a m_b T_a T_b} \cos \theta \qquad (4.123)$$

This is called two-body non-relativistic $Q$ value equation.

Quantities that are known or measurable are $m_a$, $m_X$, $m_Y$, $m_b$.

$T_a$ is the kinetic energy supplied to the projectile and hence known.

$T_b$ is the kinetic energy of product particle that is detected at angle $\theta$ and measured in the detector.

The daughter nucleus $Y$ is created and recoils at angle $\phi$ inside the target. So $T_Y$ and $\phi$ are not known—we have no control over them. They have been eliminated and hence do not occur in equation (4.123).

✓ Calculation of $Q$ value is possible from equation (4.123).

✓ The relation given by equation (4.123) is valid for any type of two-body non-relativistic nuclear reaction since this relation is independent of the mechanism of the nuclear reaction.

✓ The relation given by equation (4.123) depends on $\theta$ along which the product particle $b$ emerges out and gets detected by a detector.

• If $\theta = 90°$ (i.e. if the detector is placed at 90°) $\cos \theta = 0$. From equation (4.123) we get

$$Q = T_b \left(1 + \frac{m_b}{m_Y}\right) - T_a \left(1 - \frac{m_a}{m_Y}\right) \qquad (4.124)$$

Equation (4.123) can be rewritten as follows.

$$T_b\left(1 + \frac{m_b}{m_Y}\right) - T_a\left(1 - \frac{m_a}{m_Y}\right) - \frac{2}{m_Y}\sqrt{m_a m_b T_a T_b}\,\cos\theta - Q = 0$$

$$\left(\sqrt{T_b}\right)^2\left(1 + \frac{m_b}{m_Y}\right) - \sqrt{T_b}\frac{2}{m_Y}\sqrt{m_a m_b T_a}\,\cos\theta - \left[Q + T_a\left(1 - \frac{m_a}{m_Y}\right)\right] = 0$$

$$\left(\sqrt{T_b}\right)^2 (m_Y + m_b) - \sqrt{T_b}\,2\sqrt{m_a m_b T_a}\,\cos\theta - [Qm_Y + T_a(m_Y - m_a)] = 0 \quad (4.125)$$

This is a quadratic equation in $\sqrt{T_b}$ and can be solved using standard formula

$$\sqrt{T_b} = \frac{2\sqrt{m_a m_b T_a}\,\cos\theta \pm \sqrt{\left(2\sqrt{m_a m_b T_a}\,\cos\theta\right)^2 + 4(m_Y + m_b)[Qm_Y + T_a(m_Y - m_a)]}}{2(m_Y + m_b)}$$

$$\sqrt{T_b} = \frac{\sqrt{m_a m_b T_a}\,\cos\theta \pm \sqrt{\left(\sqrt{m_a m_b T_a}\,\cos\theta\right)^2 + (m_Y + m_b)[Qm_Y + T_a(m_Y - m_a)]}}{m_Y + m_b} \quad (4.126)$$

The condition is that kinetic energy $T_b$ should be real, positive.

The solution depends on $T_a$, $\theta$, $Q$, $m_a$, $m_b$, $m_Y$.
The solution equation (4.126) can be put in the form

$$\sqrt{T_b} = p \pm \sqrt{(p^2 + q)} \quad (4.127)$$

where

$$p = \frac{\sqrt{m_a m_b T_a}}{m_Y + m_b}\cos\theta, \quad q = \frac{Qm_Y + T_a(m_Y - m_a)}{m_Y + m_b} \quad (4.128)$$

In *exercise 4.3* we have shown how to find the energy levels of the daughter nucleus $Y$.
Based upon equations (4.126)–(4.128) let us discuss energetics of the nuclear reaction.

## 4.32 Exoergic nuclear reaction

Consider the reaction $a + X \rightarrow Y + b$.
Suppose it is exoergic.
The exoergic nuclear reaction corresponds to $Q > 0$.
Such a reaction can take place spontaneously, i.e. it may be initiated even when the kinetic energy of incident particle is zero, i.e. $T_a = 0$.
Let us study the variation of kinetic energy of the emergent particle $T_b$ and its angular distribution as the kinetic energy of incident projectile $T_a$ is varied.
✓ For zero projectile energy ($T_a = 0$)

Putting $T_a = 0$ in equation (4.128) we have

$$p = \frac{\sqrt{m_a m_b T_a}}{m_Y + m_b} \cos \theta, \quad q = \frac{Q m_Y + T_a(m_Y - m_a)}{m_Y + m_b} = \frac{Q m_Y}{m_Y + m_b} \qquad (4.129)$$

From equation (4.127) we get

$$\sqrt{T_b} = p \pm \sqrt{(p^2 + q)} = \pm \sqrt{q} \qquad (4.130)$$

$$T_b = q = \frac{Q m_Y}{m_Y + m_b} \neq T_b(\theta) \qquad (4.131)$$

Clearly $T_b$ is independent of $\theta$. So value of $T_b$ will have the same value in all directions if kinetic energy of the projectile is zero.

✓ Finite kinetic energy of projectile ($T_a \neq 0$)

We have defined in equation (4.128) that $q = \frac{Q m_Y + T_a(m_Y - m_a)}{m_Y + m_b}$. Now product nucleus has mass $m_Y$ greater than the mass of projectile $m_a$ ,i.e.

$m_Y > m_a$

$m_Y - m_a > 0$

Then $q$ is positive for all values of $T_a$.

Again, from equation (4.127)

$$\sqrt{T_b} = p \pm \sqrt{(p^2 + q)} \qquad (4.132)$$

Since $q$ is positive it follows that

$$\sqrt{(p^2 + q)} > p$$

Clearly negative sign in equation (4.132) will make $\sqrt{T_b} = -$ve which is not acceptable. Hence we accept only the positive sign in equation (4.132) to get

$$\sqrt{T_b} = p + \sqrt{(p^2 + q)} \qquad (4.133)$$

Clearly there will be only one acceptable root of the $Q$ value equation.

So in this case $T_b$ is single-valued for all possible values of incident particle energy, i.e. for all $T_a$.

Since $p = \frac{\sqrt{m_a m_b T_a}}{m_Y + m_b} \cos \theta$ [equation (4.128)] is a function of $\theta$ we see that in this case

$$T_b = T_b(\theta) \qquad (4.134)$$

For $T_a \neq 0$, when $\theta = 180°$ (i.e. when the particle emerges opposite to the incident projectile, i.e. along the backward direction), $\cos \theta = -1$ (minimum) and so

$$p = \frac{\sqrt{m_a m_b T_a}}{m_Y + m_b} \cos \theta = -\frac{\sqrt{m_a m_b T_a}}{m_Y + m_b} = -|p| \text{ This means from equation (4.133)}$$

$$\sqrt{T_b} = p + \sqrt{(p^2 + q)} = -|p| + \sqrt{(p^2 + q)} \qquad (4.135)$$

Clearly $T_b$ is minimum for $\theta = 180°$.

## 4.33 Endoergic nuclear reaction

Consider the reaction $a + X \rightarrow Y + b$.

Suppose it to be endoergic.

The endoergic nuclear reaction corresponds to $Q < 0$.

Putting $T_a = 0$, $Q = -|Q|$ in equation (4.128) we have

$$p = \frac{\sqrt{m_a m_b T_a}}{m_Y + m_b} \cos \theta = 0, \quad q = \frac{Q m_Y + T_a(m_Y - m_a)}{m_Y + m_b} = \frac{-|Q| m_Y}{m_Y + m_b} = -ve \quad (4.136)$$

From equation (4.127), i.e. $\sqrt{T_b} = p \pm \sqrt{(p^2 + q)}$ we have

$$\sqrt{T_b} = \sqrt{q} = \sqrt{-ve} = \text{imaginary quantity.}$$

This means the reaction is not possible for $T_a = 0$ (i.e. with zero kinetic energy of the projectile).

There will be some finite value of $T_a$ at which a reaction will be possible.

The smallest value of the kinetic energy of projectile $T_a$ for which a reaction will just be initiated is called threshold energy value. We shall denote threshold energy by $E_{th}$.

## 4.34 Evaluation of threshold energy value for an endoergic nuclear reaction

We are interested to find the smallest value of the kinetic energy of projectile $T_a$. Then $T_b$ will also be small.

Now the expression of $T_b$ is given by equation (4.127) viz.

$$\sqrt{T_b} = p \pm \sqrt{(p^2 + q)}$$

Let $T_a$ be such that

$$p^2 + q = 0 \quad (4.137)$$

Putting values of $p$ and $q$ from equation (4.128) we have

$$\left( \frac{\sqrt{m_a m_b T_a}}{m_Y + m_b} \cos \theta \right)^2 + \frac{Q m_Y + T_a(m_Y - m_a)}{m_Y + m_b} = 0 \quad (4.138)$$

$$\frac{m_a m_b T_a}{(m_Y + m_b)^2} \cos^2 \theta + \frac{Q m_Y + T_a(m_Y - m_a)}{m_Y + m_b} = 0$$

$$\frac{m_a m_b T_a \cos^2 \theta + [Q m_Y + T_a(m_Y - m_a)](m_Y + m_b)}{(m_Y + m_b)^2} = 0$$

$$T_a[m_a m_b \cos^2 \theta + (m_Y + m_b)(m_Y - m_a)] = -Q m_Y(m_Y + m_b) \quad (4.139)$$

$$T_a = \frac{-Q m_Y(m_Y + m_b)}{m_a m_b \cos^2 \theta + (m_Y + m_b)(m_Y - m_a)} \quad (4.140)$$

$$T_a = \frac{-Qm_Y(m_Y + m_b)}{m_a m_b \cos^2 \theta + m_Y^2 - m_Y m_a + m_Y m_b - m_a m_b}$$

$$= \frac{-Qm_Y(m_Y + m_b)}{m_Y^2 - m_Y m_a + m_Y m_b - m_a m_b(1 - \cos^2 \theta)}$$

$$T_a = \frac{-Q(m_Y + m_b)}{m_Y - m_a + m_b - \frac{m_a m_b}{m_Y} \sin^2 \theta} \tag{4.141}$$

This is the threshold value of energy.

To find the minimum value of $T_a$ we put $\theta = 0$, $\sin \theta = 0$ so that the denominator of equation (4.141) is maximum. This leads to

$$T_a \mid_{\theta=0} = \frac{-Q(m_Y + m_b)}{m_Y - m_a + m_b} = T_a = \text{minimum value of } T_a = T_a \mid_{\min} = E_{th} \quad (4.142)$$

This is the minimum energy to initiate an endoergic reaction and identified as the threshold energy $E_{th}$.

This $E_{th}$ of equation (4.142) is in terms of $m_Y$, $m_a$, $m_b$, $Q$.

Now from equation (4.61) for $Q$ value we have

$$Q = [m_X + m_a - m_Y - m_b]c^2 \tag{4.143}$$

$$m_X + m_a = m_Y + m_b + \frac{Q}{c^2} \tag{4.144}$$

$$m_X - \frac{Q}{c^2} + m_a = m_Y + m_b \tag{4.145}$$

Let us approximate $m_X \gg \frac{Q}{c^2}$ so that

$$m_X - \frac{Q}{c^2} \cong m_X$$

Then we can write equation (4.144) as

$$m_Y + m_b = m_X + m_a \tag{4.146}$$

Also

$$m_Y - m_a = m_X - m_b \tag{4.147}$$

Putting equations (4.146) and (4.147) in equation (4.142) we get

$$E \mid_{th} = \frac{-Q(m_Y + m_b)}{m_Y - m_a + m_b} = \frac{-Q(m_X + m_a)}{m_X - m_b + m_b} = \frac{-Q(m_X + m_a)}{m_X}$$

$$E \mid_{th} = -Q\left(1 + \frac{m_a}{m_X}\right) = \mid Q \mid \left(1 + \frac{m_a}{m_X}\right) \tag{4.148}$$

We have expressed the threshold energy in terms of the mass of parent $m_X$ and mass of projectile $m_a$.

✓ In the case of photodisintegration $\gamma$ serves as the projectile. Then

$$m_a \equiv m_\gamma = 0.$$

Then threshold energy will be [equation (4.148)]

$$E_{th} = -Q\left(1 + \frac{m_\gamma}{m_X}\right) = -Q = |Q| \tag{4.149}$$

So the minimum energy available from an endoergic reaction is $Q$ when $m_a = 0$.

✓ We can derive this expression of threshold energy value $E_{th}$ for an endoergic reaction more directly using conservation of momentum in L-frame, as shown in *exercise 4.1*.

We derived threshold energy of equation (4.149) taking $\theta = 0$. This means that at this threshold particles first appear in the $\theta = 0$ direction with kinetic energy $T_b$ calculated as follows from equation (4.127) viz.

$$\sqrt{T_b} = p \pm \sqrt{(p^2 + q)}$$

Using equation (4.137), i.e. $p^2 + q = 0$ we get

$$\sqrt{T_b} = p$$

$$T_b = p^2 = \left(\frac{\sqrt{m_a m_b T_a}}{m_Y + m_b} \cos\theta\right)^2 = \frac{m_a m_b T_a}{(m_Y + m_b)^2} \cos^2\theta \text{ [using equation (4.128)]}$$

Putting $\theta = 0$, $\cos\theta = 1$, $T_a = T_a|_{\theta=0} = E_{th}$ gives

$$T_b = \frac{m_a m_b E_{th}}{(m_Y + m_b)^2} \tag{4.150}$$

As the bombarding energy $T_a$ increases beyond the threshold value $E_{th}$ particles begin to appear at $\theta > 0$ directions.

• Let us find the energy $T_a$ required for emergence of a particle along $\theta = 90°$. From equation (4.128) at $\theta = 90°$

$$p = \frac{\sqrt{m_a m_b T_a}}{m_Y + m_b} \cos\theta = 0 \tag{4.151}$$

Again, threshold condition is $p^2 + q = 0$ [equation (4.137)] which means for $p = 0$

$$q = 0$$

From equation (4.128)

$$q = \frac{Qm_Y + (m_Y - m_a)T_a|_{\theta=90°}}{m_Y + m_b} = 0$$

$$Qm_Y + (m_Y - m_a)T_a|_{\theta=90°} = 0$$

$$T_a|_{\theta=90°} = -\frac{Qm_Y}{m_Y - m_a} = \frac{|Q|m_Y}{m_Y - m_a} \qquad (4.152)$$

Again from equation (4.127)

$$\sqrt{T_b} = p \pm \sqrt{(p^2 + q)} = 0$$

$$T_b = 0$$

Equation (4.152) gives the threshold energy for product particle $b$ to just emerge at $\theta = 90°$ with $T_b = 0$.

## 4.35 Resonance in nuclear reaction

The cross-section of most nuclear reactions depends on the energy of the incident projectile. Vigorous reaction occurs in some preferred regions of incident energy which give rise to peaks and are referred to as resonance. At other values of incident energy the nuclear reaction is feeble.

We give two examples.

- Consider the nuclear reaction

$$_{48}Cd^{113}(n, \gamma)\ _{48}Cd^{114} \qquad (4.153)$$

which is neutron capture by $_{48}Cd^{113}$.

The capture cross-section has been shown against neutron projectile energy in figure 4.12.

The plot shows a narrow sharp peak called resonance peak at neutron energy $T_n = 0.176\ eV$, i.e. for slow neutrons. Capture of a neutron is prominent at this energy.

For $T_n \neq 0.176\ eV$, i.e. for faster neutrons capture of a neutron is insignificant.

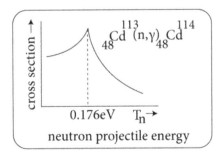

**Figure 4.12.** Resonance capture of slow neutrons by $_{48}Cd^{113}$ at $T_n = 0.176\ eV$.

We say that resonance capture occurs at this specific energy $T_n = 0.176 \, eV$ that leads to a sharp resonant peak.

    ✓ Slow neutrons are responsible for nuclear fission reactions. So if they are absorbed by cadmium then nuclear fission reaction rate diminishes. As cadmium easily absorbs slow moving neutrons it is used for control rods for nuclear reactions to control or regulate the reaction rate of nuclear fission.

   Another example
- Consider the nuclear reaction

$$_0n^1 + {}_{92}U^{238} \rightarrow {}_{92}U^{239*} \tag{4.154}$$

$$_{92}U^{239*} \rightarrow {}_{92}U^{239} + \gamma \tag{4.155}$$

Such bombardment of $_{92}U^{238}$ by neutron gives excited and unstable $_{92}U^{239*}$ that subsequently decays through emission of $\gamma$.

Momentum conservation relation in L-frame (where target $U^{238}$ is at rest) is

$$\vec{P}_n + \vec{P}_{U^{238}} = \vec{P}_{U^{239*}}$$

$$p_n = p_{U^{239*}} \; (\text{as } \vec{P}_{U^{238}} = 0) \tag{4.156}$$

Energy conservation gives

$$(T_n + m_n c^2) + m_{U^{238}} c^2 = (m_{U^{239*}} c^2 + T_{U^{239*}}) + \Delta E \tag{4.157}$$

where $\Delta E$ is the excitation energy that has been transferred by projectile neutron to $U^{238}$ and because of which we have $U^{239*}$ in an excited state now.

$$\Delta E = T_n - T_{U^{239*}} + m_n c^2 + m_{U^{238}} c^2 - m_{U^{239*}} c^2 \tag{4.158}$$

Consider

$$T_{U^{239*}} = \frac{p_{U^{239*}}^2}{2m_{U^{239*}}} = \frac{p_n^2}{2m_{U^{239*}}} \text{ since } p_n = p_{U^{239*}} \text{ [equation (4.156)]}$$

$$T_{U^{239*}} = \frac{p_n^2}{2m_n} \frac{m_n}{m_{U^{239*}}} = T_n \frac{m_n}{m_{U^{239*}}} \left( \text{as } T_n = \frac{p_n^2}{2m_n} \right)$$

With this equation (4.158) becomes

$$\Delta E = T_n - T_n \frac{m_n}{m_{U^{239*}}} + m_n c^2 + m_{U^{238}} c^2 - m_{U^{239*}} c^2 \tag{4.159}$$

Also let us use

$$m_n = 1.0087u, \; m_{U^{238}} = 238.0508u, \; m_{U^{239*}} = 239.0543u, \; 1u = 931 \, MeV$$

With these we have from equation (4.159)

$$\Delta E = T_n - T_n \frac{m_n}{m_{U^{239*}}} + (1.0087 + 238.0508 - 239.0543) \times 931 \; MeV$$

$$= T_n - T_n \frac{m_n}{m_{U^{239*}}} + 4.84 \; MeV$$

$$\Delta E = 4.84 \; MeV + T_n \left( 1 - \frac{m_n}{m_{U^{239*}}} \right) \tag{4.160}$$

We note that the factor $\left( 1 - \frac{m_n}{m_{U^{239*}}} \right)$ is constant. But we can control the energy of the neutron $T_n$. Depending on the energy possessed by the projectile neutron the reaction will proceed or not proceed as we explain now with a reference to figure 4.13.

   ✓ If almost zero kinetic energy neutron is used, i.e. $T_n = 0$, then from equation (4.160)

$$\Delta E = 4.84 \; MeV = E_1$$

which is the first discrete energy level $E_1$ of $U^{239}$. We rewrite equation (4.160) as

$$\Delta E = E_1 + T_n \left( 1 - \frac{m_n}{m_{U^{239*}}} \right) \tag{4.161}$$

We can increase the kinetic energy of the incoming neutron and study the reaction.

   If the neutron impinges with greater kinetic energy $T_n > 0$, i.e. energy is thereby made available to $U^{238}$, a reaction may or may not occur depending on the amount of neutron energy.

   ✓ If $T_n$ is such that $\Delta E = E$ where $E$ is not an energy level *of* $U^{239}$ (figure 4.13) then $U^{238}$ will not absorb energy and there will be no reaction. In other words, the probability of a nuclear reaction in that case is very very small, i.e. a reaction will not occur.

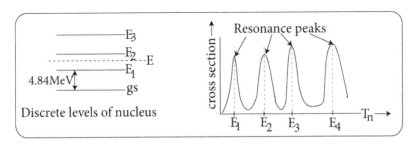

**Figure 4.13.** Resonance capture of neutrons by uranium 238.

✓ If $T_n$ is such that $\Delta E = E_1, E_2, E_3, \ldots$which are the possible discrete energy levels (excited states) of $U^{239}$ (figure 4.13) then $U^{238}$ will capture and absorb energy carried by a neutron and excite to these states. In other words, the probability of a nuclear reaction in these cases is very very large. This is referred to as a resonance reaction. Put differently, we can state this as follows.

Probability will be high for neutrons with specific energies to be captured and absorbed by $U^{238}$ and hence there exists a corresponding energy level. This is called nuclear resonance.

Due to vigorous reaction the reaction cross-section increases significantly at these certain specific projectile energies and the phenomenon is called resonance.

Clearly if the value of $T_n$ is such that $\Delta E \neq E_1, E_2, E_3, \ldots$the reaction drops drastically and so reaction cross-section falls abruptly around these energies $\Delta E = E_1, E_2, E_3, \ldots$

This resonance reaction can be pictorially represented by plotting the reaction cross-section against projectile energy as shown in figure 4.13.

We find peaks in the plot corresponding to a prominent reaction as a result of transition from $_{92}U^{238}$ to $_{92}U^{239*}$ due to resonance capture of neutrons having kinetic energy such that $\Delta E = E_1, E_2, E_3, \ldots$is ensured. Peaks obviously point to the high probability region while the off-peak zones correspond to virtually no absorption, no reaction.

Peaks correspond to the energy levels $E_1$, $E_2, E_3 \ldots$of $U^{239}$. They are discrete and well defined and that is why we have sharp peaks.

Also, we note that the energy levels may not be equally spaced and so peaks may not occur at identical separation.

We have explained the sudden sharp increase in reaction cross-section as due to the phenomenon of resonance that reflects the existence of discrete energy level structure of a nucleus.

Resonance represents the condition of energy transfer from $_{92}U^{238}$ to $_{92}U^{239*}$. No resonance means no energy transfer.

The reaction cross-section curve has narrow resonance peaks and corresponds to low projectile energies. They can be explained by the one-level dispersion formula of Breit–Wigner.

## 4.36  Lifetime and level width

- Lifetime $\tau$

    Time $\Delta t$ which a nucleus spends in a particular state before decay or transition is called lifetime $\tau$.

$$\Delta t = \tau \tag{4.162}$$

And decay constant $\lambda = \frac{1}{\tau}$ represents the probability per unit time of decay or transition.

- Level width $\Gamma$

  Level width $\Gamma$ is the spread $\Delta E$ in energy of an energy level.

$$\Delta E = \Gamma \qquad (4.163)$$

- Relation between lifetime $\tau$ and level width $\Gamma$

Lifetime $\tau$ and level width $\Gamma$ are related through Heisenberg's uncertainty principle

$$\Delta E \Delta t = \Gamma \tau \sim \hbar \qquad (4.164)$$

✓ If $\Delta t$ is small $\Delta E$ is large.

  A state of short mean lifetime or short-lived state $\Delta t$ = small corresponds to large level width, i.e. large spread in energy $\Delta E$ = large.

  In other words, the energy level is broad or diffuse and hence poorly defined in energy.

✓ If $\Delta t$ is large $\Delta E$ is small.

A state of long mean lifetime or long-lived state $\Delta t$ = large corresponds to small level width, i.e. small spread in energy $\Delta E$ = small.

In other words, the energy level is sharp or narrow and hence well-defined.

☐ Ground state

$$\tau \to \infty, \Gamma \to 0, \lambda \to 0 \qquad (4.165)$$

The nucleus remains in ground state forever unless perturbed or forced to change its state.

☐ Excited state

$$\tau \to \text{small } \Gamma \to \text{large } \lambda \to \text{large} \qquad (4.166)$$

The nucleus tries to move away from excited state and come down to ground state.

- Consider the formation of a compound nucleus

$$a + X \to C^\star$$

The projectile $a$ upon entering the target $X$ suffers a large number ($\sim$ millions) of collisions. The energy is distributed and thoroughly mixed leading to an excited compound nucleus that has completely forgotten how it derived the excitation energy. The unstable compound nucleus lives for a period of $\tau \sim 10^{-13}$ s which is its lifetime. It finds a way to de-excite as enough energy is accumulated by one or a group of nucleons through statistical fluctuation of energy. This leads to decay through one disintegration channel like

$$Y + b, \ Y' + b', \ Y'' + b'', \ \ldots.$$

Also, there is a possibility of $C^\star$ to shed its extra energy through $\gamma$ emission.

The probability of decay equals the reciprocal of the mean lifetime $\tau$ of compound nucleus. If $\Gamma$ is the width of the energy level of a state then by the uncertainty principle

$$\Gamma\tau \sim \hbar \Rightarrow \Gamma = \frac{\hbar}{\tau} \tag{4.167}$$

The width of a level is a measure of the probability of its decay.

Since decay can occur by various means we can define partial widths for each such mode of decay, e.g.

$$\Gamma_{bY}, \Gamma_{b'Y'}, \Gamma_{b''Y''}...., \Gamma_{\gamma}$$

corresponding to the decay channels $Y + b$, $Y' + b'$, $Y'' + b''$, ...., $\gamma$ respectively.

The relation of these partial widths to the total width $\Gamma$ is

$$\Gamma = \Gamma_{bY} + \Gamma_{b'Y'} + \Gamma_{b''Y''} + ...\Gamma_{\gamma} = \sum\Gamma_{partial} \tag{4.168}$$

The relative probabilities of different types of decays, i.e. channel probabilities are

$$\frac{\Gamma_{bY}}{\Gamma}, \frac{\Gamma_{b'Y'}}{\Gamma}, \frac{\Gamma_{b''Y''}}{\Gamma}, ... \frac{\Gamma_{\gamma}}{\Gamma} \tag{4.169}$$

corresponding to the decay channels $Y + b$, $Y' + b'$, $Y'' + b''$, ...., $\gamma$ respectively.

Clearly there are a number of energy levels to which decay can occur. If we consider the levels to be well separated and sharp then there will be no interference between the levels. This means that decay will occur to one level provided the projectile carries and imparts the level energy to the target nucleus. In other words, there will be only one resonance. The condition for this is

$$\text{Mean level spacing} \gg \Gamma \text{ (level width)} \tag{4.170}$$

For instance, separation between energy levels $E_1$ and $E_2$ should be greater than the level widths $\Gamma_1$, $\Gamma_2$ as in figure 4.14.

In this situation we consider the incident projectile to possess or carry excitation energy of one of the several nuclear levels leading to a single resonance. Clearly then two resonances will be widely separated and will be distinct, as shown in figure 4.14.

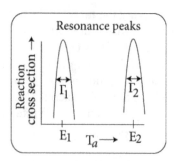

Figure 4.14. Well separated resonance peaks if mean level spacing $\gg\Gamma$.

## 4.37 Breit–Wigner one-level dispersion formula (qualitative discussion)

Breit–Wigner formula relates cross-section of a nuclear reaction to the energy of incident particle.

According to Bohr compound nucleus model

$$a + X \rightarrow C^\star$$

$$C^\star \rightarrow Y + b$$

Excitation of target nucleus by incident projectile particle can be treated as analogous to the excitation of the oscillations produced in an electric circuit by an electromagnetic wave. One expects nuclear cross-section to vary with incident energy in the same way that the energy in a forced oscillation varies with incident frequency.

The cross-section for the formation of $C^\star$ through the reaction $a + X \rightarrow C^\star$ is given by

$$\sigma_c = \pi \lambdabar^2 g \frac{\Gamma_{aX} \Gamma}{(E - E_r)^2 + \frac{\Gamma^2}{4}} \tag{4.171}$$

The cross-section for decay of $C^\star$ through the reaction $C^\star \rightarrow Y + b$ is given by

$$\sigma(a, b) = \sigma_c \frac{\Gamma_{bY}}{\Gamma} = \pi \lambdabar^2 g \frac{\Gamma_{aX} \Gamma}{(E - E_r)^2 + \frac{\Gamma^2}{4}} \frac{\Gamma_{bY}}{\Gamma}$$

$$\sigma(a, b) = \pi \lambdabar^2 g \frac{\Gamma_{aX} \Gamma_{bY}}{(E - E_r)^2 + \frac{\Gamma^2}{4}} \tag{4.172}$$

where

$$\lambdabar = \frac{\lambda}{2\pi} = \text{reduced de Broglie wavelength of incident particle}$$

$$\Gamma = \text{total level width}$$

$$\Gamma_{aX} = \text{partial level width for incident particle}$$

$$\Gamma_{bY} = \text{partial level width for decay of compound nucleus through exit channel } C^\star \rightarrow Y + b$$

$$g = \text{statistical spin factor.}$$

The value of $g = 2l + 1$ for spinless nuclei where $l$ is angular momentum of entrance channel while

$$g = \frac{2I + 1}{(2s_a + 1)(2s_X + 1)} \tag{4.173}$$

for particles with spin.

Here $s_a$, $s_X$ are spin angular momenta of projectile $a$ and target $X$, respectively and $I$ is the total angular momentum of compound nucleus $C$ i.e.

$$\vec{I} = \vec{s}_a + \vec{s}_X + \vec{l} \tag{4.174}$$

$E$ = energy of nuclear state

$E_r$ = mean energy of the resonant nuclear energy level having a spread $\Gamma$.

Equations (4.171) and (4.172) are called Breit–Wigner one-level dispersion formula.

As the Breit–Wigner result is analogous to the theory of optical dispersion the formula is referred to as dispersion formula.

✓ For elastic scattering through compound state with no other process possible

$$\Gamma_{aX} = \Gamma_{bY} = \Gamma$$

and at resonance

$$E = E_r \text{ and with } g = 2l + 1$$

$$\sigma(a, b) = \pi \lambda^2 g \frac{\Gamma_{aX}\Gamma_{bY}}{(E - E_r)^2 + \frac{\Gamma^2}{4}} = \pi \lambda^2 (2l + 1) \frac{\Gamma\Gamma}{0 + \frac{\Gamma^2}{4}}$$

$$\sigma(a, b) = 4\pi\lambda^2(2l + 1) \tag{4.175}$$

✓ For $E = E_r$

$$\sigma = \pi \lambda^2 g \frac{\Gamma_{aX}\Gamma_{bY}}{(E - E_r)^2 + \frac{\Gamma^2}{4}} = \pi \lambda^2 g \frac{\Gamma_{aX}\Gamma_{bY}}{\frac{\Gamma^2}{4}}$$

$$\sigma = 4\pi\lambda^2 g \frac{\Gamma_{aX}\Gamma_{bY}}{\Gamma^2} = \sigma_{res} \equiv \sigma_{max} \tag{4.176}$$

✓ For $E = E_r \pm \frac{\Gamma}{2}$

$$\sigma = \pi \lambda^2 g \frac{\Gamma_{aX}\Gamma_{bY}}{(E - E_r)^2 + \frac{\Gamma^2}{4}} = \pi \lambda^2 g \frac{\Gamma_{aX}\Gamma_{bY}}{\left(\pm\frac{\Gamma}{2}\right)^2 + \frac{\Gamma^2}{4}} = \pi \lambda^2 g \frac{\Gamma_{aX}\Gamma_{bY}}{\frac{\Gamma^2}{2}}$$

$$\sigma = 2\pi\lambda^2 g \frac{\Gamma_{aX}\Gamma_{bY}}{\Gamma^2} = \frac{1}{2}\sigma_{res} \equiv \frac{1}{2}\sigma_{max} \tag{4.177}$$

This implies that $\Gamma$ is full width at half maximum (figure 4.15).

✓ The factor $\lambda^2$ is a measure of the probability of forming a compound nucleus.

✓ The factor $\dfrac{1}{(E - E_r)^2 + \frac{\Gamma^2}{4}}$ is called resonance factor because at resonance $E = E_r$, denominator is minimum and so this factor becomes very large and so reaction cross-section becomes very large—resonance peak is

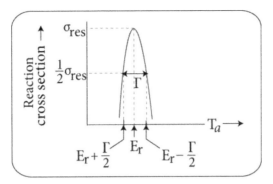

**Figure 4.15.** Cross-sections at $E = E_r$, $E = E_r \pm \frac{\Gamma}{2}$.

produced. So this factor is responsible for resonance. As $E \neq E_r$ the denominator increaseses and this factor diminishes. So reaction cross-section decreases. We have a prominent peak at $E = E_r$ as reaction cross-section drops to almost zero for $E \neq E_r$.

✓ The factor $\Gamma_{aX}\Gamma_{bY}$ gives the probability of definite type of disintegration of compound nucleus as it is expressed as a product of partial widths $\Gamma_{aX}$ and $\Gamma_{bY}$.

✓ This formula is not valid when two or more resonances are close together since resonances might overlap and distort each other.

## 4.38 Exercises

**Exercise 4.1** *Obtain the expression of threshold energy value for endoergic reaction using momentum conservation in L-frame.*

[Ans.] Consider the endoergic reaction

$$a + X \rightarrow C^* \rightarrow Y + b.$$

Apply conservation of momentum in L-frame to $a + X \rightarrow C^*$ (formation of compound nucleus)

$$\vec{p}_a + \vec{p}_X = \vec{p}_{C^*}$$

As target $X$ is at rest in L-frame $\vec{p}_X = 0$ and so

$$\vec{p}_a = \vec{p}_{C^*}$$

$$m_a v_a = (m_a + m_X)v_{C^*}$$

where

$$m_a + m_X = m_{C^*} = \text{mass of compound nucleus}$$

$\vec{v}_{C^*}$ is velocity of compound nucleus = velocity of the centre of mass. Hence

$$v_{C^*} = \frac{m_a v_a}{m_a + m_X} \tag{4.178}$$

From equation (4.83)

$$E_{th} = (T_a)_{min} = \frac{1}{2} m_a v_a^2 = |Q| + \frac{1}{2} m_{C^*} v_{C^*}^2 \qquad (4.179)$$

$$|Q| = \frac{1}{2} m_a v_a^2 - \frac{1}{2} m_{C^*} v_{C^*}^2$$

$$|Q| = \frac{1}{2} m_a v_a^2 \left( 1 - \frac{m_{C^*} v_{C^*}^2}{m_a v_a^2} \right)$$

Putting $m_a + m_X = m_{C^*}$ and using equation (4.178)

$$|Q| = \frac{1}{2} m_a v_a^2 \left[ 1 - \frac{m_a + m_X}{m_a v_a^2} \left( \frac{m_a v_a}{m_a + m_X} \right)^2 \right]$$

$$= \frac{1}{2} m_a v_a^2 \left( 1 - \frac{m_a}{m_a + m_X} \right)$$

$$|Q| = \frac{1}{2} m_a v_a^2 \frac{m_X}{m_a + m_X}$$

Using equation (4.179)

$$|Q| = E_{th} \frac{m_X}{m_a + m_X}$$

$$E_{th} = |Q| \frac{m_a + m_X}{m_X} = -Q \frac{m_a + m_X}{m_X} \text{ (in L-frame)} \qquad (4.180)$$

Since for endoergic reaction $Q < 0$ i.e. $Q = -|Q|$.

$E_{th}$ can be determined experimentally and equation (4.180) can be used to evaluate $Q$ value.

If the incident particles, i.e. $a$s are $\gamma$ rays then $m_a = 0$ and so

$$E_{th} = -Q \frac{0 + m_X}{m_X} = -Q. \qquad (4.181)$$

☐ We got in L-frame

$$E_{th} = (T_a)_{min} = \frac{1}{2} m_a v_a^2 = |Q| + \frac{1}{2} m_{C^*} v_{C^*}^2.$$

In centre of mass frame of reference centre of mass momentum is zero and so $v_C = 0$ and hence

$$E_{th}^{cm} = |Q| = -Q \qquad (4.182)$$

Threshold energy is equal to the absolute $Q$ value in the centre of mass frame of reference.

**Exercise 4.2** *The threshold energy of neutrons in the nuclear reaction* $_{11}\mathrm{Na}^{23}(n,\,\alpha)_9\mathrm{F}^{20}$ *is nearest to which of the following if Q value of the reaction is known to be* $-5.4\;MeV$ *and* $m_n = 1.008\,665u$, $m_{Na} = 22.9898u$.
(a) 4.5 MeV, (b) 5.5 MeV, (c) 6.5 MeV, (d) 7.5 MeV?
Hint: From equation (4.180)

$$E_{th} = -Q\frac{m_a + m_X}{m_X} = -Q\frac{m_n + m_{Na}}{m_{Na}}$$

$$E_{th} = -(-5.4\;MeV)\frac{1.008\,665u + 22.9898u}{22.9898u} = 5.637\;MeV$$

**Exercise 4.3** *Consider the nuclear reaction* $_8\mathrm{O}^{16}(d,\,p)_8\mathrm{O}^{17*}$. *Find the energy levels of* $_8\mathrm{O}^{17*}$ *taking* $m_dc^2 = 2$, $m_pc^2 = 1$, $m_{O^{17}}c^2 = 17$, $T_d = 10$, $Q = 1.96$ (all in MeV) and $\theta = 0.436\,111^c$ *if a detector records* $T_p = 11.69, 10.81, 8.58, 7.77$ (all in MeV).

[Ans.] Consider the nuclear reaction

$$a + X \rightarrow Y + b$$

From equation (4.120) we get

$$Q = T_b + T_Y - T_a + \Delta E = T_b - T_a + \Delta E + T_Y$$

Substituting $T_Y$ from equation (4.118) we have

$$Q = T_b - T_a + \Delta E + T_a\frac{m_a}{m_Y} + T_b\frac{m_b}{m_Y} - \frac{2}{m_Y}\sqrt{m_a m_b T_a T_b}\cos\theta$$

$$Q = T_b\left(1 + \frac{m_b}{m_Y}\right) - T_a\left(1 - \frac{m_a}{m_Y}\right) - \frac{2}{m_Y}\sqrt{m_a m_b T_a T_b}\cos\theta + \Delta E$$

$$\Delta E = Q - T_b\left(1 + \frac{m_b}{m_Y}\right) + T_a\left(1 - \frac{m_a}{m_Y}\right) + \frac{2}{m_Y}\sqrt{m_a m_b T_a T_b}\cos\theta \qquad (4.183)$$

The given nuclear reaction is

$$d + {}_8\mathrm{O}^{16} \rightarrow {}_8\mathrm{O}^{17*} + p$$

Making the substitutions viz. $a \equiv d, X \equiv {}_8\mathrm{O}^{16}$, $b \equiv p$, $Y \equiv {}_8\mathrm{O}^{17*}$

$$Q = 1.96\;MeV,\; m_ac^2 = m_dc^2 = 2\;MeV,\; m_bc^2 = m_pc^2 = 1\;MeV,$$

$$m_Yc^2 = m_{O^{17}}c^2 = 17\;MeV,\; T_a = T_d = 10\;MeV,\; T_b = T_p\text{ and}$$

$$\theta = 0.436\,111^c. = \frac{180}{\pi}0.436\,111 = 24.987\,32°$$

we have from equation (4.183)

$$\Delta E = Q - T_p\left(1 + \frac{m_p}{m_{O^{17*}}}\right) + T_d\left(1 - \frac{m_d}{m_{O^{17*}}}\right) + \frac{2}{m_{O^{17*}}}\sqrt{m_d m_p T_d T_p}\cos\theta$$

$$\Delta E = Q - T_p\left(1 + \frac{m_p c^2}{m_{O^{17}}c^2}\right) + T_d\left(1 - \frac{m_d c^2}{m_{O^{17}}c^2}\right) + \frac{2}{m_{O^{17}}c^2}\sqrt{(m_d c^2)(m_p c^2)T_d T_p}\cos\theta$$

$$\Delta E = 1.96 - T_p\left(1 + \frac{1}{17}\right) + 10\left(1 - \frac{2}{17}\right) + \frac{2}{17}\sqrt{(2)(1)(10)T_p}\cos 24.987\,32°(\,MeV)$$

$$\Delta E = 10.783\,529 - T_p(1.058\,8235) + \sqrt{T_p}(0.476\,8882)(\,MeV)$$

Putting the values of $T_p$ the energy levels $\Delta E$ of $O^{17}$ can be obtained as shown

| $T_p$ | 11.69 MeV | 10.81 MeV | 8.58 MeV | 7.77 MeV |
|---|---|---|---|---|
| $\Delta E$ | 0.0363 937 $\cong 0$ MeV | 0.905 5869 MeV $\cong 0.91$ MeV | 3.095 7069 MeV $\cong 3.1$ MeV | 3.885 7829 MeV $\cong 3.89$ MeV |

This is shown in figure 4.16.

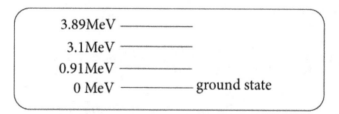

**Figure 4.16.** Energy levels of $_8O^{17}$, pertaining to *exercise 4.3*.

**Exercise 4.4** *Consider the nuclear reaction* $_1H^1 + _3Li^7 \to _4Be^7 + _0n^1$. *Calculate the threshold energy for the reaction. If the incident proton has an energy 4 MeV calculate the maximum possible velocity of the neutron. Given atomic masses of* $_3Li^7$, $_4Be^7$, p,n *to be 7.018 24u, 7.019 16u, 1.007 59u, 1.008 98u respectively and* 1u = 931 MeV.
 [Ans.] The nuclear reaction is

$$_1H^1 + _3Li^7 \to _4Be^7 + _0n^1$$

Atomic mass of $LHS = 1.007\ 59u + 7.018\ 24u = 8.025\ 83u = m_{initial}$

Atomic mass of $RHS = 7.019\ 16u + 1.008\ 98u = 8.028\ 14u = m_{final}$

$$m_{initial} - m_{final} = 8.025\ 83u - 8.028\ 14u = -0.002\ 31u$$

$$= -0.002\ 31 \times 931\ MeV$$

$$= -2.150\ 61\ MeV.$$

So the reaction cannot proceed.

The threshold energy needed for reaction to initiate is $2.150\ 61\ MeV$.
Energy pumped through $p$ is $4\ MeV$.
Energy balance is

$$4\ MeV - 2.150\ 61\ MeV = 1.849\ 39\ MeV.$$

This energy is spent as kinetic energy of the neutron $n$. So

$$\frac{1}{2}m_n v_n^2 = 1.849\ 39\ MeV$$

$$v_n = \sqrt{\frac{2(1.849\ 39\ MeV)}{m_n c^2}}\ c$$

$$= c\sqrt{\frac{2(1.849\ 39\ MeV)}{1.008\ 98u}} = c\sqrt{\frac{2(1.849\ 39\ MeV)}{1.008\ 98 \times 931\ MeV}}$$

$$v_n = 0.062\ 7499c$$

$$= 0.062\ 7499 \times 3 \times 10^8\ m\ s^{-1} \cong 19 \times 10^6\ m\ s^{-1}$$

**Exercise 4.5** *Low energy collision of a pion with deuteron results in the production of two protons. The relative orbital angular momentum of the resulting two-proton system for this reaction is*
*(a) 0, (b) 1, (c) 2, (d) 3.*
Hint: The given reaction is $\pi + d \rightarrow p + p$.
Employ parity conservation rule

$$\pi_\pi \pi_d = (-)^l \pi_p \pi_p$$

$$(-1)(+1) = (-)^l(+)(+)$$

$$(-)^l = -1$$

$$l = 1\ \text{(lowest value)}.$$

**Exercise 4.6** $_3Li^7$ *target of thickness 0.01 mm is bombarded with* $10^{13}\frac{protons}{m^2 s}$ *and as a result* $10^8\frac{neutrons}{m^2 s}$ *are generated. The cross-section for this reaction is which of the following if density of* $_3Li^7$ *is 500 kg $m^{-3}$.*
  *(a) 0.23 barn, (b) 2.3 barn, (c) 23 barn, (d) 23millibarn?*
  Hint: We use the formula of equation (4.8) viz.

$$I_b = I_i \sigma n x.$$

Using equation (4.6)

$$n = \frac{\rho N_A}{M}$$

We get

$$I_b = I_i \sigma \frac{\rho N_A}{M} x$$

$$10^8 \frac{neutrons}{m^2\ s^{-1}} = 10^{13}\frac{protons}{m^2\ s}\sigma\left(\frac{500\ kg\ m^{-3} \times 6.023 \times 10^{23}/mole}{7 \times 10^{-3}\ kg}\right)0.01\ mm$$

$$10^8 = 10^{13}\sigma\left(\frac{500 \times 6.023 \times 10^{23}}{7 \times 10^{-3}}\right)0.01 \times 10^{-3}\ m^{-2}$$

$$\sigma = \frac{10^8 \times 7 \times 10^{-3}}{10^{13} \times 500 \times 6.023 \times 10^{23} \times 0.01 \times 10^{-3}}m^2$$

$$=0.23 \times 10^{-28}m^2 = 0.23\ barn$$

**Exercise 4.7** *Consider the reaction* $_{79}Au^{197}(n, \gamma)_{79}Au^{198}$. *A beam of* $2 \times 10^{12}\frac{neutron}{m^2 s}$ *strikes* $_{79}Au^{197}$ *target of area 5 $cm^2$ and thickness 0.3 mm. What would be the number of* $_{79}Au^{198}$ *nuclei produced per second assuming reaction cross-section to be 94 barn. Take the density of* $_{79}Au^{197}$ *to be* $\rho = 19.3 \times 10^3\ kg\ m^{-3}$.
  Ans. $\sigma = 94\ barn = 94 \times 10^{-28}\ m^2$
  Number of $_{79}Au^{197}$ nuclei per unit volume is [equation (4.6)]

$$n = \frac{\rho N_A}{M}\ (N_A = \text{Avogadro's number, } M = \text{molecular weight})$$

$$n = \frac{\left(19.3 \times 10^3\frac{kg}{m^3}\right)\left(6.023 \times 10^{23}\right)}{197 \times 10^{-3}\ kg} = 5.9 \times 10^{28}\frac{1}{m^3}$$

Equation (4.5) gives the relation

$$N = N_0 e^{-n\sigma x}$$

$$N_0 = \left(2 \times 10^{12} \frac{1}{m^2 \, s}\right)(5 \, cm^2) = \left(2 \times 10^{12} \frac{1}{m^2 \, s}\right)(5 \times 10^{-4} \, m^2) = 10^9 \frac{1}{s}$$

The number of $_{79}Au^{197}$ nuclei produced per second is

$$N_0 - N = N_0 - N_0 e^{-n\sigma x}$$

$$= N_0(1 - e^{-n\sigma x})$$

$$= 10^9 \frac{1}{s}\left[1 - \exp\left(-5.9 \times 10^{28} \frac{1}{m^3} \times 94 \times 10^{-28} \, m^2 \times 0.3 \times 10^{-3} \, m\right)\right]$$

$$= 10^9(1 - e^{-0.166})\frac{1}{s}$$

$$= 1.5 \times 10^8 \frac{1}{s}$$

**Exercise 4.8** *In the reaction $_2He^4 + _5B^{11} \rightarrow _7N^{14} + _0n^1$ the incident $\alpha$ particle has a kinetic energy of 5.25 MeV towards $_5B^{11}$ which is at rest. The kinetic energies of the resultant product nuclei $_7N^{14}$ amd $_0n^1$ are 3.26 MeV and 2.139 MeV, respectively. Find the mass of the neutron. ( Mass of $_5B^{11}$ = 11.012 80u, mass of $_2He^4$ = 4.003 87u, mass of $_7N^{14}$ = 14.007 52u),1u = 931 $\frac{MeV}{c^2}$.*

$\boxed{Ans.}$ $_2He^4 + _5B^{11} \rightarrow _7N^{14} + _0n^1$.

Invoke conservation of energy

$$(m_\alpha c^2 + T_\alpha) + (m_{B^{11}}c^2 + T_{B^{11}}) = (m_{N^{14}}c^2 + T_{N^{14}}) + (m_n c^2 + T_n)$$

$$\left(m_\alpha + \frac{T_\alpha}{c^2}\right) + \left(m_{B^{11}} + \frac{T_{B^{11}}}{c^2}\right) = \left(m_{N^{14}} + \frac{T_{N^{14}}}{c^2}\right) + \left(m_n + \frac{T_n}{c^2}\right)$$

$$\left(4.003\,87u + \frac{5.25\,MeV}{c^2}\right) + (11.012\,80u + 0) = \left(14.007\,52u + \frac{3.26\,MeV}{c^2}\right) + \left(m_n + \frac{2.139\,MeV}{c^2}\right)$$

$$m_n = 4.003\,87u + 11.012\,80u - 14.007\,52u + \frac{5.25\,MeV}{c^2} - \frac{3.26\,MeV}{c^2} - \frac{2.139\,MeV}{c^2}$$

$$= 1.009\,15u - \frac{0.149\,MeV}{c^2} = 1.009\,15u - 0.149 \times \frac{1}{931}u$$

$$= 1.008\,989\,957u$$

**Exercise 4.9** *Consider α decay of $_{92}U^{238}$ to $_{90}Th^{234}$. The highest energy of emitted α is 4196 keV. Obtain mass of $_{90}Th^{234}$. Given mass of $_{92}U^{238}$ is 238.0508u and mass of α is 4.0026u.(1u = 931 MeV)*

Ans. $_{92}U^{238} \rightarrow {_{90}Th^{234}} + {_2He^4} + E_\alpha$.

Energy conservation gives

$$m_{U^{238}}c^2 = m_{Th^{234}}c^2 + m_\alpha c^2 + E_\alpha$$

$$238.0508u = m_{Th^{234}} + 4.0026u + \frac{4196 \, keV}{c^2}$$

$$m_{Th^{234}} = 238.0508u - 4.0026u - \frac{4.196 \, MeV}{c^2}$$

$$m_{Th^{234}} = 234.0482 - \frac{4.196}{931}u = 234.043\,693u$$

**Exercise 4.10** *Calculate the binding energy of the last nucleon in MeV for the nucleus $_8O^{17}$. Given $m(_8O^{16}) = 16.000\,000u$, $m(_8O^{17}) = 17.004\,530u$, $m_n = 1.008\,67u$.*

Ans. Binding energy of the least strongly bound neutron in $_8O^{17}$ nucleus

$$= m(_8O^{16}) + m_n - m(_8O^{17})$$

$$= 16.000\,000u + 1.008\,67u - 17.004\,530u$$

$$= 0.004\,14u = 0.004\,14 \times 931 \, MeV = 3.9 \, MeV$$

**Exercise 4.11** *The reaction $_5B^{10}(\alpha, p)_6C^{13}$ shows resonance for compound nucleus formation at excitation energy of 13.23 MeV. The width of this level is experimentally found to be 130 keV. Calculate the mean lifetime of the nucleus for this excitation.*

Ans. Given reaction $\alpha + {_5B^{10}} \rightarrow (_7N^{14})^* \rightarrow {_6C^{13}} + p$

Mean lifetime is given by

$$\tau = \frac{\hbar}{\Gamma} = \frac{\frac{1}{2\pi}6.626 \times 10^{-34} \, J. \, s}{130 \, keV} = \frac{6.626 \times 10^{-34} \, J. \, s}{2\pi \times 130 \times 10^3 \times 1.6 \times 10^{-19} \, J}$$

$$\tau = 5 \times 10^{-21} \, s$$

**Exercise 4.12** *Obtain the threshold energy needed to initiate the reaction $_{15}P^{31}(n, p)_{16}Si^{31}$ if it is known that $m_p = 1.008\,14u$, $m_n = 1.008\,98u$, $m_{P31} = 30.983\,56u$, $m_{Si31} = 30.985\,15u$.*

$\boxed{\text{Ans.}}$ $n + {}_{15}\text{P}^{31} \rightarrow {}_{16}\text{Si}^{31} + p$

$$Q = (m_a + m_X - m_Y - m_b)c^2 = (m_n + m_{\text{P}^{31}} - m_{\text{Si}^{31}} - m_\text{P})c^2$$

$$= (1.008\ 98 + 30.983\ 56 - 30.985\ 15 - 1.008\ 14)u = -0.000\ 75u$$

$$= -0.000\ 75 \times 931\ MeV = -0.698\ 25\ MeV$$

$$E_{th} = -Q\frac{m_a + m_X}{m_X} = -Q\frac{m_n + m_{\text{P}^{31}}}{m_{\text{P}^{31}}}$$

$$= -(-0.698\ 25\ MeV)\frac{1.008\ 98u + 30.983\ 56}{30.983\ 56}$$

$$= 0.721\ MeV$$

**Exercise 4.13** *Consider a perfectly elastic collision in L-frame. Show that analysis in L-frame involves 4 unknowns and there are 3 equations.*

$\boxed{\text{Ans.}}$ Consider a perfectly elastic collision in L-frame, as shown in figure 4.17.

Apply conservation of linear momentum

$$m_a\vec{u}_a + m_X\vec{u}_X = m_a\vec{v}_a + m_X\vec{v}_X$$

Along the $x$-axis

$$m_a u_a + m_X u_X = m_a v_a \cos\theta + m_X v_X \cos\phi \tag{4.184}$$

Along the $y$-axis

$$0 + 0 = m_a v_a \sin\theta - m_X v_X \sin\phi \tag{4.185}$$

Apply conservation of kinetic energy

$$\frac{1}{2}m_a u_a^2 + \frac{1}{2}m_X u_X^2 = \frac{1}{2}m_a v_a^2 + \frac{1}{2}m_X v_X^2 \tag{4.186}$$

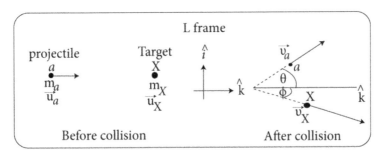

**Figure 4.17.** Perfectly elastic collision in L-frame, pertaining to *exercise 4.13*.

The three equations (4.184)–(4.186) can be solved to get values of three unknowns, e.g. $v_a$, $v_X$, $\theta$ in terms of $u_a$, $u_X$, $\phi$, $m_a m_X$.

L-frame generally involves four unknowns $v_a$, $v_X$, $\theta$, $\phi$. But we can find only three of these unknowns. Analysis in C-frame is easier.

**Exercise 4.14** *Show that two particles of equal mass undergoing perfectly elastic collision in lab frame must move along a mutually perpendicular path after collision.*

Ans. Let the mass of two particles 1,2 be $m$, $m$

Pre-collision velocities : $u_1$, $u_2 = 0$ (in lab frame)

Post-collision velocities: $v_1$, $v_2$

Consider oblique collision.

The definition of coefficient of restitution $e$ is

$$e = \frac{\text{velocity of separation}}{\text{velocity of approach}}$$

For perfectly elastic collision the coefficient of restitution is

$$e = 1.$$

Invoke momentum conservation

$$m\vec{u}_1 + m\vec{u}_2 = m\vec{v}_1 + m\vec{v}_2$$

$$\vec{u}_1 = \vec{v}_1 + \vec{v}_2 \text{ (as } \vec{u}_2 = 0 \text{ in L-frame)}$$

The $x$ component is [figure 4.18(a)]

$$u_1 \cos \alpha = v_1 \cos \theta + v_2 \tag{4.187}$$

Again

$$e = \frac{\text{velocity of separation}}{\text{velocity of approach}} = \frac{v_2 - v_1 \cos \theta}{u_1 \cos \alpha - 0} = 1 \text{ (for perfectly elastic collision)}$$

$$u_1 \cos \alpha = v_2 - v_1 \cos \theta \tag{4.188}$$

Equation 4.187—equation (4.188) gives

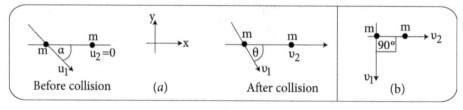

Before collision     (a)     After collision     (b)

**Figure 4.18.** Relating to *exercise 4.13*.

$$2v_1 \cos\theta = 0 \Rightarrow \cos\theta = 0$$

$$\theta = 90° \text{ (figure 4.18(b))}$$

**Exercise 4.15** *If a compound nucleus has a level width $\Gamma$ eV then its mean life would be which of the following?*

(a) $\frac{6.5 \times 10^{16}}{\Gamma} s$  (b) $\frac{6.5 \times 10^{-16}}{\Gamma} s$  (c) $\frac{1}{\Gamma} s$  (d) $\frac{66 \times 10^{-16}}{\Gamma} s.$

Hint: $\tau = \frac{\hbar}{\Gamma} = \frac{\frac{1}{2\pi} 6.626 \times 10^{-34} J.s}{\Gamma \, eV} = \frac{6.626 \times 10^{-34} J.s}{2\pi\Gamma\left(1.6 \times 10^{-19} J\right)} = \frac{6.59 \times 10^{-16}}{\Gamma} s \text{ (}\Gamma \text{ in } eV\text{)}$

**Exercise 4.16** *Find the maximum and minimum energies of emitted neutrons if 17.6 MeV deuteron undergoes the reaction given by $d + {}_1H^3 \rightarrow \alpha + n$ having Q value $=17.6$ MeV.*

Ans. The $Q$ value equation is [equation (4.123)]

$$Q = T_b\left(1 + \frac{m_b}{m_Y}\right) - T_a\left(1 - \frac{m_a}{m_Y}\right) - \frac{2}{m_Y}\sqrt{m_a m_b T_a T_b} \cos\theta$$

The given reaction is

$$d + {}_1H^3 \rightarrow \alpha + n$$

It can be compared to the reaction

$$a + X \rightarrow Y + b$$

Comparison gives

$$a \equiv d, \; X \equiv {}_1H^3, \; Y \equiv \alpha, \; b \equiv n.$$

$$T_a = 0.3 \; MeV, \; m_a = m_d = 2m_p, \; m_b = m_n - m_p$$

$$m_X = m_{H^3} = 3m_p, \; m_Y = m_\alpha = 4m_p$$

Substituting we get

$$17.6 = T_b\left(1 + \frac{m_p}{4m_p}\right) - 0.3\left(1 - \frac{2m_p}{4m_p}\right) - \frac{2}{4m_p}\sqrt{(2m_p)m_p(0.3)T_b} \cos\theta$$

$$17.6 = T_b\left(1 + \frac{1}{4}\right) - 0.3\left(1 - \frac{1}{2}\right) - \frac{1}{2}\sqrt{2(0.3)T_b} \cos\theta$$

$$17.6 = \frac{5}{4}T_b - \frac{0.3}{2} - \frac{\sqrt{0.6}}{2}\sqrt{T_b} \cos\theta \Rightarrow 17.6 = \frac{5}{4}T_b - \frac{0.3}{2} - \frac{\sqrt{0.6}}{2}\sqrt{T_b} \cos\theta$$

$$17.75 = \frac{5}{4}T_b - \frac{\sqrt{0.6}}{2}\sqrt{T_b}\cos\theta \Rightarrow 17.75 \times 4 = 5T_b - 2\sqrt{0.6}\sqrt{T_b}\cos\theta$$

$$5T_b - 1.549\sqrt{T_b}\cos\theta - 71 = 0$$

✓ When $\theta = 0°$, $\cos\theta = 1$

$5T_b - 1.549\sqrt{T_b} - 71 = 0$ . Solving $T_b = 15.4\ MeV$ (max)

✓ When $\theta = 180°$, $\cos\theta = -1$

$5T_b + 1.549\sqrt{T_b} - 71 = 0$.  Solving $T_b = 13.08\ MeV$ (min)

**Exercise 4.17** *The excitation energy of compound nuclei $_{92}U^{236}$, $_{92}U^{239}$ produced in the given reactions are which of the following?*

$$_0n^1 + {}_{92}U^{235} \rightarrow {}_{92}U^{236}$$

$$_0n^1 + {}_{92}U^{238} \rightarrow {}_{92}U^{239}$$

$$m_n = 1.0087u,\ m_{U^{235}} = 235.0439u,\ m_{U^{238}} = 238.0508u$$

$$m_{U^{236}} = 236.0456u,\ m_{U^{239}} = 239.0543u$$

*(a) 6.5 MeV,4.8 MeV (b)4.8 MeV,6.5 MeV (c)5.6 MeV,4.8 MeV (d) 4.8 MeV,4.8 MeV.*

Hint: Excitation energy for the reaction

$$_0n^1 + {}_{92}U^{235} \rightarrow {}_{92}U^{236}$$

is

$$\Delta E = m_n c^2 + m_{U^{235}}c^2 - m_{U^{236}}c^2$$

$$= 1.0087u + 235.0439u - 236.0456u$$

$$= 0.007 \times 931\ MeV = 6.5\ MeV$$

Excitation energy for the reaction

$$_0n^1 + {}_{92}U^{238} \rightarrow {}_{92}U^{239}$$

is

$$\Delta E = m_n c^2 + m_{U^{238}}c^2 - m_{U^{239}}c^2$$

$$= 1.0087u + 238.0508u - 239.0543u = 0.0052 \times 931\ MeV = 4.8\ MeV$$

**Exercise 4.18** *Which of the reactions is/are endothermic ?*

$$_3\text{Li}^7(p, \alpha)_2\text{He}^4, \quad _7\text{N}^{14}(\alpha, p)_8\text{O}^{17}$$

*Masses of $_3\text{Li}^7$, $p$, $_2\text{He}^4$, $_7\text{N}^{14}$, $_8\text{O}^{17}$ are 7.018 22u, 1.008 14u, 4.002 60u, 14.007 53u, 17.0054u respectively.*

*(a) None, (b) both, (c) (p, α) (d) (α, p).*

*The kinetic energy of α particle for the reaction $_3\text{Li}^7(p, \alpha)_2\text{He}^4$ assuming that the kinetic energy of the bombarding proton to be negligible is*

*(a)1 MeV (b) 4 MeV (c) 5 MeV (d) 10 MeV.*

Hint: ✓Let us find the $Q$ value of

$$p + _3\text{Li}^7 \rightarrow \alpha + \alpha \Leftarrow (p, \alpha) \text{ reaction}$$

$$Q = m_p c^2 + m(_3\text{Li}^7)c^2 - m(\alpha) + m(\alpha)$$

$$=1.008\ 14u + 7.018\ 22u - 4.002\ 60u - 4.002\ 60u = 0.021\ 16u.$$

$$Q = 0.021\ 16u > 0$$

The reaction is exothermic.

✓ Let us find the $Q$ value of

$$\alpha + _7\text{N}^{14} \rightarrow _8\text{O}^{17} + p \Leftarrow (\alpha, p) \text{ reaction}$$

$$Q = m_\alpha c^2 + m(_7\text{N}^{14})c^2 - m(_8\text{O}^{17}) + m(p)$$

$$= 4.002\ 60u + 14.007\ 53u - 17.0054u - 1.008\ 14u = -0.003\ 41u.$$

$$Q = -0.003\ 41u < 0$$

The reaction is endothermic.

- Energy released in $_3\text{Li}^7(p, \alpha)_2\text{He}^4$ is

$$Q = 0.021\ 16u \times 931\ MeV = 19.7\ MeV.$$

So kinetic energy of each $\alpha$ particle is

$$T_\alpha = \frac{19.7\ MeV}{2} = 9.85\ MeV.$$

**Exercise 4.19** *Consider 1D perfectly elastic collision between masses $m_1$ (projectile), $m_2$ (target). Using conservation of linear momentum, conservation of kinetic energy and Newton's collision rule obtain post-collision velocities in terms of pre-collision velocities.*

*Find the fractional kinetic energy transfer of $m_1$.*

*Find the condition of maximum energy transfer.*

*How is this result used in a nuclear reactor?*

Ans. ✓Consider 1D collision between two particles of masses $m_1$ (projectile) and $m_2$ (target). Pre-collision velocities are $u_1$ of $m_1$ and $u_2$ of $m_2$ and $u_1 > u_2$.

The post-collision velocities are $v_1$ (for $m_1$) and $v_2$ (for $m_2$).

The coefficient of restitution is

$$e = \frac{\text{velocity of separation}}{\text{velocity of approach}} = \frac{v_2 - v_1}{u_1 - u_2}$$

$$v_2 - v_1 = e(u_1 - u_2) \tag{4.189}$$

Conservation of linear momentum gives

$$m_1 v_1 + m_2 v_2 = m_1 u_1 + m_2 u_2 \tag{4.190}$$

Multiplying equation (4.189) by $m_2$ we have

$$m_2 v_2 - m_2 v_1 = m_2 e(u_1 - u_2) \tag{4.191}$$

Subtracting equation (4.191) from equation (4.190)

$$(m_1 + m_2)v_1 = u_1(m_1 - m_2 e) + m_2 u_2(1 + e)$$

$$v_1 = \frac{m_1 - m_2 e}{m_1 + m_2}u_1 + \frac{m_2(1 + e)}{m_1 + m_2}u_2 \tag{4.192}$$

Multiplying equation (4.189) by $m_1$ we have

$$m_1 v_2 - m_1 v_1 = m_1 e(u_1 - u_2) \tag{4.193}$$

Adding equations (4.190) and (4.193) we have

$$(m_1 + m_2)v_2 = u_1(m_1 + m_1 e) + u_2(m_2 - m_1 e)$$

$$v_2 = \frac{m_1(1 + e)}{m_1 + m_2}u_1 + \frac{m_2 - m_1 e}{m_1 + m_2}u_2 \tag{4.194}$$

Equations (4.192) and (4.194) express $v_1$, $v_2$ in terms of $m_1$, $m_2$, $u_1$, $u_2$, $e$.

According to Newton's collision rule $e = 1$ for perfectly elastic collision. Hence equations (4.192) and (4.194) give

$$v_1 = \frac{m_1 - m_2}{m_1 + m_2}u_1 + \frac{2m_2}{m_1 + m_2}u_2 \tag{4.195}$$

$$v_2 = \frac{2m_1}{m_1 + m_2}u_1 + \frac{m_2 - m_1}{m_1 + m_2}u_2 \tag{4.196}$$

Equations (4.195) and (4.196) give $v_1$, $v_2$ in terms of $m_1$, $m_2$, $u_1$, $u_2$.

✓ Kinetic energy transfer due to perfectly elastic collision in lab frame is

$$\frac{1}{2}m_1u_1^2 - \frac{1}{2}m_1v_1^2 = \frac{1}{2}m_1u_1^2\left(1 - \frac{v_1^2}{u_1^2}\right)$$

Fractional kinetic energy transfer is

$$\delta = \frac{\frac{1}{2}m_1u_1^2\left(1 - \frac{v_1^2}{u_1^2}\right)}{\frac{1}{2}m_1u_1^2} = 1 - \frac{v_1^2}{u_1^2} \tag{4.197}$$

In lab frame $m_2$ is at rest and so $u_2 = 0$. From equation (4.195)

$$v_1 = \frac{m_1 - m_2}{m_1 + m_2}u_1$$

$$\frac{v_1}{u_1} = \frac{m_1 - m_2}{m_1 + m_2}$$

Equation (4.197) gives

$$\delta = 1 - \frac{v_1^2}{u_1^2} = 1 - \left(\frac{m_1 - m_2}{m_1 + m_2}\right)^2 = \frac{4m_1m_2}{(m_1 + m_2)^2} \tag{4.198}$$

Condition of maximum energy transfer is obtained by maximizing equation (4.198) by varying $m_2$. This is done by setting

$$\frac{d\delta}{dm_2} = 0 \text{ i.e } \frac{d}{dm_2}\frac{4m_1m_2}{(m_1 + m_2)^2} = 0 \Rightarrow \frac{d}{dm_2}\frac{m_2}{(m_1 + m_2)^2} = 0$$

$$\frac{1}{(m_1 + m_2)^2} - \frac{2m_2}{(m_1 + m_2)^3} = 0 \Rightarrow m_1 + m_2 - 2m_2 = 0$$

$$m_1 = m_2$$

We see that for $m_1 = m_2$, $\delta = 1$ [from equation (4.198)], i.e. there is 100% transfer of kinetic energy from projectile to target.

✓ Consider the fission reaction which occurs in a nuclear reactor

$$\underset{\text{slow}}{_0n^1} + {_{92}U^{235}} \rightarrow {_{56}Ba^{144}} + {_{36}Kr^{92}} + \underset{\text{fast}}{3_0n^1} + Q$$

Slow neutron carries kinetic energy $\sim 0.026$ eV while fast neutrons carry kinetic energy $\sim MeV$.

To initiate further fission reactions, neutrons are needed to be slowed down (i.e. thermalized) by making them transfer their kinetic energy through collisions with protons of a proton-rich material like paraffin.

Since for neuton and proton $m_n \cong m_p$ it follows that due to $np$ collision the proton robs 100% of energy from the neutron and so neutrons get thermalized. The process of reducing energy, i.e. thermalizing the fast neutrons (emerging due to fission reaction) is referred to as moderation. Paraffin is the moderator referred to here.

**Exercise 4.20** *The unit of level width and its expression will be which of the following*
(a) $m, \frac{\tau}{\hbar}$   (b) barn, $\sqrt{\hbar\tau}$   (c) eV, $\frac{\hbar}{\tau}$   (d) $m^{-1}, \frac{h}{\tau}$.

**Exercise 4.21** Show that the speeds of particles are not affected due to collision in C-frame.

Ans. Consider the perfectly elastic collision $a + X \rightarrow X + a$ in C-frame.

According to conservation of linear momentum we have from equations (4.12) and (4.13)

$$\vec{P_a} = -\vec{P_X} \Rightarrow m_a U_a = -m_X U_X \tag{4.199}$$

$$\overrightarrow{p'_a} = -\overrightarrow{p'_X} \Rightarrow m_a V_a = -m_X V_X \tag{4.200}$$

where $U_a$, $U_X$ are pre-collision velocities and $V_a$, $V_X$ are post-collision velocities in C-frame (symbols defined in figure 4.4)

Use kinetic energy conservation in C-frame to get

$$\frac{1}{2}m_a U_a^2 + \frac{1}{2}m_X U_X^2 = \frac{1}{2}m_a V_a^2 + \frac{1}{2}m_X V_X^2 \tag{4.201}$$

Putting from equation (4.199) the expression $U_X = -\frac{m_a U_a}{m_X}$ and from equation (4.200) the expression $V_X = -\frac{m_a V_a}{m_X}$ into equation (4.201) we have

$$\frac{1}{2}m_a U_a^2 + \frac{1}{2}m_X \left(-\frac{m_a U_a}{m_X}\right)^2 = \frac{1}{2}m_a V_a^2 + \frac{1}{2}m_X \left(-\frac{m_a V_a}{m_X}\right)^2$$

$$m_a U_a^2 \left(1 + \frac{m_a}{m_X}\right) = m_a V_a^2 \left(1 + \frac{m_a}{m_X}\right)$$

$$U_a = V_a \tag{4.202}$$

Taking ratio of equations (4.199) and (4.200)

$$\frac{m_a U_a}{m_a V_a} = \frac{-m_X U_X}{-m_X V_X}$$

$$\frac{U_a}{V_a} = \frac{U_X}{V_X}$$

$$U_a = V_a$$

$$U_X = V_X \qquad (4.203)$$

So speeds of particles are not affected due to collision in C-frame as $U_a = V_a$, $U_X = V_X$.

**Exercise 4.22** *The Q value of a nuclear reaction $a + X \rightarrow Y + b$ is given by which of the following*
(a) $(m_Y - m_X - m_b + m_a)c^2$ (b) $(-m_Y + m_X + m_a - m_b)c^2$
(c) $(m_X - m_Y + m_b - m_a)c^2$ (d) $(-m_X + m_Y - m_a + m_b)c^2$.

**Exercise 4.23** *The lifetime of a compound nucleus is of the order of*
(a) $1\,s$, (b) $10^{-21}\,s$, (c) $10^{-10}\,s$, (d) $10^{-13}\,s$.

**Exercise 4.24** *The minimum energy available from an endoergic reaction $a + X \rightarrow Y + b$ is which of the following?*
(a) $-Q$, (b) $-Q(1 + \frac{m_a}{M_X})$, (c) $-Q(1 - \frac{m_a}{M_X})$, (d) $-Q(1 + \frac{m_a}{M_X})$.

**Exercise 4.25** *The minimum energy needed for an endoergic reaction is which of the following?*
(a) $\frac{Q}{2}$, (b) $-Q$, (c) $\frac{Q}{2}$, (d) $Q$.

**Exercise 4.26** *The process of photodisintegration is accomplished using which of the following as projectile?*
(a) alpha, (b) neutron, (c) photon, (d) proton.

**Exercise 4.27** *The partial decay widths for decay modes $x$, $y$ are $\Gamma_x$, $\Gamma_y$ respectively then the total level width is*
(a) $\frac{1}{\Gamma_x} + \frac{1}{\Gamma_y}$, (b) $\Gamma_x + \Gamma_y$, (c) $\sqrt{\Gamma_x^2 + \Gamma_y^2}$, (d) $\sqrt{\Gamma_x + \Gamma_y}$.

**Exercise 4.28** *Breit–Wigner one-level dispersion formula is which of the following?*
(a) $\pi \lambdabar^2 g \dfrac{\Gamma_{aX}\Gamma_{bY}}{(E - E_r)^2 - \frac{\Gamma^2}{4}}$, (b) $\pi \lambdabar^2 g \dfrac{\Gamma_{aX}\Gamma_{bY}}{\left(E - \frac{1}{2}E_r\right)^2 + \frac{\Gamma^2}{4}}$, (c) $\pi \lambdabar^2 g \dfrac{\Gamma_{aX}\Gamma_{bY}}{(E + E_r)^2 + \frac{\Gamma^2}{4}}$,
(d) $\pi \lambdabar^2 g \dfrac{\Gamma_{aX}\Gamma_{bY}}{(E - E_r)^2 + \frac{\Gamma^2}{4}}$.

**Exercise 4.29** *Pick up and stripping reactions are:*
*(a) compound nuclear reactions,*
*(b) direct reactions,*
*(c) occur in time $10^{-6}$ s,*
*(d) occur in time $10^{-21}$ s,*

**Exercise 4.30** *A compound nucleus:*
*(a) depends on incident channel,*
*(b) is independent of incident channel,*
*(c) decays so as to conserve energy and momentum of incident channel,*
*(d) does not conseve energy and momentum of incident channel during its decay.*

**Exercise 4.31** *Compound nucleus theory and artificial transmutation were discovered by:*
*(a) Bohr, Rutherford; (b) Rutherford, Bohr; (c) Fermi, Bohr; (d) Bohr, Einstein.*

**Exercise 4.32** *ind the minimum energy needed to disintegrate a deuteron into a proton and a neutron from the following data. Mass of proton, neutron, deuteron are, respectively,* 1.007 59*u*, 1.008 98*u*, 2.014 71*u*.
$\boxed{\text{Ans.}}$ Photodisintegration of a deuteron is

$$\gamma + d \rightarrow n + p$$

$$E_\gamma = h\nu = m_n + m_p - m_d$$

$$=1.008\ 98u + 1.007\ 59u - 2.014\ 71u = 0.001\ 86u$$

$$= 0.001\ 86 \times 931\ MeV = 1.73\ MeV$$

**Exercise 4.33** *An endoergic nuclear reaction induced with $\gamma$ rays has a threshold energy given by*
*(a) $Q$, (b) $-Q$, (c) $\frac{Q}{2}$, (d) $-Q(1 + \frac{m_a}{m_X})$.*

**Exercise 4.34** *Identify the quantity not conserved in a nuclear reaction.*
*(a) Parity, (b) total energy, (c) electric quadrupole moment, (d) angular momentum.*

**Exercise 4.35** *Which of the following is not true regarding kinetic energy in inelastic and elastic collision?*
  *(a) Not conserved, conserved,*
  *(b) conserved, not conserved,*
  *(c) both conserved,*
  *(d) both not conserved.*

**Exercise 4.36** *In elastic collision the incident particle after striking the target nucleus moves in:*
  *(a) the same direction,*
  *(b) altered direction,*
  *(c) 90° to incident direction,*
  *(d) 45° to incident diorection.*

**Exercise 4.37** *A nuclear reaction cross-section has the dimension of which of the following?*
  *(a) linear momentum, (b) angular momentum, (c) area, (d) velocity.*

**Exercise 4.38** *1 barn is*
  *(a)* $10^{-28}$ *cm*$^2$, *(b)* $10^{-24}$ *m*$^2$, *(c)* $10^{-28}$ *m*$^2$, *(d)* $10^{-24}$ *cm*$^2$.

**Exercise 4.39** *Calculate threshold energy for the endoergic reaction:* $_7\mathrm{N}^{14}(n, \alpha)_5\mathrm{B}^{11}$ *where mass of n, $\alpha$, $_7\mathrm{N}^{14}$, $_5\mathrm{B}^{11}$ are, respectively,* 1.008 987u, 4.003 879u, 14.007 550u, 11.012 811u.

$\boxed{\text{Ans.}}\ Q = (m_n + m_N - m_B - m_\alpha)c^2$

$$= 1.008\ 987u + 14.007\ 550u - 11.012\ 811u - 4.003\ 879u$$

$$= -0.000\ 153u = -0.000\ 153 \times 931\ MeV$$

$$= -0.14\ MeV$$

Threshold energy $E_{th} = -Q\left(1 + \dfrac{m_n}{M_N}\right)$

$$= -(-0.14\ MeV)\left(1 + \frac{1.008\ 987u}{14.007\ 550u}\right) = 0.15\ MeV$$

**Exercise 4.40** *Obtain mass difference between neutron and proton from the following data.*

$$Q(_1H^2 + _1H^2 \rightarrow _1H^3 + _1H^1) = 4.031 \; MeV$$

$$Q(_1H^2 + _1H^2 \rightarrow _2He^3 + _0n^1) = 3.265 \; MeV$$

$$Q(_1H^3 \rightarrow _2He^3 + \beta^-) = 0.0185 \; MeV$$

Ans. $2m_{H^2} - m_{H^3} - m_{H^1} = Q_1 = 4.031 \; MeV$        (4.204)

$$2m_{H^2} - m_{He^3} - m_n = Q_2 = 3.265 \; MeV \tag{4.205}$$

$$m_{H^3} - m_{He^3} - m_\beta = Q_3 = 0.0185 \; MeV \tag{4.206}$$

From equation (4.206)

$$m_{H^3} - m_{He^3} = Q_3 = 0.0185 \; MeV \left(\text{since } m_\beta \cong 0\right) \tag{4.207}$$

From equations (4.204), (4.205), and (4.207) we have

$$Q_1 - Q_2 + Q_3 = m_n - m_{H^1} = (4.031 - 3.265 + 0.0185) \; MeV$$

$$= 0.7845 \; MeV = \frac{0.7845}{931}u = 0.000 \; 843u$$

**Exercise 4.41** *Consider the reaction $_2He^4 + _5B^{11} \rightarrow _7N^{14} + _0n^1$. Compute the mass of $_0n^1$. The masses of $_2He^4$, $_5B^{11}$, $_7N^{14}$ are, respectively, 4.003 87u, 11.012 80u, 14.007 52u and the kinetic energy of $_2He^4$, $_5B^{11}$, $_7N^{14}$, $_0n^1$ are, respectively, 5.25 MeV, 0, 3.26 MeV, 2.139 MeV.*

Ans. From the energy conservation principle we have

$$(m_\alpha c^2 + T_\alpha) + (m_B c^2 + T_B) = (m_N c^2 + T_N) + (m_n c^2 + T_n)$$

$$\left(4.003 \; 87u + 5.25\frac{MeV}{c^2}\right) + (11.012 \; 80u + 0)$$

$$= \left(14.007 \; 52u + 3.26\frac{MeV}{c^2}\right) + \left(m_n + 2.139\frac{MeV}{c^2}\right)$$

$$m_n = 1.009 \; 15u - 0.149\frac{MeV}{c^2} = 1.009 \; 15u - \frac{0.149}{931}u$$

$$m_n = 1.008 \; 989u$$

**Exercise 4.42** *A nuclear decay process is given as* $_4\text{Be}^9(d, n)_5\text{B}^{10}$ *given that the atomic masses of* $_4\text{Be}^9$, $_5\text{B}^{10}$, $_1\text{H}^2$, $_0n^1$ *are 9.012 182u, 10.012 938u, 2.014 102u, 1.008 665u. Then the Q value of the reaction is:*

  (a) 6.5 MeV,     (b) 5.5 MeV,     (c) 4.5 MeV,     (d) 3.5 MeV,

  Hint: The reaction is $d + {}_4\text{Be}^9 \rightarrow {}_5\text{B}^{10} + n$

$$Q = m_d c^2 + m_{\text{Be}} c^2 - m_{\text{B}} c^2 - m_n c^2$$

$$= 2.014\ 102u + 9.012\ 182u - 10.012\ 938u - 1.008\ 665u$$

$$= 0.004\ 681u = 0.004\ 681 \times 931\ MeV$$

$$= 4.36\ MeV$$

**Exercise 4.43** *Obtain energy required to remove the least tightly bound neutron from* $_{20}\text{Ca}^{40}$ *if it is known that masses of* $_{20}\text{Ca}^{40}$, $_{20}\text{Ca}^{39}$, $_0n^1$ *are, respectively, 39.962 589u, 38.970 691u, 1.008 665u.*

  Ans. Consider the equation $_{20}\text{Ca}^{40} \rightarrow {}_{20}\text{Ca}^{39} + {}_0n^1$

  Let $Q$ be the energy required to remove the least tightly bound neutron from $_{20}\text{Ca}^{40}$. Then

$$Q = m_{\text{Ca}^{39}} c^2 + m_n c^2 - m_{\text{Ca}^{40}} c^2$$

$$= 38.970\ 691u + 1.008\ 665u - 39.962\ 589u$$

$$= 0.016\ 767u = 0.016\ 767u \times 931\ MeV$$

$$= 15.61\ MeV$$

**Exercise 4.44** *Identify the Q value and threshold energy for the reaction*

$$_9\text{F}^{19}(n, p)_8\text{O}^{19}$$

*Atomic masses are 18.998 404u, 1.007 825u, 19.003 577u, 1.008 665u for* $_9\text{F}^{19}$, $_1\text{H}^1$, $_8\text{O}^{19}$, $_0n^1$ *respectively.*

  (a) −4 MeV, −4 MeV, (b) −4 MeV, 4 MeV,
  (c) 4 MeV, −4 MeV, (d) 4 MeV, 4 MeV.

  Hint: The reaction is $n + {}_9\text{F}^{19} \rightarrow {}_8\text{O}^{19} + p$

$$Q = m_n c^2 + m_{\text{F}} c^2 - m_{\text{O}} c^2 - m_p c^2$$

$$=1.008\ 665u + 18.998\ 404u - 19.003\ 577u - 1.007\ 825u = -0.004\ 333u$$

$$= -0.004\ 333 \times 931\ MeV = -4.034\ MeV$$

Threshold energy

$$E_{th} = -Q\left(1 + \frac{m_n}{m_{\text{F}}}\right) = -(-4.034\ MeV)\left(1 + \frac{1.008\ 665u}{18.998\ 404u}\right) = 4.25\ MeV$$

**Exercise 4.45** *Show that kinetic energy in a perfectly elastic collision is constant.*
[Ans.] Consider the perfectly elastic collision of two masses $m_1$, $m_2$ with respective velocities $u_1$, $u_2$. The post-collision velocities are $v_1$, $v_2$ respectively.

Apply momentum conservation

$$m_1 u_1 + m_2 u_2 = m_1 v_1 + m_2 v_2$$

$$m_1(u_1 - v_1) = m_2(v_2 - u_2)$$

$$m_1(u_1 - v_1) + m_2(u_2 - v_2) = 0 \qquad (4.208)$$

We define the coefficient of restitution $e$ as

$$e = \frac{\text{velocity of separation}}{\text{velocity of approach}} = \frac{v_2 - v_1}{u_1 - u_2}$$

$$v_2 - v_1 = e(u_1 - u_2) \qquad (4.209)$$

For perfectly elastic collision the coefficient of restitution is $e = 1$. Hence

$$v_2 - v_1 = u_1 - u_2$$

$$u_1 + v_1 = u_2 + v_2 \qquad (4.210)$$

Change in kinetic energy is

$$\Delta T = \left(\frac{1}{2}m_1 u_1^2 + \frac{1}{2}m_2 u_2^2\right) - \left(\frac{1}{2}m_1 v_1^2 + \frac{1}{2}m_2 v_2^2\right)$$

$$= \frac{1}{2}m_1(u_1^2 - v_1^2) + \frac{1}{2}m_2(u_2^2 - v_2^2)$$

$$= \frac{1}{2}m_1(u_1 + v_1)(u_1 - v_1) + \frac{1}{2}m_2(u_2 + v_2)(u_2 - v_2)$$

Using equation (4.209)

$$\Delta T = \frac{1}{2}m_1(u_1 + v_1)(u_1 - v_1) + \frac{1}{2}m_2(u_1 + v_1)(u_2 - v_2) \text{ [using equation (4.209)]}$$

$$\Delta T = \frac{1}{2}(u_1 + v_1)[\, m_1(u_1 - v_1) + m_2(u_2 - v_2)]$$

Using equation (4.208)

$$\Delta T = 0$$

So no change in kinetic energy in perfectly elastic collision.

**Exercise 4.46** *Consider perfectly elastic collision. Establish that in C-frame the kinetic energy of colliding particles after collision are inversely proportional to their masses.*
[Ans.] Consider perfectly elastic collision $a + X \rightarrow X + a$

$U_a$, $U_X$ are initial velocities and $V_a$, $V_X$ are final velocities in C-frame of masses $m_a$ and $m_X$ respectively (figure 4.4 describes the symbols).

Kinetic energy of $a$ in C-frame after collision is

$$T_a = \frac{1}{2}m_a V_a^2 \tag{4.211}$$

In *exercise 4.21* we have shown that speeds do not change in C-frame. Hence

$$V_a = U_a$$

Using this in equation (4.16)

$$T_a = \frac{1}{2}m_a U_a^2 \tag{4.212}$$

Again let $u_a$, $u_X$ be the initial velocities and $v_a$, $v_X$ are final velocities in L-frame of masses $m_a$ and $m_X$ respectively. By equation (4.16)

$$U_a = \frac{m_X u_a}{m_a + m_X}. \tag{4.213}$$

Hence from equation (4.212)

$$T_a = \frac{1}{2}m_a \left( \frac{m_X u_a}{m_a + m_X} \right)^2 = \frac{1}{2} \frac{m_a m_X^2 u_a^2}{(m_a + m_X)^2} \tag{4.214}$$

Kinetic energy of $X$ in C-frame after collision is

$$T_X = \frac{1}{2}m_X V_X^2 \tag{4.215}$$

In *exercise 4.21* we have shown that speeds do not change in C-frame. Hence

$$V_X = U_X$$

Using this in equation (4.215)

$$T_X = \frac{1}{2}m_X U_X^2 \tag{4.216}$$

By equation (4.17)

$$U_X = -\frac{m_a u_a}{m_a + m_X}$$

Hence from equation (4.216)

$$T_X = \frac{1}{2}m_X \left( -\frac{m_a u_a}{m_a + m_X} \right)^2 = \frac{1}{2} \frac{m_X m_a^2 u_a^2}{(m_a + m_X)^2} \tag{4.217}$$

Taking ratio of equations (4.214) and (4.217)

$$\frac{T_X}{T_a} = \frac{m_a}{m_X}$$

Clearly in C-frame the kinetic energies of colliding particles after collision are inversely proportional to their masses.

**Exercise 4.47** *Consider a perfectly elastic collision. Express kinetic energy in L-frame in terms of scattering angle in C-frame and show that the fraction of total energy captured by the recoiling target in L-frame depends upon the masses of colliding particles and angle of scattering in C-frame.*

Ans. Consider perfectly elastic collision $a + X \to X + a$

Total kinetic energy in L-frame (before collision)

$$T_i = \frac{1}{2}m_a u_a^2 + \frac{1}{2}m_X u_X^2 \xrightarrow{u_X=0} \frac{1}{2}m_a u_a^2$$

Kinetic energy of $m_a$ in L-frame (after collision)

$$T_{af} = \frac{1}{2}m_a v_a^2$$

$$\frac{T_{af}}{T_i} = \frac{\frac{1}{2}m_a v_a^2}{\frac{1}{2}m_a u_a^2} = \frac{v_a^2}{u_a^2}$$

Apply traiangle law of vector addition to the triangle of figure 4.4 to get

$$\vec{v}_a = \vec{u}_{CM} + \vec{V}_a$$

Squaring

$$v_a^2 = u_{CM}^2 + V_a^2 + 2\vec{u}_{CM} \cdot \vec{V}_a = u_{CM}^2 + V_a^2 + 2u_{CM}V_a \cos\theta_c$$

$$\frac{T_{af}}{T_i} = \frac{v_a^2}{u_a^2} = \frac{u_{CM}^2 + V_a^2 + 2u_{CM}V_a \cos\theta_c}{u_a^2}$$

$$\frac{T_{af}}{T_i} = \left(\frac{u_{CM}}{u_a}\right)^2 + \left(\frac{V_a}{u_a}\right)^2 + 2\left(\frac{u_{CM}}{u_a}\right)\left(\frac{V_a}{u_a}\right)\cos\theta_c \qquad (4.218)$$

From equation (4.15)

$$u_{CM} = \frac{m_a u_a}{m_a + m_X}$$

$$\frac{u_{CM}}{u_a} = \frac{m_a}{m_a + m_X} \qquad (4.219)$$

In *exercise 4.21* we have shown that speeds do not change in C-frame. Hence

$$V_a = U_a \qquad (4.220)$$

From equation (4.16)

$$U_a = \frac{m_X u_a}{m_a + m_X}$$

(4.221)

Consider

$$\frac{V_a}{u_a} = \frac{U_a}{u_a} = \frac{m_X}{m_a + m_X} \text{ [using equations (4.220) and (4.221)]}$$

(4.222)

From equation (4.218), (4.219), and (4.222)

$$\frac{T_{af}}{T_i} = \left(\frac{m_a}{m_a + m_X}\right)^2 + \left(\frac{m_X}{m_a + m_X}\right)^2 + 2\left(\frac{m_a}{m_a + m_X}\right)\left(\frac{m_X}{m_a + m_X}\right)\cos\theta_c$$

$$= \frac{m_a^2 + m_X^2 + 2m_a m_X \cos\theta_c}{(m_a + m_X)^2} = \frac{(m_a + m_X)^2 - 2m_a m_X + 2m_a m_X \cos\theta_c}{(m_a + m_X)^2}$$

$$\frac{T_{af}}{T_i} = 1 - \frac{2m_a m_X(1 - \cos\theta_c)}{(m_a + m_X)^2}$$

This is the fraction of total incident energy acquired by scattered projectile $m_a$ in L-frame.

$$\frac{2m_a m_X(1 - \cos\theta_c)}{(m_a + m_X)^2} = 1 - \frac{T_{af}}{T_i} = \frac{T_i - T_{af}}{T_i}$$

(4.223)

From conservation of kinetic energy

$$T_i = T_{af} + T_{Xf}$$

where $T_{Xf}$ is the kinetic energy of $m_X$ in L-frame (after collision). Hence

$$T_{Xf} = T_i - T_{af}$$

From equation (4.223)

$$\frac{T_{Xf}}{T_i} = \frac{2m_a m_X(1 - \cos\theta_c)}{(m_a + m_X)^2}$$

This is the fraction of total energy captured by the recoiling target in L-frame. Obviously it depends upon the masses $m_a$, $m_X$, angle of scattering in C-frame $\theta_c$.

**Exercise 4.48** *Slow neutrons are captured by $_{92}U^{235}$ and resonance is seen for an excitation energy 0.29 MeV. The compound nucleus can de-excite either through $\gamma$ emission or by fission. The mean life of the compound nucleus is $4.7 \times 10^{-15}$ s and the partial width for $\gamma$ emission is $3.4 \times 10^{-2}$ eV. Calculate the partial width for fission.*

Ans. Level width and mean life are related by

$$\Gamma\tau \sim \hbar$$

The total level width $\Gamma$ is given by

$$\Gamma = \frac{\hbar}{\tau} = \frac{\frac{1}{2\pi}6.626 \times 10^{-34}\ J.\ s}{4.7 \times 10^{-15}\ s} = \frac{6.626 \times 10^{-34}}{2\pi \times 4.7 \times 10^{-15} \times (1.6 \times 10^{-19})}\ eV = 0.14\ eV$$

Let the partial level widths be $\Gamma_\gamma$ for $\gamma$ transition and $\Gamma_f$ for fission. Then

$$\Gamma = \Gamma_\gamma + \Gamma_f$$

$$\Gamma_f = \Gamma - \Gamma_\gamma = 0.14\ eV - 3.4 \times 10^{-2}\ eV = 0.14\ eV - 0.034\ eV$$

$$\Gamma_f = 0.106\ eV$$

Answers to multiple choice type questions.

4.2b, 4.5b, 4.6a, 4.15b, 4.17a, 4.18d,d, 4.20c, 4.22b, 4.23d, 4.24b, 4.25b, 4.26c, 4.27b, 4.28d, 4.29b,d, 4.30b,c,4.31a, 4.33b, 4.34c, 4.35a, 4.36b, 4.37c, 4.38c, 4.42c, 4.44b.

## 4.39 Question bank

Q4.1 How are partial cross-section and total cross-section related?

Q4.2 Explain what is meant by L-frame and C-frame.

Q4.3 Which of the frames (L-frame and C-frame) is called zero reference frame, and which is inertial reference frame? Justify.

Q4.4 What is the justification of using two frames of reference? L-frame and C-frame?

Q4.5 Establish that the total kinetic energy in C-frame is less than the total initial kinetic energy in L-frame.

Q4.6 Obtain the relation between the scattering angles in L-frame and C-frame.

Q4.7 What are the characteristics of elastic scattering? Give an example of elastic scattering.

Q4.8 What are the characteristics of inelastic scattering? Give an example of inelastic scattering.

Q4.9 How many types of inelastic collision are there?

Q4.10 In which type of collision do interacting particles stick to each other.

Q4.11 Explain why we say that kinetic energy is conserved in elastic collision but not in inelastic collision.

Q4.12 Explain with examples the following terms: reaction channel, radiative capture.

Q4.13 What is photodisintegration?

Q4.14     What do we mean by transmutation?

Q4.15     Mention the conserved quantities in a nuclear reaction. Which quantities are not conserved in a nuclear reaction?

Q4.16     What do we mean by $Q$ value of nuclear reaction? What is the significance of $Q$ value?

Q4.17     What is the significance of the $Q$ values : $Q > 0$, $Q = 0$, $Q < 0$?

Q4.18     What do we mean by exoergic and endoergic reactions? Give examples.

Q4.19     Which of the reactions (exoergic, endoergic) needs to be initiated externally and why?

Q4.20     What is threshold energy for endoergic reaction? How is it used up?

Q4.21     Differentiate between compound nuclear reaction and direct nuclear reaction.

Q4.22     Classify various types of direct nuclear reaction.

Q4.23     Explain with examples the following nuclear reactions: transfer reaction, knock out reaction, pick up reaction.

Q4.24     Explain Bohr independence hypothesis and its use to analyse a compound nuclear reaction.

Q4.25     Discuss the experimental verification of Bohr independence hypothesis of a compound nuclear reaction.

Q4.26     Show in a diagram the angular distributuion of product particles in a compound nuclear reaction and a direct nuclear reaction.

Q4.27     Obtain the two-body non-relativistic $Q$ value equation. Obtain a solution of the equation.

Q4.28     Obtain the expression of threshold energy for an endoergic reaction.

Q4.29     What do we mean by resonance in nuclear reaction? Explain with examples.

Q4.30     What do we mean by lifetime and level width? How are they related?

Q4.31     Give a qualitative brief description of Breit–Wigner one-level dispersion formula.

# Further reading

[1] Krane S K 1988 *Introductory Nuclear Physics* (New York: Wiley)

[2] Tayal D C 2009 *Nuclear Physics* (Mumbai: Himalaya Publishing House)

[3] Satya P 2005 *Nuclear Physics and Particle Physics* (New Delhi: Sultan Chand & Sons)

[4] Guha J 2019 *Quantum Mechanics: Theory, Problems and Solutions* 3rd edn (Kolkata: Books and Allied (P) Ltd))

[5] Lim Y-K 2002 *Problems and Solutions on Atomic, Nuclear and Particle Physics* (Singapore: World Scientific)

**IOP** Publishing

Nuclear and Particle Physics with Cosmology, Volume 1
Nuclear physics
**Jyotirmoy Guha**

# Chapter 5

## Nuclear fission and fusion

## 5.1 Introduction

The energy released in nuclear fission is around 200 $MeV$ per fission. This is huge—enormously large compared to a conventional chemical reaction. Nuclear power plants and nuclear reactors run on the principle of nuclear fission.

Fission means breaking of a heavy mass nucleus $X$ of mass $m$ into two middle mass nuclei $Y$, $Z$ of respective masses $m_1, m_2$ as

$$\underset{\text{slow}}{_0n^1} + X \rightarrow Y + Z + \underset{\text{fast}}{v\,_0n^1} + \text{Energy} \quad (v = 2 \text{ or } 3) \tag{5.1}$$

and the mass deficit

$$m - (m_1 + m_2) \tag{5.2}$$

is converted into energy.

$Y$ and $Z$ are neutron rich product nuclei of intermediate mass and called fission fragments. Two or three fast neutrons are emitted during the process of fission. Each fission gives off a large amount of energy.

- First observed fission

In 1939 Otto Hahn and Fritz Strassmann observed nuclear fission of $_{92}U^{235}$ on being struck with a thermal neutron $_0n^1$. The reaction was

$$\underset{\text{slow}}{_0n^1} + {_{92}U^{235}_{143}} \rightarrow {_{92}U^{236*}_{144}} \rightarrow {_{56}Ba^{141}_{85}} + {_{36}Kr^{92}_{56}} + \underset{\text{fast}}{3\,_0n^1} + \text{Energy} \tag{5.3}$$

## 5.2 Characteristics of fission products: size of fission fragments

The $_{92}U^{235}_{143}$ nucleus does not always split up into those of $_{56}Ba^{141}_{85}$ and $_{36}Kr^{92}_{56}$. They may divide into different pairs of nuclei of elements lying in the central region of the periodic table with unequal nuclear masses. In addition to the decay mode shown in equation (5.3), some other alternative modes of decay are given as follows

$$_{92}U_{143}^{235} + {}_0n^1 \rightarrow {}_{92}U_{144}^{236*} \rightarrow {}_{40}Zr_{58}^{98} + {}_{52}Te_{83}^{135} + 3 \, {}_0n^1 + \text{Energy} \qquad (5.4)$$

$$_{92}U_{144}^{236*} \rightarrow {}_{54}Xe_{86}^{140} + {}_{38}Sr_{56}^{94} + 2 \, {}_0n^1 + \text{Energy} \qquad (5.5)$$

$$_{92}U_{144}^{236*} \rightarrow {}_{56}Ba_{83}^{139} + {}_{36}Kr_{58}^{94} + 3 \, {}_0n^1 + \text{Energy} \qquad (5.6)$$

$$_{92}U_{144}^{236*} \rightarrow {}_{55}Cs_{86}^{141} + {}_{37}Rb_{56}^{93} + 2 \, {}_0n^1 + \text{Energy} \qquad (5.7)$$

$$_{92}U_{144}^{236*} \rightarrow {}_{53}I_{84}^{137} + {}_{39}Y_{58}^{97} + 2 \, {}_0n^1 + \text{Energy} \qquad (5.8)$$

etc. Clearly $_{92}U_{144}^{236*}$ splits into various combinations of fission fragments and there is no unique pair.

A study of the mass numbers of the fission fragments of $_{92}U^{236*}$ due to fission by slow neutron in equation (5.3) to equation (5.8) shows how the 236 nucleons are split up amongst two daughter nuclei. Let us write the mass number combinations.

$$236 = 141 + 92 + 3 \text{ [equation (5.3)]}$$

$$236 = 98 + 135 + 3 \text{ [equation (5.4)]}$$

$$236 = 140 + 94 + 2 \text{ [equation (5.5)]}$$

$$236 = 139 + 94 + 3 \text{ [equation (5.6)]}$$

$$236 = 141 + 93 + 2 \text{ [equation (5.7)]}$$

$$236 = 137 + 97 + 2 \text{ [equation (5.8)]}$$

Obviously the fragments are of unequal size and have widely different mass numbers. Hence this fission is called asymmetric fission.

The mass distribution of fission products is represented by plotting a quantity called fission yield $Y(A)$ defined as follows.

$$Y(A) = \frac{\text{Number of nuclei of mass number A formed in fission}}{\text{Total number of fission}} = \frac{N_A}{N_0}$$

Percentage fission yield is

$$Y(A)\% = \frac{N_A}{N_0} \times 100\%. \qquad (5.9)$$

A plot of percentage fission yield against mass number is shown in figure 5.1(a).

Percentage yield is maximum around $A \sim 95$ and $A \sim 139$. In other words, the probability of getting one fragment with $A \sim 139$ and another fragment with $A \sim 95$ is large. So the probability of asymmetric fission is large.

So fission fragments are not of equal mass. The distribution of fission products induced by slow neutrons is highly asymmetric.

✓ If $_{92}U^{236}$ was split into two fragments of equal size ($236 = 117 + 117 + 2$) then such splitting would have been called symmetric fission.

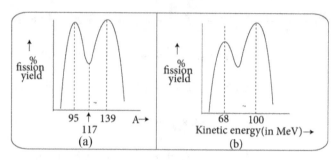

**Figure 5.1.** (a) Nuclear fission yield is minimum at $A = 117$ and maximum at $A = 95, 139$, (b) kinetic energy is maximum at around 68, 100 $MeV$.

Fission yield for $A = 117$ is very small, i.e. the probability of symmetric fission is low. This is evident from the fact that the yield is minimum at $A = 117$.

This dominance of asymmetric fission is not explained by the liquid drop model of a nucleus. However, we can explain it considering shell structure of nuclei from the shell model.

Mass number $A = 117$ corresponds to atomic number $Z{\sim}50$ which is a magic number and corresponds to strong nuclear binding and filled nucleonic shells. This means low probability of splitting of the nuclei, i.e. a small number of nuclei undergo splitting. So at $Z{\sim}50$, $A = 117$ fission yield is minimum. Symmetric splitting is thus less probable.

(We note that in equation (5.1) the symbols $X$, $Y$, $Z$ represent nuclei)

On the other hand, for values of mass numbers around $A \sim 95, 139$ the atomic numbers $Z \sim 40, 52, 54, 56$ etc, are possible. As these numbers are not magic numbers, binding is not so strong and fission into various unequal two or three fragments like $_{40}Zr^{98}$, $_{52}Te^{135}$, $_{54}Xe^{140}$, $_{56}Ba^{139}$ is possible. Asymmetric fission is more probable or favoured.

We also note that the ratio of mass numbers, i.e. $A$ s of the larger fragment $Z$ and smaller fragment $Y$ is (equation (5.1): $n + X \rightarrow Y + Z + 3n$)

$$\frac{A(\text{larger fragment})}{A(\text{smaller fragment})} = \frac{m_Z}{m_Y} = \frac{139}{95} \cong 1.5 \qquad (5.10)$$

- Let us find the kinetic energy ratio of the fragments.

    The nucleus undergoing nuclear fission can be considered to be at rest. Also, for simplicity of calculation let us overlook the mass of the neutron as it is negligibly small. So the reaction of equation (5.1) is approximately written as

$$X \rightarrow Y + Z \qquad (5.11)$$

Apply conservation of linear momentum to get

$$\vec{p}_X = 0$$

$$\vec{p}_Y + \vec{p}_Z = 0$$

$$\vec{p}_Y = -\vec{p}_Z \tag{5.12}$$

Fission fragments $Y$, $Z$ have equal and opposite momentum. Hence

$$p_Y = p_Z \text{ (magnitude same)} \tag{5.13}$$

If $m_Y$, $m_Z$ are masses of $Y$ and $Z$ respectively, then the ratio of their kinetic energies are

$$\frac{T_Y}{T_Z} = \frac{p_Y^2/2m_Y}{p_Z^2/2m_Z} = \frac{m_Z}{m_Y} \text{ [using equation (5.13)]} \tag{5.14}$$

Equation (5.14) shows that kinetic energies of the fission fragments are inversely proportional to their masses. As mass is proportional to mass number we expect from equations (5.10) and (5.14) that

$$\frac{T_Y}{T_Z} = \frac{m_Z}{m_Y} = 1.5 \tag{5.15}$$

The experimental plot of fission yield against kinetic energy of fragments is shown in figure 5.1(b).

It follows that the ratio of kinetic energies of the fragments is

$$\frac{T_Y}{T_Z} = \frac{\text{kinetic energy of lighter fragment}}{\text{kinetic energy of heavier fragment}} = \frac{100 \; MeV}{68 \; MeV} \cong 1.5 \tag{5.16}$$

Thus the experimental result of equation (5.16) corroborates the theoretical result of equation (5.15). This also shows the asymmetry of the nuclear fission process.

• Neutrons released in fission process.

✓ As shown in the fission reactions equation (5.3) to equation (5.8) neutrons are produced.

✓ Either two neutrons or three neutrons are generated in the process of one fission. This we wrote in equation (5.1) as $\nu = 2$ or $3$. The average number of neutrons produced is 2.5 per fission.

✓ Neutrons emitted in the fission reaction are fast neutrons. The energy of neutrons emitted varies from 0.05 to 17 $MeV$ with an average of 2 $MeV$.

✓ Neutrons emitted in the process of fission are of two classes: prompt neutron and delayed neutron.

• Neutron emission pattern

Consider the fission reaction of equation (5.1) viz. $n + X \rightarrow Y + Z + 3n$.

For heavy nuclei, the number of neutrons exceeds the number of protons, e.g.

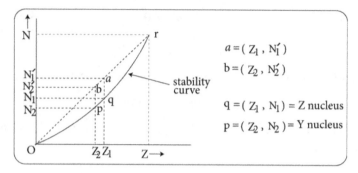

**Figure 5.2.** Explanation of neutron emission from fission fragments by stability curve.

$$_{92}U_{143}^{235} \text{ has } \frac{N}{Z} = \frac{143}{92} = 1.55.$$

When a heavy nucleus $_ZX_N^A$ splits into two parts we assume that the $\frac{N}{Z}$ ratio is also maintained in the products $_{Z_1}Y_{N_1}^{A_1}$ and $_{Z_2}Z_{N_2}^{A_2}$ i.e. $\frac{N_1}{Z_1}, \frac{N_2}{Z_2}, \frac{N}{Z}$ are more or less of the same value. In other words, the ratio $\frac{N}{Z}$ should be the same in the nuclei $X$, $Y$, $Z$. We thus expect the product fragments to lie on the line $Obar$ in the $N$ versus $Z$ plot of figure 5.2 at points $a(Z_1, N_1')$, $b(Z_2, N_2')$.

However, for stability, the fragments are allowed to possess a lower number of neutrons, say $N_1 < N_1'$ and $N_2 < N_2'$.

Clearly at the time of creation the fragments $Y$ and $Z$ are overloaded with neutrons, i.e. they have neutron excess. They therefore get rid of some neutrons to fall in the line of stability (curve $Opqr$ of figure 5.2). Hence on the average two to three neutrons (2.5 on the average) are emitted per fission of $_{92}U_{143}^{235}$.

The fragments $Y$ and $Z$ thus settle to the limit of stability, i.e. occupy a position on (or near) the stability curve, say at points $q \equiv (Z_1N_1)$ which is the nuclear fragment $Z$ and $p \equiv (Z_2N_2)$ which is the nuclear fragment $Y$.

## 5.3 Prompt neutrons

✓ Neutrons that are emitted instantaneously during the fission process, i.e. within a time of $10^{-14}$ s are called prompt neutrons.

✓ Most of the neutrons, i.e. 99% of the fission neutrons produced in fission, are prompt neutrons.

☐ Explanation

When a thermal slow neutron $_0n^1$ strikes $_{92}U^{235}$ it gets absorbed and a compound nucleus $_{92}U^{236*}$ in excited state is formed. It splits into two fragments which are initially unstable. The fragments have excess neutrons and also some excess energy around 6 $MeV$ that makes them unstable. This excess energy is released through neutron emission. So the excited unstable nuclear fragments eject two or three neutrons for attainment of stability within a very short period of time ($\sim 10^{-14}$ s) after their formation and these are prompt neutrons. And the nuclear fragments $Y$, $Z$ result [equation (5.1): $n + X \rightarrow Y + Z + 3n$].

☐ Example:

In the fission process

$$_{92}U^{236*}_{144} \rightarrow _{54}Xe^{140}_{86} + _{38}Sr^{94}_{56} + 2_{0}n^1$$

the two neutrons emitted are prompt neutrons. And the nuclear fragments are $_{54}Xe^{140}_{86}$ and $_{38}Sr^{94}_{56}$.

On the average, 2.5 prompt neutrons are emitted per fission.

• Beta chain

After fission the products $Y, Z$ along with 2.5 prompt neutrons (on the average) are created. However, often it so happens that such fission products $(Y, Z)$ are so neutron rich that they are not stable—they do not lie on the stability curve in the $N$ versus $Z$ plot. The $\frac{N}{Z}$ ratio is high for them.

Further neutron emission may be energetically unfavorable and so beta decay is preferred.

To attain stability they are $\beta^-$ active ($n$ converts to $p$) in an effort to reduce the $\frac{N}{Z}$ ratio. Thus beta chain follows after fission.

For the fission

$$_{92}U^{236*}_{144} \rightarrow _{54}Xe^{140}_{86} + _{38}Sr^{94}_{56} + 2_{0}n^1$$

two beta chains follow as $_{54}Xe^{140}_{86}$ is radioactive and becomes stable after four beta decays, while $_{38}Sr^{94}_{56}$ is radioactive too and becomes stable after two beta decays. We indicate this now.

$$_{54}Xe^{140}_{86} \xrightarrow{\beta^-} _{55}Cs^{140}_{85} \xrightarrow{\beta^-} _{56}Ba^{140}_{84} \xrightarrow{\beta^-} _{57}La^{140}_{83} \xrightarrow{\beta^-} _{58}Ce^{140}_{82} \text{ (stable)}$$

$$_{38}Sr^{94}_{56} \xrightarrow{\beta^-} _{39}Y^{94}_{55} \xrightarrow{\beta^-} _{40}Zr^{94}_{54} \text{ (stable)}$$

## 5.4 Delayed neutron

✓ Neutrons that are emitted with gradually decreasing intensity with a time lag or time delay of several seconds to more than one minute after the fission process are called delayed neutrons since neutron emission occurs after a delay.

✓ Delayed neutrons constitute only 0.64% of the total neutrons.

Average number of delayed neutrons per fission of $_{92}U^{235}_{143}$ is 0.2.

✓ Delayed neutrons play important role in the control of a nuclear reactor. As the number of delayed neutrons is low they prevent rapid changes in neutron density when the condition of operation is close to critical.

✓ Explanation

In the process of nuclear fission the nuclear fragments produced are neutron rich and hence may be unstable. They try to attain stability through

$\beta$ decay. After $\beta$ decay the product nucleus may find itself still in an excited state with energy greater than the binding energy of neutron in that nucleus. This might lead to emission of neutron which is a delayed neutron.

✓ We give a few examples.

☐ During the fission

$$_{92}U^{236*}_{144} \rightarrow {}_{53}I^{137}_{84} + {}_{39}Y^{97}_{58} + 2{}_0n^1$$

Two prompt neutrons get emitted.

After this fission two beta chains follow and then again there is neutron emission.

One beta chain is because $_{39}Y^{97}_{58}$ is radioactive and becomes stable after three beta decays. We indicate these transformations in the following.

$$_{39}Y^{97}_{58} \xrightarrow{\beta^-} {}_{40}Zr^{97}_{57} \xrightarrow{\beta^-} {}_{41}Nb^{97}_{56} \xrightarrow{\beta^-} {}_{42}Mo^{97}_{55} \text{ (stable)}$$

Another beta chain is because $_{53}I^{137}_{84}$ is radioactive too and becomes stable after one beta decay and one delayed neutron emission after 24 $s$.

$$_{53}I^{137} \xrightarrow{\beta^-} {}_{54}Xe^{137} \xrightarrow{\text{delayed n}} {}_{54}Xe^{136} \text{ (stable)}$$

This neutron has not been emitted promptly but after a delay of 24 $s$ and hence called delayed neutron.

☐ Consider a fission fragment $_{35}Br^{87}$ which suffers $\beta^-$ decay followed by a 55 $s$ delayed neutron emission.

$$_{35}Br^{87} \xrightarrow{\beta^-} {}_{36}Kr^{87} \xrightarrow{\text{delayed n}} {}_{36}Kr^{86} \text{ (55 } s\text{)}$$

☐ Consider a fission fragment $_{37}Rb^{93}$ which suffers $\beta^-$ decay followed by a 6 $s$ delayed neutron emission.

$$_{37}Rb^{93} \xrightarrow{\beta^-} {}_{38}Sr^{93} \xrightarrow{\text{delayed } n} {}_{38}Sr^{92} \text{ (6 } s\text{)}$$

- Gamma emission: prompt and delayed gammas

    The fragments created in the process of fission are in higher energy state. And $\gamma$ emission is another process through which they can release their extra energy. So just after or during fission some $\gamma$ rays are emitted called prompt $\gamma$ photons.

    Also, in the beta chain the daughters that are produced at various intermediate steps may be in excited state and may release their extra energy through $\gamma$ photon emission. These $\gamma$ photons are obviously emitted much after the prompt $\gamma$ emission. So the $\gamma$ photons that are emitted with a delay after beta decays are called delayed $\gamma$ photons.

- Antineutrino emission

There will be emission of antineutrinos which are invariably associated with $\beta^-$ in the beta chain.

## 5.5 Estimation of the energy release during nuclear fission from binding energy curve

The energy released in fission can be estimated in a number of ways.

Let us investigate whether it is energetically possible to convert a heavy mass nucleus $_Z X_N^A$ with

$$\frac{B}{A} = 7.5 \frac{MeV}{\text{nucleon}} \tag{5.17}$$

into two nuclei $_{Z_1} Y_{N_1}^{A_1}$, $_{Z_2} Z_{N_2}^{A_2}$ of intermediate mass with

$$\frac{B}{A} = 8.5 \frac{MeV}{\text{nucleon}}. \tag{5.18}$$

This is shown in figure 5.3. Here

$$N_1 + Z_1 = A_1, \ N_2 + Z_2 = A_2, \ N + Z = A \tag{5.19}$$

$$N_1 + N_2 = N, \ Z_1 + Z_2 = Z, \ A_1 + A_2 = A \tag{5.20}$$

Mass of a nucleus $M(Z, N, A)$ is related to its binding energy $B$ through the formula

$$M(Z, N, A)c^2 = (Z m_p + N m_n)c^2 - B(Z, N, A) \tag{5.21}$$

Increase of $B$ means $M$ decreases.

The $Q$ value of the reaction $X \rightarrow Y + Z$ is

$$Q = (m_{\text{initial}} - m_{\text{final}})c^2 = (m_X - m_Y - m_Z)c^2$$

$$Q = m_X c^2 - m_Y c^2 - m_Z c^2 \tag{5.22}$$

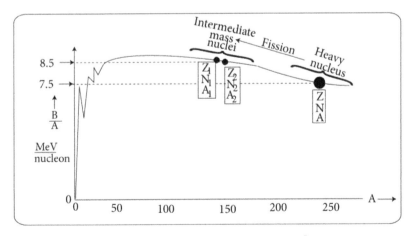

**Figure 5.3.** Fission energy from plot of binding energy per nucleon $\frac{B}{A}$ versus mass number $A$.

Using equation (5.21) with equations (5.17) and (5.18) we can write

$$m_X c^2 = \left(Z m_p + N m_n\right)c^2 - (7.5\ \text{MeV})A$$

$$m_Y c^2 = \left(Z_1 m_p + N_1 m_n\right)c^2 - (8.5\ \text{MeV})A_1$$

$$m_Z c^2 = \left(Z_2 m_p + N_2 m_n\right)c^2 - (8.5\ \text{MeV})A_2$$

Equation (5.22) thus gives

$$Q = [\,Z - (Z_1 + Z_2)]m_p c^2 + [\,N - (N_1 + N_2)]m_n c^2 - 7.5\ MeV A + 8.5\ MeV (A_1 + A_2)$$

Using equations (5.19) and (5.20) we get

$$Q = 8.5\ MeV A - 7.5\ MeV A$$

$$= A MeV$$

For $A = 220$ we have

$$Q = 220\ MeV = +ve \tag{5.23}$$

$Q$ is positive means that such spontaneous splitting or fission of a heavy nucleus $X$ into two nuclei of intermediate mass is energetically favourable and the fission fragments $Y$ and $Z$ carry this released energy of $220\ MeV$ as their kinetic energy. Such process causes decrease in rest mass as evident from equation (5.2) in which

$$m > m_1 + m_2 \tag{5.24}$$

Higher mass $X$ goes to lower mass combination $Y + Z$, the mass deficit $m - m_1 - m_2$ being converted to energy that is released through the process of nuclear fission.

## 5.6 Estimation of the energy release during nuclear fission from mass of fission fragments

Consider the fission process of equation (5.3)

$$_0 n^1 + {}_{92}U^{235}_{143} \rightarrow {}_{92}U^{236*}_{144} \rightarrow {}_{56}Ba^{141}_{85} + {}_{36}Kr^{92}_{56} + 3{}_0 n^1 + Q$$

$$Q = m_{\text{initial}} - m_{\text{final}}$$

$$Q = m({}_{92}U^{235}_{143}) + m({}_0 n^1) - m({}_{56}Ba^{141}_{85}) - m({}_{36}Kr^{92}_{56}) - m({}_0 n^1) \tag{5.25}$$

Masses of ${}_{92}U^{235}$, ${}_{56}Ba^{141}$, ${}_{36}Kr^{92}$, ${}_0 n^1$ are $235.113\,92u$, $140.9177u$, $91.926\,156u$, $1.008\,665u$ respectively. Hence from equation (5.25) we have

$$Q = 235.113\,92u + 1.008\,665u - 140.9177u - 91.926\,156u - 3(1.008\,665u)$$

$$= 0.252\,734u = 0.252\,734 \times 931\ MeV$$

$$Q = 235\ MeV\ \text{(per fission)} \tag{5.26}$$

☐ We can find the energy released in *watt. hr* from 1 *g* of $_{92}U^{235}_{143}$.

235 *g* of $_{92}U^{235}_{143}$ have Avogadro's number of atoms, i.e. $6.023 \times 10^{23}$ atoms.

1 *g* will have $\frac{6.023 \times 10^{23}}{235}$ atoms.

Assume each atom undergoes nuclear fission.

The energy released by 1 *g* of $_{92}U^{235}_{143}$ will be

(Number of atoms in 1 *g*)(energy per fission)

$$= \frac{6.023 \times 10^{23}}{235} \times 235 \; MeV \; \text{[using equation (5.26)]}$$

$$= 6.023 \times 10^{23} \, \text{MeV} = 6.023 \times 10^{23} \times 10^{6} \times (1.6 \times 10^{-19}) J$$

$$= 9.64 \times 10^{10} \frac{J}{s}. \; s = 9.64 \times 10^{10} \; watt. \; \frac{hr}{60 \times 60} = 2.67 \times 10^{7} \; watt. \; hr \quad (5.27)$$

## 5.7 Spontaneous fission

✓ Spontaneous fission occurs automatically or spontaneously and is not induced or triggered.

✓ Nuclei with $Z > 82$ suffer spontaneous fission.

✓ Nuclei with $Z < 82$ do not undergo spontaneous fission.

✓ Explanation

The number of protons is large in a heavy nucleus. As there are a large number of protons in the nucleus, $Z$ is large, which means the mass number $A$ is also large and hence it follows from the formula for nuclear radius, namely $R = R_0 A^{1/3}$, that radius is large. So it is a big nucleus. As population of protons is large, electrostatic repulsion between them is significantly large. For two diametrically opposite protons *p–p*, as shown in figure 5.4, the nuclear force is small since it is a short-range force but electrostatic repulsive force is large. In other words, electrostatic repulsion dominates over nuclear binding force, leading to a split in the nucleus.

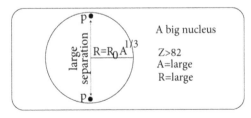

**Figure 5.4.** Nucleus with $Z > 82$ undergoes spontaneous fission.

## 5.8 Induced fission

- ✓ If fission does not take place automatically but is induced or triggered by an external particle, i.e. by a projectile, we call it induced fission.
- ✓ A projectile strikes a heavy nucleus and in the process energy is supplied to it. The nucleus becomes unstable and breaks into two pieces. Neutrons are emitted and energy is released.
- ✓ The energy needed to activate fission is called fission activation energy $E_A$.
- ✓ Fission activation energy $E_A$ can also be supplied to the nucleus by particles other than neutrons such as protons, deuterons, $\alpha$ particles, $\gamma$ rays, $\pi^-$ etc.
- ✓ The fission induced by $\gamma$ rays is called photofission. In photofission a heavy nucleus $X$ is struck with a $\gamma$ photon of requisite energy $E_A$ and fission is induced.

$\gamma + X \rightarrow X^* \rightarrow$ fission occurs

This process helps to find the fission activation energy $E_A$ which is the energy required by a nucleus for it to split apart.

- ✓ Neutron is the most appropriate projectile. A positively charged projectile requires sufficient energy to overcome the nuclear potential barrier of target. Neutrons are neutral. They are neither affected by the presence of orbital electrons nor by the positive charge of the nucleus. Hence the neutron is the most effective or appropriate projectile to induce fission.
- ✓ Neutron-induced fission

$n + X \rightarrow X^* \rightarrow$ fission occurs

Power plants and nuclear reactors run on the principle of neutron-induced nuclear fission.

A heavy nucleus $X$ is struck by neutrons—and if it can provide the necessary activation energy $E_A$, fission is induced.

- ☐ Heavy nuclei $_{92}U^{235}_{143}$ having even $Z = 92$ and odd $N = 143$ undergo fission by slow or thermal neutrons.
- ☐ Heavy nuclei $_{92}U^{238}_{146}$ having even $Z = 92$ and even $N = 146$ undergo fission by fast neutrons having kinetic energy greater than 1 $MeV$.

## 5.9 Fissile material

- ✓ Materials that are capable of undergoing fission by absorbing low energy neutron (thermal slow neutron) are fissile materials. They see no fission barrier and readily break apart.
- ✓ $_{92}U^{235}_{143}$, $_{94}Pu^{239}_{145}$, $_{92}U^{233}_{141}$ are fissile materials. They can be fissioned with thermal neutrons. The fission reactions are

$$_0n^1 + {}_{92}U^{235}_{143} \rightarrow {}_{56}Ba^{141}_{85} + {}_{36}Kr^{92}_{56} + 3{}_0n^1$$

$$_0n^1 + {}_{94}Pu^{239}_{145} \rightarrow {}_{54}Xe^{134}_{80} + {}_{40}Zr^{103}_{63} + 3{}_0n^1$$

$$_0n^1 + {}_{92}U^{233}_{141} \rightarrow {}_{54}Xe^{137}_{83} + {}_{38}Sr^{94}_{56} + 3{}_0n^1$$

✓ The only material that is naturally occurring and fissile (i.e. can be fissioned with thermal neutron) is $_{92}U_{143}^{235}$.

Only 0.72% of the naturally occurring uranium is $_{92}U_{143}^{235}$.

✓ $_{94}Pu_{145}^{239}$, $_{92}U_{141}^{233}$ do not occur in nature and can be produced by the interaction of neutrons with the fertile materials namely $_{92}U_{146}^{238}$, $_{90}Th_{142}^{232}$, respectively.

## 5.10 Fertile or fissionable material and breeding reaction

✓ $_{92}U_{146}^{238}$, $_{90}Th_{142}^{232}$ are called fertile or fissionable materials.
✓ They are not fissile materials, i.e. they are not capable of undergoing fission with a low energy thermal neutron.
✓ They can be used as raw materials for the production of fissile materials.
✓ The nuclear reactions which convert these fertile materials into fissile materials are called breeding reactions. They are

$$_{0}n^{1} + {}_{92}U_{146}^{238} \rightarrow {}_{92}U_{147}^{239} + \gamma$$

$$_{92}U_{147}^{239} \rightarrow {}_{93}Np_{146}^{239} + {}_{-1}e^{0} + \bar{\nu}_{e} \left(\beta^{-}\text{decay}\right)$$

$$_{93}Np_{146}^{239} \rightarrow {}_{94}Pu_{145}^{239} + {}_{-1}e^{0} + \bar{\nu}_{e} \left(\beta^{-}\text{decay}\right),$$

$$_{94}Pu_{145}^{239} \text{ is fissile material.}$$

Again

$$_{0}n^{1} + {}_{90}Th_{142}^{232} \rightarrow {}_{90}Th_{143}^{233} + \gamma$$

$$_{90}Th_{143}^{233} \rightarrow {}_{91}Pa_{142}^{233} + {}_{-1}e^{0} + \bar{\nu}_{e} \ (\beta^{-}\text{decay})$$

$$_{91}Pa_{142}^{233} \rightarrow {}_{92}U_{141}^{233} + {}_{-1}e^{0} + \bar{\nu}_{e} \ (\beta^{-}\text{decay})$$

$$_{92}U_{141}^{233} \text{ is a fissile material.}$$

## 5.11 Minimum value of mass number $A$ for fission to be energetically favourable

Consider spontaneous fission of a nucleus $_{Z}X_{N}^{A}$ as

$$_{Z}X_{N}^{A} \rightarrow {}_{Z_{1}}Y_{N_{1}}^{A_{1}} + {}_{Z_{2}}Z_{N_{2}}^{A_{2}} \tag{5.28}$$

where

$$Z_1 + Z_2 = Z, \; A_1 + A_2 = A \tag{5.29}$$

$Z_1$, $Z_2$ are of comparable values; $A_1$, $A_2$ are of comparable values also.

For simplicity of calculation let us consider symmetric fission of $_ZX_N^A$ so that

$$Z_1 = Z_2 = \frac{Z}{2}, \; N_1 = N_2 = \frac{N}{2}, \; A_1 = A_2 = \frac{A}{2} \tag{5.30}$$

$$_ZX_N^A \rightarrow \; _{Z/2}Y_{N/2}^{A/2} + \; _{Z/2}Y_{N/2}^{A/2} = 2\,_{Z/2}Y_{N/2}^{A/2} \tag{5.31}$$

The $Q$ value of the nuclear fission reaction (called fission energy) is

$$Q = (m_{\text{initial}} - m_{\text{final}})c^2 = (m_X - 2m_Y)c^2 \tag{5.32}$$

$$Q = \left[ M(Z, N, A) - 2M\left(\frac{Z}{2}, \frac{N}{2}, \frac{A}{2}\right) \right]c^2 \tag{5.33}$$

Using equation (5.21) we can write

$$Q = \left[ (Zm_p + Nm_n)c^2 - B(Z, N, A) \right] - 2\left[ \left(\frac{Z}{2}m_p + \frac{N}{2}m_n\right)c^2 - B\left(\frac{Z}{2}, \frac{N}{2}, \frac{A}{2}\right) \right]$$

$$Q = 2B\left(\frac{Z}{2}, \frac{N}{2}, \frac{A}{2}\right) - B(Z, N, A) \tag{5.34}$$

Use the semi-empirical mass formula

$$B(Z, N, A) = a_V A - a_S A^{2/3} - a_C \frac{Z^2}{A^{1/3}} - a_{\text{asym}}\frac{(A - 2Z)^2}{A} + \frac{\delta(Z, N, A)}{A^{3/4}} \tag{5.35}$$

$$B\left(\frac{Z}{2}, \frac{N}{2}, \frac{A}{2}\right) = a_V \frac{A}{2} - a_S(A/2)^{2/3} - a_C\frac{(Z/2)^2}{(A/2)^{1/3}} - a_{\text{asym}}\frac{\left(\frac{A}{2} - 2\frac{Z}{2}\right)^2}{A/2} + \frac{\delta\left(\frac{Z}{2}, \frac{N}{2}, \frac{A}{2}\right)}{(A/2)^{3/4}}$$

$$2B\left(\frac{Z}{2}, \frac{N}{2}, \frac{A}{2}\right) = a_V A - 2a_S(A/2)^{2/3} - 2a_C\frac{(Z/2)^2}{(A/2)^{1/3}} - a_{\text{asym}}\frac{(A - 2Z)^2}{A} + 2\frac{\delta\left(\frac{Z}{2}, \frac{N}{2}, \frac{A}{2}\right)}{(A/2)^{3/4}} \tag{5.36}$$

From equations (5.34)–(5.36) we have

$$Q = a_S A^{\frac{2}{3}}\left(1 - 2^{1-\frac{2}{3}}\right) - a_C\frac{Z^2}{A^{1/3}}\left(1 - 2^{-1+\frac{1}{3}}\right) + 2\frac{\delta\left(\frac{Z}{2}, \frac{N}{2}, \frac{A}{2}\right)}{(A/2)^{3/4}} - \frac{\delta(Z, N, A)}{A^{3/4}} \tag{5.37}$$

$Q = -0.26 a_S A^{2/3} + 0.37 a_C \frac{Z^2}{A^{1/3}}$+pairing energy difference contribution

As pairing energy difference contribution is small we neglect it. Hence

$$Q = -0.26 a_S A^{2/3} + 0.37 a_C \frac{Z^2}{A^{1/3}} \tag{5.38}$$

This analysis shows that, in the splitting process, volume energy and asymmetry energy do not play any active role. An active role is played only by surface energy term and Coulomb energy term which occur with opposite sign and thus try to cancel each other out. Surface energy prevents splitting of nucleus while Coulomb energy promotes splitting.

Using $a_S = 17.8 \; MeV$, $a_C = 0.71 \; MeV$ we have

$$Q = -0.26(17.8 \; MeV)A^{2/3} + 0.37(0.71 \; MeV)\frac{Z^2}{A^{1/3}}$$

$$Q = \left(-4.628A^{2/3} + 0.2627\frac{Z^2}{A^{1/3}}\right) MeV \tag{5.39}$$

✓ As an example let us find the energy released in the symmetric fission of a nucleus with $A = 200$, $Z = 80$.

From equation (5.39) we have

$$Q = \left(-4.628(200)^{2/3} + 0.2627\frac{(80)^2}{(200)^{1/3}}\right) MeV$$

$$Q = 129 \; MeV$$

- For symmetric fission to occur spontaneously $Q$ value has to be positive, i.e.

$$Q \geqslant 0 \tag{5.40}$$

Hence from equation (5.39) we have

$$\left(-4.628A^{2/3} + 0.2627\frac{Z^2}{A^{1/3}}\right) MeV \geqslant 0 \tag{5.41}$$

$$0.2627\frac{Z^2}{A^{1/3}} \geqslant 4.628A^{2/3}$$

$$\frac{Z^2}{A^{\frac{1}{3}+\frac{2}{3}}} \geqslant \frac{4.628}{0.2627}$$

$$\frac{Z^2}{A} \geqslant 17.6 \tag{5.42}$$

This is the condition for symmetric spontaneous fission to be energetically possible.

✓ For instance if $Z = 40$, $A = 90$ we note that

$$\frac{Z^2}{A} = \frac{40^2}{90} = 17.77$$

So the condition for spontaneous fission is obeyed for nuclei with $A > 90$, $Z > 40$.

☐ In reality it is observed that even though energetically allowed, nuclei with $A > 90$, $Z > 40$ are not all spontaneously fissile—rather they are very stable (they do not break on their own to reduce their rest mass in a bid to go to a stable region with larger binding energy).

☐ To explain the discrepancy, Bohr and Wheeler considered Coulomb potential barrier $E_0$ of the two fragments at the instant of separation.

Existence of this Coulomb barrier $E_0$ prevents fragmentation of the nucleus. Fragmentation will occur only if the Coulomb barrier $E_0$ is surmounted. In other words if $Q > E_0$ spontaneous fission occurs while if $Q < E_0$ spontaneous fission does not take place.

Let us calculate the value of the Coulomb barrier $E_0$.

The Coulomb potential energy between two symmetric fragments each of atomic number $\frac{Z}{2}$, mass number $\frac{A}{2}$ and radius $R = R_0\left(\frac{A}{2}\right)^{1/3}$, when they are just in contact with each other, is

$$E_0 = \frac{1}{4\pi\varepsilon_0} \frac{\left(\frac{z}{2}|e|\right)^2}{2R} = \frac{1}{4\pi\varepsilon_0} \frac{1}{8} \frac{(Z|e|)^2}{R_0(A/2)^{1/3}} = \frac{1}{4\pi\varepsilon_0} \frac{|e|^2}{8R_0} \frac{Z^2 2^{1/3}}{A^{1/3}}$$

$$= \frac{1}{4\pi\frac{10^{-9}}{36\pi}\frac{F}{m}} \frac{(1.6 \times 10^{-19}\ C)^2}{8(1.5 \times 10^{-15}\ m)} \frac{Z^2 2^{1/3}}{A^{1/3}} \quad (\text{using } R_0 = 1.5 \times 10^{-15}\ m)$$

$$= 2.4 \times 10^{-14} \frac{Z^2}{A^{1/3}} J = \frac{2.4 \times 10^{-14}}{1.6 \times 10^{-19}} \frac{Z^2}{A^{1/3}}\ eV = \frac{150 \times 10^3}{10^6} \frac{Z^2}{A^{1/3}}\ MeV$$

$$E_0 = 0.15 \frac{Z^2}{A^{1/3}}\ MeV$$

For spontaneous fission to occur, instead of equation (5.40) we should have

$$Q \geqslant E_0 \tag{5.43}$$

Using equation (5.39)

$$\left(-4.628 A^{2/3} + 0.2627 \frac{Z^2}{A^{1/3}}\right) MeV \geqslant 0.15 \frac{Z^2}{A^{1/3}}\ MeV \tag{5.44}$$

$$-4.628 A^{2/3} + 0.2627 \frac{Z^2}{A^{1/3}} - 0.15 \frac{Z^2}{A^{1/3}} \geqslant 0$$

$$0.1 \frac{Z^2}{A^{1/3}} \geqslant 4.628 A^{2/3}$$

$$\frac{Z^2}{A} \geqslant \frac{4.628}{0.1}$$

$$\frac{Z^2}{A} \geqslant 46 \quad \text{(for spontaneous fission)} \tag{5.45}$$

$$\frac{Z^2}{A} < 46 \quad \text{(for no spontaneous fission)} \tag{5.46}$$

- The difference $E_0 - Q$ is referred to as fission activation energy $E_A$ i.e.

$$E_0 - Q = E_A \tag{5.47}$$

This $E_A$ serves as a fission barrier.

Obviously if a nucleus possesses energy $Q$ spontaneous fission will not occur if $Q < E_0$.

Only if nucleus possesses a minimum energy of

$$Q + E_A = E_0$$

or energy greater than $E_0$ then spontaneous fission will occur.

The quantities $E_0$, $E_A$, $Q$ are shown in figure 5.6.

✓ Example

Consider $_{92}U^{235}$ nucleus. For it

$$\frac{Z^2}{A} = \frac{92^2}{235} = 36.$$

Since condition $\frac{Z^2}{A} < 46$ is obeyed it follows from equation (5.46) that spontaneous fission does not occur for a $_{92}U^{235}$ nucleus. It is stable against spontaneous fission.

A better and a more rigorous estimation was done in Bohr and Wheeler's theory on the basis of the liquid drop model that we explain now.

- Let us discuss whether we can build a nucleus with mass number $A > 300$.

Let the potential energy of interaction be $E$ which is surface energy + Coulomb energy. It depends upon the amount of deformation of nucleus. Amount of deformation is measured by a parameter $r$ called deformation parameter. It is the separation between the two lobes or fragments into which the nucleus is going to be split. The energy of deformation for each stage or shape can be evaluated using semi-empirical mass formula, as shown in figure 5.7 and also in the $E$ versus $r$ diagram in figure 5.6.

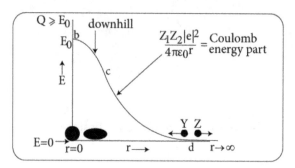

**Figure 5.5.** Nuclear interaction energy $E$ against deformation parameter $r$ for $A > 300$. Here $Q \geqslant E_0$ and fission barrier is absent so that nucleus runs downhill in the energy diagram to suffer spontaneous fission.

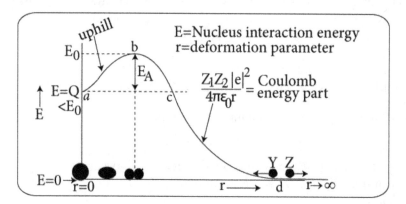

**Figure 5.6.** Nuclear interaction energy as a function of deformation parameter.

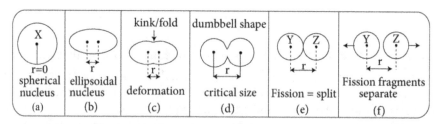

**Figure 5.7.** Splitting of nucleus as per Bohr–Wheeler model.

Consider a very large nucleus. For a very large nucleus with a very large mass number the fission barrier is extremely small or absent. We have shown in figure 5.5 the $E$ versus $r$ plot. The nucleus has initially a spherical shape and has energy $Q \geqslant E_0$.

So it starts right from the top (maximum or peak point $b$) and simply falls downhill along curve $bcd$ to get split into lower mass number nuclei $Y$ and $Z$ at point $d$.

The large energy $Q \geqslant E_0$ possessed by the nucleus might cause it to deform slightly. A small deformation into ellipsoidal shape (say) triggers more deformation and the system starts a downhill journey along *bcd* towards disrupture and soon splits. Such nuclei are in unstable equilibrium.

Any disturbance or perturbation aggravates and pushes the nucleus from unstable equilibrium state and enhances its disrupture. Such nuclei are spontaneously fissile (no activation needed). This means we cannot indefinitely go on building nuclei with a gradually increasing number of nucleons. It is not possible to build nuclei with $A > 300$.

In other words, such a large nucleus for which spontaneous fission is possible without any activation only moves downhill in the $E$ versus $r$ plot of figure 5.5 and gets fissioned. It will not need any energy to move uphill because it is already sitting at the topmost spot (point *b*) of the potential energy hill.

- In figure 5.6 we show the $E$ versus $r$ plot for a nucleus corresponding to energy $E = Q < E_0$.

At point *a*, i.e. r = 0, the nucleus is undeformed, i.e. it has spherical shape.

For fission to occur the nucleus should move downhill along curve portion *cd*. But to reach point *c* it has to first move uphill along the curve portion *ab* and then fall downhill along the curve portion *bcd*. In other words the nucleus has to overcome the fission barrier represented by the curve portion *abc* which again corresponds to activation energy $E_A$.

Activation energy or fission barrier height is defined from equation (5.47) as

$$E_A = E_0 - Q \tag{5.48}$$

In other words, the nucleus has to overcome the fission barrier by getting activation energy $E_A$, moving uphill to reach the peak point *b*, i.e. point of irreversible deformation, and then splitting through fission (i.e. going downhill).

As the nucleus gets deformed, the shape deviates from sphericity more and more and so area increases. So surface energy increases. Again, deformation means the nucleus gives way to two lobes—the distance between them increasing. Hence Coulomb repulsion decreases with increase in separation of the two lobes.

The two forces—attractive surface energy and repulsive Coulomb energy are always in competition.

When deformation begins energy $E$ increases from $E = Q$ (at point *a*) to become maximum $E = E_0$ (at point *b*) and then decreases as

$$E = \frac{Z_1 Z_2 \mid e \mid^2}{4\pi\varepsilon_0 r} \propto \frac{1}{r} \text{ (Coulomb energy of repulsion)} \tag{5.49}$$

When $r \rightarrow \infty$, i.e. when the fragments are well separated, there is no interaction between the split nuclei and so $E = 0$.

Clearly for spontaneous fission the energy of interaction of the nucleus has to pass through, i.e. overcome the barrier height $E_0$ and it will be able to do that (and split itself) only if it manages to derive or get the activation energy $E_A$. In other words, the nucleus $X$ will be able to activate itself (for fission) only if activation energy is received.

Mere possession of extra rest mass energy ($Q$ value) cannot cause it to break. This explains stability of high $A$ nuclei which has a $+ve$ $Q$ value. So Bohr–Wheeler's theory explains why high mass number nuclei are not spontaneously fissile.

This activation energy $E_A$ (that is needed for attaining the irreversible deformation —i.e. to move from spherical undeformed shape to the deformed shape of a dumbbell) is dependent on mass number $A$.

For instance for $A \sim 240$, $E_A \sim 5.5\ MeV - 6.5\ MeV$.

## 5.12 Explanation of nuclear fission by the liquid drop model

Nucleus behaves as a liquid drop. Assume that nucleus is a spherical or nearly spherical drop (initially). It is uniformly charged. It is incompressible, i.e. corresponds to constant volume, mass, density. It has a well defined surface.

The shape of the nuclear drop is governed by two forces.

   ✓ Cohesive short-range nuclear force which is similar to surface tension force that tries to make the nuclear drop spherical.
   ✓ Electrostatic repulsive forces between protons within the nucleus that tends to distort the spherical shape.

Normally the disruptive repulsive electrostatic force is overshadowed by a 100 times stronger binding nuclear strong force and so the spherical shape is nearly retained.

The spontaneous fission (which is breaking of the nucleus $X$ into two lobes $Y$ and $Z$) can be thought of as a journey of the nucleus from a spherical (or nearly spherical) unbroken state to two fragments or lobes. We can describe the transition in terms of a deformation parameter $r$, which we define as follows.

$r$ = separation between the centres of the lobes.

Increase of deformation of nucleus is characterized by deformation parameter $r$.

   □ The idea is to start with a spherical or nearly spherical nucleus $X$ (of constant volume).

No deformation is there. If we consider it to be made up of two lobes it follows that the two lobes have the same centre $r = 0$ and occupy the same volume. This is shown in figure 5.7(a).

   □ Due to excess rest mass energy the spherical (or nearly spherical) shape gets deformed to an ellipsoidal shape (of the same constant volume). This spontaneously occurs to achieve stability. The deformation parameter shows up as $r \neq 0$. This is shown in figure 5.7(b).
   □ The ellipsoid narrows and a sharp kink (or twist) is developed. This corresponds to more deformation and $r$ increases. This is shown in figure 5.7(c).
   □ The nucleus is irreversibly deformed to a dumbbell shape as if two spheres or lobes are held together, i.e. attached at a point. The excitation energy spreads and makes the nucleus drop to oscillate—there are alternate elongations and

contractions and the kink or fold becomes prominent with clear sign of impending disintegration into two lobes. This is shown in figure 5.7(d).

☐ The nucleus finally splits into two spherical fission fragments $Y$ and $Z$. Nuclear fission occurs. This is shown in figure 5.7(e).

☐ The fission fragments (positively charged nuclei $Y$ and $Z$) move away from each other due to Coulomb repulsion. This is shown in figure 5.7(f).

• Explanation of nuclear fission of $_{92}U^{235}$

The fissile nucleus $_{92}U^{235}$ is normally maintained in equilibrium under the combined action of two forces.

One force is the short-range nuclear force which is of attractive nature and tends to maintain the shape, i.e. spherical symmetry of the nucleus. The other force is the Coulomb repulsion among the protons of nucleus that tries to disrupt the nucleus and tries to distort the shape of the nucleus.

$_{92}U^{235}$ can be assumed to be like a liquid drop—actually a drop of nuclear fluid having a spherically symmetric shape. It is struck by a slow thermal neutron projectile. This is shown as stage $a$ in figure 5.8.

The neutron $_0n^1$ is captured by $_{92}U^{235}$ and a compound nucleus $_{92}U^{236}$ is produced. Energy is imparted to $_{92}U^{235}$ by $_0n^1$. So $_{92}U^{236}$ is in excited state and it starts oscillating. The force of surface tension tries to retain spherical shape while the excitation energy induced oscillation distorts the shape of $_{92}U^{236}$ and forces it to assume an ellipsoidal shape. This is shown as stage $b$ of figure 5.8.

The vigorous oscillation of nucleus due to the excitation energy it absorbed from the neutron soon distorts the ellipsoid further and a kink develops, as shown in stage $c$ of figure 5.8.

The kink deepens and two lobes or fragments are formed. Each lobe contains positive charge. The two positive charges repel each other and the lobes get extended. We have a dumbbell shaped nucleus as shown in stage $d$ of figure 5.8. It appears that the two fragments are joined through a tenuous link. It is clear that the nuclear attractive force still succeeds in overcoming the Coulomb repulsion and holds onto the two fragments as parts of the same nucleus.

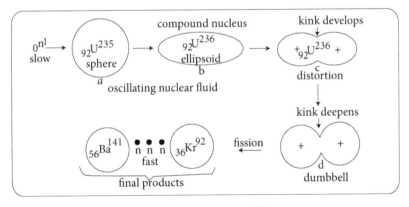

**Figure 5.8.** Fission of $_{92}U^{235}$.

As the Coulomb repulsion exceeds the nuclear attraction the two lobes get separated into $_{56}Ba^{141}$ and $_{36}Kr^{92}$ along with the ejection of three fast neutrons $_0n^1$.

This was how Bohr and Wheeler explained qualitatively nuclear fission of $_{92}U^{235}$ by thermal neutrons.

## 5.13 Explanation of nuclear fission by the theory of Bohr and Wheeler

To explain nuclear fission theoretically the nucleus is treated to be initially spherical, made up of incompressible nuclear fluid having a well defined surface and constant volume.

On account of incompressibility the vibrations set up due to absorption of neutron are confined to the surface of the drop—they cannot penetrate deep within.

The deformation produced in the nucleus may be of any type. For mathematical simplicity it is assumed that the drop maintains axial symmetry in the lowest mode of excitation. The axis of symmetry is chosen as the polar axis $\hat{z}$ and spherical polar coordinate system $r$, $\theta$, $\phi$ used.

Let $R = R_0A^{1/3}$ be the radius of undeformed spherical nucleus drop, $A$ being mass number and let $r$ be the radial coordinate of the distorted surface of the nucleus.

For no deformation

$$r = R$$

The deformed nucleus can be described by expressing $r$ as an expansion or a series in terms of Legendre poynomials $P_l(\cos\theta)$ and the expansion coefficients are $\alpha_l$ that serve as deformation parameters.

$$r = R[1 + \sum_{l=0}^{\infty}\alpha_l P_l(\cos\theta)] \tag{5.50}$$

We mention that though deformed the volume should be constant and the coefficient $\alpha_0$ should be chosen accordingly.

We choose $\alpha_0 = 0$.

We have assumed that the centre of mass of the drop remans unchanged despite deformation. The coefficient of $P_1(\cos\theta)$ viz. $\alpha_1$ gives the distance through which centre of the spherical drop moves and for pure deformation without any translation we must have

$$\alpha_1 = 0.$$

Hence from equation (5.50) we have the expanded form excluding the $l = 1$ term.

$$r = R[1 + \alpha_2 P_2(\cos\theta) + \alpha_3 P_3(\cos\theta) + \ldots] = r(\theta) \tag{5.51}$$

Using the expression of Legendre polynomials

$$P_2(\cos\theta) = \frac{1}{2}(3\cos^2\theta - 1), \quad P_3(\cos\theta) = \frac{1}{2}(5\cos^3\theta - 3\cos\theta)$$

etc, in equation (5.51) we have

$$r = R[1 + \alpha_2 \frac{1}{2}(3\cos^2\theta - 1) + \alpha_3 \frac{1}{2}(5\cos^3\theta - 3\cos\theta) + ...] = r(\theta) \qquad (5.52)$$

Surface energy of the liquid drop is related to the surface tension $S$ of the liquid as

$$\text{Surface energy} = S.\text{area} \qquad (5.53)$$

For the undeformed liquid drop the surface energy is given by

$$E_{s0} = S. 4\pi R^2 \qquad (5.54)$$

$$E_{s0} = S. 4\pi(R_0 A^{1/3})^2 = S4\pi R_0^2 A^{2/3} = a_s A^{2/3} \qquad (5.55)$$

where $a_s = 17.8 \, MeV$ is surface energy constant as used in semi-empirical mass formula.

For the deformed liquid drop the surface energy is given from equations (5.53) and (5.52)

$$E_s = S. 4\pi r^2$$

$$= S. 4\pi R^2 \left[1 + \alpha_2 \frac{1}{2}(3\cos^2\theta - 1) + \alpha_3 \frac{1}{2}(5\cos^3\theta - 3\cos\theta) + ...\right]^2 \qquad (5.56)$$

Using equation (5.54) and on calculation we have

$$E_s = E_{s0}\left(1 + \frac{2}{5}\alpha_2^2 + \frac{5}{7}\alpha_3^2 + ...\right) \qquad (5.57)$$

The Coulomb energy for an undeformed liquid drop is given by

$$E_{C0} = \frac{1}{4\pi\varepsilon_0}\frac{3}{5}\frac{(Z|e|)^2}{R} \qquad (5.58)$$

$$E_{C0} = \frac{1}{4\pi\varepsilon_0}\frac{3}{5}\frac{(Z|e|)^2}{R_0 A^{1/3}} = a_C \frac{Z^2}{A^{1/3}} \qquad (5.59)$$

where $a_C = 0.71 \, MeV$ is Coulomb energy constant as used in the semi-empirical mass formula.

$Z$ is the number of protons, $Z|e|$ is the total positive charge of the nucleus.

The Coulomb energy for the deformed liquid drop is given by, upon using equation (5.52)

$$E_C = \frac{1}{4\pi\varepsilon_0}\frac{3}{5}\frac{(Z|e|)^2}{r}$$

$$= \frac{1}{4\pi\varepsilon_0}\frac{3}{5}\frac{(Z|e|)^2}{R}\left[1 + \alpha_2 \frac{1}{2}(3\cos^2\theta - 1) + \alpha_3 \frac{1}{2}(5\cos^3\theta - 3\cos\theta)...\right]^{-1} \qquad (5.60)$$

Using equation (5.58) and on calculation

$$E_C = E_{C0}\left(1 - \frac{1}{5}\alpha_2^2 - \frac{10}{49}\alpha_3^2 - .....\right)$$ (5.61)

The energy change due to deformation of the nucleus is given by

$$\Delta E = E_{\text{initial}} - E_{\text{final}} = (E_s + E_C) - (E_{s0} + E_{C0})$$

From equations (5.57) and (5.61) we have

$$\Delta E = E_{s0}\left(1 + \frac{2}{5}\alpha_2^2 + \frac{5}{7}\alpha_3^2 + ...\right) + E_{C0}\left(1 - \frac{1}{5}\alpha_2^2 - \frac{10}{49}\alpha_3^2 - .....\right) - (E_{s0} + E_{C0})$$

$$\Delta E = E_{s0}\left(\frac{2}{5}\alpha_2^2 + \frac{5}{7}\alpha_3^2 + ...\right) + E_{C0}\left(-\frac{1}{5}\alpha_2^2 - \frac{10}{49}\alpha_3^2 - .....\right)$$ (5.62)

We consider small distortions. So we neglect higher order deformation parameters. Keeping only $\alpha_2^2$ we have

$$\Delta E = E_{s0}\frac{2}{5}\alpha_2^2 - E_{C0}\frac{1}{5}\alpha_2^2$$

$$\Delta E = \frac{1}{5}\alpha_2^2(2E_{s0} - E_{C0})$$ (5.63)

The surface energy term $E_{s0}$ occurs with a $+ve$ sign while Coulomb energy term $E_{C0}$ occurs with a $-ve$ sign. So sign of $\Delta E$ depends on relative strengths of these terms.

$$\checkmark \ \Delta E = +ve \ \text{ if } \ 2E_{s0} > E_{C0}$$ (5.64)

then the drop, i.e., the nucleus is stable to small distortions and there is no spontaneous fission. (Surface energy term dominates.)

$$\checkmark \ \Delta E = -ve \ \text{ if } \ 2E_{s0} < E_{C0}$$ (5.65)

then the drop i.e., nucleus is unstable to small distortions and spontaneous fission may occur. (Coulomb energy term dominates.)

$$\Delta E = 0 \ \text{ if } \ 2E_{s0} = E_{C0}$$ (5.66)

This is the critical condition.

Putting the expresions from equations (5.55) and (5.59) in equation (5.66) we have

$$2a_s A^{2/3} = a_C \frac{Z^2}{A^{1/3}}$$

$$\frac{2a_S}{a_C} = \frac{Z^2}{A}$$ (5.67)

Putting the values of $a_s = 17.8 \, MeV$ and $a_C = 0.71 \, MeV$ the critical value is

$$\frac{Z^2}{A} = \frac{2(17.8 \, MeV)}{0.71 \, MeV}$$

$$\frac{Z^2}{A} \equiv \left(\frac{Z^2}{A}\right)_{cr} = 50 \qquad (5.68)$$

The critical value that decides whether spontaneous fission would occur for a nucleus of mass number $A$ or not is obtained.

Let us define a quantity called fission parameter $\chi$ as

$$\chi = \frac{Z^2/A}{(Z^2/A)_{cr}} = \frac{Z^2/A}{50} \qquad (5.69)$$

$\chi$ decides if a nucleus is stable or unstable against spontaneous fission. When

$$\chi < 1 \text{ i.e. } \frac{Z^2}{A} < 50 \text{ i.e. } 2E_{s0} > E_{C0} \qquad (5.70)$$

then the nucleus is stable against spontaneous fission.

$$\chi > 1 \text{ i.e. } \frac{Z^2}{A} > 50 \text{ i.e. } 2E_{s0} < E_{C0} \qquad (5.71)$$

then the nucleus is unstable against spontaneous fission.

Suppose as a rough estimate we take $Z = \frac{41}{100}A$ (i.e. for 41 of the 100 nucleons are protons). Then the condition for spontaneous fission equation (5.71) viz. $\frac{Z^2}{A} > 50$ becomes

$$\left(\frac{41}{100}A\right)^2 \frac{1}{A} > 50$$

$$A > 297 \qquad (5.72)$$

Roughly any $A > 300$ nuclei will suffer spontaneous fission (i.e. fission by itself with no external excitation) just after their formation and so we will not find them to stay in nature.

- Let us check if $_{92}U_{143}^{235}$ is stable against spontaneous instantaneous fission.

  Nuclei with $\frac{Z^2}{A} > 50$ are unstable and suffer spontaneous instantaneous fission.

  For $_{92}U_{143}^{235}$

$$\frac{Z^2}{A} = \frac{92^2}{235} = 36 < 50 \qquad (5.73)$$

and hence $_{92}U_{143}^{235}$ will not suffer spontaneous instantaneous fission. In fact, the condition of stability against spontaneous fission is $\frac{Z^2}{A} < 50$.

- Let us check the stability w.r.t. spontaneous fission of the following elements $Z = 117$, $A = 270$ and $Z = 117$, $A = 280$.

For $Z = 117$, $A = 270$, $\frac{Z^2}{A} = \frac{117^2}{270} = 50.7 > 50 \Leftarrow$ unstable

For $Z = 117$, $A = 280$, $\frac{Z^2}{A} = \frac{117^2}{280} = 49 < 50 \Leftarrow$ stable

- Consider a heavy nucleus $_ZX^A$ satisfying the condition $\frac{Z^2}{A} < 50$ which means it is stable and will not suffer spontaneous fission due to its incapability to surmount the fission barrier $E_0$. It lacks the activation energy $E_A = E_0 - Q$ necessary for the purpose. In other words it cannot fission by itself.

We can trigger or induce fission in the system $(_ZX_N^A)$ by pumping requisite energy (greater than the fission barrier energy).

## 5.14 Fission of $_{92}U_{143}^{235}$ by slow neutron

It was shown in equation (5.73) that $_{92}U_{143}^{235}$ cannot fission spontaneously (as $\frac{Z^2}{A} = 36 < 50$).

A slow thermal neutron (carrying energy of $0.026\ eV$, *excerise 5.4*) can act as a trigger to cause fission of $_{92}U_{143}^{235}$ as we show now.

Consider capture of a slow thermal neutron by a $_{92}U_{143}^{235}$ nucleus.

$$_0n^1 + {}_{92}U_{143}^{235} \rightarrow {}_{92}U_{144}^{236*} \tag{5.74}$$

Bohr and Wheeler calculated the activation energy for fission of $_{92}U_{144}^{236}$ to be $E_A = 5.7\ MeV$ as shown in figure 5.9(a) in the form of a fission barrier $abc$ of height $5.7\ MeV$.

If energy imparted by slow thermal neutron can make $_{92}U_{144}^{236}$ cross the fission barrier $abc$ then fission will take place.

The $Q$ value of the interaction of equation (5.74) is obtained by putting the masses of reactants and products as follows

$$Q = (m_{\text{initial}} - m_{\text{final}})c^2 = m(_0n^1) + m(_{92}U_{143}^{235}) - m(_{92}U_{144}^{236*})$$

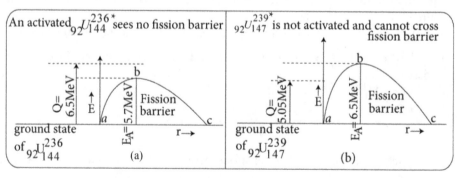

**Figure 5.9.** (a) Fission activation energy of $_{92}U_{143}^{235}$ is 5.7 *MeV*. (b) Fission activation energy of $_{92}U_{146}^{238}$ is 6.5 *MeV*.

$$= 1.008\,665u + 235.113\,92u - 236.115\,59u$$

$$= 0.006\,995u$$

$$= 0.006\,995 \times 931\ MeV$$

$$Q = 6.5\ MeV \tag{5.75}$$

When a slow/thermal neutron is captured or absorbed by $_{92}U^{235}_{143}$ an excited spherical compound nucleus $_{92}U^{236}_{144}$ is formed. We can take the neutron to have zero kinetic energy and the target $_{92}U^{235}_{143}$ to be at rest. So the compound nucleus $_{92}U^{236}_{144}{}^{*}$also has zero linear momentum, zero kinetic energy. Thus all the energy received upon absorption of the neutron goes to its internal excitation. As the excitation energy $6.5\ MeV$ exceeds the required fission activation energy of $5.7\ MeV$ the nucleus $_{92}U^{236}_{144}{}^{*}$ will find itself above the fission barrier and fission will occur.

In other words due to excitation $_{92}U^{236}_{144}{}^{*}$ moves uphill (figure 5.9(a)) from point $a$ to $b$ and then falls downhill along $bc$ and fission occurs. Hence slow thermal neutron can induce fission in $_{92}U^{235}_{143}$.

✓ If it is a non-fission capture then fission barrier is not crossed and $_{92}U^{236}_{144}{}^{*}$ may shed off the extra energy by emitting a $\gamma$ i.e. suffer radiative decay.

## 5.15 Fission of $_{92}U^{238}_{146}$

$_{92}U^{238}_{146}$ cannot fission spontaneously since

$$\frac{Z^2}{A} = \frac{92^2}{238} = 36 < 50 \tag{5.76}$$

- Let us study whether a slow thermal neutron (having energy $\sim 0.026\ eV$, *excerise 5.4*) can induce fission in $_{92}U^{238}_{146}$.

Let us study capture of thermal neutron by $_{92}U^{238}_{146}$

$$_{0}n^1 + {}_{92}U^{238}_{146} \rightarrow {}_{92}U^{239}_{147}{}^{*} \tag{5.77}$$

Bohr and Wheeler calculated the activation energy for fission of $_{92}U^{239}_{147}$ to be $E_A = 6.5\ MeV$ as shown in figure 5.9(b) in the form of a fission barrier $abc$ of height $6.5\ MeV$.

Let us check if energy imparted by slow thermal neutron can make $_{92}U^{239}_{147}$ cross the fission barrier $abc$.

The $Q$ value of the interaction of equation (5.77) is obtained by putting the masses of reactants and products as follows

$$Q = (m_{\text{initial}} - m_{\text{final}})c^2 = m(_0n^1) + m(_{92}U^{238}_{146}) - m(_{92}U^{239}_{147}{}^{*})$$

$$= 1.008\,665u + 238.050\,786u - 239.054\,0293u$$

$$= 0.005\,4217u$$

$$= 0.005\,4217 \times 931 \; MeV$$

$$= 5.05 \; MeV \tag{5.78}$$

This $Q$ is the reduction in rest mass energy and it is less than $E_A = 6.5 \; MeV$, which is the height of the fission barrier as shown in figure 5.9(b).

Clearly excitation energy 5.05 $MeV$ supplied by a slow thermal neutron cannot make $_{92}U_{147}^{239}{}^{*}$ to cross the fission barrier $abc$ of height 6.5 $MeV$. So slow neutrons cannot induce fission in $_{92}U_{147}^{239}{}^{*}$. In other words, a thermal neutron cannot drive $_{92}U_{147}^{239}$ uphill from point $a$ to point $b$ and therefore no question of it running downhill along the curve $bc$. Hence slow thermal neutron cannot induce fission in $_{92}U_{146}^{238}$.

✓ Fast energetic neutrons may help $_{92}U_{146}^{238}$ cross the fission barrier and cause fission.

- Let us make a gist.

| Nucleus | Reaction | Excitation energy received by thermal slow neutron | Activation energy needed to overcome fission barrier | Does fission occur |
|---|---|---|---|---|
| $_{92}U_{143}^{235}$ | $_0n^1 + {}_{92}U_{143}^{235} \rightarrow {}_{92}U_{144}^{236}{}^{*}$ | 6.5 $MeV$ | 5.7 $MeV$ | Yes |
| $_{92}U_{146}^{238}$ | $_0n^1 + {}_{92}U_{146}^{238} \rightarrow {}_{92}U_{147}^{239}{}^{*}$ | 5.05 $MeV$ | 6.5 $MeV$ | No |

## 5.16 Classification of neutrons

According to energy $E$ of neutron we can make the following classification.
- Neutrons for which $0 < E < 10^3 \; eV$ are slow neutrons. They fall in various ranges.

✓ Neutrons with $E < 0.002 \; eV$ are cold neutrons. They can penetrate crystalline and poly crystalline materials.
✓ Neutrons with $E = 0.026 \; eV$ are thermal neutrons (*excerise 5.4*).

These neutrons are thermalized meaning thereby that they are in thermal equilibrium with the medium. Their velocity distribution is Maxwellian in character.

Materials rich in hydrogen (i.e. proton rich materials like paraffin) are very efficient in slowing neutrons and thermalizing them. The process of slowing down of neutrons is referred to as moderation. Paraffin, water, heavy water are examples of moderator (*excerise 4.19*).

    ✓ Neutrons with $E \geqslant 0.5 \, eV$ are called epithermal neutrons.

    ✓ Neutrons with $1 \, eV < E < 100 \, eV$ are called resonance neutrons. They produce sharp resonances in heavy nuclei.

- $1 \, keV < E < 500 \, keV$ are called intermediate neutrons.
- $0.5 \, MeV < E < 10 \, MeV$ are called fast neutrons.
- $10 \, MeV < E < 50 \, MeV$ are called very fast neutrons.
- $E > 50 \, MeV$ are called ultra fast neutrons.
- $E > 10^4 \, MeV$ are called relativistic neutrons.

## 5.17 Interaction with a neutron

- Fission capture of a neutron by a nucleus

In fission capture $n$ is absorbed by a nucleus followed by fission of the nucleus.

Neutrons capable of inducing fission in $_{92}U^{235}_{143}$ are of thermal energy, i.e. of very low energy. The excitation energy imparted is greater than the fission barrier and so fission is activated. Fission thus occurs and the fission cross-section is large.

The fission cross-section depends on the speed of neutrons.

☐ Dependence of fission cross-section on neutron speed. The $\frac{1}{v}$ effect.

If speed is small the neutron has greater probability to stay close to or in the vicinity of $_{92}U^{235}_{143}$ and so there is greater probability of the neutron getting captured and then fission occurs. On the other hand, a high-speed neutron spends little time in contact with $_{92}U^{235}_{143}$ and so the probability of capture and fission will be less. So if neutron velocity $v$ and hence neutron kinetic energy is raised, the fission cross-section falls. This is referred to as $\frac{1}{v}$ effect. In other words, the fission cross-section will not increase if high energy neutrons strike $_{92}U^{235}_{143}$.

Figure 5.10(a) shows a plot of fission cross-section against neutron energy.

The fission cross-section of $_{92}U^{235}_{143}$ drops from 600 barn (at thermal energy) to 1 barn (at $MeV$ energy). In other words, this means that the lower the $n$ energy the larger the fission of $_{92}U^{235}_{143}$.

    ✓ The fission cross-section for $_{92}U^{238}_{146}$ is zero at thermal energies and above 1 $MeV$, i.e. for fast neutrons fission occurs and the fission cross-section value is around 1 barn.

- Non-fission radiative capture of a neutron by nucleus

In non-fission, radiative capture neutron is absorbed by a nucleus followed by emission of a $\gamma$ photon (hence the process is called radiative) and no fission occurs.

    ✓ For $_{92}U^{235}_{143}$

        Non-fission radiative capture of thermal neutrons is very low for $_{92}U^{235}$. Thermal energy neutrons suffer radiative capture of cross-section $\sim$97 barns.

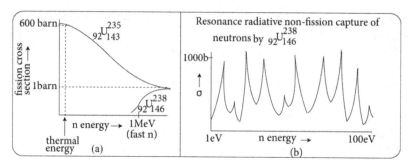

**Figure 5.10.** (a) Fission cross-section for various neutron energies. (b) Non-fission capture of neutrons.

This is very low compared to the fission capture cross-section of 600 barn by $_{92}U^{235}$ for thermal neutrons.

✓ For $_{92}U^{238}_{146}$

Neutrons of particular energy between 1 $eV$ and 100 $eV$ get absorbed—cross-section peaking up to ~1000barn followed by emission of γ photon. These are radiative resonance capture and represented by the resonance peaks shown in figure 5.10(b).

Clearly
- For $_{92}U^{235}_{143}$ cross-sections are

✓ 600 barn for fission capture of thermal $n$.
✓ 1 barn for fission capture of fast 1 $MeV$ $n$.
✓ 97 barn for radiative capture of thermal neutrons.

- For $_{92}U^{238}_{146}$ cross-sections are

✓ 0 barn for fission capture of thermal $n$.
✓ Less than 1 barn for fission capture of fast 1 $MeV$ $n$.
✓ 1000 barn peak for radiative capture of 1 $eV$–100 $eV$ $n$.

- Elastic scattering of neutrons by the nucleus
   Neutrons may be elastically scattered by nucleus. Kinetic energy will be conserved in the process. The scattering cross-section is ~1 to 10 barn.
- Inelastic scattering of neutrons by the nucleus
   Neutrons may be scattered by the nucleus inelastically—some energy goes into the nucleus as internal excitation. Threshold neutron energy for inelastic scattering is 14 $keV$ for $_{92}U^{235}_{143}$ and 44 $keV$ for $_{92}U^{238}_{146}$ and the scattering cross-section is ~1 to 10 barns.

## 5.18 Nuclear reactor based on nuclear fission

A nuclear reactor is a device to initiate and control a fission chain reaction.

- First nuclear reactor

  The first nuclear reactor is known as Chicago pile 1 and was created in 1942 at the university of Chicago by a group of scientists led by Fermi.

- A schematic diagram of a nuclear fission reactor is shown in figure 5.11. The components are

  ✓ Fuel rods. These are nuclear fuel, i.e. uranium enriched in $_{92}U^{235}$.

  ✓ Moderator

  ✓ Control rods (Cd rods)

  ✓ Reflector

  ✓ Coolant

Natural uranium has three major isotopes (all $\alpha$ radioactive). Let us list them.

| Isotope name | $_{92}U^{238}_{146}$ | $_{92}U^{235}_{143}$ | $_{92}U^{234}_{142}$ |
|---|---|---|---|
| Natural abundance | 99.27% | 0.72% | 0.01% |
| Half-life | $4.5 \times 10^9$ yr | $7 \times 10^8$ yr | $2.5 \times 10^5$ yr |
| Fission (by thermal $n$) | Non-fissile (c.s =0 b) | Fissile (c.s =600 barn) | |
| Fission (by fast $n$) | ~1 barn (low c.s) | ~1 barn | |

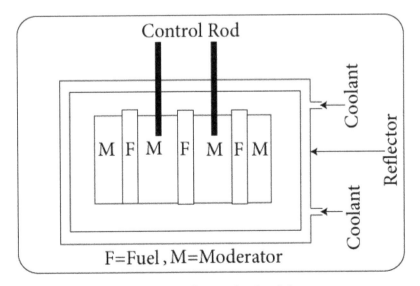

**Figure 5.11.** Schematic diagram of nuclear fission reactor.

☐ Enriched uranium is used as fuel

As majority of natural uranium is the non-fissile $_{92}U^{238}_{146}$ the only way to achieve fission process is to increase the percentage of the fissile $_{92}U^{235}_{143}$. This is called enrichment of uranium. Enrichment of $_{92}U^{235}_{143}$ of about 3% to 4% is done.

Low enriched uranium (LEU) 0.72% is the primary fuel for nuclear reactors.

High enriched uranium (HEU) greater than 20% is used primarily in weapons

☐ Moderator

Water ($H_2O$), graphite (carbon), heavy water ($D_2O$), beryllium are all used as moderators.

Fission cross-section of $_{92}U^{235}_{143}$ is large (about 600 barn) for thermalized neutron, and very small (around 1 barn) for fast 1 $MeV$ neutron. So we have to rely on a thermal neutron for carrying out fission of $_{92}U^{235}_{143}$.

The fission process generates, on the average 2.5 fast neutrons per fission having energy 1 $MeV$ to 2 $MeV$ and they are thermalized by moderators. These thermalized neutrons cause further fission of $_{92}U^{235}_{143}$.

Moderators are materials that thermalize the fast neutrons produced in fission. After one head-on elastic collision with a moderator nucleus of mass $M$, a fast neutron of mass $m$ transfers energy of amount

$$\delta = \frac{4mM}{(m + M)^2} \text{ [excerise 4.19, equation (4.198)].}$$

Clearly for $m = M$ (i.e. neutron mass $\cong$ moderator nucleus mass) a head-on elastic collision of neutron with a moderator nucleus will cause maximum (100%) energy transfer and neutron will be quickly thermalized.

So water (that contains hydrogen, i.e. $_1H^1_0$ or proton having same mass as that of neutron) can be used as moderator.

The drawback of using water or hydrogen-rich materials as moderator is that in the process of ( ~ 100%) energy transfer from $n$ to $p$, deuterons are produced ($n$ gets stuck to $p$ to form $d = {_1H^2_1}$)

$$n + {_1H^1_0} \rightarrow {_1H^2_1}.$$

So neutrons are lost, i.e. removed as a result of absorption by moderator which is not a desirable feature.

- Moderation is needed outside the fuel rods.

    As neutron energy reduces from MeV (for fast $n$) to 0.026 $eV$ (for thermalized $n$) a large number of neutrons might be having energy ~1 $eV$ to 100 $eV$. Now $_{92}U^{238}_{146}$ of the fuel rod has large (1000 barn peak) non-fission radiative capture cross-section for these neutrons. So if moderation occurs in the fuel rod a significant number of slow neutrons will be lost. Design of nuclear fission reactor is such that the fast neutron, upon generation, should quickly come out of the fuel rod for moderation outside the fuel rod.

- Moderator characteristics:

    A moderator will have good performance if it satisfies the following criteria.

✓ It should have low *n* absorption probability or cross-section.

✓ It should be a low *Z* material that readily receives the bulk of neutron energy.

✓ It should be easily available.

✓ It should be a cost-effective material.

✓ Design of the nuclear fission reactor should be such that the fission-generated fast neutron should quickly come out of the fuel rod for moderation outside the fuel rod. After thermalization the thermalized neurons should move into the fuel rod for causing fission.

☐ Control rods

Cadmium rods are used as contol rods.

Control rods are rods made of materials that absorb neutrons with high probabilities.

✓ If the nuclear reaction goes faster than what is desired or designed, the Cd rods are inserted deeper into the reactor. Due to neutron absorption by Cd rod, the number of fission reactions are reduced.

✓ If nuclear reaction goes slower than what is desired or designed, the Cd rods are lifted up from within the reactor. Due to less neutron absorption by Cd rod, the number of fission reactions are increased.

So with Cd rods one can keep control of the nuclear fission reaction. Resonant neutron absorption by Cd is shown in figure 4.12.

☐ Reflector

Reflects back the escaping neutrons so that they remain in the system.

☐ Coolant

Fission reaction generates heat and cooling is necessary to keep the reactor core cool. Coolant goes into the reactor core, gets heated and in this way heat is taken away.

## 5.19 Uses, advantages and disadvantages of a nuclear reactor

Major uses of a nuclear reactor are:

✓ Power generation, i.e. production of electrical energy

✓ Nuclear marine propulsion

✓ Production of radio isotopes

✓ Scientific reasearch using a neutron beam

Let us mention the major advantages of a nuclear power reactor.

☐ Nuclear power plant requires less space compared to other power plants.

☐ Energy from a nuclear power plant is cost-effective.

☐ A nuclear power plant uses much less fuel than a fossil fuel plant.

Let us mention the major disadvantages of a nuclear power reactor.

☐ There is the problem of radioactive waste that should be disposed of carefully.

☐ Any radioactive leak leads to disastrous consequences to human life as well as to the environment.

☐ Maintenance cost of the plant is high.

☐ Radioactive materials associated have long half-lives and so any radioactive leak or mismanagement or errors will affect the region and the surrounding population or society for many many years.

☐ Proper safety measures are essential.

## 5.20 Fission chain reaction

Due to fission capture of a thermal neutron $_0n^1$ by $a_{92}U^{235}_{143}$ nucleus, around 2.5 prompt fast neutrons are produced. These fast neutrons are slowed down (i.e. thermalized) by moderators. The thermalized neutrons can now induce further fission. This is referred to as a chain reaction. In other words, once started, a fission process can be made to continue automatically and the energy derived from the fission process can be used.

Fission involves the following processes in competition.

- Fission capture of thermal $n$ by $_{92}U^{235}_{143}$ followed by generation of 2.5 prompt fast $n$.
- Non-fission capture of $n$ by uranium nucleus. This leads to loss of thermal neutron.
- Non-fission capture of $n$ by nuclei other than uranium, e.g. moderator nuclei. This leads to loss of thermal neutron.
- Escape of $n$ without being captured. This leads to loss of thermal neutron.

The state or status of a nuclear fission reactor is determined by the rate of fission reactions and loss of thermal neutrons, i.e. how many neutrons replace one neutron of the previous generation. This is measured by a quantity called reproduction factor or multiplication factor of nuclear fission reactor.

Figure 5.12 shows how one neutron named $n0$ generates nine second generation neutrons named $n2$ through fission of $U \equiv {}_{92}U^{235}_{143}$.

## 5.21 Neutron multiplication factor or neutron reproduction factor $k$

Neutron reproduction factor $k$ is defined as follows

$$k = \frac{\text{Number of thermal neutrons in the (n + 1)th generation}}{\text{Number of thermal neutrons in the nth generation}} = \frac{N_{th}^{(n+1)}}{N_{th}^{(n)}} \quad (5.79)$$

The value of $k$ may be $k = 1$, $k < 1$, $k > 1$. We mention the significance of having these values.

- $k = 1$ (critical condition)
    - ✓ $k = 1$ corresponds to a controlled, self sustained chain reaction.
    - ✓ This condition ($k = 1$) is maintained in a nuclear reactor.
    - ✓ Thermal neutron population is stabilized. Thermal neutron density is constant on the average (hence controllable).

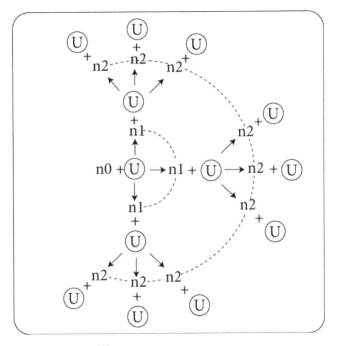

**Figure 5.12.** Chain reaction. $U \equiv {}_{92}U^{235}_{143}$. $n0$ refers to the thermal neutron that initiates fission. $n1$ is the first generation neutron. $n2$ is the second generation neutron. Clearly one neutron generates nine second generation neutrons.

  ✓ Each thermal neutron is succeeded by one thermal neutron on the average. This indicates a control of population—neither too few nor too many thermal neutrons would be there to push the situation out of control.
  ✓ Thermal neutron population of $(n + 1)th$ generation is equal to the population of $n$th generation.
  ✓ This is a sustained (controlled) chain reaction—the chain reaction is maintained and controlled.
  ✓ The system is said to have a critical size and in the steady state.
  ✓ Leads to release of energy that is designed and expected. In other words, the energy release is controlled.
• $k < 1$ (Sub-critical condition)
  ✓ $k < 1$ corresponds to a decaying chain reaction that would soon stop.
  ✓ This condition ($k < 1$) is not maintained in a nuclear reactor.
  ✓ Loss of thermal neutrons occurs so that thermal neutron density drops out.
  ✓ Each thermal neutron is succeeded by less than one thermal neutron on the average. This indicates population extinction. So the reaction would soon stop as there will soon be a situation when there will be an insufficient number of thermal neutrons to carry on the process of fission.
  ✓ Thermal neutron population of $(n + 1)th$ generation is less than the population of $n$th generation.

✓ This is a convergent chain reaction—the chain reaction becomes weaker and stops.

✓ The system is called sub-critical.

✓ Energy release diminishes and soon falls to zero.

- $k > 1$ (Super-critical condition)

    ✓ $k > 1$ corresponds to an uncontrolled chain reaction.

    ✓ This condition ($k > 1$) occurs during explosion of an atomic bomb or nuclear bomb and is not maintained in a nuclear reactor.

    ✓ Build-up of thermal neutrons occurs so that thermal neutron density increases uncontrollably.

    ✓ Each thermal neutron is succeeded by more than one thermal neutron on the average. This indicates population explosion since there will soon be too many thermal neutrons to cause fission and the situation will go out of control.

    ✓ Thermal neutron population of $(n + 1)th$ generation is more than the population of $nth$ generation.

    ✓ This is a divergent (uncontrollable) chain reaction—the chain reaction becomes uncontrollably vigorous.

    ✓ The system is called super-critical.

    ✓ It leads to enormous release of energy in a short time which is called an explosion. In other words there is uncontrolled release of energy.

## 5.22 Power output of a nuclear reactor

We calculate the power output of a nuclear reactor. Let

$m$ = number of fissile $U^{235}$ nuclei present in the reactor per unit volume $\left(\frac{1}{m^3}\right)$

$n$ = number of thermal neutrons available for fission per unit volume $\left(\frac{1}{m^3}\right)$

$v$ = average velocity of thermal neutrons $\left(\frac{m}{s}\right) \Leftarrow \frac{\text{length}}{\text{time}}$

Hence

$$nv = \frac{\text{number of thermal neutrons}}{\text{volume}} \times \frac{\text{length}}{\text{time}}$$

$$nv = \frac{\text{number of thermal neutrons}}{\text{area. time}} \left(\frac{1}{m^2 s}\right) \tag{5.80}$$

So $nv$ is neutron flux. Let

$\sigma_f$ = fission cross-section ($m^2$). So

$$(nv)m\sigma_f = \frac{\text{number of thermal neutrons}}{\text{area. time}} \times \frac{\text{number of } U^{235}}{\text{volume}} \times m^2$$

$$nvm\sigma_f = \frac{\text{number of nuclei undergoing thermal fission}}{m^3 \, s} = N \tag{5.81}$$

So in a volume $V(m^3)$ of reactor, number of nuclei undergoing fission per second is

$$NV = nvm\sigma_f V = \frac{\text{number of nuclei undergoing fission}}{s} \qquad (5.82)$$

Each fission of $_{92}U_{143}^{235}$ generates 235 $MeV$ of energy [equation (5.26)].

So $NV$ fissions will generate $(235\ MeV)(NV)$ energy per second, which is the power output.

Power output of the reactor will thus be

$$P = (235 \times 10^6 \times 1.6 \times 10^{-19}\ J)NV\frac{1}{s} = 3.76 \times 10^{-11}NV\frac{J}{s}$$

$$P = 3.76 \times 10^{-11}\ nvm_f\sigma_f V\ watt \qquad (5.83)$$

## 5.23 Four factor formula

Consider the cycle of processes that occur in a nuclear reactor.

Let at an instant there be $N_{th}$ thermal neutrons absorbed by the fuel (Uranium). This is the first generation population of thermal neutrons and we denote this as

$$N_{th}^{(1)} = N_{th}$$

We wish to evaluate the number of thermal neutrons of the second generation.

Some of $N_{th}$ thermal neutrons are fission captured by $_{92}U_{143}^{235}$. And so fission occurs. Some fast neutrons are produced. The number of fast neutrons produced is say $\eta N_{th}$.

(Others are non-fission captured producing $_{92}U_{144}^{236}$, $_{92}U_{147}^{239}$).

$\eta$ = thermal fission neutron yield.

$$\eta = \frac{\text{number of fast neutrons produced due to thermal fission}}{\text{number of thermal neutrons absorbed by fuel}}. \qquad (5.84)$$

These $\eta N_{th}$ fast neutrons will cause fission in $_{92}U_{146}^{238}$ and some more fast neutrons are generated. The number of fast neutrons due to fast fission is say $\varepsilon(\eta N_{th}) = \varepsilon\eta N_{th}$.

$\varepsilon$ = fast fission factor

$$\varepsilon = \frac{\text{number of fast neutrons produced due to fast fission}}{\text{number of fast neutrons absorbed by fuel}} \qquad (5.85)$$

These $\varepsilon\eta N_{th}$ fast neutrons are passed through moderator. Some are thermalized (escaping absorption). The number of thermalized neutrons is say $p(\varepsilon\eta N_{th}) = p\varepsilon\eta N_{th}$.

[Others are absorbed by moderator. Their number is $(1 - p)\varepsilon\eta N_{th}$

$p$ = Resonance escape probability

$$p = \frac{\text{number thermalized neutrons}}{\text{number of fast neutrons passed through moderator}} \qquad (5.86)$$

Some of the $pe\eta N_{th}$ thermalized neutrons cannot leak out and are retained. The number of thermalized neutrons that do not leak out but get retained is say $P(pe\eta N_{th}) = Ppe\eta N_{th}$.

[Others leak out. Their number is $(1 - P)pe\eta N_{th}$]

$P$ = Non-leakage factor

$$P = \frac{\text{number of thermalized neutrons that do not leak out but are retained}}{\text{number of thermalized neutrons}} \tag{5.87}$$

$(1 - P$ = leakage factor)

Some of these $Ppe\eta N_{th}$ thermalized neutrons get absorbed in the fuel and produce further thermal fission. The number of thermalized neutrons absorbed in the fuel and produce further thermal fission is, say, $(Ppe\eta N_{th}) = fPpe\eta N_{th} = N_{th}^{(2)}$.

(Others are lost and left unutilized)

$f$ = thermal utilization factor

$$f = \frac{\text{number of thermalized neutrons absorbed in the fuel that will produce further thermal fission}}{\text{number of thermalized neutrons}} \tag{5.88}$$

We thus arrive at the second generation of thermal neutrons whose population is

$$N_{th}^{(2)} = fPpe\eta N_{th}^{(1)}$$

starting from $N_{th}^{(1)}$ which is the first generation population of thermal neutrons.

The reproduction factor or multiplication factor is

$$k = \frac{N_{th}^{(2)}}{N_{th}^{(1)}} = fPpe\eta \tag{5.89}$$

This result holds for a finite reactor or medium. The various stages have been represented in figure 5.13.

- Infinitely extended medium

For an infinitely extended medium there would be no leakage. Hence we set the leakage factor $1 - P = 0$ and non-leakage factor

$$P = 1 \tag{5.90}$$

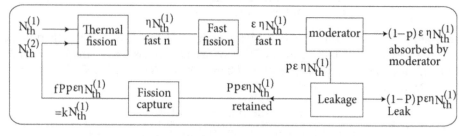

**Figure 5.13.** Schematic representation of four-factor formula.

For an infinitely extended medium let us denote the reproduction factor or multiplication factor by

$$k \equiv k_\infty.$$

We thus have for an infinitely extended medium using equations (5.89) and (5.90)

$$k_\infty = fpe\eta \tag{5.91}$$

This is called four-factor formula.

☐ Statement of four-factor formula:

The reproduction factor for a large reactor $k_\infty$ is the product of four factors

Fission neutron yield $\eta$

Fast fission factor $\varepsilon$

Resonance escape probability $p$

Thermal utilization factor $f$ i.e. $k_\infty = fpe\eta$

$$\bullet\ k = Pk_\infty (P = \text{Non leakage factor}). \tag{5.92}$$

## 5.24 Two-factor formula

Suppose the reactor contains only $_{92}U_{143}^{235}$ as fuel, i.e. no $_{92}U_{146}^{238}$ are present. This means no further fast neutrons will be produced due to fast fission of $_{92}U_{146}^{238}$. Then fast fission factor

$$\varepsilon = 1 \tag{5.93}$$

Suppose all the fast neutrons passed through the moderator are thermalized and none is absorbed, i.e. all escape absorption. Hence resonance escape probability is

$$p = 1. \tag{5.94}$$

Then from equation (5.89)

$$k = fP\eta \tag{5.95}$$

For a large reactor, equation (5.90) gives $P = 1$ and we arrive at the two-factor formula

$$k_\infty = f\eta. \tag{5.96}$$

• Reactivity

Reactivity $\rho$ of a nuclear fission reactor is defined as

$$\rho = \frac{k-1}{k} \tag{5.97}$$

Since $k = 1$ is critical condition it follows that $\rho$ is relative deviation of the multiplication factor from unity, i.e. relative deviation from criticality.

• $k < 1$, $\rho < 0$ Sub-critical

- $k = 1$, $\rho = 0$ Critical
- $k > 1$, $\rho > 0$ Super-critical

## 5.25 Time elapsed between two successive generations of thermal neutrons

Let us find the time elapsed between two successive generations of thermal neutrons.

Fission of $U^{235}$ by thermal neutrons of $nth$ generation occurs in a time $\sim 10^{-14}\,s$ (i.e. almost instantaneously).

Fission of $U^{238}$ by fast neutrons occurs in a time $\sim 10^{-8}\,s$.

The fast neutrons move into the moderator and lose energy through collisions. Time taken for thermalization, scattering and wandering within the moderator is $\sim 10^{-3}$–$10^{-4}\,s$. So after this time the thermalized neutron of $(n + 1)th$ generation go into the fuel rod.

Clearly neutrons of successive generations are separated by generation time scale $\tau \cong 10^{-3}\,s$. In other words $N$ thermal neutrons become $Nk$ thermal neutrons in a time of 1 millisecond.

If $k < 1$ then in every millisecond the number of thermal neutrons is going down by the factor $k$, reactor energy diminishes and the rate of reaction will be extremely low so that the chain reaction stops.

If $k > 1$ then in every millisecond the number of thermal neutrons will keep on increasing by the factor $k$, reactor energy enhances and in a short time the rate of reaction will be extremely high so that the chain reaction becomes uncontrollable. This will lead to explosion.

It is essential that $k$ is kept exactly at unity. In this situation one thermal neutron causes fission and such fission will generate several neutrons. But in every millisecond only one thermal neutron would be ready to initiate fission. So in every millisecond one thermal neutron is relaced by one thermal neutron. This means reaction rate is under control. It is sustained. Reaction will neither die down nor swell.

## 5.26 Reaction rate

Let us write the amount of deviation of the reproduction factor $k$ from unity, i.e. deviation from criticality needed for sustained and controlled chain reaction as $\varepsilon$, i.e.

$$k = 1 + \varepsilon \tag{5.98}$$

where $\varepsilon$ is a small quantity.

$\tau$ is time for one generation to pass.

So in time $t$ number of generations that would pass is $\frac{t}{\tau}$.

Number of thermal neutrons is increasing in each generation by the factor.

So for $\frac{t}{\tau}$ generations in time $t$ the number of thermal neutrons will be

$$k^{t/\tau} = (1 + \varepsilon)^{t/\tau} \tag{5.99}$$

If at time $t = 0$ the rate of reaction is $R(0)$ then at time $t$ the rate of reaction will be

$$R(t) = R(0)(1 + \varepsilon)^{t/\tau} \tag{5.100}$$

Taking $\log_e \equiv \ln$ of equation (5.100) we get

$$\ln R(t) = \ln R(0) + \frac{t}{\tau} \ln (1 + \varepsilon) \tag{5.101}$$

Expansion of $\ln (1 + \varepsilon)$ gives

$$\ln (1 + \varepsilon) = \varepsilon - \frac{\varepsilon^2}{2} + \frac{\varepsilon^3}{3} - \cdots$$

Using the approximation $\varepsilon \ll 1$ we have

$$\ln (1 + \varepsilon) = \varepsilon - \frac{\varepsilon^2}{2} + \frac{\varepsilon^3}{3} - \cdots \cong \varepsilon$$

and so

$$\ln R(t) \cong \ln R(0) + \frac{t}{\tau}\varepsilon \tag{5.102}$$

$$R(t) = R(0)e^{\varepsilon t/\tau} \tag{5.103}$$

Clearly rate changes exponentially.

If $\varepsilon$ is positive then rate increases, while if $\varepsilon$ is negative then the rate goes down.

Let us get an estimate.

Suppose $k = 1.01 = 1 + 0.01 = 1 + \varepsilon$

$$\varepsilon = 0.01$$

Using $\tau = 10^{-3}s$ the reaction rate in $t = 1$ s will be

$$R(1 \text{ s}) = R(0)e^{(0.01)(1 \text{ s})/10^{-3}s} = R(0)e^{10} \cong 22\,000R(0)$$

Clearly the rate of reaction increases by 22 000 in 1 s if $k$ differs from unity by 0.01. Evidently a very small increase in $k$ can lead to extremely high energy output in a short time. And if this happens it will be impossible to handle such high energy output in such a short span of time. This means that the reactor would go out of control. It follows therefore that $k$ should be maintained at unity.

- Maintaing $k = 1$

  To maintain $k = 1$ cadmium control rods are used.

  Fission occurs in fuel rods. The fuel rods are kept separated.

  In a $k > 1$ situation the control rods are pushed and inserted in the core so as to absorb more thermal neutrons. So the reaction rate goes down.

  In a $k < 1$ situation the control rods are lifted so that only a small portion of the rods is inside the core. This means thermal neutrons are less absorbed. So the reaction rate enhances.

- Movement of control rods

    We have evaluated the time scale in which one generation is replaced by the following generation and it is $\tau = 10^{-3}\,s$. Again movement of control rods is to be done to ensure that the reactor does not become sub-critical ($k < 1$) or super-critical ($k > 1$). Mechanical movement of control rods during this time scale of $10^{-3}\,s$ is not possible.

    The design of the reactor is done in such a way that the reactor is slightly sub-critical ($k < 1$) when working with prompt neutrons that are emitted in a time of $10^{-13}\,s$. But when delayed neutrons are taken care of the reactor becomes critical. In other words, the delayed neutrons are needed to make the reactor critical. And the delayed neutrons are emitted after a time of several seconds to a minute. This means we get sufficient time to control and adjust the control rods, i.e. enough time is obtained to make mechanical movement of the control rods. This is how the nuclear reactor is maintained in the critical state through mechanical adjustment and movement of control rods depending upon the delayed neutrons. This is how a sustained chain reaction can be accomplished.

- Temperature dependence of $k$

    The reproduction factor $k$ depends upon the reaction cross-section, absorption cross-section etc, all of which depend on temperature. So it is evident that $k$ depends on temperature.

    Suppose temperature increases due to choice of design, materials, geometry or any other factor. Then if

$$\frac{dk}{dT} > 0 \qquad (5.104)$$

then increase of temperature will lead to increase in $k$ which means increase in the rate of reaction. And if rate of reaction increases then temperature will further increase. The process is cumulative and leads to an uncontrolled reaction which is not desired.

    For reaction stability the condition

$$\frac{dk}{dT} < 0 \qquad (5.105)$$

is to be ensured. Choice of materials, choice of design etc should be such that if temperature increases due to some factors then $k$ should go down, i.e. rate of reaction should diminish. It is a kind of self-correction to ensure reactor stability.

- Form of output energy from a nuclear reactor

    The output of a nuclear reactor is in the form of thermal energy generated. Due to fission, kinetic energy of fission fragments is produced and neutrons emitted. All these are absorbed in the reactor core and this increases temperature of the nuclear reactor. Heat or thermal energy is generated. This is the primary output of an active nuclear reactor due to fission operation.

This energy generated is taken out and used for power or electricity generation, driving submarines, scientific research etc.

## 5.27 Typical design of a nuclear reactor

Figures 5.14 and 5.15 show a schematic diagram of the following nuclear reactors.
- ✓ Pressurised Water Reactor (PWR)
- ✓ Pressurised Heavy Water Reactor (PHWR)

☐ Water $H_2O$ or heavy water $D_2O$ acts as a coolant as well as a moderator in PWR and PHWR, respectively.

☐ The heat generated in the reactor core is taken up by water which is kept at a very high pressure of 100 *atm* so that on acceptance of heat its boiling point goes up and reaches a temperature around 250°C. This means water remains in liquid form and carries away the heat. This is how it acts as a coolant.

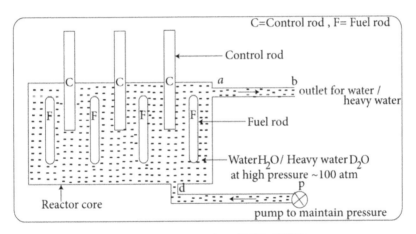

**Figure 5.14.** Typical design of PWR, PHWR.

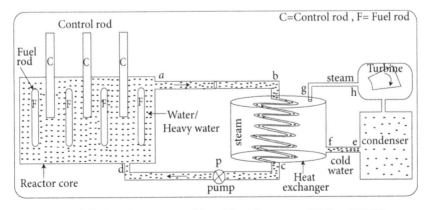

**Figure 5.15.** Running a steam turbine with nuclear reactor PWH or PHWH.

☐ Water also acts as a moderator. In fact, the fuel rods are surrounded by water moderator, as shown in figure 5.14.

Hot water at very high pressure gets expelled from the outlet end along the pipe *ab*. This heat energy can be used for steam generation to run a turbine. This we show in figure 5.15 in a schematic diagram.

The hot water from the outlet (point *b*) is made to pass through a suitably designed heat exchanger.

Within the heat exchanger, hot water (~250°C) flows through pipes at high pressure and it is recycled, i.e. enters along *cd* into the reactor core through the pump as shown. In other words, it flows along the path *cPd*.

Cold water is made to enter along the tube *ef* into the heat exchanger where it gets heated up due to the hot water flowing internally along the pipe *bc*. Since cold water at normal pressure introduced into the heat exchanger has boiling point 100°C, water starts to boil and converts into steam. This steam flows out along the path *gh* and is introduced into a turbine chamber to run a turbine. From the turbine, steam can move on to a condenser where it is converted, i.e. condensed into water. This cold water can be recycled by making it flow along path *ef* into the heat exchanger.

- The moderator has to be a low $Z$ material as then it can reduce the kinetic energy of a neutron more effectively. So light water that contains hydrogen would be preferred on this ground. Further, neutron absorption aspect is also a very important aspect. Proton or hydrogen nucleus can capture the neutron to produce deuteron—the cross-section for such interaction is significant.

So neutron absorption is significant if one uses light water. In this case to keep the reaction proceeding at a reasonable rate one has to use enriched (3% or 5%) uranium. So in this type of reactor $U^{235}$ is used as fuel.

If heavy water is used as moderator then the absorption cross-section for the neutron is small. Also, deuteron upon absorption of neutron makes tritium which is β radioactive. But with heavy water as moderator and coolant, natural uranium (without enrichment) can be used as fuel. This is a great advantage. However, heavy water is produced from light water and is costly.

## 5.28 Breeder reactor

In a breeder reactor one breeds fissile nuclei, i.e. a fertile, non-fissionable material is converted to a fissile material or fuel.

✓ For instance fissile $_{94}\text{Pu}_{145}^{239}$ is bred starting from $_{92}\text{U}_{146}^{238}$.

The reactor core is surrounded by a blanket of fertile material $_{92}\text{U}_{146}^{238}$. Resonance radiative capture of epithermal neutrons by $_{92}\text{U}_{146}^{238}$ occurs and ultimately fissile fuel $_{94}\text{Pu}_{145}^{239}$ is produced or bred. Hence the name breeder reactor.

The steps in the breeding process are

$$_0n^1 + {}_{92}\text{U}_{146}^{238} \rightarrow {}_{92}\text{U}_{147}^{239} + \gamma$$

$$_{92}U^{239}_{147} \rightarrow {}_{93}Np^{239}_{146} + {}_{-1}e^0 + \bar{\nu}_e \ \left(\beta^- \text{ decay}\right)$$

$$_{93}Np^{239}_{146} \rightarrow {}_{94}Pu^{239}_{145} + {}_{-1}e^0 + \bar{\nu}_e \ \left(\beta^- \text{decay}\right)$$

✓ Another instance is that fissile $_{92}U^{233}_{141}$ is bred starting from $_{90}Th^{232}_{142}$ and the breeding reactions are

$$_0n^1 + {}_{90}Th^{232}_{142} \rightarrow {}_{90}Th^{233}_{143} + \gamma$$

$$_{90}Th^{233}_{143} \rightarrow {}_{91}Pa^{233}_{142} + {}_{-1}e^0 + \bar{\nu}_e \ \left(\beta^- \text{ decay}\right)$$

$$_{91}Pa^{233}_{142} \rightarrow {}_{92}U^{233}_{141} + {}_{-1}e^0 + \bar{\nu}_e \ \left(\beta^- \text{ decay}\right).$$

We note some differences between breeder reactor and conventional nuclear reactor.

✓ A conventional nuclear reactor uses only $_{92}U^{235}$ as fuel which is very scarce as its amount in natural uranium is only 0.72%.

A breeder reactor uses $_{92}U^{238}_{146}$ or $_{90}Th^{232}_{142}$ as fertile material which are readily available.

✓ In breeder reactors fission occurs with fast neutrons. Hence they are called fast breeder reactors (FBRs).

In conventional nulear reactors fission by thermal neutrons is done. So they are thermal fission reactors.

✓ In breeder reactors as fast neutrons cause fission no moderation is required.

In conventional nuclear reactors fast neutrons are produced during fission by thermal neutrons. These fast neutrons do not cause fission as efficiently as slow thermal neutrons and hence have to be slowed down by the process of moderation.

✓ In breeder reactors liquid sodium is used as coolant.

In a conventional nuclear reactor water or heavy water is used as coolant (as in PWR, PHWR).

✓ A breeder reactor uses a coolant (eg. liquid sodium) that is not an efficient moderator so that fast neutrons are not thermalized. In fact, moderation is not needed in a breeder reactor.

In a conventional nuclear reactor the moderator can serve as coolant also (eg. water or heavy water).

## 5.29 Homogeneous and heterogeneous nuclear reactor

• In a homogeneous nuclear reactor the nuclear fuel is mixed with coolant and the moderator to form the same physical state. Hence it is called homogeneous.

The first aqueous homogeneous reactor was built in 1952 at Oak Ridge National Laboratory (ORNL) United States.

- In a heterogeneous nuclear reactor the fissionable material and the moderator are arranged in a geometric pattern or in a lattice-like pattern.

## 5.30  Safety measures

☐ Nuclear reactor design should be such that in the event of a failure (say power failure) the system should function in a way that will cause no harm to the surrounding population and environment.

☐ A nuclear power plant is constructed away from human habitation.

☐ Regular periodic vigilance or checks are made.

☐ Materials used for construction should be of high standard.

☐ When nuclear waste is disposed of it is ensured that it does not contaminate rivers or the sea.

☐ There should be proper foolproof safety measures.

☐ Provisions are made to shut down the plant when not required.

☐ Waste water should be purified.

☐ Solid, liquid and gaseous wastes should have their own methods of disposal.

☐ Proper arrangements should be taken in case of a natural calamity (like flood, earthquake, lightning, storm) and war.

## 5.31  Reactor accidents

A reactor accident is defined as an event that leads to disastrous consequences in and around the location of a nuclear reactor such as loss of life and loss of environmental purity.

A reactor accident occurs due to release of radioactivity from radioactive isotopes produced in the reactor, or meltdown of the reactor core. Again, such things happen due to human error, unprecedented technical faults, breakdown or malfunctioning of instruments, maintenance problems, natural calamity.

- World's first reactor accident occurred in Chalk River Laboratories, Ontario, Canada in 1952.
- Some of nuclear reactor accidents.

✓ Three mile island disaster occurred in Pennsylvania, United States in 1979. Human errors and technical flaws led to partial meltdown of reactor core, radioactive contamination and release of radioactive gas into the atmosphere.

✓ Chernobyl disaster, occurred in 1986 in Ukraine. This led to several direct deaths, numerous cancer deaths, alarmimg increase in radiation levels and radioactive contamination in the vicinity. Disastor occurred due to flawed reactor design, inadequate safety measures, sudden power surge that damaged fuel rods causing explosion and meltdown leading to release of 5% of reactor core into the atmosphere.

✓ The Fukushima disaster, occurred in 2011 in Japan. Disaster was triggered by a tsunami that flooded and made the cooling system inoperative leading to reactor meltdown and radioactive contamination due to overheating.

✓ The Marcoule nuclear disaster occurred in 2011 in France. Disaster occurred due to a blast that occurred in a furnace used to melt metallic waste.

## 5.32 Slowing down of thermal neutrons

Neutrons emitted in a nuclear fission reaction are fast and possess energy of the order of $MeV$. As neutrons are electrically neutral they will not loose energy by ionization. They slow down by successive collision in matter until they reach thermal equilibrium with the surroundings.

Let us calculate the energy loss of a neutron due to a single collision. Let us consider the collision

$$a + X \to X + a \qquad (5.106)$$

where $a$ is neutron and $X$ is target moderator nucleus. They have masses $m_a$, $m_X$ respectively.

Their velocities before collision are $u_a$, $u_X = 0$ in L-frame and $v_a$, $v_X$ in C-frame, respectively. And after collision their velocities are $U_a$, $U_X$ in L-frame and $V_a$, $V_X$ in C-frame, respectively. Define

$$\frac{m_X}{m_a} = A \qquad (5.107)$$

Let velocity of centre of mass of $a$ and $X$ be $u_{CM}$ given by equation (4.15)

$$u_{CM} = \frac{m_a u_a}{m_a + m_X} = \frac{u_a}{1 + \frac{m_X}{m_a}} = \frac{u_a}{A + 1} \qquad (5.108)$$

In C-frame the centre of mass is at rest and we have to add $-u_{CM}$ to the velocities in L-frame. In view of equations (4.16) and (4.17) we have using equation (5.107)

$$u_a - u_{CM} = U_a = \frac{m_X u_a}{m_a + m_X} = \frac{u_a}{1 + \frac{m_X}{m_a}} \frac{m_X}{m_a} = \frac{u_a A}{A + 1} \qquad (5.109)$$

$$u_X - u_{CM} = U_X = -\frac{m_a u_a}{m_a + m_X} = -\frac{u_a}{1 + \frac{m_X}{m_a}} = -\frac{u_a}{A + 1} \qquad (5.110)$$

The relation between L-frame and C-frame has been depicted in figure 4.4 through a triangle formed by $\vec{v}_a$, $\vec{V}_a$, $\vec{u}_{CM}$. From triangle law we have

$$\vec{v}_a = \vec{u}_{CM} + \vec{V}_a \qquad (5.111)$$

$$v_a^2 = u_{CM}^2 + V_a^2 + 2u_{CM} V_a \cos \theta_c \qquad (5.112)$$

In *excerise 4.21* we have shown that the speeds of particles are not affected due to collision in C-frame. Hence by equation (4.202) we have

$$U_a = V_a \qquad (5.113)$$

So we rewrite equation (5.112) as

$$v_a^2 = u_{CM}^2 + U_a^2 + 2u_{CM}U_a \cos \theta_c \tag{5.114}$$

Using equations (5.108) and (5.109) we get

$$v_a^2 = \left(\frac{u_a}{A+1}\right)^2 + \left(\frac{u_a A}{A+1}\right)^2 + 2\left(\frac{u_a}{A+1}\right)\left(\frac{u_a A}{A+1}\right)\cos \theta_c$$

$$\frac{v_a^2}{u_a^2} = \frac{1 + A^2 + 2A \cos \theta_c}{(A+1)^2} \tag{5.115}$$

The ratio of kinetic energy of the incident neutron $E_2 = \frac{1}{2}m_a v_a^2$ after collision and the kinetic energy of neutron $E_1 = \frac{1}{2}m_a u_a^2$ before collision is

$$\frac{E_2}{E_1} = \frac{\frac{1}{2}m_a v_a^2}{\frac{1}{2}m_a u_a^2} = \frac{v_a^2}{u_a^2} \tag{5.116}$$

From equation (5.115)

$$\frac{E_2}{E_1} = \frac{1 + A^2 + 2A \cos \theta_c}{(1+A)^2} \tag{5.117}$$

$$= \frac{1 + A^2 - 2A + 2A + 2A \cos \theta_c}{(A+1)^2} = \frac{(A-1)^2 + 2A(1 + \cos \theta_c)}{(A+1)^2}$$

$$\frac{E_2}{E_1} = \frac{(A-1)^2}{(A+1)^2} + \frac{2A}{(A+1)^2}(1 + \cos \theta_c) \tag{5.118}$$

Putting

$$\alpha = \frac{(A-1)^2}{(A+1)^2} \tag{5.119}$$

we rewrite equation (5.118)

$$\frac{E_2}{E_1} = \alpha + \frac{2A}{(A+1)^2}(1 + \cos \theta_c) \tag{5.120}$$

*Excerise 5.32* shows an alternate form of this formula.
    Hence fractional energy loss of neutron is

$$\frac{E_1 - E_2}{E_1} = 1 - \frac{E_2}{E_1} = 1 - \alpha - \frac{2A}{(A+1)^2}(1 + \cos \theta_c) \tag{5.121}$$

For grazing collision $\theta_c = 0$, $\cos \theta_c = 1$.

From equation (5.121)

$$\frac{E_1 - E_2}{E_1} = 1 - \alpha - \frac{2A}{(A+1)^2}(2)$$

Using equation (5.119) we have

$$\frac{E_1 - E_2}{E_1} = 1 - \frac{(A-1)^2}{(A+1)^2} - \frac{4A}{(A+1)^2} = \frac{(A+1)^2 - (A-1)^2}{(A+1)^2} - \frac{4A}{(A+1)^2} = 0 \quad (5.122)$$

This means no energy is transferred from neutron to moderator nucleus.

Also, from equation (5.120) with $\theta_c = 0$, $\cos \theta_c = 1$ we have

$$\frac{E_2}{E_1} = \alpha + \frac{4A}{(A+1)^2} = \frac{(A-1)^2}{(A+1)^2} + \frac{4A}{(A+1)^2} = \frac{(A-1)^2 + 4A}{(A+1)^2} = \frac{(A+1)^2}{(A+1)^2} = 1$$

$$E_2 = E_1 \qquad (5.123)$$

For back scattering $\theta_c = 180°$, $\cos \theta_c = -1$.

From equation (5.121)

$$\frac{E_1 - E_2}{E_1} = 1 - \alpha - \frac{2A}{(A+1)^2}(0) = 1 - \alpha \qquad (5.124)$$

Hence maximum energy is transferred from neutron to moderator.

Also, from equation (5.120) with $\theta_c = 180°$, $\cos \theta_c = -1$

$$\frac{E_2}{E_1} = \alpha = \frac{(A-1)^2}{(A+1)^2}$$

$$E_2 = \alpha E_1 \qquad (5.125)$$

since $\alpha$ depends on $A$. So the maximum loss of kinetic energy depends on the mass number of the nucleus acting as moderator.

For instance, for hydrogen ($X \equiv p$)

$$A = \frac{m_X}{m_a} = 1 \text{ [equation (5.107)]}, \alpha = \frac{(A-1)^2}{(A+1)^2} = 0 \text{ and so from equation (5.124)}$$

$$\frac{E_1 - E_2}{E_1} = 1 \Rightarrow E_1 - E_2 = E_1$$

$$E_2 = 0$$

i.e. maximum loss of kinetic energy occurs. Neutron loses all energy in a single collision.

For larger values of $A$ we have

$$\alpha = \frac{(A-1)^2}{(A+1)^2} = \frac{\left(1 - \frac{1}{A}\right)^2}{\left(1 + \frac{1}{A}\right)^2} = \left(1 - \frac{1}{A}\right)^2\left(1 + \frac{1}{A}\right)^{-2} = \left[\left(1 - \frac{1}{A}\right)\left(1 + \frac{1}{A}\right)^{-1}\right]^2$$

$$= \left[ \left( 1 - \frac{1}{A} \right)\left( 1 - \frac{1}{A} + \frac{1}{A^2} - \frac{1}{A^3} + \dots \right) \right]^2 = \left( 1 - \frac{1}{A} + \frac{1}{A^2} - \frac{1}{A^3} + \dots - \frac{1}{A} + \frac{1}{A^2} - \frac{1}{A^3} + \dots \right)^2$$

$$= \left( 1 - \frac{2}{A} + \frac{2}{A^2} - \frac{2}{A^3} + \dots \right)^2 = 1 + \frac{4}{A^2} + \dots - \frac{4}{A} + \frac{4}{A^2} + \dots$$

$$\alpha = 1 - \frac{4}{A} + \frac{8}{A^2} + \dots \tag{5.126}$$

The maximum fractional energy loss is from equation (5.124)

$$\frac{E_1 - E_2}{E_1} \Big|_{\max} = 1 - \alpha = 1 - \left( 1 - \frac{4}{A} + \frac{8}{A^2} + \dots \right) = \frac{4}{A} - \frac{8}{A^2} + \dots \cong \frac{4}{A}$$

$$(E_1 - E_2)_{\max} = \frac{4}{A} E_1 \tag{5.127}$$

For $A$ small $\frac{E_1 - E_2}{E_1} \Big|_{\max}$ is large. This means that light nuclei are more effective moderators than heavy nuclei and the maximum loss of energy $(E_1 - E_2)_{\max}$ is proportional to the initial energy $E_1$.

- For carbon moderator $A = 12$ and for a head-on collision $\theta = \pi$ equation (5.125) gives

$$\frac{E_2}{E_1} = \frac{(A-1)^2}{(A+1)^2} = \frac{(12-1)^2}{(12+1)^2} = \frac{11^2}{13^2} = 0.72$$

## 5.33 Scattering angles in L-frame and C-frame

Equation (4.22) viz. $v_a \cos \theta = u_{CM} + V_a \cos \theta_c$ inter-relates the velocities of the projectile neutron along the direction of incidence before and after collision.

Also, using equation (4.28), i.e. $U_a = V_a$ we can rewrite

$$v_a \cos \theta = u_{CM} + U_a \cos \theta_c \tag{5.128}$$

Substituting the values of $u_{CM}$, $U_a$ from equations (5.108) and (5.109) we have

$$v_a \cos \theta = \frac{u_a}{A+1} + \frac{u_a A}{A+1} \cos \theta_c$$

$$\frac{v_a}{u_a} \cos \theta = \frac{1 + A \cos \theta_c}{A + 1} \tag{5.129}$$

Again, from equation (5.115) we have

$$\frac{v_a}{u_a} = \frac{\sqrt{1 + A^2 + 2A \cos \theta_c}}{A + 1} \tag{5.130}$$

Putting equation (5.130) in equation (5.129) we get

$$\frac{\sqrt{1 + A^2 + 2A \cos \theta_c}}{A + 1} \cos \theta = \frac{1 + A \cos \theta_c}{A + 1}$$

$$\cos \theta = \frac{1 + A \cos \theta_c}{\sqrt{1 + A^2 + 2A \cos \theta_c}} \tag{5.131}$$

This is the relation between the scattering angles $\theta$ in L-frame and $\theta_c$ in C-frame.

When the collision is against a heavy nucleus $A \gg 1$ then $A^2$ is large and we rewrite equation (5.131) as

$$\cos \theta = \frac{1 + A \cos \theta_c}{\sqrt{A^2}} = \frac{1 + A \cos \theta_c}{A}$$

$$\cos \theta = \cos \theta_c + \frac{1}{A} \tag{5.132}$$

If $A \gg 1$, $\frac{1}{A} \cong 0$.

From equation (5.132) it follows that
$\cos \theta = \cos \theta_c$

$$\theta = \theta_c \tag{5.133}$$

So the scattering angles are the same in the two frames for very high $A$.

- However, it is evident from figure 4.4 that the scattering angles in L-frame are smaller than the scattering angle in C-frame.
- Neutrons show a preferential forward scattering in the L-frame although the scattering is spherically symmetrical (isotropic) in the C-frame.

## 5.34 Average logarithmic energy decrement

A moderator should have the following characteristics: large scattering cross-section, small absorption cross-section, large energy loss per collision.

To study the behavior of slowed down neutrons after many collisions it is convenient to use the logarithmic scale of energy.

Define average logarithmic energy decrement per collision as the average decrease per collision in the logarithm of neutron energy

$$\xi = <\ln \frac{E_1}{E_2}> = \frac{\int_{E_2=\alpha E_1}^{E_2=E_1} \ln \frac{E_1}{E_2} P(E_2) dE_2}{\int_{E_2=\alpha E_1}^{E_2=E_1} P(E_2) dE_2} \tag{5.134}$$

where $E_1$, $E_2$ are, respectively, the average initial and final energy of neutron.

$P(E_2)$ is the probability for the neutron to be scattered into the solid angle lying between $\theta$ and $\theta + d\theta$ with energy $E_2$. So the integration variable is $E_2$.

The limits of integration correspond to equation (5.125), i.e. $\theta_c = \pi$, $E_2 = \alpha E_1$ and equation (5.123), i.e. $\theta_c = 0$, $E_2 = E_1$.

And $P(E_2)dE_2$ is the probability that a neutron of energy $E_2$ is scattered within solid angle

$$d\Omega = \int_0^{2\pi} d\phi \int \sin\theta_c d\theta_c = 2\pi \sin\theta_c d\theta_c. \tag{5.135}$$

Hence

$$P(E_2)dE_2 = \frac{d\Omega}{4\pi} = \frac{2\pi \sin\theta_c d\theta_c}{4\pi} = \frac{1}{2}\sin\theta_c d\theta_c \tag{5.136}$$

In *excerise 5.32* we have shown that

$$\frac{E_2}{E_1} = \frac{1}{2}[1 + \alpha + (1-\alpha)\cos\theta_c]. \tag{5.137}$$

This gives

$$dE_2 = -\frac{E_1}{2}(1-\alpha)\sin\theta_c d\theta_c$$

$$\sin\theta_c d\theta_c = -\frac{2dE_2}{E_1(1-\alpha)}$$

From equation (5.136)

$$P(E_2)dE_2 = \frac{1}{2}\left(-\frac{2dE_2}{E_1(1-\alpha)}\right) = -\frac{dE_2}{E_1(1-\alpha)} \rightarrow \frac{dE_2}{E_1(1-\alpha)} = \frac{1}{1-\alpha}d\left(\frac{E_2}{E_1}\right) \tag{5.138}$$

as $E_1$ is constant and $dE_2$ is negative.

Again, the sum of the probabilities over all values of energy is unity, i.e.

$$\int_{E_2=\alpha E_1}^{E_2=E_1} P(E_2)dE_2 = 1 \text{ (total probability)} \tag{5.139}$$

It follows from equations (5.134) and (5.139) that

$$\xi = \int_{E_2=\alpha E_1}^{E_2=E_1} \ln\frac{E_1}{E_2} P(E_2)dE_2$$

Using equation (5.138) we get

$$\xi = -\frac{1}{1-\alpha}\int_{E_2=\alpha E_1}^{E_2=E_1} \ln\frac{E_2}{E_1}d\left(\frac{E_2}{E_1}\right) \tag{5.140}$$

Putting $x = \frac{E_2}{E_1}$ and using $\int \ln x\, dx = x\ln x - x$ we have

$$\xi = -\frac{1}{1-\alpha}\left[\left(\frac{E_2}{E_1}\ln\frac{E_2}{E_1} - \frac{E_2}{E_1}\right)\Big|_{E_2=E_1} - \left(\frac{E_2}{E_1}\ln\frac{E_2}{E_1} - \frac{E_2}{E_1}\right)\Big|_{E_2=\alpha E_1}\right]$$

$$= -\frac{1}{1-\alpha}[0 - 1 - (\alpha \ln \alpha - \alpha)] = \frac{1 - \alpha + \alpha \ln \alpha}{1 - \alpha}$$

$$\xi = 1 + \frac{\alpha}{1 - \alpha} \ln \alpha \qquad (5.141)$$

Putting $A$ from equation (5.119) viz. $\alpha = \frac{(A-1)^2}{(A+1)^2}$ we have

$$\xi = 1 + \frac{\frac{(A-1)^2}{(A+1)^2}}{1 - \frac{(A-1)^2}{(A+1)^2}} \ln \frac{(A-1)^2}{(A+1)^2} = 1 + \frac{(A-1)^2}{2A} \ln \frac{A-1}{A+1} \qquad (5.142)$$

✓ Clearly fractional energy loss due to scattering does not depend on the initial energy of the neutron but depends only on the mass number ratio $A = \frac{m_X}{m_a}$ [equation (5.107)].

✓ For a hydrogen target (i.e. $X \equiv p$ and $a$ is neutron)

$$A = \frac{m_X}{m_a} = 1, \alpha = 0$$

Equation (5.142) gives

$$\xi = <\ln \frac{E_1}{E_2}> = 1 \Rightarrow \frac{E_1}{E_2} = e \qquad (5.143)$$

$$E_2 = \frac{1}{e}E_1 \qquad (5.144)$$

Average energy of a scatterered neutron is $\frac{1}{e}$ times the initial energy of the neutron.

✓ For $A > 10$ an alternate approximate handy expression is used viz.

$$\xi \cong \frac{2}{A + \frac{2}{3}} \qquad (5.145)$$

This shows that for heavy nuclei $A$ is large

$$\xi = 0 \qquad (5.146)$$

This means that there is practically no loss of energy due to collision of a neutron with heavy nuclei.

Thus we conclude that only light nuclei like hydrogen, heavy water, carbon etc, can serve as good moderators.

$$✓ \ \xi = <\ln \frac{E_1}{E_2}> \Rightarrow <\frac{E_1}{E_2}> = e^\xi$$

$$< \frac{E_2}{E_1} > = e^{-\xi} \tag{5.147}$$

$$1 - < \frac{E_2}{E_1} > = 1 - e^{-\xi}$$

$$< \frac{E_1 - E_2}{E_1} > = 1 - e^{-\xi} \tag{5.148}$$

This is the fractional energy loss for neutrons which have made one collision with a moderator nucleus.

- For carbon $\xi = 0.158$

$$< \frac{E_1 - E_2}{E_1} > = 1 - e^{-\xi} = 1 - e^{-0.158} = 0.1462.$$

This is the fractional energy loss for neutrons which have made one collision with carbon nuclei.

✓ If neutrons start out with average energy $E_1$ and are slowed down to average energy $E_2$ then the total logarithmic energy loss is $\ln \frac{E_1}{E_2}$. The average number of collisions is given by

$$n = \frac{\text{logarithmic interval of average energy}}{\text{average logarithmic decrement per collision}} = \frac{\ln E_1 - \ln E_2}{\xi} = \frac{1}{\xi} \ln \frac{E_1}{E_2} \tag{5.149}$$

☐ Suppose a neutron slows down from 2 $MeV$ to 0.026 $eV$ (thermal energy). The average number of collisions will be

$$n = \frac{1}{\xi} \ln \frac{E_1}{E_2} = \frac{1}{\xi} \ln \frac{2\ MeV}{0.026\ eV} = \frac{18.2}{\xi}$$

☐ For carbon $\xi = 0.158$

$$n = \frac{18.2}{\xi} = \frac{18.2}{0.158} = 115$$

Now from equation (5.149) we write

$$\frac{E_1}{E_2} = e^{n\xi} \Rightarrow \frac{E_2}{E_1} = e^{-n\xi}$$

$$E_2 = E_1 e^{-n\xi} \tag{5.150}$$

☐ Suppose neutron makes 115 collisions with carbon moderator nuclei starting with initial energy of 2 $MeV$. The final energy of neutrons will be

$$E_2 = E_1 e^{-n\xi} = 2\ MeV\ e^{-(115)(0.158)} = 0.026\ eV$$

- Let us calculate the average energy loss of the scattered neutron after one collision. This is given by, using equation (5.139)

$$<E_1 - E_2> = \frac{\int_{E_2=\alpha E_1}^{E_2=E_1} (E_1 - E_2)P(E_2)dE_2}{\int_{E_2=\alpha E_1}^{E_2=E_1} P(E_2)dE_2} = \int_{E_2=\alpha E_1}^{E_2=E_1} (E_1 - E_2)P(E_2)dE_2 \quad (5.151)$$

Using equation (5.138) viz. $P(E_2)dE_2 = \frac{dE_2}{E_1(1-\alpha)}$

$$<E_1 - E_2> = \int_{E_2=\alpha E_1}^{E_2=E_1} (E_1 - E_2)\frac{dE_2}{E_1(1-\alpha)} = \frac{1}{E_1(1-\alpha)} \int_{E_2=\alpha E_1}^{E_2=E_1} (E_1 - E_2)dE_2$$

$$= -\frac{1}{E_1(1-\alpha)} \int_{E_2=\alpha E_1}^{E_2=E_1} (E_1 - E_2)d(E_1 - E_2) = -\frac{1}{E_1(1-\alpha)}\frac{1}{2}(E_1 - E_2)^2 \Big|_{E_2=\alpha E_1}^{E_2=E_1}$$

$$= -\frac{1}{E_1(1-\alpha)}\frac{1}{2}[(E_1 - E_1)^2 - (E_1 - \alpha E_1)^2]$$

$$= -\frac{1}{E_1(1-\alpha)}\frac{1}{2}[-E_1^2(1-\alpha)^2]$$

$$<E_1 - E_2> = \frac{1}{2}E_1(1-\alpha) \quad (5.152)$$

So average fractional energy loss is

$$<\frac{E_1 - E_2}{E_1}> = \frac{1}{2}(1-\alpha) \quad (5.153)$$

It follows that the same fraction of neutron energy is transferred to the moderator nucleus in each successive collision.

## 5.35 Slowing down power (sdp)

The ability of a moderator to slow down neutrons is represented by a quantity called slowing down power (sdp). Slowing down power (sdp) is the product of the following three quantities:
- ✓ Average logarithmic energy decrement per collision $\xi$;
- ✓ Number of scattering centres (atoms) per unit volume $N_a$;
- ✓ Scattering cross-section $\sigma_s$.

Hence

$$sdp = \xi N_a \sigma_s \quad (5.154)$$

For efficient slowing down we require large $\xi$, large $\sigma_s$, large $N_a$ and small absorption cross-section, otherwise too many neutrons would be lost due to absorption.

Define

$$N_a\sigma_s = \Sigma_s \tag{5.155}$$

as the mascroscopic cross-section for scattering and

$$\Sigma_s = \frac{1}{\lambda_s} \tag{5.156}$$

where $\lambda_s$ is the scattering mean free path. So

$$sdp = \xi N_a\sigma_s = \xi\Sigma_s = \frac{\xi}{\lambda_s} \tag{5.157}$$

Slowing down power (sdp) is thus interpreted as the average loss in the logarithm of the energy per unit distance of travel in the moderator.

So a large $\xi$ and a large $\Sigma_s$ make a good moderator.

A good moderator should have a large sdp and a small absorption cross-section.

## 5.36 Moderating ratio

An overall criterion that decides the effectiveness of a material as a moderator is the ratio of sdp and macroscopic absorption crosss-section $\Sigma_a$

$$\text{Moderating ratio} = \frac{sdp}{\Sigma_a} \tag{5.158}$$

As $\Sigma_a = \frac{1}{\lambda_a}$ where $\lambda_a$ is the absorption mean free path we have using equation (5.157)

$$\text{Moderating ratio} = \frac{sdp}{\Sigma_a} = sdp\lambda_a = \frac{\xi}{\lambda_s}\lambda_a = \xi\frac{\lambda_a}{\lambda_s} = \xi\frac{\Sigma_s}{\Sigma_a} \tag{5.159}$$

Heavy water has moderating ratio of 12 000, graphite has moderating ratio of 170, water has moderating ratio of 72.

## 5.37 Slowing down density

The rate at which neutrons slow down at a particular energy $E$ per unit volume of a moderator is called slowing down density at that energy.

Let us denote slowing down density as $q(E)$.

The process of slowing down of neutrons is a discontinuous process since neutron energy gets reduced in discrete steps after each collision with moderator nuclei.

For heavy moderator nuclei, the average loss of energy at each collision is small. Hence for heavy moderator nuclei the slowing down process may be treated as continuous. This simplifies the problems greatly.

Suppose $Q$ is the source density. In other words, $Q$ is the rate of production of neutrons per unit volume per second and the neutrons produced have initial high energy $E_0 =$ constant. Also, suppose that no neutrons are lost either by leakage, i.e. escape or by absorption before being thermalized. Under this condition the slowing down density $q(E)$ and the source density $Q$ can be taken to be equal, i.e.

$$Q = q(E) \tag{5.160}$$

This is the case of an ideal moderator.

Unit of $Q$, $q \rightarrow \dfrac{1}{m^3 s}$

## 5.38 Neutron current density, microscopic and macroscopic cross-section, mean free path

If neutron density varies in a given volume, neutrons move from region of higher density to region of lower density. This process is called neutron diffusion. When neutron density becomes uniform the net flow of neutrons stops.

The net number of neutrons travelling per second through a unit area normal to the direction of flow is called neutron current density $\vec{J}$.

If $n$ is the neutron density, $v$ is neutron velocity then $nv = \phi$ is the number of neutrons per unit area per second, called neutron flux.

□ Fick's law

Neutron current density $J$ and space rate of change of neutron flux $\frac{d\phi}{dx}$ (considering movement along $x$) are proportional to each other, i.e.

$$J \propto -\frac{d\phi}{dx}.$$

The negative sign signifies that neutron current is in the direction of decreasing neutron density. Hence

$$J = -D\frac{d\phi}{dx} \tag{5.161}$$

where $D$ is called diffusion coefficient. The diffusion coefficient is a characteristic of the medium which relates the movement of neutrons to the spatial distribution of neutron density $n$. In three dimensions the relation becomes

$$\vec{J} = -D\vec{\nabla}\phi \tag{5.162}$$

✓ Unit of $J = \dfrac{number}{m^2 \, s}$

Unit of $\phi = nv = \dfrac{number \ of \ neutrons}{volume} \cdot \dfrac{length}{time} = \dfrac{number}{m^2 \, s}$

From $J = -D\frac{d\phi}{dx} \Rightarrow \dfrac{number}{m^2 \, s} = D\dfrac{number}{m^2 \, s} \cdot \dfrac{1}{m}$

Clearly the unit of $D$ is that of length, i.e. $m$.

- Microscopic cross-section

  We can refer to the probability of a particular nuclear reaction by specifying the target area presented by an individual atom to the incoming neutron. This area is the microscopic cross-section $\sigma$ and is measured in $m^2$ or *barn*.

- Macroscopic cross-section

    We can refer to the probability of a particular nuclear reaction by specifying the total target area per unit volume presented to the incoming neutron. This area per unit volume is the macroscopic cross-section $\Sigma$ and is measured in $\frac{m^2}{m^3} = \frac{1}{m}$.

    The total target area in unit volume is just the number of atoms per unit volume say $N$ times the cross-section $\sigma$. Hence

$$\Sigma = N\sigma = \frac{\rho N_A}{A}\sigma \tag{5.163}$$

where $N_A$ is Avogadro's number, $\rho$ is density, i.e. mass per unit volume, $A$ is atomic weight. The unit of $\Sigma = N\sigma \rightarrow \frac{\text{number}}{m^3} m^2 \sim \frac{1}{m}$

- Mean free path $\lambda$

    Mean free path $\lambda$ designates the average distance travelled by a neutron before undergoing a particular reaction.

    $\lambda$ is related to $\Sigma$ as

$$\lambda = \frac{1}{\Sigma} \text{ (excerise 5.33)} \tag{5.164}$$

    If the macroscopic cross-section is large this means amount of interaction is large and this means mean free path is small so that interaction occurs more.

- Scattering mean free path $\lambda_s$

    Scattering mean free path $\lambda_s$ is the average distance travelled between scattering collisions.

    It is related to the macroscopic scattering cross-section as $\lambda_s = \frac{1}{\Sigma_s}$.

- Transport mean free path $\lambda_{tr}$

If neutrons are preferentially scattered in the forward direction as seen in the L-frame, the neutrons will travel further than $\lambda_s$ before being deflected. This greater distance is called transport mean free path $\lambda_{tr}$. In fact, forward scattering results in an enhanced migration of neutrons from their point of appearance in the medium. Their free paths are effectively longer and to describe them $\lambda_{tr}$ is used instead of $\lambda_s$.

We can define the transport mean free path $\lambda_{tr}$ as the average distance travelled by a neutron in its original direction after a large number (infinite number) of scattering collisions. This is explained in figure 5.16.

Suppose a neutron eneters a medium at point $p$ and travels a distance $pq = \lambda_s$ freely without any collision or scattering. It suffers a scattering encounter at point $q$ and gets scattered, i.e. deflected by an angle $\bar{\theta}$ which is the average angle of scattering of neutron. The neutron travels along this deflected direction $qr = \lambda_s$. Its projection along the initial direction is $q'r' = \lambda_s < \cos\theta > = \lambda_s\bar{\mu}$ where $\bar{\mu} = <\cos\theta>$. It again gets deflected at point $r$ by the average angle $\bar{\theta}$, travels an amount $rm = \lambda_s$ and an amount $r'm' = \lambda_s\bar{\mu} < \cos\theta > = \lambda_s\bar{\mu}^2$ along the initial direction and so on. We do our calculation w.r.t. the initial direction of motion. Hence

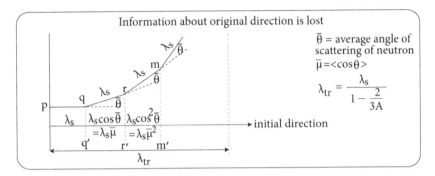

**Figure 5.16.** Scattering mean free path and transport mean free path.

$$\lambda_{tr} = \lambda_s + \lambda_s\bar{\mu} + \lambda_s\bar{\mu}^2 + \ldots\infty = \lambda_s(1 + \bar{\mu} + \bar{\mu}^2 + \ldots\infty)$$

$$\lambda_{tr} = \lambda_s\frac{1}{1 - \bar{\mu}} = \frac{\lambda_s}{1 - <\cos\theta>} \tag{5.165}$$

$\lambda_{tr}$ and $\lambda_s$ are related through the average cosine of the scattering angle $\theta$. Since

$$<\cos\theta> = \frac{2}{3A} \text{ (can be shown)} \tag{5.166}$$

$$\lambda_{tr} = \frac{\lambda_s}{1 - <\cos\theta>} = \frac{\lambda_s}{1 - \frac{2}{3A}} \tag{5.167}$$

This is the relation between scattering angle of neutron in L-frame and mass number of target nucleus.

For hydrogen $A = 1$, $<\cos\theta> = \frac{2}{3}$ meaning that the neutron tends to be scattered in the forward direction.

For a uranium nucleus which is heavy, $A$ is large $<\cos\theta> \simeq 0$ meaning that scattering is almost equally likely in all directions, i.e. isotropic.

In fact, in L-frame the forward scattering is predominant in the collision with lighter nuclei. But for large $A$ the average forward component becomes very small and scattering is almost isotropic.

We note that $\lambda_{tr}$ is larger than $\lambda_s$.

✓ For carbon $A = 12$, $\lambda_s = 2.5 \; cm$, $\lambda_{tr} = \frac{\lambda_s}{1 - \frac{2}{3A}} = \frac{2.5 \; cm}{1 - \frac{2}{3(12)}} = 2.7 \; cm$

• $D$ is related to $\lambda_{tr}$ as

$$D = \frac{\lambda_{tr}}{3} \tag{5.168}$$

• Absorption mean free path $\lambda_a$

The average distance travelled by the neutron before it gets absorbed is called absorption mean free path $\lambda_a$ and it is related to the macroscopic absorption cross-section as $\lambda_a = \frac{1}{\Sigma_a}$.

## 5.39 Diffusion equation for monoenergetic neutrons

We consider neutron diffusion inside a nuclear reactor. Study of spatial distribution of neutrons during slowing down process is required in calculations relating to the determination of critical size of a reactor.

Inside a nuclear reactor the number of neutrons is very large and the neutron assembly inside the reactor can be treated as a neutron gas. So we can apply the results of kinetic theory of gas to describe neutron diffusion.

The equation that governs the movement of monoenergetic thermal neutron in a medium is

$$\begin{pmatrix} \text{Rate of change} \\ \text{of neutron density} \end{pmatrix} = \begin{pmatrix} \text{Production} \\ \text{rate} \end{pmatrix} - \begin{pmatrix} \text{absorption} \\ \text{rate} \end{pmatrix} - \begin{pmatrix} \text{Leakage} \\ \text{rate} \end{pmatrix} \quad (5.169)$$

$$\int_V \frac{\partial n}{\partial t} dV = \int_V Q dV - \int_V \Sigma_a \phi dV - \int_V \vec{\nabla} \cdot \vec{J} dV \quad \text{(each term has unit} \rightarrow \frac{1}{s}\text{)} \quad (5.170)$$

In view of equation (5.162) viz. $\vec{J} = -D\vec{\nabla}\phi$ we write the third integral on the RHS of equation (5.170)

$$\int_V \vec{\nabla} \cdot \vec{J} \, dV = \int_V \vec{\nabla} \cdot (-D\vec{\nabla}\phi) \, dV = -D \int_V \nabla^2 \phi \, dV$$

$$= -\frac{\lambda_{tr}}{3} \int_V \nabla^2 \phi dV \quad \text{[using equation (5.168)]} \quad (5.171)$$

Hence from equation (5.170)

$$\int_V \frac{\partial n}{\partial t} dV = \int_V Q dV - \int_V \Sigma_a \phi dV + \frac{\lambda_{tr}}{3} \int_V \nabla^2 \phi dV \quad (5.172)$$

Removing the volume integral

$$\frac{\partial n}{\partial t} = Q - \Sigma_a \phi + \frac{\lambda_{tr}}{3} \nabla^2 \phi \quad (5.173)$$

This is the general form of diffusion equation for monoenergetic neutrons.

If the reactor is operating at a steady state then

$$\frac{\partial n}{\partial t} = 0$$

i.e. neutron flux and power will be constant. Then
production = leakage + absorption.

Hence

$$Q - \Sigma_a\phi + \frac{\lambda_{tr}}{3}\nabla^2\phi = 0 \qquad (5.174)$$

Boundary conditions imposed are
- ✓ Neutron flux is finite and non-negative in the region where diffusion takes place. This condition does not hold near to the source. (Neutron flux is infinite at a point source of neutrons.)
- ✓ The neutron flux is continuous.
- ✓ The neutron current component normal to the boundary between two media is continuous.

Production of thermal neutrons is mainly due to slowing down of fast neutrons to thermal energies so that by equation (5.160) $Q = q$ (the slowing down density). Also, $\Sigma_a = \frac{1}{\lambda_a}$. Hence from equation (5.174)

$$q - \frac{\phi}{\lambda_a} + \frac{\lambda_{tr}}{3}\nabla^2\phi = 0$$

$$\nabla^2\phi - \frac{3\phi}{\lambda_a\lambda_{tr}} + \frac{3q}{\lambda_{tr}} = 0 \qquad (5.175)$$

- Case of a point source

Production of neutrons takes place only at the location of the point source. The neutron production in all regions that excludes this point source is zero. Hence in the space excluding the point source we put $q = 0$. The differential equation therefore reduces to (for all points excluding the source point)

$$\nabla^2\phi - \frac{3\phi}{\lambda_a\lambda_{tr}} = 0 \qquad (5.176)$$

✓ Thermal diffusion length

Thermal diffusion length $L$ is a measure of the average direct flight distance travelled by a thermal neutron before capture. In other words it is a measure of the average distance a thermal neutron travels from the point of its formation to the point where it is absorbed. Let us define

$$\frac{1}{L} = \sqrt{\frac{3}{\lambda_a\lambda_{tr}}} \text{ i.e.}$$

$$L = \sqrt{\frac{\lambda_a\lambda_{tr}}{3}} = \text{Thermal diffusion length (unit} \rightarrow m) \qquad (5.177)$$

$$\nabla^2\phi - \frac{1}{L^2}\phi = 0 \qquad (5.178)$$

Point source corresponds to spherical symmetry and we thus take $\phi = \phi(r)$, $r$ being the radial coordinate. So we use spherical polar coordinate system that gives

$$\frac{1}{r^2}\frac{d}{dr}\left(r^2\frac{d\phi}{dr}\right) - \frac{1}{L^2}\phi = 0 \tag{5.179}$$

Substitute

$$r\phi = R$$

$$rd\phi + \phi dr = dR \Rightarrow r\frac{d\phi}{dr} + \phi = \frac{dR}{dr}$$

$$r^2\frac{d\phi}{dr} + r\phi = r\frac{dR}{dr} \Rightarrow r^2\frac{d\phi}{dr} + R = r\frac{dR}{dr}$$

$$r^2\frac{d\phi}{dr} = r\frac{dR}{dr} - R$$

From equation (5.179)

$$\frac{1}{r^2}\frac{d}{dr}\left(r\frac{dR}{dr} - R\right) - \frac{1}{L^2}\phi = 0 \Rightarrow \frac{1}{r^2}\left(r\frac{d^2R}{dr^2} + \frac{dR}{dr} - \frac{dR}{dr}\right) - \frac{1}{L^2}\phi = 0$$

$$\frac{1}{r}\frac{d^2R}{dr^2} - \frac{1}{L^2}\phi = 0 \Rightarrow \frac{d^2R}{dr^2} - \frac{1}{L^2}r\phi = 0$$

$$\frac{d^2R}{dr^2} - \frac{1}{L^2}R = 0 \tag{5.180}$$

General solution of this equation is

$$R = Ae^{r/L} + Be^{-r/L} \tag{5.181}$$

$$\phi = \frac{R}{r} = \frac{A}{r}e^{r/L} + \frac{B}{r}e^{-r/L} \tag{5.182}$$

At $r = \infty$, the term $\frac{A}{r}e^{r/L} \sim e^{\infty}$ makes $\phi$ diverge. To prevent divergence we set $A = 0$. Hence the solution in equation (5.182) becomes

$$\phi = \frac{B}{r}e^{-r/L} \tag{5.183}$$

We have to evaluate the constant $B$.

The medium is infinitely extended and so there is no leakage. Hence it follows that

Total production rate or source rate = total neutron absorption rate for the medium

Now total neutron absorption rate for the medium is $\int_0^\infty \Sigma_a\phi dV$

The total production rate or source rate is $\int_0^\infty QdV = Q_0$. Hence

$$Q_0 = \int_0^\infty \Sigma_a \phi \, dV = \int_0^\infty \Sigma_a \phi 4\pi r^2 dr$$

$$= \int_0^\infty \Sigma_a \frac{B}{r} e^{-r/L} 4\pi r^2 dr = 4\pi B \Sigma_a \int_0^\infty e^{-r/L} \, r \, dr$$

$$= 4\pi B \Sigma_a [ r \int e^{-r/L} dr \, |_0^\infty - \int_0^\infty \frac{d}{dr} r \left( \int e^{-r/L} dr \right) dr ] \text{ (partially integrating)}$$

$$= 4\pi B \Sigma_a [ 0 - \int_0^\infty \frac{e^{-r/L}}{-1/L} \, dr ]$$

$$= 4\pi B \sum_a L \int_0^\infty e^{-r/L} dr = 4\pi B \sum_a L \frac{e^{-r/L}}{-1/L} \, |_0^\infty$$

$$Q_0 = 4\pi B \sum_a L^2 \tag{5.184}$$

$$B = \frac{Q_0}{4\pi \sum_a L^2} = \frac{Q_0}{4\pi \sum_a} \frac{3}{\lambda_a \lambda_{tr}} \text{ [using equation (5.177)]}$$

$$B = \frac{3Q_0}{4\pi \lambda_{tr}} \left( \text{as } \sum_a = 1/\lambda_a \right) \tag{5.185}$$

So the solution in equation (5.183) becomes

$$\phi = \frac{B}{r} e^{-r/L} = \frac{3Q_0}{4\pi r \lambda_{tr}} e^{-r/L} \tag{5.186}$$

This is the solution of the diffusion equation.

## 5.40 Diffusion of fast neutrons and Fermi age equation

Consider diffusion of neutrons during slowing down stage. We investigate neutron balance for a moderator in which there is no neutron absorption and so $\Sigma_a = 0$, $\lambda_a = \infty$ and no production of neutrons, i.e. $Q = 0$. Neutrons diffuse through the moderator before getting thermalized.

The rate at which neutron density increases with time at a given point is equal to the net rate at which neutrons diffuse to that point. This also follows from the neutron balance equation (5.173)

$$\frac{\partial n}{\partial t} = Q - \Sigma_a \phi + \frac{\lambda_{tr}}{3} \nabla^2 \phi$$

With $\Sigma_a = 0$, $Q = 0$

$$\frac{\partial n}{\partial t} = \frac{\lambda_{tr}}{3} \nabla^2 \phi \tag{5.187}$$

Since $\phi = nv$, $\frac{\partial \phi}{\partial t} = v \frac{\partial n}{\partial t} \Rightarrow \frac{\partial n}{\partial t} = \frac{1}{v} \frac{\partial \phi}{\partial t}$.

With this equation (5.187) becomes

$$\frac{1}{v}\frac{\partial \phi}{\partial t} = \frac{\lambda_{tr}}{3}\nabla^2 \phi \qquad (5.188)$$

The rate at which neutrons slow down at a particular energy $E$ per unit volume of a moderator, i.e. the slowing down density at that energy $E$ is

$$q(E) = \begin{pmatrix} \text{Number of neutrons per unit volume} \\ \text{with energy between } E \text{ and } E + dE \end{pmatrix}\begin{pmatrix} \text{number of} \\ \text{collisions per second} \end{pmatrix}$$

$$= n(E)dE.\,\frac{v}{\lambda_s} \qquad (5.189)$$

Since

$$\phi(E) = n(E)v \qquad (5.190)$$

$$q(E) = \frac{\phi(E)dE}{\lambda_s} \qquad (5.191)$$

If a neutron of energy $E$ loses energy of amount $dE$ per collision then we identify

$$\frac{dE}{E} = \xi \Rightarrow dE = \xi E \qquad (5.192)$$

$$q(E) = \frac{\phi(E)\xi E}{\lambda_s} = \phi(E)\xi E\, \Sigma_s \left( \text{as } \frac{1}{\lambda_s} = \Sigma_s \right)$$

$$\phi(E) = \frac{q(E)}{\xi E \Sigma_s} \qquad (5.193)$$

$$\nabla^2 \phi(E) = \frac{1}{\xi E \Sigma_s}\nabla^2 q(E) \qquad (5.194)$$

$$\frac{\partial}{\partial t}\phi(E) = \frac{1}{\xi E \Sigma_s}\frac{\partial}{\partial t}q(E) \qquad (5.195)$$

Equation (5.188) with equations (5.194) and (5.195) becomes

$$\frac{1}{v}\frac{1}{\xi E \Sigma_s}\frac{\partial}{\partial t}q(E) = \frac{\lambda_{tr}}{3}\frac{1}{\xi E \Sigma_s}\nabla^2 q(E)$$

$$\frac{1}{v}\frac{\partial}{\partial t}q(E) = \frac{\lambda_{tr}}{3}\nabla^2 q(E)$$

$$\frac{1}{v}\frac{\partial q(E)}{\partial E}\frac{dE}{dt} = \frac{\lambda_{tr}}{3}\nabla^2 q(E) \qquad (5.196)$$

Let us take into account the number of collisions in time $dt$ which is

$$\text{(number of collisions per second)}dt = \frac{v}{\lambda_s}dt \qquad (5.197)$$

Equation (5.192) thus modifies to (associating negative sign since it is decrease in energy)

$$-\frac{dE}{E} = \xi \frac{v}{\lambda_s} dt$$

$$-\frac{dE}{dt} = \frac{v e \xi}{\lambda_s} \tag{5.198}$$

Equation (5.196) thus becomes

$$\frac{1}{v} \frac{\partial q(E)}{\partial E} \left(-\frac{v e \xi}{\lambda_s}\right) = \frac{\lambda_{tr}}{3} \nabla^2 q(E)$$

$$\nabla^2 q = -\frac{3 E \xi}{\lambda_s \lambda_{tr}} \frac{\partial q}{\partial E} \tag{5.199}$$

Introduce a new variable $\tau$ defined as

$$d\tau = -\frac{\lambda_s \lambda_{tr}}{3 E \xi} dE \tag{5.200}$$

Then equation (5.199) becomes

$$\nabla^2 q = \frac{\partial q}{\partial \tau} \Rightarrow \nabla^2 q - \frac{\partial q}{\partial \tau} = 0 \tag{5.201}$$

This equation is known as the Fermi age equation. The variable $\tau$ is called Fermi age or neutron age.

✓ Dimension of $\tau$ is $m^2$ (not of time) [following from equation (5.200)]
✓ The Fermi age equation does not contain time explicitly and hence it is a steady state equation.
✓ From equation (5.200) it follows that

$$\tau = \int_{E_1}^{E_2} d\tau = \int_{E_1}^{E_2} -\frac{\lambda_s \lambda_{tr}}{3 E \xi} dE = -\frac{\lambda_s \lambda_{tr}}{3 \xi} \int_{E_1}^{E_2} \frac{dE}{E}$$

$$= -\frac{\lambda_s \lambda_{tr}}{3 \xi} \ln \frac{E_2}{E_1}$$

$$\tau = \frac{\lambda_s \lambda_{tr}}{3 \xi} \ln \frac{E_1}{E_2} \tag{5.202}$$

✓ Since $\frac{1}{\xi} \ln \frac{E_1}{E_2} = n_0$ (call it so) represents the average number of collisions suffered by a neutron with moderator nuclei to change energy from $E_1$ to $E_2$ [equation (5.149)] we can write

$$\tau = \frac{n_0 \lambda_s \lambda_{tr}}{3} = \frac{\lambda_{tr} \Lambda_s}{3} \tag{5.203}$$

where $\Lambda_s = n_0\lambda_s$ represents the total zigzag path length of a neutron starting from the beginning of slowing down and attaining thermal energy.

The quantity $\sqrt{\frac{\lambda_{tr}\Lambda_s}{3}} = L_f$ is called fast diffusion length.

## 5.41 Critical size of nuclear reactor

When a nuclear reactor operates such that there is exact balance between neutron production and neutron losses due to their absorption and leakage the reactor is said to operate under critical condition. And the size of the nuclear reactor is said to be of critical size. The equation that expresses the relationship between the geometrical properties of the reactor and its material properties is called the critical equation.

Take the source of thermal neutron to be represented by the slowing down density $q$ to be a function of position $\vec{r}$ and neutron age $\tau$ i.e.

$$q = q(\vec{r}, \tau) \qquad (5.204)$$

Fermi age is given by equation (5.201) viz. $\nabla^2 q = \frac{\partial q}{\partial \tau}$

Apply the method of separation of variables

$$q = R((\vec{r})T(\tau) \qquad (5.205)$$

Putting in the Fermi age equation we get

$$\nabla^2 RT = \frac{\partial}{\partial \tau}RT$$

$$T\nabla^2 R = R\frac{dT}{d\tau}$$

$$\frac{1}{R}\nabla^2 R = \frac{1}{T}\frac{dT}{d\tau} = -B^2 \qquad (5.206)$$

LHS is a function of $r$ only, RHS is a function of $\tau$ only. This is possible if each side is a constant called separation constant $-B^2$.

Spatial equation is

$$\frac{1}{R}\nabla^2 R = -B^2 \Rightarrow \nabla^2 R + B^2 R = 0 \qquad (5.207)$$

The $\tau$ equation is

$$\frac{1}{T}\frac{dT}{d\tau} = -B^2 \Rightarrow \frac{dT}{T} = -B^2 d\tau \Rightarrow dlnT = -B^2 d\tau$$

$$\int dlnT = -B^2 \int d\tau + C$$

$$ln\,T = -B^2\tau + C$$

$$\text{At } \tau = 0, \ T = T_0, \ C = \ln T_0$$

$$\ln T = -B^2\tau + \ln T_0 \Rightarrow \ln \frac{T}{T_0} = -B^2\tau$$

$$T = T_0 e^{-B^2\tau} \tag{5.208}$$

From equations (5.204) and (5.205) we write

$$\tau = 0, \ q = q_0 = RT_0 \tag{5.209}$$

Rate of production of a fission neutron in a reactor core per thermal neutron absorbed is given by the four-factor formula equation (5.91) $k_\infty = p\varepsilon\eta f$.

Number of thermal neutrons absorbed per unit area per second is $\phi\Sigma_a$.

Hence the rate at which thermal neutrons are produced per unit area is

$$p\varepsilon\eta f. \ \phi\Sigma_a = k_\infty\phi\Sigma_a$$

This is also the rate at which fast neutrons become available for slowing down. So it is the initial slowing down density $q_0$. Hence

$$q_0 = k_\infty\phi\Sigma_a = RT_0 \ [\text{from equation (5.209)}]$$

Now from equations (5.205), (5.208), and (5.209)

$$q = R(\vec{r})T(\tau) = RT_0 e^{-B^2\tau} = q_0 e^{-B^2\tau}$$

$$q = k_\infty\phi\Sigma_a e^{-B^2\tau} = p\varepsilon\eta f\phi\Sigma_a e^{-B^2\tau} \tag{5.210}$$

This is the solution of the Fermi age equation.

For a reactor operating at steady power the neutron balance equation is equation (5.175) viz.

$\nabla^2\phi - \dfrac{3\phi}{\lambda_a\lambda_{tr}} + \dfrac{3q}{\lambda_{tr}} = 0$ which becomes

$$\nabla^2\phi - \frac{3\phi}{\lambda_a\lambda_{tr}} + \frac{3k_\infty\phi\Sigma_a e^{-B^2\tau}}{\lambda_{tr}} = 0$$

$$\nabla^2\phi - \frac{3\phi}{\lambda_a\lambda_{tr}} + \frac{3k_\infty\phi e^{-B^2\tau}}{\lambda_a\lambda_{tr}} = 0 \ \left(\text{as } \Sigma_a = \frac{1}{\lambda_a}\right)$$

$$\nabla^2\phi + \frac{3}{\lambda_a\lambda_{tr}}\frac{k_\infty\phi e^{-B^2\tau} - \phi}{\lambda_a\lambda_{tr}} = 0$$

Using thermal diffusion length $L = \sqrt{\dfrac{\lambda_a\lambda_{tr}}{3}}$ we have

$$\nabla^2\phi + \frac{1}{L^2}(k_\infty e^{-B^2\tau} - 1)\phi = 0 \tag{5.211}$$

This is the neutron balance equation for a thermal reactor.

From equation (5.210) $q = k_\infty \phi \Sigma_a e^{-B^2 \tau}$

Let us calculate

$$\nabla^2 q = k_\infty \Sigma_a e^{-B^2 \tau} \nabla^2 \phi$$

and

$$\frac{\partial q}{\partial \tau} = k_\infty \phi \Sigma_a \frac{\partial}{\partial \tau} e^{-B^2 \tau} = k_\infty \phi \Sigma_a (-B^2) e^{-B^2 \tau}$$

With this the Fermi age equation (5.201) viz. $\nabla^2 q - \frac{\partial q}{\partial \tau} = 0$ can be written as

$$k_\infty \Sigma_a e^{-B^2 \tau} \nabla^2 \phi - k_\infty \phi \Sigma_a (-B^2) e^{-B^2 \tau} = 0$$

$$\nabla^2 \phi + B^2 \phi = 0 \qquad (5.212)$$

Comparison of equation (5.212) with (5.211) gives

$$B^2 = \frac{1}{L^2}(k_\infty e^{-B^2 \tau} - 1) \Rightarrow L^2 B^2 = k_\infty e^{-B^2 \tau} - 1$$

$$1 + L^2 B^2 = k_\infty e^{-B^2 \tau}$$

$$\frac{k_\infty e^{-B^2 \tau}}{1 + L^2 B^2} = 1 \Rightarrow \frac{1 + L^2 B^2}{k_\infty} = e^{-B^2 \tau} \qquad (5.213)$$

This equation is known as the critical equation.

✓ $B^2$ has unit of $m^{-2}$.

✓ The value of $B^2$ that follows from equation (5.212) is in terms of shape, size and geometry of the nuclear reactor and is called geometrical buckling $B_g$ of the system. It is a function of geometry.

✓ The value of $B^2$ that follows from equation (5.213) is in terms of physical properties of reactor material $k_\infty$, $\tau$, $L^2$ and is called material buckling $B_m$ of the system. It is a function of material composition.

✓ If $B_g = B_m$ then the reactor is critical. If $B_g > B_m$ then the reactor is sub-critical. If $B_g < B_m$ then the reactor is super-critical

## 5.42 Basics of fusion

• Definition of nuclear fusion

Fusion means joining of two low mass nuclei $X$, $Y$ of masses $m_1$, $m_2$, respectively, to form a larger mass nucleus $Z$ of mass $m$ as

$$X + Y \rightarrow Z \qquad (5.214)$$

that is

$$m_1 + m_2 \rightarrow m \qquad (5.215)$$

and the mass deficit $(m_1 + m_2) - m$ is converted into energy.

For instance, two low mass nuclei $_{10}\text{Ne}_{10}^{20}$, $_{10}\text{Ne}_{10}^{20}$ can join or fuse to form a larger mass nucleus $_{20}\text{Ca}_{20}^{40}$.

Another example is joining or fusion of two low mass nuclei $_1\text{H}_1^2$, $_1\text{H}_1^2$ to form a larger mass nucleus $_2\text{He}_2^4$.

## 5.43 Calculation of the energy release during nuclear fusion from binding energy curve

Let us investigate whether it is energetically possible to fuse two low mass nuclei $_{Z_1}X_{N_1}^{A_1}$ and $_{Z_2}X_{N_2}^{A_2}$ to form $_Z Z_N^A$ by determining $Q$ value. Here

$$N_1 + Z_1 = A_1, \; N_2 + Z_2 = A_2, \; N + Z = A$$

$$N_1 + N_2 = N, \; Z_1 + Z_2 = Z, \; A_1 + A_2 = A \tag{5.216}$$

A nucleus of mass $M(Z, N, A)$ is related to its binding energy $B$ through the formula:

$$M(Z, N, A)c^2 = \left(Zm_p + Nm_n\right)c^2 - B(Z, N, A) \tag{5.217}$$

which shows that increase of $B$ means $M$ decreases.

The $Q$ value of the reaction of equation (5.214) viz. $X + Y \rightarrow Z$ is

$$Q = (M_{\text{initial}} - M_{\text{final}})c^2$$

$$Q = (M_X + M_Y - M_Z)c^2 \tag{5.218}$$

Construct relations similar to equation (5.217)

$$M_X c^2 = \left(Z_1 m_p + N_1 m_n\right)c^2 - \left(\frac{B}{A}\right)_X A_1$$

$$M_Y c^2 = \left(Z_2 m_p + N_2 m_n\right)c^2 - \left(\frac{B}{A}\right)_Y A_2$$

$$M_Z c^2 = \left(Z m_p + N m_n\right)c^2 - \left(\frac{B}{A}\right)_Z A$$

Putting them in equation (5.218) we get

$$Q = (Z_1 + Z_2 - Z)m_p c^2 + (N_1 + N_2 - N)m_n c^2 - \left(\frac{B}{A}\right)_X A_1 - \left(\frac{B}{A}\right)_Y A_2 + \left(\frac{B}{A}\right)_Z A$$

Using equation (5.216) we have

$$Q = \left(\frac{B}{A}\right)_Z A - \left(\frac{B}{A}\right)_X A_1 - \left(\frac{B}{A}\right)_Y A_2 \tag{5.219}$$

✓ Let us give an example. Consider the reaction

$$_1H_1^2 + {}_1H_1^2 \rightarrow {}_2He_2^4 + Q \ ( \text{i.e } d + d \rightarrow \alpha + Q)$$

Comparing with equation (5.214) viz. $X + Y \rightarrow Z$ we have
$X = Y = d, \ Z = \alpha$. Also $A_1 = A_2$. So from equation (5.219) we can write

$$Q = \left(\frac{B}{A}\right)_\alpha A - 2\left(\frac{B}{A}\right)_d A_1 \tag{5.220}$$

Using $\left(\frac{B}{A}\right)_\alpha = 7.7\frac{MeV}{\text{nucleon}}, \left(\frac{B}{A}\right)_d = 1.11\frac{MeV}{\text{nucleon}}$ from figure 5.17(a) and that $A = 4$ for $\alpha$, $A_1 = 2$ for $d$ we have from equation (5.220)

$$Q = 7.7\frac{MeV}{\text{nucleon}} \times 4 \text{ nucleon} - 2 \times 1.11\frac{MeV}{\text{nucleon}} \times 2 \text{ nucleon}$$

$$= 26.36 \ MeV \tag{5.221}$$

✓ Let us consider another example

$$_{10}Ne_{10}^{20} + {}_{10}Ne_{10}^{20} \rightarrow {}_{20}Ca_{20}^{40} + Q$$

Comparing with equation (5.214) viz. $X + Y \rightarrow Z$ we have
$X = Y = {}_{10}Ne_{10}^{20}, \ Z = {}_{20}Ca_{20}^{40}$. Also, $A_1 = A_2$. So from equation (5.219) we can
write

$$Q = \left(\frac{B}{A}\right)_{Ca} A - 2\left(\frac{B}{A}\right)_{Ne} A_1 \tag{5.222}$$

Using $\left(\frac{B}{A}\right)_{Ca} = 8.3\frac{MeV}{\text{nucleon}}, \left(\frac{B}{A}\right)_{Ne} = 7.78\frac{MeV}{\text{nucleon}}$ from figure 5.17(b), and that $A = 40$
for $_{20}Ca_{20}^{40}$, $A_1 = 20$ for $_{10}Ne_{10}^{20}$ we have from equation (5.222)

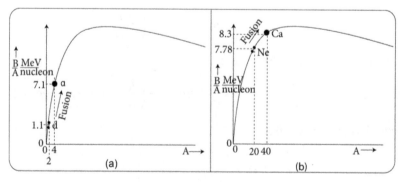

**Figure 5.17.** Fusion energy from plot of binding energy $\frac{B}{A}$ versus mass number $A$. $(a) \frac{B}{A}$ for $d$ and $\alpha$, $(b) \frac{B}{A}$ for Ne and Ca.

$$Q = 8.3 \frac{MeV}{nucleon} \times 40 \text{ nucleon} - 2 \times 7.78 \frac{MeV}{nucleon} \times 20 \text{ nucleon}$$

$$= 20.8 \; MeV \tag{5.223}$$

$Q$ is positive means that such fusion of two low mass nuclei into a heavier mass nucleus is energetically favourable and this process is associated with release of energy of about 24 $MeV$ per fusion. Such process causes decrease in rest mass (higher mass state of $X + Y$ goes to lower mass state of $Z$, the mass deficit being converted to energy that is released).

## 5.44 Coulomb barrier preventing fusion

Fusion means combination of two different positively charged nuclei.

Let $r$ be the distance of separation between the two nuclei of radii $R_1$, $R_2$.

✓ When they are apart by a distance $r > R_1 + R_2$ the two nuclei interact via Coulomb repulsion interaction potential energy given by (figure 5.18)

$$V_C = \frac{1}{4\pi\varepsilon_0} \frac{(Z_1|e|)(Z_2|e|)}{r} = \frac{1}{4\pi\varepsilon_0} \frac{Z_1 Z_2 \mid e\mid^2}{r} \tag{5.224}$$

$$rV_C = \text{constant} \tag{5.225}$$

As $r$ falls $V_C$ increases. So the $V_C$ versus $r$ plot is hyperbolic. In this state the two nuclei are separate.

✓ At $r < R_1 + R_2$ Coulomb repulsive interaction is overshadowed by 100 times stronger nuclear interaction (or strong interaction) which is attractive (represented by say $-V_{\text{strong}}$). In this state the two nuclei have built the new system by fusing into each other.

✓ At $r = R_1 + R_2$ the two nuclei are just in contact (and about to form the combined system). The Coulomb repulsive potential energy is maximum

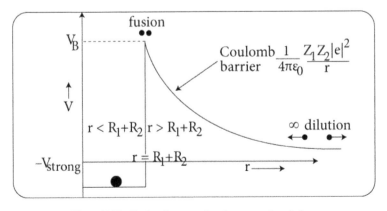

**Figure 5.18.** Coulomb energy barrier preventing fusion.

$V_B = \frac{1}{4\pi\varepsilon_0} \frac{Z_1 Z_2 | e |^2}{R_1 + R_2}$ and is overtaken by the strong potential energy of interaction $-V_{\text{strong}}$.

Clearly the two nuclei (initially in infinite dilution) have built the combined nuclei because it is able to overcome the Coulomb repulsive potential energy barrier $V_B$.

• Let us calculate the order of this Coulomb barrier height for the interaction

$$_{10}\text{Ne}_{10}^{20} + {}_{10}\text{Ne}_{10}^{20} \rightarrow 20\text{Ca}_{20}^{40} + Q \tag{5.226}$$

$$V_B = \frac{1}{4\pi\varepsilon_0} \frac{Z_1 Z_2 | e |^2}{R_1 + R_2} = \frac{1}{4\pi\left(\frac{10^{-9}}{36\pi} \frac{F}{m}\right)} \frac{(10)(10)(1.6 \times 10^{-19} \, C)^2}{2(1.2 \times 10^{-15} \, m)(20)^{1/3}}$$

$$= 3.5367 \times 10^{-12} \, J = \frac{3.5367 \times 10^{-12}}{1.6 \times 10^{-19}} eV$$

$$V_B = 22 \, MeV \tag{5.227}$$

For two protons fusing to form a deuteron in the reaction ($Z_1 = Z_2 = 1$)

$$p + p \rightarrow d + e^+ + \nu_e \tag{5.228}$$

$$V_B = \frac{1}{4\pi\varepsilon_0} \frac{| e |^2}{R_1 + R_2} = \frac{1}{4\pi\left(\frac{10^{-9}}{36\pi} \frac{F}{m}\right)} \frac{(1.6 \times 10^{-19} \, C)^2}{2(10^{-15} \, m)}$$

$$= 1.152 \times 10^{-13} \, J = \frac{1.152 \times 10^{-13}}{1.6 \times 10^{-19}} eV$$

$$V_B = 720 \, keV \tag{5.229}$$

This is the minimum Coulomb potential energy barrier that the two protons have to surmount.

## 5.45 Thermonuclear fusion

Accelerators can provide such energy for instance $720 \, keV$ for two protons [equation (5.229)] or $22 \, MeV$ for two $_{10}\text{Ne}_{10}^{20}$ [equation (5.227)] and punch them to fuse.

The energy (i.e. the $Q$ value) released through the process due to reduction of rest mass $m_1 + m_2 - m$ [equation (5.215)] is of the same order of the energy they absorb from accelerators.

As our main aim is to derive significant energy from the process of fusion it follows that the method of supplying energy through accelerators will not be a profitable one. We can then get a very small energy output through fusion. So we think of thermonuclear fusion process.

We discuss the possibility of two particles or nuclei approaching each other with their thermal kinetic energy, overcoming the Coulomb repulsion barrier in order to fuse.

Let us evaluate the temperature that is needed to overcome or surmount the Coulomb barrier and combine.

At room temperature $T = 300\ K$ energy scale is

$$kT = (1.38 \times 10^{-23}\ J\ K^{-1})(300\ K)$$

$$kT = 4.14 \times 10^{-21}\ J = \frac{4.14 \times 10^{-21}}{1.6 \times 10^{-19}} eV = 0.026\ eV \qquad (5.230)$$

Clearly nuclei at room temperature $T = 300\ K$ cannot fuse as they are far below the Coulomb barrier.

In a thermal system in equilibrium at certain temperature $T$, different particles will have different kinetic energies, i.e. there will be a distribution of velocity, momentum, kinetic energy. We can take the distribution to be a Maxwell–Boltzmann distribution which has average kinetic energy $\frac{3}{2}kT$.

For $T = 300\ K$, $kT = 0.026\ eV$ we have

$$\frac{3}{2}kT = \frac{3}{2}(0.026\ eV) - 0.039\ eV \qquad (5.231)$$

This kinetic energy is very small compared to the Coulomb barrier height $\sim 720\ keV$ for two protons fusing to form a deuteron [equation (5.229)].

Let us estimate the temperature $T$ at which kinetic energy will be comparable to the Coulomb barrier height $V_B$—the order of which we take to be $100\ keV$. Equating the average kinetic energy $\frac{3}{2}kT$ to $V_B = 100\ keV$ we have

$$\frac{3}{2}kT = V_B = 100\ keV$$

$$T = \frac{2V_B}{3k} = \frac{2 \times 100\ keV}{3 \times 1.38 \times 10^{-23}\ J\ K^{-1}} = \frac{2 \times 100 \times 10^3 \times 1.6 \times 10^{-19}\ J}{3 \times 1.38 \times 10^{-23}\ J\ K^{-1}} \cong 1 \times 10^9\ K \qquad (5.232)$$

Fusion can occur between two protons and they will form deuteron if the two protons approach each other with a thermal kinetic energy in an environment where temperature is $\sim 10^9\ K$. Hence the fusion process is called thermonuclear fusion reaction.

- In the Sun the core temperature is $\sim 1.4 \times 10^7\ K$. This means that the interaction

$$p + p \rightarrow d + e^+ + \nu_e$$

occurs in the Sun's core at a lower temperature because we have shown in equation (5.232) that thermonuclear fusion needs a much higher temperature $\sim 10^9\ K$.

The reaction in the Sun's core occurs at a reduced temperature because of gigantic pressure ~200 billion atm. in the core.

- The essential condition of thermonuclear fusion is that the temperature has to be very high to get a reasonable barrier penetration.
- Plasma

When temperature of a material is raised, its state changes from solid to liquid and then to gas. If temperature is elevated further, gas atoms ionize and there are nuclei and electrons and such a state is called plasma.

Plasma is a superheated soup of positively charged particles (ions) and negatively charged particles (electrons).

At the temperature needed for fusion all matter will be in the form of plasma.

## 5.46 Reaction rate for fusion

Consider the fusion reaction of two elements $X$ and $Y$.

$$X + Y \rightarrow Z$$

Number of fusion reactions per unit time is called reaction rate and depends upon concentration and temperature.

Let $n_X$, $n_Y$ be the concentration, i.e. $\frac{\text{number}}{\text{volume}}$ of $X$, $Y$ respectively. $v$ is the relative velocity of $X$ w.r.t. $Y$.

Let $\sigma$ represent the area presented by $X$ nucleus also called the cross-section (figure 5.19). $\sigma$ is related to the probability of fusion reaction. If $Y$ nucleus hits that area a fusion reaction occurs. The number of reactions is determined by the number density of $Y$ nucleus $n_Y$ within the cylinder of length $vdt$ striking the area $\sigma$ presented by one $X$ nucleus, i.e.

$n_Y \sigma v dt$

and for $n_X$ nuclei the number of reactions will be

$n_X n_Y \sigma v dt$

And number of reactions per second will be

$n_X n_Y \sigma v$.

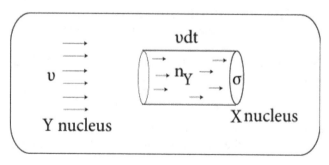

**Figure 5.19.** Reaction rate is $n_X n_Y < \sigma v >$.

In thermonuclear fusion there is no solid target as everything is in plasma state—the nuclei $X$, $Y$ move randomly in all directions and with all velocities and not in a particular direction, as shown in figure 5.19.

For simplicity, just to make an estimate let us assume that the velocities of the nuclei are distributed according to Maxwell–Boltzmann distribution. So we have to integrate over all velocities involved. In other words we have to take average over all velocities.

If $\sigma$ is large, probability is large and the reaction rate is large. As amount of reaction depends on velocity $v$ and the cross-section $\sigma$ i.e. on $\sigma v$ we have to average over $\sigma v$. Hence the reaction rate is

$$R = n_X n_Y < \sigma v > \tag{5.233}$$

The angular bracket denotes average of $\sigma v$.

The fusion reaction has to go through a barrier penetration process. When two nuclei are fusing, the two positive charges approach each other overcoming the Coulomb repulsion and when they are within nuclear range nuclear force takes over and they fuse. If we assume that the Maxwell velocity distribution is followed then there will be nuclei moving with kinetic energy greater than $\frac{3}{2}kT$ and they will have higher barrier penetrability and greater probability of fusion. The fusion cross-section $\sigma$ involves tunneling probability $T$ and hence

$$\sigma \propto T$$

Using the result we derived in *excerise 2.25* viz. $T = e^{-G} = e^{-\frac{(Z-2)|e|^2}{\hbar v \varepsilon_0}}$ we can take

$$\sigma \propto e^{-\frac{(Z-2)|e|^2}{\hbar v \varepsilon_0}} \propto e^{-\frac{C'}{v}} \tag{5.234}$$

where $C'$ is a constant.

If relative speed $v$ is high, kinetic energy of the particle is high and probability of penetration $T$ increases and hence rate of fusion reaction $R$ increases.

The relative angular momentum is

$$\vec{l} = \vec{r} \times \vec{p} \sim mvb$$

where $b$ is impact parameter.

Figure 5.20 shows the area presented by $X$ nucleus for a nuclear reaction with angular momentum quantum numbers $l$ and $l+1$ corresponding to impact parameters $b_l$ and $b_{l+1}$ respectively. Now angular momentum is quantized as

$$mvb_l = l\hbar$$

$$mvb_{l+1} = (l+1)\hbar$$

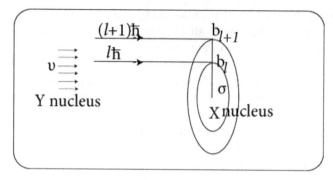

**Figure 5.20.** Area presented by $X$ nucleus for nuclear reaction with angular momentum quantum numbers $l$ and $l + 1$ corresponding to impact parameters $b_l$ and $b_{l+1}$ respectively.

$$b_l = \frac{l\hbar}{mv}, \; b_{l+1} = \frac{(l + 1)\hbar}{mv} \tag{5.235}$$

Maximum $l$ is given by nuclear range of interaction $R$ as

$$mvR = l_{\max}\hbar \Rightarrow l_{\max} = \frac{mvR}{\hbar} \tag{5.236}$$

The elemental area magnitude is

$$\pi b_{l+1}^2 - \pi b_l^2 = \pi\left(\frac{(l + 1)\hbar}{mv}\right)^2 - \pi\left(\frac{l\hbar}{mv}\right)^2 = [(l + 1)^2 - l^2]\frac{\pi\hbar^2}{m^2v^2}$$

$$= (2l + 1)\frac{\pi\hbar^2}{m^2v^2} \tag{5.237}$$

The total cross-section is obtained by summing over area from $l = 0$ to $l = l_{\max}$ as

$$\sigma = \sum_{l=0}^{l_{\max}}(2l + 1)\frac{\pi\hbar^2}{m^2v^2} = \frac{\pi\hbar^2}{m^2v^2}\sum_{l=0}^{l_{\max}}(2l + 1)$$

$$\sigma = \frac{\pi\hbar^2}{m^2v^2}(1 + 3 + 5 + 7 + \ldots + l_{\max}) = \frac{\pi\hbar^2}{m^2v^2}(l_{\max} + 1)^2$$

Using equation (5.236) we have

$$\sigma = \frac{\pi\hbar^2}{m^2v^2}\left(\frac{mvR}{\hbar} + 1\right)^2$$

$$\sigma = \pi\left(R + \frac{\hbar}{mv}\right)^2 \tag{5.238}$$

Now consider the quantity

$$\left(\frac{\hbar}{mv}\right)^2 = \frac{(\hbar c)^2}{mc^2 mv^2} \tag{5.239}$$

For the $d + d$ reaction (fusion of two deuterons) since $d = 2m_p$, reduced mass of two deuterons is

$\frac{(2m_p)(2m_p)}{2m_p + 2m_p} = m_p$ and hence we take

$$mc^2 \sim m_p c^2 = 938 \; MeV.$$

Also, kinetic energy $= \frac{1}{2}mv^2$ and we take $mv^2$ to be of the order of $keV$ since temperature is around $T \sim 10^7 \; K$ for fusion. Hence from equation (5.239) we have with *excerise 1.2* viz. $\hbar c = 200 \; MeV. \; fm$

$$\left(\frac{\hbar}{mv}\right)^2 = \frac{(197 \; MeV. \; fm)^2}{(938 \; MeV)(keV)} = \frac{197^2 \; MeV^2 fm^2}{938 \; MeV. \; keV} = 41 \times 10^3 \; fm^2$$

$$\frac{\hbar}{mv} \sim 200 \; fm.$$

This term dominates in equation (5.238) since $R \sim 2 \; fm$. So from equation (5.238)

$$\sigma = \pi\left(R + \frac{\hbar}{mv}\right)^2 \cong \pi\left(\frac{\hbar}{mv}\right)^2 = \frac{\pi \hbar^2}{m^2 v^2} \tag{5.240}$$

Incorporating the barrier penetration probability factor [equation (5.234)] viz. $\sigma \propto e^{-C'/v}$ we rewrite

$$\sigma = \frac{\pi \hbar^2}{m^2 v^2} e^{-C'/v} = \frac{C_1}{v^2} e^{-C'/v} \; (C_1, \; C' \; \text{are constants}) \tag{5.241}$$

We now proceed to evaluate $<\sigma v>$ using the Maxwell–Boltzmann probability distribution function

$$P(v)dv = \frac{C_2}{T^{3/2}} v^2 e^{-\frac{mv^2}{2kT}} dv \; (C_2 \; \text{is constant}) \tag{5.242}$$

$$<\sigma v> = \int_0^\infty \sigma v \frac{C_2}{T^{3/2}} v^2 e^{-\frac{mv^2}{2kT}} dv$$

$$= \int_0^\infty \frac{C_1}{v^2} e^{-C'/v} \; v \frac{C_2}{T^{3/2}} v^2 e^{-\frac{mv^2}{2kT}} dv$$

$$<\sigma v> = C \int_0^\infty e^{-C'/v} \frac{v}{T^{3/2}} e^{-\frac{mv^2}{2kT}} dv \; (C \; \text{is constant})$$

$$<\sigma v> = \frac{C}{T^{3/2}} \int_0^\infty e^{-C'/v} \; e^{-\frac{mv^2}{2kT}} v dv \tag{5.243}$$

Putting

$$E = \frac{1}{2}mv^2, \ dE = mvbv, \ vdv = \frac{dE}{m},$$

$$v = \sqrt{\frac{2E}{m}} \propto \sqrt{E}$$

From equation (5.243) we get

$$<\sigma v> = \frac{C}{T^{3/2}} \int_0^\infty e^{-c'/\sqrt{E}} \ e^{-E/kT} \ \frac{dE}{m} \ \text{(integration limits remains same 0 to } \infty\text{)} \quad (5.244)$$

$e^{-c'/\sqrt{E}}$ is the barrier contribution—it rises with energy increase and $e^{-E/kT}$ is the contribution of the distribution—it falls with energy increase. Their product corresponds to a common region that contributes.

So reaction rate can be calculated.

- Possible reactions in a fusion reactor.

✓ Consider the reaction

$$p + p \rightarrow d + e^+ + \nu_e \ (\text{H} + \text{H} \rightarrow \text{D} + \beta^+ + \nu_e) \quad (\text{D} \rightarrow \text{deuteron d}) \quad (5.245)$$

This reaction is not a popular fusion reaction. Two protons cannot combine to form a stable bound di-proton system because such $pp$ system does not exist in nature. Protons interact and a deuteron $d \equiv {}_1\text{H}^2 \equiv np$ is formed. Clearly a proton is converted into a neutron and this occurs through weak interaction. In a nuclear time scale (which is very short) when two protons come close to each other to fuse, the probability of β decay to occur is low.

✓ Consider the reaction D–T reaction (deuteron–tritium)

$${}_1\text{H}^2 + {}_1\text{H}^3 \rightarrow {}_2\text{He}^4 + {}_0n^1 + 17.6 \ MeV \ (\text{D} + \text{T} \rightarrow \text{He} + n + 17.6 \ MeV) \quad (5.246)$$

Such a reaction occurs at a temperature of around $10^8 \ K$ and at such a high temperature nuclei and electrons move and scatter with high velocity and acceleration. Again acceleration of charged particle means emission of radiation—called bremsstrahlung radiation and the system loses energy. Most of the radiation loss comes from electrons as their mass is small. The loss of energy per unit volume per unit time is proportional to

concentration of the positive ions $n$

concentration of electrons $n_e = n$ (for fusion between light elements)

$Z^2$ and

the relative velocity, i.e. $(kT)^{1/2}$.

In other words, loss $L$ per unit volume per unit time is given by

$$L = nn_e Z^2 (kT)^{1/2} \quad (5.247)$$

We demand that the rate at which the energy is produced should be larger than the rate of loss of energy.

Let energy emitted per fusion reaction be $Q$. So energy emitted from fusion per unit time per unit volume is

$$P = \text{Reaction rate} \times Q = RQ$$

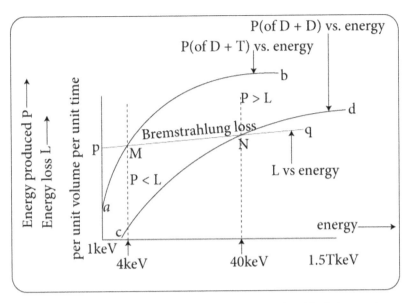

**Figure 5.21.** Energy loss should be smaller than energy produced by fusion reaction.

$$P = n_X n_Y <\sigma v> Q \tag{5.248}$$

Clearly the operating temperature has to be chosen in such a way that

$$P > L \tag{5.249}$$

and this is shown in figure 5.21.

In figure 5.21 the abscissa is an energy scale. Along the ordinate energy production rate is plotted that gives the curve $ab$ for D + T fusion reaction [equation (5.246)] and the curve $cd$ for D + D fusion reaction ($_1H^2 + _1H^2 \rightarrow _2He^4 +$ energy i.e. D + D $\rightarrow _2He^4 +$ energy). The energy loss rate is shown in the plot as the curve $pq$.

- ✓ At low temperature loss is large and so production is small. But at high temperature production is larger.
- ✓ Production rate of energy per unit volume is larger for D + T as evident from figure 5.21.
- ✓ The production rate and loss rate matches at point $M$ (4 $keV$ for D + T) and $N$ (40 $keV$ for D + D).
- ✓ The operating temperature should be chosen large enough such that the production rate is larger than the bremstrahlung loss rate.
- ✓ $L \propto Z^2$ which means if we use heavy elements then loss will be very large. So lighter elements are used for fusion.

## 5.47 Lawson criterion

Suppose plasma is confined at a high enough temperature $T$ so that the bremstrahling loss is small and overcome. To take plasma to that temperature it has to be heated—energy has to be given from outside. To confine plasma in a small volume at high temperature for a long period is difficult because plasma would leak out and get cooled. Generally, confinement is done in a pulsed manner for a time say $\tau$ called confinememt time which is of the order of seconds). Let

$$n_X = n_Y = \frac{n}{2} \text{ (say equal amounts of D and T)} \qquad (5.250)$$

where $n$ is the total number of nuclei per unit volume.

Energy supplied at temperature $T$ is $\frac{3}{2}nkT$ for nuclei and $\frac{3}{2}nkT$ for electrons, i.e. the total energy per unit volume given to heat up the plasma is

$$\frac{3}{2}nkT + \frac{3}{2}nkT = 3nkT \qquad (5.251)$$

Fusion energy produced per unit volume in confinement time $\tau$ (in one pulse) is the product of the reaction rate and the confinement time, i.e.

$$n_X n_Y < \sigma v > Q\tau \qquad (5.252)$$

Neglecting bremstrahlung loss of energy radiation the energy produced per unit volume should be greater than the energy supplied per unit volume. Hence

$$n_X n_Y < \sigma v > Q\tau \geqslant 3nkT \qquad (5.253)$$

Using equation (5.250) we have

$$\frac{n}{2}\frac{n}{2} < \sigma v > Q\tau \geqslant 3nkT$$

$$n\tau \geqslant \frac{12kT}{Q < \sigma v>} \qquad (5.254)$$

This is Lawson criterion. Reactor design should be such that the product of concentration and confinement time should be greater than $\frac{12kT}{Q<\sigma v>}$ to get greater output than the input energy.

✓ For D–T reaction the $T$ is given by $kT = 20\ keV$ and $n \sim 20\ \frac{nuclei}{m^3}$.

## 5.48 Problems of fusion reactor

The main problem for a fusion reactor is the confinement problem.

High temperature plasma ($\sim 10^8\ K$) has to be confined in a small volume to get the large concentration. Again, no material boundary can be used to confine plasma since the material container cannot sustain such high temperature. So magnetic confinement is done.

- Magnetic confinement principle

Suppose we have a cylindrical container, as shown in figure 5.22, with axis $PQ$ and gas particles are moving in it along random directions with random speeds. The component of velocity along $LM$ will make the particle hit the wall which is not desired.

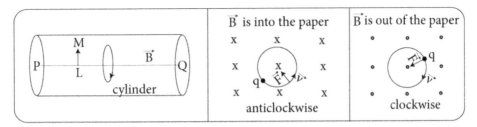

**Figure 5.22.** Confinement of plasma by applying a magnetic field. The motion of the charged particle is anticlockwise or clockwise depending upon whether $\vec{B}$ is into the paper or out of the paper.

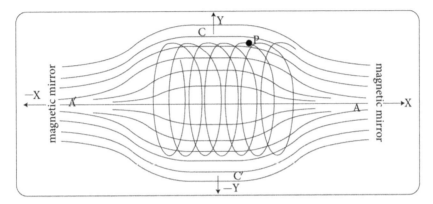

**Figure 5.23.** Magnetic mirror, magnetic bottle, magnetic confinement.

Let us apply an axial magnetic induction $\vec{B} = B\widehat{PQ}$. This magnetic field will bend the charged particle before it hits the wall. So the particle turns around and moves along a helix/circle avoiding any contact with the wall. This follows from the Lorentz force experienced by charge $q$ namely

$$\vec{F} = q\,\vec{v} \times \vec{B}.$$

This is the principle of magnetic confinement.

- Magnetic mirror or magnetic trap

    Properly designed magnetic fields can contain or retain charged particles in a volume.

    Consider a non-uniform magnetic field $\vec{B}$, as shown in figure 5.23. Consider the motion of a charged particle moving in this region. The $\vec{B}$ lines condense or crowd at region $A$, $A'$ and thin down at region $C$, $C'$. Clearly $\vec{B}$ field is stronger at region $A$, $A'$ and weak at region $C$, $C'$. Suppose the charged particle is at point $P$ and is acted upon by the magnetic field

$$\vec{B} = B_x\hat{i} - B_y\hat{j}$$

and has velocity components along $X$ as well as along $Z$ as

$$\vec{v} = v_x\hat{i} - v_z\hat{k}.$$

This means the charged particle has a velocity component along $X$ namely $v_x\hat{i}$.

The charged particle of charge $q$ is acted upon by Lorentz force given by

$$\vec{F} = q\vec{v} \times \vec{B} = q \begin{vmatrix} \hat{i} & \hat{j} & \hat{k} \\ v_x & 0 & -v_z \\ B_x & -B_y & 0 \end{vmatrix}$$

$$\vec{F} = \hat{i}(-qv_zB_y) + \hat{j}(-qv_zB_x) + \hat{k}(-qv_xB_y)$$

$$= -\hat{i}qv_zB_y - \hat{j}qv_zB_x - \hat{k}-qv_xB_y \tag{5.255}$$

Clearly the charged particle is acted upon by a force component $-\hat{i}qv_zB_y$ i.e. the force acts along $-X$. This means the particle will decelerate. The stronger magnetic field at region $A$ pushes it back. So region $A$ acts like a magnetic mirror. The same thing is repeated at $A'$ which is the location of a strong magnetic field and from here the charged particle is decelerated. In other words it is reflected back along $X$. The region $A'$ also acts as a magnetic mirror. Obviously the charged particle, so to say gets trapped in this region between $AA'$ and this region is called magnetic bottle.

- Tokamak design

  Consider a cylinder to be bent axially and the ends are joined to form a toroid, as shown in figure 5.24. Copper coils are wound over the toroid and current through it generates a magnetic field within the toroid. The concentration of the coils is larger on the inner side than the outer side because of the difference in radius. So the magnetic field inside will not be uniform. The magnetic field is stronger in the inner

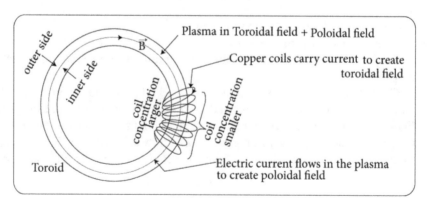

**Figure 5.24.** Tokamak design.

side and weaker in the outer side. The stronger magnetic field will push the charged particles to the outer wall due to magnetic mirror effect. To prevent this effect and to confine plasma another magnetic field is applied called poloidal field by sending current along the length of the toroid that would push the charged particles towards the centre and away from the walls. So now the charged particles would travel in the toroidal region in a helical path. A combination of toridal field and poloidal field is thus able to confine plasma for some reasonable period of time.

## 5.49 Physics of the sun—an introduction

Mass of the sun is $M = 2 \times 10^{30} \, kg = 3.33 \times 10^5 \, m_{earth}$

The radius of the sun is $R = 7 \times 10^8 \, m = 109 R_{earth}$

Mean distance of the sun from the earth is $1.5 \times 10^{11} \, m$

Layers of the sun are shown in figure 5.25.

The four layers corona, transition zone, chromosphere and photosphere are the outer layers, while convective zone, radiative zone and core are the inner layers.

Photosophere is the outer layer or surface of the sun that is visible from earth during day time. It is 500 $km$ thick. The thickness of this outer layer of the sun is actually very small compared to the radius of the sun. So it is actually a thin layer.

Composition of the photosphere by mass of the most abundant elements is the following:

Hydrogen (73.46%), helium (24.85%), oxygen (0.77%), carbon (0.29%), iron (0.16%).

Composition of the photosphere by atoms is hydrogen (90%), helium (9%) others (1%).

Photosphere temperature is 5780 $K$ (low). Pressure is 0.1–0.001 atm (low).

The sun is not a solid and so the layers are difficult to demarcate. Most of the sun is made up of hydrogen in the form of a glowing hot gas called plasma.

Core is of radius 0.25$R$. Fusion occurs and energy is produced here and gets transported to the surface of the sun. Temperature in the core is $1.4 \times 10^7 \, K$ (huge). Pressure is $200 \times 10^9 atm$ (huge). Density is 160 times the density of water, i.e. 160 $g \, cm^{-3}$ (huge). Mass here is made of protons, electrons, helium in plasma state.

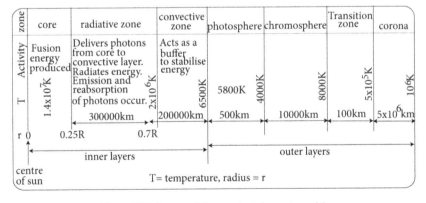

**Figure 5.25.** Layers of the sun (not drawn to scale).

## 5.50 Energy release in stars (Sun)

Study of various data point to the fact that the temperature of the sun has been same for at least $10^9$ *yrs*. Amount of energy radiated by the sun is $3.8 \times 10^{26}$ *J* $s^{-1}$.

The origin of such a fantastically huge amount of energy coming out from the sun and other stars is basically due to the nuclear fusion of four protons ($_1H^1 \equiv p$) to form alpha ($_2He^4 \equiv \alpha$) along with two positrons ($e^+$) plus release of ~24 *MeV* of energy per cycle of fusion process.

There are a few models that prescribe such a fusion cycle which end with this outcome viz.

$$4p \rightarrow \alpha + 2e^+ + 2\nu_e + Q \tag{5.256}$$

Let us mention the models.

- *pp*-1 chain or proton–proton cycle (1)

First step
This cycle starts with the initial reaction

$$_1H^1 + {}_1H^1 \rightarrow {}_1H^2 + {}_1e^0 + \nu_e \text{ or } p + p \rightarrow d + e^+ + \nu_e) \tag{5.257}$$

Consider two protons moving with thermal velocity (at the temperature of the sun) and coming sufficiently close so that the strong force becomes operative and binds them. This *pp* system or di-proton system is not stable. The only stable two-nucleon system is a deuteron $d = np = {}_1H_1^2$ system. This necessitates that one proton has to change to a neutron during the process of fusion of two protons through the reaction

$$n \rightarrow p + e^+ + \nu_e.$$

In other words we would have

$$p + p \rightarrow np + e^+ + \nu_e.$$

This means two processes should occur simultaneously for this reaction to be a reality.

- (a) strong interaction between protons that occur in a time period ~$10^{-21}$ *s*.
- (b) conversion of proton to neutron and generation of $e^+ + \nu_e$ which is a weak interaction characterized by a long time period. This cannot occur in such a short time ~$10^{-21}$ *s*.

As the two parts of the process $p + p \rightarrow d + e^+ + \nu_e$ are incompatible (there being a violent mismatch of lifetimes and cross-sections of the strong and weak inearctions) we expect that the probability of such a process, i.e. weak interaction driven $e^+$ emission and strong interaction driven $d$ formation to occur simultaneously and directly is extremely small.

This is referred to as deuteron bottleneck.

Probability of such interaction is so small that a proton has to wait on average $10^{10}$ *yrs* to find another proton with which it can fuse to form a deuteron. In other words, out of $10^{18}$ protons only one will fuse to form a deuteron through this interaction.

But we note that the total number of protons present is so enormous (pressure = 200 billion *atm* in the core) that despite such low probability around $6 \times 10^{11}$ *kg* of hydrogen, or $3.7 \times 10^{38}$ protons fuse per second inside the core of the Sun.

$$\text{Second step: } {}_1H^2 + {}_1H^1 \rightarrow {}_2He^3 \text{ (or } d + p \rightarrow {}_2He^3) \tag{5.258}$$

$$\text{Third step: } {}_2He^3 + {}_2He^3 \rightarrow {}_2He^4 + 2{}_1H^1 \text{ (or } {}_2He^3 + {}_2He^3 \rightarrow \alpha + 2p) \tag{5.259}$$

Let us suitably add the intermediate steps as follows

2 (equation (5.257)) + 2 (equation (5.258)) + equation (5.259) to get for the *pp-1* chain or proton–proton cycle (1)

$$2{}_1H^1 + 2{}_1H^1 \rightarrow 2{}_1H^2 + 2{}_1e^0 + 2\nu_e \text{ (or } 2p + 2p \rightarrow 2d + 2e^+ + 2\nu_e)$$

$$2{}_1H^2 + 2{}_1H^1 \rightarrow 2{}_2He^3 \text{ (or } 2d + 2p \rightarrow 2{}_2He^3)$$

$${}_2He^3 + {}_2He^3 \rightarrow {}_2He^4 + 2{}_1H^1 \text{ (or } {}_2He^3 + {}_2He^3 \rightarrow \alpha + 2p)$$

Addition gives

$$4{}_1H^1 \rightarrow {}_2He^4 + 2{}_1e^0 + 2\nu_e + Q \text{ (or } 4p \rightarrow \alpha + 2e^+ + 2\nu_e + Q)$$

So equation (5.256) is reproduced. This occurs 85% of the time, as data analysis indicates.

Let us discuss some other fusion cycles leading to the same final reaction.

- pp-2 chain or proton–proton cycle (2)

$${}_1H^1 + {}_1H^1 \rightarrow {}_1H^2 + {}_1e^0 + \nu_e$$

$${}_1H^2 + {}_1H^1 \rightarrow {}_2He^3$$

$${}_2He^3 + {}_2He^4 \rightarrow {}_4Be^7$$

$${}_4Be^7 + {}_{-1}e^0 \rightarrow {}_3Li^7 + \nu_e \tag{5.260}$$

$${}_3Li^7 + {}_1H^1 \rightarrow 2{}_2He^4$$

Addition gives

$$4{}_1H^1 \rightarrow {}_2He^4 + 2{}_1e^0 + 2\nu_e + Q$$

So equation (5.256) is reproduced. This occurs 14.9% of the time as data analysis indicates.

- pp-3 chain or proton–proton cycle (3)

$${}_1H^1 + {}_1H^1 \rightarrow {}_1H^2 + {}_1e^0 + \nu_e$$

$${}_1H^2 + {}_1H^1 \rightarrow {}_2He^3$$

$$_2\text{He}^3 + {}_2\text{He}^4 \rightarrow {}_4\text{Be}^7$$

$$_4\text{Be}^7 + {}_1\text{H}^1 \rightarrow {}_5\text{B}^8$$

$$_5\text{B}^8 + {}_{-1}e^0 \rightarrow 2{}_2\text{He}^4 + \nu_e \qquad (5.261)$$

Addition gives

$$4_1\text{H}^1 \rightarrow {}_2\text{He}^4 + 2_1e^0 + 2\nu_e + Q$$

So equation (5.256) is reproduced. This occurs 0.1% of the time as data analysis indicates.

- Carbon–Nitrogen–Oxygen cycle or CNO cycle

$$_6\text{C}^{12} + {}_1\text{H}^1 \rightarrow {}_7\text{N}^{13} + Q$$

$$_7\text{N}^{13} \rightarrow {}_6\text{C}^{13} + {}_1e^0 + \nu_e$$

$$_6\text{C}^{13} + {}_1\text{H}^1 \rightarrow {}_7\text{N}^{14} + Q$$

$$_7\text{N}^{14} + {}_1\text{H}^1 \rightarrow {}_8\text{O}^{15} + Q$$

$$_8\text{O}^{15} \rightarrow {}_7\text{N}^{15} + {}_1e^0 + \nu_e$$

$$_7\text{N}^{15} + {}_1\text{H}^1 \rightarrow {}_6\text{C}^{12} + {}_2\text{He}^4$$

Addition gives

$$4_1\text{H}^1 \rightarrow {}_2\text{He}^4 + 2_1e^0 + 2\nu_e + Q$$

So equation (5.256) is reproduced. This occurs 2% of the time as data analysis indicates.

Energy released in the cycle can be evaluated by considering the end net result of the reaction cycle given by equation (5.256) namely $4_1\text{H}^1 \rightarrow {}_2\text{He}^4 + 2_1e^0 + 2\nu_e + Q$.

$$4m(p) - m(\alpha) - 2m(e^+) - 2m(\nu_e)$$

$$= 4(1.007\,276u) - (4.002\,604) - 2(0.000\,549u) - 2(0) = 0.025\,402u$$

$$= 0.025\,402 \times 931 \; MeV = 23.65 \; MeV \cong 24 \; MeV \text{ (per fusion cycle)}$$

The conversion of hydrogen to helium would continue till the whole of hydrogen in the star is completely used up. The time for complete burn-out of fuel in the sun is $3 \times 10^{10} \; yrs$.

- Information regarding the core of the sun

The models predict release of energy due to fusion which moves from the core and passes through the various layers like radiative zone, convective zone and then comes to the photosphere. For instance, once photon enters convective zone it may take $1.7 \times 10^5 \; yrs$ for it to reach the photosphere. The photosphere is transparent to electromagnetic waves. Photons from the photosphere cross outer layers and come to the earth.

Clearly to cross various layers and move from the core to the photosphere it takes millions of years. The energy we are receiving today may have been created millions of years before due to fusion inside the core. Also, a multitude of processes of absorption, emission, re-absorption and re-emission of photons take place in the intermediate layers. This means it is difficult to relate the energy we are receiving directly to the fusion in the sun's core.

Direct experimental evidence of the models proposed comes from the two neutrinos that the models predict to be created in the sun's core. The neutrinos travel very nearly with the speed of light and almost do not interact with matter. Hence they directly escape from the core of the sun, suffering no interaction in the intermediate layers. These neutrinos reach the earth in copious amounts. These neutrinos carry direct information of events that they underwent in the core of the sun. In other words, the neutrinos carry information regarding the phenomena occurring in the core of the sun. We can get the information if it is possible to study these neutrinos. And we have to wait just a few seconds (8 *min* delay) to get an update of events occuring within the sun.

Let us study the energy distribution of the neutrinos. Neutrinos have different energies in different cycles.

Neutrinos are produced in the $pp1$ chain in the reaction [equation (5.257)]

$$_1H^1 + {}_1H^1 \rightarrow {}_1H^2 + {}_1e^0 + \nu_e$$

The energy of the products is 0.42 *MeV*, which is shared between $_1e^0$ and $\nu_e$ (neglecting the negligible recoil energy of $_1H^2$). It is a continuous spectrum. So energy of $\nu_e$ is $\leqslant 0.42$ *MeV*.

Neutrinos are produced in the $pp2$ chain in the reaction [equation (5.260)]

$$_4Be^7 + {}_{-1}e^0 \rightarrow {}_3Li^7 + \nu_e$$

The energy of the products is 0.861 *MeV*, which is taken away by $\nu_e$ (neglecting the recoil energy of $_3Li^7$). It is a discrete spectrum. So energy of $\nu_e$ is 0.861 *MeV*.

Neutrinos are produced in the $pp3$ chain in the reaction [equation (5.261)]

$$_5B^8 + {}_{-1}e^0 \rightarrow 2{}_2He^4 + \nu_e$$

The energy of the products is 14.03 *MeV*, which is taken away by $\nu_e$ (neglecting the recoil energy of $_2He^4$). It is a discrete spectrum. So energy of $\nu_e$ is 14.03 *MeV*.

Solar neutrino detection was done by Raymond Davies Jr. in 1968 based on the reactions

$$p \rightarrow n + e^+ + \nu_e$$

$$n + \nu_e \rightarrow p + e^-$$

$$_{17}Cl^{37} + \nu_e \rightarrow {}_{18}Ar^{37} + {}_{-1}e^0$$

The theory predicted that an incoming solar neutrino would interact with chlorine nucleus and produce argon. This was successfully detected by Davies proving existence of solar neutrinos. However, the experimentally observed number of solar neutrinos detected was one third the expected number. Subsequent experiments at neutrino observatory at Kamiokande, Japan, SAGE (Soviet American Gallium Experiment), GALLEX (Gallium solar neutrino experimets) all observed that the flux of solar neutrinos was much smaller than expected.

As per the models discussed, the sun can only make electron neutrinos $\nu_e$. But there are three neutrino flavours viz. electron neutrino $\nu_e$ (the electron flavour), muon neutrino $\nu_\mu$, tauon neutrino $\nu_\tau$ (these are non-electron flavours).

The experimental set up of Davies could detect only electron neutrinos. In 2001, SNO (Sudbury Neutrino Observatory), Canada observed that two thirds of solar electron neutrinos produced in the sun had converted to non-electron flavours. This was confirmed by the neutrino detection experiment called Super Kamiokande at the neutrino observatory at Kamiokande, Japan. Such flavour conversion of two thirds of $\nu_e$ to $\nu_\mu$, $\nu_\tau$ (non-electron flavours) happened as the $\nu_e$ produced in fusion in the core of the sun were in transit from the sun to earth.

The conversion of flavours between the three types of neutrinos is called neutrino oscillation and was discovered by Super Kamiokande and SNO. Also, such flavour conversion can occur if neutrino masses are non-zero. Possession of mass by the neutrinos requires revision of our basic model of particles and fields and has a bearing on our understanding of the structure of the universe.

## 5.51 Exercises

**Exercise 5.1** *An amount of energy of 200 MeV is released when a $U^{235}$ nucleus undergoes fission. How many fissions per second are needed to generate power of 1 watt? How much energy is released due to fission of 1 g of $U^{235}$. Express the result in J and also in kwatt. hr.*

[Ans.] Amount of power to be generated $= 1 \ watt = 1\frac{J}{s}$

Energy per fission $= 200 \ MeV$.

Number of fissions needed per second to generate 1 *watt* is

$$\frac{1 \ J/s}{200 \ MeV} = \frac{1 \ J/s}{200 \times 10^6 \times 1.6 \times 10^{-19} \ J} = 3.125 \times 10^{10}\frac{1}{s}$$

$$235 \ g \ U^{235} \text{ contains } 6.023 \times 10^{23} \text{ atoms}$$

$$1 \ g \ U^{235} \text{ contains } \frac{6.023 \times 10^{23}}{235} \text{ atoms}$$

Energy released by 1 $g$ $U^{235}$ due to fission is
(number of atoms that suffer fission)(energy per fission)

$$= \left(\frac{6.023 \times 10^{23}}{235}\right)(200 \; MeV) = 5.126 \times 10^{23} \; MeV$$

$$= 5.126 \times 10^{23} \times 10^6 \times 1.6 \times 10^{-19} \; J = 8.2 \times 10^{10} \; J$$

$$= 8.2 \times 10^{10} \frac{J}{s}. \; s = 8.2 \times 10^{10} \; watt. \; s$$

$$= 8.2 \times 10^{10} \; watt. \; \frac{1}{3600} h = \frac{8.2 \times 10^{10}}{10^3 \times 3600} \; kwatt. \; hr$$

$$= 22 \; 778 \; kwatt. \; hr$$

**Exercise 5.2** *Find the mass of uranium which must undergo fission to produce the same energy as is produced by* $10^5$ *kg of coal that produces energy of* $3.5 \times 10^7 \; J \; kg^{-1}$.
[Ans.] Energy produced by $10^5$ *kg* of coal is

$$10^5 \times 3.5 \times 10^7 \frac{J}{kg} = 3.5 \times 10^{12} \; J$$

$$= \frac{3.5 \times 10^{12}}{1.6 \times 10^{-19}} eV = 2.18 \times 10^{31} \; eV$$

Energy released per fission = 200 *MeV* (take)

$$= 200 \times 10^6 \; eV = 2 \times 10^8 \; eV$$

Number of fissions required $= \frac{2.18 \times 10^{31} \; eV}{2 \times 10^8 \; eV} = 1.09 \times 10^{23} =$ number of uranium atoms needed

$$6.023 \times 10^{23} \text{ atoms are in 235 } g \text{ of uranium}$$

$1.09 \times 10^{23}$ atoms are in $\frac{235}{6.023 \times 10^{23}} 1.09 \times 10^{23} \; g = 42.5 g$ of uranium.

**Exercise 5.3** *Why is there no prompt and delayed proton emission in fission reaction?*
[Ans.] The fission fragments have protons—but their number is not in excess of what is required for stability of the fragment. So the question of emission of prompt and delayed protons does not arise.

**Exercise 5.4** *Show that energy of thermal neutron is* 0.026 *eV*.
[Ans.] Consider a neutron moving through a medium whose temperature is *T*.

The thermal energy that it possesses is $kT$ where $k$ is the Boltzmann constant having value $1.38 \times 10^{-23}\frac{J}{K}$.

For $T = 300\ K$ the energy of thermal neutron will be

$$kT = 1.38 \times 10^{-23}\frac{J}{K} \times 300\ K = 4.14 \times 10^{-21}\ J = \frac{4.14 \times 10^{-21}}{1.6 \times 10^{-19}}eV = 0.026\ eV$$

Neutrons that are in thermal equilibrium with the surroundings have this energy.

**Exercise 5.5** *The Q value of energy released in nuclear fission of* $U^{235}$ *is 200 MeV. Calculate the amount of energy released from 1 g of* $U^{235}$. *If 1 kg of coal, when burnt releases 8.93 kwatt. hr of energy, how much of it is required to burn to get the same amount of energy obtained from 1 g of* $U^{235}$?

Ans. 235 g of $U^{235}$ has $6.023 \times 10^{23}$ atoms.

So 1 g of $U^{235}$ will have $\frac{6.023 \times 10^{23}}{235}$ atoms.

Energy released by 1 g of $U^{235}$ is

$$\frac{6.023 \times 10^{23}}{235} \times 200\ MeV = 5.13 \times 10^{23}\ MeV$$

$$= 5.13 \times 10^{23} \times 10^{6} \times 1.6 \times 10^{-19}\ J = 8.2 \times 10^{10}\ J.$$

1 kg of coal supplies 8.93 kwatt. hr energy

$$= 8.93 \times 10^{3}J\ s^{-1} \times 60 \times 60\ s = 32 \times 10^{6}\ J.$$

So $32 \times 10^{6}\ J$ is obtained from 1 kg of coal.

Hence $8.2 \times 10^{10}\ J$ will be obtained from $\frac{1}{32 \times 10^{6}\ J}8.2 \times 10^{10}\ J = 2563\ kg.$

**Exercise 5.6** *Calculate in kWh how much energy is generated when 0.1 kg of* $_3Li^7$ *is converted to* $_2He^4$ *by proton bombardment. Atomic masses are 7.0183 amu for* $_3Li^7$, *4.0040 amu for* $_2He^4$ *and 1.0081 amu for* $_1H^1$.

Ans. The concerned reaction is given by

$$_3Li^7 + {}_1H^1 \rightarrow 2{}_2He^4 + Q$$

$$Q = m(_3Li^7) + m(_1H^1) - m(2{}_2He^4)$$

$$Q = 7.0183u + 1.0081u - 2(4.0040u) = 0.0184u$$

$$= 0.0184 \times 931\ MeV = 17.13\ MeV = 17.13 \times 10^{6} \times 1.6 \times 10^{-19}\ J$$

$$Q = 2.74 \times 10^{-12}\ J$$

This is the heat energy generated by 7.0183 amu of $_3Li^7$.

$$7.0183u = 7.0183 \times 1.67 \times 10^{-27} \, kg = 1.17 \times 10^{-26} \, kg.$$

So we can write that

$$1.17 \times 10^{-26} \, kg \text{ of } {}_3Li^7 \text{ generates } 2.74 \times 10^{-12} \, J \text{ of heat.}$$

$$0.1 \, kg \text{ of } {}_3Li^7 \text{ generates } \frac{2.74 \times 10^{-12} \, J}{1.17 \times 10^{-26} \, kg} \times 0.1 \, kg = 2.34 \times 10^{13} \, J$$

$$= 2.34 \times 10^{13} \frac{J}{s} \, s = 2.34 \times 10^{13} \, watt. \, \frac{1}{3600} hr = 6.5 \times 10^9 \, watt. \, hr = 6.5 \times 10^6 \, kWh$$

**Exercise 5.7** *A reactor is known to develop nuclear energy of 32 000 kW. How many nuclei of* $U^{235}$ *have to undergo fission per second? If a reactor is active for 1000 hrs find how many kg of* $U^{235}$ *would be used up? Assume energy per fission to be* $2 \times 10^8 \, eV$.

Ans.    Reactor    power    $32\,000 \, kW = 32\,000 \times 10^3 \, J \, s^{-1} = \frac{32\,000 \times 10^3}{1.6 \times 10^{-19}} eVs^{-1}$
$= 2 \times 10^{26} \, eV. \, s^{-1}$

Energy per fission $2 \times 10^8 \, eV$.

Number of fissions $\frac{2 \times 10^{26} \, eV/s}{2 \times 10^8 \, eV} = 10^{18} \frac{1}{s}$

This is the number of $U^{235}$ nuclei that undergo fission per second.

Number of fissions, i.e. number of $U^{235}$ nuclei fissioned in 1000 hrs is

$$10^{18} \frac{1}{s} \times 1000 \, hrs = 10^{18} \frac{1}{s} \times 1000 \times 3600 \, s = 3.6 \times 10^{24}$$

$$6.023 \times 10^{23} \text{ atoms are there in } 235 \, g \text{ of } U^{235}$$

$$3.6 \times 10^{24} \text{ atoms are there in } \frac{235}{6.023 \times 10^{23}} \times 3.6 \times 10^{24} \, g \text{ of } U^{235} = 1.41 \times 10^3 \, g = 1.41 \, kg$$

So 1.41 $kg$ of $U^{235}$ will be used up.

**Exercise 5.8** *If energy released per fission of* $U^{235}$ *is 200 MeV then calculate the fission rate of* $U^{235}$ *that produces 2 W. Also, find the amount of energy that is liberated in a complete fission of 500 g of* $U^{235}$.

Ans. Power needed $2 \, W = 2 \frac{J}{s} = \frac{2}{1.6 \times 10^{-19}} \frac{eV}{s} = 1.25 \times 10^{19} \frac{eV}{s}$

Energy released per fission 200 $MeV$

Number of fissions per second is

$$\frac{1.25 \times 10^{19} \, eV/s}{200 \, MeV} = \frac{1.25 \times 10^{19} \, eV/s}{200 \times 10^6 \, eV} = 6.25 \times 10^{10} \frac{1}{s} = \text{Fission rate}$$

$$235 \ g \ \text{of} \ U^{235} \ \text{has} \ 6.023 \times 10^{23} \ \text{atoms}$$

$$500 \ g \ \text{of} \ U^{235} \ \text{has} \ \frac{6.023 \times 10^{23}}{235 \ g} \times 500 \ g = 1.2815 \times 10^{24} \ \text{atoms}$$

Energy liberated due to fission of $1.2815 \times 10^{24}$ atoms is

$$1.2815 \times 10^{24} \times 200 \ MeV = 2.563 \times 10^{26} \ MeV$$

**Exercise 5.9** *Fission of $_{92}U^{235}$ (235.0439u) occurs by slow neutrons (1.0087u) and the fission fragments are $_{36}Kr^{92}$(91.8973u) and $_{56}Ba^{141}$(140.9139u). Write down the complete fission reaction. Find the energy produced in one fission. If 1 g of $_{92}U^{235}$ is fissioned calculate the total energy produced in kWhr.*

Given fission reaction is

$$_0n^1 + {}_{92}U^{235} \rightarrow {}_{36}Kr^{92} + {}_{56}Ba^{141} + 3{}_0n^1 + Q$$

$$Q = m(_0n^1) + m(_{92}U^{235}) - m(_{36}Kr^{92}) - m(_{56}Ba^{141}) - m(3_0n^1)$$

$$Q = 1.0087u + 235.0439u - 91.8973u - 140.9139u - m(3 \times 1.0087u)$$

$$= 0.2153u = 0.2153 \times 931 \ MeV = 200.44 \ MeV$$

This is the energy produced in one fission.

$$235 \ g \ \text{of} \ U^{235} \ \text{contains} \ 6.023 \times 10^{23} \ \text{atoms.}$$

$$1 \ g \ \text{of} \ U^{235} \ \text{contains} \ \frac{6.023 \times 10^{23}}{235 \ g} \times 1 \ g = 2.56 \times 10^{21} \ \text{atoms}$$

Energy produced is

$$200.44 \ MeV \ \text{per atom} \times 2.56 \times 10^{21} \ \text{atoms}$$

$$= 5.13 \times 10^{23} \ MeV = 5.13 \times 10^{23} \times 10^6 \times 1.6 \times 10^{-19} \ J$$

$$= 8.2 \times 10^{10} \ J$$

$$= 8.2 \times 10^{10} \frac{J}{s} s = 8.2 \times 10^{10} \ watt \frac{1}{3600} \ hr$$

$$= 2.28 \times 10^7 \ watt \ hr = 2.28 \times 10^4 \ kw \ hr$$

**Exercise 5.10** *0.1% mass loss occurs in a fission process. The energy released due to fission of 1 kg of the substance will be which of the following?*
   *(a) $9 \times 10^{13} \ J$, (b) $5.6 \times 10^{26} \ MeV$, (c) $2.5 \times 10^7 \ kwhr$, (d) $5.6 \times 10^{32} \ eV$, (e) All, (f) None.*

Hint: Loss of mass is 0.1% of $1\ kg = \frac{0.1}{100} \times 1\ kg = 0.001\ kg$

$$E = mc^2 = (0.001\ kg)\left(3 \times 10^8 \frac{m}{s}\right)^2 = 9 \times 10^{13}\ J$$

$$= 9 \times 10^{13} \frac{J}{s} s = 9 \times 10^{13}\ watt \frac{1}{3600} hr = 2.5 \times 10^{10}\ watt\ hr = 2.5 \times 10^7\ kwhr$$

$$= \frac{9 \times 10^{13}}{1.6 \times 10^{-19}}\ eV = 5.6 \times 10^{32}\ eV = 5.6 \times 10^{26}\ MeV$$

**Exercise 5.11** *The fission of 0.001 g of* $U^{235}$ *leads to X amount of energy which is y times less than the energy derived due to fission of 1 kg of* $U^{235}$*. Which of the following correctly gives* (*X, y*)*?*
(*a*) $5.12 \times 10^{20}\ MeV$, $10^4$,       (*b*) $8.192 \times 10^3\ J$, $10^6$,
(*c*) $2.28 \times 10^4\ watt\ hr$, $10^4$,    (*d*) $22.8\ kWhr$, $10^6$.
Hint: ✓ For 0.001 g of $U^{235}$

$$235\ g\ \text{of}\ U^{235}\ \text{contains}\ 6.023 \times 10^{23}\ \text{atoms}$$

$$0.001\ g\ \text{of}\ U^{235}\ \text{contains}\ \frac{6.023 \times 10^{23}}{235\ g} \times 0.001\ g\ \text{atoms} = 2.56 \times 10^{18}\ \text{atoms}$$

Per fission 200 $MeV$ energy is liberated (take).
So 0.001 g $U^{235}$ will liberate energy

$$2.56 \times 10^{18} \times 200\ MeV = 5.12 \times 10^{20}\ MeV = X.$$

$$= 5.12 \times 10^{20} \times 10^6 \times 1.6 \times 10^{-19}\ J = 8.192 \times 10^7\ J = X$$

$$= 8.192 \times 10^7 \frac{J}{s} s = 8.192 \times 10^7\ watt \frac{1}{3600}\ hr = 2.28 \times 10^4\ watt\ hr = X$$

$$= 22.8\ kWh = X$$

✓ For 1 $kg = 1000$ g of $U^{235}$

$$235\ g\ \text{of}\ U^{235}\ \text{contains}\ 6.023 \times 10^{23}\ \text{atoms}$$

$$1000\ g\ \text{of}\ U^{235}\ \text{contains}\ \frac{6.023 \times 10^{23}}{235\ g} \times 1000\ g\ \text{atoms} = 2.56 \times 10^{24}\ \text{atoms}$$

Per fission 200 $MeV$ energy is liberated (take).

So 1 $kg$ $U^{235}$ will liberate energy $2.56 \times 10^{24} \times 200$ $MeV = 5.12 \times 10^{26}$ $MeV$.

$$= 5.12 \times 10^{26} \times 10^6 \times 1.6 \times 10^{-19} J = 8.192 \times 10^{13} J$$

$$= 8.192 \times 10^{13} \frac{J}{s} s = 8.192 \times 10^{13} \, watt \frac{1}{3600} hr$$

$$= 2.28 \times 10^{10} \, watt \, hr = 2.28 \times 10^7 \, kw. \, hr = yx$$

Clearly $y = \dfrac{5.12 \times 10^{26} \, MeV}{5.12 \times 10^{20} \, MeV} = 10^6$; $y = \dfrac{8.192 \times 10^{13} \, J}{8.192 \times 10^7 \, J} = 10^6$

$y = \dfrac{2.28 \times 10^{10} \, watt \, hr}{2.28 \times 10^4 \, watt \, hr} = 10^6$; $y = \dfrac{2.28 \times 10^7 \, kW \, hr}{22.8 \, kW \, hr} = 10^6$

**Exercise 5.12** *Which of the following materials can be used as a fuel in fission process?*
(a) $_{92}U^{234}$, (b) $_{92}U^{235}$, (c) $_{93}U^{236}$, (d) $_{94}U^{238}$.

**Exercise 5.13** *Which of the following materials can be used as a fuel in fission process?*
(a) $_{92}U^{235}$, (b) $_{36}Kr^{86}$, (c) $_{7}N^{14} + {}_{8}O^{16}$, (d) $_{1}D^2 + {}_{1}T^3$.

**Exercise 5.14** *In thermal fission the fission fragments are seen to be $\beta$ active because of excess*
(a) proton, (b) neutron, (c) neutrino, (d) electron.

**Exercise 5.15** *Slow neutron can cause fission in which of the following?*
(a) $_{92}U^{234}$, (b) $_{92}U^{235}$, (c) $_{92}U^{238}$, (d) $_{92}U^{239}$.

**Exercise 5.16** *If the multiplication factor k corresponding to an assembly of finite size has a value $k = 1$ then the size of the assembly is*
(a) not critical, (b) sub-critical, (c) critical, (d) super-critical.

**Exercise 5.17** *Which of the following is fissionable by fast neutron?*
(a) $_{92}U^{234}$, (b) $_{92}U^{235}$, (c) $_{92}U^{238}$, (d) $_{92}U^{239}$.

**Exercise 5.18** *Which of the following materials is/are fertile?*
*(a)* $_{92}U^{238}$, *(b)* $_{92}U^{235}$, *(c)* $_{90}Th^{232}$, *(d)* $_{92}U^{239}_{147}$.

**Exercise 5.19** *Thermal fission occurs in which of the following?*
*(a) PWR, (b) PHWR, (c) breeder reactor, (d) none.*

**Exercise 5.20** *What is used as moderator in a conventional nuclear reactor?*
*(a) Water, (b) heavy water, (c) liquid sodium, (d) cadmium.*

**Exercise 5.21** *What is used as moderator in a breeder reactor?*
*(a) Water, (b) heavy water, (c) liquid sodium, (d) moderator not used.*

**Exercise 5.22** *What is used as coolant in a breeder reactor?*
*(a) Water, (b) heavy water, (c) liquid sodium, (d) cadmium.*

**Exercise 5.23** *A fast fission reactor is which of the following?*
*(a) PWR, (b) PHWR, (c) breeder reactor, (d) none.*

**Exercise 5.24** *Which of the following materials is fissile?*
*(a)* $_{92}U^{238}$, *(b)* $_{92}U^{235}$, *(c)* $_{90}Th^{232}$, *(d)* $_{92}U^{239}_{147}$, *(e)* $_{94}Pu^{239}_{145}$.

**Exercise 5.25** *The de Broglie wavelength of thermal neutron is which of the following?*
*(a) 1.8 nm, (b) 1.8Å, (c) 0.18 pm, (d) 0.18 nm.*
Hint: Energy of a thermal neutron is $0.026 \, eV = 0.026 \times 1.6 \times 10^{-19} \, J$.
The de Broglie wavelength is given by

$$\lambda = \frac{h}{p} = \frac{h}{\sqrt{2mE}}$$

$$= \frac{6.626 \times 10^{-34} \, J \, s}{\sqrt{2(1.675 \times 10^{-27} \, kg)(0.026 \, eV)}} = \frac{6.626 \times 10^{-34} \, J \, s}{\sqrt{2(1.675 \times 10^{-27} \, kg)(0.026 \times 1.6 \times 10^{-19} \, J)}}$$

$$\lambda = 0.18 \times 10^{-9} \, m = 0.18 \, nm$$

**Exercise 5.26** *Choose the most appropriate condition regarding occurrence of spontaneous fission.*

(a) $\frac{Z^2}{A} > 50$, (b) $\frac{Z^2}{A} < 50$, (c) $\frac{Z^2}{A} < 17$, (d) $\frac{Z^2}{A} < 50$.

**Exercise 5.27** *Which material is used as control rod in a nuclear reactor?*
(a) Water, (b) heavy water, (c) liquid sodium, (d) cadmium.

**Exercise 5.28** *Fission of* $U^{235}$ *yields two fragments of mass numbers 95, 140. If the fission products are emitted with equal and opposite momentum, the energy distribution of the fragments will be which of the following?*
(a) 28: 19, (b) $28^2$: $19^2$, (c) $\sqrt{19} : \sqrt{28}$, (d) 19: 28.
Hint. Fission fragments have equal and opposite momentum. Hence

$$P_{95} = P_{140}$$

Energy ratios will be

$$\frac{E_{95}}{E_{140}} = \frac{p_{95}^2/2m_{95}}{p_{140}^2/2m_{140}} = \frac{m_{140}}{m_{95}} = \frac{140}{95} = \frac{28}{19}$$

**Exercise 5.29** *Suppose fission process begins with 100 neutrons. Multiplication factor is 1.05. The number of neutrons in the 100th generation will approximately be which of the following?*
(a) 1000, (b) 9000, (c) 13 000, (d) 130 000.
Hint: Number of nuclei in the first generation is say $N_0$. This is the initial number of nuclei.
Number of nuclei in the second generation is $N_0 k = N_0 k^{2-1}$.
Number of nuclei in the third generation is $(N_0 k)k = N_0 k^2 = N_0 k^{3-1}$.
Number of nuclei in the $n$th generation is $N_0 k^{n-1}$.
Hence number of nuclei in the 100th generation is

$$N_0 k^{100-1} = 100 \times 1.05^{100-1} = 1.25 \times 10^4 \cong 13\,000$$

**Exercise 5.30** *A nuclear power station has a capacity of 60 000 kW. 20% of energy generated is converted to electricity. If relative abundance of* $U^{235}$ *is 0.7% then the amount of natural uranium spent per year is given by*
(a) 1600 kg, (b) 67 000 kg, (c) 16 500 kg, (d) 25 700 kg.
Hint: 235 g of $U^{235}$ has $6.023 \times 10^{23}$ atoms

$$1 \text{ kg of } U^{235} \text{ has } \frac{6.023 \times 10^{23}}{235 \text{ g}} \times 1000 \text{ g} = 2.56 \times 10^{24} \text{ atoms}$$

Energy per fission is 200 *MeV*.

Energy from 1 $kg$ of uranium is $2.56 \times 10^{24} \times 200 \, MeV$

$$= 2.56 \times 10^{24} \times 200 \times 10^6 \times 1.6 \times 10^{-19} \, J = 8.192 \times 10^{13} \, J$$

$$= 8.192 \times 10^{13} \frac{J}{s} s = 8.192 \times 10^{13} \, watt \frac{1}{3600} \, hr$$

$$= 2.276 \times 10^{10} \, watt \, hr = 2.276 \times 10^7 \, kWhr$$

Energy for electricity from 1 $kg$ of uranium is 20% of $2.276 \times 10^7 \, kWhr$

$$= \frac{20}{100} \times 2.276 \times 10^7 \, kWhr = 4.55 \times 10^6 \, kWhr.$$

Amount of $U^{235}$ spent per hour$\frac{60\,000 \, kW}{4.55 \times 10^6 \, kWhr} = 13.1868 \times 10^{-3} \, kg \, hr^{-1}$

Consumption of $U^{235}$ per year is $13.1868 \times 10^{-3}$ $kg. \, hr^{-1} \times 365 \times 24 \, hr$ $= 115.5 \, kg$

Consumption of natural uranium per year $= \frac{100}{0.7} \times 115.5 \, kg = 16\,500 \, kg$

**Exercise 5.31** *The power of a nuclear reactor is $10^9$ W. How many free neutrons are present in the reactor if the average time lapse between emission of a prompt neutron and its fission capture to initiate the next generation is $10^{-3}$ s?*

(a) $10^{10}$, (b) $10^{12}$, (c) $10^{14}$, (d) $10^{16}$.

Hint: Energy per fission by one neutron is

$$200 \, MeV = 200 \times 10^6 \times 1.6 \times 10^{-19} \, J = 3.2 \times 10^{-11} \, J$$

One neutron produces power $= \frac{energy}{time} = \frac{3.2 \times 10^{-11} \, J}{10^{-3} \, s} = 3.2 \times 10^{-8} \, W$

Number of free neutrons in the reactor is $\frac{10^9 \, W}{3.2 \times 10^{-8} \, W} = 3.125 \times 10^{16}$

**Exercise 5.32** *Show that the ratio of final to initial kinetic energy of a neutron upon collision with a moderator nucleus of mass number A can be written as*

$\frac{E_2}{E_1} = \frac{1}{2}[1 + \alpha + (1 - \alpha)\cos\theta_c]$

*where* $= \frac{(A-1)^2}{(A+1)^2}$, $\theta_c =$ *angle of scatterer in C-frame*

*Hence find the amount of energy transfer for $\theta_c = 0°$, $180°$.*

**Ans.** From equation (5.117) viz. $\frac{E_2}{E_1} = \frac{1 + A^2 + 2A\cos\theta_c}{(1+A)^2}$

Multiplying numerator and denominator by 2 we have

$$\frac{E_2}{E_1} = \frac{2 + 2A^2 + 4A\cos\theta_c}{2(1+A)^2} = \frac{(1 + A^2 + 2A) + (1 + A^2 - 2A) + 4A\cos\theta_c}{2(1+A)^2}$$

$$= \frac{1}{2}\left[\frac{(1+A)^2 + (1-A)^2 + 4A\cos\theta_c}{(1+A)^2}\right]$$

$$= \frac{1}{2}\left[1 + \frac{(1-A)^2}{(1+A)^2} + \frac{4A\cos\theta_c}{(1+A)^2}\right] = \frac{1}{2}[1 + \alpha + \frac{(1+A)^2 - (1-A)^2}{(1+A)^2}\cos\theta_c]$$

$$= \frac{1}{2}[1 + \alpha + \left(1 - \frac{(1-A)^2}{(1+A)^2}\right)\cos\theta_c]$$

$$= \frac{1}{2}[1 + \alpha + (1-\alpha)\cos\theta_c]$$

For $\theta_c = 0°$, $\cos\theta_c = 1$, $\frac{E_2}{E_1} = \frac{1}{2}[1 + \alpha + (1-\alpha)] = 1$, i.e.

$E_2 = E_1 \Leftarrow$ amount of energy transfer for $\theta_c = 0°$.

For $\theta_c = 180°$, $\cos\theta_c = -1$, $\frac{E_2}{E_1} = \frac{1}{2}[1 + \alpha - (1-\alpha)] = \alpha$, i.e.

$E_2 = \alpha E_1 \Leftarrow$ amount of energy transfer for $\theta_c = 180°$.

**Exercise 5.33** *Derive the survival equation in neutron diffusion. Show that mean free path and macroscopic cross-sections are related as* $\lambda = \frac{1}{\Sigma}$.

Ans. Consider that as a neutron moves forward along the $x$ direction, neutron flux

$$\phi = \frac{\text{number of neutrons}}{m^2\ s}$$

diminishes because of interaction suffered by the neutron.

Let $dx$ be the elemental length travelled by neutron leading to reduction of neutron flux $-d\phi$

$\sigma$ is microscopic cross-section ($m^2$)

Nuclear density of moderator is

$$N = \frac{\text{number of nuclei}}{m^3}$$

Rate of interaction is

$$\sigma\phi NAdx \sim m^2 \cdot \frac{1}{m^2\ s} \cdot \frac{\text{number of nuclei}}{m^3} \cdot m^2 \cdot m = \frac{1}{s}$$

Also,

$$Ad\phi \sim m^2 \frac{1}{m^2\ s} = \frac{1}{s}$$

So the change of neutron flux is governed by the equation

$$\sigma\phi NAdx = -Ad\phi \tag{5.262}$$

where negative sign indicates decrease in neutron flux

$$\sigma\phi Ndx = -d\phi$$

Using $N\sigma = \Sigma$

$$\Sigma dx = -\frac{d\phi}{\phi} = -dln\phi$$

Integration gives, taking $\phi(x = 0) = \phi_0$

$$\int_{x=0}^{x} \Sigma dx = -\int_{\phi_0}^{\phi} dln\phi$$

$$-\Sigma x = ln\frac{\phi}{\phi_0}$$

$$\phi(x) = \phi_0 e^{-\Sigma x}$$

This equation characterizes how many neutrons survive collision with moderator nuclei and is called survival equation.

Let us find the average distance travelled before interaction occurs which is called mean free path $\lambda$ and defined as

$$\lambda = \frac{\int_0^\infty x d\phi}{\int_0^\infty d\phi} = \frac{1}{\phi_0} \int_0^\infty x d\phi$$

$$= \frac{1}{\phi_0} \int_0^\infty x d\phi_0 e^{-\Sigma x} = \int_0^\infty x de^{-\Sigma x}$$

$$= -\Sigma \int_0^\infty x e^{-\Sigma x} dx = -\Sigma[x \int e^{-\Sigma x} dx \mid_0^\infty - \int_0^\infty \frac{d}{dx} x \left( \int e^{-\Sigma x} dx \right) dx]$$

$$= -\Sigma[x \frac{e^{-\Sigma x}}{-\Sigma} \mid_0^\infty - \int_0^\infty \frac{e^{-\Sigma x}}{-\Sigma} dx]$$

$$= -\Sigma \left[ 0 - \int_0^\infty \frac{e^{-\Sigma x}}{-\Sigma} dx \right] = -\int_0^\infty e^{-\Sigma x} dx = -\frac{e^{-\Sigma x}}{-\Sigma} \mid_0^\infty = \frac{1}{\Sigma}$$

(we neglect negative sign in the final step as it indicates reduction in neutron population with increase in distance).

**Exercise 5.34** *Find energy released in the fusion process* D + D → He *if binding energy of deuteron nucleus is* 1.1 MeV. nucleon$^{-1}$ *and that of helium is* 7 MeV. nucleon$^{-1}$.

Ans. $_1H^2 + _1H^2 \rightarrow _2He^4$

Energy required to break two nucleons in $_1H^2$ is 2 × 1.1 MeV = 2.2 MeV
Energy required to break four nucleons in two $_1H^2$ is 2 × 2.2 MeV = 4.4 MeV
Energy released to form $_2He^4$ is 4 × 7 MeV = 28 MeV
Hence energy released in the fusion process D + D → He is

$$28 \; MeV - 4.4 \; MeV = 23.6 \; MeV$$

**Exercise 5.35** *Complete conversion of all hydrogen to helium atoms of the sun occurs in time*
(*a*) $3 \times 10^{20}$ *yrs,* (*b*) $3 \times 10^{15}$ *yrs,* (*c*) $2 \times 10^8$ *yrs,* (*d*) $3 \times 10^{10}$ *yrs.*

**Exercise 5.36** *The origin of energy of sun is which of the following?*
(*a*) *thermonuclear reaction,* (*b*) *nuclear fusion,* (*c*) *nuclear fission,* (*d*) *nucleosynthesis.*

**Exercise 5.37** *The energy source of the sun is due to the formation of helium nucleus with four hydrogen nuclei. Calculate the energy released when 1 g of hydrogen is fused into helium. Also, estimate how much hydrogen is to be converted to helium in the sun per second. Take mass of hydrogen nucleus* $1.008\ 13u$*, mass of He nucleus* $4.003\ 86u$*. Also, the solar constant is* $1.35\ kW.\ m^{-2}$*, earth to sun distance* $1.5 \times 10^8\ km$*.*

Ans. ✓4 hydrogen nuclei fuse to form one helium nucleus.
Energy released is

$$(4m_{\mathrm{H}} - m_{\mathrm{He}})c^2 = 4 \times 1.008\ 13u - 4.003\ 86u = 0.028\ 66u$$

$$= 0.028\ 66 \times 931\ MeV = 26.68\ MeV$$

This is the energy released because of creation of one helium nuclei due to fusion of four hydrogen nuclei.

1 $g$ contains $6.023 \times 10^{23}$ hydrogen atoms.

Number of sets of four hydrogen atoms will be $\frac{6.023 \times 10^{23}}{4}$

Hence energy released due to fusion of 1 $g$ of hydrogen is

$$26.68\ MeV \times \frac{6.023 \times 10^{23}}{4} = 4 \times 10^{24}\ MeV.$$

✓ 4 hydrogen, upon fusion produce energy $26.68\ MeV$.

One hydrogen corresponds to energy amount

$$\frac{26.68}{4}\ MeV = \frac{26.68}{4} \times 10^6 \times 1.6 \times 10^{-19}\ J = 1.07 \times 10^{-12}\ J$$

$$\text{Solar constant} = 1.35\ kW.\ m^{-2}$$

Energy radiated per sec per unit area from sun is $1.35 \times 10^3\ J\ s^{-1}\ m^{-2}$
Energy radiated per sec from sun in a sphere of radius $1.5 \times 10^8\ km$ is

$$1.35 \times 10^3 J\ s^{-1}\ m^{-2} \times 4\pi(1.5 \times 10^{11})^2\ m^2 = 3.8 \times 10^{26} J\ s^{-1}$$

$$\text{Number of hydrogen atoms required is } \frac{3.8 \times 10^{26}\frac{J}{s}}{1.07 \times 10^{-12}\ J} = 3.55 \times 10^{38}\frac{1}{s}$$

Mass of hydrogen needed is

$$3.55 \times 10^{38} \frac{1}{s} \times m_p = 3.55 \times 10^{38} \frac{1}{s} \times 1.67 \times 10^{-27} \, kg$$

$$= 5.9 \times 10^{11} \, kg$$

**Exercise 5.38** *The temperature of the core and the photosphere of the Sun is which of the following?*
   *(a)* $10^5 \, K, 10^3 \, K$, *(b)* $10^7 \, K$, $6000 \, K$, *(c)* $10^9 \, K \, 8000 \, K$, *(d)* $10^{11} \, K, 10^5 \, K$.

| Answer to Mulitiple Choice Type Questions |
|---|

5.10 *abcd*, 5.11*d*, 5.12*b*, 5.13*a*, 5.14*b*, 5.15*b*, 5.16*c*, 5.17*c*, 5.18*ac*, 5.19*ab*, 5.20*ab*, 5.21*d*, 5.22*c*, 5.23*c*, 5.24*be*, 5.25*d*, 5.26*a*, 5.27*d*, 5.28*a*, 5.29*c*, 5.30*c*, 5.31*d*, 5.35*d*, 5.36*b*, 5.38*b*

## 5.52 Question bank

Q5.1   What amount of energy is released in one fission, one fusion?
Q5.2   Mention a few devices that run on nuclear fission.
Q5.3   Write the reactions describing nuclear fission, fusion and explain the energy conversions.
Q5.4   What is the speed (fast or slow) of neutrons released during nuclear fission?
Q5.5   Mention the size (equal or unequal) of nuclear fragments produced during fission by a slow neutron.
Q5.6   What do we mean by symmetric and asymmetric fission?
Q5.7   Show in a plot the fission yield against mass number, against kinetic energy of fission fragments.
Q5.8   What are prompt neutrons?
Q5.9   What are delayed neutrons?
Q5.10  Estimate the amount of energy release during nuclear fission from the binding energy curve.
Q5.11  Estimate the amount of energy release during nuclear fission from mass of fission fragments.
Q5.12  What do we mean by spontaneous fission?
Q5.13  What is induced fission?
Q5.14  What do we mean by fissile material? Give an example.
Q5.15  What do we mean by fertile material? Give an example.
Q5.16  Write down the nuclear reactions that convert fertile materials into fissile materials.

Q5.17    What are breeding reactions?

Q5.18    Find the minimum value of mass number for fission to be energetically possible.

Q5.19    Find the condition for symmetric spontaneous fission to be energetically possible.

Q5.20    What is fission activation energy?

Q5.21    Outline Bohr–Wheeler's theory on the basis of the liquid drop model.

Q5.22    Discuss whether we can build nucleus with mass number greater than 300.

Q5.23    Explain nuclear fission by the liquid drop model.

Q5.24    What is fission parameter? What is its significance?

Q5.25    Check stability w.r.t. spontaneous fission of a $Z = 117$, $A = 270$ nucleus and a $Z = 117$, $A = 280$ nucleus.

Q5.26    Explain fission of the following elements: $_{92}U^{235}$ and $_{92}U^{238}$.

Q5.27    Classify neutrons in terms of their speed.

Q5.28    What is $\frac{1}{v}$ effect in relation to nuclear fission?

Q5.29    What is a nuclear reactor? What are its components?

Q5.30    Draw a schematic diagram of a nuclear reactor.

Q5.31    Why are moderators used in a nuclear reactor?

Q5.32    What is enriched uranium? Mention uses of low enriched uranium and high enriched uranium.

Q5.33    What is the utility of control rods in a nuclear reactor?

Q5.34    What is fission chain reaction? Under what condition is it a controlled chain reaction? What happens if it is uncontrolled?

Q5.35    Define and explain the following terms in relation to nuclear fission in a nuclear reactor: neutron reproduction factor, critical condition, super-critical condition, sub-critical condition.

Q5.36    Which of the following conditions are favoured in a nuclear reactor: critical condition, super-critical condition, sub-critical condition?

Q5.37    Calculate the power output of a nuclear reactor.

Q5.38    Obtain the four-factor formula.

Q5.39    Under what condition does the four-factor formula reduce to two-factor formula?

Q5.40    What do we mean by reactivity of a nuclear reactor?

Q5.41    Establish that the rate of nuclear fission reaction increases drastically if reproduction factor differs slightly from unity.

Q5.42    Show in a neat diagram a schematic design of a pressurised water reactor and a pressurised heavy water reactor.

Q5.43    How can one run a turbine with a nuclear reactor? Show in a schematic diagram the design of such a nuclear reactor.

Q5.44    What is a breeder reactor?

Q5.45    What is bred in a breeder reactor? Mention some differences between a breeder reactor and a conventional reactor.

Q5.46    What do we mean by a homogeneous and a heterogeneous reactor?

Q5.47     What safety measures are expected in a nuclear reactor?

Q5.48     Prove that light nuclei are more effective moderators than heavy nuclei. Show that the maximum loss of energy is proportional to the initial energy.

Q5.49     Find the amount of maximum energy transferred from neutron to moderator and under what condition.

Q5.50     Inter-relate the scattering angles in L-frame and C-frame. Show that they are the same for nuclei of very high mass number.

Q5.51     Prove that $\cos\theta = \cos\theta_c + \frac{1}{A}$ where $A$ is mass number, $\theta$, $\theta_c$ are scattering angles in L-frame and C-frame, respectively.

Q5.52     What is average logarithmic energy decrement? How is it related to the average number of collisions?

Q5.53     Show that the same fraction of neutron energy is transferred to the moderator nucleus in each successive collision.

Q5.54     Define and explain slowing down power, moderating ratio, slowing down density.

Q5.55     What is Fick's law?

Q5.56     Define the following terms: microscopic and macroscopic cross-section, scattering mean free path, transport mean free path, absorption mean free path.

Q5.57     Obtain the general form of diffusion equation for monoenergic neutrons. Imposing suitable boundary conditions, reduce it for the case of a point source and obtain a solution for neutron flux.

Q5.58     What is thermal diffusion length? What does it represent?

Q5.59     What is fast diffusion length? What does it represent?

Q5.60     Obtain the Fermi age equation. What does the Fermi age equation represent? Define Fermi age or neutron age. Establish that it does not represent a time.

Q5.61     Obtain a solution of the Fermi age equation. Calculate the energy realsed during nuclear fusion from a binding energy curve.

Q5.62     What is the expression of the Coulomb barrier that prevents fusion of two nuclei? What is the order of its value?

Q5.63     What is thermonuclear fusion? Mention the condition for it to occur.

Q5.64     Obtain a rough expression for the reaction rate of fusion.

Q5.65     Mention some possible reactions in a fusion reactor. Identify the problems of a fusion reactor.

Q5.66     What is the Lawson criterion for nuclear fusion?

Q5.67     What is plasma? Explain the following terms: magnetic mirror, magnetic bottle, magnetic confinement.

Q5.68     Briefly sketch the Tokamak design.

Q5.69     How many layers does the sun consist of? Show in a diagram mentioning their size and temperature. What is the temperature of the photosphere? What is the composition of the photosphere?

Q5.70     What is the source of energy release in the sun?

Q5.71   Describe the proton–proton cycle, carbon–nitrogen cycle in relation to energy release from the sun.

Q5.72   How is information obtained from the core of the sun?

Q5.73   What is neutrino oscillation?

Q5.74   Can a neutrino of one flavour be converted to a neutrino of another flavour?

## Further reading

[1] Krane S K 1988 *Introductory Nuclear Physics* (New York: Wiley)

[2] Tayal D C 2009 *Nuclear Physics* (Mumbai: Himalaya Publishing House)

[3] Satya P 2005 *Nuclear Physics and Particle Physics* (New Delhi: Sultan Chand & Sons)

[4] Guha J 2019 *Quantum Mechanics: Theory, Problems and Solutions* 3rd edn (Kolkata: Books and Allied (P) Ltd))

[5] Lim Y-K 2002 *Problems and Solutions on Atomic, Nuclear and Particle Physics* (Singapore: World Scientific)

**IOP** Publishing

# Nuclear and Particle Physics with Cosmology, Volume 1
### Nuclear physics
**Jyotirmoy Guha**

# Chapter 6

## Deuteron problem

## 6.1 Introduction

Nucleus is made of protons and neutrons which are collectively called nucleons. We are interested to study the force between the nucleons that hold them together and build the nucleus.

Deuteron $_1\text{H}_1^2 \equiv d \equiv np$ is the only stable bound system consisting of two nucleons (neutron and proton). It is the simplest system upon which investigation can be done to know the nature of nuclear force. Deuteron is an ideal system to investigate nucleon–nucleon potential.

Nature does not provide any bound system of two nucleons other than deuteron which is an $np$ system built with a specific spin combination—that is preferred by nature.

We can think of a di-proton system, i.e. a system of two protons or a $pp$ system. But such a stable bound system does not exist.

We can also think of a di-neutron system, i.e. a system of two neutrons or an $nn$ system. But such a stable bound system does not exist.

- We consider a deuteron system as an ideal system for investigating the nature of nuclear force because of the following two reasons.

✓ In the case of nuclei with mass number $A > 2$ there are more than two nucleons in the nucleus and so such a nucleus has to be treated as a many-body problem. And it is very difficult to handle a many-body problem.
✓ A nucleus with mass number $A > 2$ will contain multiple protons and there would come into play Coulomb force of repulsion between them inside the nucleus. This means that in a larger nucleus pure nuclear force does not operate. In the nucleus there will be present nuclear force as well as Coulomb force—though Coulomb force is of smaller magnitude compared to a force of nuclear origin.

However, in deuteron there is only one proton and so it cannot exert any Coulomb force. The neutron is uncharged and so proton cannot electromagnetically

doi:10.1088/978-0-7503-5027-3ch6

interact with it. The proton interacts with the neutron only through nuclear force. So there is no Coulomb force inside a deuteron. Force between $n$ and $p$ inside the deuteron nucleus will be a force only of nuclear origin.

- It is evident that a system of just two nucleons—neutron and proton—is a very simple system and would be easy to handle and we can gather a lot of information about nuclear force from it.

## 6.2 Description of nucleons through quarks

The structure of proton and neutron, i.e. a nucleon, can be studied by sending high energy projectile particles like electrons of energy $\sim GeV$, having de Broglie wavelength $\ll 10^{-15}\,m \equiv fm$ and analyzing the angular distribution of scattered particles.

Electron is a structureless particle.

But nucleons have an internal structure—called quark structure. They are built out of point-like particles called quarks. In other words, nucleons are made of quarks.

Matter in the universe is made of two kinds of particles—quarks and leptons.

There are six varieties of leptons. They are:

electron $e^-$

negative muon $\mu^-$

negative tauon $\tau^-$

electron neutrino $\nu_e$

muon neutrino $\nu_\mu$

tauon neutrino $\nu_\tau$

and their respective six anti-leptons. They are:

positron $e^+$

positive muon $\mu^+$

positive tauon $\tau^+$

electron anti neutrino $\overline{\nu}_e$

muon anti neutrino $\overline{\nu}_\mu$

tauon anti neutrino $\overline{\nu}_\tau$.

Quarks are of six varieties or flavours. The flavours are

up quark $u$

down quark $d$

strange quark $s$

charm quark $c$

bottom quark $b$

top quark $t$

and there are their six respective antiquarks. The flavours of antiquarks are

anti-up quark $\overline{u}$

anti-down quark $\overline{d}$

anti-strange quark $\overline{s}$

anti-charm quark $\overline{c}$

anti-bottom quark $\overline{b}$

anti-top quark $\overline{t}$.

✓ The quarks have fractional charge. And also

$Q$(quark) $= -Q$(antiquark).
Let us mention the charges.

$$Q(u, c, t) = \frac{2}{3} \mid e \mid, \ Q(d, s, b) = -\frac{1}{3} \mid e \mid \text{ and}$$

$$Q(\bar{u}, \bar{c}, \bar{t}) = -\frac{2}{3} \mid e \mid, \ Q(\bar{d}, \bar{s}, \bar{b}) = \frac{1}{3} \mid e \mid$$

- According to quark structure a proton is built of three quarks, namely

$$p = uud.$$

Let us evaluate the charge of a proton from quark structure.

$$Q(p) = Q(uud) = Q(u) + Q(u) + Q(d)$$

$$= \frac{2}{3} \mid e \mid + \frac{2}{3} \mid e \mid - \frac{1}{3} \mid e \mid = \mid e \mid.$$

This means that the unit of positive charge carried by proton comes from the fractional charges carried by the constituent quarks.
- According to quark structure a neutron is built of three quarks, namely

$$n = ddu.$$

Let us evaluate the charge of a neutron from quark structure.

$$Q(n) = Q(duu) = Q(d) + Q(d) + Q(u)$$

$$= -\frac{1}{3} \mid e \mid - \frac{1}{3} \mid e \mid + \frac{2}{3} \mid e \mid = 0.$$

This means that though neutron is overall neutral (carries no net charge), it is made up of charged particles called quarks. In other words, charged quarks are the constituents of the neutral neutron. So neutron is built out of charged particles though it is overall uncharged.
- The forces that originate between nucleons are essentially the forces between the quarks. The nature of force or interaction is strong force.

It is known that gravitational interaction originates from the intrinsic property called mass.
Electromagnetic interaction comes from the intrinsic property called charge.
The strong interaction comes from an intrinsic property called colour charge.
We say that quarks have colour charge.
Colour charge can be of three types—red, green, blue. In other words, a quark can have a colour charge either red colour charge, or green colour charge or blue colour charge.

When quarks constitute a nucleon, the sum of colour charges of the constituent quarks should be zero so that we have a nucleon carrying no net colour charge. For instance, a red $u$, green $u$ and a blue $d$ can produce a proton while a red $d$, green $d$ and a blue $u$ can produce a neutron. And neutrons and protons are colourless, i.e. do not carry a net colour charge.

## 6.3 Binding force and energy of a deuteron system

✓ Experimentally determined binding energy of deuteron is

$$B = 2.225 \ MeV \tag{6.1}$$

Since deuteron mass number is $A = 2$ there are two constituent nucleons $n$ and $p$ and so the binding energy per nucleon will be

$$\frac{B}{A} = \frac{2.225 \ MeV}{2 \ \text{nucleon}} = 1.1125\frac{MeV}{\text{nucleon}}. \tag{6.2}$$

This value is very small compared to the average binding energy per nucleon

$$\frac{B}{A} = 8\frac{Mev}{\text{nucleon}}. \tag{6.3}$$

So deuteron is a weakly bound system. The ground state of deuteron has energy

$$E_{gs} \equiv E = -B = -2.225 \ MeV \tag{6.4}$$

This is the only bound state and there is no excited state. If we pump energy $B = 2.225 \ MeV$ into the deuteron system, its energy would be from equations (6.1) and (6.4)

$$B + E = 2.225 \ MeV - 2.225 \ MeV = 0$$

which means deuteron would disintegrate, i.e. break up into its constituent $n$ and $p$ that would separate or move away from each other.

Due to small $\frac{B}{A}$ the deuteron system is not so stable—it is a loosely bound system, i.e. the constituent nucleons are weakly bound to each other. Clearly from equations (6.1) and (6.4)

$$|E| = B = 2.225 \ MeV, \ E = -|E| \tag{6.5}$$

In *exercise 6.24* we discuss how binding energy of deuteron is measured or determined.

✓ Since neutron is overall neutral and proton is charged, the binding force between them cannot be of electrical or electromagnetic in nature.

✓ Since both neutron and proton have mass $m_n = 1.675 \times 10^{-27} \ kg$, $m_p = 1.6 \times 10^{-27} \ kg$ there is gravitational attractive interaction between them. But gravitational interaction between these microparticles is very small in magnitude and can be neglected. So binding force between neutron and proton cannot be of gravitational nature.

✓ The binding force between neutron and proton has nuclear origin and is called nuclear force—that we wish to investigate.

## 6.4 Range and strength of nuclear force

✓ Nuclear range is $r_0 = 2.1\ fm$.
✓ Since deuteron is a loosely bound system, the nucleons can interact and stay bound even when their separation is greater than $r_0$.

We define the distance over which the nucleons of deuteron can relax and still stay bound as the deuteron radius denoted by $\rho$. We shall find its order of magnitude (*exercises 6.10 and 6.11*).
✓ Strength of deuteron potential

Since deuteron is a very simple system consisting of two nucleons $n$ and $p$ we can approximate the nucleon–nucleon interaction potential by a very simple form that would be easy to handle mathematically.

☐ We choose a central potential to represent deuteron $V = V(\vec{r}) = V(r)$.
☐ As deuteron is a bound system holding together neutron and proton, the interaction between them is obviously attractive and so the deuteron potential $V = V(r)$ is negative.
☐ The strength of interaction is measured by the depth of the potential—the deeper the potential the stronger it is.

We can assume a simple form by taking the deuteron potential to be a 3D square well potential, as shown in figure 6.1(a), and expressed as

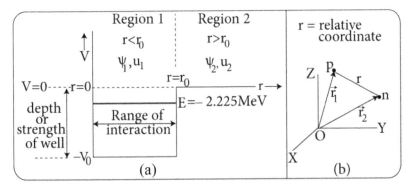

**Figure 6.1.** (a) Deuteron potential well, $r_0$ is range, $V_0$ is depth. (b) Relative coordinate of neutron and proton is $r$.

$$V = V(r) = \begin{cases} V = 0 & \text{for } r > r_0 \\ V = -V_0 & \text{for } r < r_0 \end{cases} \tag{6.6}$$

where $r$ is the separation between the nucleons, i.e. separation between $n$ and $p$. In other words $r$ is relative coordinate, shown in figure 6.1(b).

$-V_0$ in equation (6.6) represents the depth or strength of deuteron potential.

As $r$ increases radially from the $r = 0$ point there is attractive potential—$V_0$ in all directions, i.e. within the spherical volume of radius $r = r_0$.

Our task is to find a relation between the range $r_0$ and depth $V_0$ of deuteron potential—that characterizes the nuclear force operating within deuteron.

## 6.5 Deuteron wave function from solution of the Schrödinger equation

To investigate the relative motion between neutron and proton in a deuteron nucleus let us write down the time independent Schrödinger wave equation

$$H\psi\,(\vec{r}) = E\psi\,(\vec{r}) \tag{6.7}$$

where $\psi(\vec{r})$ is the deuteron wave function, $E$ is the nucleon–nucleon interaction energy given by equation (6.4). The Hamiltonian is

$$H = -\frac{\hbar^2}{2\mu}\nabla^2 + V(\vec{r}) = -\frac{\hbar^2}{2\mu}\nabla^2 + V(r) \tag{6.8}$$

where potential is spherically symmetric $V(\vec{r}) = V(r)$ and $\mu$ is the reduced mass of proton and neutron, i.e.

$$\mu = \frac{m_p m_n}{m_p + m_n} \tag{6.9}$$

As $m_p \cong m_n = m$ we have

$$\mu = \frac{m}{2} \tag{6.10}$$

Writing the Schrödinger equation (6.7) explicitly we get

$$\left[ -\frac{\hbar^2}{2\mu}\nabla^2 + V(r) \right]\psi(\vec{r}) = E\psi\,(\vec{r})$$

$$\nabla^2\psi(\vec{r}) + \frac{2\mu}{\hbar^2}[E - V(r)]\psi(\vec{r}) = 0 \tag{6.11}$$

We choose equation (6.6) to be the deuteron potential $= V(r)$.

Let us use method of separation of variables. Let us write

$\psi(\vec{r}) = R(r)Y(\theta, \phi)$

where $R(r)$ is radial part and $Y(\theta, \phi)$ is angular part.

Since deuteron potential $V = V(r)$ is central and spherically symmetric, the angular part $Y(\theta, \phi)$ will be spherical harmonic, i.e.

$$Y(\theta, \phi) = Y_{lm_l}(\theta, \phi) \qquad (6.12)$$

The solution thus takes the form

$$\psi(\vec{r}) = R(r) Y_{lm_l}(\theta, \phi) \qquad (6.13)$$

where $l = 0, 1, 2, 3, \ldots$ and $m_l = 0, \pm 1, \pm 2, \ldots$ are the allowed values.

The radial equation as derived in *exercise 6.4* is

$$\frac{d^2}{dr^2}R(r) + \frac{2}{r}\frac{d}{dr}R(r) + \frac{2\mu}{\hbar^2}\left[E - V(r) - \frac{l(l+1)\hbar^2}{2\mu r^2}\right]R(r) = 0 \qquad (6.14)$$

Simplification is achieved as shown in *exercise 6.5* by substituting

$$R(r) = \frac{u(r)}{r}$$

$$u(r) = rR(r) \qquad (6.15)$$

The simplified equation is

$$\frac{d^2}{dr^2}u(r) + \frac{2\mu}{\hbar^2}\left[E - V(r) - \frac{l(l+1)\hbar^2}{2\mu r^2}\right]u(r) = 0 \qquad (6.16)$$

This resembles a 1D equation in $u$.

Deuteron is a loosely bound just stable system. It has only one energy state. So we can identify this energy state as the lowest energy state.

The centrifugal potential energy term $\frac{l(l+1)\hbar^2}{2\mu r^2}$ makes a positive contribution to energy for $l \neq 0$. Clearly thus lowest energy or the ground state energy corresponds to $l = 0$.

According to our discussion therefore deuteron corresponds to a single $l$ value namely $l = 0$. This means that $l$ is conserved or fixed. This is consistent with the fact that we are working with a central potential $V(r)$ which leads to conservation of angular momentum, as shown in *exercise 6.2*.

With $l = 0$, $m_l = 0$ the angular part namely equation (6.12) is

$$Y_{lm_l}(\theta, \phi) \xrightarrow{\quad l = 0,\ m_l = 0 \quad} Y_{00} = \frac{1}{\sqrt{4\pi}} = \text{constant and so equation (6.13) gives with}$$

equation (6.15)

$$\psi(\vec{r}) = RY_{00} = R(r)\frac{1}{\sqrt{4\pi}} = \frac{u(r)}{r}\frac{1}{\sqrt{4\pi}} = \frac{u(r)}{\sqrt{4\pi}\,r} \qquad (6.17)$$

where $u(r)$ satisfies equation (6.16).

For $l = 0$ we call it the $s$ wave radial equation, since $l = 0$ state is spectroscopically denoted as $S$ state. So with $l = 0$ equation (6.16) becomes

$$\frac{d^2}{dr^2}u(r) + \frac{2\mu}{\hbar^2}[E - V(r)]u(r) = 0$$

Using $\mu = \frac{m}{2}$ from equation (6.10) we have

$$\frac{d^2}{dr^2}u(r) + \frac{m}{\hbar^2}[E - V(r)]u(r) = 0 \tag{6.18}$$

As shown in figure 6.1(a) there are two regions.

- In the region 1, $r < r_0$ we have $V = V(r) = -V_0$ from equation (6.6). The wave function is $u = u_1$, $\psi = \psi_1$
  and so from equation (6.18) we have

$$\frac{d^2}{dr^2}u_1 + \frac{m}{\hbar^2}(E + V_0)u_1 = 0$$

$$\frac{d^2}{dr^2}u_1 + k^2u_1 = 0 \tag{6.19}$$

where

$$k^2 = \frac{m}{\hbar^2}(E + V_0) \tag{6.20}$$

Using $E = -|E|$ from equation (6.5) we rewrite equation (6.20)

$$k^2 = \frac{m}{\hbar^2}(V_0 - |E|)$$

$$k = \sqrt{\frac{m}{\hbar^2}(V_0 - |E|)} \xrightarrow{V_0 > |E|} \text{real quantity} \tag{6.21}$$

Equation (6.19) is a simple harmonic equation and the solution is

$$u_1 = A \sin kr + B \cos kr \quad A, B \rightarrow \text{constants} \tag{6.22}$$

From equation (6.17) it follows that the solution is

$$\psi_1 = \frac{u_1(r)}{\sqrt{4\pi r}} = \frac{A \sin kr + B \cos kr}{\sqrt{4\pi r}} \tag{6.23}$$

At $r = 0$, $\psi_1 |_{r=0} = \dfrac{A \sin kr + B \cos kr}{\sqrt{4\pi r}}\bigg|_{r=0} = \dfrac{B}{0} \rightarrow \infty.$

Clearly to keep $\psi_1$ finite we need to set $B = 0$. Hence

$$u_1 = A \sin kr \tag{6.24}$$

$$\psi_1 = \frac{u_1}{\sqrt{4\pi r}} = \frac{A \sin kr}{\sqrt{4\pi r}} \xrightarrow[\frac{A}{\sqrt{4\pi}} \equiv A]{\text{redefine}} A\frac{\sin kr}{r} \tag{6.25}$$

So solution of equation (6.19) in the region $r < r_0$ is oscillatory in nature.

- In the region 2, $r > r_0$ it follows from equation (6.6) that $V = 0$. The wave function is

$$u = u_2, \psi = \psi_2$$

and so from equation (6.18) we have

$$\frac{d^2}{dr^2}u_2 + \frac{m}{\hbar^2}Eu_2 = 0 \tag{6.26}$$

$$\frac{d^2}{dr^2}u_2 - \beta^2 u_2 = 0 \tag{6.27}$$

where

$$\beta^2 = -\frac{m}{\hbar^2}E \tag{6.28}$$

Using equation (6.5) viz. $E = -|E|$ we have

$$\beta^2 = \frac{m}{\hbar^2}\,|\,E\,|$$

$$\beta = \sqrt{\frac{m}{\hbar^2}\,|\,E\,|} = \text{real quantity} \tag{6.29}$$

The solution of equation (6.26) is

$$u_2 = Ce^{-\beta r} + De^{\beta r} \quad C, D \rightarrow \text{constants}$$

$$\psi_2 = \frac{u_2}{\sqrt{4\pi}\,r} = \frac{Ce^{-\beta r} + De^{\beta r}}{\sqrt{4\pi}\,r}$$

At $r \rightarrow \infty$, $\quad \psi_2\,|_{r=\infty} = \frac{Ce^{-\beta r} + De^{\beta r}}{\sqrt{4\pi}\,r}\,|_{r=\infty} = \frac{De^{\beta\infty}}{\infty} \rightarrow \infty$.

Clearly to keep $\psi_2$ finite we need to set $D = 0$. Hence

$$u_2 = Ce^{-\beta r} \tag{6.30}$$

$$\psi_2 = \frac{Ce^{-\beta r}}{\sqrt{4\pi}\,r} \xrightarrow[\frac{C}{\sqrt{4\pi}} \equiv C]{\text{redefine}} C\frac{e^{-\beta r}}{r} \tag{6.31}$$

So solution in the region $r > r_0$ is exponentially damped in nature.

The acceptable solution of Schrödinger equation (6.18) for a deuteron problem is with equations (6.24) and (6.30)

$$u = \begin{cases} A \sin kr = u_1 \;(r < r_0) & \text{oscillatory} \\ Ce^{-\beta r} = u_2 \;(r > r_0) & \text{exponentially damped} \end{cases} \tag{6.32}$$

or with equations (6.25) and (6.31)

$$\psi = \begin{cases} A \dfrac{\sin kr}{r} & (r < r_0) \\[2ex] C \dfrac{e^{-\beta r}}{r} & (r > r_0) \end{cases} \tag{6.33}$$

The condition which a well behaved wave function $\psi$ obeys is that $\psi$ and its derivative $\frac{d}{dr}\psi$ have to be finite, single-valued, continuous.

Normalization constants have been evaluated in *exercise 6.12*.

Let us impose continuity requirement on $u$. So $u$ is continuous across boundary at $r = r_0$, i.e.

$$u_1 \big|_{r_0} = u_2 \big|_{r_0} \tag{6.34}$$

$$A \sin kr_0 = C e^{-\beta r_0} \tag{6.35}$$

Also, $\frac{du}{dr}$ = continuous at $r = r_0$, i.e.

$$\frac{du_1}{dr} \bigg|_{r_0} = \frac{du_2}{dr} \bigg|_{r_0}$$

$$\frac{d}{dr} A \sin kr \big|_{r_0} = \frac{d}{dr} C e^{-\beta r} \big|_{r_0} \tag{6.36}$$

$$Ak \cos kr_0 = -C\beta \, e^{-\beta r_0} \tag{6.37}$$

Dividing equation (6.35) by equation (6.37) we have

$$\frac{A \sin kr_0}{Ak \cos kr_0} = \frac{C e^{-\beta r_0}}{-C\beta \, e^{-\beta r_0}} \Rightarrow \frac{\tan kr_0}{k} = -\frac{1}{\beta}$$

$$\tan kr_0 = -\frac{k}{\beta} \tag{6.38}$$

$$\cot kr_0 = -\frac{\beta}{k} \Rightarrow k \cot kr_0 = -\beta \tag{6.39}$$

Equation (6.39) is a transcendental equation giving a relationship between $V_0$ and $r_0$.

$kr_0$ lies in the second quadrant ($\frac{\pi}{2} \leqslant kr_0 \leqslant \pi$) or fourth quadrant ($\frac{3\pi}{2} \leqslant kr_0 \leqslant 2\pi$), as shown in figure 6.2, as evident from the points of intersection $a$, $b$ and $c$, respectively.

Consider

$$\operatorname{cosec}^2 kr_0 = 1 + \cot^2 kr_0$$

$$= 1 + \frac{1}{\tan^2 kr_0} = 1 + \frac{1}{\left(-\frac{k}{\beta}\right)^2} \quad \text{[using equation (6.38)]}$$

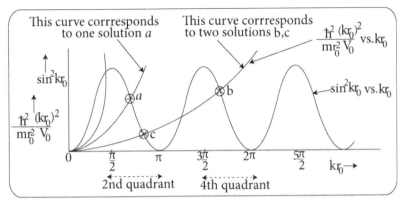

**Figure 6.2.** Plot of $\sin^2 kr_0$ versus $\sin^2 kr_0$ and $\frac{\hbar^2}{mr_0^2 V_0}(kr_0)^2$ versus $kr_0$ yields solutions $a$, $b$, $c$.

$$\csc^2 kr_0 = 1 + \frac{\beta^2}{k^2} = \frac{k^2 + \beta^2}{k^2}$$

$$\sin^2 kr_0 = \frac{1}{\csc^2 kr_0} = \frac{k^2}{k^2 + \beta^2} \tag{6.40}$$

Using equations (6.21) and (6.29) we have

$$\sin^2 kr_0 = \frac{k^2}{\frac{m}{\hbar^2}(V_0 - |E|) + \frac{m}{\hbar^2}|E|} = \frac{\hbar^2 k^2}{mV_0}$$

$$\sin^2 kr_0 = \frac{\hbar^2}{mV_0 r_0^2}(kr_0)^2 \tag{6.41}$$

The LHS $\sin^2 kr_0$ of equation (6.41) is a function of $kr_0$. So let us make a plot of $\sin^2 kr_0$ versus $kr_0$. This is shown in figure 6.2.

The RHS $\frac{\hbar^2}{mV_0 r_0^2}(kr_0)^2$ of equation (6.41) is a function of $kr_0$. So let us make a plot

of $\frac{\hbar^2}{mV_0 r_0^2}(kr_0)^2$ versus $kr_0$ which is a parabola. This is also shown in figure 6.2.

It follows from equation (6.41) that mathematically acceptable solution for $kr_0$ corresponds to points of intersection of the parabola and sine square curve lying in the second and fourth quadrants. The points of intersection are $a$, $b$, $c$.

Again the curve containing the solutions $b$ and $c$ is not acceptable since deuteron has a single bound state ($E = -2.225\ MeV$) which is the ground state, there being no excited state. We cannot thus accept a curve that corresponds to two solutions.

So the point of intersection $a$ is the only acceptable solution, which corresponds to

$$\frac{\pi}{2} \leqslant kr_0 \leqslant \pi \text{(curve with one solution)} \tag{6.42}$$

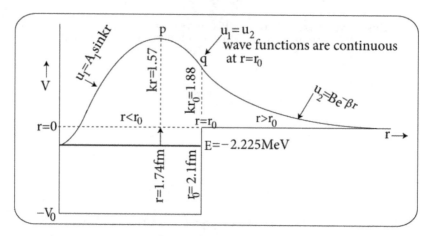

**Figure 6.3.** Deuteron wave function.

We show the plot of deuteron wave function $u(r)$ versus $r$ in figure 6.3 where $r$ is the inter-nucleon distance, i.e. separation between neutron and proton of the deuteron system.

In the $r < r_0$ region the sine wave function $u_1(r) = A \sin kr$ [equation (6.32)] reaches its peak at point $p$ where $kr = \frac{\pi}{2} = 1.57$. Then it turns over and goes down. It joins the exponentially decaying function $u_2(r) = e^{-\beta r}$ [equation (6.32)] at $r = r_0$ at point $q$ to preserve the continuity condition requirement.

We mention that if the wave function did not turn over inside the well, i.e. within the span $0 \leqslant r \leqslant r_0$, it would not be possible for it to connect smoothly to a decaying exponential (that has a negative slope) and in that case there would be no bound state. Actually, deuteron is very close to the top of the well. If nucleon–nucleon force were a just bit weaker, the deuteron bound state would not exist. The weak binding means that $u(r)$ is just barely able to turn over in the well so as to connect at point $q$ of figure 6.3 with the negative slope of the decaying exponential. Let us find an estimate of $kr_0$.

- Estimate of $kr_0$

It follows from equation (6.42) that $kr_0$ is slightly greater than $\frac{\pi}{2}$ and less than $\pi$. Let us take

$$kr_0 = \frac{\pi}{2} + \varepsilon, \; \varepsilon \to \text{ small quantity} \tag{6.43}$$

Now from equation (6.38) $\tan kr_0 = -\frac{k}{\beta} \Rightarrow \tan kr_0 = -\frac{kr_0}{\beta r_0}$

$$\beta r_0 = -kr_0 \cot kr_0 \tag{6.44}$$

$$\beta r_0 = -\left(\frac{\pi}{2} + \varepsilon\right) \cot \left(\frac{\pi}{2} + \varepsilon\right) = \left(\frac{\pi}{2} + \varepsilon\right) \tan \varepsilon$$

Taking $\tan \varepsilon \cong \varepsilon$ we get

$$\beta r_0 = \left( \frac{\pi}{2} + \varepsilon \right) \varepsilon = \frac{\pi}{2} \varepsilon$$

$$\varepsilon = \frac{2\beta r_0}{\pi} \tag{6.45}$$

Hence from equation (6.43)

$$k r_0 = \frac{\pi}{2} + \frac{2\beta r_0}{\pi} \tag{6.46}$$

Using $\beta$ from equation (6.29) viz. $= \sqrt{\frac{m}{\hbar^2} |E|}, |E| = 2.225 \ MeV$ [equation (6.5)], $r_0 = 2.1 \ fm$ and $\frac{\hbar^2}{m} = 41.37 \ MeV. fm^2$ (exercise 6.1) we write equation (6.46) as

$$k r_0 = \frac{\pi}{2} + \frac{2\beta r_0}{\pi} = \frac{\pi}{2} + \frac{2}{\pi} \sqrt{\frac{m}{\hbar^2} |E|} r_0 = \frac{\pi}{2} + \frac{2}{\pi} \sqrt{\frac{2.225 \ MeV}{41.37 \ MeV. fm^2}} 2.1 \ fm$$

$$k r_0 = 1.88 \tag{6.47}$$

Clearly the internal wave function $u_1(r < r_0)$ joins with the external wave function $u_2(r > r_0)$ at $k r_0 = 1.88$

## 6.6 Relation between range $r_0$ and strength $V_0$ of deuteron potential

Let us consider the ratio $\frac{k}{\beta}$ and employ equation (6.46).

$$\frac{k}{\beta} = \frac{k r_0}{\beta r_0} = \frac{\frac{\pi}{2} + \frac{2\beta r_0}{\pi}}{\beta r_0} = \frac{\pi}{2\beta r_0} + \frac{2}{\pi} \tag{6.48}$$

Using $k$ and $\beta$ from equations (6.21) and (6.29) we have

$$\frac{\sqrt{\frac{m}{\hbar^2}(V_0 - |E|)}}{\sqrt{\frac{m}{\hbar^2} |E|}} = \frac{\pi}{2 r_0 \sqrt{\frac{m}{\hbar^2} |E|}} + \frac{2}{\pi}$$

$$\left( \frac{V_0}{|E|} - 1 \right)^{1/2} = \frac{\pi \hbar}{2 r_0 \sqrt{m |E|}} + \frac{2}{\pi}$$

$$\left( \frac{V_0}{|E|} - 1 \right)^{1/2} = \frac{\pi}{2 r_0} \sqrt{\frac{\hbar^2}{m |E|}} + \frac{2}{\pi} \tag{6.49}$$

Using $\frac{\hbar^2}{m} = 41.37 \ MeV. fm^2$ (exercise 6.1) and $|E| = 2.225 \ MeV$ [equation (6.5)] we write

$$\left(\frac{V_0}{2.225\ MeV} - 1\right)^{1/2} = \frac{1.57}{r_0}\sqrt{\frac{41.37\ MeV.fm^2}{2.225\ MeV}} + 0.64$$

$$\left(\frac{V_0}{2.225\ MeV} - 1\right)^{1/2} = \frac{6.77\ fm}{r_0} + 0.64 \tag{6.50}$$

This is the required relation between range $r_0$ and depth $V_0$.

From equation (6.50) we can estimate

✓ the range of nuclear force $r_0$ if strength of deuteron potential $V_0$ is given say $V_0 = 36\ MeV$ (*exercise 6.6*) or

✓ the strength of deuteron potential $V_0$ if range $r_0$ is given say $r_0 = 2.1\ fm$ (*exercise 6.7*).

• *Exercise 6.8* gives a rough estimate of the relation between $V_0$ and $r_0$ taking $kr_0 = \frac{\pi}{2} = 1.57$ instead of $kr_0 = 1.88$ (figure 6.3).

## 6.7 Deuteron wave function parameters

Let us proceed with the values

$$V_0 = 36\ MeV,\ r_0 = 2.1\ fm,\ |E| = 2.225\ MeV \tag{6.51}$$

Now by equation (6.21) we have, using $\frac{\hbar^2}{m} = 41.37\ MeV.fm^2$ (*exercise 6.1*)

$$k = \sqrt{\frac{m}{\hbar^2}(V_0 - |E|)} = \sqrt{\frac{36 MeV - 2.225\ MeV}{41.37\ MeV.fm^2}} = 0.9\ fm^{-1}$$

This also follows from equation (6.47) namely $kr_0 = 1.88$

$$k = \frac{1.88}{r_0} = \frac{1.88}{2.1\ fm} = 0.9\ fm^{-1} \tag{6.52}$$

Again equation (6.29) gives

$$\beta = \sqrt{\frac{m}{\hbar^2}|E|} = \sqrt{\frac{2.225 MeV}{41.37\ MeV.fm^2}} = 0.23\ fm^{-1} \tag{6.53}$$

We note that in the internal region $r < r_0, u_1 = A\sin kr$ has a peak at

$$kr = \frac{\pi}{2} = 1.57\ i.e\ r = \frac{1.57}{k} = \frac{1.57}{0.9\ fm^{-1}} = 1.74\ fm \tag{6.54}$$

and matches the external wave function $u_2 = Ce^{-\beta r}$ at

$$kr_0 = 1.88\ i.e\ at\ r_0 = \frac{1.88}{k} = \frac{1.88}{0.9\ fm^{-1}} = 2.1\ fm \tag{6.55}$$

as evident from figure 6.3.

## 6.8 Deuteron radius

The range of nuclear potential is $r_0 = 2.1 \, fm$. But we have shown in *exercise 6.8* that the nucleons of the deuteron have much greater probability to lie outside the potential well—$n$ and $p$ show less interest to lie within.

This reflects the fact that deuteron is a loosely bound system as the nucleons are located most of the time at the edges of the well. Deuteron has an open structure with the nucleons relaxing most of the time outside the well but still maintaining the binding.

This state of affairs, i.e. such behavaviour of nucleons of deuteron can be better visualized if we consider the Woods–Saxon potential well, as shown in figure 6.4(a).

Since neutrons and protons prefer to stay outside the well for a considerable amount of time, it follows that the size of deuteron or the deuteron radius $\rho$ will be greater than the range $r_0$. The deuteron has an extended and relaxed structure.

The exterior function of deuteron is $u_2(r > r_0) = Ce^{-\beta r}$. This exponential function suggests existence of nucleons outside the deuteron potential well. It suggests the fact that nucleons can relax over a large distance—spend most of the time at the edges or outside the well. Since the range $r_0$ is smaller than this relaxation length $\rho$ and since mathematically $u_2 \to 0$ at $r \to \infty$ one can treat the exponential part of the wave function as a fairly close approximation for the entire range of variation of $r$. Actually $u_2$ is non-vanishing over a much longer range compared to range $r_0$. This is called zero range approximation and defined in *exercise 6.13*.

In *exercises 6.10 and 6.11* we have defined

$$\frac{1}{\beta} = \rho = 4.35 \, fm \tag{6.56}$$

as the deuteron radius. This can be compared to the nuclear range $r_0 = 2.1 \, fm$.

Since $\rho > r_0$ it turns out that the radius of deuteron is larger than the range of nuclear interaction.

The result $\rho > r_0$ is consistent with the fact that nuclear force is of short-range nature.

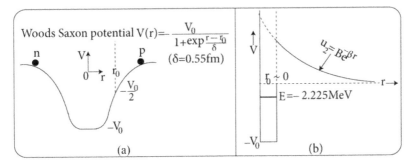

**Figure 6.4.** (a) Woods–Saxon potential. (b) Zero range approximation (*exercise 6.13*).

- We emphasize that discussion on deuteron has been made assuming that angular momentum *l* is conserved and corresponds to a fixed value of orbital angular momentum *l*. The fixed value of *l* considered is $l = 0$. In other words, only central force is assumed to be operative in the analysis.
- Experimentally measured value of total angular momentum (also called spin) of deuteron is $J = 1$.
- Also, it will be shown in section 6.9 that deuteron, which is a composition of two spin $\frac{1}{2}$ fermions (*n* and *p*) has spin value $s = 1$ (deuteron is a spin triplet).
- Let us write the spectroscopic symbol of the state of deuteron.

We assumed deuteron to be in the state $|\ l = 0>$. The spectroscopic symbol for $l = 0$ state is $S$ state. So deuteron state that we discussed is represented by the spectroscopic notation $^{2S+1}(l = 0)_J \equiv\ ^{2.1+1}S_{J=1} =\ ^3S_1$.

- We shall show in section 6.9 that deuteron is not entirely in the $^3S_1$ or $|l = 0>$ state. Actually, deuteron is 96% $l = 0$ or $^3S_1$ state and there is 4% mixture of $l = 2$ state. The spectroscopic symbol for $l = 2$ state is $D$ state. So spectroscopic notation of $D$ state is

$$^{2S+1}(l = 2)_J \equiv\ ^{2.1+1}D_{J=1} =\ ^3D_1.$$

As both $l = 0, 2$ values occur, it is evident that $l$ is not conserved and hence potential is not central. So deuteron is predominantly in $l = 0$ or $^3S_1$ state as there is slight presence of a non-central part.

The spectroscopic notation of deuteron is thus

$$^3S_1,\ ^3D_1$$

## 6.9 Angular momentum of deuteron

Angular momentum of nucleon are of three types
- ✓ Orbital angular momentum $\vec{l}$
- ✓ Spin angular momentum $\vec{s}$ and
- ✓ Total angular momentum $\vec{j} = \vec{l} + \vec{s}$.

For a single particle small letters $l, s, j$ are used as symbols and for a multi-particle system capital letters $L, S, J$ are generally used.

- Nuclear spin

When we add all the angular momenta $\vec{j_i}\ s$ of the constituent nucleons we get the final or total angular momentum of the nucleus. The total angular momentum of nucleus $\vec{J}$ is also called nuclear spin in literature and it can be experimentally measured.

Let the space part of the wave function be

$$\psi(\vec{r}) = \psi(r, \theta, \phi).$$

The behaviour of wave function under mirror reflection is expressed through an operator called parity operator, denoted by $\pi$. It is defined through the operation that changes $\vec{r}$ to $-\vec{r}$

$$\pi \mid \vec{r} > = \mid -\vec{r}>$$

that is

$$(x, y, z) \xrightarrow{\pi} (-x, -y, -z)$$

$$(r, \theta, \phi) \xrightarrow{\pi} (r, \pi - \theta, \pi + \phi)$$

In the nucleus the space part of wave function $\psi(\vec{r})$ is either symmetric or anti-symmetric and we say that parity is definite.

If the relation

$$\psi(\vec{r}) = \psi(-\vec{r})$$

is satisfied, wave function is symmetric, parity is even, parity eigenvalue is $\pi = +ve$.

If the relation

$$\psi(\vec{r}) = -\psi(-\vec{r})$$

is satisfied, wave function is anti-symmetric, parity is odd, parity eigenvalue is $\pi = -ve$.

For any central spherically symmetric potential $V(\vec{r}) = V(r)$, $V \neq V(\theta, \phi)$ the angular part of wave function [equations (6.12) and (6.13)] is the spherical harmonic $Y_{lm_l}(\theta, \phi)$ where $l$ is the orbital angular momentum quantum number and $m_l$ is its projection along $z$ direction (the field direction). This $l$ is directly related to parity. For

$$l = 0, 2, 4, \ldots(even)\ \pi = +ve\ ,\ \text{while for}$$

$$l = 1, 3, 5, \ldots(odd)\quad \pi = -ve.\ \text{So}$$

$$\pi = (-1)^l \tag{6.57}$$

A quantum system will have definite parity if the Hamiltonian $H$ of the system commutes with the parity operator $\pi$. All nuclei have wave function with definite parity (either even or odd) because the nuclear Hamiltonian commutes with the parity operator

$$[H, \pi] = 0.$$

The spin parity of a nucleus is denoted by $J^\pi$.

• For a deuteron system experimentally measured spin parity is

$$J^\pi = 1^+\ \text{i.e.}\ J = 1,\ \pi = +ve \tag{6.58}$$

## 6.10 Spin dependence of nuclear force

✓ Deuteron system is an $np$ system. Nucleons are spin $\frac{1}{2}$ fermions. Proton has spin $s_p = \frac{1}{2}$. Neutron has spin $s_n = \frac{1}{2}$.

The square of intrinsic spin angular momentum $s^2$ has eigenvalue $s(s + 1)\hbar^2$. So for a proton

$$s_p^2 \text{ has eigenvalue } s_p(s_p + 1)\hbar^2 = \frac{1}{2}\left(\frac{1}{2} + 1\right)\hbar^2 = \frac{3}{4}\hbar^2 \text{ and}$$

for neutron

$$s_n^2 \text{ has eigenvalue } s_n(s_n + 1)\hbar^2 = \frac{1}{2}\left(\frac{1}{2} + 1\right)\hbar^2 = \frac{3}{4}\hbar^2$$

The projection of angular momentum $s_z$ has eigenvalue $m_s\hbar$ where $m_s = -s$ to $+s$ in unit steps. If $s = \frac{1}{2}$ then $m_s$ is $-\frac{1}{2}, +\frac{1}{2}$ i.e. $\pm\frac{1}{2}$.

For proton

$$s_{zp} \text{ has eigenvalue } m_{sp}\hbar = \pm\frac{\hbar}{2} \text{ where } m_{sp} = \pm\frac{1}{2}$$

For neutron

$$s_{zn} \text{ has eigenvalue } m_{sn}\hbar = \pm\frac{\hbar}{2} \text{ where } m_{sn} = \pm\frac{1}{2}$$

We often denote

$$m_s = \frac{1}{2} \text{ as up state and write it as } |\uparrow> = |up> \text{ and}$$

$$m_s = -\frac{1}{2} \text{ as down state and write it as } |\downarrow> = |down>.$$

- Spin state of a deuteron system can be represented by specifying the spin state of the constituent nucleons as follows.

Proton states

$$|proton> = |s_p m_{sp}> = \begin{cases} |\frac{1}{2}\frac{1}{2}> \equiv |up> & \equiv |\uparrow> \\ |\frac{1}{2} -\frac{1}{2}> \equiv |down> \equiv |\downarrow> \end{cases} \text{2 possibilities}$$

Neutron states

$$|neutron> = |s_n m_{sn}> = \begin{cases} |\frac{1}{2}\frac{1}{2}> \equiv |up> & \equiv |\uparrow> \\ |\frac{1}{2} -\frac{1}{2}> \equiv |down> \equiv |\downarrow> \end{cases} \text{2 possibilities}$$

Clearly there are four possibilities and with them we can define four deuteron states since deuteron is an *np* system. We write the states as

$$| s_p s_n m_{sp} m_{sn} > = | \frac{1}{2} \frac{1}{2} m_{sp} m_{sn} > \equiv | m_{sp} m_{sn} > = \begin{cases} | \frac{1}{2} \frac{1}{2} > & \equiv | \uparrow \uparrow > \\ | \frac{1}{2} -\frac{1}{2} > & \equiv | \uparrow \downarrow > \\ | -\frac{1}{2} \frac{1}{2} > & \equiv | \downarrow \uparrow > \\ | -\frac{1}{2} -\frac{1}{2} > \equiv | \downarrow \downarrow > \end{cases} \quad (6.59)$$

These four deuteron states $| m_{sp} m_{sn} >$ are called uncoupled basis or states since we are not performing any addition of angular momentum of the constituents of deuteron—the neutron and proton. This is one way to represent the deuteron system through $| m_{sp} m_{sn} >$ representation.

- The total angular momentum is the sum of orbital and spin angular momentum, i.e.

$$\vec{J} = \vec{l} + \vec{s}$$

We take $l = 0$ for which the spectroscopic symbol is $S$. So deuteron is considered as $S$ state.

$$\vec{J} = \vec{0} + \vec{s} = \vec{s}$$

Again $\vec{s}$ being the total spin angular momentum of deuteron is the sum of the contribution from proton and neutron intrinsic spin angular momenta, namely $s_p = \frac{1}{2}$, $s_n = \frac{1}{2}$. Hence

$$\vec{s} = \vec{s}_p + \vec{s}_n = \frac{\vec{1}}{2} + \frac{\vec{1}}{2} = \begin{cases} 0 \\ 1 \end{cases}$$

$s = 0 \Rightarrow m_s = 0$ and we denote this one state by $|sm_s > \rightarrow | 00>$ called singlet state.

$s = 1 \Rightarrow m_s = 1, 0, -1$ and we denote the three states by $| sm_s > \rightarrow | 11 >$, $| 10 >$, $| 1 - 1>$ called triplet states.

Clearly we have four possibilities and with them we can define four deuteron states as

$$| s_p s_n sm_s > = | \frac{1}{2} \frac{1}{2} sm_s > \equiv | sm_s > = \begin{cases} | 00 > \\ | 11 > \\ | 10 > \\ | 1 - 1 > \end{cases} \quad (6.60)$$

These four deuteron states $|sm_s>$ can be called coupled basis or states since we have performed addition of angular momentum of the constituents of deuteron—the neutron and proton. This is another way to represent the deuteron system through $|sm_s>$ representation.

- Relation between the bases $|m_{sp}m_{sn}>$ and $|sm_s>$

We can inter-relate the two representations, i.e. establish a connection between the two sets of basis—the uncoupled basis $|m_{sp}m_{sn}>$ and the coupled basis $|sm_s>$. Let us use the coupling relation between the projections of $\vec{s}$, $\vec{s}_p$, $\vec{s}_n$, namely

$$m_s = m_{sp} + m_{sn} \tag{6.61}$$

☐ Since $m_s = 0$ (corresponding to $s = 0$) is obtained through addition of

$$m_{sp} = \frac{1}{2} \quad \text{and } m_{sn} = -\frac{1}{2} (\text{as } m_{sp} + m_{sn} = \frac{1}{2} - \frac{1}{2} = 0 = m_s) \text{ and}$$

$$m_{sp} = -\frac{1}{2} \text{ and } m_{sn} = \frac{1}{2} (\text{as } m_{sp} + m_{sn} = -\frac{1}{2} + \frac{1}{2} = 0 = m_s)$$

it follows that $|sm_s> = |00>$ can be obtained from a linear combination of the states, namely

$$| m_{sp}m_{sn} > = |\frac{1}{2} - \frac{1}{2}> \equiv| \uparrow \downarrow > \text{ and}$$

$$| m_{sp}m_{sn} > = |-\frac{1}{2}\frac{1}{2}> \equiv| \downarrow \uparrow > .$$

Hence we take (incorporating normalization constant $\frac{1}{\sqrt{2}}$ and proper symmetry requirement that $s = 0$ state has symmetry defined as $(-1)^{1+s} = (-1)^{1+0} = -ve$, i.e. spin singlet state is anti-symmetric, as explained in figure 8.5)

$$| 00 > = \frac{1}{\sqrt{2}}(| \uparrow \downarrow > -| \downarrow \uparrow > ) \tag{6.62}$$

☐ Since $m_s = 1$ (corresponding to $s = 1$) is obtained through addition of

$$m_{sp} = \frac{1}{2} \text{ and } m_{sn} = \frac{1}{2} (\text{as } m_{sp} + m_{sn} = \frac{1}{2} + \frac{1}{2} = 1 = m_s)$$

it follows that $| sm_s > = | 11>$ can be obtained from
$$| m_{sp}m_{sn}> = |\frac{1}{2}\frac{1}{2}> \equiv| \uparrow \uparrow > \text{ as}$$

$$| 11 > = | \uparrow \uparrow > \tag{6.63}$$

☐ Since $m_s = 0$ (corresponding to $s = 1$) is obtained through addition of

$$m_{sp} = \frac{1}{2} \text{ and } m_{sn} = -\frac{1}{2} \left(\text{as } m_{sp} + m_{sn} = \frac{1}{2} - \frac{1}{2} = 0 = m_s\right) \quad \text{and}$$

$$m_{sp} = -\frac{1}{2} \text{ and } m_{sn} = \frac{1}{2} \left(\text{as } m_{sp} + m_{sn} = -\frac{1}{2} + \frac{1}{2} = 0 = m_s\right)$$

it follows that $|sm_s> = |10>$ can be obtained from a linear combination of states, namely

$$| m_{sp}m_{sn} > = | \frac{1}{2} - \frac{1}{2} > \equiv | \uparrow \downarrow > \text{ and}$$

$$| m_{sp}m_{sn} > = | - \frac{1}{2}\frac{1}{2} > \equiv | \downarrow \uparrow > .$$

Hence we take (incorporating normalization constant $\frac{1}{\sqrt{2}}$ and proper symmetry requirement that $s = 1$ state has symmetry defined as $(-1)^{1+s} = (-1)^{1+1} = +ve$, i.e. spin triplet state is symmetric as explained in figure 8.5).

$$| 10 > = \frac{1}{\sqrt{2}}(| \uparrow \downarrow > +| \downarrow \uparrow >) \tag{6.64}$$

☐ Since $m_s = -1$ (corresponding to $s = 1$) is obtained through addition of
$m_{sp} = -\frac{1}{2}$ and $m_{sn} = -\frac{1}{2}$ (as $m_{sp} + m_{sn} = -\frac{1}{2} - \frac{1}{2} = -1 = m_s$)
it follows that $| sm_s > = | 1 - 1>$ can be obtained from
$| m_{sp}m_{sn} > = | - \frac{1}{2} - \frac{1}{2} > \equiv | \downarrow \downarrow >$ as

$$| 1 - 1 > = | \downarrow \downarrow > \tag{6.65}$$

The relation between the two sets of basis   the uncoupled basis $| m_{sp}m_{sn}>$ and the coupled basis $|sm_s>$ is thus

$$| 00 > = \frac{1}{\sqrt{2}}(| \uparrow \downarrow > -| \downarrow \uparrow >)$$

$$| 11 > = | \uparrow \uparrow >$$

$$| 10 > = \frac{1}{\sqrt{2}}(| \uparrow \downarrow > +| \downarrow \uparrow > )$$

$$| 1 - 1 > = | \downarrow \downarrow > \tag{6.66}$$

✓ If nuclear interaction had been spin independent then there should not be any preference in the coupling of proton and neutron spins. In that case the four configurations of equation (6.66) would be equally probable and hence probability of each configuratuion would be $\frac{1}{4}$.

Since there is one state corresponding to $s = 0$ its probability would be $\frac{1}{4}$, i.e. 25% and since there are three states corresponding to $s = 1$ its probability would be $\frac{3}{4}$, i.e. 75%. In other words spin singlet state probability is expected to be 25% and spin triplet state probability is expected to be 75%.

Now the measured value of deuteron total angular momentum, called spin, is $J = 1$. Again

$$\vec{J} = \vec{l} + \vec{s}.$$

For a central potential and with $l = 0$ we have

$$\vec{J} = \vec{l} + \vec{s} = \vec{0} + \vec{s} = \vec{s}$$

$$\vec{s} = \vec{J} = \vec{1}$$

$$s = 1$$

Clearly deuteron bound state should correspond to $s = 1$ if potential is central and if $l = 0$. Even if there is a mixture of $l \neq 0$ part (i.e. for non-central potential) it is found that $s = 1$ for deuteron. Clearly nuclear interaction prefers a particular or specific combination of constituent neutron and proton that generates a spin triplet $s = 1$ for deuteron bound state.

In other words deuteron state is a spin triplet state and not a spin singlet state. This clearly indicates that nuclear force is spin dependent. In fact, nuclear force is stronger for parallel ↑↑ spin orientation, i.e. in the triplet state.
$$_{n\,p}$$

## 6.11 Possible $l$ values of deuteron

We discussed in section 6.5 deuteron system in state $l = 0$ with a central potential [equation (6.6)].

We also mentioned that the total angular momentum, called spin, of deuteron is experimentally determined to be $J = 1$. Based on this finding, let us focus on the $l$ value that the deuteron system can assume.

The angular momentum addition relation is

$$\vec{J} = \vec{l} + \vec{s}.$$

For deuteron $J = 1$ and so

$$\vec{l} + \vec{s} = \vec{J} = \vec{1}$$

$J = 1$ can come from the following combinations of $l$ and $s$.

$$\vec{l} + \vec{s} = \begin{cases} \vec{0} + \vec{1} \rightarrow 1 \\ \vec{1} + \vec{0} \rightarrow 1 \\ \vec{1} + \vec{1} \rightarrow 2, 1, 0 \\ \vec{2} + \vec{1} \rightarrow 3, 2, 1 \end{cases}$$

This leads to the following spectroscopic notations $^{2s+1}(l)_{J=1}$.

$$\begin{cases} l = 0 \Rightarrow \text{S state; s} = 1 \text{ means triplet, } 2s + 1 = 2.1 + 1 = 3, \ J = 1 \Rightarrow \quad ^3S_1 \text{ state} \\ l = 1 \Rightarrow \text{P state; s} = 0 \text{ means singlet, } 2s + 1 = 2.0 + 1 = 1, \ J = 1 \Rightarrow \quad ^1P_1 \text{ state} \\ l = 1 \Rightarrow \text{P state; s} = 1 \text{ means triplet, } 2s + 1 = 2.1 + 1 = 3, \ J = 1 \Rightarrow \quad ^3P_1 \text{ state} \\ l = 2 \Rightarrow \text{D state; s} = 1 \text{ means triplet, } 2s + 1 = 2.1 + 1 = 3, \ J = 1 \Rightarrow \quad ^3D_1 \text{ state} \end{cases}$$

Again $l = 1$ leads to parity value $(-1)^l = (-1)^1 = -ve$. But deuteron parity is $\pi = +ve$ [equation (6.58)].

So $l = 1$ states, i.e. $^1P_1$, $^3P_1$ states are not acceptable because of parity mismatch. Acceptable combinations are

$$\vec{l} + \vec{s} = \begin{cases} \vec{0} + \vec{1} \to 1 \\ \vec{2} + \vec{1} \to 3, 2, 1 \end{cases} \Rightarrow \begin{cases} ^3S_1 \text{ state } (l = 0, \ s = 1, \text{ triplet}) \\ ^3D_1 \text{ state } (l = 2, \ s = 1, \text{ triplet}) \end{cases} \tag{6.67}$$

✓ Clearly with $s = 0$ (singlet) we cannot get $J = 1$. Actually $s$ has to be 1 (triplet). Nuclear interaction chooses $s = 1$ (spin triplet state) and hence nuclear interaction is spin dependent.

    Clearly $n$ and $p$ can form a stable combination only when spin of the nucleons are parallel, i.e. only in triplet state ($s = 1$).

    In other words, the state in which spins are antiparallel, i.e. the singlet state ($s = 0$) is not a bound state.

✓ We see that $l = 0$ and $l = 2$ states lead to $J = 1$. So ground state of deuteron is an admixture of $l = 0$ and $l = 2$ states, i.e. a mixture of $^3S_1$, $^3D_1$ states is allowed.

✓ Presence of a single $l$ value means $l$ is conserved and potential is central. Having different values of $l$ hints at the fact that potential is non-central (*exercise 6.2*).

✓ So the observed spin parity $J^\pi = 1^+$ of deuteron is possible with $s = 1$ and a mixture of $l = 0$ and $l = 2$.

Spin dependence of nuclear force means the nuclear Hamiltonian should be spin dependent. In other words, the nuclear Hamiltonian should involve a term that takes care of this spin dependence. Such a term that represents spin dependence is

$$V_s(r)\vec{s}_p \cdot \vec{s}_n \tag{6.68}$$

and it should be added to the central potential term.

In *exercise 6.16* we indicate how the term $\vec{s}_p \cdot \vec{s}_n$ assumes different values for $s = 0$ and $s = 1$ and hence is spin dependent. So we can include $V_s(r)\vec{s}_p \cdot \vec{s}_n$ in the nuclear Hamiltonian to take care of the spin dependence of nuclear force. In other words, in *exercise 6.16* we show the justification of including $V_s(r)\vec{s}_p \cdot \vec{s}_n$ in the nuclear Hamiltonian.

We can alternatively use the operator $P_\sigma$ expressed in terms of Pauli spin operators to take care of spin dependence of nuclear force in the nuclear Hamiltonian.

$$P_\sigma = \frac{1}{2}\left(1 + \frac{4}{\hbar^2}\vec{s}_p \cdot \vec{s}_n\right) = \frac{1}{2}\left(1 + \frac{2}{\hbar}\vec{s}_p \cdot \frac{2}{\hbar}\vec{s}_n\right)$$

Using $\vec{s}_p = \frac{\hbar}{2}\vec{\sigma}_p$, $\vec{s}_n = \frac{\hbar}{2}\vec{\sigma}_n$ we have

$$P_\sigma = \frac{1}{2}\left(1 + \vec{\sigma}_p \cdot \vec{\sigma}_n\right) \tag{6.69}$$

- The spin dependent nuclear Hamiltonian containing $\sim V(r)\vec{s}_p \cdot \vec{s}_n$ is more attractive for the triplet $s = 1$ state and so we have the deuteron bound state corresponding to $s = 1$. The singlet $s = 0$ state cannot keep deuteron bound.
- Existence of $l = 2$, i.e. $^3D_1$ part along with $l = 0$ $^3S_1$ part is experimentally supported from the value of deuteron magnetic moment $\mu_d \sim \mu_{dz}$ and the non-vanishing quadrupole moment $Q_d \neq 0$. Mixture of the $l = 2$ part in deuteron implies that a part of the nuclear force in deuteron nucleus is non-central or tensor force. This we discuss now.

## 6.12 Magnetic moment of proton, neutron and gyromagnetic ratio

✓ The orbital motion of a proton (which is charged) gives rise to current which in turn produces magnetic field and hence there is magnetic moment due to orbital motion of proton.

✓ As the neutron is uncharged, its orbital motion will not create any current, no magnetic field is created and hence there is no magnetic moment due to orbital motion of neutron.

✓ Both proton and neutron possess intrinsic spin angular momentum due to their internal structures and hence will possess magnetic moments due to their spins $s_p = \frac{1}{2}$, $s_n = \frac{1}{2}$.

To measure magnetic moment we have to apply a magnetic field— the direction of which is defined as the $\hat{z}$ axis. Experimentally observed magnetic moment of deuteron is

$$\mu_{dz} = 0.8574\mu_N \equiv \langle \mu_{dz} \rangle \tag{6.70}$$

where

$\mu_N$ is called nuclear magneton defined as

$$\mu_N = \frac{|e|\hbar}{2m_p} = \frac{(1.6 \times 10^{-19}\ C)\left(\frac{1}{2\pi}6.626 \times 10^{-34}\ J.\ s\right)}{2(1.67 \times 10^{-27}\ kg)} = 5 \times 10^{-27}\ \frac{J.\ C.\ s}{kg}$$

Let us simplify the units. The Lorentz force relation

$$\overrightarrow{F} = q\,\overrightarrow{v}\times\overrightarrow{B} \text{ suggests } B \sim \frac{F}{qv} = \frac{MLT^{-2}}{CLT^{-1}} = \frac{M}{CT} \to \frac{kg}{C.\,s} = \text{Tesla} \equiv T.$$

Using this we have

$$\mu_N = \frac{|e|\,\hbar}{2m_p} = 5\times 10^{-27}\frac{J}{T} \text{ or } Am^2 \tag{6.71}$$

Nuclear magneton is the natural unit used to measure the magnetic dipole moment of nuclei.

✓ For atoms the atomic magnetic moments are measured in units of Bohr magneton

$$\mu_B = \frac{|e|\hbar}{2m_e} = 9.27\times 10^{-24}\frac{J}{T} \text{ and we note that}$$

$$\frac{\mu_N}{\mu_B} = \frac{|e|\,\hbar}{2m_p}\cdot\frac{2m_e}{|e|\,\hbar} = \frac{m_e}{m_p} = \frac{1}{1836}$$

$$\mu_N - \frac{1}{1836}\mu_B \tag{6.72}$$

- We show how magnetic moment for a charged particle is related to angular momemtum.

Consider a charged particle of charge $q$, mass $m$ revolving with velocity $v$ in a circle of radius $r$ with time period $T = \frac{2\pi r}{v}$ as shown in figure 6.5. The circle encloses area

$$\overrightarrow{A} = \pi r^2 \hat{n}\,.$$

This generates current

$$i = \frac{q}{T} = \frac{qv}{2\pi r}.$$

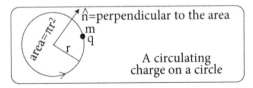

Figure 6.5. Magnetic moment of a circulating charged particle.

Magnetic moment of the circulating charged particle is

$$\vec{\mu} = i\vec{A} = \frac{qv}{2\pi r}\pi r^2\hat{n} = \frac{qvr}{2}\hat{n} = \frac{q}{2m}mvr\hat{n}$$

$$\vec{\mu} = \frac{q\vec{l}}{2m} \tag{6.73}$$

where $\vec{l} = mvr\hat{n}$ is the orbital angular momentum of the particle.
- If magnetic field is applied along $\hat{z}$ axis then the magnetic moment $\vec{\mu}$ takes maximum value along $\hat{z}$ axis and this is what is measured experimentally. In other words, $\vec{\mu} \cdot \hat{z} = \mu_z$ is experimentally measured. This has been referred to in equation (6.70).
- Orbital angular momentum gives rise to the following magnetic moments
✓ For electron $q = -|e|$, $m = m_e$ the contribution to magnetic moment due to its orbital angular momentum $\vec{l}$ as per equation (6.73) is given by

$$\vec{\mu}_{le} = -\frac{|e|\vec{l}}{2m_e} = -\frac{|e|\hbar}{2m_e}\frac{\vec{l}}{\hbar} = -\mu_B\frac{\vec{l}}{\hbar}$$

The magnetic moment $\vec{\mu}_{le}$ is measured in units of nuclear magneton $\mu_N$ while the angular momentum $\vec{l}$ is measured in units of $\hbar$. It follows that the ratio $|\frac{\mu_{le}/\mu_B}{l/\hbar}| = 1$ in this case. However, this ratio is not always unity if we consider other magnetic moments say $\vec{\mu}_{se}$, $\vec{\mu}_{sp}$, $\vec{\mu}_{sn}$. So it is customary to denote the ratio by a quantity called gyromagnetic ratio or g-factor. Hence we define

$$\left|\frac{\mu_{le}/\mu_B}{l/\hbar}\right| = g_{le} = 1$$

where $g_{le}$ is called orbital gyromagnetic ratio or orbital g-factor for electron. Accordingly

$$\vec{\mu}_{le} = -g_{le}\mu_B\frac{\vec{l}}{\hbar} \tag{6.74}$$

✓ For proton $q = |e|$, $m = m_p$ the contribution to magnetic moment due to its orbital angular momentum $\vec{l}$ as per equation (6.73) is given by

$$\vec{\mu}_{lp} = \frac{|e|\vec{l}}{2m_p} = \frac{|e|\hbar}{2m_p}\frac{\vec{l}}{\hbar} = g_{lp}\mu_N\frac{\vec{l}}{\hbar} \tag{6.75}$$

The quantity $g_{lp}$ is the ratio $\frac{\mu_{lp}/\mu_N}{l/\hbar}$ and is called orbital gyromagnetic ratio or orbital g-factor for proton. Its value is $g_{lp} = 1$.

The $z$ component of proton magnetic moment due to orbital angular momentum $\vec{l}$ of proton is obtained by considering projection of $\vec{l}$ viz. $l_z$. Hence we have

$$\mu_{lp_z} = g_{lp}\,\mu_N \frac{l_z}{\hbar} \tag{6.76}$$

✓ For neutron $q = 0$, $m = m_n$ the contribution to magnetic moment as per equation (6.73) is given by

$$\vec{\mu}_{ln} = \frac{(0)\vec{l}}{2m_n} = 0,\ \mu_{ln_z} = 0 \tag{6.77}$$

- Spin angular momentum gives rise to the following magnetic moments. Being a magnetic moment we present it in the same form as prescribed in equation (6.73).
✓ For an electron the magnetic moment due to intrinsic spin $\vec{s}_e$ is

$$\vec{\mu}_{se} = -g_{se}\,\mu_B \frac{\vec{s}_e}{\hbar} \tag{6.78}$$

The quantity $g_{se}$ is called spin gyromagnetic ratio or spin $g$-factor for electron and its value is

$$g_{se} = 2 \tag{6.79}$$

✓ For proton the magnetic moment due to intrinsic spin $\vec{s}_p$ is

$$\vec{\mu}_{sp} = g_{sp}\,\mu_N \frac{\vec{s}_p}{\hbar} \tag{6.80}$$

The $z$ component of magnetic moment due to intrinsic spin of proton is

$$\mu_{sp_z} = g_{sp}\,\mu_N \frac{s_{pz}}{\hbar} \tag{6.81}$$

The quantity $g_{sp}$ is called spin gyromagnetic ratio or spin $g$-factor for proton and its value is

$$g_{sp} = 5.585\ 691 \tag{6.82}$$

✓ For a neutron the magnetic moment due to intrinsic spin $\vec{s}_n$ is

$$\vec{\mu}_{sn} = g_{sn}\,\mu_N \frac{\vec{s}_n}{\hbar} \tag{6.83}$$

The $z$ component of magnetic moment due to spin of neutron is

$$\mu_{snz} = g_{sn} \mu_N \frac{S_{nz}}{\hbar} \tag{6.84}$$

The quantity $g_{sn}$ is called spin gyromagnetic ratio or spin $g$-factor for neutron and its value is

$$g_{sn} = -3.826\,084 \tag{6.85}$$

✓ We note that electron is a fundamental particle. It is structureless, i.e. does not have an internal structure and is a point particle. It carries charge $-|e|$. The value $g_{se} = 2$ [equation (6.79)] follows from Dirac's electron theory.

• Let us explain the values of spin gyromagnetic ratios, namely

$$g_{sp} = 5.585\,691, \; g_{sn} = -3.826\,084$$

from the quark structure of nucleons.
✓ Proton and neutron are not fundamental particles. They have an internal structure. They are made up of three quarks

$$p \equiv uud \text{ and } n \equiv udd$$

The quarks are bound or confined within nucleons due to strong interaction described in terms of exchange of gluons.

Quarks are charged and so there is a charge distribution inside nucleons that gives rise to magnetic moment.
Proton is charged—its constituents $u, u, d$ are charged viz.

$$Q(u) = \frac{2}{3} \, | \, e \, |, \; Q(d) = -\frac{1}{3} \, | \, e \, |$$

Hence the spin $g$-factor of a proton, i.e. $g_{sp} = 5.585\,691$ is different from the spin $g$-factor of an electron, i.e. $g_{se} = 2$ though both carry a charge of the same magnitude $| \, e \, | = 1.6 \times 10^{-19} \, C$.
Neutron is uncharged but its constituents $d, u, u$ are charged. In other words neutron has an internal charge distribution since $Q(u) = \frac{2}{3} \, | \, e \, |$, $Q(d) = -\frac{1}{3} |e|$. This internal charge distribution of quarks inside neutron gives it a spin $g$-factor $g_{sn} = -3.826\,084$.

## 6.13 Magnetic moment of deuteron

Magnetic moment of deuteron $\vec{\mu}_d$ is the vector sum of the orbital and spin magnetic moments of proton and neutron. Hence

$$\vec{\mu}_d = \vec{\mu}_{lp} + \vec{\mu}_{ln} + \vec{\mu}_{sp} + \vec{\mu}_{sn} \tag{6.86}$$

The $z$ component of magnetic moment of deuteron is measured and its expression is

$$\mu_{dz} = \mu_{lp_z} + \mu_{ln_z} + \mu_{sp_z} + \mu_{snz} \tag{6.87}$$

Using equations (6.76), (6.77), (6.81) and (6.84) we have

$$\mu_{dz} = g_{lp}\,\mu_N \frac{l_z}{\hbar} + 0 + g_{sp}\,\mu_N \frac{S_{pz}}{\hbar} + g_{sn}\,\mu_N \frac{S_{nz}}{\hbar}$$

$$\mu_{dz} = g_{lp}\,\mu_N \frac{l_z}{\hbar} + g_{sp}\,\mu_N \frac{S_{pz}}{\hbar} + g_{sn}\,\mu_N \frac{S_{nz}}{\hbar} = \mu_{lz} + \mu_{sp_z} + \mu_{snz} \tag{6.88}$$

In a centre of mass system the angular momentum of a proton is half of the total. So we have to put $\frac{l_z}{2}$ instead of $l_z$ in the equation (6.88) of $\mu_{l_z}$. Hence we rewrite the magnetic moment of deuteron of equation (6.88) as

$$\mu_{dz} = \frac{g_{lp}\,\mu_N}{\hbar} \frac{l_z}{2} + \frac{g_{sp}\,\mu_N}{\hbar} S_{pz} + \frac{g_{sn}\,\mu_N}{\hbar} S_{nz} \tag{6.89}$$

The observed value is given by the expectation value

$$<\mu_{dz}> = <\psi \,|\, \mu_{dz} \,|\, \psi> \tag{6.90}$$

where $|\,\psi>$ is the bound state wave function of the deuteron.

$$<\mu_{dz}> = <\psi\,|\, \frac{g_{lp}\,\mu_N}{\hbar}\frac{l_z}{2} + \frac{g_{sp}\,\mu_N}{\hbar}S_{pz} + \frac{g_{sn}\,\mu_N}{\hbar}S_{nz} \,|\, \psi> \tag{6.91}$$

$$= <\psi\,|\, \frac{g_{lp}\,\mu_N}{\hbar}\frac{l_z}{2}\,|\,\psi> + <\psi\,|\,\frac{g_{sp}\,\mu_N}{\hbar}S_{pz}\,|\,\psi> + <\psi\,|\,\frac{g_{sn}\,\mu_N}{\hbar}S_{nz}\,|\,\psi> \tag{6.92}$$

Assuming central potential and value of angular momentum $l = 0$, $l_z = 0$ for ground state of deuteron

$$<\psi\,|\,\frac{g_{lp}\mu_N}{\hbar}\frac{l_z}{2}\,|\,\psi> = 0 \tag{6.93}$$

Hence from equation (6.92) we get

$$<\mu_{dz}> = <\psi\,|\,\frac{g_{sp}\,\mu_N}{\hbar}S_{pz}\,|\,\psi> + <\psi\,|\,\frac{g_{sn}\,\mu_N}{\hbar}S_{nz}\,|\,\psi> \tag{6.94}$$

For deuteron $J = 1$, $m_J = 1, 0, -1$.

Quantum mechanically the maximum value of $\mu_{dz}$ is compared with the experimental value. The maximum value of $\mu_{dz}$ is obtained by using the wave function corresponding to $m_J = 1$. Thus the state we consider is

$|\,\psi> = |\,J = 1, m_J = 1, l = 0, s = 1> \equiv |\,J = 1, m_J = 1>$

and we evaluate

$$<\mu_{dz}> = <J = 1, m_J = 1 \,|\, \mu_{dz} \,|\, J = 1, m_J = 1> \qquad (6.95)$$

Now with $J = 1$, $m_J = 1$, $l = 0$, $m_l = 0$ the relation $m_J = m_l + m_s$ becomes $m_s = 1$.

Again since $\vec{s} = \vec{s}_p + \vec{s}_n$ their projections are inter-related as $m_s = m_{sp} + m_{sn}$. Clearly $m_s = 1$ comes from $m_{sp} = \frac{1}{2}$, $m_{sn} = \frac{1}{2}$. This means both proton and neutron are *up* states, i.e. spin-ups ↑↑. We can thus denote the state as

$|\psi> = |J = 1, m_J = 1> = |\uparrow\uparrow>$.

So we can write equation (6.95) as

$$<\mu_{dz}> = <\uparrow\uparrow\,|\, \mu_{dz} \,|\, \uparrow\uparrow> \qquad (6.96)$$

And equation (6.94) can be recast as

$$<\mu_{dz}> = <\uparrow\uparrow\,|\, \frac{g_{sp}\,\mu_N}{\hbar} s_{pz} \,|\, \uparrow\uparrow> + <\uparrow\uparrow\,|\, \frac{g_{sn}\,\mu_N}{\hbar} s_{nz} \,|\, \uparrow\uparrow>$$

$$= \frac{g_{sp}\,\mu_N}{\hbar} <\uparrow\uparrow\,|\, s_{pz} \,|\, \uparrow\uparrow> + \frac{g_{sn}\,\mu_N}{\hbar} <\uparrow\uparrow\,|\, s_{nz} \,|\, \uparrow\uparrow> \qquad (6.97)$$

In the state $|\uparrow\uparrow>$ the first arrow represents proton spin eigenstate and so

$$s_{pz} \,|\, \uparrow\uparrow> = \frac{\hbar}{2} \,|\, \uparrow\uparrow>$$

and the second arrow in $|\uparrow\uparrow>$ represents neutron eigenstate and so

$$s_{nz} \,|\, \uparrow\uparrow> = \frac{\hbar}{2} \,|\, \uparrow\uparrow> .$$

We thus have from equations (6.96) and (6.97)

$$<\mu_{dz}> = <\uparrow\uparrow\,|\, \mu_z \,|\, \uparrow\uparrow> = \frac{g_{sp}\,\mu_N}{\hbar}\frac{\hbar}{2} <\uparrow\uparrow\,|\,\uparrow\uparrow> + \frac{g_{sn}\,\mu_N}{\hbar}\frac{\hbar}{2} <\uparrow\uparrow\,|\,\uparrow\uparrow>$$

$$<\mu_{dz}> = \frac{\mu_N}{2}(g_{sp} + g_{sn}) \qquad (6.98)$$

Using equations (6.82) and (6.85) we have

$$<\mu_{dz}> = \frac{\mu_N}{2}(5.585\ 691 - 3.826\ 084)$$

$$<\mu_{dz}> = 0.8798\mu_N = <\mu_{dz}>_{theo}^{l=0} \qquad (6.99)$$

This value disagrees with the experimental value given in equation (6.70) viz.

$$<\mu_{dz}>_{\text{expt}} = 0.8574\mu_N$$

The theoretical value of magnetic moment of deuteron nucleus was obtained assuming it to be described by a central nuclear potential in the $l = 0$ state, i.e. $^3S_1$ state). Hence the disagreement.

- Let us consider a slight admixture of $l = 2$ state $^3D_1$ state in the ground state orbital motion of deuteron.

    Evaluation of $< \mu_z >_{\text{theo}}^{l=2}$ has been done in *exercise 6.17*. The value obtained is given by

$$< \mu_z >_{\text{theo}}^{l=2} = \frac{3 - g_{sp} - g_{sn}}{4} \mu_N = \frac{3 - 5.585\,691 - (-3.826\,084)}{4} \mu_N = 0.3101 \mu_N \quad (6.100)$$

We claim that a particular admixture of $l = 0$ state and $l = 2$ state will generate the observed magnetic dipole moment $< \mu_z >_{\text{expt}} = 0.8574 \mu_N$ of deuteron.

Let the probability of finding deuteron in $l = 0$, i.e. $^3S_1$ state be $| a_s |^2$ and the probability of finding deuteron in $l = 2$, i.e. $^3D_1$ state be $| a_D |^2$ so that total probability is

$$| a_S |^2 + | a_D |^2 = 1 \quad (6.101)$$

Let us set

$$| \psi > = a_s | l = 0 > + a_D | l = 2>$$

where the equation (6.101) is the normalization relation. Accordingly

$$< \mu_{dz} >_{\text{observed}} = | a_S |^2 < \mu_{dz} >_{l=0} + | a_D |^2 < \mu_{dz} >_{l=2} \quad (6.102)$$

Putting values from equations (6.70), (6.99), and (6.100) we have the relation

$$0.8574 \mu_N = | a_S |^2 (0.8798 \mu_N) + | a_D |^2 (0.3101 \mu_N)$$

$$0.8574 \mu_N = | a_S |^2 (0.8798 \mu_N) + (1 - | a_S |^2)(0.3101 \mu_N) \text{ [using equation (6.101)]}$$

$$| a_S |^2 = \frac{0.8574 - 0.3101}{0.8798 - 0.3101} = 0.96 \text{ and}$$

$$| a_D |^2 = 1 - | a_S |^2 = 1 - 0.96 = 0.04 \quad (6.103)$$

It is clear that the observed magnetic moment suggests that deuteron ground state is an admixture of 96% $l = 0$, i.e. $^3S_1$ state and 4% $l = 2$, i.e. $^3D_1$ state. So there is a small non-central part. Obviously nuclear potential is slightly non-central.

- The experimental value of magnetic moment of deuteron is $\mu_d \sim \mu_{dz} = 0.857\,36 \mu_N$.

Again the value of magnetic moment of the proton is

$$\mu_p = 2.792\,81\mu_N \tag{6.104}$$

and the value of magnetic moment of the neutron is

$$\mu_n = -1.9131\,48\mu_N. \tag{6.105}$$

In deuteron ground state the spin of proton and neutron are parallel and so the net magnetic moment is obtained through a simple addition as

$$\mu_p + \mu_n = 2.792\,81\mu_N - 1.913\,148\mu_N$$

$$= 0.879\,662\mu_N \neq \mu_d = 0.857\,36\mu_N \text{ (experimental value). Clearly}$$

$$\mu_d \neq \mu_p + \mu_n \tag{6.106}$$

The difference or discrepancy is because of the fact that we have ignored magnetic moment due to orbital angular momemtum through equation (6.93), i.e. we assumed that deuteron is in state $l = 0$ in the ground state and took $\mu_{l_z} = 0$. This is evident from the actual relation given by equation (6.88). Actually, we have to consider contribution from $l = 2$ also since deuteron ground state is a combination of $l = 0$, i.e. $^3S_1$ state (96%) and $l = 2$, i.e. $^3D_1$ state (4%).

## 6.14 Quadrupole moment of deuteron

The electric quadrupole moment is a measure of departure of the nucleus from spherical symmetry. The quadrupole moment is zero for a spherically symmetric nucleus, as shown in figure 6.6(a).

The quadrupole moment of a nucleus arises from deformation of its charge distribution from spherical shape, as shown in figure 6.6(b),(c).

Electrical quadrupole moment $Q$ of a nucleus is defined as

$$Q = \frac{1}{|e|} \int_\tau (3z^2 - r^2)\rho(\vec{r})d\tau \tag{6.107}$$

where $\rho(\vec{r})$ is charge density of a nucleus.

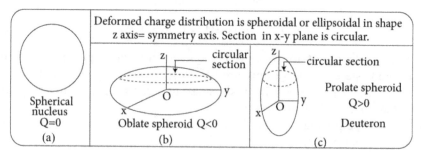

**Figure 6.6.** (a) Spherical nucleus being undistorted has zero quadrupole moment. (b) Oblate spheroid. (c) Prolate spheroid of deuteron.

If charge distribution is spherically symmetric

$$V(\vec{r}) = V(r) \neq V(\theta, \phi)$$

then charge density is the same everywhere on the spherical surface, i.e.

$$\rho(\vec{r}) = \rho(r).$$

So

$$Q = \frac{1}{|e|} \int_\tau (3z^2 - r^2)\rho(r)d\tau \qquad (6.108)$$

Since spherical symmetry implies that all directions $x$, $y$, $z$ are equivalent hence

$$\int x^2 \rho(r)d\tau = \int y^2 \rho(r)d\tau = \int z^2 \rho(r)d\tau$$

$$= \frac{1}{3} \int (x^2 + y^2 + z^2)\rho(r)d\tau$$

$$\int z^2 \rho(r)d\tau = \frac{1}{3} \int r^2 \rho(r)d\tau$$

$$\int 3z^2 \rho(r)d\tau = \int r^2 \rho(r)d\tau$$

Clearly thus for spherically symmetric charge distribution

$$Q = \frac{1}{|e|} \int_\tau (3z^2 - r^2)\rho(r)d\tau = 0 \qquad (6.109)$$

If deuteron is described by a purely central potential corresponding to $l = 0$, i.e. $^3S_1$ state then its wave function would have been spherically symmetric and quadrupole moment would have vanished, i.e we would get $Q = 0$.

But the observed value of quadrupole moment of deuteron is

$$Q = 2.82 \times 10^{-31}\,m^2 = 2.82 \times 10^{-3}\,barn = 2.82\,mbarn \neq 0 \qquad (6.110)$$

So deuteron charge distribution is spherically asymmetric—there is slight deviation from spherical symmetry.

✓ This spherically asymmetric charge distribution arises because of slight ($\sim 4\%$) admixture of $l = 2$, i.e. $^3D_1$ state along with 96% $l = 0$, i.e. $^3S_1$ state. In other words, 96% of the time deuteron stays in $^3S_1$ state and only 4% of the time deuteron stays in $^3D_1$ state. We can write the deuteron state as

$$| \text{deuteron} > = \sqrt{0.96} |\ ^3S_1 > + \sqrt{0.04} |\ ^3D_1 > \qquad (6.111)$$

So nuclear potential is non-central. Orbital angular momentum $l$ is not conserved.

• If charge distribution is stretched perpendicular to the $z$ direction it is an oblate spheroid or disc shaped as depicted in figure 6.6(b). Then

$$\int 3z^2\rho(r)d\tau \; < \; \int r^2\rho(r)d\tau \text{ and so}$$

$$Q < 0 \tag{6.112}$$

- If charge distribution is stretched or elongated in the $z$ direction, i.e. it is a prolate spheroid or cigar shaped as depicted in figure 6.6(c) then

$$\int 3z^2\rho(r)d\tau \; > \; \int r^2\rho(r)d\tau \text{ and so}$$

$$Q > 0 \tag{6.113}$$

Obviously the positive value of quadrupole moment namely $Q = 2.82 \; mbarn > 0$ signifies that deuteron has a prolate spheroidal charge distribution as depicted in figure 6.6(c). It is stretched along the direction of angular momentum (taken as $z$-axis).

## 6.15 Tensor force

The deuteron problem was dealt with assuming that the neutron–proton force is a central force. Then the ground state of deuteron must be a $^3S_1$ state. The magnetic moment would then be the algebraic sum of the magnetic moments of proton $\vec{\mu}_p$ and neutron $\vec{\mu}_n$, i.e.

$$\vec{\mu}_d = \vec{\mu}_l + \vec{\mu}_p + \vec{\mu}_n = \vec{\mu}_p + \vec{\mu}_n \text{ since for } l = 0, \; \vec{\mu}_l = 0.$$

And the quadrupole moment would have been zero, i.e.

$$Q = 0$$

But these facts are not corroborated in experiments. The magnetic moment is not the sum of proton and neutron magnetic moments, i.e. $\vec{\mu}_d \neq \vec{\mu}_p + \vec{\mu}_n$. And the quadrupole moment of the deuteron system is non-zero as $Q = 2.82 \times 10^{-31} m^2 \neq 0$.

Nucleon–nucleon interaction is partly ($\sim$96%) due to central force and partly (4%) due to non-central force and so $l$ is not unique and hence $l$ is not conserved.

We say that there is no tensor force for $l = 0$ ($S$ state) since there is spherical symmetry and all directions are equivalent.

It is said that tensor force is responsible for the non-central part of interaction.
It is spin dependent.
It gives a non-zero value of quadrupole moment.
Tensor force part is smaller than central force part.

In other words, the non-central part of the nucleon–nucleon potential can be described through a term like the tensor force with the help of which one can explain the non-zero quadrupole moment of deuteron.

- An example of interaction through non-central potential is shown in figure 6.7(a), where two magnetic dipole moments $\bar{m}_1$, $\bar{m}_2$ exert force upon each other. We note that the distance $r$ between the dipoles $\bar{m}_1$ and $\bar{m}_2$ is the

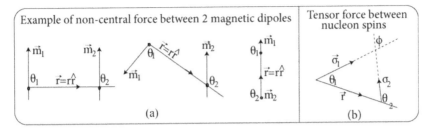

**Figure 6.7.** (a) Example of non-central force between two magnetic moments. (b) Tensor force between nucleons depends on their spin orientation, relative orientation and orientation of spins w.r.t. relative position vector.

same in all the three cases shown but the force of interaction will be different since the orientation of the dipoles w.r.t. the line joining them is different.

The magnetic potential energy between the two magnetic dipoles has been calculated in *exercise 6.18* which is

$$U = \frac{\mu_0}{4\pi} \frac{\vec{m}_1 . \vec{m}_2 - 3(\vec{m}_1 . \hat{r})(\vec{m}_2 . \hat{r})}{r^3} \tag{6.114}$$

Magnetic potential energy depends upon magnetic dipole moments $\vec{m}_1$, $\vec{m}_2$ and their relative orientation as evident from the term $\vec{m}_1 . \vec{m}_2$; the relative separation between dipoles $r$; relative orientation of the magnetic dipoles $\vec{m}_1$, $\vec{m}_2$ w.r.t. the relative position vector $\vec{r}$ as evident from the terms $(\vec{m}_1 . \hat{r})$, $(\vec{m}_2 . \hat{r})$.

Magnetic potential energy is lowered when line joining the dipoles is parallel to their spin direction compared to other orientations. This is shown in *exercise 6.19.*

- Tensor force resembles the magnetic interaction between two magnetic dipoles [figure 6.7(b)] which is non-central. Since tensor force part should be included in the nucleon–nucleon system let us try to predict the terms that the tensor force part would include.

  Suppose the spins of two nucleons is $\vec{\sigma}_p \equiv \vec{\sigma}_1$, $\vec{\sigma}_n \equiv \vec{\sigma}_2$

  Tensor force would depend on the relative spin orientation of $\vec{\sigma}_1$, $\vec{\sigma}_2$ of the two nucleons and this is taken care of by the term

$$\vec{\sigma}_1 . \vec{\sigma}_2.$$

  Tensor force would depend upon their relative separation $\vec{r}$ as well as on the orientation of spins w.r.t. their relative separation. This means the tensor force would include the terms

$$(\vec{\sigma}_1 . \hat{r}), (\vec{\sigma}_2 . \hat{r})$$

- Though nuclear force is non-central we can still assume that it is conservative and is derivable from a potential $V$. The difference from the previously assumed potential is that now potential $V$ depends not only on inter-nucleon distance $r$ but also on spins and spin orientations.

Since the nuclear force should be derivable from a potential it is required that the corresponding potential is needed to be a scalar which means that it should be invariant under rotation and reflection of the coordinate system used to describe the relative motion of particles. Such a potential that is scalar, conserves parity as required in the description of nuclear forces.

We note that if potential is taken to be pseudoscalar—it would change sign under reflection and would lead to non-conservation of parity. This is not acceptable for the description of nuclear force.

- The general form of potential describing nucleon–nucleon potential will be

$$V = V_1(r) + V_2(r)\vec{\sigma}_1 \cdot \vec{\sigma}_2 + V_3(r)S_{12} \tag{6.115}$$

$V$ depends on position of nucleons and their spins.

$V_1(r)$ is the ordinary central potential part, i.e. corresponds to central force as it depends only on $r$ and is the major part of nucleon–nucleon potential.

$V_2(r)\vec{\sigma}_1 \cdot \vec{\sigma}_2$ is the spin dependent central force part. It is central since it depends on $r$. It is spin dependent since it depends on $\vec{\sigma}_1, \vec{\sigma}_2$.

$V_3(r)S_{12}$ is the non-central angle dependent part called the tensor potential part. It is the spin–spin interaction potential and depends on angular orientation of spins $\vec{\sigma}_1$ and $\vec{\sigma}_2$ w.r.t. the line joining the nucleons. This angle dependent non-central nuclear force is called tensor force.

$S_{12}$ is called tensor force operator and is defined as the following scalar operator

$$S_{12} = \frac{3(\vec{\sigma}_1 \cdot \vec{r})(\vec{\sigma}_2 \cdot \vec{r})}{r^2} - \vec{\sigma}_1 \cdot \vec{\sigma}_2 = 3(\vec{\sigma}_1 \cdot \hat{r})(\vec{\sigma}_2 \cdot \hat{r}) - \vec{\sigma}_1 \cdot \vec{\sigma}_2 \tag{6.116}$$

The first term on the RHS of equation (6.116) viz. $\frac{3(\vec{\sigma}_1 \cdot \vec{r})(\vec{\sigma}_2 \cdot \vec{r})}{r^2} = 3(\vec{\sigma}_1 \cdot \hat{r})(\vec{\sigma}_2 \cdot \hat{r})$ gives dependence of nucleon–nucleon interaction upon the angles $\theta_1$, $\theta_2$ between the nucleon spins $\vec{\sigma}_1$, $\vec{\sigma}_2$ and the relative position vector $\vec{r}$ respectively.

The second term on the RHS of equation (6.116) is $\vec{\sigma}_1 \cdot \vec{\sigma}_2$ and depends on angle between the spins of nucleons. It is subtracted so that the average value of $S_{12}$ over all directions of $\vec{r}$ is zero.

The quantities $V_1$, $V_2$, $V_3$ are functions of nucleon–nucleon distance.

☐ Let us find the form of tensor force operator $S_{12} = \frac{3(\vec{\sigma}_1 \cdot \vec{r})(\vec{\sigma}_2 \cdot \vec{r})}{r^2} - \vec{\sigma}_1 \cdot \vec{\sigma}_2$ for the four cases (i) to (iv), as shown in figure 6.8.

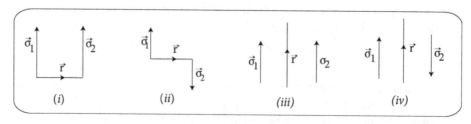

**Figure 6.8.** We find $S_{12}$ for these configurations of $\vec{\sigma}_1$, $\vec{\sigma}_2$.

✓ Refer to figure 6.8($i$)

Spins $\vec{\sigma}_1, \vec{\sigma}_2$ are parallel ↑ ↑ . So it is triplet state $s = 1$. And $\vec{\sigma}_1, \vec{\sigma}_2$ are perpendicular to $\vec{r}$. Hence

$$\vec{\sigma}_1. \vec{r} = 0$$

$$\vec{\sigma}_2. \vec{r} = 0.$$

And by *exercise 6.22* $\vec{\sigma}_1. \vec{\sigma}_2 = 1$ for triplet state. Hence by equation (6.116)

$$S_{12} = \frac{3(\vec{\sigma}_1. \vec{r})(\vec{\sigma}_2. \vec{r})}{r^2} - \vec{\sigma}_1. \vec{\sigma}_2 = \frac{3.0.0}{r^2} - 1$$

$$S_{12} = -1$$

✓ Refer to figure 6.8($ii$)

Spins $\vec{\sigma}_1, \vec{\sigma}_2$ are antiparallel ↑ ↓ . So it is singlet state $s = 0$. And $\vec{\sigma}_1, \vec{\sigma}_2$ are perpendicular to $\vec{r}$. Hence

$$\vec{\sigma}_1. \vec{r} = 0$$

$$\vec{\sigma}_2. \vec{r} = 0.$$

And by *exercise 6.22* $\vec{\sigma}_1. \vec{\sigma}_2 = -3$ for singlet state. Hence by equation (6.116)

$$S_{12} = \frac{3(\vec{\sigma}_1. \vec{r})(\vec{\sigma}_2. \vec{r})}{r^2} - \vec{\sigma}_1. \vec{\sigma}_2 = \frac{3.0.0}{r^2} - (-3)$$

$$S_{12} = 3$$

✓ Refer to figure 6.8($iii$)

Spins $\vec{\sigma}_1, \vec{\sigma}_2$ are parallel ↑ ↑ . So it is triplet state $s = 1$. And $\vec{\sigma}_1, \vec{\sigma}_2$ are along $\vec{r}$. Hence

$$\vec{\sigma}_1. \vec{r} = \sigma_1 r$$

$$\vec{\sigma}_2. \vec{r} = \sigma_2 r.$$

Also

$$\vec{\sigma}_1. \vec{\sigma}_2 = \sigma_1 \sigma_2.$$

And by *exercise 6.22* $\vec{\sigma}_1. \vec{\sigma}_2 = 1$ for triplet state. Also

$$\sigma_1^2 = 3, \ \sigma_2^2 = 3.$$

Hence by equation (6.116)

$$S_{12} = \frac{3(\vec{\sigma}_1 \cdot \vec{r})(\vec{\sigma}_2 \cdot \vec{r})}{r^2} - \vec{\sigma}_1 \cdot \vec{\sigma}_2 = \frac{3 \cdot \sigma_1 r \cdot \sigma_2 r}{r^2} - 1 = 3\sigma_1\sigma_2 - 1$$

$$= 3\sqrt{\sigma_1^2\sigma_2^2} - 1 = 3\sqrt{3.3} - 1$$

$$S_{12} = 8$$

✓ Refer to figure 6.8(*iv*)

Spins $\vec{\sigma}_1$, $\vec{\sigma}_2$ are antiparallel ↑ ↓ . So it is singlet state $s = 0$. And $\vec{\sigma}_1$ and $\vec{r}$ are parallel. Hence

$$\vec{\sigma}_1 \cdot \vec{r} = \sigma_1 r$$

Also, $\vec{\sigma}_2$ and $\vec{r}$ are antiparallel. So

$$\vec{\sigma}_2 \cdot \vec{r} = -\sigma_2 r.$$

Also,

$$\vec{\sigma}_1 \cdot \vec{\sigma}_2 = -\sigma_1\sigma_2.$$

And by *exercise 6.22* $\vec{\sigma}_1 \cdot \vec{\sigma}_2 = -3$ for singlet state. Also

$$\sigma_1^2 = 3, \ \sigma_2^2 = 3.$$

Hence by equation (6.116)

$$S_{12} = \frac{3(\vec{\sigma}_1 \cdot \vec{r})(\vec{\sigma}_2 \cdot \vec{r})}{r^2} - \vec{\sigma}_1 \cdot \vec{\sigma}_2 = \frac{3(\sigma_1 r)(-\sigma_2 r)}{r^2} - (-3)$$

$$= -3\sigma_1\sigma_2 + 3 = -3\sqrt{\sigma_1^2\sigma_2^2} + 3 = -3\sqrt{3.3} + 3$$

$$S_{12} = -6$$

☐ In *exercise 6.25* we calculate the tensor force operator $S_{12}$ for deuteron.
- Let us find if tensor force operates in spin singlet state.

   In singlet spin state there is no preferred direction for the spin orientation. So total spin angular momentum should be zero, i.e. for the singlet spin state we can write

$$\vec{s} = 0$$

$$\vec{s}_p + \vec{s}_n = 0$$

$$\frac{\hbar}{2}\vec{\sigma}_1 + \frac{\hbar}{2}\vec{\sigma}_2 = 0 \ \Rightarrow \ \frac{\hbar}{2}(\vec{\sigma}_1 + \vec{\sigma}_2) = 0 \quad (\vec{\sigma}_p = \vec{\sigma}_1, \vec{\sigma}_n = \vec{\sigma}_2)$$

$$\vec{\sigma}_1 = -\vec{\sigma}_2$$

So in singlet state the orientation of $\vec{\sigma}_1$ and $\vec{\sigma}_2$ is equal and opposite. For singlet spin state the tensor force operator [equation (6.116)] will thus be

$$S_{12} = \frac{3(\vec{\sigma}_1.\,\vec{r})(\vec{\sigma}_2.\,\vec{r})}{r^2} - \vec{\sigma}_1.\,\vec{\sigma}_2 = \frac{-3(\vec{\sigma}_1.\,\vec{r})(\vec{\sigma}_1.\,\vec{r})}{r^2} - \vec{\sigma}_1.\,\vec{\sigma}_2$$

$$= -3(\vec{\sigma}_1.\,\hat{r})(\vec{\sigma}_1.\,\hat{r}) - \vec{\sigma}_1.\,\vec{\sigma}_2 \tag{6.117}$$

In *exercise 6.22* we have shown that for $s = 0$ (singlet state)

$$\vec{\sigma}_1.\,\vec{\sigma}_2 = -3$$

And also we use the result of *exercise 6.20* namely

$$(\vec{\sigma}.\,\vec{A})(\vec{\sigma}.\,\vec{B}) = \vec{A}.\,\vec{B} + i\ \vec{\sigma}.\ \vec{A} \times \vec{B}$$

to get

$$(\vec{\sigma}_1.\,\hat{r})(\vec{\sigma}_1.\,\hat{r}) = \hat{r}.\,\hat{r} + i\ \vec{\sigma}_1.\,\hat{r} \times \hat{r} = 1 + 0 = 1 \tag{6.118}$$

Hence from equations (6.117) and (6.118) we have

$$S_{12} = -3(\vec{\sigma}_1.\,\hat{r})(\vec{\sigma}_1.\,\hat{r}) - \vec{\sigma}_1.\,\vec{\sigma}_2 = -3(1) - (-3) = -3 + 3 = 0 \tag{6.119}$$

Clearly in equation (6.115) the term $V_3(r)S_{12} = 0$ in case of singlet spin state in which there is no preferred direction. So no tensor force is present in singlet spin state. In singlet spin state the nuclear force is purely a central force.

• If spin state is not a pure singlet then tensor force arises. So in a mixture of singlet plus triplet state tensor force will operate.

• Some properties of tensor force are

✓ Tensor force potential $V_3(r)S_{12}$ is a scalar. Hence the total angular momentum $J$ and parity $\pi$ are constants of motion.

✓ Tensor force is a non-central force. So tensor force is not invariant under rotation of space coordinate or spin coordinate separately. Hence orbital angular momentum $\vec{l}$ and spin angular momentum $\vec{s}$ are not constants of motion.

✓ Nucleon–nucleon potential has a central force part that is actually the force between two nucleons and it corresponds to singlet spin state of the two nucleons. It has also a tensor force that corresponds to triplet spin state. And tensor force part is actually a minor part of the nuclear force—the major part being the central force. Tensor force part is weaker and has a longer range compared to the central force part.

✓ Central force between nucleons is spin independent but the tensor force or the non-central force is spin dependent. Actually, the spin dependence of nuclear force arises entirely from the tensor force. This is evident since tensor force is zero in the singlet spin state and contributes in the triplet spin state.

✓ The existence of one bound spin triplet state of the nucleon–nucleon system (the *n–p* system or deuteron system) is due to the tensor force. The singlet state has only the central force and no tensor force and it fails to create a bound state.

- In *exercise 8.8* we show that deuteron state is an isospin singlet.

## 6.16 Exercises

**Exercise 6.1** *Show that* $\frac{\hbar^2}{m} = 41.37\ MeV.\ fm^2$ *where* $m = m_p =$ *proton mass.*
$\boxed{Ans}$ $\frac{\hbar^2}{m} = \frac{(\hbar c)^2}{mc^2} = \frac{(197\ MeV.fm)^2}{938\ MeV} = 41.37\ MeV.\ fm^2$ (using *exercise 6.1*)

**Exercise 6.2** *Show that central potential leads to conservation of angular momentum.*
$\boxed{Ans}$ Relation between angular momentum and torque is

$$\frac{d}{dt}\vec{l} = \vec{N} = \vec{r} \times \vec{F}$$

Being a central force

$$\vec{F} = \hat{r}F(r)$$

and so

$$\frac{d}{dt}\vec{l} = \vec{r} \times \hat{r} \quad F(r) = r\hat{r} \times \hat{r} \quad F(r) = 0$$

$\vec{l}$ = conserved.
Central potential leads to conservation of angular momentum.

**Exercise 6.3** *(a) Two molecules in a substance interact although the molecules are neutral. Explain if there is any anomaly.*
*(b) Two nucleons inside a nucleus interact through strong interaction which depends on colour charge. Again nucleons are colourless. Explain if there is any anomaly.*
$\boxed{Ans}$ (a) A molecule is by itself electrally neutral on the whole. Each molecule of a substance contains atoms, nuclei and electrons which are charged particles and charged particles interact electromagnetically. In a substance, therefore, molecules are not indifferent to one another but interact electromagnetically. This keeps the molecules bound together forming the substance.

We note that two charges can interact irrespective of their intervening separation. But two neutral molecules virtually do not interact if their separation is large compared to their sizes. This is because at large distance the molecules do not see or feel each other's internal structure. In fact, at a large distance it behaves as a point particle of zero charge since molecules are neutral. So they fail to influence each other and there is virtually no interaction. But when the two molecules are brought closer to each other, i.e. when their separation is smaller or of the same order of their sizes then they can feel or see each other's internal structure and hence interaction is possible at very small distance. So despite the fact that the total molecular charge is zero there is molecular force which is obviously short range and the strength is $\sim eV$.

(b) The total colour charge on each nucleon is zero. But if the separation of nucleons is small and is of the same order as the distribution inside the nucleon then there will be colour interaction. So nucleon–nucleon force originates from here. Nuclear force is thus short range and occurs between colour-neutral entities (nucleons) and the strength is $\sim MeV$.

**Exercise 6.4** *Write down* $\nabla^2$ *in spherical polar coordinates. Use it in the time independent Schrödinger equation describing deuteron system and obtain the radial equation (6.14).*

$\boxed{Ans}$ ✓ In spherical polar coordinates

$$\nabla^2 = \frac{1}{r^2}\frac{\partial}{\partial r}\left(r^2\frac{\partial}{\partial r}\right) + \frac{1}{r^2 \sin\theta}\frac{\partial}{\partial\theta}\left(\sin\theta\frac{\partial}{\partial\theta}\right) + \frac{1}{r^2 \sin^2\theta}\frac{\partial^2}{\partial\phi^2} \tag{6.120}$$

✓ We can use it in the time independent Schrödinger equation describing deuteron system and obtain the radial equation as follows.
The square of orbital angular momentum operator $\hat{l}^2$ is

$$\hat{l}^2 = -\hbar^2\left[\frac{1}{\sin\theta}\frac{\partial}{\partial\theta}\left(\sin\theta\frac{\partial}{\partial\theta}\right) + \frac{1}{\sin^2\theta}\frac{\partial^2}{\partial\phi^2}\right] \tag{6.121}$$

Hence from equations (6.120) and (6.121) we have

$$\nabla^2 = \frac{1}{r^2}\frac{\partial}{\partial r}\left(r^2\frac{\partial}{\partial r}\right) + \frac{1}{r^2}\left[\frac{1}{\sin\theta}\frac{\partial}{\partial\theta}\left(\sin\theta\frac{\partial}{\partial\theta}\right) + \frac{1}{\sin^2\theta}\frac{\partial^2}{\partial\phi^2}\right]$$

$$= \frac{1}{r^2}\frac{\partial}{\partial r}\left(r^2\frac{\partial}{\partial r}\right) + \frac{1}{r^2}[-\frac{\hat{l}^2}{\hbar^2}]$$

$$\nabla^2 = \frac{1}{r^2}\frac{\partial}{\partial r}\left(r^2\frac{\partial}{\partial r}\right) - \frac{1}{\hbar^2 r^2}\hat{l}^2 \tag{6.122}$$

The time independent Schrödinger equation (6.11) for the deuteron problem is

$$\nabla^2\psi(\vec{r}) + \frac{2\mu}{\hbar^2}[E - V(\vec{r})]\psi(\vec{r}) = 0 \tag{6.123}$$

We use method of separation of variables. Let us write

$$\psi(\vec{r}) = R(r)Y(\theta, \phi) \tag{6.124}$$

where $R(r)$ is radial part and $Y(\theta, \phi)$ is angular part.

The deuteron potential is given by equation (6.6) to be

$$V = -V_0 \text{ for } r < r_0 \text{ and}$$

$$V = 0 \quad \text{for } r > r_0.$$

Since deuteron potential $V = V(r)$ is central, spherically symmetric the angular part $Y(\theta, \phi)$ will be [equation (6.12)] spherical harmonic, i.e. $Y(\theta, \phi) = Y_{lm_l}(\theta, \phi)$ where $l = 0, 1, 2, 3, \ldots$ and $m_l = 0, \pm1, \pm2, \ldots$ are the allowed values.

With equation (6.122), equation (6.124) the Schrödinger equation (6.123) becomes

$$\left[ \frac{1}{r^2} \frac{\partial}{\partial r} \left( r^2 \frac{\partial}{\partial r} \right) - \frac{1}{\hbar^2 r^2} \hat{l}^2 \right] R(r) Y(\theta, \phi) + \frac{2\mu}{\hbar^2} [E - V(r)] R(r) Y(\theta, \phi) = 0$$

$$Y(\theta, \phi) \frac{1}{r^2} \frac{d}{dr} \left( r^2 \frac{d}{dr} \right) R(r) - R(r) \frac{1}{\hbar^2 r^2} \hat{l}^2 Y(\theta, \phi) + \frac{2\mu}{\hbar^2} [E - V(r)] R(r) Y(\theta, \phi) = 0$$

Dividing by $R(r) Y(\theta, \phi)$ we get

$$\frac{1}{R(r)} \frac{1}{r^2} \frac{d}{dr} \left( r^2 \frac{d}{dr} \right) R(r) - \frac{1}{Y(\theta, \phi)} \frac{1}{\hbar^2 r^2} \hat{l}^2 Y(\theta, \phi) + \frac{2\mu}{\hbar^2} [E - V(r)] = 0$$

Since $Y(\theta, \phi) = Y_{lm_l}(\theta, \phi)$ we use the eigenvalue equation of $\hat{l}^2$ viz.

$$\hat{l}^2 Y_{lm_l}(\theta, \phi) = l(l + 1)\hbar^2 Y_{lm_l}(\theta, \phi)$$

to get

$$\frac{1}{R(r)} \frac{1}{r^2} \frac{d}{dr} \left( r^2 \frac{d}{dr} \right) R(r) - \frac{1}{Y_{lm_l}(\theta, \phi)} \frac{1}{\hbar^2 r^2} l(l + 1)\hbar^2 Y_{lm_l}(\theta, \phi) + \frac{2\mu}{\hbar^2} [E - V(r)] = 0$$

$$\frac{1}{R(r)} \frac{1}{r^2} \frac{d}{dr} \left( r^2 \frac{d}{dr} \right) R(r) + \frac{2\mu}{\hbar^2} \left[ E - V(r) - \frac{l(l + 1)\hbar^2}{2\mu r^2} \right] = 0$$

$$\frac{1}{r^2} \frac{d}{dr} \left( r^2 \frac{d}{dr} \right) R(r) + \frac{2\mu}{\hbar^2} \left[ E - V(r) - \frac{l(l + 1)\hbar^2}{2\mu r^2} \right] R(r) = 0$$

$$\frac{1}{r^2} [r^2 \frac{d^2}{dr^2} + 2r \frac{d}{dr}] R(r) + \frac{2\mu}{\hbar^2} \left[ E - V(r) - \frac{l(l + 1)\hbar^2}{2\mu r^2} \right] R(r) = 0$$

$$\frac{d^2}{dr^2} R(r) + \frac{2}{r} \frac{d}{dr} R(r) + \frac{2\mu}{\hbar^2} \left[ E - V(r) - \frac{l(l + 1)\hbar^2}{2\mu r^2} \right] R(r) = 0$$

This is the radial equation (6.14) for the deuteron problem.

**Exercise 6.5** *Show that the substitution $R(r) = \frac{u(r)}{r}$ in radial equation (6.14) describing deuteron problem leads to a radial equation resembling 1D equation describing 1D motion of a particle in a potential $V_{\text{eff}} = V(r) + \frac{l(l+1)}{2\mu r^2}$.*

[Ans] The radial equation for the deuteron problem is by equation (6.14)

$$\frac{d^2}{dr^2}R(r) + \frac{2}{r}\frac{d}{dr}R(r) + \frac{2\mu}{\hbar^2}\left[E - V(r) - \frac{l(l+1)\hbar^2}{2\mu r^2}\right]R(r) = 0 \qquad (6.125)$$

Substitute $R(r) = \frac{u(r)}{r}$

$$\frac{dR}{dr} = \frac{d}{dr}\frac{u}{r} = \frac{1}{r}\frac{du}{dr} - \frac{u}{r^2} \quad \text{and}$$

$$\frac{d^2R}{dr^2} = \frac{d}{dr}\left[\frac{1}{r}\frac{du}{dr} - \frac{u}{r^2}\right] = \frac{1}{r}\frac{d^2u}{dr^2} + \frac{du}{dr}\left(-\frac{1}{r^2}\right) - \left[\frac{1}{r^2}\frac{du}{dr} - \frac{2u}{r^3}\right]$$

$$= \frac{1}{r}\frac{d^2u}{dr^2} - \frac{2}{r^2}\frac{du}{dr} + \frac{2u}{r^3}$$

Substituting $\frac{dR}{dr}$ and $\frac{d^2R}{dr^2}$ in equation (6.125) we have

$$\left[\frac{1}{r}\frac{d^2u}{dr^2} - \frac{2}{r^2}\frac{du}{dr} + \frac{2u}{r^3}\right] + \frac{2}{r}\left[\frac{1}{r}\frac{du}{dr} - \frac{u}{r^2}\right] + \frac{2\mu}{\hbar^2}\left[E - V(r) - \frac{l(l+1)\hbar^2}{2\mu r^2}\right]\frac{u}{r} = 0$$

$$\frac{1}{r}\frac{d^2u}{dr^2} + \frac{2\mu}{\hbar^2}\left[E - V(r) - \frac{l(l+1)\hbar^2}{2\mu r^2}\right]\frac{u}{r} = 0$$

$$\frac{d^2u}{dr^2} + \frac{2\mu}{\hbar^2}\left[E - V(r) - \frac{l(l+1)\hbar^2}{2\mu r^2}\right]u = 0$$

This is the radial equation (6.16) for the deuteron problem.

$$\frac{d^2u}{dr^2} + \frac{2\mu}{\hbar^2}[E - V_{\text{eff}}(r)]u = 0 \qquad (6.126)$$

with

$$V_{\text{eff}} = V(r) + \frac{l(l+1)\hbar^2}{2\mu r^2}$$

**Exercise 6.6** *Estimate range of nucleon–nucleon interaction in deuteron if strength of deuteron potential is $V_0 = 36\ MeV$.*

[Ans] The relation between strength of deuteron potential and range is given by equation (6.50)

$$\left(\frac{V_0}{2.225\ MeV} - 1\right)^{\frac{1}{2}} = \frac{6.77\ fm}{r_0} + 0.64$$

$$\left(\frac{36\ MeV}{2.225\ MeV} - 1\right)^{\frac{1}{2}} = \frac{6.77\ fm}{r_0} + 0.64$$

$$r_0 = 2.1\ fm$$

**Exercise 6.7** *Estimate strength of deuteron potential if the range of nucleon–nucleon interaction in deuteron is $r_0 = 2.1\ fm$.*

[Ans] The relation between strength of deuteron potential and range is given by equation (6.50).

$$\left(\frac{V_0}{2.225\ MeV} - 1\right)^{\frac{1}{2}} = \frac{6.77\ fm}{r_0} + 0.64$$

$$\left(\frac{V_0}{2.225\ MeV} - 1\right)^{\frac{1}{2}} = \frac{6.77\ fm}{2.1\ fm} + 0.64$$

$$V_0 = 35.44\ MeV$$

**Exercise 6.8** *If the range of deuteron potential is 2 fm estimate the probability for the nucleons in deuteron to be found within this separation.*

[Ans] The probability for the nucleons in deuteron to be found within a separation of $r_0$

$$P(0 \leqslant r \leqslant r_0) = \int_{r=0}^{r_0} \int_{\theta=0}^{\pi} \int_{\phi=0}^{2\pi} |\psi|^2\ d\tau$$

where $\psi$ is given by equation (6.33). Since

$$\psi(0 \leqslant r \leqslant r_0) = \frac{A \sin kr}{r}$$

we write

$$P(0 \leqslant r \leqslant r_0) = \int_{r=0}^{r_0} \int_{\theta=0}^{\pi} \int_{\phi=0}^{2\pi} \left(\frac{A \sin kr}{r}\right)^2 (r^2 dr \sin\theta d\theta d\phi)$$

$$= A^2 \int_{r=0}^{r_0} \sin^2 kr dr \int_{\theta=0}^{\pi} \sin\theta d\theta \int_{\phi=0}^{2\pi} d\phi = A^2.4\pi \int_{r=0}^{r_0} \sin^2 kr dr$$

$$P(0 \leqslant r \leqslant r_0) = A^2.\ 2\pi(r_0 - \frac{\sin 2kr_0}{2k})$$

From exercise *6.12* [equation (6.131)] $A = \sqrt{\frac{\beta}{2\pi}}(1 + \beta r_0)^{-1/2}$ so

$$P(0 \leqslant r \leqslant r_0) = \frac{\beta}{2\pi}(1 + \beta r_0)^{-1} 2\pi(r_0 - \frac{\sin 2kr_0}{2k})$$

$$= \frac{\beta}{1 + \beta r_0}\left(r_0 - \frac{\sin kr_0 \cos kr_0}{k}\right) = \frac{\beta}{1 + \beta r_0}\left(r_0 - \frac{\tan kr_0 \cos^2 kr_0}{k}\right)$$

Use from equation (6.38) $\tan kr_0 = -\frac{k}{\beta}$ to get

$$P(0 \leqslant r \leqslant r_0) = \frac{\beta}{1 + \beta r_0}\left(r_0 - \frac{-\frac{k}{\beta}\cos^2 kr_0}{k}\right) = \frac{\beta}{1 + \beta r_0}\left(r_0 + \frac{1}{\beta}\cos^2 kr_0\right)$$

$$= \frac{\beta r_0 + \cos^2 kr_0}{1 + \beta r_0} \tag{6.127}$$

Using equations (6.47) and (6.53) we have with $r_0 = 2.1\ fm$

$$P(0 \leqslant r \leqslant r_0) = \frac{(0.23\ fm^{-1})(2.1\ fm) + \cos^2\left(\frac{180}{\pi}1.88\right)}{1 + (0.23\ fm^{-1})(2.1\ fm)} = 0.388 \tag{6.128}$$

$\Rightarrow$ Small probability of the nuclcons of dcuteron to lie inside the well.

$$P(r > r_0) = 1 - P(0 \leqslant r \leqslant r_0) = 1 - 0.388 = 0.612 \tag{6.129}$$

$\Rightarrow$ Much greater probability of the nucleons of deuteron to lie outside the well.

**Exercise 6.9** *Consider the deuteron problem. Show that* $\cot kr_0 = -\sqrt{\frac{|E|}{V_0|}}$ *where k is given by* equation (6.21). *Hence find how strength* $V_0$ *and range* $r_0$ *are related. Hence find* $V_0$ *if* $r_0 = 2.1\ fm$.

$\boxed{Ans}$✓ By equation (6.38) $\tan kr_0 = -\frac{k}{\beta}$

$$\cot kr_0 = -\frac{\beta}{k}$$

Using equations (6.21) and (6.29) we can write

$$\cot kr_0 = -\frac{\sqrt{\frac{m}{\hbar^2}|E|}}{\sqrt{\frac{m}{\hbar^2}(V_0 - |E|)}} = -\sqrt{\frac{|E|}{V_0 - |E|}} \overset{V_0 \gg |E|}{\rightarrow} -\sqrt{\frac{|E|}{V_0}}$$

✓ As $V_0 \gg |E|$ we can write $\frac{|E|}{V_0} \cong 0$ and so

$$\cot kr_0 \cong 0$$

$$kr_0 \cong \frac{\pi}{2}, \frac{3\pi}{2}, \ldots \tag{6.130}$$

$$k \cong \frac{\pi}{2r_0}, \frac{3\pi}{2r_0}, \ldots \tag{6.131}$$

Again, use of equation (6.21) gives

$$k = \sqrt{\frac{m}{\hbar^2}(V_0 - |E|)} \xrightarrow{V_0 \gg |E|} \sqrt{\frac{mV_0}{\hbar^2}} \tag{6.132}$$

Hence from equations (6.131) and (6.132)

$$k = \sqrt{\frac{mV_0}{\hbar^2}} \cong \frac{\pi}{2r_0}, \frac{3\pi}{2r_0}, \ldots$$

Upon squaring we get

$$\frac{mV_0}{\hbar^2} = \frac{\pi^2}{4r_0^2}, \frac{9\pi^2}{4r_0^2}, \ldots$$

$$V_0 r_0^2 \cong \frac{\pi^2 \hbar^2}{4m}, \frac{9\pi^2 \hbar^2}{4m}, \ldots \tag{6.133}$$

Let us investigate the correct value of $V_0 r_0^2$ from simple arguments because $V_0 r_0^2$ cannot correspond to multiple values as deuteron has a single bound state.

Deuteron is in ground state and has no excited state. Again, ground state wave function is characterized by the fact that it will have no node since it corresponds to lowest energy—insufficient for making the wave function cross the $r$ axis.

Now the internal wave function given by equation (6.32) viz. $u_1(r) = A \sin kr$ goes to zero at $kr = 0$, $kr = \pi$. This implies the ground state wave function will be within this range of $0 \leqslant kr \leqslant \pi$. In other words, the range $r_0$ will be such that $kr_0$ falls within $0 \leqslant kr \leqslant \pi$. This is shown in figure 6.9(a).

From equation (6.130) it follows that the choice $kr_0 = \frac{3\pi}{2} = 1.5\pi > \pi$ would mean that then the internal wave function $u_1(r) = A \sin kr$ will have a node inside the well which is unacceptable. This is shown in figure 6.9(b).

Again, the choice $kr_0 = \frac{\pi}{2} < \pi$ means that the internal wave function will have no node inside the well. This is shown in figure 6.9(c).

There can be matching of internal wave function $u_1(r) = A \sin kr$ with the external wave function $u_2(r) = e^{-\beta r}$ at $kr_0 = \frac{\pi}{2} = 1.57$ that corresponds to $r = 1.74 \, fm$ [equation (6.54)].

(We earlier took the matching at $kr_0 = 1.88$ [equations (6.47) and (6.55)], i.e. at $r_0 = 2.1 \, fm$ as shown in figure 6.3.)

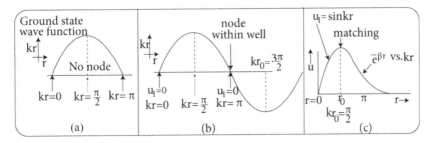

**Figure 6.9.** (a) Ground state wave function cannot become zero at any intermediate value of $r$. (b) Unacceptable wave function as it has a node within well. (c) Acceptable deuteron wave function where matching occurs at $kr_0 = \frac{\pi}{2}$.

Hence $kr_0 = \frac{\pi}{2}$ is acceptable. So

$$V_0 r_0^2 = \frac{\pi^2 \hbar^2}{4m} \tag{6.134}$$

can be accepted as a rough estimate. This is how $V_0$ and $r_0$ are related.

This is a rough estimate of how the depth or strength of deuteron potential $V_0$ is related to range $r_0$ of nuclear potential. Equation (6.50) gives a better estimate.

Clearly the quantity $V_0 r_0^2$ is a measure of the strength of the potential.

✓ If $r_0 = 2.1\,fm$ then

$$V_0 = \frac{\pi^2 \hbar^2}{4mr_0^2} = \frac{\pi^2 \hbar^2}{4m(2.1\,fm)^2} = \frac{\pi^2(41.37\,MeV.fm^2)}{4(2.1\,fm)^2} = 23\,MeV$$

where we have used the result of *exercise 6.1*.

**Exercise 6.10** *Define deuteron radius. Estimate its value.*

Ans Deuteron is a loosely bound system and there is greater probability of nucleons to stay outside the well, i.e. at the edges of the well [equation (6.129)].

So deuteron nucleus can be thought to have an extended structure.

The wave function outside the well decays exponentially as [equation (6.32)] viz.

$$u_2(r \geqslant r_0) = Ce^{-\beta r} \tag{6.135}$$

where $r_0$ is the range of of nuclear interaction in deuteron and where external wave function matches the internal wave function.

We can define a relaxation length $\rho$ of this external part of the wave function as the length at which wave function $u_2$ reduces to $\frac{1}{e}$ times its value at $r_0$. It is this relaxation length $\rho$ over which the nucleons $n$ and $p$ can be located with sufficient probability and is defined as deuteron radius.

✓ Estimation of $\rho$ (relaxation length or deuteron radius)

At $r = r_0$, equation (6.135) gives $u_2(r_0) = Ce^{-\beta r_0}$.
As $r_0 \cong 0$ (zero range approximation of *exercise 6.13*) we can rewrite

$$u_2(r_0) = C$$

$$\text{Thus } \frac{1}{e}u_2(r_0) = \frac{C}{e} = Ce^{-1} \tag{6.136}$$

By definition deuteron radius $\rho$ is such that [from equations (6.135) and (6.136)]

$$u_2(\rho) = \frac{1}{e}u_2(r_0)$$

$$Ce^{-\beta r}\big|_{r=\rho} = \frac{1}{e}C$$

$$Ce^{-\beta\rho} = Ce^{-1}$$

$$\beta\rho = 1$$

$$\rho = \frac{1}{\beta}$$

Using from equation (6.53), $\beta = \sqrt{\frac{m}{\hbar^2}|E|} = 0.23\,fm^{-1}$ we have

$$\rho = \frac{1}{\beta} = \sqrt{\frac{\hbar^2}{m|E|}} = \frac{1}{0.23}\,fm = 4.35\,fm \tag{6.137}$$

As $\rho > r_0 = 2.1\,fm$ it follows that deuteron is not at all a tightly bound system. This is also evident from the fact that the deuteron ground state energy level is sitting just below the top of the well—actually only at a depth of *2.225* MeV from the top.

**Exercise 6.11** *Show that deuteron radius is* $\rho = \frac{2r_0}{\pi}\left(\frac{V_0}{|E|} - 1\right)^{1/2} - \frac{4}{\pi^2}r_0$

[Ans] In *exercise 6.10* the deuteron radius was obtained to be [equation (6.137)]

$$\rho = \sqrt{\frac{\hbar^2}{m\,|E|}}.$$

Again from equation (6.49) we have the relation between $V_0$ and range $r_0$ to be

$$\left(\frac{V_0}{|E|} - 1\right)^{1/2} = \frac{\pi}{2r_0}\sqrt{\frac{\hbar^2}{m\,|E|}} + \frac{2}{\pi}$$

$$= \frac{\pi\rho}{2r_0} + \frac{2}{\pi}$$

$$\frac{\pi\rho}{2r_0} = \left(\frac{V_0}{|E|} - 1\right)^{1/2} - \frac{2}{\pi}$$

$$\rho = \frac{2r_0}{\pi}\left[\left(\frac{V_0}{|E|} - 1\right)^{1/2} - \frac{2}{\pi}\right]$$

$$\rho = \frac{2r_0}{\pi}\left(\frac{V_0}{|E|} - 1\right)^{1/2} - \frac{4}{\pi^2}r_0$$

Taking $V_0 = 36\ MeV$ we have

$$\rho = \frac{2(2.1\ fm)}{\pi}\left(\frac{36\ MeV}{2.225\ MeV} - 1\right)^{1/2} - \frac{4}{\pi^2}(2.1\ fm) = 4.35\ fm$$

**Exercise 6.12** *Normalise the deuteron wave function* $\psi = \begin{cases} A\frac{\sin kr}{r} & (r < r_0) \\ C\frac{e^{-\beta r}}{r} & (r > r_0) \end{cases}$

$\boxed{Ans}$ Normalization condition is

$$\int_{\text{all space}} |\psi|^2\ d\tau = 1 \Rightarrow \int_{\text{inside}} |\psi|^2\ d\tau + \int_{\text{outside}} |\psi|^2\ d\tau = 1$$

Using equation (6.33) we have

$$\int_{r=0}^{r_0}\int_{\theta=0}^{\pi}\int_{\phi=0}^{2\pi}\left(A\frac{\sin kr}{r}\right)^2(r^2 dr\ \sin\theta d\theta d\phi) + \int_{r=r_0}^{\infty}\int_{\theta=0}^{\pi}\int_{\phi=0}^{2\pi}\left(C\frac{e^{-\beta r}}{r}\right)^2(r^2 dr\ \sin\theta d\theta d\phi) = 1$$

$$A^2\int_{r=0}^{r_0}\sin^2 kr dr\int_{\theta=0}^{\pi}\int_{\phi=0}^{2\pi}\sin\theta d\theta d\phi + C^2\int_{r=r_0}^{\infty}e^{-2\beta r}dr\int_{\theta=0}^{\pi}\int_{\phi=0}^{2\pi}\sin\theta d\theta d\phi = 1$$

Using

$$\int_{\theta=0}^{\pi}\int_{\phi=0}^{2\pi}\sin\theta d\theta d\phi = \int_{\theta=0}^{\pi}\sin\theta d\theta\int_{\phi=0}^{2\pi}d\phi = 2.2\pi = 4\pi$$

we have

$$A^2 4\pi\int_{r=0}^{r_0}\sin^2 kr dr + C^2 4\pi\int_{r=r_0}^{\infty}e^{-2\beta r}dr = 1$$

$$A^2 2\pi\int_{r=0}^{r_0}(1 - \cos 2kr)dr + C^2 4\pi\int_{r=r_0}^{\infty}e^{-2\beta r}dr = 1$$

$$A^2 2\pi\left(r_0 - \frac{\sin 2kr_0}{2k}\right) + C^2 4\pi\frac{e^{-2\beta r_0}}{2\beta} = 1$$

Use the continuity condition given by equation (6.35) viz. $A \sin kr_0 = Ce^{-\beta r_0}$ we get

$$C = \frac{A \sin kr_0}{e^{-\beta r_0}} \tag{6.138}$$

This gives

$$A^2 2\pi \left( r_0 - \frac{\sin 2kr_0}{2k} \right) + \left( \frac{A \sin kr_0}{e^{-\beta r_0}} \right)^2 4\pi \frac{e^{-2\beta r_0}}{2\beta} = 1$$

$$\frac{A^2 2\pi}{k} \left( kr_0 - \frac{\sin 2kr_0}{2} + \frac{k \sin^2 kr_0}{\beta} \right) = 1$$

$$\frac{A^2 2\pi}{k} \left( kr_0 - \sin kr_0 \cos kr_0 + \frac{k}{\beta} \sin^2 kr_0 \right) = 1$$

Use equation (6.38) $\tan kr_0 = -\frac{k}{\beta}$ in the third term on the LHS to get

$$\frac{A^2 2\pi}{k} (kr_0 - \tan kr_0 \cos^2 kr_0 - \tan kr_0 \sin^2 kr_0) = 1$$

$$\frac{A^2 2\pi}{k} (kr_0 - \tan kr_0) = 1$$

Using $\tan kr_0 = -\frac{k}{\beta}$ again we get

$$\frac{A^2 2\pi}{k} \left( kr_0 + \frac{k}{\beta} \right) = 1$$

$$A^2 = \frac{\beta}{2\pi(1 + \beta r_0)}$$

$$A = \sqrt{\frac{\beta}{2\pi}} (1 + \beta r_0)^{-1/2} \tag{6.139}$$

As $\beta r_0 \ll 1$ we approximate and write

$$A \cong \sqrt{\frac{\beta}{2\pi}} \left( 1 - \frac{1}{2}\beta r_0 \right) \tag{6.140}$$

From equation (6.138) we get

$$C = \frac{A \sin kr_0}{e^{-\beta r_0}} = \sqrt{\frac{\beta}{2\pi}} \left( 1 - \frac{1}{2}\beta r_0 \right) \frac{\sin kr_0}{e^{-\beta r_0}}$$

Using equation (6.40), i.e.

$$\sin^2 k r_0 = \frac{k^2}{k^2 + \beta^2} = \left(1 + \frac{\beta^2}{k^2}\right)^{-1}$$

we have

$$C = \sqrt{\frac{\beta}{2\pi}}\left(1 - \frac{1}{2}\beta r_0\right)e^{\beta r_0}\left(1 + \frac{\beta^2}{k^2}\right)^{-1/2}$$

As $\dfrac{\beta}{k} \ll 1$, $\dfrac{\beta^2}{k^2} \ll < 1$, $\left(1 + \dfrac{\beta^2}{k^2}\right)^{-1/2} \cong 1$. So

$$C \cong \sqrt{\frac{\beta}{2\pi}}\left(1 - \frac{1}{2}\beta r_0\right)e^{\beta r_0}$$

$$C \cong \sqrt{\frac{\beta}{2\pi}}\left(1 - \frac{1}{2}\beta r_0\right)(1 + \beta r_0) \cong \sqrt{\frac{\beta}{2\pi}}\left(1 + \beta r_0 - \frac{1}{2}\beta r_0\right)$$

$$C = \sqrt{\frac{\beta}{2\pi}}\left(1 + \frac{1}{2}\beta r_0\right) \tag{6.141}$$

Hence deuteron wave function of equation (6.33) along with equations (6.140) and (6.141) is

$$\psi = \begin{cases} \sqrt{\dfrac{\beta}{2\pi}}\left(1 - \dfrac{1}{2}\beta r_0\right)\dfrac{\sin kr}{r} \equiv \psi_1(r < r_0) \\[3mm] \sqrt{\dfrac{\beta}{2\pi}}\left(1 + \dfrac{1}{2}\beta r_0\right)\dfrac{e^{-\beta r}}{r} \equiv \psi_2(r > r_0) \end{cases} \tag{6.142}$$

This is the normalized deuteron wave function.

**Exercise 6.13** *Write down deuteron wave function for zero range approximation.*
⏍*Ans* For the purpose of rough calculation we can make zero range approximation, i.e. set

$$r_0 \cong 0$$

We then would neglect the inner part of the wave function $\psi_1$ of equation (6.142).
The exterior wave function then extends from 0 to $\infty$. From equation (6.142) we can write

$$\psi(r) = \sqrt{\frac{\beta}{2\pi}}\frac{e^{-\beta r}}{r} \equiv \psi_2 \quad \text{(figure 6.4(b))} \tag{6.143}$$

This form resembles the Yukawa potential form

$$V(r) = -V_0 \frac{e^{-r/r_0}}{r/r_0} \tag{6.144}$$

**Exercise 6.14** . *Find the minimum depth of a rectangular potential well to get a just bound state in it for $r_0 = 2.1$ fm.*

Ans From equation (6.38) we have

$$\tan kr_0 = -\frac{k}{\beta}$$

$$\cot kr_0 = -\frac{\beta}{k}$$

Using equations (6.21) and (6.29) we get

$$\cot \sqrt{\frac{m}{\hbar^2}(V_0 - |E|)}\, r_0 = \frac{\sqrt{\frac{m}{\hbar^2}|E|}}{\sqrt{\frac{m}{\hbar^2}(V_0 - |E|)}}$$

Taking $E = 0$ for just bound state

$$\cot \sqrt{\frac{m}{\hbar^2} V_0}\, r_0 = 0 = \cot \frac{\pi}{2}$$

$$\sqrt{\frac{m}{\hbar^2} V_0}\, r_0 = \frac{\pi}{2}$$

Squaring we have

$$\frac{m}{\hbar^2} V_0 r_0^2 = \frac{\pi^2}{4}$$

$$V_0 = \frac{\pi^2 \hbar^2}{4m r_0^2} = \frac{\pi^2 (41.37\ MeV.fm^2)}{4(2.1\ fm)^2} = 23\ MeV \text{ (using exercise 6.1)}$$

In figure 6.10 we compare the deuteron wells $-V_0 = -23\ MeV$ for just bound state $E = 0$ and $-V_0 = -36\ MeV$ for actual ground state $E = -2.225\ MeV$ corresponding to range $r_0 = 2.1\ fm$.

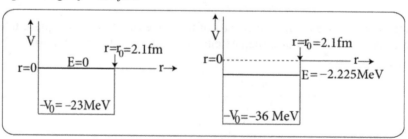

**Figure 6.10.** Deuteron potential well for $r_0 = 2.1$ *fm* for $E = 0$ and $E = -2.225\ MeV$.

**Exercise 6.15** . *Solve the radial wave equation of deuteron problem for $l = 1$. Calculate the well depth needed to bind two nucleons together with $E = -2.225\ MeV$ and range $r_0 = 2\ fm$.*

[Ans] Wave equation determining the relative motion of the deuteron system is by equation (6.11)

$$\nabla^2 \psi + \frac{2\mu}{\hbar^2}(E - V)\psi = 0 \text{ with } \psi(r, \theta, \phi) = R(r)Y(\theta, \phi)$$

The deuteron potential is spherically symmetric $V = V(r)$ and so the angular part is spherical harmonic. So

$$Y(\theta, \phi) = Y_{lm_l}(\theta, \phi).$$

The radial part is by *exercise 6.4*

$$\frac{d^2}{dr^2}R(r) + \frac{2}{r}\frac{d}{dr}R(r) + \frac{2\mu}{\hbar^2}\left[E - V(r) - \frac{l(l+1)\hbar^2}{2\mu r^2}\right]R(r) = 0$$

and with

$$R(r) = \frac{u(r)}{r} \Rightarrow u(r) = rR(r)$$

we have by *exercise 6.5*

$$\frac{d^2}{dr^2}u(r) + \frac{2\mu}{\hbar^2}\left[E - V(r) - \frac{l(l+1)\hbar^2}{2\mu r^2}\right]u(r) = 0 \tag{6.145}$$

This is 1D equation in $u = rR$.

As $l$ increases the repulsive centrifugal potential energy $\frac{l(l+1)\hbar^2}{2\mu r^2}$ increases and so binding energy decreases since if system is excited it will be easier for it to disintegrate. In other words the binding energy $B'$ will now be less, i.e.

$$B' < B$$

where $B = 2.225\ MeV$ is the binding energy for the ground state [equation (6.1)].

Figure 6.11(a) depicts the level of excited state $E = E'$ if it exists. Also, $E = -2.225\ MeV$ is the ground state energy, $\mu = \frac{m_p m_n}{m_p + m_n} = \frac{m}{2}$ where $m_p = m_n = m$. So $2\mu = m$

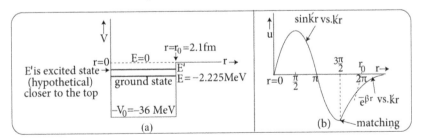

(a)          (b)  matching

**Figure 6.11.** (a) For an excited state to occur it should sit closer to the top $E = 0$ above the ground state $E = -2.225\ MeV$. (b) To accommodate a node for $l = 1$ state $\sin k'r$ completes $\frac{3}{4}$ of its wavelength at or before $r = r_0$. We take $k'r_0 = \frac{3\pi}{2}$. Matching between internal and external wave functions should occur at this point.

For $l = 1$ the radial equation (6.145) in $u(r)$ becomes

$$\frac{d^2}{dr^2}u(r) + \frac{m}{\hbar^2}\left[E - V(r) - \frac{1(1+1)\hbar^2}{2\mu r^2}\right]u(r) = 0$$

As we are considering bound states $E = -|E|$

$$\frac{d^2}{dr^2}u(r) + \frac{m}{\hbar^2}\left[-|E| - V(r) - \frac{2\hbar^2}{mr^2}\right]u(r) = 0 \tag{6.146}$$

For $r < r_0$, $V(r) = -V_0'$ (say), $u = u_1(r)$ and so from equation (6.146)

$$\frac{d^2}{dr^2}u_1(r) + \frac{m}{\hbar^2}\left[-|E| + V_0' - \frac{2\hbar^2}{mr^2}\right]u_1(r) = 0$$

$$\frac{d^2}{dr^2}u_1(r) + \left(\frac{m(V_0' - |E|)}{\hbar^2} - \frac{2}{r^2}\right)u_1(r) = 0$$

Defining $k'^2 = \dfrac{m(V_0' - |E|)}{\hbar^2}$ \hfill (6.147)

$$\frac{d^2}{dr^2}u_1(r) + \left(k'^2 - \frac{2}{r^2}\right)u_1(r) = 0 \tag{6.148}$$

For $r > r_0$, $V(r) = 0$, $u = u_2(r)$ and so from equation (6.146) we have

$$\frac{d^2}{dr^2}u_2(r) - \frac{m}{\hbar^2}\left[|E| + \frac{2\hbar^2}{mr^2}\right]u_2(r) = 0$$

With $\beta^2 = \dfrac{m|E|}{\hbar^2}$ \hfill (6.149)

$$\frac{d^2}{dr^2}u_2(r) - \left(\beta^2 + \frac{2}{r^2}\right)u_2(r) = 0 \tag{6.150}$$

The least well depth just required to produce bound state is the one for which the binding energy $E = 0$. So from equation (6.147)

$$k'^2 = \frac{mV_0'}{\hbar^2} \tag{6.151}$$

and from equation (6.149)

$$\beta = 0 \tag{6.152}$$

Hence with equations (6.151) and (6.152) we have from equations (6.148) and (6.150)

$$\frac{d^2}{dr^2}u_1(r) + \left(k'^2 - \frac{2}{r^2}\right)u_1(r) = 0 \quad \text{for } r < r_0 \text{ and}$$

$$\frac{d^2}{dr^2}u_2(r) - \frac{2}{r^2}u_2(r) = 0 \quad \text{for } r > r_0$$

Rewriting in terms of $k'r$ we have

$$\frac{d^2}{d(k'r)^2}u_1(r) + \left(1 - \frac{2}{(k'r)^2}\right)u_1(r) = 0 \quad \text{for } r < r_0 \text{ and}$$

$$\frac{d^2}{d(k'r)^2}u_2(r) - \frac{2}{(k'r)^2}u_2(r) = 0 \quad \text{for } r > r_0$$

Put $k'r = x$, $k'r_0 = x_0$ and rewrite the equations

$$\frac{d^2}{dx^2}u_1(x) + \left(1 - \frac{2}{x^2}\right)u_1(x) = 0 \quad \text{for } x < x_0 \tag{6.153}$$

$$\frac{d^2}{dx^2}u_2(x) - \frac{2}{x^2}u_2(x) = 0 \quad \text{for } x > x_0 \tag{6.154}$$

Put $u_1 = \frac{v}{x}$ in equation (6.153) to get

$$\frac{d^2}{dx^2}\frac{v}{x} + \left(1 - \frac{2}{x^2}\right)\frac{v}{x} = 0 \Rightarrow \frac{d}{dx}\frac{d}{dx}\frac{v}{x} + \frac{v}{x} - \frac{2v}{x^3} = 0$$

$$\frac{d}{dx}\left(\frac{1}{x}\frac{dv}{dx} - \frac{v}{x^2}\right) + \frac{v}{x} - \frac{2v}{x^3} = 0 \Rightarrow \frac{d}{dx}\left(\frac{1}{x}\frac{dv}{dx}\right) - \frac{d}{dx}\frac{v}{x^2} + \frac{v}{x} - \frac{2v}{x^3} = 0$$

$$\left(\frac{1}{x}\frac{d^2v}{dx^2} - \frac{1}{x^2}\frac{dv}{dx}\right) - \left(\frac{1}{x^2}\frac{dv}{dx} - \frac{2v}{x^3}\right) + \frac{v}{x} - \frac{2v}{x^3} = 0$$

$$\frac{d^2v}{dx^2} - \frac{2}{x}\frac{dv}{dx} + v = 0$$

Differentiate w.r.t. $x$ to get

$$\frac{d}{dx}\left(\frac{d^2v}{dx^2} - \frac{2}{x}\frac{dv}{dx} + v\right) = 0$$

$$\frac{d^3v}{dx^3} - \frac{2}{x}\frac{d^2v}{dx^2} + \frac{2}{x^2}\frac{dv}{dx} + \frac{dv}{dx} = 0$$

Dividing by $x$ throughout we have

$$\frac{1}{x}\frac{d^3v}{dx^3} - \frac{2}{x^2}\frac{d^2v}{dx^2} + \frac{2}{x^3}\frac{dv}{dx} + \frac{1}{x}\frac{dv}{dx} = 0 \tag{6.155}$$

Use

$$\frac{d^2}{dx^2}\left(\frac{1}{x}\frac{dv}{dx}\right) = \frac{d}{dx}\frac{d}{dx}\left(\frac{1}{x}\frac{dv}{dx}\right) = \frac{d}{dx}\left(\frac{1}{x}\frac{d^2v}{dx^2} - \frac{1}{x^2}\frac{dv}{dx}\right) = \frac{1}{x}\frac{d^3v}{dx^3} - \frac{2}{x^2}\frac{d^2v}{dx^2} + \frac{2}{x^3}\frac{dv}{dx}$$

in equation (6.155) to have

$$\frac{d^2}{dx^2}\left(\frac{1}{x}\frac{dv}{dx}\right) + \frac{1}{x}\frac{dv}{dx} = 0 \text{ for } x < x_0 \tag{6.156}$$

Putting $\frac{1}{x}\frac{dv}{dx} = \xi$ in equation (6.156) we get

$$\frac{d^2}{dx^2}\xi + \xi = 0$$

The solution of this equation is

$$\xi = A \sin x$$

$$\frac{1}{x}\frac{dv}{dx} = A \sin x$$

$$\frac{dv}{dx} = Ax \sin x$$

$$v = A \int x \sin x dx$$

$$v = A[x \int \sin x dx - \int \frac{d}{dx}x \left(\int \sin x dx\right)dx] \text{ (integrating by parts)}$$

$$v = A\left[-x \cos x - \int -\cos x dx\right] = A(-x \cos x + \sin x)$$

$$u_1 = \frac{v}{x} = A\left(\frac{\sin x}{x} - \cos x\right)$$

Hence

$$u_1(r) = A\left(\frac{\sin k'r}{k'r} - \cos k'r\right) \quad \text{for } r < r_0.$$

The solution of equation (6.154), i.e. $\frac{d^2}{dx^2}u_2(x) - \frac{2}{x^2}u_2(x) = 0$ for $x > x_0$ is

$$u_2(x) = \frac{B}{x}$$

$$u_2(r) = \frac{B}{k'r} \text{ (check by substitution)}$$

Impose continuity condition at the boundary $r = r_0$ i.e. $x = x_0$ which is

$$\frac{d}{dx} x u_1(x) \big|_{x=x_0} = \frac{d}{dx} x u_2(x) \big|_{x=x_0}$$

$$\frac{d}{dx} x A \left( \frac{\sin x}{x} - \cos x \right) \big|_{x=x_0} = \frac{d}{dx} x \frac{B}{x} \big|_{x=x_0}$$

$$\frac{d}{dx} A (\sin x - x \cos x) \big|_{x=x_0} = \frac{d}{dx} B \big|_{x=x_0}$$

$$(\cos x + x \sin x - \cos x) \big|_{x=x_0} = 0$$

$$x_0 \sin x_0 = 0 \Rightarrow \sin x_0 = 0$$

$$\sin k' r_0 = 0 = \sin \pi$$

$$k' r_0 = \pi.$$

This is the smallest value of $k' r_0$ that corresponds to state $l = 1$. The corresponding deuteron potential well depth $V_0'$ is calculated as follows using equation (6.151).

$$k' r_0 = \sqrt{\frac{m V_0'}{\hbar^2}} \, r_0 = \pi$$

$$V_0' = \frac{\pi^2 \hbar^2}{m r_0^2} = \frac{\pi^2 (41.37 \, MeV. \, fm^2)}{(2.1 \, fm)^2} = 93 \, MeV \text{ where we took help of exercise 6.1.}$$

This depth of deuteron well that is needed for existence of $l = 1$ bound state does not match the actual or physical deuteron well $V_0 \sim 36 \, MeV$. This indicates that deuteron has got no bound excited state for $l = 1$.

We give a few other arguments to demonstrate non-existence of excited state of deuteron.

- Even the ground state of deuteron is not strongly bound as both the nucleons relax up to a large distance. They stay most of the time outside the deuteron well. This qualitatively explains why an excited state does not exist.
- The magnitude of the centrifugal potential energy term is

$$V_{cf} = \frac{l(l+1)\hbar^2}{2\mu r^2}$$

Taking $\mu \cong \frac{m}{2}$, $r \cong \rho$ we get for $l = 1$ state

$$V_{cf} = \frac{1(1+1)\hbar^2}{m\rho^2} = \frac{2\hbar^2}{m\rho^2} = \frac{2(41.37 \, MeV. \, fm^2)}{(4.35 \, fm)^2} = 4.4 \, MeV \text{ (where we used exercise 6.1, 6.10)}$$

Clearly $V_{cf} > E$. This centrifugal repulsive potential energy term will not allow a stable $l = 1$ bound state. To compensate the centrifugal barrier a much deeper well is required.

- Lowest energy state corresponds to the state with lowest number of nodes. The $l = 0$ ground state has no node (at $r < r_0$). (figure 6.3)

On the other hand, excited states correspond to larger momentum, energy and hence larger number of oscillations and so larger number of nodes. The excited $l = 1$ state would have at least one node, i.e. $\sin k'r$ would go to zero at least once (figure 6.11(b)). So, to accommodate a node $\sin k'r$ completes $\frac{3}{4}$ of its wavelength at or before $r = r_0$, i.e. the required condition for $l = 1$ excited state is

$$\frac{3\pi}{2} \leqslant k'r_0 \leqslant 2\pi \text{ say } k'r_0 = \frac{3\pi}{2} + \delta.$$

Let us take $\delta = 0$. Then

$$k'r_0 = \frac{3\pi}{2}$$

Using equation (6.151) we have

$$k'r_0 = \sqrt{\frac{mV_0'}{\hbar^2}}\, r_0 = \frac{3\pi}{2}$$

$$V_0' = \frac{9\pi^2\hbar^2}{4mr_0^2} = \frac{9\pi^2(41.37 \; MeV.\, fm^2)}{4(2.1 \, fm)^2} = 208 \; MeV \text{ (where we used exercise 6.1).}$$

This is considerably deeper than the real deuteron potential well $V_0 \sim 36 \; MeV$ and hence unacceptable. Since this value is a complete misfit with experimental data, it follows that it is impossible to have $r_0$ and $V_0$ coordinated in such a way that the excited state corresonding to $l = 1$ exists. So deuteron has no excited state.

**Exercise 6.16** *How is the term $\vec{s}_p \cdot \vec{s}_n$ spin dependent? Establish that this term can distinguish between spin singlet state and spin triplet state.*

$\boxed{Ans}$ Consider the deuteron system of neutron and proton. The total spin angular momentum is

$$\vec{s} = \vec{s}_p + \vec{s}_n$$

$$s^2 = s_p^2 + s_n^2 + 2\vec{s}_p \cdot \vec{s}_n$$

$$\vec{s}_p \cdot \vec{s}_n = \frac{1}{2}(s^2 - s_p^2 - s_n^2)$$

Replacing by eigenvalues we have

$$\vec{s}_p \cdot \vec{s}_n \to \frac{s(s+1)\hbar^2 - s_p(s_p+1)\hbar^2 - s_n(s_n+1)\hbar^2}{2}$$

Put $s_p = \frac{1}{2}$, $s_n = \frac{1}{2}$ to get

$$\vec{s}_p \cdot \vec{s}_n \rightarrow \frac{s(s+1)\hbar^2 - \frac{1}{2}\left(\frac{1}{2}+1\right)\hbar^2 - \frac{1}{2}\left(\frac{1}{2}+1\right)\hbar^2}{2} = \frac{s(s+1)\hbar^2 - \frac{3}{2}\hbar^2}{2}$$

Clearly

For singlet state $s = 0$ and we have

$$\vec{s}_p \cdot \vec{s}_n \rightarrow \frac{0(0+1)\hbar^2 - \frac{3}{2}\hbar^2}{2} = -\frac{3}{4}\hbar^2 \qquad (6.157)$$

For triplet state $s = 1$ and we have

$$\vec{s}_p \cdot \vec{s}_n \rightarrow \frac{1(1+1)\hbar^2 - \frac{3}{2}\hbar^2}{2} = \frac{1}{4}\hbar^2. \qquad (6.158)$$

Evidently the term $\vec{s}_p \cdot \vec{s}_n$ assumes different values for $s = 0$ and $s = 1$ and hence is spin dependent. This is the justification of including the term $V_s(r)\vec{s}_p \cdot \vec{s}_n$ in nuclear Hamiltonian.

**Exercise 6.17** *Calculate the magnetic dipole moment of deuteron in the $l = 2$ state.*
$\boxed{Ans}$ For deuteron $J = 1$, $m_J = 1, 0, -1$.

Quantum mechanically, the maximum value of $\mu_z$ is compared with the experimental value. The maximum value of $\mu_z$ is obtained by using the wave function corresponding to $m_J = 1$.

Also, we are considering $l = 2$ and deuteron intrinsic spin is $s = 1$. Thus the state we consider is

$$| J = 1, m_J = 1, l = 2, s = 1 > \text{ and we evaluate } < \mu_z >_{l=2} .$$

The relation between the projections is

$$m_J = m_l + m_s$$

With $m_J = 1$ it becomes

$$m_l + m_s = 1$$

This leads to three possibilities viz.

$$(m_l, m_s) = (2, -1), (1, 0), (0, 1).$$

Also, $m_s = m_{sp} + m_{sn}$ has to be obeyed.

✓ For $(m_l, m_s) = (2, -1)$

$l = 2$, $m_l = 2$ corresponds to the spherical harmonic $Y_{lm_l} = Y_{22}$.
$m_s = -1$ can come from $m_{sp} = -\frac{1}{2}($ proton down state ↓$),m_{sn} = -\frac{1}{2}$
(neutron down state ↓). This is expressed as $| \downarrow \downarrow >$. The corresponding state is

$$Y_{22} | \downarrow \downarrow > \qquad (6.159)$$

✓ For $(m_l, m_s) = (1, 0)$.

$l = 2$, $m_l = 1$ corresponds to the spherical harmonic $Y_{lm_l} = Y_{21}$.

$m_s = 0$ can come from

$$m_{sp} = \frac{1}{2}(\text{proton up state } \uparrow), \ m_{sn} = -\frac{1}{2}(\text{neutron down state } \downarrow) \text{ and}$$

$$m_{sp} = -\frac{1}{2}(\text{proton down state } \downarrow), \quad m_{sn} = \frac{1}{2}(\text{neutron down state } \uparrow)$$

This is expressed by taking a linear combination of

$$m_{sp} = \frac{1}{2}, \ m_{sn} = -\frac{1}{2}(\text{state } \uparrow \downarrow) \text{ and } m_{sp} = -\frac{1}{2}, \ m_{sn} = \frac{1}{2}(\text{state } \downarrow \uparrow) \text{ as}$$

$$\frac{1}{\sqrt{2}}(|\uparrow \downarrow > + |\downarrow \uparrow >). \tag{6.160}$$

The corresponding state is

$$Y_{21}\frac{1}{\sqrt{2}}(|\uparrow \downarrow > + |\downarrow \uparrow >) \tag{6.161}$$

✓ For $(m_l, m_s) = (0, 1)$

$l = 2$, $m_l = 0$ corresponds to the spherical harmonic $Y_{lm_l} = Y_{20}$.

$m_s = 1$ can come from $m_{sp} = \frac{1}{2}(\text{proton up state } \uparrow)$, $m_{sn} = \frac{1}{2}(\text{neutron up state } \downarrow)$.

This is expressed as $|\uparrow \uparrow >$ . The corresponding state is

$$Y_{20}|\uparrow \uparrow > \tag{6.162}$$

With these states we construct the wave function after attaching appropriate coefficients to equations (6.159), (6.161), and (6.162)

$$|\psi> = \sqrt{\frac{3}{5}} \, Y_{22}|\downarrow \downarrow > + \sqrt{\frac{3}{10}} \, Y_{21}\frac{1}{\sqrt{2}}(|\uparrow \downarrow > + |\downarrow \uparrow >) + \sqrt{\frac{1}{10}} \, Y_{20}|\uparrow \uparrow > \tag{6.163}$$

where $\left(\sqrt{\frac{3}{5}}\right)^2 + \left(\sqrt{\frac{3}{10}}\right)^2 + \left(\sqrt{\frac{1}{10}}\right)^2 = 1$.

In the ket, the first position is for proton and the second position is for neutron.

Magnetic moment of deuteron is given in equation (6.89) to be

$$\mu_{dz} = \frac{g_{lp}\,\mu_N}{\hbar}\frac{l_z}{2} + \frac{g_{sp}\,\mu_N}{\hbar}s_{pz} + \frac{g_{sn}\,\mu_N}{\hbar}s_{nz}$$

The observed value (expected value) is given by equations (6.90) and (6.91) to be

$$<\mu_{dz}> \; = \; <\psi \mid \mu_z \mid \psi>$$

$$= \; <\psi \left| \frac{g_{lp}\mu_N}{\hbar} \frac{l_z}{2} + \frac{g_{sp}\mu_N}{\hbar} S_{pz} + \frac{g_{sn}\mu_N}{\hbar} S_{nz} \right| \psi>$$

$$<\mu_{dz}> \; = \; <\psi \mid \frac{g_{lp}\mu_N}{\hbar} \frac{l_z}{2} \mid \psi> + <\psi \mid \frac{g_{sp}\mu_N}{\hbar} S_{pz} \mid \psi> + <\psi \mid \frac{g_{sn}\mu_N}{\hbar} S_{nz} \mid \psi> \quad (6.164)$$

Using the eigenvalue equation for $l_z$ operator viz. $l_z Y_{lm_l} = m_l \hbar Y_{lm_l}$ we get using equation (6.163)

$$l_z \mid \psi> \; = l_z \left[ \sqrt{\frac{3}{5}} Y_{22} \mid \downarrow \downarrow > + \sqrt{\frac{3}{10}} Y_{21} \frac{1}{\sqrt{2}} (\mid \uparrow \downarrow > + \mid \downarrow \uparrow >) + \sqrt{\frac{1}{10}} Y_{20} \mid \uparrow \uparrow > \right]$$

$$= \sqrt{\frac{3}{5}} (2\hbar) Y_{22} \mid \downarrow \downarrow > + \sqrt{\frac{3}{10}} (\hbar) Y_{21} \frac{1}{\sqrt{2}} (\mid \uparrow \downarrow > + \mid \downarrow \uparrow >) + \sqrt{\frac{1}{10}} (0) Y_{20} \mid \uparrow \uparrow >$$

$$l_z \mid \psi> \; = \sqrt{\frac{3}{5}} (2\hbar) Y_{22} \mid \downarrow \downarrow > + \sqrt{\frac{3}{10}} (\hbar) Y_{21} \frac{1}{\sqrt{2}} (\mid \uparrow \downarrow > + \mid \downarrow \uparrow >) \quad (6.165)$$

Consider the first term of the RHS of equation (6.164) and use $<\psi \mid$ from equations (6.163) and (6.165) to get

$$<\psi \mid \frac{g_{lp}\mu_N}{\hbar} \frac{\hat{l}_z}{2} \mid \psi>$$

$$= \frac{g_{lp}\mu_N}{2\hbar} [ \sqrt{\frac{3}{5}} Y_{22} < \downarrow \downarrow \mid + \sqrt{\frac{3}{10}} Y_{21} \frac{1}{\sqrt{2}} (< \uparrow \downarrow \mid + < \downarrow \uparrow \mid)$$

$$+ \sqrt{\frac{1}{10}} Y_{20} < \uparrow \uparrow \mid][ \sqrt{\frac{3}{5}} (2\hbar) Y_{22} \mid \downarrow \downarrow > + \sqrt{\frac{3}{10}} (\hbar) Y_{21} \frac{1}{\sqrt{2}} (\mid \uparrow \downarrow > + \mid \downarrow \uparrow >)]$$

$$= \frac{g_{lp}\mu_N}{2\hbar} \left( \frac{3}{5} 2\hbar + \frac{3}{10}\hbar \right) = g_{lp}\mu_N \frac{15}{20} \quad \text{(using orthonormality relation of } Y_{lm_l})$$

As $g_{lp} = 1$

$$<\psi \mid \frac{g_{lp}\mu_N}{\hbar} \frac{\hat{l}_z}{2} \mid \psi> \; = \frac{3}{4}\mu_N \quad (6.166)$$

Now we shall use the following results of operation of $s_{pz}$ operator.

$$s_{pz} \mid \downarrow \downarrow > \; = -\frac{\hbar}{2} \mid \downarrow \downarrow >$$

$$S_{pz}\frac{1}{\sqrt{2}}(|\uparrow\downarrow> + |\downarrow\uparrow>) = \frac{1}{\sqrt{2}}\left(\frac{\hbar}{2}|\uparrow\downarrow> - \frac{\hbar}{2}|\downarrow\uparrow>\right)$$

$$S_{pz}|\uparrow\uparrow> = \frac{\hbar}{2}|\uparrow\uparrow>$$

Consider the second term of equation (6.164) and use equation (6.163).

$$<\psi|\frac{g_{sp}\mu_N}{\hbar}S_{pz}|\psi> = \frac{g_{sp}\mu_N}{\hbar}<\psi|S_{pz}|\psi>$$

$$= \frac{g_{sp}\mu_N}{\hbar}<\psi|S_{pz}|\left[\sqrt{\frac{3}{5}}Y_{22}|\downarrow\downarrow> + \sqrt{\frac{3}{10}}Y_{21}\frac{1}{\sqrt{2}}(|\uparrow\downarrow> + |\downarrow\uparrow>) + \sqrt{\frac{1}{10}}Y_{20}|\uparrow\uparrow>\right]$$

$$= \frac{g_{sp}\mu_N}{\hbar}<\psi|\left[\sqrt{\frac{3}{5}}Y_{22}\left(-\frac{\hbar}{2}\right)|\downarrow\downarrow> + \sqrt{\frac{3}{10}}Y_{21}\frac{1}{\sqrt{2}}\frac{\hbar}{2}(|\uparrow\downarrow> - |\downarrow\uparrow>) + \sqrt{\frac{1}{10}}Y_{20}\frac{\hbar}{2}|\uparrow\uparrow>\right]$$

$$= \frac{g_{sp}\mu_N}{\hbar}\left[\sqrt{\frac{3}{5}}Y_{22}<\downarrow\downarrow| + \sqrt{\frac{3}{10}}Y_{21}\frac{1}{\sqrt{2}}(<\uparrow\downarrow| + <\downarrow\uparrow|) + \sqrt{\frac{1}{10}}Y_{20}<\uparrow\uparrow|\right]\left[\sqrt{\frac{3}{5}}Y_{22}\left(-\frac{\hbar}{2}\right)|\downarrow\downarrow>\right.$$

$$\left.+ \sqrt{\frac{3}{10}}Y_{21}\frac{1}{\sqrt{2}}\frac{\hbar}{2}(|\uparrow\downarrow> - |\downarrow\uparrow>) + \sqrt{\frac{1}{10}}Y_{20}\frac{\hbar}{2}|\uparrow\uparrow>\right]$$

$$= \frac{g_{sp}\mu_N}{\hbar}\left[\frac{3}{5}\cdot\left(-\frac{\hbar}{2}\right) + \frac{1}{10}\left(\frac{\hbar}{2}\right)\right]$$

$$= g_{sp}\mu_N\left(-\frac{3}{10} + \frac{1}{20}\right)$$

$$<\psi|\frac{g_{sp}\mu_N}{\hbar}S_{pz}|\psi> = -\frac{g_{sp}\mu_N}{4} \tag{6.167}$$

Similarly the third term of equation (6.164) upon using equation (6.163) gives

$$<\psi|\frac{g_{sp}\mu_N}{\hbar}S_{nz}|\psi> = -\frac{g_{sn}\mu_N}{4} \tag{6.168}$$

Hence from equation (6.164) we have using equations (6.166)–(6.168)

$$<\mu_{dz}> = \frac{3}{4}\mu_N - \frac{g_{sp}}{4}\mu_N - \frac{g_{sn}}{4}\mu_N = \left(\frac{3 - g_{sp} - g_{sn}}{4}\right)\mu_N$$

$$= \frac{3 - 5.585\,691 - (-3.826\,084)}{4}\mu_N$$

$$<\mu_{dz}>_{l=2} = 0.3101\mu_N$$

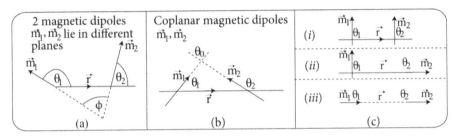

**Figure 6.12.** (a) Two magnetic dipoles $\bar{m}_1$, $\bar{m}_2$ of arbitrary orientation. (b) Two magnetic dipoles lying in the same plane. (c) Various orientations of two coplanar magnetic dipoles $\bar{m}_1$, $\bar{m}_2$ relative to their separation.

**Exercise 6.18** *Find out the magnetic energy in the case of interaction between two magnetic dipoles $\bar{m}_1$, $\bar{m}_2$.*

[Ans] Consider two magnetic dipoles $\bar{m}_1$, $\bar{m}_2$. The position vector of $\bar{m}_2$ w.r.t. $\bar{m}_1$ is $\vec{r}$ as depicted in figure 6.12(a). The magnetic induction at $\bar{m}_2$ due to $\bar{m}_1$ is given by

$$\vec{B} = \frac{\mu_0}{4\pi} \frac{3\hat{r}(\bar{m}_1 \cdot \hat{r}) - \bar{m}_1}{r^3}$$

$\bar{m}_2$ is placed in this field $\vec{B}$.

Potential energy of interaction is

$$U = -\bar{m}_2 \cdot \vec{B} = -\bar{m}_2 \cdot \frac{\mu_0}{4\pi} \frac{3\hat{r}(\bar{m}_1 \cdot \hat{r}) - \bar{m}_1}{r^3}$$

$$= -\frac{\mu_0}{4\pi} \frac{3(\bar{m}_1 \cdot \hat{r})(\bar{m}_2 \cdot \hat{r}) - \bar{m}_1 \cdot \bar{m}_2}{r^3}$$

$$U = \frac{\mu_0}{4\pi} \frac{\bar{m}_1 \cdot \bar{m}_2 - 3(\bar{m}_1 \cdot \hat{r})(\bar{m}_2 \cdot \hat{r})}{r^3} \tag{6.169}$$

If $\angle \bar{m}_1, \bar{m}_2 = \phi$, $\angle \bar{m}_1, \vec{r} = \theta_1$, $\angle \bar{m}_2, \vec{r} = \theta_2$

$$U = \frac{\mu_0}{4\pi} \frac{m_1 m_2 \cos\phi - 3m_1 \cos\theta_1 m_2 \cos\theta_2}{r^3}$$

$$U = \frac{\mu_0}{4\pi} \frac{m_1 m_2}{r^3} (\cos\phi - 3\cos\theta_1 \cos\theta_2)$$

**Exercise 6.19** . *Show the orientation between two coplanar magnetic dipoles w.r.t. their relative separation for minimum magnetic energy.*

[Ans] The magnetic energy for two magnetic dipoles is by equation (6.169)

$$U = \frac{\mu_0}{4\pi} \frac{\bar{m}_1 \cdot \bar{m}_2 - 3(\bar{m}_1 \cdot \hat{r})(\bar{m}_2 \cdot \hat{r})}{r^3}$$

For coplanar dipoles as shown in figure 6.12(b)

$$\angle \vec{m}_1, \vec{m}_2 = \theta_0, \ \angle \vec{m}_1, \vec{r} = \theta_1, \ \angle \vec{m}_2, \vec{r} = \theta_2, \ \theta_2 = \theta_0 + \theta_1 \Rightarrow \theta_2 - \theta_1 = \theta_0$$

$$U = \frac{\mu_0}{4\pi} \frac{m_1 m_2 \cos\theta_0 - 3 m_1 \cos\theta_1 m_2 \cos\theta_2}{r^3}$$

$$= \frac{\mu_0}{4\pi} \frac{m_1 m_2}{r^3} [\cos(\theta_2 - \theta_1) - 3\cos\theta_1 \cos\theta_2]$$

$$= \frac{\mu_0}{4\pi} \frac{m_1 m_2}{r^3} [\cos\theta_1 \cos\theta_2 + \sin\theta_1 \sin\theta_2 - 3\cos\theta_1 \cos\theta_2]$$

$$U = \frac{\mu_0}{4\pi} \frac{m_1 m_2}{r^3} [\sin\theta_1 \sin\theta_2 - 2\cos\theta_1 \cos\theta_2]$$

Let us find the magnetic interaction energy for different orientations shown in figure 6.12(c).

✓ For parallel dipoles $(\vec{m}_1 \| \vec{m}_2)$ shown in figure 6.12(c)(i)

$$\theta_1 = 90°, \ \theta_2 = 90°$$

$$U = \frac{\mu_0}{4\pi} \frac{m_1 m_2}{r^3} [\sin\theta_1 \sin\theta_2 - 2\cos\theta_1 \cos\theta_2] = \frac{\mu_0}{4\pi} \frac{m_1 m_2}{r^3}$$

✓ For perpendicular dipoles $(\vec{m}_1 \perp \vec{m}_2)$ shown in figure 6.12(c)(ii)

$$\theta_1 = 90°, \ \theta_2 = 0°$$

$$U = \frac{\mu_0}{4\pi} \frac{m_1 m_2}{r^3} [\sin\theta_1 \sin\theta_2 - 2\cos\theta_1 \cos\theta_2] = 0$$

✓ For parallel dipoles oriented along their relative position vector $(\vec{m}_1 \| \vec{m}_2 \| \vec{r})$ shown in figure 6.12(c)(iii)

$$\theta_1 = 0°, \ \theta_2 = 0°$$

$$U = \frac{\mu_0}{4\pi} \frac{m_1 m_2}{r^3} [\sin\theta_1 \sin\theta_2 - 2\cos\theta_1 \cos\theta_2] = -\frac{\mu_0}{4\pi} \frac{2 m_1 m_2}{r^3}$$

This orientation of figure 6.12(c)(iii) corresponds to minimum energy.

**Exercise 6.20** *Prove that*

$$(\vec{\sigma} \cdot \vec{A})(\vec{\sigma} \cdot \vec{B}) = \vec{A} \cdot \vec{B} + i \ \vec{\sigma} \cdot (\vec{A} \times \vec{B}),$$

$\vec{\sigma}$ *is Pauli spin matrix,* $\vec{A}, \vec{B}$ *are two vectors.*

$\boxed{Ans}$ $(\vec{\sigma} \cdot \vec{A})(\vec{\sigma} \cdot \vec{B})$

$$= (\sigma_x A_x + \sigma_y A_y + \sigma_z A_z)(\sigma_x B_x + \sigma_y B_y + \sigma_z B_z)$$

$$= \sigma_x^2 A_x B_x + \sigma_y^2 A_y B_y + \sigma_z^2 A_z B_z + \sigma_x \sigma_y A_x B_y + \sigma_x \sigma_z A_x B_z$$

$$+ \sigma_y \sigma_x A_y B_x + \sigma_y \sigma_z A_y B_z + \sigma_z \sigma_x A_z B_x + \sigma_z \sigma_y A_z B_y$$

Using $\sigma_x^2 = \sigma_y^2 = \sigma_z^2 = I$

$\sigma_y \sigma_x = -\sigma_x \sigma_y$ ; $\sigma_z \sigma_y = -\sigma_y \sigma_z$; $\sigma_x \sigma_z = -\sigma_z \sigma_x$ (as Pauli spin matrices anti-commute)

we have

$$(\vec{\sigma}.\,\vec{A})(\vec{\sigma}.\,\vec{B}) = (A_x B_x + A_y B_y + A_Z B_z) + \sigma_x \sigma_y (A_x B_y - A_y B_x)$$

$$+ \sigma_y \sigma_z (A_y B_z - A_z B_y) + \sigma_z \sigma_x (A_z B_x - A_x B_z)$$

Since $\sigma_x \sigma_y = i\sigma_z$ ; $\sigma_y \sigma_z = i\sigma_x$ ; $\sigma_z \sigma_x = i\sigma_y$

$$(\vec{\sigma}.\,\vec{A})(\vec{\sigma}.\,\vec{B}) = \vec{A}.\,\vec{B} + i\sigma_z \quad (\vec{A} \times \vec{B})_z + i\sigma_x \quad (\vec{A} \times \vec{B})_x + i\sigma_y \quad (\vec{A} \times \vec{B})_y$$

$$(\vec{\sigma}.\,\vec{A})(\vec{\sigma}.\,\vec{B}) = \vec{A}.\,\vec{B} + i \quad \vec{\sigma}.\quad \vec{A} \times \vec{B}$$

**Exercise 6.21** *Consider a state of a two-nucleon system that is a linear combination of orbital angular momentum quantum numbers $l = J - 1, J, J + 1$ which form a triplet of states. Show that the even parity $J = 1$ state is a mixture of $^3S_1$, $^3D_1$ states while the odd parity $J = 1$ state is a pure $^3P_1$ state.*

$\boxed{Ans}$ Given $J = 1$. So $l$ values are

✓ $l = J - 1 = 1 - 1 = 0$. This $l = 0$ state has spectroscopic notation $S$.
  Parity is $(-)^l = (-)^0 = +ve$ or even.
✓ $l = J = 1$. This $l = 1$ state has spectroscopic notation $P$.
  Parity is $(-)^l = (-)^1 = -ve$ or odd.
✓ $l = J + 1 = 1 + 1 = 2$. This $l = 2$ state has spectroscopic notation $D$.
  Parity is $(-)^l = (-)^2 = +ve$ or even.
  We are discussing a triplet of states. So multiplicity is $2s + 1 = 3$.
  So now we can write the states in spectroscopic notation
  $^{2s+1=3}(l)_{J=1}$, i.e. $^3S_1$ (even), $^3P_1$ (odd) $^3D_1$ (even). Hence we can
state the following.
  Even parity states that $J = 1$ corresponds to is a mixture of $^3S_1$ and $^3D_1$
states.
  Odd parity state that $J = 1$ corresponds to is a pure $^3P_1$ state.

**Exercise 6.22** *For a system of two nucleons (np system) evaluate $\vec{\sigma}_1.\,\vec{\sigma}_2$ for singlet and triplet states and show that nuclear potential is spin dependent.*

$\boxed{Ans}$ Define for the two nucleons (proton and neutron)

$$\vec{S}_p = \frac{\hbar}{2}\vec{\sigma}_1, \ \vec{S}_n = \frac{\hbar}{2}\vec{\sigma}_2$$

$$\vec{s} = \vec{S}_p + \vec{S}_n = \frac{1}{2} + \frac{1}{2} = \begin{cases} 0 \ (\text{singlet}) \\ 1 \ (\text{triplet}) \end{cases}$$

$$\vec{s} = \frac{\hbar}{2}\vec{\sigma}_1 + \frac{\hbar}{2}\vec{\sigma}_2 = \frac{\hbar}{2}(\vec{\sigma}_1 + \vec{\sigma}_2)$$

$$\vec{\sigma}_1 + \vec{\sigma}_2 = \frac{2}{\hbar}\vec{s}$$

$$(\vec{\sigma}_1 + \vec{\sigma}_2)^2 = \frac{4}{\hbar^2}s^2$$

Replace by eigenvalue to get

$$(\vec{\sigma}_1 + \vec{\sigma}_2)^2 = \frac{4}{\hbar^2}s(s+1)\hbar^2 = 4s(s+1) \tag{6.170}$$

Also

$$(\vec{\sigma}_1 + \vec{\sigma}_2)^2 = \sigma_1^2 + \sigma_2^2 + 2\vec{\sigma}_1 \cdot \vec{\sigma}_2$$

$$\vec{\sigma}_1 \cdot \vec{\sigma}_2 = \frac{1}{2}[(\vec{\sigma}_1 + \vec{\sigma}_2)^2 - \sigma_1^2 - \sigma_2^2] \tag{6.171}$$

Using the property of Pauli spin matrix that

$$\sigma_x^2 = 1, \quad \sigma_y^2 = 1, \ \sigma_z^2 = 1$$

and so

$$\sigma_1^2 = \sigma_{1x}^2 + \sigma_{1y}^2 + \sigma_{1z}^2 = 1 + 1 + 1 = 3$$

$$\sigma_2^2 = \sigma_{2x}^2 + \sigma_{2y}^2 + \sigma_{2z}^2 = 1 + 1 + 1 = 3$$

From equation (6.171) it follows that

$$\vec{\sigma}_1 \cdot \vec{\sigma}_2 = \frac{1}{2}[(\vec{\sigma}_1 + \vec{\sigma}_2)^2 - 3 - 3] = \frac{1}{2}[(\vec{\sigma}_1 + \vec{\sigma}_2)^2 - 6] \tag{6.172}$$

For singlet $s = 0$. From equation (6.170) we have

$$(\vec{\sigma}_1 + \vec{\sigma}_2)^2 = 4.0(0 + 1) = 0$$

Hence equation (6.172) gives

$$\vec{\sigma}_1 \cdot \vec{\sigma}_2 = \frac{1}{2}[(\vec{\sigma}_1 + \vec{\sigma}_2)^2 - 6] = \frac{1}{2}[0 - 6] = -3 \tag{6.173}$$

For triplet $s = 1$. From equation (6.170) we have

$$(\vec{\sigma}_1 + \vec{\sigma}_2)^2 = 4.1(1 + 1) = 8$$

Hence equation (6.172) gives

$$\vec{\sigma}_1 \cdot \vec{\sigma}_2 = \frac{1}{2}[(\vec{\sigma}_1 + \vec{\sigma}_2)^2 - 6] = \frac{1}{2}[8 - 6] = 1 \qquad (6.174)$$

The nuclear potential is given by equation (6.115). Ignoring the tensor force part we have

$$V(r) = V_1(r) + V_2(r) \quad \vec{\sigma}_1 \cdot \vec{\sigma}_2$$

For singlet $s = 0$ case, $\vec{\sigma}_1 \cdot \vec{\sigma}_2 = -3$ and so

$$V(r) = V_1(r) - 3V_2(r)$$

For triplet $s = 1$ case, $\vec{\sigma}_1 \cdot \vec{\sigma}_2 = 1$ and we have

$$V(r) = V_1(r) + V_2(r)$$

Nuclear potential $V(r)$ is different for different spin. Clearly nuclear force is spin dependent.

**Exercise 6.23** *Show that the tensor force operator is a scalar operator.*
[Ans] The tensor force operator $S_{12}$ is given by equation (6.116) to be

$$S_{12} = \frac{3(\vec{\sigma}_1 \cdot \vec{r})(\vec{\sigma}_2 \cdot \vec{r})}{r^2} - \vec{\sigma}_1 \cdot \vec{\sigma}_2$$

Under reflection of coordinates, behaviour of $\vec{r}$ and $\vec{\sigma}$ is as follows

$$\vec{r} \xrightarrow{\text{reflection about origin}} -\vec{r} \text{ as } \vec{r} \text{ is a polar vector}$$

$$\vec{\sigma} \xrightarrow{\text{reflection about origin}} \vec{\sigma} \text{ as } \vec{\sigma} \text{ is a pseudo vector}$$

Hence under reflection of coordinates behavior of $S_{12}$ is as follows.

$$S_{12} = \frac{3(\vec{\sigma}_1 \cdot \vec{r})(\vec{\sigma}_2 \cdot \vec{r})}{r^2} - \vec{\sigma}_1 \cdot \vec{\sigma}_2$$

$$\xrightarrow{\text{reflection about origin}} \frac{3[\vec{\sigma}_1 \cdot (-\vec{r})][\vec{\sigma}_2 \cdot (-\vec{r})]}{r^2} - \vec{\sigma}_1 \cdot \vec{\sigma}_2$$

$$= \frac{3(\vec{\sigma}_1 \cdot \vec{r})(\vec{\sigma}_2 \cdot \vec{r})}{r^2} - \vec{\sigma}_1 \cdot \vec{\sigma}_2 = S_{12}$$

So $S_{12}$ is a scalar operator.

**Exercise 6.24** *Discuss how one measures the binding energy of deuteron.*

$\boxed{Ans}$ ✓ One can determine the binding energy of deuteron directly by bringing a proton $p$ and a neutron $n$ together to form a deuteron $d \equiv {}_1H^2$ as per the reaction

$$n + p \rightarrow d + \gamma$$

and measuring the energy of the $\gamma$ photon that is emitted.

    ✓ One can also use the reverse reaction to determine the binding energy of deuteron by studying the photo diusintegration of deuteron as per the reaction

$$\gamma + d \rightarrow p + n$$

in which a $\gamma$ ray photon tears apart the deuteron. The minimum $\gamma$ ray energy that accomplishes this process is identified as the binding energy of deuteron.

Binding energy of $d$ as measured from experiment is $B = 2.225\ MeV$ that proves equation (6.1).

**Exercise 6.25** *Find the tensor force operator for deuteron.*

$\boxed{Ans}$ The tensor force operator as given in equation (6.116) is

$$S_{12} = \frac{3(\vec{\sigma}_1 \cdot \vec{r})(\vec{\sigma}_2 \cdot \vec{r})}{r^2} - \vec{\sigma}_1 \cdot \vec{\sigma}_2$$

Consider $d \equiv np$, i.e. deuteron system. It is a spin triplet state with parallel spins of proton and neutron, as shown in figure 6.13(a),(b). We consider two configurations.

Refer to figure 6.13(a) where spins of proton and neutron are parallel to $\vec{r}$. Then

$$\vec{\sigma}_1 \cdot \vec{r} = \sigma_1 r, \ \vec{\sigma}_2 \cdot \vec{r} = \sigma_2 r, \ \vec{\sigma}_1 \cdot \vec{\sigma}_2 = \sigma_1 \sigma_2 = 1$$

(by *exercise 6.22* for $s = 1$ spin triplet state)

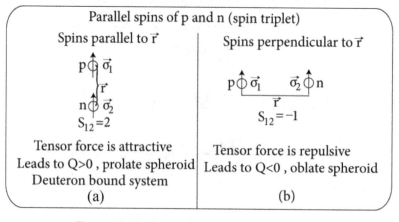

Figure 6.13. Finding tensor force operator for deuteron

$$S_{12} = \frac{3(\vec{\sigma}_1 \cdot \vec{r})(\vec{\sigma}_2 \cdot \vec{r})}{r^2} - \vec{\sigma}_1 \cdot \vec{\sigma}_2 = \frac{3\sigma_1 r \sigma_2 r}{r^2} - 1 = 3 - 1 = 2$$

When spins of proton and neutron are parallel to $\vec{r}$, $S_{12} = 2$ and the tensor force is attractive leading to deuteron bound state. It is stretched along the angular momentum direction resulting in prolate shape giving $Q_d = +ve$.

Refer to figure 6.13(b) where spins of proton and neutron are perpendicular to $\vec{r}$. Then

$$\vec{\sigma}_1 \cdot \vec{r} = 0, \ \vec{\sigma}_2 \cdot \vec{r} = 0, \ \vec{\sigma}_1 \cdot \vec{\sigma}_2 = 1 \text{ (by exercise 6.22 for } s = 1 \text{ spin triplet state)}$$

$$S_{12} = \frac{3(\vec{\sigma}_1 \cdot \vec{r})(\vec{\sigma}_2 \cdot \vec{r})}{r^2} - \vec{\sigma}_1 \cdot \vec{\sigma}_2 = \frac{3(0)(0)}{r^2} - 1 = -1$$

When spins of proton and neutron are perpendicular to $\vec{r}$, $S_{12} = -1$ and the tensor force is repulsive. It corresponds to oblate shape. So this configuration does not represent deuteron bound state.

**Exercise 6.26** *Write down the possible spectroscopic terms for deuteron following the L–S coupling scheme.*

$\boxed{Ans}$ In the L–S coupling scheme we add $l$ values to get a resultant $L$, we add $s$ values to get a resultant $S$. Then we add the $L$ and $S$ to get the resultant $J$ value.

✓ The $l$ values

      Let us take $l \rightarrow 0, 1, 2 \Rightarrow S, P, D$ states

✓ Addition of $s$ values

$$\vec{s}_p + \vec{s}_n = \frac{\overline{1}}{2} + \frac{\overline{1}}{2} \rightarrow \begin{cases} \frac{1}{2} - \frac{1}{2} = 0 \text{ (singlet): } 2s + 1 = 2.0 + 1 = 1 \\ \frac{1}{2} + \frac{1}{2} = 1 \text{ (triplet): } 2s + 1 = 2.1 + 1 = 3 \end{cases}$$

Possibilities for $J$ have been depicted in figure 6.14.

| $l$ | + | $s$ | = | $J$ | | | $l$ | + | $s$ | = | $J$ | |
|---|---|---|---|---|---|---|---|---|---|---|---|---|
| $0$ | | | | $0$ | $^1S_0$ | | $0$ | | | | $1$ | $^3S_1$ |
| $1$ | $+ 0$ | | $=$ | $1$ | $^1P_1$ | | $1$ | $+ 1$ | | $=$ | $2,1,0$ | $^3P_{2,1,0}$ |
| $2$ | | | | $2$ | $^1D_2$ | | $2$ | | | | $3,2,1$ | $^3D_{3,2,1}$ |

For $J=1$ allowed spectroscopic terms are $^1P_1$, $^3S_1$, $^3P_1$, $^3D_1$

Parity –ve +ve –ve +ve

For deuteron allowed states are $^3S_1$, $^3D_1$

**Figure 6.14.** Spectroscopic terms for deuteron.

Let us extract out $J = 1$ cases since $J = 1$ for deuteron. They are

$$^1P_1, \,^3S_1, \,^3P_1, \,^3D_1.$$

Again, deuteron is a spin triplet having $s = 1$ and parity $(-1)^{l+s} = (-1)^{l+1} = +ve$ (figure 8.5). So the allowed deuteron states are

$$^3S_1, \,^3D_1$$

**Exercise 6.27** *To explain the observed magnetic moment of deuteron $0.8574\mu_N$ , its ground state wave function is taken to be an admixture of S and D states. The expectation values of the z component of the magnetic moment in pure S and pure D states are $0.8797\mu_N$ and $0.3101\mu_N$ respectively. The contribution of the D state to the mixed ground state is approximately*

*(a) 40 %, (b) 4 %, (c) 0.4 %, (d) 0.04 %.*

Hint: $\mid d > \,= \sqrt{0.96} \mid ^3S_1> + \sqrt{0.04} \mid ^3D_1>$

**Exercise 6.28** *The ground state wave function of deuteron is in a superposition of S and D states. Which of the following is not true as a consequence?*
  *(a) It has a non-zero quadrupole moment.*
  *(b) The neutron–proton potential is non-central.*
  *(c) The orbital wave function is not spherically symmetric.*
  *(d) The Hamiltonian does not conserve total angular momentum.*

**Exercise 6.29** *The neutron and proton form a deuteron bound state which is stable while there is no bound state for two neutrons because:*
  *(a) nuclear forces are saturated,*
  *(b) nuclear forces are spin dependent,*
  *(c) nuclear forces are charge independent,*
  *(d) nuclear forces depend on magnetic moment.*

**Exercise 6.30** *The ground state of deuteron is*
  *(a) pure $^3S_1$ state,*
  *(b) pure $^3P_1$ state,*
  *(c) mixture of $^3S_1$ and $^3P_1$ states,*
  *(d) mixture of $^3S_1$ and $^3D_1$ states.*

**Exercise 6.31** *Deuteron has only one bound state with spin parity $1^+$ , isospin 0 , electric dipole moment 2.82 milli barn. These data suggest that the nuclear forces are having:*
  *(a) only spin and isospin dependence,*
  *(b) no spin dependence and no tensor component,*
  *(c) spin dependence but no tensor components,*
  *(d) spin dependence along with tensor components.*

**Exercise 6.32** *With reference to nuclear forces which of the following is not true?*
(a) *Short range,*
(b) *charge independent,*
(c) *velocity dependent,*
(d) *spin independent.*

**Exercise 6.33** . *Deuteron in ground state has spin parity* $1^+$ . *Which of the following combinations is correct?*
(a) $l = 0$, $s = 1$ and $l = 2$, $s = 0$,
(b) $l = 0$, $s = 1$ and $l = 1$, $s = 0$,
(c) $l = 0$, $s = 1$ and $l = 2$, $s = 1$,
(d) $l = 1$, $s = 1$ and $l = 2$, $s = 1$.

**Exercise 6.34** . *An admissible potential between the proton and the neutron in a deuteron is:*
(a) *Coulomb potential,*
(b) *harmonic oscillator,*
(c) *finite square well,*
(d) *Infinite square well.*

**Exercise 6.35** *The binding energy per nucleon of helium nucleus is 7 MeV and that of deuteron is 1 MeV. In view of this statement which option is correct?*
(a) *Helium nucleus is more stable.*
(b) *Deuteron nucleus is nore stable.*
(c) *Deuteron nucleus has binding energy per nucleon 2.225 MeV.*
(d) *Deuteron nucleus has binding energy per nucleon –2.225 MeV.*

**Exercise 6.36** *The binding energy of deuteron is*
(a) *2.225 MeV, (b) –2.225 McV, (c) 8 MeV, (d) –8 MeV.*

**Exercise 6.37** *The existence of quadrupole moment of deuteron reveals the existence of:*
(a) *central force, (b) non-central force,*
(c) *strong force, (d) electromagnetic forces.*

**Exercise 6.38** *The magnetic moment of deuteron is*
(a) *equal to the sum of the magnetic moments of proton and neutron,*
(b) *less than the sum of magnetic moments of proton and neutron,*
(c) *greater than the sum of magnetic moments of proton and neutron,*
(d) *equal to the magnetic moment deuteron.*

[Hint. Equation (6.106)]

**Exercise 6.39** *If we consider only the central forces then magnetic moment of deuteron is:*

    *(a) equal to the sum of the magnetic moments of proton and neutron,*

    *(b) less than the sum of magnetic moments of proton and neutron,*

    *(c) greater than the sum of magnetic moments of proton and neutron,*

    *(d) equal to the magnetic moment of deuteron.*

Ans to Multiple Choice Questions

6.27 *b*, 6.28 *d*, 6.29 *b*, 6.30 *d*, 6.31 *d*, 6.32 *d*, 6.33 *c*, 6.34 *c*, 6.35 *a*, 6.36 *a*, 6.37 *b*, 6.38 *b*, 6.39 *a*.

## 6.17 Question bank

Q6.1    Why is deuteron considered as an ideal system to study about nuclear force?

Q6.2    What is the difficulty in studying a nucleus with $A > 2$ to investigate the nature of nuclear force?

Q6.3    Is electromagnetic force operative within deuteron?

Q6.4    Write down the quark structure of neutron and proton.

Q6.5    How is the structure of neutron and proton studied experimentally?

Q6.6    What should be the energy and de Broglie wavelength of the agency probing neutron and proton?

Q6.7    What is the role of quarks and leptons in the Universe?

Q6.8    How many varieties of leptons are there? Mention them.

Q6.9    How many varieties of quarks are there? Mention them.

Q6.10    Write down the charges of quarks and antiquarks.

Q6.11    Find out the charge of proton from its quark structure.

Q6.12    Find out the charge of neutron from its quark structure.

Q6.13    Justify the statement that neutron is built out of charged particles though it is overall uncharged.

Q6.14    The strong interaction comes from an intrinsic property called colour charge. How many types of colour charge are there?

Q6.15    Justify that neutron and proton are colourless.

Q6.16    Why is it claimed that deuteron is a loosely or weakly bound system?

Q6.17    Explain what we mean by deuteron radius.

Q6.18    What is the depth of deuteron potential well?

Q6.19    Obtain deuteron wave function from solution of the Schrödinger equation. Show deuteron wave function in a plot against distance.

Q6.20    What is the relation between depth of deuteron potential well and nuclear range?

Q6.21    Can you describe the deuteron well with Woods–Saxon potential well?

Q6.22    How do you explain the fact that deuteron is not a pure $l = 0$ state?

Q6.23    Write down the spectroscopic notation of deuteron ground state.

Q6.24    What do we mean by the term nuclear spin in nuclear physics?

Q6.25    What is the experimentally confirmed spin parity of deuteron?

Q6.26    Deuteron is an $np$ system. With the help of two spin-half particles construct a coupled basis $|sm_s\rangle$ and an uncoupled basis $|m_{sp}m_{sn}\rangle$. Express $|sm_s\rangle$ in terms of $|m_{sp}m_{sn}\rangle$.

Q6.27    Explain the fact that deuteron bound state is a spin triplet and not a spin singlet.

Q6.28    Show that magnetic moment of a charged particle is related to angular momentum.

Q6.29    What is nuclear magneton? Obtain its value in SI. Express deuteron magnetic moment in units of nuclear magneton. Can you express deuteron magnetic moment in units of Bohr magneton?

Q6.30    Obtain the ratio of nuclear magneton and Bohr magneton. Which is larger?

Q6.31    What is the implication of the fact that deuteron has magnetic moment $0.857\ 36\mu_N$ and a non-vanishing quadrupole moment.

Q6.32    What is gyromagnetic ratio?

Q6.33    Write down the magnetic moments of electron, proton and neutron for their orbital motion and for their intrinsic spin.

Q6.34    Explain the spin gyromagnetic ratios of electron (2), proton (5.585 691) and neutron ($-3.826\ 084$).

Q6.35    How is it that the neutral neutron has a non-vanishing spin gyromagnetic ratio.

Q6.36    Show that if we consider deuteron in $l = 0$ state the vector sum of orbital and spin magnetic moments of proton and neutron fails to generate the observed magnetic moment. What is the result if we consider that deuteron has an admixture of $l = 2$ state? Obtain the amount of admixture of $l = 2$ state.

Q6.37    Write down the expression of quadrupole moment of a nucleus. What would have been the value of deuteron quadrupole moment if it existed in a pure $l = 0$ state?

Q6.38    Justify the shape of deuteron nucleus: prolate, oblate or spherically symmetric.

Q6.39    How do you account for the non-central part of interaction within the deuteron nucleus?

Q6.40    What is tensor force? Why is it necessary to explain nuclear force in a deuteron nucleus?

Q6.41    Write down the general form of potential describing nucleon–nucleon potential in a deuteron.

Q6.42    What is tensor force operator? Find its value for two magnetic dipoles that are parallel, antiparallel.

Q6.43    Obtain the tensor force operator for deuteron.

Q6.44    What are the properties of tensor force?
Q6.45    Is tensor force present in singlet spin state?
Q6.46    Is tensor force a central force? Justify your answer.

## Further reading

[1] Krane S K 1988 *Introductory Nuclear Physics* (New York: Wiley)
[2] Tayal D C 2009 *Nuclear Physics* (Mumbai: Himalaya Publishing House)
[3] Satya P 2005 *Nuclear Physics and Particle Physics* (New Delhi: Sultan Chand & Sons)
[4] Guha J 2019 *Quantum Mechanics: Theory, Problems and Solutions* 3rd edn (Kolkata: Books and Allied (P) Ltd))
[5] Lim Y-K 2002 *Problems and Solutions on Atomic, Nuclear and Particle Physics* (Singapore: World Scientific)

**IOP** Publishing

Nuclear and Particle Physics with Cosmology, Volume 1
Nuclear physics
**Jyotirmoy Guha**

# Chapter 7

## Scattering

## 7.1 Introduction

Nucleon–nucleon scattering is a powerful technique used to know about many features of nuclear force like its range, strength, spin dependence etc.

Study of the deuteron problem gave us some information about nucleon–nucleon interaction. But we had to be satisfied with the fact that deuteron bound state, which is the ground state, has no excited state and corresponds to $l = 0$, parallel spins (triplet) and nuclear range 2 *fm*. So study of the deuteron problem fails to give a complete quantitative nuclear force law.

If excited states were present in deuteron it might correspond to different $l$ values and different spin orientations (singlet plus triplet).

To study the nucleon–nucleon interaction in different configurations we can perform nucleon–nucleon scattering experiments—in which an incident beam of nucleons is scattered from a target of nucleons.

Scattering corresponds to unbound state of nucleons.

If we choose a nucleus as target then since the nucleus contains many nucleons, there will be several target nucleons within the range of nuclear potential of the incident nucleon. Such a case will be difficult to analyse as it would involve multiple encounters and their complicated effects.

Therefore, a hydrogen nucleus or proton can be considered as a target and it can scatter incident particles. In such a case scattering occurs first from one proton then from another proton—situated far from the first proton on the scale of nuclear dimensions. The probability of a single encounter is small and the probability for multiple encounters will thus be negligibly small.

Nucleon–nucleon scattering can be studied both at low projectile energy as well as high projectile energy.

To study nucleon–nucleon interaction in different configurations we can investigate nucleon–nucleon scattering where a projectile nucleon would be scattered by a target nucleon. We can consider *np* scattering, *pp* scaterring, *nn* scattering.

We are interested to know the nature of nuclear interaction when one nucleon scatters another nucleon.

- Difference between study of scattering problem and study of bound state problem.

   ✓ In the scattering problem the emphasis is on wave function in contrast to bound state problem where focus is mainly on energy eigenvalues.

   ✓ In the scattering problem we are concerned with the positive part of energy spectrum ($E > 0$) in contrast to bound state problem where we deal with the negative part of the energy spectrum ($E < 0$).

## 7.2 Classical idea of scattering cross-section

In figure 7.1 we show the scattering process of a monoenergetic, uniformly distributed nucleon beam—so to say, an ensemble of particles—approaching the target. The target is a set of nucleons (say hydrogen target consisting of protons) which are scattering centres. The incident beam of nucleons gets scattered or deflected by the target nucleons. So basically it is scattering of a projectile nucleon by target nucleon. The scattering centre of target is represented by a potential $V$. We say that the projectile is scattered, i.e. gets deviated by the potential $V$ in all possible directions—some are undeviated or forward scattered at $\theta = 0°$, some suffer back scattering at $\theta = 180°$.

The aim of the scattering experiment is to find out the intensity of scattered beam along a particular direction $\theta$, $\phi$ which also is a measure of the probability of scattering along that direction.

Suppose a detector is at angular position $(r, \theta, \phi)$. The detector window subtends an elemental solid angle $d\Omega = \sin\theta d\theta d\phi$ at the target. The detector records flux of particles that get scattered along the particular direction ($\hat{r} \equiv \theta, \phi$).

Let $dN$ be the number of scattered particles into solid angle $d\Omega$ per unit time and $dN$ is also the number of particles received by the detector per unit time.

Let $J_{inc}$ be the incident flux along $\hat{z}$, i.e. the number of incident particles proceeding along $\hat{z}$ crossing unit area (which is the $xy$ plane perpendicular to $\hat{z}$) per unit time.

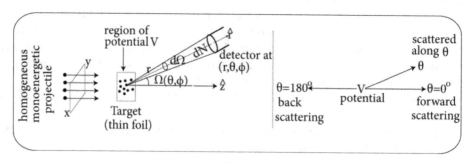

Figure 7.1. The scattering experiment.

The larger the $J_{\text{inc}}$ the more is $dN$. Hence

$$dN \propto J_{\text{inc}}$$

Let $n$ be the number of scattering centres in the target. The more $n$ is, the more is $dN$. Hence

$$dN \propto n$$

Also, the wider the $d\Omega$ the more the $dN$. Hence

$$dN \propto d\Omega$$

In other words, experimental observation suggests that

$$dN \propto n, \; dN \propto J_{\text{inc}}, \; dN \propto d\Omega$$

Hence we write

$$dN \propto nJ_{\text{inc}}d\Omega \tag{7.1}$$

- Such a proportionality relation is valid if the following conditions are obeyed.

  ✓ There is no mutual interaction between particles in the incident beam. And this occurs if incident particles are significantly separated from each other during their journey to the target. And this means that the number of particles per unit volume (i.e. particle density or intensity) of the incident beam has to be low.
  ✓ The incident particle is to be scattered by a scattering centre. Hence the size of the wave packet associated with the incident particles should be smaller than the distance between the scattering centres. For this to occur we expect that the de Broglie wavelength $\lambda = \frac{h}{p}$ of the incident pariticles has to be smaller than the average separation $d$ between the scattering centres, i.e.

$$\lambda \ll d.$$

Again, to keep $\lambda$ small, the momentum of the incident particles $p = \frac{h}{\lambda}$ should be large.

Assuming that the above conditions are satisfied we have from equation (7.1)

$$dN = (\text{constant})nJ_{\text{inc}}d\Omega$$

This constant is written as $\frac{d\sigma}{d\Omega}$ and is called differential scattering cross-section. So

$$dN = \frac{d\sigma}{d\Omega}nJ_{\text{inc}}d\Omega \tag{7.2}$$

  ✓ Let us find the unit of $\frac{d\sigma}{d\Omega}$

From equation (7.2) we have ($d\Omega$ has no dimension)

$$\left(\frac{\text{number}}{\text{time}}\right) = \frac{d\sigma}{d\Omega}(\text{number})\left(\frac{\text{number}}{\text{area. time}}\right)$$

$$\frac{d\sigma}{d\Omega} \rightarrow \text{area}$$

We have shown that the differential scattering cross-section has unit of area. $\frac{d\sigma}{d\Omega}$ is measured in $m^2$.

Since value of $\frac{d\sigma}{d\Omega}$ in the case of nuclear interactions is of the order of $10^{-28}\ m^2$ we measure the differential scattering cross-section in unit of $10^{-28}\ m^2$. We define

$$10^{-28}\ m^2 = 1\ barn$$

Let $J_s(\Omega)$ be the scattered flux in the direction ($\hat{r} \equiv \theta, \phi$) which is the number of scattered particles crossing unit area perpendicular to $\hat{r}$ per unit time.

The detector is placed at a distance $r$ from the target and

$$r \gg \text{dimension of target.}$$

The elemental volume of detector placed at ($r, \theta, \phi$) is

$$dV = r^2 dr \sin\theta d\theta d\phi = r^2 dr d\Omega$$

Now

$$J_s(\Omega) = \frac{\text{number}}{\text{area. time}}$$

Multiplying numerator and denominator by length we have

$$J_s(\Omega) = \frac{\text{number}}{\text{area. time}}\frac{\text{length}}{\text{length}} = \frac{\text{number}}{\text{time}}\cdot\frac{\text{length}}{\text{volume}} = dN\frac{dr}{dV} = dN\frac{.dr}{r^2 dr d\Omega}$$

$$J_s(\Omega) = \frac{dN}{r^2 d\Omega}$$

$$dN = J_s(\Omega)r^2 d\Omega \tag{7.3}$$

Comparing (7.2) and (7.3) we can write

$$dN = \frac{d\sigma}{d\Omega}nJ_{\text{inc}}d\Omega = J_s(\Omega)r^2 d\Omega.$$

This leads to

$$\frac{d\sigma}{d\Omega} = \frac{r^2 J_s(\Omega)}{nJ_{\text{inc}}} \tag{7.4}$$

In general the differential scattering cross-section $\frac{d\sigma}{d\Omega}$ depends on $\theta$, $\phi$, i.e.

$$\frac{d\sigma}{d\Omega} = \frac{d\sigma}{d\Omega}(\theta, \phi).$$

Physically the total scattering cross-section represents the cross-section of the incident beam that is traversed by as many particles as are scattered in all directions by each scattering centre.

The total scattering cross-section $\sigma$ is obtained from the differential scattering cross-section as

$$\sigma = \int \frac{d\sigma}{d\Omega} d\Omega = \int_0^\pi \int_0^{2\pi} \frac{d\sigma}{d\Omega}(\theta, \phi)\sin\theta d\theta d\phi \tag{7.5}$$

For spherically symmetric or isotropic potential

$$V(\vec{r}) = V(r) \neq V(\theta, \phi)$$

the differential scattering cross-section $\frac{d\sigma}{d\Omega}$ is independent of $\phi$ and only depends on $\theta$, i.e. in that case

$$\frac{d\sigma}{d\Omega} = \frac{d\sigma}{d\Omega}(\theta)$$

and so from equation (7.5)

$$\sigma = \int_0^{2\pi} d\phi \int_0^\pi \frac{d\sigma}{d\Omega}(\theta)\sin\theta d\theta = 2\pi \int_0^\pi \frac{d\sigma}{d\Omega}(\theta)\sin\theta d\theta \tag{7.6}$$

The differential scattering cross-section $\frac{d\sigma}{d\Omega}$ as well as the total scattering cross-section $\sigma$ give information regarding the nuclear force or interaction responsible for the scattering process.

We have decribed the classical concept of scattering cross-section.

## 7.3 Types of scattering

We mention four types of scattering.

- Elastic scattering.

    Let us consider scattering by a potential created by scattering centre. This is also called potential scattering.

    Here there is transfer of momentum between projectile and target but no transfer of energy. No energy is absorbed by the target—there being no internal excitation. Kinetic energy is conserved in the process.

    Looked from centre of mass frame the velocity of particles change due to change in direction but there is no change in speed (magnitude of velocity remains constant), i.e. momentum magnitude remains constant and hence kinetic energy is also constant.

- Inelastic scattering.

    In inelastic scattering there is exchange of energy between projectile and target. Target absorbs energy and gets internally excited.

- Radiative capture.

  In radiative capture projectile is completely absorbed or captured by target and there is no emission from target. This justifies the name capture.
- Nuclear reaction.

  In nuclear reaction incident and emergent particles are generally different. This was discussed in detail in chapter 4.

## 7.4 Scattering amplitude

Scattering cross-section is determined through experiment and provides information regarding forces responsible for the process of scattering. So in the study of nucleon–nucleon scattering the scattering cross-section will furnish information about the nature of nuclear force. This is achieved by relating scattering cross-section and wave function.

The wave function is obtained by solving the Schrödinger time independent equation the Hamiltonian of which incorporates the interaction under consideration.

We shall discuss nucleon–nucleon elastic scattering.

Refer to figure 7.2(a) that decribes elastic scattering in lab frame.

A projectile of mass $m_1$ moves with velocity $\vec{v}_1$ and strikes the target of mass $m_2$. Origin is taken on the target, i.e. we consider a stationary or fixed target, i.e. target velocity $\vec{v}_2 = 0$.

The projectile is scattered by a potential $V(\vec{r})$ associated with the target.

After being hit by the projectile the target recoils. Projectile velocity is now changed from $\vec{v}_1$ to $\vec{v}_1'$ while the recoil velocity of target changes from $\vec{v}_2 = 0$ to $\vec{v}_2'$.

When this interaction is viewed from centre of mass system in the centre of mass frame of reference (figure 7.2(b)) we take the centre of mass (the origin in this case) of the two masses $m_1$ and $m_2$ to be stationary.

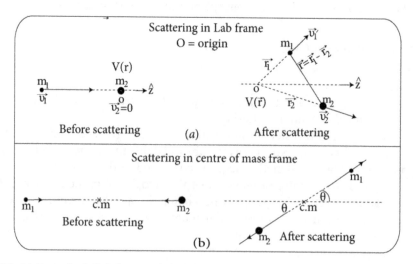

**Figure 7.2.** (a). Scattering in Lab frame. $m_1$ is projectile, $m_2$ is target which is initially at rest. (b) Scattering in centre of mass frame in which centre of mass is at rest.

We note that in lab frame the centre of mass is moving and the velocity of the centre of mass in lab frame is

$$\vec{v} = \frac{m_1 \vec{v_1} + m_2 \vec{v_2}}{m_1 + m_2} \xrightarrow{\vec{v_2}=0} \frac{m_1 \vec{v_1}}{m_1 + m_2} \quad (7.7)$$

Also, if $\vec{r_1}$ and $\vec{r_2}$ be the position vectors of masses $m_1$ and $m_2$ respectively then the relative position vector of $m_1$ w.r.t. $m_2$ will be

$$\vec{r} = \vec{r_1} - \vec{r_2}.$$

When interaction depends only on the relative coordinate $r$ then the time independent Schrödinger equation for the scattering problem can be separated into two parts
   ✓ one part describes the uniform motion of the centre of mass of projectile–target system
   ✓ another part describes the relative motion of the particles that appears as a motion of a single particle of reduced mass

$$\mu = \frac{m_1 m_2}{m_1 + m_2}$$

which moves with relative velocity

$$\vec{v} = \vec{v_1} - \vec{v_2}$$

which is the velocity of $m_1$ w.r.t. $m_2$ under a potential $V(\vec{r})$.

Scattering affects only the relative motion which is independent of the velocity of the centre of mass. So the theory of scattering of a particle by a potential is also the theory of two-particle scattering.

We can reduce this problem to a one-body problem by taking the mass to be the reduced mass.

The process of scattering consists of two situations, as described in figure 7.3.
   ✓ Before scattering
      The projectile was projected a long time ago say at time $t = -\infty$ from a very large distance that we denote as $r \to \infty$ towards target.

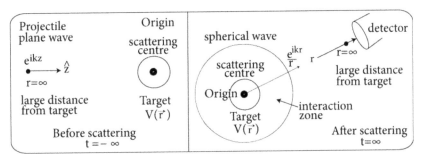

**Figure 7.3.** At $t = -\infty$ (before interaction or scattering) distance between projectile and scattering centre is ∞. At $t = +\infty$ (after interaction or scattering) the distance between the scattering centre and scattered particle is again ∞. Origin is chosen on the target.

Origin is taken on target. The target is described by potential $V(\vec{r})$ which is of finite or short range (not Coulombic potential, which is long range).

So $V(\vec{r})$ is effective only around a small region around the location of target and goes to zero faster than $\frac{1}{r}$ (in contrast to Coulombic variation $\propto \frac{1}{r}$) as $r \to \infty$. Clearly the projectile starts from $\infty$ distance from target or scattering centre. Interaction occurs only when the projectile enters into the field of the target $V(\vec{r})$, say at time $t = 0$.

✓ After scattering

The scattered particle gets scattered or deflected along a direction $\hat{r}$ and proceeds to the detector kept at a large distance from the target that we designate as $r \to \infty$ at time say $t = \infty$, the target being at the origin.

This means the purpose of scattering problem involves quantum description of how a wave packet, representing incident projectile at $r \to \infty$ at time $t = -\infty$ behaves, as it passes the short-range potential $V(\vec{r})$, i.e. the manner in which it gets perturbed in its journey across the scattering centre (target) and reaches the detector at time $t = \infty$ located at a large distance $r \to \infty$. In the actual experiment there are many particles that pass the perturbation region of potential and get scattered.

Nuclear scattering resembles optical diffraction by an obstacle. In *exercise 7.5* we mention some features of optical diffraction.

Let us analyse the process of scattering.

The time independent Schrödinger equation for the process of scattering is

$$-\frac{\hbar^2}{2\mu}\nabla^2\psi(\vec{r}) + V(\vec{r})\psi(\vec{r}) = E(\vec{r})\psi(\vec{r}) \tag{7.8}$$

where

$$E = \frac{p^2}{2\mu} = \frac{\hbar^2 k^2}{2\mu}$$

is kinetic energy of monoenergetic projectile which are mutually non-interacting. The momentum of the incident particle is $\vec{p} = \hbar\vec{k}$ and $\mu$ is the reduced mass of projectile–target particles. The potential energy $V(\vec{r})$ is responsible for scattering.

✓ We are interested in solving the time independent Schrödinger equation long before scattering event (at $t = -\infty$) as well as long after the scattering event (at time $t \to \infty$), as shown in figure 7.3. In other words, we are interested in the solution in two regions.

✓ We also note that while writing down the solution we can ignore the normalization factors since we are interested in obtaining relative probabilities like $\frac{dN}{N}$ in which the normlization constants cancel out. ($N$ is the total number of incident particles)

• Long before scattering the incident projectile beam is far away from the target and can be considered as a parallel beam of free particles travelling along $\hat{z}$

towards scattering centre with kinetic energy $E = \frac{p^2}{2\mu}$, momentum $\vec{p} = \hbar\vec{k} = \hbar k\hat{z}$ in a potential free region $V = 0$. It can be represented at $r \to \infty$ by a plane wave $e^{i\,\vec{k}.\vec{r}}$ at $t = -\infty$. Hence we write

$$\psi_{\text{initial}} \equiv \psi_k^i(\vec{r}) \xrightarrow{r \to \infty} e^{i\,\vec{k}.\vec{r}} = e^{i\,k\hat{z}.(\hat{i}x+\hat{j}y+\hat{z}z)} = e^{ikz} \tag{7.9}$$

We can find normalization constants through the standard techniques such as Dirac delta normalization or box normalization. Dirac delta normalization factor is $(2\pi)^{-3/2}$. However, we ignore the normalization factor as it will cancel out in our analysis.

The incident plane wave is the solution of equation (7.8) at $r = \infty$ at $t = -\infty$.

- After interaction, due to the potential $V = V(\vec{r})$ particles are deflected or scattered in all directions emerging out from the scattering centre. So, long after scattering, i.e. at time $t = \infty$ and at $r = \infty$, the scattered beam can be considered as outgoing plane spherical waves emanating from the target region $V(\vec{r})$ in all directions as a result of interaction or scattering. Again, energy is conserved as it propagates. The energy of spherical wave front is spread out over the spherical surface area $4\pi r^2$. Therefore, energy per unit area of an expanding spherical wave varies as

$$\frac{\text{Energy}}{\text{area}} = \frac{\text{Energy}}{4\pi r^2} \propto \frac{1}{r^2}.$$

Again energy $\propto$ | amplitude |$^2$

As energy is proportional to amplitude square it follows that the amplitude of spherical wave will be $\propto \frac{1}{r}$. Hence the outgoing plane spherical waves will be denoted by

$$\frac{e^{ikr}}{r}.$$

The strength of the outgoing spherical waves in different directions or orientations $\Omega \equiv \theta, \phi$ is different, characterized by the amount of interaction or scattering suffered by the incident beam. It depends upon a quantity defined as scattering amplitude and denoted by

$$f_k(\Omega) = f_k(\theta, \phi)$$

Hence the spherical scattered wave at $t \to \infty$ as detected in the detector at $r = \infty$ will be

$$\psi_{\text{scatt}} = \psi_k^s(\vec{r}) \xrightarrow{r \to \infty} f_k(\Omega)\frac{e^{i\,kr}}{r} \tag{7.10}$$

where normalization factor has been ignored.

This scattered spherical wave is the solution of equation (7.8) at $r \to \infty$ at time $t \to \infty$.

The scattering amplitude $f_k(\Omega) \equiv f_k(\theta, \phi)$ has the following characteristics

✓ It is independent of $r$ and depends on $\theta, \phi$.

✓ It describes how a particle is scattered in various directions.

✓ In the scattering phenomena the scattering amplitude represents the probability amplitude of the outgoing spherical wave relative to the incoming plane wave.

✓ Scattering amplitude $f_k(\theta, \phi)$ has dimension of length. This is because $e^{ikr}$ in $f_k(\Omega)\dfrac{e^{ikr}}{r}$ is dimensionless.

✓ In the case of spherically symmetric potential

$$f_k(\Omega) \equiv f_k(\theta, \phi) = f_k(\theta)$$

i.e. scattering amplitude becomes independent of $r, \phi$ and depends on the angle of scattering $\theta$ only.

The asymptotic solution of the time independent Schrödinger equation (7.8) will be the combination of the solutions (7.9) and (7.10). Hence the final asymptotic solution for the wave function will be

$$\Psi_k(\vec{r}) = \Psi_k^i(\vec{r}) + \Psi_k^s(\vec{r})$$

$$\xrightarrow{r \to \infty} e^{ikz} + f_k(\Omega)\frac{e^{ikr}}{r} \tag{7.11}$$

$$= \begin{pmatrix} \text{incident} \\ \text{plane wave} \end{pmatrix} + \begin{pmatrix} \text{scattered} \\ \text{spherical wave} \end{pmatrix}$$

What one observes far away from the scattering site is the incident plane wave $e^{ikz}$ plus the scattered spherical wave $f_k(\Omega)\dfrac{e^{ikr}}{r}$.

This is pictorially indicated in figure 7.4.

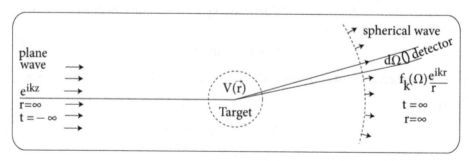

**Figure 7.4.** What one observes far away from the scattering site are the incident plane wave $e^{ikz}$ and the scattered spherical wave $f_k(\Omega)\dfrac{e^{ikr}}{r}$.

## 7.5 Differential scattering cross-section in terms of scattering amplitude

Let us find the expression of differential scattering cross-section as given by equation (7.4) viz. $\frac{d\sigma}{d\Omega} = \frac{r^2 J_s(\Omega)}{n J_{inc}}$ and try to relate it with scattering amplitude.

Density of particles in the incident beam is defined as

$$\rho_{inc} = \left| \psi_k^i(\vec{r}) \right|^2 = \psi_k^{i*}(\vec{r}) \psi_k^i(\vec{r})$$

From equation (7.9) we get

$$\rho_{inc} = (e^{ikz})^* (e^{ikz}) = e^{-ikz} e^{ikz} = 1 \tag{7.12}$$

The incident flux is defined as

$$J_{inc} = \frac{number}{area \cdot time} = \frac{number}{volume} \cdot \frac{length}{time} = density \cdot velocity$$

$$= \rho_{inc} v_i = 1. \, v_i = v_i = \frac{p}{\mu}$$

[as $\rho_{inc} = 1$ by equation (7.12)]

$$J_{inc} = \frac{\hbar k}{\mu} (as \, p = \hbar k) \tag{7.13}$$

Density of particles in the scattered beam in the direction $(\theta, \phi)$ is defined as

$$\rho_{scatt} = \left| \psi_k^s(\vec{r}) \right|^2 = \psi_k^{s*}(\vec{r}) \psi_k^s(\vec{r})$$

From equation (7.10) we get

$$\rho_{scatt} = \left( f_k(\Omega) \frac{e^{ikr}}{r} \right)^* \left( f_k(\Omega) \frac{e^{ikr}}{r} \right) = f_k^*(\Omega) \frac{e^{-ikr}}{r} f_k(\Omega) \frac{e^{ikr}}{r}$$

$$\rho_{scatt} = \frac{|f_k(\Omega)|^2}{r^2} \tag{7.14}$$

In elastic scattering, event velocity of scattered particles $v_s$ and velocity of incident particles $v_i$ are the same, i.e.

$$v_s = v_i = \frac{\hbar k}{\mu} \tag{7.15}$$

The scattered flux along $(\theta, \phi)$ is defined as

$$J_s(\Omega) = \rho_{scatt} v_s = \frac{|f_k(\Omega)|^2}{r^2} \frac{\hbar k}{\mu} \tag{7.16}$$

So from equation (7.4) we get using equations (7.13) and (7.16)

$$\frac{d\sigma}{d\Omega} = \frac{r^2 J_s(\Omega)}{n J_{inc}} = r^2 \cdot \frac{|f_k(\Omega)|^2}{r^2} \frac{\hbar k}{\mu} \cdot \frac{1}{n \frac{\hbar k}{\mu}}$$

$$\frac{d\sigma}{d\Omega} = \frac{\left| f_k(\Omega) \right|^2}{n} \tag{7.17}$$

If the potential $V(\vec{r})$ responsible for scattering phenomenon corresponds to a single scattering centre then we take $n = 1$ and so

$$\frac{d\sigma}{d\Omega} = \left| f_k(\Omega) \right|^2 \tag{7.18}$$

The total scattering cross-section is

$$\sigma = \int \frac{d\sigma}{d\Omega} d\Omega = \int_0^\pi \int_0^{2\pi} \left| f_k(\Omega) \right|^2 \sin\theta d\theta d\phi = \int_\Omega \left| f_k(\Omega) \right|^2 d\Omega \tag{7.19}$$

It follows that if in the scattering experiment the scattering amplitude $f_k(\Omega)$ is evaluated, then differential scattering cross-section $\frac{d\sigma}{d\Omega} = \left| f_k(\Omega) \right|^2$ and the total scattering cross-section, namely $\sigma = \int_\Omega \left| f_k(\Omega) \right|^2 d\Omega$ are known.

We have discussed the quantum mechanical idea of scattering cross-section.

## 7.6 Neutron proton scattering at low energy

We wish to find out the scattering cross-section in case of neutron–proton scattering process. This scattering cross-section depends on incident particle energy.

We shall study neutron–proton scattering both at low projectile energy as well as high projectile energy.

In the case of low energy $n$–$p$ scattering the projectile neutron has energy $E < 10\ MeV$.

The experimental curve for low energy $n$–$p$ scattering is depicted in figure 7.5. The plot shows dependence of scattering cross-section $\sigma$ on projectile neutron energy $E$.

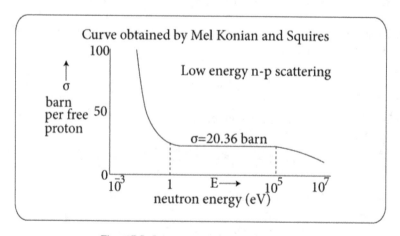

**Figure 7.5.** Low energy $n$–$p$ scattering curve.

It will be shown that for low projectile neutron energy $E < 10 \, MeV$ scattering is due to orbital angular momentum $l = 0$ ($s$ wave) that corresponds to spherically symmetric potential $V = V(r)$. In other words, neutrons suffer isotropic scattering.

The $n{-}p$ scattering is an elastic scattering that is studied in the centre of mass system for convenience.

The two-body problem of $np$ scattering can equivalently be treated using a single particle having mass

$$\mu = \frac{m_p m_n}{m_p + m_n} \xrightarrow{m_p \cong m_n = m(say)} \frac{m}{2} \text{ called the reduced mass of the } np \text{ system.}$$

We would use partial wave analysis to explain things.

We consider the projectile neutron to be incident on a proton which is the target. An actual target is comprised of a multitude of protons and they mutually interact through Coulomb force. Again, the energy that binds the protons to target material is small. In fact, the molecular binding energy of a proton in a molecule is $\sim 0.1 \, eV$. So to the incident projectile neutron having energy $\sim MeV$ target protons can be treated as good as free and Coulomb force is negligibly small. So we can consider that incident projectile neutron is being scattered by a free proton acting as target and Coulomb interaction avoided. (In the actual experiment the striking neutron should not thus have energy less than $1 \, eV$.)

Let us choose the square well potential as shown in figure 7.6.

$$V = \begin{cases} -V_0 \text{ for } r < r_0 \\ 0 \quad \text{ for } r > r_0 \end{cases} \tag{7.20}$$

as the spherically symmetric potential, $r_0$ being the range of the potential.

The only difference of this scattering problem from the deuteron problem is that in the scattering problem we are concerned with free incident particles having positive energy $E > 0$.

The time independent Schrödinger equation (7.8) with the spherically symmetric potential $V(r)$ is

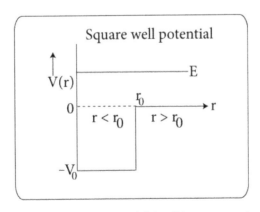

**Figure 7.6.** Square well potential describing $n{-}p$ scattering.

$$\nabla^2\psi(\vec{r}) + \frac{2\mu}{\hbar^2}[E - V(r)]\psi(\vec{r}) = 0$$

$$\nabla^2\psi(\vec{r}) + \frac{m}{\hbar^2}[E - V(r)]\psi(\vec{r}) = 0 \quad \text{(as } \mu = \frac{m}{2}) \tag{7.21}$$

The boundary conditions for this problem are the following:
  ✓ at $t \to -\infty$ (long before scattering) the incident wave should be plane wave, and
  ✓ at $t \to \infty$ (long after scattering) the scattered wave is spherical wave emerging from the scattering centre.

In other words, we will be concerned with asymptotic solutions.
The solution of the equation (7.21) is

$$\psi = \psi_{\text{inc}} + \psi_{\text{scatt}}$$

At $t \to -\infty$, $V(r) = 0$ (as there is no potential at $r \to \infty$), $\psi = \psi_{\text{inc}}$ as $\psi_{\text{scatt}} = 0$ (as no scattering occurs at $t \to -\infty$,).
From equation (7.21) we have

$$\nabla^2\psi(\vec{r}) + \frac{mE}{\hbar^2}\psi(\vec{r}) = 0$$

Define

$$k^2 = \frac{mE}{\hbar^2} \Rightarrow k = \pm\sqrt{\frac{mE}{\hbar^2}} \tag{7.22}$$

We thus have

$$\nabla^2\psi(\vec{r}) + k^2\psi(\vec{r}) = 0 \tag{7.23}$$

Supposing incident wave moves along $\hat{z}$ having wave vector $\vec{k} = k\hat{z}$ we have the plane wave solution of (7.23) to be

$$\psi_{\text{inc}} = e^{ikz} = e^{i\vec{k}.\vec{r}} = e^{ik\hat{z}.r\hat{r}} = e^{ikr\cos\theta} \tag{7.24}$$

written without normalization factor.
The probability density of incident beam is

$$|\psi_{\text{inc}}|^2 = \psi_{\text{inc}}^*\psi_{\text{inc}} = e^{-ikz}. e^{ikz} = 1 \tag{7.25}$$

i.e. wave function $\psi_{\text{inc}}$ represents one particle per unit volume.
This is the solution of equation (7.21) at $r = \infty$, $t = -\infty$ when the scattering potential has not been turned on.
After scattering by the potential the scattered wave is a spherical wave function written as

$$\psi_{\text{scatt}} = f(\theta)\frac{e^{ikr}}{r} \tag{7.26}$$

where $f(\theta)$ is the amplitude of the scattered wave in the $\theta$, $\phi$ direction. Actually, the scattering amplitude is

$$f(\Omega) = f(\theta, \phi) = f(\theta)$$

since potential is spherically symmetric.

Also, since $\phi$ does not contribute and only $r$, $\theta$ contributes, it is a 2D scattering.

- Probability of finding a scattered particle is

$$|\psi_{scatt}|^2 = \psi_{scatt}^* \psi_{scatt}$$

$$= f^*(\theta) \frac{e^{-ikr}}{r} f(\theta) \frac{e^{ikr}}{r} = \frac{|f(\theta)|^2}{r^2} \tag{7.27}$$

Let $N_{scatt}$ be the number of scattered particles. We note that the number of scattered particles found per unit volume within a spherical annulus between radii $r$ and $r + dr$ will be

$$\frac{N_{scatt}}{\text{annular volume}} = \frac{N_{scatt}}{4\pi r^2 dr} \propto \frac{1}{r^2}. \tag{7.28}$$

Clearly this wave function $\psi_{scatt}$ leads to a particle density that is inversely proportional to $r^2$.

Total wave function in the asymptotic region is

$$\psi = \psi_{inc} + \psi_{scatt}$$

$$= e^{ikz} + f(\theta) \frac{e^{ikr}}{r} \tag{7.29}$$

Let us expand the incident plane wave $e^{ikz} = e^{ikr\cos\theta}$ in terms of Legendre polynomial $P_l(\cos\theta)$ as follows, where $l$ is the integer representing various partial waves

$$e^{ikz} = e^{ikr\cos\theta} = \sum_{l=0}^{\infty} B_l(r) P_l(\cos\theta) \tag{7.30}$$

where

$B_l(r)$ is the radial wave function defined in terms of spherical Bessel function $j_l(kr)$ as

$$B_l(r) = i^l (2l + 1) j_l(kr) \tag{7.31}$$

And the definition of spherical Bessel function is

$$j_l(kr) = (-)^l (kr)^l \left( \frac{1}{kr} \frac{d}{d(kr)} \right)^l \left( \frac{\sin kr}{kr} \right) \tag{7.32}$$

Let us find the value of $j_l(kr)$ for $l = 0, 1$

$$j_0(kr) = (-)^0(kr)^0\left(\frac{1}{kr}\frac{d}{d(kr)}\right)^0\left(\frac{\sin kr}{kr}\right) = \frac{\sin kr}{kr} \tag{7.33}$$

$$j_1(kr) = (-)^1(kr)^1\left(\frac{1}{kr}\frac{d}{d(kr)}\right)^1\left(\frac{\sin kr}{kr}\right) = -kr\left[\frac{1}{kr}\frac{d}{d(kr)}\right]\left(\frac{\sin kr}{kr}\right)$$

$$= -\frac{d}{d(kr)}\left(\frac{\sin kr}{kr}\right) = -\left[-\frac{\sin kr}{(kr)^2} + \frac{\cos kr}{kr}\right]$$

$$j_1(kr) = \frac{\sin kr}{(kr)^2} - \frac{\cos kr}{kr} \tag{7.34}$$

The asymptotic value (i.e. value at $r = \infty$) of spherical Bessel function would be

$$j_l(kr) \xrightarrow{r \to \infty} \frac{\sin\left(kr - \frac{l\pi}{2}\right)}{kr} \tag{7.35}$$

From equation (7.31) we get the value of $B_l$ for $r \to \infty$ to be using equation (7.35)

$$B_l(r) = i^l(2l + 1)j_l(kr) \xrightarrow{r \to \infty} i^l(2l + 1)\frac{\sin\left(kr - \frac{l\pi}{2}\right)}{kr}$$

$$= i^l\frac{2l + 1}{kr}\left[\frac{e^{i\left(kr-\frac{l\pi}{2}\right)} - e^{-i\left(kr-\frac{l\pi}{2}\right)}}{2i}\right] \tag{7.36}$$

We can similarly expand the scattering amplitude $f(\theta)$ in terms of the Legendre polynomial as

$$f(\theta) = \sum_{l=0}^{\infty} f_l P_l(\cos\theta) \tag{7.37}$$

So from equation (7.29) we have using equations (7.30) and (7.37)

$$\psi = \psi_{\text{inc}} + \psi_{\text{scatt}}$$

$$= e^{ikz} + f(\theta)\frac{e^{ikr}}{r}$$

$$= \sum_{l=0}^{\infty} B_l(r)P_l(\cos\theta) + \sum_{l=0}^{\infty} f_l P_l(\cos\theta)\frac{e^{ikr}}{r} \tag{7.38}$$

Using equation (7.31)

$$\psi = \psi_{\text{inc}} + \psi_{\text{scatt}}$$

$$= \sum_{l=0}^{\infty}\left[i^l(2l + 1)j_l(kr) + f_l\frac{e^{ikr}}{r}\right]P_l(\cos\theta) \tag{7.39}$$

The RHS of equation (7.39) involves summation over $l = 0$ to $\infty$. Each term in the sum corresponds to a particular orbital angular momentum quantum number $l$ and represents the solution of the Schrödinger equation (7.21) in spherical polar coordinates with a constant potential.

The sum over $l$, i.e. the expansion classifies the paritcles of the beam according to their angular momentum $l$ called partial waves. The sum indicates the number of partial waves superposed.

- For low energy $E < 10\ MeV$ scattering by $l = 0$ will predoiminate as we investigate in the following.

We estimate the kinetic energy of a nucleon projectile for $l = 0$ interaction in *exercise 7.6.*

Let us calculate $B_l$ from equation (7.31) viz. $B_l(r) = i^l(2l + 1)j_l(kr)$ for $l = 0, 1$.

$$B_0(r) = i^0(2.0 + 1)j_0(kr) = j_0(kr)$$

Using equation (7.33) we have

$$B_0(r) = j_0(kr) = \frac{\sin kr}{kr} \tag{7.40}$$

Expanding $\sin kr$ we have

$$\sin kr \cong kr - \frac{1}{6}(kr)^3 + \ldots$$

Hence

$$B_0(r) = \frac{\sin kr}{kr} = \frac{kr - \frac{1}{6}(kr)^3 + \ldots}{kr} \cong 1 - \frac{(kr)^2}{6}$$

$$B_0(r) \cong 1 \tag{7.41}$$

Again

$$B_1(r) = i^1(2.1 + 1)j_1(kr) = 3ij_1(kr) \tag{7.42}$$

Using equation (7.34) we have

$$B_1(r) = 3ij_1(kr) = 3i\left(\frac{\sin kr}{(kr)^2} - \frac{\cos kr}{kr}\right) \tag{7.43}$$

Expanding

$$\sin kr \cong kr - \frac{1}{6}(kr)^3 + \ldots \text{ and } \cos kr \cong 1 - \frac{1}{2}(kr)^2 - \ldots$$

we get

$$B_1(r) = 3i\left(\frac{\sin kr}{(kr)^2} - \frac{\cos kr}{kr}\right) = 3i\left[\frac{kr - \frac{1}{6}(kr)^3 + \cdots}{(kr)^2} - \frac{1 - \frac{1}{2}(kr)^2 - \cdots}{kr}\right]$$

$$B_1(r) \cong 3i\left[\frac{1}{kr} - \frac{kr}{6} - \frac{1}{kr} + \frac{kr}{2}\right] = 3i\frac{kr}{3}$$

$$B_1(r) \cong ikr \qquad (7.44)$$

Since probability density involves $|B_l(r)|^2$ let us take ratio of squares of $B_0$ and $B_1$. Hence from equations (7.41) and (7.44) we get

$$\left|\frac{B_1(r)}{B_0(r)}\right|^2 = (kr)^2 \qquad (7.45)$$

Suppose the kinetic energy of the striking neutron of mass $m$ is $E = 1\ MeV$ in lab system. So kinetic energy of the neutron in centre of mass system where things are described in terms of reduced mass $\mu = \frac{m}{2}$ will be $\frac{E}{2} = \frac{1\ MeV}{2} = 0.5\ MeV$.

In centre of mass system the magnitude of momentum of neutron will be

$$p = \sqrt{2m\frac{E}{2}} = \sqrt{mE} = \sqrt{(1.675 \times 10^{-27}\ kg)(1\ MeV)}$$

$$= \sqrt{(1.675 \times 10^{-27}\ kg)(10^6 \times 1.6 \times 10^{-19}\ J)}$$

$$p = 1.6 \times 10^{-20}\ kg.\ ms^{-1}.$$

The corresponding wave number is

$$k = \frac{p}{\hbar} = \frac{1.6 \times 10^{-20}\ kg.\ ms^{-1}}{\frac{1}{2\pi}6.626 \times 10^{-34}\ J.\ s} = 1.5 \times 10^{14}\ m^{-1}.$$

Taking range of nuclear force $r_0 \sim 2\ fm = 2 \times 10^{-15}\ m$ gives from equation (7.45)

$$\left|\frac{B_1(r)}{B_0(r)}\right|^2 = (kr)^2 \sim (kr_0)^2 = (1.5 \times 10^{14}\ m^{-1} \times 2 \times 10^{-15}\ m)^2$$

$$\left|\frac{B_1(r)}{B_0(r)}\right|^2 = 0.09$$

$$\frac{B_1}{B_0} = \sqrt{0.09} = 0.3 = \frac{0.3}{1}$$

$$B_0: B_1 = 1: 0.3$$

It is clear that $B_1(r) \ll B_0(r)$ implying that $l = 0$ term contributes predominantly.

Contrbution of $l = 1$ term in the scattering of $n$ by $p$ at neutron energy of $1\ MeV$ is only around 1% while 99% scattering is by $l = 0$, i.e. $S$ wave term. Clearly $l = 0$ scattering dominates in the low energy region.

If we take the kinetic energy of the striking neutron of mass $m$ to be $E = 10\ MeV$ in lab system then kinetic energy of the neutron in centre of mass system will be $\frac{E}{2} = \frac{10\ MeV}{2} = 5\ MeV$.

In centre of mass system the magnitude of momentum of neutron will be

$$p = \sqrt{2m\frac{E}{2}} = \sqrt{mE} = \sqrt{(1.675 \times 10^{-27}\ kg)(10\ MeV)}$$

$$= \sqrt{(1.675 \times 10^{-27}\ kg)(10 \times 10^6 \times 1.6 \times 10^{-19}\ J)}$$

$$= 5.2 \times 10^{-20}\ kg.\ ms^{-1},$$

$$k = \frac{p}{\hbar} = \frac{5.2 \times 10^{-20}\ kg.\ ms^{-1}}{\frac{1}{2\pi}6.626 \times 10^{-34}\ J.\ s} = 4.9 \times 10^{14}\ m^{-1}$$

Hence

$$\left|\frac{B_1(r)}{B_0(r)}\right|^2 \sim (kr_0)^2 = (4.9 \times 10^{14}\ m^{-1} \times 2 \times 10^{-15}\ m)^2.$$

$$\left|\frac{B_1(r)}{B_0(r)}\right|^2 = 0.96$$

$$\frac{B_1}{B_0} = \sqrt{0.96} = 0.98 = \frac{0.98}{1}$$

$$B_0: B_1 = 1: 0.98$$

This means for neutron energy of 10 $MeV$ the $l = 1$ scattering increases significantly to 96%. Clearly $l = 0$ and $l = 1$ scatterings become comparable. Obviously, as neutron energy increases, the probability of $l = 1$ scattering increases.

So for $S$ wave scattering, incident neutron energy should be much less than 10 $MeV$—it should be around $\sim 1\ MeV$.

We have seen that for low energy $n$–$p$ scattering the first partial wave $l = 0$ plays a dominant role.

$\psi_{\text{inc}}$ from equation (7.38) is

$$\psi_{\text{inc}} = \sum_{l=0}^{\infty} B_l(r)P_l(\cos\theta).$$

For low energy scattering $l = 0$ and so we have

$$\psi_{\text{inc}}^{l=0} = B_0(r)P_0(\cos\theta)$$

Using equation (7.40) for $B_0(r) = \frac{\sin kr}{kr}$ and $P_0(\cos \theta) = 1$ we write

$$\psi_{\text{inc}}^{l=0} = \frac{\sin kr}{kr} \cdot 1 = \frac{\sin kr}{kr} = \frac{1}{kr} \frac{e^{ikr} - e^{-ikr}}{2i}$$

$$\psi_{\text{inc}}^{l=0} = \frac{e^{ikr} - e^{-ikr}}{2ikr} \tag{7.46}$$

The time evolution of wave function $\psi_{\text{inc}}^{l=0}$ at a later time $t$ is obtained by multiplying with $e^{-iwt}$ as follows

$$\psi_{\text{inc}}^{l=0}(r, t) = e^{-iwt} \frac{e^{ikr} - e^{-ikr}}{2ikr} = \frac{1}{2ikr} [e^{i(kr-wt)} - e^{-i(kr+wt)}] \tag{7.47}$$

As $e^{-ikr}$ represents radially incoming spherical waves (moving towards the scattering centre) and $e^{ikr}$ represents radially outgoing spherical waves (moving away from the scattering centre) we can state that for $l = 0$ low energy incident $S$-wave is a combination of two spherical waves. The minus sign between the two terms keeps $\psi_{\text{inc}}^{l=0}(r, t)$ finite. This part of the solution corresponds to $V = 0$.

Let us now switch on the scattering potential $V(r)$ to calculate the scattering cross-section.

The outgoing part of the wave function is to be affected by the potential $V(r)$.

The number of particles emerging out after scattering should be the same as the number of particles moving in—since it is an elastic scattering and there is no absorption or emission of particles. In other words, the amplitude of the outgoing wave should remain unchanged. The effect of the scattering potential will be to change the phase of the wave and there is no effect on the amplitude of the wave.

Let the phase change suffered by the outgoing wave $e^{ikr}$ for $l = 0$ due to scattering potential be $2\delta_0$ after scattering (the factor 2 is taken for future convenience).

For $l = 0$ we had, from equation (7.46), for $V = 0$, $\psi_{\text{inc}}^{l=0} = \frac{e^{ikr} - e^{-ikr}}{2ikr}$

and now when $V$ is switched on we have scattering, the effect of which is taken care of by adding a phase shift $2\delta_0$ to the outgoing wave part $e^{ikr}$.

Scattering does not create or destroy particles. Scattering cannot change the amplitudes of $e^{ikr}$ or $e^{-ikr}$ since the square of amplitudes give the probability to detect incoming or outgoing particles. The sole effect of scattering is a change in phase of the outgoing wave. So for $l = 0$ a part of the plane wave can be written as

$$\psi_0 = \frac{e^{i(kr+2\delta_0)} - e^{-ikr}}{2ikr} \tag{7.48}$$

$$= \frac{e^{i\delta_0}}{kr} \frac{e^{i(kr+\delta_0)} - e^{-i(kr+\delta_0)}}{2i}$$

$$\psi_0 = e^{e^{i\delta_0}} \frac{\sin (kr + \delta_0)}{kr} \tag{7.49}$$

This is the wave function outside the range of the nuclear force.

This $\psi_0$ consists of two parts.

One part is the incident wave function and the other part is the scattered wave function—the expression of which we are trying to obtain. So we write

$$\psi_0 = \psi_{\text{inc}}^{l=0} + \psi_{\text{scatt}}^{l=0}$$

$$\psi_{\text{scatt}}^{l=0} = \psi_0 - \psi_{\text{inc}}^{l=0}$$

Substituting their values from equations (7.46) and (7.48) we have

$$\psi_{\text{scatt}}^{l=0} = \frac{e^{i(kr+2\delta_0)} - e^{-ikr}}{2ikr} - \frac{e^{ikr} - e^{-ikr}}{2ikr}$$

$$= \frac{e^{i(kr+2\delta_0)} - e^{ikr}}{2ikr}$$

$$\psi_{\text{scatt}}^{l=0} = \frac{e^{ikr}}{2ikr}(e^{i2\delta_0} - 1) \tag{7.50}$$

We note the following

✓ Expression for $\psi_{\text{scatt}}^{l=0}$ contains $e^{ikr}$ only and hence represents outgoing wave.

✓ Also, if we switch off potential term, i.e. if $V(r) = 0$ then there is no interaction and hence no phase shift, i.e. $2\delta_0 = 0$ and so

$$\psi_{\text{scatt}}^{l=0} = \frac{e^{ikr}}{2ikr}(e^{i2\delta_0} - 1) = \frac{e^{ikr}}{2ikr}(e^{i(0)} - 1) = 0$$

implying no outgoing wave and hence no scattering.

Again, the scattered wave function is expressed in terms of scattering amplitude as [equation (7.26)]

$$\psi_{\text{scatt}} = f(\theta)\frac{e^{ikr}}{r} \tag{7.51}$$

Comparison of equations (7.50) and (7.51) gives for $l = 0$ the scattering amplitude to be

$$f(\theta)_{l=0} = \frac{e^{i2\delta_0} - 1}{2ik}$$

$$= \frac{e^{i\delta_0}}{k}\frac{e^{i\delta_0} - e^{-i\delta_0}}{2i}$$

$$f(\theta)_{l=0} = \frac{e^{i\delta_0}}{k}\sin \delta_0 \tag{7.52}$$

The differential scattering cross-section will be for $S$ wave, $l = 0$

$$\frac{d\sigma}{d\Omega} = |f(\theta)|^2 = f^*(\theta)_{l=0} f(\theta)_{l=0}$$

$$= \frac{e^{-i\delta_0}}{k} \sin\delta_0 \cdot \frac{e^{i\delta_0}}{k} \sin\delta_0$$

$$\frac{d\sigma}{d\Omega} = \frac{\sin^2\delta_0}{k^2} \tag{7.53}$$

The total scattering cross-section for $l = 0$ will be

$$\sigma_{l=0} = \int \frac{d\sigma}{d\Omega} d\Omega = \int \frac{\sin^2\delta_0}{k^2} \sin\theta d\theta d\phi$$

$$= \frac{\sin^2\delta_0}{k^2} \int_0^\pi \sin\theta d\theta \int_0^{2\pi} d\phi = \frac{\sin^2\delta_0}{k^2} \cdot 2.2\pi$$

$$\sigma_{l=0} = \frac{4\pi \sin^2\delta_0}{k^2} \tag{7.54}$$

Using $\lambdabar = \frac{\lambda}{2\pi} = \frac{1}{k}$ we have

$$\sigma_{l=0} = 4\pi\lambdabar^2 \sin^2\delta_0 \tag{7.55}$$

We note that $\frac{d\sigma}{d\Omega} = \frac{\sin^2\delta_0}{k^2}$ [equation (7.53)] and $\sigma_{l=0} = 4\pi\lambdabar^2 \sin^2\delta_0$ [equation (7.55)] are independent of $\theta$. They depend on phase shift $\delta_0$ for $l = 0$. So the scattering is isotropic or spherically symmetric. Intensity of scattered beam measured by detector is same for various $\theta$, $\phi$ for same $r$, i.e. if distance from scattering centre is fixed the intensity will be same even if angular position of detector is varied arbitrarily.
- We have discussed low energy scattering (with projectile neutron energy $E < 10\ MeV$) for $S$ wave or $l = 0$ which predominates at such low energies say 1 $MeV$, 2 $MeV$...).

But if neutron beam strikes proton with higher energy (but less than 10 $MeV$) then contribution from higher $l$, say $l = 1, 2, \ldots$ will become increasingly significant.

We then have to take into account the contribution from all the partial waves. Obviously we have to sum over the partial waves. Let us mention the differential scattering cross-section if a neutron beam strikes proton with higher energy.

$$\frac{d\sigma}{d\Omega} = \frac{1}{k^2} \left| \sum_{l=0}^{\infty} (2l+1)\, e^{i\delta_l} \sin\delta_l P_l(\cos\theta) \right|^2 \tag{7.56}$$

$$\xrightarrow{l=0} \frac{\sin^2\delta_0}{k^2} = \frac{d\sigma}{d\Omega}\bigg|_{l=0} \quad \text{[Equation (7.53)]}$$

and the total scattering cross-section will be

$$\sigma = \frac{4\pi}{k^2} \sum_{l=0}^{\infty} (2l + 1) \sin^2 \delta_l \tag{7.57}$$

$$\xrightarrow[l=0]{} \frac{4\pi \sin^2 \delta_0}{k^2} = \sigma_{l=0} \text{ [equation (7.54)]}$$

Here $k^2$ represents energy, $2\delta_l$ is the phase shift and $\delta_l$ is half phase shift. If $\delta_l$ is known then $\frac{d\sigma}{d\Omega}$ can be found out from equation (7.56) and $\sigma$ can be found out from equation (7.57).

If $\delta_l = (2n + 1)\frac{\pi}{2}$ with $n = 0, \pm1, \pm2, \ldots$ then $\sigma$ is maximum.

✓ We note that expression of $\sigma$ and $\frac{d\sigma}{d\Omega}$ obtained for low energy scattering does not involve any parameter describing nuclear force or potential and hence fails to provide much detail of nuclear force.

This can be explained from the de Broglie wavelength of low energy neutrons given by

$$\lambda = \frac{h}{\sqrt{2mE}}$$

If energy of neutron $E$ is small then $\lambda$ is large and if $\lambda > r_0$, which is the range of nuclear force, then the low energy neutron beam will fail to probe the nucleus. Thus in this case for low energy scattering, the scattering cross-section obtained does not reveal much information regarding scattering potential.

## 7.7 Scattering length $a$

The wave function outside the range of the nuclear force is from equation (7.49)

$$\psi_0 = e^{i\delta_0} \frac{\sin(kr + \delta_0)}{kr}$$

$$u(r) = r\psi_0 = re^{i\delta_0} \frac{\sin(kr + \delta_0)}{kr} = e^{i\delta_0} \frac{\sin(kr + \delta_0)}{k} \tag{7.58}$$

where from equation (7.22) we have $k = \sqrt{\frac{mE}{\hbar^2}}$.

The dimension of $k$ is

$$k \sim \sqrt{\frac{M \cdot J}{(J \cdot s)^2}} = \sqrt{\frac{M}{J \cdot s^2}} = \sqrt{\frac{M}{(MLT^{-2} \cdot L)T^2}} = \sqrt{\frac{1}{L^2}} = \frac{1}{L} \text{ i.e. } k \text{ has dimension of length}$$

inverse.

Clearly for $E \to 0$, $k \to 0$.

Scattering length $a$ is defined as

$$a = \mathop{Lt}_{k\to 0} - \frac{\sin \delta_0}{k} \tag{7.59}$$

Clearly $a$ has dimension of length as $a \sim \frac{1}{k} = \frac{1}{1/L} = L \rightarrow$ length.

In the limiting condition $E \rightarrow 0$, $k \rightarrow 0$ the total scattering cross-section is from equation (7.54)

$$\underset{k \rightarrow 0}{Lt}\, \sigma_{scatt} = \underset{k \rightarrow 0}{Lt}\, \frac{4\pi \sin^2 \delta_0}{k^2} = 4\pi \left[ \underset{k \rightarrow 0}{Lt}\, - \frac{\sin \delta_0}{k} \right]^2 = 4\pi a^2 \tag{7.60}$$

So in the limiting condition when $k \rightarrow 0$, the total scattering cross-section $\sigma_{scatt}$ is $4\pi a^2$ which is the surface area of a sphere of radius $a$. This means that in the limit $k \rightarrow 0$, $E \rightarrow 0$ scattering is taking place from a hard impenetrable sphere of radius $a$ around the scattering centre.

This quantum mechanical result that cross-section is a surface area

$$\sigma_{scatt} = 4\pi a^2$$

is in contrast to the classical scattering cross-section

$$\sigma_{classical} = \pi a^2$$

which is the area of a disc of radius $a$.

This implies that diffraction taking place at the boundaries gives rise to a large amount of small angle scattering.

Let us rewrite equation (7.54) as

$$\sigma_{scatt} = \frac{4\pi}{k^2} \sin^2 \delta_0$$

If $k \rightarrow 0$ and $\delta_0 \neq 0$ or $\pi$ then $\sin \delta_0 \neq 0$ then

$$\sigma_{scatt} = \frac{4\pi \sin^2 \delta_0}{0} = \infty.$$

This result is unacceptable since this implies the unphysical phenomena that scattering takes place from everewhere. This suggests that if $k \rightarrow 0$ then $\delta_0$ has to be 0 or $\pi$ so that $\sin \delta_0 = 0$.

Consider the situation for $k \rightarrow 0$, $\delta_0 \rightarrow 0$ then the expression of scattering length becomes

$$a = \underset{k \rightarrow 0}{Lt}\, - \frac{\sin \delta_0}{k} = -\frac{\delta_0}{k} \tag{7.61}$$

In this limiting condition of very low energy, i.e. $k \rightarrow 0$ the wave function in the region $r > r_0$ (i.e. outside the square well of figure 7.6) is [using equation (7.58)]

$$\underset{k \rightarrow 0}{Lt}\, u(r) = \underset{k \rightarrow 0}{Lt}\, (r\psi_0) = \underset{k \rightarrow 0}{Lt}\, e^{i\delta_0} \frac{\sin(kr + \delta_0)}{k} \tag{7.62}$$

If we take $\delta_0 \rightarrow 0$ then from equation (7.62) we get

$$\underset{\substack{k \rightarrow 0 \\ \delta_0 \rightarrow 0}}{Lt}\, u(r) = \underset{\substack{k \rightarrow 0 \\ \delta_0 \rightarrow 0}}{Lt}\, e^{i\delta_0} \frac{\sin(kr + \delta_0)}{k} = \underset{\delta_0 \rightarrow 0}{Lt}\, e^{i\delta_0} \underset{\substack{k \rightarrow 0 \\ \delta_0 \rightarrow 0}}{Lt}\, \frac{\sin(kr + \delta_0)}{k}$$

$$= \underset{\substack{k \to 0 \\ \delta_0 \to 0}}{Lt} \frac{\sin (kr + \delta_0)}{k} = \frac{kr + \delta_0}{k}.$$

$$\underset{k \to 0}{Lt}\, u(r) = r + \frac{\delta_0}{k}$$

Using equation (7.61), i.e. $a = -\frac{\delta_0}{k}$ we have

$$\underset{k \to 0}{Lt}\, u(r) = r + \frac{\delta_0}{k} = r - \left(-\frac{\delta_0}{k}\right) = r - a \qquad (7.63)$$

If we take $\delta_0 \to \pi$ then we have from equation (7.62)

$$\underset{\substack{k \to 0 \\ \delta_0 \to \pi}}{Lt}\, u(r) = \underset{\substack{k \to 0 \\ \delta_0 \to \pi}}{Lt}\, e^{i\delta_0}\frac{\sin (kr + \delta_0)}{k} = \underset{\delta_0 \to \pi}{Lt}\, e^{i\delta_0}\, \underset{\substack{k \to 0 \\ \delta_0 \to \pi}}{Lt}\, \frac{\sin (kr + \delta_0)}{k}$$

$$= \underset{\substack{\| \\ -1}}{e^{i\pi}}\frac{kr + \delta_0}{k} = -\frac{kr + \delta_0}{k} = -r - \frac{\delta_0}{k}$$

Using equation (7.61), i.e. $a = -\frac{\delta_0}{k}$ we have

$$\underset{k \to 0}{Lt}\, u(r) = -r - \frac{\delta_0}{k} = -r + a \qquad (7.64)$$

Combining the two results of equations (7.63) and (7.64) we write

$$\underset{k \to 0}{Lt}\, u(r) = \pm r \mp a = \pm(r - a) \qquad (7.65)$$

We write

$$u = \pm(r - a)$$

It follows that the plot of $u(r)$ versus $r$ outside the well, i.e. in the region $r > r_0$ is a straight line.

Inside the square well $r < r_0$ the wave function is sinusoidal or oscillatory. We have plotted $u$ versus $r$ in figure 7.7(b),(c).

The intercept of the straight line on the $r$ axis is defined as the scattering length. The intercept can be positive (figure 7.7(b)) or negative (figure 7.7(c)). So we are able to give a geometrical interpretation of scattering length.

The physical significance of positive and negative scattering length is explained in the following.

- In the case of the bound state of an $n$–$p$ system, i.e. for deuteron bound state (figure 7.7(a)) the internal sinusoidal wave function attains a maximum at point $p$ and then turns over near the edge $r = r_0$ to become a decreasing wave function in the portion $pe$. At point $e$ it matches with an exponentially decaying wave function as $u(r) = Ce^{-\beta r}$ outside the square well $r > r_0$ where

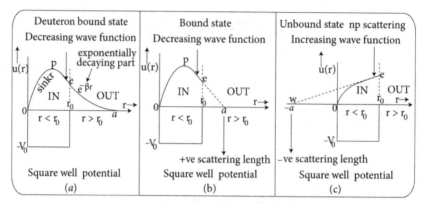

**Figure 7.7.** (a) $u(r)$ versus $r$ plot showing deuteron bound state. (b) $u(r)$ versus $r$ plot for bound state having positive scattering length. (c) $u(r)$ versus $r$ plot for unbound state ($np$ scattering) having negative scattering length.

$\beta = \sqrt{\dfrac{m|E|}{\hbar^2}}$ = positive constant. For a bound state we cannot expect wave function to relax too much outside the well and we see that the exponentially decaying part of the wave function in the region $r > r_0$ soon reduces to zero at $r = a$ (say).

Let us now discuss figure 7.7(b),(c) with reference to figure 7.7(a).
 ✓ +ve scattering length indicates a bound state (figure 7.7(b))
    In figure 7.7(b) we have a situation similar to what we got in figure 7.7(a). To represent a bound state the curve representing wave function should be a decreasing function $pe$ such that the extrapolation $ea$ of the wave function outside the nuclear range ($r > r_0$) intersects the $r$ axis at a point. This positive intercept on $r$ axis is called the positive scattering length. Hence positive scattering length represents a bound state. In fact the wave function has to die down for it to represent a bound state. This is the significance of positive scattering length $a$.
    So for positive scattering length wave function for the inner region can be matched to the exponentially decreasing function in the outer region so as to create a bound state.
 ✓ −ve scattering length indicates an unbound state (figure 7.7(c)). Bound state is impossible.

In unbound state ($n$–$p$ scattering problem) the wave function should not decrease at the edge $r = r_0$ because then the extrapolation would give a positive intercept on the $r$ axis leading to a bound state. The wave function should increase at the edge $r = r_0$ such that an extrapolation $ew$ would lead to a negative intercept $-a$ on the $r$ axis. This is called negative scattering length $-a$. This means that the wave function will not die down at $r > r_0$ and will correspond to an unbound state.

So for negative scattering length the wave function in the inner region behaves in such a way that it is impossible to match it with any exponentially decreasing function in the outer region, i.e. a bound state of the system is impossible in this case.

- We have brought in the concept of scattering length for $k \to 0$, $E \to 0$. We can extend the concept of scattering length to situations other than $E = 0$.

  For a general $k$ (zero or otherwise) the scattering length is a function of $k$, i.e. energy of striking neutron and is defined as

$$a(k) = -\frac{\tan \delta_0}{k} \qquad (7.66)$$

This definition is also valid for $k \to 0$ since then

$$a(0) = \underset{k \to 0}{Lt} -\frac{\tan \delta_0}{k}$$

Again when $k \to 0$, $\delta_0 \to 0$ we can write

$\tan \delta_0 \cong \delta_0$.

Thus

$$a(0) = \underset{k \to 0}{Lt} -\frac{\tan \delta_0}{k} = -\frac{\delta_0}{k}$$

as defined in equation (7.61) previously. So we have successfully defined scattering length for all $k$.

- The total scattering cross-section for all $k$ can be calculated from its value for $l = 0$ which is given by equation (7.54) to be

$$\sigma_{\text{scatt}} = \frac{4\pi \sin^2 \delta_0}{k^2} = \frac{4\pi}{k^2} \frac{1}{\text{cosec}^2 \delta_0}$$

$$= \frac{4\pi}{k^2} \frac{1}{1 + \cot^2 \delta_0} = \frac{4\pi}{k^2} \frac{1}{1 + \frac{1}{\tan^2 \delta_0}}$$

Using

$$a(k) = -\frac{\tan \delta_0}{k}$$

$$\tan \delta_0 = -ka(k)$$

we have

$$\sigma_{\text{scatt}} = \frac{4\pi}{k^2} \frac{1}{1 + \frac{1}{k^2 a^2}} = \frac{4\pi}{k^2 + \frac{1}{a^2}}$$

$$\sigma_{\text{scatt}} = \frac{4\pi}{k^2 + \frac{1}{a^2}} \qquad (7.67)$$

Clearly the total cross-section is energy dependent, i.e. $k$ dependent. If we know scattering length $a = a(k)$ we can find the scattering cross-section $\sigma_{scatt}$ from equation (7.67).

Let us check if the result we obtained [equation (7.67)] can reproduce the result for $k \to 0$.

For $k \to 0$ we have

$$\underset{k \to 0}{Lt}\, \sigma_{scatt} = \frac{4\pi}{k^2 + \frac{1}{a^2}} \xrightarrow{r \to 0} \frac{4\pi}{\frac{1}{a^2}} = 4\pi a^2$$

a result we derived previously in equation (7.60).

Knowledge of a single parameter namely the scattering length $a(k)$ is sufficient to describe $s$-wave scattering, i.e. low energy $n$–$p$ scattering.

## 7.8 Wigner's explanation of scattering cross-section for low energy $np$ scattering through spin dependence of nuclear force

Experimental value of scattering cross-section in the case of low energy $n$–$p$ scattering for $l = 0$ state is $\sigma_{scatt}^{l=0} = 20\ barn$. This has been shown in figure 7.5. However, the theoretical estimate is 2.3 $barn$. So there is a large discrepancy between the theoretical and experimental values.

Wigner explained this discrepancy by considering that nuclear force is spin dependent.

The radial part of the time independent Schrödinger equation for $S$ state ($l = 0$) of a nucleon–nucleon system is

$$\frac{1}{r^2} \frac{d}{dr} \left[ r^2 \frac{d}{dr} \psi(r) \right] + \frac{2\mu}{\hbar^2} [E - V(r)] \psi(r) = 0$$

No centrifugal term is here as $l = 0$.

$$\frac{d^2}{dr^2} \psi(r) + \frac{2}{r} \frac{d}{dr} \psi(r) + \frac{2\mu}{\hbar^2} [E - V(r)] \psi(r) = 0 \ [exercise\ 6.14] \qquad (7.68)$$

where reduced mass of the $n$–$p$ system is $\mu = \frac{m_n m_p}{m_n + m_p} = \frac{m_p}{2} = \frac{m}{2}$ where $m_p = m_n = m$

The solution of equation (7.68) can be greatly simplified through the substitution $\psi(r) = \frac{u(r)}{r}$ as shown in *exercise 6.5*. We thus get

$$\frac{d^2}{dr^2} u(r) + \frac{m}{\hbar^2} [E - V(r)] u(r) = 0 \qquad (7.69)$$

We now consider a square well potential given by equation (7.20) and shown in figure 7.6.

$$V(r) = \begin{cases} - V_0 \text{ for } r < r_0 \\ 0 \quad \text{ for } r > r_0 \end{cases}$$

Unlike the deuteron problem here the energy $E$ is positive. Also $E < 10\ MeV$, small compared to $V_0$.

Let $u_1$ be inside wave function, $u_2$ be outside wave function.

In the region $r < r_0$, $u = u_1$, $V = -V_0$.

The Schrödinger equation (7.69) becomes

$$\frac{d^2}{dr^2} u_1(r) + \frac{m}{\hbar^2}(E + V_0)u_1(r) = 0$$

$$\frac{d^2}{dr^2} u_1(r) + k^2 u_1(r) = 0 \tag{7.70}$$

$$\text{with } k^2 = \frac{m}{\hbar^2}(E + V_0) \tag{7.71}$$

In the region $r > r_0$, $u = u_2$, $V = 0$.

The Schrödinger equation (7.69) becomes

$$\frac{d^2}{dr^2} u_2(r) + \frac{m}{\hbar^2} E u_2(r) = 0$$

$$\frac{d^2}{dr^2} u_2(r) + \gamma^2 u_2(r) = 0 \tag{7.72}$$

with

$$\gamma^2 = \frac{m}{\hbar^2} E \tag{7.73}$$

Let the solution of the equations be

$$u_1 = A \sin kr, u_2 = B \sin(\gamma r + \delta_0) \tag{7.74}$$

where we have introduced the $l = 0$ phase shift $\delta_0$ for $n$–$p$ scattering.

Boundary conditions or continuity conditions satisfied are

✓ Continuity of wave function

$$u_1]_{r=r_0} = u_2]_{r=r_0}$$

Using equation (7.74) we get

$$A \sin kr]_{r=r_0} = B \sin(\gamma r + \delta_0)]_{r=r_0}$$

$$A \sin kr_0 = B \sin(\gamma r_0 + \delta_0) \tag{7.75}$$

✓ Continuity of derivative of wave function

$$\frac{du_1}{dr}\bigg]_{r=r_0} = \frac{du_2}{dr}\bigg]_{r=r_0}$$

Using equation (7.74) we get

$$\frac{d}{dr}A \sin kr]_{r=r_0} = \frac{d}{dr}B \sin(\gamma r + \delta_0)]_{r=r_0}$$

$$-Ak \cos kr]_{r=r_0} = B\gamma \cos(\gamma r + \delta_0)]_{r=r_0}$$

$$Ak \cos kr_0 = B\gamma \cos(\gamma r_0 + \delta_0) \qquad (7.76)$$

Taking the ratio of equations (7.75) and (7.76) we get

$$\frac{Ak \cos kr_0}{A \sin kr_0} = \frac{B\gamma \cos(\gamma r_0 + \delta_0)}{B \sin(\gamma r_0 + \delta_0)}$$

$$k \cot kr_0 = \gamma \cot(\gamma r_0 + \delta_0) \qquad (7.77)$$

This result equation (7.77) may be compared with the continuity condition given by equation (6.39) viz. $k \cot kr_0 = -\beta$ where $\beta = \sqrt{\frac{m}{\hbar^2}|E|}$ [equation (6.29)] for the ground state of the deuteron problem. Accordingly we write equation (7.77) as

$$\gamma \cot(\gamma r_0 + \delta_0) = -\beta$$

$$\tan(\gamma r_0 + \delta_0) = -\frac{\gamma}{\beta} \qquad (7.78)$$

In the experiment neutron energy is very small $\sim$ few $keV$. So $\gamma r_0$, $\delta_0$ are very small, i.e.

$$\gamma r_0 \ll 1, \delta_0 \ll 1 \text{ and so}$$
$$k^2 \ll \beta^2$$

Then

$$\gamma r_0 + \delta_0 \rightarrow \text{small.}$$

Hence

$$\tan(\gamma r_0 + \delta_0) \cong \gamma r_0 + \delta_0$$

and so equation (7.78) suggests that

$$\gamma r_0 + \delta_0 = -\frac{\gamma}{\beta}$$

$$\delta_0 \cong -\frac{\gamma}{\beta} - \gamma r_0 \qquad (7.79)$$

We can find total scattering cross-section at low energy $np$ scattering for $l = 0$ using equation (7.57) and equations (7.22) and (7.73)

$$\sigma_{sc}^{l=0} = \frac{4\pi}{\gamma^2} \sin^2 \delta_0 \tag{7.80}$$

For small $\delta_0$

$$\sin \delta_0 \cong \delta_0$$

and so using equation (7.79)

$$\sigma_{sc}^{l=0} = \frac{4\pi}{\gamma^2} \delta_0^2$$

$$= \frac{4\pi}{\gamma^2} \left( -\frac{\gamma}{\beta} - \gamma r_0 \right)^2$$

$$\sigma_{sc}^{l=0} = 4\pi \left( \frac{1}{\beta} + r_0 \right)^2 \tag{7.81}$$

Again from equation (6.56) the deuteron radius is $\frac{1}{\beta} = \rho = 4.35\,fm$ and $r_0 = 2.1\,fm$ is nuclear range. Then equation (7.81) gives

$$\sigma_{sc}^{l=0} = 4\pi(4.35\,fm + 2.1\,fm)^2 = 5.2 \times 10^{-28}\,m^2$$

$$\sigma_{sc}^{l=0} = 5.2\,barn \tag{7.82}$$

Let us make another assumption.

As $\gamma r_0$ is very small let us neglect $\gamma r_0$ in equation (7.79) to get

$$\delta_0 = -\frac{\gamma}{\beta} - \gamma r_0 \cong -\frac{\gamma}{\beta} \tag{7.83}$$

Now the scattering cross-section given in equation (7.80) is

$$\sigma_{sc}^{l=0} = \frac{4\pi}{\gamma^2} \sin^2 \delta_0$$

As $\delta_0 =$ small, $\sin \delta_0 \cong \delta_0$ and so with equation (7.83)

$$\sigma_{sc}^{l=0} = \frac{4\pi}{\gamma^2} \delta_0^2 = \frac{4\pi}{\gamma^2} \left( -\frac{\gamma}{\beta} \right)^2$$

$$\sigma_{sc}^{l=0} = \frac{4\pi}{\beta^2} \tag{7.84}$$

Using equation (6.56), i.e. $\frac{1}{\beta} = \rho = 4.35\,fm$ we have

$$\sigma_{sc}^{l=0} = 4\pi(4.35\,fm)^2 = 2.38 \times 10^{-28}\,m^2$$

$$\sigma_{sc}^{l=0} = 2.38 \; barn \tag{7.85}$$

The values $\sigma_{sc}^{l=0} = 5.2 \; barn$ [equation (7.82)] or $\sigma_{sc}^{l=0} = 2.38 \; barn$ [equation (7.85)] are the theoretical values of scattering cross-section for low energy $np$ scattering based on a square well potential and using the result of deuteron ground state where $np$ is bound in spin triplet state with parallel spin.

- Explanation of the discrepancy of theoretical and experimental value was done by Wigner.

The discrepancy between theoretical and experimental value can be explained as being due to the spin dependence of nuclear force.

In deuteron bound state $n$ and $p$ are bound with parallel spins $| \uparrow \uparrow >$ in a spin

$\phantom{xxxxxxxxxxxxxxxxxxxxxxxxxxxxxxxxxxxxxxxxxxxxxxxx} n \; p$

triplet state, i.e. spins are correlated to one another. Deuteron state is [equation (6.111)]

$$| d > = a_s | \, ^3S_1 > + a_D | \, ^3D_1 >$$

In a general $np$ scattering experiment, spin of neutron and spin of proton are not correlated—we cannot say that spins are parallel or anti-parallel.

Actually in $np$ scattering the neutron in the neutron beam has spins oriented randomly—both singlet $^1S_0$ and triplet $^3S_1$ orientations are possible. So scattering occurs neither from a pure singlet state nor from a pure triplet state.

The probability of finding the system in triplet state or singlet state is equal to its statistical weight factor.

As there are three spin orientations for triplet state

$$| sm_s > \; \rightarrow | \, 11 > , \; | \, 10 > , \; | \, 1 - 1 >$$

and only one spin orientation for singlet state

$$| sm_s > \; \rightarrow | \, 00 >$$

the statistical weight factor for triplet state is $\frac{3}{4}$ and statistical weight factor for singlet state is $\frac{1}{4}$.

Let $\sigma_{triplet}^0$ be the cross-section for scattering in triplet state and $\sigma_{singlet}^0$ be the cross-section for scattering in singlet state. Then experimentally measured cross-section is

$$\sigma^0 = \frac{3}{4}\sigma_{triplet}^0 + \frac{1}{4}\sigma_{singlet}^0 \tag{7.86}$$

Now we found theoretically that $\sigma_{triplet}^0$ lies between 2.38 $barn$ and 5 $barn$ [equations (7.82) and (7.85)].

If we put $\sigma_{triplet}^0 = 2.38 \; barn$ and $\sigma^0 = 20 \; barn$ in equation (7.86) we get

$$20 \; barn = \frac{3}{4} 2.38 \; barn + \frac{1}{4}\sigma_{singlet}^0$$

$$\sigma_{\text{singlet}}^0 = 73 \; barn$$

If we put $\sigma_{\text{triplet}}^0 = 5.2 \; barn$ and $\sigma^0 = 20 \; barn$ in equation (7.86) we get

$$20 \; barn = \frac{3}{4} 5.2 \; barn + \frac{1}{4}\sigma_{\text{singlet}}^0$$

$$\sigma_{\text{singlet}}^0 = 64 \; barn$$

So we see that $\sigma_{\text{singlet}}^0 = 6 \; barn$ is very large in comparison to $\sigma_{\text{triplet}}^0 = 5.2 \; barn$ in low energy $np$ scattering. Singlet scattering contributes only $\frac{1}{4}$th of the time to the scattering process but has a major contribution to scattering cross-section. The enormous difference between the cross-section in the singlet and triplet state suggests that nuclear force must be spin dependent.

We mention that $\sigma_{\text{singlet}}^0$, $\sigma_{\text{triplet}}^0$ are not measured in the experiment since during scattering $np$ is neither purely in singlet state nor purely in triplet state. The experimentally measured quantity is $\sigma^0 = 20 \; barn$.

Due to different spin states the $np$ scattering corresponds to different scattering cross-sections. So nuclear force strongly depends on spin orientation. In the case of anti-parallel spin orientation the force is stronger than in the case of parallel spin orientation.

We can verify our conclusion about singlet and triplet cross-sections in a variety of ways. One method that we discuss now is to scatter very low energy neutrons from hydrogen molecules.

## 7.9 Coherent scattering of slow neutrons

Hydrogen is a diatomic homonuclear molecule. The nuclear part of its wave function obeys Pauli exclusion principle since protons obey Fermi–Dirac statistics. So the symmetry of the space part and the spin part are opposite.

Molecular hydrogen can exist in two states.
☐ Ortho hydrogen
In ortho hydrogen molecules proton spins are parallel ↑↑. So total spin is
$s_H = \frac{1}{2} + \frac{1}{2} = 1$
Statistical weight is

$$2s_H + 1 = 2.1 + 1 = 3.$$

It is a triplet state (ortho triplet).

Spin part is symmetric since $(-1)^{1+s} = (-1)^{1+1} = +ve$, figure 8.5 and space wave function is antisymmetric under exchange of two protons.
☐ Para hydrogen
In para hydrogen molecules proton spins are anti-parallel ↑↓. So total spin is

$$s_H = \frac{1}{2} - \frac{1}{2} = 0.$$

Statistical weight is

$$2s_H + 1 = 2.0 + 1 = 1.$$

It is a singlet state (para singlet).

Spin part is anti-symmetric since $(-1)^{1+s} = (-1)^{1+0} = -ve$, figure 8.5 and space wave function is symmetric under exchange of two protons.

Since the space wave functions of ortho and para hydrogen have opposite symmetry character they will have different energy levels.

Teller and Schwinger suggested performing an experiment to test spin dependence of $np$ interaction. It was done through an experimental comparison of coherent scattering from ortho and para hydrogen. A difference between the neutron scattering cross-sections of ortho and para hydrogen will be evidence of the spin dependent part of nucleon–nucleon force.

Hydrogen molecules are randomly oriented in a hydrogen target and so the scattering of neutrons by different molecules is incoherent. On the other hand, scattering amplitudes of two protons in each molecule coherently add up with each other.

Coherent scattering occurs when low energy neutrons strike hydrogen molecules. Very low energy neutrons with $E < 0.01 \, eV$ have a de Broglie wavelength larger than 0.5 Å, which is greater than the separation of the two protons in a hydrogen molecule. So the wave packet of the incident neutron overlaps simultaneously with both protons in a hydrogen molecule, even though the range of nuclear force of the individual $np$ interactions is $\sim 1 \, fm$. The scattered neutron waves $\psi_1$ and $\psi_2$ from the two protons will combine coherently, i.e. they will interfere, and the cross-section would depend on $|\psi_1 + \psi_2|^2$ and not on $|\psi_1|^2 + |\psi_2|^2$. We thus cannot simply add the cross-sections from the two individual scatterings.

At higher energy the de Broglie wavelength is smaller than the proton–proton separation. The scattered waves would not interfere and we can add the cross-sections directly.

The reason to work at very low energy is twofold.

☐ To observe interference effect, i.e. coherence.

☐ To prevent neutron from transferring energy to the hydrogen molecule that might start rotating. This would introduce additional levels and complicate the analysis. The minimum rotational energy is of the order of $0.015 \, eV$ and hence neutrons with energies in the range $0.01 \, eV$ will not excite rotational states of the molecule.

The coherent scattering experiment reveals that nuclear force in $np$ interaction is spin dependent and that singlet state scattering length is negative.

Equation (7.60) namely $\sigma_{scatt} = 4\pi a^2$ involves $a^2$ and hence it follows that knowing $\sigma_{scatt}$ we can find the magnitude but not the sign of scattering length $a$.

Let us compare coherent and incoherent scattering.

✓ When projectile neutrons strike randomly distributed protons then there is no definite phase relationship between spins of neighbouring protons and so scattering is incoherent.

When projectile neutrons strike protons bound in molecules or crystal then there is a definite phase relationship between spins of neighbouring protons and so scattering is coherent.

✓ If the de Broglie wavelength of incident neutrons is less than interatomic distances $\sim 10^{-10}$ $m$ then there is thermal agitation and coherence is destroyed.

If the de Broglie wavelength of incident neutrons is greater than interatomic distances $\sim 10^{-10}$ $m$ then there is coherent scattering. This is the condition of coherent scattering.

✓ In incoherent scattering intensities are added up.

In coherent scattering amplitudes are added up.
Total spin of $np$ system is

$$\vec{s} = \vec{s}_n + \vec{s}_p \tag{7.87}$$

where

$$\vec{s}_n = \frac{\hbar}{2}\vec{\sigma}_n$$

is spin of neutron, $\vec{\sigma}_n$ is Pauli spin matrix for a neutron
and

$$\vec{s}_p = \frac{\hbar}{2}\vec{\sigma}_p$$

is spin of proton, $\vec{\sigma}_p$ is Pauli spin matrix for a proton

Now the square of spin of $np$ system is

$$\vec{s}^2 = \left(\vec{s}_n + \vec{s}_p\right)^2 = \left(\vec{s}_n + \vec{s}_p\right) \cdot \left(\vec{s}_n + \vec{s}_p\right)$$

$$= s_n^2 + s_p^2 + \vec{s}_n \cdot \vec{s}_p + \vec{s}_p \vec{s}_n$$

*As* $\vec{s}_n$ *and* $\vec{s}_p$ *are independent they commute, i.e.* $\vec{s}_n \cdot \vec{s}_p = \vec{s}_p \cdot \vec{s}_n$. Hence

$$\vec{s}^2 = s_n^2 + s_p^2 + 2\vec{s}_n \cdot \vec{s}_p \tag{7.88}$$

Taking $\hbar = 1$ the eigenvalue of $s^2$ is

$$s(s + 1) = \begin{cases} 0(0 + 1) = 0 \text{ for singlet state } s = 0 \\ 1(1 + 1) = 2 \text{ for triplet state } s = 1 \end{cases} \tag{7.89}$$

Also, eigenvalue of $s_n^2$ is (taking $\hbar = 1$)

$$s_n(s_n + 1) = \frac{1}{2}\left(\frac{1}{2} + 1\right) = \frac{3}{4} \text{ as } s_n = \frac{1}{2} \text{ for neutron.} \tag{7.90}$$

And eigenvalue of $s_p^2$ is

$$s_p(s_p + 1) = \frac{1}{2}\left(\frac{1}{2} + 1\right) = \frac{3}{4} \text{ as } s_p = \frac{1}{2} \text{ for proton} \tag{7.91}$$

Consider from equation (7.88)

$$\vec{s}_n \cdot \vec{s}_p = \frac{1}{2}\left(s^2 - s_n^2 - s_p^2\right)$$

Using the eigenvalues $s(s+1)$, $s_n^2 \rightarrow \frac{3}{4}$ and $s_p^2 \rightarrow \frac{3}{4}$ we get

$$\vec{s}_n \cdot \vec{s}_p = \frac{1}{2}\left[s(s+1) - \frac{3}{4} - \frac{3}{4}\right]$$

$$\vec{s}_n \cdot \vec{s}_p = \frac{1}{2}\left[s(s+1) - \frac{3}{2}\right] \qquad (7.92)$$

For singlet $s = 0$ we have from equations (7.89) and (7.92)

$$\vec{s}_n \cdot \vec{s}_p = \frac{1}{2}\left(0 - \frac{3}{2}\right) = -\frac{3}{4}$$

For triplet $s = 1$ we have from equations (7.89) and (7.92)

$$\vec{s}_n \cdot \vec{s}_p = \frac{1}{2}\left(2 - \frac{3}{2}\right) = \frac{1}{4}$$

Let us construct projection operators $P_s$, $P_t$ as follows

$$P_s = \frac{1}{4} - \vec{s}_n \cdot \vec{s}_p$$

$$P_t = \frac{3}{4} + \vec{s}_n \cdot \vec{s}_p$$

Let us check the operation of $P_s$ and $P_t$ on singlet state for which $\vec{s}_n \cdot \vec{s}_p = -\frac{3}{4}$

$$P_s = \frac{1}{4} - \vec{s}_n \cdot \vec{s}_p = \frac{1}{4} - \left(-\frac{3}{4}\right) = 1$$

$$P_t = \frac{3}{4} + \vec{s}_n \cdot \vec{s}_p = \frac{3}{4} + \left(-\frac{3}{4}\right) = 0$$

So $P_s$ operator projects out singlet state.

Let us check the operation of $P_s$ and $P_t$ on triplet state $\vec{s}_n \cdot \vec{s}_p = \frac{1}{4}$

$$P_s = \frac{1}{4} - \vec{s}_n \cdot \vec{s}_p = \frac{1}{4} - \left(\frac{1}{4}\right) = 0$$

$$P_t = \frac{3}{4} + \vec{s}_n \cdot \vec{s}_p = \frac{3}{4} + \left(\frac{1}{4}\right) = 1$$

So $P_t$ operator projects out triplet state.

Let $a_1$ and $a_2$ be, respectively, the scattering lengths for scattering of neutron by two protons 1 and 2 of a hydrogen molecule—which can be defined in terms of $a_s$ and $a_t$ as follows.

$$a_1 = a_s P_{s1} + a_t P_{t1}$$

$$= a_s\left(\frac{1}{4} - \vec{s}_n \cdot \vec{s}_{p1}\right) + a_t\left(\frac{3}{4} + \vec{s}_n \cdot \vec{s}_{p1}\right)$$

$$a_2 = a_s P_{s2} + a_t P_{t2}$$

$$= a_s\left(\frac{1}{4} - \vec{s}_n \cdot \vec{s}_{p2}\right) + a_t\left(\frac{3}{4} + \vec{s}_n \cdot \vec{s}_{p2}\right)$$

Coherent scattering occurs when the distance between two protons is less than the de Broglie wavelength of a neutron.

In coherent scattering the amplitudes are added and so we write

$$a = a_1 + a_2$$

$$= a_s\left(\frac{1}{4} - \vec{s}_n \cdot \vec{s}_{p1}\right) + a_t\left(\frac{3}{4} + \vec{s}_n \cdot \vec{s}_{p1}\right) + a_s\left(\frac{1}{4} - \vec{s}_n \cdot \vec{s}_{p2}\right) + a_t\left(\frac{3}{4} + \vec{s}_n \cdot \vec{s}_{p2}\right)$$

$$= \frac{1}{2}a_s + \frac{3}{2}a_t - a_s\left(\vec{s}_n \cdot \vec{s}_{p1} + \vec{s}_n \cdot \vec{s}_{p2}\right) + a_t\left(\vec{s}_n \cdot \vec{s}_{p1} + \vec{s}_n \cdot \vec{s}_{p2}\right)$$

$$= \frac{1}{2}a_s + \frac{3}{2}a_t - a_s\vec{s}_n \cdot \left(\vec{s}_{p1} + \vec{s}_{p2}\right) + a_t\vec{s}_n \cdot \left(\vec{s}_{p1} + \vec{s}_{p2}\right)$$

$$= \frac{1}{2}a_s + \frac{3}{2}a_t + (a_t - a_s)\vec{s}_n \cdot \left(\vec{s}_{p1} + \vec{s}_{p2}\right)$$

Defining $\vec{s}_H = \vec{s}_{p1} + \vec{s}_{p2}$

$$a = \frac{1}{2}a_s + \frac{3}{2}a_t + (a_t - a_s)\vec{s}_n \cdot \vec{s}_H \qquad (7.93)$$

For para hydrogen the proton spins are anti-parallel ↑↑ and so

$$s_H = \frac{1}{2} - \frac{1}{2} = 0$$

Equation (7.93) gives

$$a \equiv a_p = \frac{1}{2}a_s + \frac{3}{2}a_t$$

$$a_p = \frac{1}{2}(a_s + 3a_t)$$

$$\sigma_p = 4\pi a_p^2 = 4\pi\left(\frac{a_s + 3a_t}{2}\right)^2 \qquad (7.94)$$

For ortho hydrogen the proton spins are parallel ↑↑ and we have

$$s_H = \frac{1}{2} + \frac{1}{2} = 1$$

Equation (7.93) gives

$$a \equiv a_o = \frac{1}{2}(a_s + 3a_t) + (a_t - a_s)\vec{s}_n. \ \vec{s}_H \text{ with } s_H = 1$$

Scattering cross-section for each ortho hydrogen molecule is

$$\sigma_o = 4\pi a_o^2$$

$$= 4\pi[\frac{1}{2}(a_s + 3a_t) + (a_t - a_s)\vec{s}_n. \ \vec{s}_H]^2 \qquad (7.95)$$

$$\sigma_o = 4\pi[\left(\frac{a_s + 3a_t}{2}\right)^2 + (a_t - a_s)^2 \ | \ \vec{s}_n. \ \vec{s}_H \ |^2 + 2. \ \frac{a_s + 3a_t}{2}(a_t - a_s) \ | \ \vec{s}_n. \ \vec{s}_H \ |] \qquad (7.96)$$

Since incident neutrons are unpolarized this expression must be averaged over all possible polarizations of the incident beam.

We focus on the factor $\vec{s}_n. \ \vec{s}_H$. Let us consider the average of its square.

$$|\vec{s}_n. \ \vec{s}_H|^2_{avg} = (s_{nx}s_{Hx} + s_{ny}s_{Hy} + s_{nz}s_{Hz})^2_{avg}$$

$$= \left(s_{nx}^2 s_{Hx}^2 + s_{ny}^2 s_{Hy}^2 + s_{nz}^2 s_{Hz}^2\right)_{avg} + 2(s_{nx}s_{Hx}s_{ny}s_{Hy} + s_{nx}s_{Hx}s_{nz}s_{Hz})_{avg}$$

$$+ 2(s_{ny}s_{Hy}s_{nx}s_{Hx} + s_{ny}s_{Hy}s_{nz}s_{Hz})_{avg} + 2(s_{nz}s_{Hz}s_{nx}s_{Hx} + s_{nz}s_{Hz}s_{ny}s_{Hy})_{avg}$$

As Pauli spin matrices anticommute we have

$$s_{nx}s_{ny} + s_{ny}s_{nx} \sim \sigma_x\sigma_y + \sigma_y\sigma_x = 0 \text{ etc.}$$

And hence all the last six terms average out to zero. So we are left with

$$|\vec{s}_n. \ \vec{s}_H|^2_{avg} = \left(s_{nx}^2 s_{Hx}^2 + s_{ny}^2 s_{Hy}^2 + s_{nz}^2 s_{Hz}^2\right)_{avg} \qquad (7.97)$$

As $s_n^2$ has eigenvalue $\frac{3}{4}$ [equation (7.90)] we write

$$s_n^2 = s_{nx}^2 + s_{ny}^2 + s_{nz}^2 = \frac{3}{4}$$

$$(s_n^2)_{avg} = (s_{nx}^2)_{avg} + (s_{ny}^2)_{avg} + (s_{nz}^2)_{avg} = \frac{3}{4}$$

This means that

$$\left(s_{nx}^2\right)_{\text{avg}} = \left(s_{ny}^2\right)_{\text{avg}} = \left(s_{nz}^2\right)_{\text{avg}} = \frac{1}{4}$$

With this we have from equation (7.97)

$$|\vec{s}_n \cdot \vec{s}_H|_{\text{avg}}^2 = \frac{1}{4}\left(s_{Hx}^2 + s_{Hy}^2 + s_{Hz}^2\right)_{\text{avg}} = \frac{1}{4}(s_H^2)_{\text{avg}}$$

Replacing $(s_H^2)_{\text{avg}}$ by its eigenvalue we have

$$| \vec{s}_n \cdot \vec{s}_H |_{\text{avg}}^2 = \frac{1}{4}s_H(s_H + 1) \qquad (7.98)$$

As $s_H = 1$(triplet) we have from equation (7.98)

$$| \vec{s}_n \cdot \vec{s}_H |_{\text{avg}}^2 = \frac{1}{4}1(1 + 1) = \frac{1}{2} \qquad (7.99)$$

We shall now consider average of equation (7.96) and in it we have to put $| \vec{s}_n \cdot \vec{s}_H |_{\text{avg}}^2 = \frac{1}{2}$ [equation (7.99)] and $| \vec{s}_n \cdot \vec{s}_H |_{\text{avg}} = 0$. This gives from equation (7.96)

$$\sigma_o = 4\pi\left[\left(\frac{a_s + 3a_t}{2}\right)^2 + (a_t - a_s)^2\, | \vec{s}_n \cdot \vec{s}_H |_{\text{avg}}^2 + 2.\,\frac{a_s + 3a_t}{2}(a_t - a_s)\, | \vec{s}_n \cdot \vec{s}_{II} |_{\text{avg}}\right]$$

$$= 4\pi\left[\left(\frac{a_s + 3a_t}{2}\right)^2 + (a_t - a_s)^2\frac{1}{2}\right]$$

$$= 4\pi\left(\frac{a_s + 3a_t}{2}\right)^2 + 2\pi(a_t - a_s)^2 \qquad (7.100)$$

Using equation (7.94) we rewrite equation (7.100)

$$\sigma_o = \sigma_p + 2\pi(a_t - a_s)^2 \qquad (7.101)$$

Let us take the ratio of $\sigma_p$ [equation (7.94)] and $\sigma_o$ (7.101) to get

$$\frac{\sigma_o}{\sigma_p} = \frac{\sigma_p + 2\pi(a_t - a_s)^2}{\sigma_p}$$

$$= 1 + \frac{2\pi(a_t - a_s)^2}{\sigma_p}$$

$$= 1 + \frac{2\pi(a_t - a_s)^2}{4\pi\left(\frac{a_s + 3a_t}{2}\right)^2} \quad \text{[using equation (7.94)]}$$

$$\frac{\sigma_o}{\sigma_p} = 1 + 2\frac{(a_t - a_s)^2}{(a_s + 3a_t)^2} > 1 \tag{7.102}$$

- This is the relation between scattering cross-sections for ortho and para hydrogen.
- If $np$ forces were independent of spin then we would expect the same scattering cross-section for both, i.e. one would expect the result

$$\frac{\sigma_o}{\sigma_p} = 1.$$

But the experimental result is that

$$\sigma_p = 4 \; barn, \; \sigma_o = 120 \; barn$$

leading to

$$\frac{\sigma_o}{\sigma_p} = \frac{120}{4} = 30.$$

- That singlet $np$ scattering length is negative follows as explained in the following.

From equation (7.102) it follows that $\sigma_o > \sigma_p$.

Hence the second term in equation (7.102) should be large.

Now if $a_s > 0$ then the denominator gets larger and the numerator gets smaller. This will not make the second term larger.

But if $a_s < 0$ then the denominator is smaller and the numerator is larger. So the second term will be large.

We thus conclude that the fact that $a_s = -ve$ is consistent with experimental data.

Experimental values of $a_s$ and $a_t$ are

$$a_s = -23.7 \; fm, \; a_t = 5.4 \; fm.$$

Singlet $np$ scattering length is negative; this means that the singlet $np$ state is unbound.

## 7.10 Effective range theory of low energy $np$ scattering

Scattering cross-section is energy dependent, i.e. it depends on the energy of the striking neutron. We expressed the scattering cross-section in terms of the scattering length $a$.

We reformulate the theory of scattering such that the range over which the nuclear potential is effective occurs explicitly in the expression of scattering cross-section along with phase shift and scattering length. The theory is applicable only for low energies.

In other words, we will express the scattering cross-section in terms of the scattering length $a$ and another parameter called effective range denoted by $R_0$ whose value is of the order of the range of nuclear force.

Since the scattering cross-section is expressed in terms of scattering length $a$ and the effective range $R_0$ this theory is called effective range theory.

At low energy the phase shift is $\delta_0$. The effective range theory predicts that the phase shift will be a function of energy. This energy dependence of phase shift was calculated by Schwinger and Bethe.

We shall consider the low energy $s$ wave scattering.

The process of scattering is described by wave functions. Let $u_1$ and $u_2$ be the radial wave functions describing scattering when neutrons strike protons with energies say $E_1$ and $E_2$ respectively. Here

$$u_1 = r\psi_1 \text{ and } u_2 = r\psi_2 \tag{7.103}$$

The Schrödinger time independent equations are

$$\frac{d^2 u_1}{dr^2} + \frac{m}{\hbar^2}(E_1 - V)u_1 = 0$$

$$\frac{d^2 u_1}{dr^2} + \left(k_1^2 - \frac{mV}{\hbar^2}\right)u_1 = 0, \; k_1^2 = \frac{mE_1}{\hbar^2} \tag{7.104}$$

$$\frac{d^2 u_2}{dr^2} + \frac{m}{\hbar^2}(E_2 - V)u_2 = 0$$

$$\frac{d^2 u_2}{dr^2} + \left(k_2^2 - \frac{mV}{\hbar^2}\right)u_2 = 0, \; k_2^2 = \frac{mE_2}{\hbar^2} \tag{7.105}$$

where $m$ is average mass of a nucleon.

Let us multiply equation (7.104) by $u_2$ and equation (7.105) by $u_1$ to get

$$u_2\frac{d^2 u_1}{dr^2} + \left(k_1^2 - \frac{mV}{\hbar^2}\right)u_1 u_2 = 0 \tag{7.106}$$

$$u_1\frac{d^2 u_2}{dr^2} + \left(k_2^2 - \frac{mV}{\hbar^2}\right)u_1 u_2 = 0 \tag{7.107}$$

Now take the difference of equation (7.106) and equation (7.107) to get

$$u_2\frac{d^2 u_1}{dr^2} - u_1\frac{d^2 u_2}{dr^2} + (k_1^2 - k_2^2)u_1 u_2 = 0$$

$$u_2\frac{d^2 u_1}{dr^2} - u_1\frac{d^2 u_2}{dr^2} = (k_2^2 - k_1^2)u_1 u_2 \tag{7.108}$$

Since

$$\frac{d}{dr}\left(u_2\frac{du_1}{dr} - u_1\frac{du_2}{dr}\right) = u_2\frac{d^2u_1}{dr^2} - \frac{du_2}{dr}\frac{du_1}{dr} - \frac{du_1}{dr}\frac{du_2}{dr} - u_1\frac{d^2u_2}{dr^2}$$

$$= u_2\frac{d^2u_1}{dr^2} - u_1\frac{d^2u_2}{dr^2}$$

we rewrite equation (7.108) as

$$\frac{d}{dr}\left(u_2\frac{du_1}{dr} - u_1\frac{du_2}{dr}\right) = (k_2^2 - k_1^2)u_1u_2 \tag{7.109}$$

Integrate from $r = 0$ to $r = \infty$ to get

$$\int_{r=0}^{\infty}\frac{d}{dr}\left(u_2\frac{du_1}{dr} - u_1\frac{du_2}{dr}\right)dr = \int_{r=0}^{\infty}(k_2^2 - k_1^2)u_1u_2dr$$

$$\left(u_2\frac{du_1}{dr} - u_1\frac{du_2}{dr}\right)_0^{\infty} = (k_2^2 - k_1^2)\int_{r=0}^{\infty}u_1u_2dr \tag{7.110}$$

Boundary condition satisfied by $u_1$ and $u_2$ are as follows.

At $r = 0$ the wave function starts from zero (figure 6.3) and so we take

$$u_1(r = 0) = u_2(r = 0) = 0$$

Let us introduce auxiliary wave functions $v_1$ and $v_2$ which represent, respectively, the asymptotic behaviour of $u_1$ and $u_2$, i.e. behavior at large distances $r > r_0$ and also at $r \to \infty$ (outside the range of nuclear force). So

$$u_1(r) \xrightarrow{r\to\infty} v_1(r)$$

$$u_2(r) \xrightarrow{r\to\infty} v_2(r)$$

Again at $r > r_0, V = 0$.

So $v_1$ and $v_2$ are, respectively, the solutions of Schrödinger equations (7.104) and (7.105) with $V = 0$ and the equations are

$$\frac{d^2v_1}{dr^2} + k_1^2v_1 = 0 \tag{7.111}$$

$$\frac{d^2v_2}{dr^2} + k_2^2v_2 = 0 \tag{7.112}$$

We note that in the inside region $r \leqslant r_0$, $u_1$ and $v_1$ are different, $u_2$ and $v_2$ are also different. But in the outside region $r \geqslant r_0$, $u_1$ and $v_1$ are same, $u_2$ and $v_2$ are same too.

The solution of equation (7.111) is

$$v_1 = B_1 \sin(k_1r + \delta_1) \tag{7.113}$$

$B_1$ is normalization factor.

We choose $B_1$ in such a manner as to fix the amplitude of the external wave function $v_1$ when extended back.

We set $v_1 = 1$ at $r = 0$.

Hence from equation (7.113)

$$1 = B_1 \sin(k_1 . 0 + \delta_1)$$

$$B_1 = \frac{1}{\sin \delta_1}$$

Hence from equation (7.113) we have

$$v_1 = \frac{\sin(k_1 r + \delta_1)}{\sin \delta_1} = v_1(r) \tag{7.114}$$

Similarly

$$v_2 = \frac{\sin(k_2 r + \delta_2)}{\sin \delta_2} = v_2(r) \tag{7.115}$$

where we set $v_2 = 1$ at $r = 0$.

Clearly from equations (7.114) and (7.115) the asymptotic behavior of $u$ is evident.

$$u(r) \xrightarrow{r \to \infty} v(r) = \frac{\sin(kr + \delta)}{\sin \delta} \tag{7.116}$$

Multiply equation (7.111) by $v_2$ and equation (7.112) by $v_1$

$$v_2 \frac{d^2 v_1}{dr^2} + k_1^2 v_1 v_2 = 0 \tag{7.117}$$

$$v_1 \frac{d^2 v_2}{dr^2} + k_2^2 v_1 v_2 = 0 \tag{7.118}$$

Now take the difference of equations (7.117) and (7.118) to get

$$\left( v_2 \frac{d^2 v_1}{dr^2} - v_1 \frac{d^2 v_2}{dr^2} + k_1^2 - k_2^2 \right) v_1 v_2 = 0$$

$$\frac{d}{dr}\left( v_2 \frac{dv_1}{dr} - v_1 \frac{dv_2}{dr} \right) = (k_2^2 - k_1^2) v_1 v_2 \tag{7.119}$$

Integrate from $r = 0$ to $r = \infty$

$$\int_{r=0}^{\infty} \frac{d}{dr}\left( v_2 \frac{dv_1}{dr} - v_1 \frac{dv_2}{dr} \right) dr = \int_{r=0}^{\infty} (k_2^2 - k_1^2) v_1 v_2 dr$$

$$\left( v_2 \frac{dv_1}{dr} - v_1 \frac{dv_2}{dr} \right)\Big|_0^{\infty} = (k_2^2 - k_1^2) \int_{r=0}^{\infty} v_1 v_2 dr \tag{7.120}$$

Equations (7.120) and (7.110) are similar. Take their difference to get

$$\left(u_2\frac{du_1}{dr} - u_1\frac{du_2}{dr} - v_2\frac{dv_1}{dr} + v_1\frac{dv_2}{dr}\right)_0^{\infty} = (k_2^2 - k_1^{\,2})\int_{r=0}^{\infty}(u_1u_2 - v_1v_2)dr \qquad (7.121)$$

At $r \to \infty$, $u_1 = v_1 \to 0$, $u_2 = v_2 \to 0$.

At $r = 0$, $u_1 = u_2 = 0$ ($v_1 \neq 0$, $v_2 \neq 0$).

With this we have from equation (7.121)

$$0 - \left(-v_2\frac{dv_1}{dr} + v_1\frac{dv_2}{dr}\right)_{r=0} = (k_2^2 - k_1^{\,2})\int_{r=0}^{\infty}(u_1u_2 - v_1v_2)dr$$

$$\left(v_1\frac{dv_2}{dr} - v_2\frac{dv_1}{dr}\right)_{r=0} = (k_2^2 - k_1^{\,2})\int_{r=0}^{\infty}(v_1v_2 - u_1u_2)dr \qquad (7.122)$$

Again, we took the normalization constant to be such that

$$v_1 \,|_{r=0} = v_2 \,|_{r=0} = 1$$

With this we have from equation (7.122)

$$\left(\frac{dv_2}{dr} - \frac{dv_1}{dr}\right)_{r=0} = (k_2^2 - k_1^{\,2})\int_{r=0}^{\infty}(v_1v_2 - u_1u_2)dr$$

Using equations (7.114) and (7.115) we have

$$\left(\frac{d}{dr}\frac{\sin(k_2r + \delta_2)}{\sin \delta_2} - \frac{d}{dr}\frac{\sin(k_1r + \delta_1)}{\sin \delta_1}\right)_{r=0} = (k_2^2 - k_1^{\,2})\int_{r=0}^{\infty}(v_1v_2 - u_1u_2)dr$$

$$\left[\frac{k_2\cos(k_2r + \delta_2)}{\sin \delta_2} - \frac{k_1\cos(k_1r + \delta_1)}{\sin \delta_1}\right]_{r=0} = (k_2^2 - k_1^{\,2})\int_{r=0}^{\infty}(v_1v_2 - u_1u_2)dr$$

$$k_2\frac{\cos \delta_2}{\sin \delta_2} - k_1\frac{\cos \delta_1}{\sin \delta_1} = (k_2^2 - k_1^{\,2})\int_{r=0}^{\infty}(v_1v_2 - u_1u_2)dr$$

$$k_2\cot \delta_2 - k_1\cot \delta_1 = (k_2^2 - k_1^{\,2})\int_{r=0}^{\infty}(v_1v_2 - u_1u_2)dr \qquad (7.123)$$

We are considering low energy $np$ scattering. Let the striking neutron have zero energy, i.e. $E_1 \to 0$. Hence

$$k_1 = \sqrt{\frac{mE_1}{\hbar^2}} \to 0 \text{ then } \delta_1 \to 0 \text{ and so}$$

$$\sin \delta_1 \to \delta_1, \cos \delta_1 \to 1.$$

So

$$k_1 \cot \delta_1 \xrightarrow{k_1 \to 0} k_1 \frac{\cos \delta_1}{\sin \delta_1} \cong k_1 \frac{1}{\delta_1}$$

Since scattering length can be defined as

$$a = -\frac{\delta}{k} \Rightarrow \frac{1}{a} = -\frac{k}{\delta}$$

we write

$$k_1 \cot \delta_1 \xrightarrow{k_1 \to 0} -\frac{1}{a} \tag{7.124}$$

Hence equation (7.123) becomes with $k_1 = 0$ and using equation (7.124)

$$k_2 \cot \delta_2 - (-\tfrac{1}{a}) = k_2^2 \int_{r=0}^{\infty} (v_1 v_2 - u_1 u_2) dr$$

Dropping the suffix 2 to make things simple and writing

$$u_1 \equiv u_0, \; v_1 \equiv v_0, \; k = \sqrt{\frac{mE}{\hbar^2}} \tag{7.125}$$

we have

$$k \cot \delta + \frac{1}{a} = k^2 \int_{r=0}^{\infty} (v_0 v - u_0 u) dr \tag{7.126}$$

This is the fundamental equation for effective range theory and is the basis of an approximation method to find the scattering cross-section in the case of low energy $np$ scattering. The approximation method is called shape independent approximation meaning that the shape of the potential does not matter.

A significant contribution to the integral in the RHS of equation (7.126) comes from the region within the range of nuclear force, i.e. $r \leqslant r_0$.

In the region $r \leqslant r_0$, $u$ and $u_0$ differ from $v$ and $v_0$.

Outside the region of nuclear force they do not differ.

In the region $r \leqslant r_0$ their dependence on energy is small and the potential is also large, i.e. $E \ll V$. So dependence of the integral on $E$ is very small.

$u$ and $v$ can be replaced in the region $r \leqslant r_0$ by their zero energy forms, i.e. by $u_0$ and $v_0$.

So in zero energy approximation $u$ will be replaced by $u_0$ and $v$ by $v_0$.

Hence in zero energy approximation we have equation (7.126) to be

$$k \cot \delta + \frac{1}{a} = k^2 \int_{r=0}^{\infty} (v_0^2 - u_0^2) dr$$

Define

$$R_0 = 2 \int_{r=0}^{\infty} (v_0^2 - u_0^2) dr \tag{7.127}$$

called effective range. With this

$$k \cot \delta + \frac{1}{a} = \frac{k^2}{2} R_0 \qquad (7.128)$$

We can express the scattering cross-section in terms of the effective range $R_0$ and scattering length $a$. From equation (7.128)

$$k \cot \delta = \frac{k^2}{2} R_0 - \frac{1}{a}$$

$$\cot \delta = \frac{k R_0}{2} - \frac{1}{ka} \qquad (7.129)$$

This is the expression of $\cot \delta$ in terms of the effective range $R_0$ and scattering length $a$.

This result of equation (7.129) expresses the phase shift $\delta$ in the so-called shape independent approximation. Since this result does not depend on the form of the potential it is called shape independent.

Scattering cross-section in the case of low energy $np$ scattering is (dropping the subscript zero in equation (7.54))

$$\sigma_{sc} = \frac{4\pi}{k^2} \sin^2 \delta = \frac{4\pi}{k^2} \frac{1}{\cosec^2 \delta} = \frac{4\pi}{k^2} \frac{1}{1 + \cot^2 \delta} \qquad (7.130)$$

$$\sigma_{sc} = \frac{4\pi}{k^2 + k^2 \cot^2 \delta}$$

Using equation (7.129)

$$\sigma_{sc} = \frac{4\pi}{k^2 + k^2 \left( \frac{k R_0}{2} - \frac{1}{ka} \right)^2}$$

$$\sigma_{sc} = \frac{4\pi}{k^2 + \left( \frac{k^2 R_0}{2} - \frac{1}{a} \right)^2} \qquad (7.131)$$

This is the expression of the total scattering cross-section in case of low energy $np$ scattering in terms of the parameters, namely scattering length $a$ and the effective range $R_0$ and energy $k$ of a striking neutron.

So measuring $R_0$ and $a$, we can find the cross-section $\sigma_{sc}$.

This result is independent of the shape of the potential well. This is why the method is called shape independent approximation.

- Determination of scattering length $a$ and effective range $R_0$

Since $np$ scattering is spin dependent actually we should consider four parameters while dealing with $\sigma_{sc}$. They are:

✓ $a_S$ which is the singlet scattering length (since the $np$ system can be in the singlet state with probability $\frac{1}{4}$).

✓ $a_t$ which is the triplet scattering length (since the $np$ system can be in the triplet state with probability $\frac{3}{4}$).

✓ $R_{0s}$ which is the singlet effective range.

✓ $R_{0t}$ which is the triplet effective range.

Let $\sigma_{\text{triplet}}^0$ be the scattering cross-section in triplet state and $\sigma_{\text{singlet}}^0$ be the scattering cross-section in singlet state. The probability of finding the $np$ system in singlet state is $\frac{1}{4}$ and probability of finding the $np$ system in triplet state is $\frac{3}{4}$.

Expression of total scattering cross-section taking spin dependence into account is

$$\sigma^0 = \frac{3}{4}\sigma_{\text{triplet}}^0 + \frac{1}{4}\sigma_{\text{singlet}}^0 \tag{7.132}$$

In view of equation (7.131) we rewrite equation (7.132) as

$$\sigma^0 = \frac{3}{4}\frac{4\pi}{k^2 + \left(\frac{k^2 R_{0t}}{2} - \frac{1}{a_t}\right)^2} + \frac{1}{4}\frac{4\pi}{k^2 + \left(\frac{k^2 R_{0s}}{2} - \frac{1}{a_s}\right)^2}$$

$$\sigma^0 = \frac{3\pi}{k^2 + \left(\frac{k^2 R_{0t}}{2} - \frac{1}{a_t}\right)^2} + \frac{\pi}{k^2 + \left(\frac{k^2 R_{0s}}{2} - \frac{1}{a_s}\right)^2} \tag{7.133}$$

This is the total scattering cross-section in terms of the four parameters $a_S$, $a_t$, $R_{0s}$, $R_{0t}$ and if these are measured from very low energy $np$ scattering experiment we can calculate $\sigma^0$. Actually, the two parameters $a_t$, $R_{0t}$ are also measured from the knowledge of deuteron ground state.

In the case of very low or zero energy approximation we replace $u$ by the internal wave function of the deuteron $u_{1d}$—the radial wave function and $v$ by the external wave function of deuteron $u_{2d}$, i.e.

$$u \rightarrow u_{1d}$$

$$v \rightarrow u_{2d}.$$

The energy $E$ is replaced by binding energy of deuteron viz. $B = -|E|$. In this case $k^2$ is replaced by $-\beta^2$ where $\beta^2 = \sqrt{\frac{m|E|}{\hbar^2}}$. Also, $k \cot \delta = -\beta$.

With these changes the fundamental equation (7.126) viz.

$$k \cot \delta + \frac{1}{a} = k^2 \int_{r=0}^{\infty} (v_0 v - u_0 u) dr$$

becomes

$$-\beta + \frac{1}{a_t} = -\beta^2 \int_{r=0}^{\infty} (v_0 u_{2d} - u_0 u_{1d}) dr \tag{7.134}$$

In zero energy approximation $u$ will be replaced by $u_0$ and $v$ by $v_0$. Hence in zero energy approximation equation (7.126) gives

$$-\beta + \frac{1}{a_t} = -\beta^2 \int_{r=0}^{\infty} (v_0^2 - u_0^2)dr = -\frac{\beta^2}{2} R_{0t} \tag{7.135}$$

where we define $R_{0t}$ as the effective range in triplet state. So

$$R_{0t} = -\frac{2}{\beta^2}\left(-\beta + \frac{1}{a_t}\right)$$

$$R_{0t} = \frac{2}{\beta^2}\left(\beta - \frac{1}{a_t}\right) \tag{7.136}$$

If we measure $R_{0t}$, $a_t$ from experiment we can herefrom find $\beta$.

With this $\beta$ we can find the binding energy of deuteron since $\beta^2 = \frac{m|E|}{\hbar^2}$ and $B = -|E|$. So binding energy of deuteron can be measured by measuring $R_{0t}$ and $a_t$ from experiment.

In the extreme limiting condition we take $R_{0t} \cong 0$ then equation (7.136) yields

$$0 = \frac{2}{\beta^2}\left(\beta - \frac{1}{a_t}\right)$$

$$a_t \cong \frac{1}{\beta}$$

Again $\frac{1}{\beta} = \rho$ is deuteron radius (*exercise 6.10*)

So in the limit of zero effective range scattering takes place from a hard sphere of radius $\rho = \frac{1}{\beta}$. But this result is valid only when we consider the limiting case of zero effective range.

The experimentally obtained best fit values of low energy ($l = 0$) $np$ scattering parameters are

$$a_t = 5.4\,fm,\ R_{0t} = 1.7\,fm,\ a_s = -23.7\,fm,\ R_{0s} = 2.7\,fm \tag{7.137}$$

## 7.11 Essential difference between *np* and *pp* scattering

In nucleon–nucleon scattering we discussed low energy $np$ scattering. Now we shall discuss $pp$ scattering at low energy.

In order to obtain quantitative information about the nuclear force acting between two charged particles we perform the $pp$ scattering experiment.

Low enrgy $pp$ scattering gives information about spin dependence of nuclear force and charge independence of nuclear force.

  • Superiority of $pp$ scattering over the $np$ scattering experiment.

  ✓ Protons are easily available over a wide range of energies.
  ✓ Protons can be made monoenergetic.

✓ A well collimated beam of protons can be easily produced but it is very difficult to prepare a well collimated beam with neutrons.

✓ Protons are easily detected by their ionizing property. Neutrons being neutral particles have less ionizing power and so are not easily detected.

✓ Protons undergo Coulomb scattering along with nuclear scattering. There is increase in sensitivity in case one of the scattering probabilities is small. There are also phase shifts resulting from nuclear scattering.

✓ Use of FD statistics to *pp* combination in the case of *pp* scattering simplifies the analysis.

- In *pp* scattering, in addition to nuclear forces, Coulomb forces are present since Coulomb force operates between the two protons as they are charged particles. A pure unscreened Coulomb field has an infinite range, i.e. only when $r \to \infty$, Coulomb force $\to 0$.

  At low energy, i.e. at $100\ keV$ proton energy, the two interacting protons will not come very close to each other due to mutual Coulomb repulsion and so $r > r_0$, i.e. the two protons stay outside the nuclear range of interaction. Under this situation of low energy $E < 100\ keV$ the *pp* scattering will predominantly be due to Coulomb repulsive force.

  In the case of large energy, the protons approach sufficiently close $r \leqslant r_0$ and interact through nuclear force.

  So in the case of *pp* scattering the interaction potential will consist of both nuclear and Coulomb potentials. The Coulomb potential appreciably distorts the incident wave function.

  In contrast, in the case of *np* scattering process no Coulomb force acts. And so potential is purely nuclear potential. There is no question of the incident wave function getting distorted by Coulomb force.

- In *pp* the scattering projectile is a proton and the target is also proton, i.e. target and projectile are identical particles. But in the case of *np* scattering projectile is neutron and the target is proton, i.e. target and projectile are different particles.

  In *pp* scattering the identical projectile and target nucleons must be described by a common wave function.

- The fact that the nature of projectile and target in *pp* scattering are identical can be addressed through the use of quantum mechanics—an issue that was not present in the case of *np* scattering, as in *np* scattering projectile and target were not identical.

- Protons are fermions obeying Pauli exclusion principle. The wave function describing the *pp* system must be antisymmetric w.r.t. exchange of protons. This requires the spatial part of the wave function to have symmetry opposite to that of the spin part of the wave function.

In singlet state the spin part has parity $(-1)^{1+s} = (-1)^{1+0} = -ve$ and is antisymmetric (figure 8.5). So the spatial part of the wave function is symmetric w.r.t. the interchange of protons.

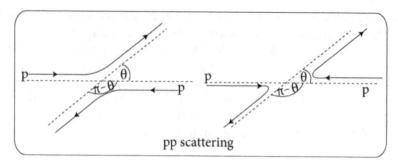

pp scattering

**Figure 7.8.** Scattering of identical particles ($pp$) in centre of mass system. Exchange of the coordinates of the two protons means exchanging $\theta$ with $\pi - \theta$.

In triplet state the spin part has parity $(-1)^{1+s} = (-1)^{1+1} = +ve$ and is symmetric (figure 8.5). So the spatial part of the wave function is antisymmetric w.r.t. the interchange of protons.

Now at low energy below 10 $MeV$ there is only $l = 0$, i.e. $s$ wave scattering. The wave function has parity $(-1)^l = (-1)^0 = +ve$, i.e. symmetric. So the spin state will be antisymmetric. It will thus be singlet state scattering. So the information regarding interaction is restricted only to the singlet state.

It is clear from figure 7.8 that one particle emerges at angle $\theta$ and the other at $\pi - \theta$. Since particles are identical there is no way to tell which of the particles— projectile or target—emerges at which angle. In other words, since particles are identical we cannot experimentally distinguish between the two cases shown in the figure. The scattered wave function must therefore include both the contributions for scattering at $\theta$ as well as scattering at $\pi - \theta$. So when we square the scattered wave function to calculate cross-section there will be a term to take care of interference between the parts of the wave function that give scattering at $\theta$ and scattering at $\pi - \theta$. The phenomenon of interference is purely quantum mechanical and has no classical analogue.

We shall discuss the $pp$ scattering focussing on the $s$ wave scattering.

- The theory of $pp$ scattering is more complicated than the $np$ scattering.

We give an otline of the theory.

## 7.12 The $pp$ scattering

We shall find the differential scattering cross-section using partial wave analysis. The same problem can be dealt with using effective range theory also.

For particles incident along the $z$-axis the total wave function, i.e. the sum of the incident wave function plus the scattered wave function is

$$\psi(r, \theta) = \sum_l C_l \frac{u_l}{r} P_l(\cos \theta) \tag{7.138}$$

where $P_l(\cos\theta)$ is the Legendre polynomial, $u_l(r)$ is the radial wave function—it being the solution of the radial wave equation for the proton–proton system in the case of $pp$ scattering, namely

$$\frac{d^2u}{dr^2} + \frac{m}{\hbar^2}\left[E - V(r) - \frac{|e|^2}{4\pi\varepsilon_0 r} - \frac{l(l+1)\hbar^2}{mr^2}\right]u = 0 \qquad (7.139)$$

$$\frac{d^2u}{dr^2} + \left[k^2 - \frac{mV}{\hbar^2} - \frac{m}{\hbar^2}\frac{|e|^2}{4\pi\varepsilon_0 r} - \frac{l(l+1)}{r^2}\right]u = 0$$

where $k^2 = \frac{mE}{\hbar^2}$

The term $V(r)$ is the nuclear potential (short range) and the term $\frac{|e|^2}{4\pi\varepsilon_0 r}$ represents the Coulomb potential (long or infinite range). This term was not present in the radial wave equation (7.69) describing the $np$ scattering case.

Clearly $V(r) \to 0$ at $r \to \infty$ and the solution at large distances would correspond to a pure Coulomb potential. Also, $\frac{l(l+1)\hbar^2}{mr^2}$ is the centrifugal potential energy term.

For $s$ wave scattering $l = 0$, no centrifugal term will be there.

Also, $V(r)$ is taken to be the square well nuclear potential given by equation (7.20) and shown in figure 7.6. And reduced mass is

$$\mu = \frac{m_p m_p}{m_p + m_p} = \frac{m_p}{2} = \frac{m}{2} \text{ where } m_p = m_n = m.$$

For low energies nuclear potential will affect only $l = 0$ partial waves. The Coulombian potential produces higher $l$ scattering.

For low energy for which $l = 0$, the $pp$ scattering can only be in the singlet state. For high energy there will be higher $l$ values and scattering can be from singlet state or triplet state, decided by whether $l$ is even or odd.

In the centre of mass system the probability that either proton will be scattered through an angle $\theta$ will be

$$|f(\theta)|^2 + |f(\pi - \theta)|^2$$

since exchange of the spatial coordinates of the two protons means exchanging $\theta$ with $\pi - \theta$ (figure 7.8).

At higher energies, scattering involves $l = 1$ which means spatial wave function is antisymmetric w.r.t. interchange of spatial coordinate (since $(-1)^l = (-1)^1 = -ve$) and so the spin part will be symmetric w.r.t. interchange of spin coordinate—so that the total wave function (spatial part times spin part) is antisymmetric.

For identical particles, their waves interfere and we have to add their amplitudes and then take the square of the modulus because the particles are identical. So eigenfunctions containing only spatial coordinates without spin will be

$$f(\theta) \pm f(\pi - \theta)$$

For +*ve* sign the wave function is symmetric and for −*ve* sign the wave function is antisymmetric w.r.t. exchange of spatial coordinates of the two protons.

Again, spin wave function is symmetric in the triplet state and antisymmetric in the singlet state. Since for proton system the total wave function must be antisymmetrical hence the resultant scattering amplitude will have a symmetric space part for the singlet state, i.e.

$$f_s(\theta) = f(\theta) + f(\pi - \theta) \tag{7.140}$$

and an antisymmetric space part for the triplet state, i.e.

$$f_t(\theta) = f(\theta) - f(\pi - \theta) \tag{7.141}$$

The protons of target material as well as the projectile protons have randomly oriented spins and the cross-section for *pp* scattering is obtained after taking into account the proper statistical weights $\frac{1}{4}$ for the singlet state and $\frac{3}{4}$ for the triplet state.

The differential scattering cross-section is defined as

$$\frac{d\sigma}{d\Omega} = \frac{1}{4}|f_s(\theta)|^2 + \frac{3}{4}|f_t(\theta)|^2 \tag{7.142}$$

Using equations (7.140) and (7.141) we write

$$\frac{d\sigma}{d\Omega} = \frac{1}{4}|f(\theta) + f(\pi - \theta)|^2 + \frac{3}{4}|f(\theta) - f(\pi - \theta)|^2$$

$$= \frac{1}{4}[f(\theta) + f(\pi - \theta)]^*[f(\theta) + f(\pi - \theta)] + \frac{3}{4}[f(\theta) - f(\pi - \theta)]^*[f(\theta) - f(\pi - \theta)]$$

$$= \frac{1}{4}[f^*(\theta) + f^*(\pi - \theta)][f(\theta) + f(\pi - \theta)] + \frac{3}{4}[f^*(\theta) - f^*(\pi - \theta)][f(\theta) - f(\pi - \theta)]$$

$$= \frac{1}{4}[f^*(\theta)f(\theta) + f^*(\pi - \theta)f(\pi - \theta) + f^*(\theta)f(\pi - \theta) + f^*(\pi - \theta)f(\theta)]$$

$$+ \frac{3}{4}[f^*(\theta)f(\theta) + f^*(\pi - \theta)f(\pi - \theta) - f^*(\theta)f(\pi - \theta) - f^*(\pi - \theta)f(\theta)]$$

$$= \frac{1}{4}[|f(\theta)|^2 + |f(\pi - \theta)|^2 + f^*(\theta)f(\pi - \theta) + f^*(\pi - \theta)f(\theta)]$$

$$+ \frac{3}{4}[|f(\theta)|^2 + |f(\pi - \theta)|^2 - f^*(\theta)f(\pi - \theta) - f^*(\pi - \theta)f(\theta)]$$

Hence

$$\frac{d\sigma}{d\Omega} = |f(\theta)|^2 + |f(\pi - \theta)|^2 - \frac{1}{2}[f^*(\theta)f(\pi - \theta) + f^*(\pi - \theta)f(\theta)]$$

This can be expressed as

$$\frac{d\sigma}{d\Omega} = |f(\theta)|^2 + |f(\pi - \theta)|^2 - \mathrm{Re}\, f^*(\theta)f(\pi - \theta) \tag{7.143}$$

We show in the following how the third term of equation (7.143) follows.
    Let us put

$$f(\theta) = x + iy, f(\pi - \theta) = u + iv$$

Then

$$f^*(\theta)f(\pi - \theta) + f^*(\pi - \theta)f(\theta)$$

$$= (x - iy)(u + iv) + (u - iv)(x + iy)$$

$$= xu + ixv - iyu + yv + xu + iyu - ixv + yv$$

$$= 2(xu + yv). \tag{7.144}$$

Again

$$Re[f^*(\theta)f(\pi - \theta)] = Re[(x - iy)(u + iv)]$$

$$= Re[xu + ixv - iyu + yv]$$

$$= xu + yv \tag{7.145}$$

Hence from equations (7.144) and (7.145) we can write

$$f^*(\theta)f(\pi - \theta) + f^*(\pi - \theta)f(\theta) = 2(xu + yv) = 2Re[f^*(\theta)f(\pi - \theta)]$$

$$\frac{1}{2}f^*(\theta)f(\pi - \theta) + f^*(\pi - \theta)f(\theta)] = Re[f^*(\theta)f(\pi - \theta)]$$

which explains the third term of equation (7.143).
    The first term on the RHS of equation (7.143), i.e. $|f(\theta)|^2$ refers to scattering, the second term $|f(\pi - \theta)|^2$ refers to exchange scattering and the third term $Re\, f^*(\theta)f(\pi - \theta)$ refers to interference between direct and exchange scattering waves.
    We now refer to the radial equation (7.139).
    The nuclear potential $V(r)$ is effective within $r \leqslant r_0$. The Coulomb potential has a long range and is effective at all $r$. We mention the solution of equation (7.139) which is

$$u(r) = \sin\left[kr - \eta \ln 2kr - \frac{l\pi}{2} + \zeta_l + \delta_l\right] \tag{7.146}$$

where
    $\delta_l$ is the change in phase shift due to the nuclear potential $V(r)$ in presence of Coulomb force

$$\eta = \frac{e^2}{4\pi\varepsilon_0 \hbar v}$$

$$\zeta_l = \arg\Gamma(l + 1 + i\eta)$$

$v$ is the relative velocity in the centre of mass framework.

Presence of Coulomb force complicates this problem.

If Coulomb force were not present the plane wave could be expressed as $e^{ikz}$. But due to the presence of Coulomb potential the plane wave is distorted. So the plane wave after distortion in the presence of Coulomb force will be

$$e^{i[kz+\eta lnk(r-z)]} = \sum_{l=0}^{\infty}(2l+1)i^l(kr)^{-1}\sin\left(kr - \eta ln2kr - \frac{1}{2}l\pi\right)P_l(\cos\theta) \qquad (7.147)$$

where we put

$$C_l = (2l+1)k^{-1}i^l e^{i(\zeta_l+\delta_l)} \qquad (7.148)$$

in equation (7.138)

The wave function will have the form

$$\Psi(r,\theta) = e^{i[kz+\eta lnk\,(r-z)]} - f(\theta)\frac{1}{r}e^{i(kr-\eta ln2kr)} \qquad (7.149)$$

Here

$$f(\theta) = -\frac{1}{2ik}\sum_{l=0}^{\infty}(2l+1)[e^{2i(\zeta_l+\delta_l)} - 1]P_l(\cos\theta) \qquad (7.150)$$

The complete solution written in equation (7.149) may be expressed as

$$\psi(r,\theta) = I(r,\theta) + f(r,\theta)S(r) \qquad (7.151)$$

where

$$I(r,\theta) = e^{i[kz+\eta lnk(r-z)]} \qquad (7.152)$$

may be regarded as the incident plane wave modified by the Coulomb force.

$$S(r) = \frac{1}{r}e^{i(kr-\eta ln2kr)} \qquad (7.153)$$

may be regarded as the scattered wave modified by the Coulomb force.

For pure Coulombian scattering $\delta_l = 0$

$f(\theta)$ is the amplitude of the scattered wave.

So $f(\theta)$ will become $f_c(\theta)$, which is the contribution due to Coulomb interaction and has the form

$$f_c(\theta) = -\frac{1}{4k^2}\frac{|e|^2}{4\pi\varepsilon_0}\cosec^2\frac{\theta}{2}e^{-2i\eta lnsin\frac{\theta}{2}} + i\pi + 2i\zeta_0 \qquad (7.154)$$

Now

$$f(\theta) = f_C(\theta) + f_N(\theta) \qquad (7.155)$$

where

$$f_c(\theta) = -\frac{1}{2ik} \sum_{l=0}^{\infty} (2l + 1)(e^{2i\zeta_l} - 1)P_l(\cos\theta) \tag{7.156}$$

(due to Coulombian potential)

$f_c(\theta)$ can be regarded as Coulomb scattering amplitude.

$$f_N(\theta) = -\frac{1}{2ik} \sum_{l=0}^{\infty} (2l + 1)\, e^{2i\zeta_l}(e^{2i\delta_l} - 1)P_l(\cos\theta) \tag{7.157}$$

(due to nuclear potential)

$f_N(\theta)$ can be regarded as nuclear scattering amplitude. The phase shift $\delta_l$ is due to nuclear scattering and $\zeta_l$ corresponds to Coulomb scattering.

Substituting the value of $f_c(\theta)$, only for Coulomb force from (7.154) in (7.143) we get for the differential scattering cross-section due to the Coulmb force only—also called Mott scattering cross-section (all calculations being done in the centre of mass system)

$$\frac{d\sigma}{d\Omega}\Big|_{\text{Mott}} = \left(\frac{|e|^2}{8\pi\varepsilon_0 E}\right)^2 \left[\operatorname{cosec}^4\frac{\theta}{2} + \sec^4\frac{\theta}{2} - \operatorname{cosec}^2\frac{\theta}{2}\sec^2\frac{\theta}{2}\cos\left(\eta\, ln\, \tan^2\frac{\theta}{2}\right)\right]$$

$$= \left(\frac{|e|^2}{8\pi\varepsilon_0 E}\right)^2 \left[\frac{1}{\sin^4\frac{\theta}{4}} + \frac{1}{\cos^4\frac{\theta}{4}} - \frac{\cos\left(\eta\, ln\, \tan^2\frac{\theta}{2}\right)}{\sin^2\frac{\theta}{4}\cos^2\frac{\theta}{4}}\right] \tag{7.158}$$

The simplified form of this equation is with

$$\cos\left(\eta\, ln\, \tan^2\frac{\theta}{2}\right) \cong 1 \text{ if } \eta \text{ is very small}$$

$$\frac{d\sigma}{d\Omega} = \left(\frac{|e|^2}{8\pi\varepsilon_0 E}\right)^2 \left[\frac{1}{\sin^4\frac{\theta}{2}} + \frac{1}{\cos^4\frac{\theta}{2}} - \frac{1}{\sin^2\frac{\theta}{2}\cos^2\frac{\theta}{2}}\right] \tag{7.159}$$

The first term on the RHS of equation (7.159) viz.

$$\left(\frac{|e|^2}{8\pi\varepsilon_0 E}\right)^2 \frac{1}{\sin^4\frac{\theta}{2}}$$

is a characteristic of Coulomb scattering and is identical to the classical result of Rutherford scattering.

The second term on the RHS of equation (7.159) viz.

$$\left(\frac{|e|^2}{8\pi\varepsilon_0 E}\right)^2 \frac{1}{\cos^4\frac{\theta}{2}}$$

can be obtained from the first term $\left(\frac{|e|^2}{8\pi\varepsilon_0 E}\right)^2 \frac{1}{\sin^4\frac{\theta}{2}}$ by exchanging $\theta$ with $\pi - \theta$ because

$$\sin^4\frac{\pi - \theta}{2} = \cos^4\frac{\theta}{2}.$$

Clearly the second term represents exchange of protons. Since two protons are identical we cannot tell whether the incident proton comes out at $\theta$ and the target proton comes out at $\pi - \theta$ in the centre of mass system or whether incident proton comes out at $\pi - \theta$ and the target proton comes out at $\theta$. This term is clearly a characteristic Coulomb term representing Rutherford scattering.

In the third term on the RHS of equation (7.159) viz.

$$\left(\frac{|e|^2}{8\pi\varepsilon_0 E}\right)^2 \frac{1}{\sin^2\frac{\theta}{2}\cos^2\frac{\theta}{2}}$$

both the factors $\sin^2\frac{\theta}{2}$ and $\cos^2\frac{\theta}{2}$ are present. Clearly the third term represents the interference between the proton wave functions as the two protons might get exchanged after Coulomb scattering at $\theta$ and $\pi - \theta$ scattering.

As the Coulomb force has a long range at low energy, high angular momenta are less frequent and the effect of nuclear interaction is considered through $s$ wave scattering that produces a phase shift $\delta_0$ due to nuclear scattering.

The total differential scattering cross-section for $s$ wave scattering is given by

$$\frac{d\sigma}{d\Omega}\Big|_{total} = \left(\frac{|e|^2}{8\pi\varepsilon_0 E}\right)^2 \left[\frac{1}{\sin^4\frac{\theta}{2}} + \frac{1}{\cos^4\frac{\theta}{2}} - \frac{\cos\left(\eta \ln \tan^2\frac{\theta}{2}\right)}{\sin^2\frac{\theta}{2}\cos^2\frac{\theta}{2}}\right]$$

$$-\frac{2\sin\delta_0}{\eta}\left[\frac{\cos\left(\delta_0 + \eta \ln \sin^2\frac{\theta}{2}\right)}{\sin^2\frac{\theta}{2}} + \frac{\cos\left(\delta_0 + \eta \ln \cos^2\frac{\theta}{2}\right)}{\cos^2\frac{\theta}{2}}\right] + \frac{4}{\eta^2}\sin^2\delta_0 \quad (7.160)$$

The fourth and fifth terms on the RHS of equation (7.160) viz.

$$-\frac{2\sin\delta_0}{\eta}\left[\frac{\cos\left(\delta_0 + \eta \ln \sin^2\frac{\theta}{2}\right)}{\sin^2\frac{\theta}{2}} + \frac{\cos\left(\delta_0 + \eta \ln \cos^2\frac{\theta}{2}\right)}{\cos^2\frac{\theta}{2}}\right.$$

is the interference term and represents interference between Coulomb and nuclear scattering as is evident from occurrence of $\sin^2\frac{\theta}{2}$ and $\cos^2\frac{\theta}{2}$.

The sixth term on the RHS of equation (7.160) viz.

$$\frac{4}{\eta^2}\sin^2\delta_0$$

is the pure nuclear scattering.

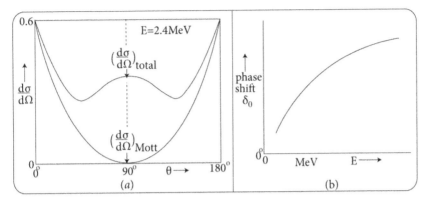

**Figure 7.9.** (a) Differential scattering cross-section $\frac{d\sigma}{d\Omega}$ against angle of scattering $\theta$ for $pp$ scattering. (b) $S$ wave phase shift for $pp$ scattering.

This expression (7.160) reduces to Coulomb scattering cross-ection of equation (7.159) if we take nuclear phase shift $\delta_0 = 0$, i.e. when there is no nuclear scattering effect.

Variation of differential scattering cross-section against $\theta$ is shown in figure 7.9(a). Dependence of phase shift $\delta_0$ on energy $E$ is shown in figure 7.9(b).

In the limit $|e| \to 0$ equation (7.159) reduces to

$$\frac{d\sigma}{d\Omega}\Big|_{|e|\to 0} = \frac{4}{\eta^2}\sin^2 \delta_0$$

which is the sixth term of equation (7.160) representing pure nuclear scattering.

- Facts from low energy $pp$ scattering data

✓ Nuclear forces are charge independent. Forces in the spin singlet states of $nn$, $np$ and $pp$ are identical.
✓ Nuclear forces are spin dependent.
✓ Nuclear forces are independent of the shape of nuclear potential.

The $pp$ scattering length and effective range values are

$$a = -7.8\,fm,\ R_0 = 2.8\,fm.$$

The effective range value of $R_0^{pp} = 2.8\,fm$ is consistent with the singlet $np$ effective range value of $R_{0s}^{np} = 2.7\,fm$ [equation (7.137)].

The scattering length values

$$a^{pp} = -7.8\,fm \text{ and } a_s^{np} = -23.7\,fm \text{ [equation (7.137)]}$$

do not match and actually we cannot compare them.

The scattering length measures the strength of interaction and in $pp$ scattering it includes effect of Coulomb as well as nuclear effects in contrast to the $np$ case where Coulomb effect is not present.

However both $a^{pp}$ and $a_s^{np}$ are negative. The negative value of $a^{pp}$ suggests that there is no $pp$ bound state. In other words, a di-proton bound system does not exist because of spin dependence of nuclear interaction (and the nonexistence is not linked to Coulomb repulsion) (*exercise 7.7*).

## 7.13 Neutron–neutron scattering

Just like $np$ scattering and unlike $pp$ scattering, the $nn$ scattering will not involve Coulomb scattering effects. So $nn$ scattering is expected to be not as complicated as the $pp$ scattering that involves Coulomb plus nuclear interaction.

In $nn$ scattering the difficulty is related to the experimental process itself. Here the projectile is a neutron, the target is also a neutron. While a neutron beam acting as a projectile is available, targets of free neutrons are not available.

One can use a nuclear reaction to create two neutrons in relative motion within the range of each other's nuclear force. The two neutrons separate due to scattering and this can be analysed as an $nn$ scattering experiment. However, such a reaction creates a third particle which will have interactions with both of the neutrons. So necessary corrections are to be made. Some reactions studied are

$$\pi^- + d \rightarrow 2n + \gamma$$

$$n + d \rightarrow 2n + p$$

Analysis of $nn$ scattering leads to the following values of scattering length and effective range

$$a = -16.6\,fm, \quad R_0 = 2.66\,fm \tag{7.161}$$

The negative scattering length shows that two neutrons do not form a stable bound state. In other words, a di-neutron bound system does not exist due to spin dependence of nuclear interaction (*exercise 7.7*).

## 7.14 Nucleon–nucleon scattering at high energy

✓ Valuable information regarding the nature of nuclear forces is obtained in low energy scattering analysis. However, nothing concrete is obtained about the distance dependence of nuclear force.

✓ Also, in low energy scattering analysis one deals with two states only, namely one singlet $^1S$ state and one triplet $^3S$ state. The experiments do not involve states with higher angular momenta.

More information regarding nuclear forces is obtained through high energy scattering experiments. This is because for a high energy incident nucleon the de Broglie wavelength is shorter than range of nuclear force (since $\lambda = \frac{h}{\sqrt{2mE}}$ and as $E$ increases $\lambda$ decreases). Hence a small $\lambda$ will be able to probe the nucleus in greater detail and give information regarding nuclear force.

The problems involved in high energy scattering experiments are as follows.

✓ A monoenergetic neutron beam at higher energies is not available. Energy dependence of cross-section will be unclear the more the spread in energy.

✓ At high energy larger angular momentum states will be present. This results in a large number of parameters and the data will be difficult to interpret.

## 7.15 The *np* scattering above 10 *MeV*

Figure 7.10(a) shows the experimental plot of *np* scattering at high energy. It shows a plot of differential scattering cross-section $\frac{d\sigma}{d\Omega}$ versus scattering angle $\theta$ at high neutron energies.

✓ All the curves show minimum near 90°. This effect is associated with the exchange character of nuclear force and not with higher angular momentum states.

✓ All curves show forward maximum at 0°. This can be explained as being due to absorption and diffraction of neutrons in the scattering medium.

✓ All curves show backward maximum at 180°. This is due to Majorana exchange force (figure 8.5) that interchanges position coordinates of neutron and proton.

The high energy *np* scattering is due to a mixture of ordinary and exchange forces. The observed angular distribution has maximum at 0° due to pure ordinary force and at 180° due to pure exchange force.

• Serber force

The experimental data points to the fact that the nuclear force is an equal mixture of ordinary and exchange forces and this mixture is called Serber force.

Serber force is represented by the operator

$$\rho_{\text{Serber}} = \frac{1}{2}(1 + \rho_M)$$

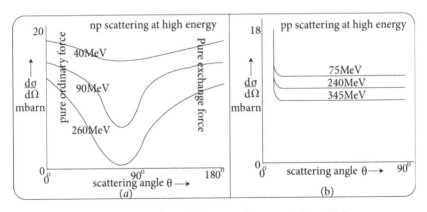

**Figure 7.10.** (a) *np* scattering at high energy (b) *pp* scattering at high energy.

where $\rho_M$ is the Majorana exchange operator [equation (8.125)] which is $+1$ for $l = 0, 2, 4, \ldots$ (attractive force) and $-1$ for $l = 1, 3, 5, \ldots$ (repulsive force).

Clearly for odd states $l = 1, 3, 5, \ldots$ i.e. for $P, F, \ldots$ states

$$\rho_{\text{Serber}} = 0$$

and so it cannot account for the saturation.

The effective range for high energy $np$ scattering depends on shape of potential well.

The high energy $np$ scattering shows the presence of exchange forces.

## 7.16 The $pp$ scattering above 10 $MeV$

Figure 7.10(b) shows the experimental plot of $pp$ scattering at high energy. It shows a plot of differential scattering cross-section $\frac{d\sigma}{d\Omega}$ versus scattering angle $\theta$ at high proton energies.

✓ High energy $pp$ scattering data is easier to interpret than $np$ scattering data since the number of phase shifts reduces to half for $pp$ scattering to satisfy the requirement of Pauli exclusion principle.

✓ For each proton scattered at $\theta$ in centre of mass frame, the other proton is scattered at angle $\pi - \theta$. So the differential cross-section curve for a given incident proton energy is symmetrical about $\theta = 90°$.

✓ As protons are identical particles the coherence between scattered amplitudes gives

$$|f|^2 = |f(\theta) + f(\pi - \theta)|^2$$

This makes the singlet and triplet $pp$ cross-sections four times larger than the corresponding $np$ cross-sections.

✓ Pauli exclusion principle excludes all odd $l$ singlet states and all even $l$ triplet states. So $pp$ scattering can take place from ${}^1S, \; {}^3P, \; {}^1D, \; {}^3F, \ldots$ states.

✓ At small angles $\theta \leqslant 20°$ scattering is dominated by Coulomb interaction and so $\frac{d\sigma}{d\Omega}$ is large.

✓ At larger angles $\theta \geqslant 20°$ $\frac{d\sigma}{d\Omega}$ is nearly independent of scattering angle, i.e. scattering is isotropic (isotropy being a characteristic of $s$ wave scattering).

✓ The differential scattering cross-section is nearly independent of energy in the range 150 to 400 $MeV$ and increases thereafter due to pion creation.

- Jastrow's interpretation

The isotropy of $pp$ differential cross-section was interpreted by Jastrow by assuming a short-range repulsive core or hard sphere of radius $\sim$0.4 $fm$.

The higher $l$ states (other than the $l = 0$ $s$ wave) are kept away from the repulsive core by centrifugal forces.

The $s$ wave phase shift is assumed to be negative, which corresponds to a repulsive potential.

As the incident proton energy increases the $S$ wave will be affected more by the repulsive hard core potential and less by the outer attractive potential so that $S$ wave phase shift will change from positive to negative.

The higher $l$ states are affected by attractive potential and have positive phase shifts.

So at higher proton energies nuclear force is described by a potential that is attractive at larger distances but has a strongly repulsive hard spherical core.

The hypothesis of repulsive core at short distance helps explain the saturation character of nuclear force.

## 7.17 Exercises

**Exercise 7.1** *Suppose in a centre of mass system incident neutron energy is 4 MeV. Calculate the total scattering cross-section for np scattering if it is known that the singlet scattering length is 24 fm, triplet scattering length is 5 fm, singlet effective range is 2.4 fm, triplet effective range is 1.6 fm.*

$\boxed{Ans}$ Given $E = 4\ MeV = 4 \times 10^{6} \times 1.6 \times 10^{-19}\ J = 6.4 \times 10^{-13}\ J$

$$a_s = 24\ fm,\ a_t = 5\ fm,\ R_{0s} = 2.4\ fm,\ R_{0t} = 1.6\ fm$$

From equation (7.125)

$$k^2 = \frac{mE}{\hbar^2} = \frac{(1.675 \times 10^{-27}\ kg)(6.4 \times 10^{-13}\ J)}{\left(\frac{1}{2\pi}6.626 \times 10^{-34}\ J.\ S\right)^2} = 9.6 \times 10^{28}\ m^{-2}$$

From equation (7.133)

$$\sigma^0 = \frac{3\pi}{k^2 + \left(\frac{k^2 R_{0t}}{2} - \frac{1}{a_t}\right)^2} + \frac{\pi}{k^2 + \left(\frac{k^2 R_{0s}}{2} - \frac{1}{a_s}\right)^2}$$

$$= \frac{3\pi}{9.6 \times 10^{28}\ m^{-2} + \left(\frac{\left(9.6 \times 10^{28}\ m^{-2}\right)(1.6\ fm)}{2} - \frac{1}{5\ fm}\right)^2} + \frac{\pi}{9.6 \times 10^{28}\ m^{-2} + \left(\frac{\left(9.6 \times 10^{28}\ m^{-2}\right)(2.4\ fm)}{2} - \frac{1}{24\ fm}\right)^2}$$

$$= \frac{3\pi}{9.6 \times 10^{28} + \left(\frac{1.536 \times 10^{14}}{2} - \frac{1}{5 \times 10^{-15}}\right)^2}m^2 + \frac{\pi}{9.6 \times 10^{28} + \left(\frac{2.304 \times 10^{14}}{2} - \frac{1}{24 \times 10^{-15}}\right)^2}m^2$$

$$= 9.25 \times 10^{-29}\ m^2 + 3.09 \times 10^{-29} m^2$$

$$= 1.234 \times 10^{-28}\ m^2$$

$$= 1.234\ barn$$

**Exercise 7.2** *Construct all states of a two-nucleon system corresponding to $J \leqslant 2$.*

$\boxed{Ans}$ $\vec{J} = \vec{l} + \vec{s}$

Spectroscopic notation $^{2s+1}(l)_J$

Nucleons are spin $\frac{1}{2}$ particles $s_1 = \frac{1}{2}$, $s_2 = \frac{1}{2}$.

$$\vec{s} = \vec{s_1} + \vec{s_2} = \frac{1}{2} + \frac{1}{2} = \begin{cases} \frac{1}{2} - \frac{1}{2} = 0: s = 0 \Rightarrow \text{singlet } 2s + 1 = 2.0 + 1 = 1 \\ \frac{1}{2} + \frac{1}{2} = 1: s = 1 \Rightarrow \text{triplet } 2s + 1 = 2.1 + 1 = 3 \end{cases}$$

- For singlet $s = 0$, $\vec{J} = \vec{l} + \vec{0} = \vec{l}$, overall parity $(-1)^{l+s+1}$
- ✓ $l = 0$, $S$ state, $J = l = 0$, $^1S_0$, parity $(-1)^{l+s+1} = (-1)^{0+0+1} = -ve$, antisymmetric
- ✓ $l = 1$, $P$ state, $J = l = 1$, $^1P_1$, parity $(-1)^{l+s+1} = (-1)^{1+0+1} = +ve$, symmetric
- ✓ $l = 2$, $D$ state, $J = l = 2$, $^1D_2$, parity $(-1)^{l+s+1} = (-1)^{2+0+1} = -ve$, antisymmetric

- For triplet $s = 1$, $\vec{J} = \vec{l} + \vec{1}$, overall parity $(-1)^{l+s+1}$
- ✓ $l = 0$, $S$ state, $J = \vec{0} + \vec{1} \rightarrow 1$, $^3S_1$, parity $(-1)^{l+s+1} = (-1)^{0+1+1} = +ve$, symmetric
- ✓ $l = 1$, $P$ state, $J = \vec{1} + \vec{1} \rightarrow 2, 1, 0$, $^3P_{2,1,0}$, parity $(-1)^{l+s+1} = (-1)^{1+1+1} = -ve$, antisymmetric
- ✓ $l = 2$, $D$ state, $J = \vec{2} + \vec{1} \rightarrow 3, 2, 1$, $^3D_{3,2,1}$, Ignore $J = 3$ as $J \leqslant 2$ is to be considered. So we consider $^3D_{2,1}$, parity $(-1)^{l+s+1} = (-1)^{2+1+1} = +ve$, symmetric.
- ✓ $l = 3$, $F$ state, $J = \vec{3} + \vec{1} \rightarrow 4, 3, 2$, $^3F_{4,3,2}$, ignore $J = 4, 3$ as $J \leqslant 2$ is to be considered. So we consider $^3F_2$, parity $(-1)^{l+s+1} = (-1)^{3+1+1} = -ve$, antisymmetric.

The two-nucleon system must be antisymmetric w.r.t. exchange of particle coordinates as determined from the parity factor $(-1)^{l+s+1}$. So the spectroscopic notation for the two-nucleon states will be

$$^1S_0, \quad ^1D_2, \quad ^3P_{2,1,0}, \quad ^3F_2,$$

**Exercise 7.3** *The singlet scattering length is*
*(a) positive, (b) negative, (c) zero, (d) half the range of nuclear force.*

**Exercise 7.4** *The ratio of scattering cross-sections for ortho and para states in hydrogen is nearly*
*(a) 1, (b) 2, (c) 30, (d) 120.*

**Exercise 7.5** *Explain why optical diffraction is analogous to the scattering of nucleons.*

[*Ans*] Some features of optical diffraction resemble scattering of nucleons.

✓ Incident wave is represented by plane wave.

✓ Far from the obstacle the scattered wave fronts are spherical

✓ The total energy content of any expanding spherical wave front is conserved. So its intensity per unit area decreases as $\frac{1}{r^2}$ and so amplitude decreases as $\frac{1}{r}$.

✓ Intensity depends on angular coordinates $\theta$, $\phi$. There is variation of intensity along the surface of spherical scattered wave front due to diffraction.

✓ Any detector placed at a far away point from the obstacle would record both incident and scattered waves.

**Exercise 7.6** *Obtain a rough estimate of the kinetic energy of neutron projectile for $l = 0$ interaction with a target nucleon.*

[*Ans*] Consider an incident nucleon striking a target nucleon—the range of interaction being $r \sim 1\ fm$. The orbital angular momentum of a projectile nucleon relative to target is $mvr$ where $m$ is nucleon mass, $mc^2 \sim 939\ MeV$ being the energy, $v$ is the velocity of incident nucleon (neutron).

The relative orbital angular momentum between the nucleons must be quantized in units of $\hbar$ i.e.

$$mvr = l\hbar$$

$$\text{If } mvr \ll \hbar$$

then only $l = 0$ interaction is likely to occur. Thus

$$v \ll \frac{\hbar}{mr}$$

The corresponding kinetic energy is

$$T = \frac{1}{2}mv^2 \ll \frac{1}{2}m\left(\frac{\hbar}{mr}\right)^2 = \frac{\hbar^2}{2mr^2} = \frac{\hbar^2 c^2}{2(mc^2)r^2} = \frac{(197\ MeV.\,fm)^2}{2(939\ MeV)(1\ fm)^2} = 20.66\ MeV$$

where we used *exercise 1.2*.

If the incident energy is $T \ll 20\ MeV$ the $l = 0$ interaction predominates. In other words, for low energy nucleon–nucleon scattering, $l = 0$ interaction will be predominant if projectile nucleon energy is much less than $20\ MeV$.

**Exercise 7.7** *From a study of $l = 0$ two-nucleon state argue that pp and nn bound states do not exist.*

[*Ans*] Study of $l = 0$ two-nucleon state means the spatial wave function has parity $(-1)^l = (-1)^0 = +ve$, i.e. symmetric. Again, nucleons are fermions and system of two nucleons has total wave function antisymmetric. This means the spin part should be antisymmetric, i.e. singlet spin state. Again, singlet spin state corresponds to negative scattering lengths and hence are unbound states.

**Exercise 7.8** *Identify the correct statement from the following propositions.*
  *(a) Space part of wave function for para hydrogen is symmetric.*
  *(b) Space part of wave function for para hydrogen is antisymmetric.*
  *(c) Space part of wave function for ortho hydrogen is symmetric.*
  *(d) Space part of wave function for ortho hydrogen is antisymmetric.*

(Hint: Para hydrogen is in singlet state and has antisymmetric spin part, the space part is symmetric.

Ortho hydrogen is in triplet state and has symmetric spin part, the space part is antisymmetric.

**Exercise 7.9** *Identify the correct statement.*
  *(a) Positive scattering length corresponds to attractive potential giving rise to scattering from unbound state.*
  *(b) Positive scattering length corresponds to attractive potential giving rise to scattering from bound state.*
  *(c) Positive scattering length corresponds to repulsive potential giving rise to scattering from unbound state.*
  *(d) Positive scattering length corresponds to repulsive potential giving rise to scattering from bound state.*

**Exercise 7.10** *Identify the correct statement.*
  *(a) Negative scattering length corresponds to attractive potential giving rise to scattering from unbound state.*
  *(b) Negative scattering length corresponds to attractive potential giving rise to scattering from bound state.*
  *(c) Negative scattering length corresponds to repulsive potential giving rise to scattering from unbound state.*
  *(d) Negative scattering length corresponds to repulsive potential giving rise to scattering from bound state.*

**Exercise 7.11** *Which partial waves are present in* $f(\theta) = \frac{4 + 3\cos 2\theta}{2ik}$?

$$\boxed{Ans}\, f(\theta) = \frac{4 + 3\cos 2\theta}{2ik}$$

$$= \frac{4 + 3(2\cos^2\theta - 1)}{2ik} = \frac{1 + 2.3\cos^2\theta}{2ik}$$

$$= \frac{1 + 2(3\cos^2\theta - 1) + 2}{2ik} = \frac{3 + 4\frac{1}{2}(3\cos^2\theta - 1)}{2ik}$$

$$f(\theta) = \frac{3}{2ik} + \frac{4}{2ik}\frac{1}{2}(3\cos^2\theta - 1)$$

Using $P_0(\cos\theta) = 1$, $P_1(\cos\theta) = \frac{1}{2}(3\cos^2\theta - 1)$

$$f(\theta) = \frac{3}{2ik}P_0(\cos\theta) + \frac{4}{2ik}P_2(\cos\theta)$$

So the expression $f(\theta)$ contains partial waves $l = 0, 2$.

$$\boxed{\text{Ans to Multiple Choice Type Questions}}$$

7.3 *b*, 7.4 *c*, 7.8 *a*, *d*, 7.9 *b*, 7.10 *c*

## 7.18 Question bank

Q7.1 Why cannot all information regarding nuclear force be obtained from deuteron?

Q7.2 What is the difference between study of the scattering problem and study of the deuteron problem?

Q7.3 Define scattering cross-section. What is its physical significance? What is its unit and dimension?

Q7.4 Define differential scattering cross-section. What is its unit and dimension? What is the effect on it for isotropic scattering.

Q7.5 What is elastic scattering?

Q7.6 What is inelastic scattering?

Q7.7 What do you mean by radiative capture?

Q7.8 Define scattering amplitude. How is it related to differential scattering cross-section. Write down the total scattering cross-section in terms of scattering amplitude.

Q7.9 Establish that the asymptotic solution to the scattering problem has the form

$$\psi_k = e^{ikz} + f_k(\Omega)\frac{e^{ikr}}{r}$$

Q7.10 Show the experimental curve of low energy *np* scattering as obtained by Mel, Konian and Squires. How can it be explained theoretically?

Q7.11 Define scattering length. What is its significance?

Q7.12 How can you explain that positive scattering length refers to a bound state?

Q7.13    How can you explain that negative scattering length refers to an unbound state?

Q7.14    Why is the knowledge of scattering length sufficient to describe $s$ wave $np$ scattering?

Q7.15    Why was spin dependence of nuclear force needed to explain scattering cross-section for low energy $np$ scattering?

Q7.16    How was Wigner able to explain scattering cross-section for low energy $np$ scattering assuming that nuclear force is spin dependent?

Q7.17    How was spin dependence of nuclear force inferred from study of low energy $np$ scattering?

Q7.18    Molecular hydrogen can exist in two states. What are they?

Q7.19    What is ortho hydrogen?

Q7.20    What is para hydrogen?

Q7.21    How do you infer spin dependence of nuclear force from a study of coherent scattering from ortho and para hydrogen?

Q7.22    What do you mean by coherent scattering from ortho and para hydrogen?

Q7.23    Compare coherent and incoherent scattering.

Q7.24    Obtain the relation between scattering cross-sections for ortho and para hydrogen.

Q7.25    Write down the expected ratio of $\frac{\sigma_{ortho}}{\sigma_{para}}$ if (a) $np$ forces were independent of spin, (b) $np$ forces are spin dependent.

Q7.26    How do you explain that singlet state $np$ scattering length is negative from a study of coherent scattering of ortho and para hydrogen?

Q7.27    What is the significance of the fact that the singlet state $np$ scattering length is negative?

Q7.28    Outline briefly the effective range theory of low energy $np$ scattering.

Q7.29    Obtain total $np$ scattering cross-section in terms of scattering length and effective range for low energy scattering. Write down the experimentally obtained best fit values of low energy $np$ scattering parameters like triplet and singlet scattering lengths and effective ranges.

Q7.30    What are the essential differences between $np$ and $pp$ scattering?

Q7.31    Why do think that $pp$ scattering theory is more complicated compared to $np$ scattering theory.

Q7.32    In $pp$ scattering identical particles are involved. What effect does it have on the scattering analysis? Does it complicate or simplify the problem?

Q7.33    Give a brief outline of $pp$ scattering.

Q7.34    Write down the differential scattering cross-section for $s$ wave $pp$ scattering. Herefrom explain the following terms: Mott scattering, nuclear scattering, interference between Coulomb and nuclear scattering, identifying the terms they represent.

Q7.35    Why is $nn$ scattering considered to be simpler than $pp$ scattering? What are the difficulties associated with $nn$ scattering?

Q7.36 The scattering length and of *nn* scattering is $-16.6\,fm$. What does the negative sign signify?

Q7.37 Nonexistence of stable bound di-neutron system is connected with spin dependence of nuclear force. Justify.

Q7.38 What are the problems involved in nucleon–nucleon scattering at high energy?

Q7.39 Sketch the experimental plot of (a) *np* scattering at high energy, (b) *pp* scattering at high energy. Compare them.

Q7.40 What is Serber force?

Q7.41 What was Jastrow's interpretation regarding *pp* scattering?

Q7.42 Explain the hypothesis of existence of a repulsive core regarding nuclear force? Justify.

Q7.43 Identify the region of isotropic *pp* scattering. Justify.

## Further reading

[1] Krane S Kenneth 1988 *Introductory Nuclear Physics* (New York: Wiley)

[2] Tayal D C 2009 *Nuclear Physics* (Mumbai: Himalaya Publishing House)

[3] Satya Prakash 2005 *Nuclear Physics and Particle Physics* (New Delhi: Sultan Chand & Sons)

[4] Guha J 2019 *Quantum Mechanics: Theory, Problems and Solutions* 3rd edn (Kolkata: Books and Allied (P) Ltd))

[5] Lim Y-K 2002 *Problems and Solutions on Atomic, Nuclear and Particle Physics* (Singapore: World Scientific)

**IOP** Publishing

**Nuclear and Particle Physics with Cosmology, Volume 1**
Nuclear physics
**Jyotirmoy Guha**

# Chapter 8

## Nature of nuclear force

## 8.1 Theory of nuclear force

During β decay, the nuclear interaction involved, changes the charge state of nucleons.

✓ If number of neutrons is greater than that required for stability of nucleus then neutron changes to proton and electron emission occurs, e.g.

$$n \rightarrow p + e^- + \bar{\nu}_e \text{ called } \beta^- \text{decay.}$$

✓ If number of protons is greater than that required for stability of nucleus then proton changes to neutron and positron emission occurs, e.g.

$$p \rightarrow n + e^+ + \bar{\nu}_e \text{ called } \beta^+ \text{decay.}$$

Beta decay is a weak interaction and we can get an estimate of the strength of interaction from the Fermi theory of beta decay. The coupling constant of beta decay is of small magnitude $\sim 10^{-6}$. Interaction of such a small magnitude cannot be responsible for nuclear force which is a strong force and corresponds to nuclear binding energy $\sim 8 \frac{MeV}{nucleon}$. To account for such binding energy one needs a strong interaction characterized by a large coupling constant $\sim$ unity.

Heisenberg made a hypothesis which said that nuclear force possesses exchange character. Meson theory of exchange force, also called meson theory of nuclear force, was proposed by Yukawa on the basis of meson hypothesis.

## 8.2 Force generation mechanism

Force is generated when particles interact. When we say that particles are exchanging forces we mean that they exert force on each other. We can think of various mechanisms of such interaction.

doi:10.1088/978-0-7503-5027-3ch8

✓ Concept of particle–particle direct interaction or action at a distance.

Newton's law of gravitation and Coulomb's law of electrostatics are based upon the concept of action at a distance.

We say that particle 1 directly exerts force or influences particle 2 though they are not in contact but are separated. Again by Newton's third law of motion particle 2 will similarly exert equal and opposite direct force on particle 1. Clearly here we are referring to forces without contact. There is no idea of field involved here.

✓ Concept of field

Particle 1 does not exert force directly on particle 2 but creates a field around it. This field exerts force on particle 2. Similarly particle 2 also creates a field around it and this field exerts force on particle 1. In other words, the two particles interact through the field they establish.

✓ Exchange theory—the basis of Yukawa's theory.

The quantum field theory supposes that the fields are carried by quanta. Instead of setting up a field it is said that the particle emits quanta of the field and another particle absorbs this quanta. According to quantum field theory, force or interaction is mediated by some virtual particle that is exchanged between the interacting particles.

The particle exchanged is virtual. Virtual because during interaction when particle 1 shoots off the exchange particle towards particle 2 (or vice versa) the exchange particle cannot be detected. This again follows from Heisenberg's uncertainty principle

$\Delta E \Delta t \sim \hbar$

Actually, during interaction mediated by the exchange particle one cannot detect whether there is energy violation $\Delta E$ over a time $\Delta t$ given by

$$\Delta t = \frac{\hbar}{\Delta E} \tag{8.1}$$

since energy is uncertain by an amount $\Delta E$ over a time $\Delta t$. This throws up the possibility of violation of energy conservation during that period of time $\Delta t$. Clearly even if energy violation occurs in time $\Delta t$ we fail to detect that. In other words energy violation is allowed in time $\Delta t$ given by equation (8.1).

So during mediation of interaction over the time $\Delta t$ the exchange particles cannot be detected—they are beyond the reach of any physical measurement.

However, these particles do exist and have been detected in various physical measurement processes.

Each type of field has its own characteristic particle. For instance, a photon mediates electromagnetic interaction.

Pion mediates strong interaction between nucleons according to Yukawa theory. Actually, it is the gluons that mediate strong interaction between quarks which are the constituents of nucleons.

$W$, $Z^{\pm}$ bosons mediate weak interaction.

Graviton is said to mediate gravitational interaction—though it is not detected in experiment yet.

- Electromagnetic interaction

    Electromagnetic interaction occurs when two charges, say $q_1$ and $q_2$, interact. The virtual particle exchanged between the interacting charges is photon $\gamma$. In other words, electromagnetic interaction between two particles can be viewed in terms of emission and absorption of photons which are quanta of electromagnetic field.

    A photon has zero rest mass and moves with speed $c = 3 \times 10^8 \, m \, s^{-1}$ in free space. The photon quanta, so to say carries the interaction. In other words the two charges electromagnetically interact through an incessant and continuous give and take of photons. We say that a photon is the quantum of electromagnetic field or electromagnetic wave. Electromagnetic force acts due to exchange of photons between charges. So a photon is a mediator of electromagnetic interaction.

- Graviational interaction

    Gravitationl interaction between two masses is propagated or mediated by field quanta called gravitons. We say that graviton is the quantum of gravitational field or gravitational wave.

    Graviton has zero rest mass and moves with speed $c = 3 \times 10^8 \, m \, s^{-1}$ in free space. Gravitational force acts due to exchange of gravitons between masses. So a graviton is a mediator of gravitational interaction.

- Nuclear interaction

Yukawa explained short-range nuclear interaction in terms of field quanta that was named $\pi$ meson. Let us discuss Yukawa's meson theory of nuclear force in the following.

## 8.3 Yukawa's meson theory of nuclear force

$\pi$ meson or pion was thought to mediate nuclear interaction. In other words, $\pi$ meson is the quantum of nuclear interaction.

The mediator of strong interaction—the $\pi$ meson was not discovered when Yukawa made the hypothesis in 1935. Later, in 1947, the pion was discovered by Powell during investigation of cosmic radiation.

Nucleon was regarded as being surrounded by a cloud of virtual pions—that are continuously absorbed and emitted.

Mass of pion was estimated assuming that there was exchange of virtual pions between nucleons in a nucleus.

Exchange of pion implies two processes taking place.

✓ Emission of pion

$$N_1 \rightarrow N_1 + \text{pion}$$

and

✓ Absorption of pion

$$N_2 + \text{pion} \rightarrow N_2.$$

Here $N_1$, $N_2$ represent two nucleons, say $n$, $p$ or $n$, $n$ or $p$, $p$.

The $n \leftrightarrows p$ transformation suggests that pions can be charged and can have mass while $p \leftrightarrows p$ transformation and $n \leftrightarrows n$ transformation suggest that pions can be neutral.

Clearly there can be three types of pions:

+*vely* charged, –*vely* charged, neutral.

Later, three types of pions were indeed discovered, namely

$\pi^+$ (+ve pion)

$\pi^-$ (–ve pion)

$\pi^0$ (neutral pion).

Let us write the interactions $p \leftrightarrows p$, $n \leftrightarrows n$, $n \leftrightarrows p$ mediated through three types of pion exchanges $\pi^+$, $\pi^-$, $\pi^0$.

$$n \rightarrow n + \pi^0 \Rightarrow n \text{ emits } \pi^0$$
$$p + \pi^0 \rightarrow p \Rightarrow p \text{ absorbs } \pi^0$$
$$p \rightarrow p + \pi^0 \Rightarrow p \text{ emits } \pi^0$$
$$n + \pi^0 \rightarrow n \Rightarrow n \text{ absorbs } \pi^0$$
$$n \rightarrow p + \pi^- \Rightarrow n \text{ emits } \pi^- \text{ and converts to } p$$
$$p + \pi^- \rightarrow n \Rightarrow p \text{ absorbs } \pi^- \text{ and converts to } n$$
$$p \rightarrow n + \pi^+ \Rightarrow p \text{ emits } \pi^+ \text{ and converts to } n$$
$$n + \pi^+ \rightarrow p \Rightarrow n \text{ absorbs } \pi^+ \text{ and converts to } p$$

So a nucleon is transformed to another nucleon and in the process a pion is either created or absorbed, i.e. destroyed.

Since pions have non-zero rest mass this process implies that in the above transformations mass is either created or destroyed. And this suggests violation of energy conservation principle.

Let us illustrate.

☐ Consider the exchange interaction

$$n \rightarrow n + \pi^0$$

A neutron emits a $\pi^0$ and still remains a neutron. We have $n$ on both sides —so $m_n c^2$ occurs on both sides and hence we fail to account for the pion mass viz. $m_{\pi^0} c^2$ since we see that a neutron remains a neutron but a pion has been created. So mass $\Delta E = m_{\pi^0} c^2$ has been created out of nothing. This suggests violation of energy conservation principle during the exchange interaction process.

☐ Consider the exchange interaction

$$p + \pi^0 \rightarrow p$$

A proton absorbs a $\pi^0$ and still remains a proton. We have $p$ on both sides —so $m_p c^2$ occurs on both sides and hence we fail to account for the pion mass viz. $m_{\pi^0} c^2$ since we see that a proton remains a proton but a pion has been absorbed. So the energy $\Delta E = m_{\pi^0} c^2$ seems to get destroyed. This suggests violation of the energy conservation principle during the exchange interaction process.

Such violation of energy conservation principle by amount of energy $\Delta E = m_\pi c^2$ is permitted by Heisenberg's uncertainty principle provided it occurs within a time

$$\Delta t = \frac{\hbar}{\Delta E} = \frac{\hbar}{m_\pi c^2} \qquad (8.2)$$

In other words, pions can be created or destroyed. Pions can play the role of mediator of strong interaction between nucleons over a time $\Delta t$.

Again, a pion travels from one nucleon to another nucleon with speed say $v$. Suppose the concerned nucleons are sitting at the edges of the nucleus and let us take the inter-nucleon distance to be 1.4 *fm*, which is thus the range of pion within nucleus. And the pion has to travel this distance in time $\Delta t$. Hence

$$v \Delta t = 1.4 \, fm$$

By equation (8.2)

$$v \frac{\hbar}{m_\pi c^2} = 1.4 \, fm.$$

For getting an estimate we take $v \cong c$ where $c$ is the speed of light in free space. So

$$c \frac{\hbar}{m_\pi c^2} = 1.4 \, fm$$

$$m_\pi c^2 = \frac{\hbar c}{1.4 \, fm}$$

Using *exercise 1.2*

$$m_\pi c^2 = \frac{197 \, MeV. \, fm}{1.4 \, fm} = 141 \, MeV \qquad (8.3)$$

So the pion mass as predicted is

$$m_\pi \sim 141 \frac{MeV}{c^2}$$

compared to proton mass $m_p = 938 \frac{MeV}{c^2}$, neutron mass $m_n = 939 \frac{MeV}{c^2}$.

Clearly pions are massive particles. Later experiments showed that

$$m_{\pi^0} = 135 \frac{MeV}{c^2}$$

$$m_{\pi^\pm} = 140 \frac{MeV}{c^2}$$

Experimental agreement with pion masses suggests that range of nuclear force is ~1.4 *fm* as suggested by Yukawa's theory.

Let us express pion mass in terms of electron mass $m_e = 0.51\frac{MeV}{c^2}$. Consider

$$m_{\pi^0} = \frac{m_{\pi^0}}{m_e}m_e = \frac{135}{0.51}m_e = 265m_e$$

$$m_{\pi^\pm} = \frac{m_{\pi^\pm}}{m_e}m_e = \frac{140}{0.51}m_e = 275m_e$$

The mass of pions as predicted by Yukawa's theory and the mass of pions that was experimentally detected turned out to be in very good agreement and this proved the success of Yukawa's theory and meson hypothesis.

We can find the range of electromagnetic and gravitational interaction along similar arguments as shown in *exercise 8.1*.

✓ In *exercise 8.30* we find the time interval during which a nucleon can emit and absorb a pion without detection of any violation of energy conservation. We also find the range $r_0$ or the distance between two nucleons within which the nucleons would interact strongly through exchange of pions and beyond which pion exchange will not occur and they will not interact strongly.

Let us define what is meant by exchange force.

* Exchange force or exchange character of nuclear force

Force accomplished through exchange of quanta between two nucleons is called exchange force. Continuous give and take, i.e. exchange of quanta binds two nucleons through a force which is exchange force. In other words generation of nuclear force due to exchange of particles between nucleons is referred to as the exchange character of nuclear force.

## 8.4 One-pion exchange potential (OPEP) or Yukawa potential

Assume for simplicity that interactions are accomplished through exchange of one pion.

Let us find the corresponding potential called Yukawa potential or one-pion exchange potential, abbreviated as OPEP.

In equation (8.3) we took the speed of pions to be $v \sim c$ and this means we have to use relativistic formula. Total energy of pion is given by the well-known formula of special theory of relativity as

$$E^2 = p^2c^2 + m_{0\pi}^2c^4 \tag{8.4}$$

where $p$ is pion momentum, $m_{0\pi}$ is pion rest mass.

Invoke Hermitian operators for energy

$$E \rightarrow i\hbar\frac{\partial}{\partial t}$$

and momentum

$$\vec{p} = -i\hbar\vec{\nabla}$$

Operator form of equation (8.4) is

$$\left(i\hbar\frac{\partial}{\partial t}\right)\left(i\hbar\frac{\partial}{\partial t}\right) = (-i\hbar\vec{\nabla}) \cdot (-i\hbar\vec{\nabla})c^2 + m_{0\pi}^2 c^4$$

$$-\hbar^2\frac{\partial^2}{\partial t^2} = -c^2\hbar^2\nabla^2 + m_{0\pi}^2 c^4$$

We introduce a scalar pion wave function $\phi\,(\vec{r}, t)$ describing the pion field and using it we develop wave equation

$$-\hbar^2\frac{\partial^2\phi}{\partial t^2} = -c^2\hbar^2\nabla^2\phi + m_{0\pi}^2 c^4\phi$$

$$\nabla^2\phi - \frac{1}{c^2}\frac{\partial^2\phi}{\partial t^2} - \frac{m_{0\pi}^2 c^2}{\hbar^2}\phi = 0 \tag{8.5}$$

This is the Klein–Gordon equation which is the relativistic wave equation applicable for spin zero particle.

In the case of quantum of electromagnetic field, i.e. zero rest mass photon we have to replace $m_\pi$ by $m_\gamma = 0$ and $\phi$ will be the wave function of electromagnetic field. Then equation (8.5) reduces to

$$\nabla^2\phi - \frac{1}{c^2}\frac{\partial^2\phi}{\partial t^2} = 0 \tag{8.6}$$

which is the well-known wave equation for quanta of electromagnetic wave. It is Maxwell's electromagnetic wave equation.

Again, time independent electromagnetic field is called electrostatic field which is the simplest electromagnetic field.

Electrostatic field $\phi$ is not an explicit function of time $t$ and so

$$\frac{\partial\phi}{\partial t} = 0$$

The equation (8.6) reduces to Laplace equation that is valid in case of electromagnetic field in the absence of charges, i.e. in charge-free space viz.

$$\nabla^2\phi = 0 \tag{8.7}$$

Now in the pion field we can similarly obtain an equation that resembles Laplace equation by removing in equation (8.5) the time dependent part $\frac{1}{c^2}\frac{\partial^2\phi}{\partial t^2}$ to get

$$\nabla^2\phi - \frac{m_{0\pi}^2 c^2}{\hbar^2}\phi = 0 \tag{8.8}$$

Define

$$k = \frac{m_{0\pi}c}{\hbar} = \frac{1}{\hbar/m_{0\pi}c} = \frac{1}{r_0} \tag{8.9}$$

As $r_0$ is range of nuclear force clearly $k$ is the reciprocal of the range of nuclear force.

We rewrite equation (8.8) as

$\nabla^2\phi - k^2\phi = 0$

$$(\nabla^2 - k^2)\phi = 0 \tag{8.10}$$

Again, we note that if charges are present we cannot use the Laplace equation. We have to take into account the charge density given by

$$\rho(\vec{r}) = q\delta(\vec{r})$$

for a point charge $q$ located at $\vec{r} = 0$, where $\delta(\vec{r})$= Dirac delta function.

In the presence of charge we have to employ the Poisson equation

$$\nabla^2\phi = -\frac{\rho(\vec{r})}{\varepsilon_0} = -\frac{q\delta(\vec{r})}{\varepsilon_0} \tag{8.11}$$

Solution of the Poisson equation is

$$\phi = -\frac{1}{4\pi\varepsilon_0}\frac{q}{r} \text{ (attractive solution)} \tag{8.12}$$

Similarly, the nucleus is not a charge-free region.

And for a single point nucleon at origin $\vec{r} = 0$ we have to use an equation analogous to the Poisson equation (8.11) to solve our problem.

Let us rewrite equation (8.10) by analogy to equation (8.11)

$$(\nabla^2 - k^2)\phi = 4\pi g\delta(\vec{r}) \tag{8.13}$$

where $g$ is referred to as coupling constant and plays the same role as charge in the electrostatic case.

The solution of equation (8.13) as assumed by Yukawa was

$$\phi = -g\frac{e^{-kr}}{r} \tag{8.14}$$

In *exercise 8.2* it is shown that $\phi = -g\frac{e^{-kr}}{r}$ is a solution of equation (8.13).

In the electrostatic case potential energy at a location of a charge is

$$V = \text{(charge)(potential at that point)} = q\phi. \tag{8.15}$$

Similarly, in the pion field within nucleus one can also obtain the potential energy to be

$$V = g\phi$$

From equation (8.14) we have

$$V = g\left(-g\frac{e^{-kr}}{r}\right)$$

$$V = -g^2\frac{e^{-kr}}{r} \tag{8.16}$$

Using equation (8.9), i.e. $k = \frac{1}{r_0}$ we have the Yukawa potential energy, also called Yukawa potential or one-pion exchange potential (OPEP) energy given by

$$V(r) = -g^2\frac{e^{-r/r_0}}{r} = -\frac{g^2}{r}e^{-r/r_0} \tag{8.17}$$

This represents the strong interaction between two elementary particles, i.e. nucleons.

At $r \to 0$, $V \to \infty$ and at

$$r \to \infty, \ V \to 0.$$

Plot of Yukawa potential energy versus $r$ is shown in figure 8.1.
Coulombian attractive potential energy

$$V(r) = -\frac{q^2}{4\pi\varepsilon_0 r} \equiv -\frac{g^2}{r} \quad \left(g^2 = \frac{q^2}{4\pi\varepsilon_0}\right) \tag{8.18}$$

is also characterized by the fact that at $r \to 0$, $V \to \infty$ and at $r \to \infty$, $V \to 0$.

We have shown a plot of the Coulomb potential in the same figure 8.1 for comparison purpose.

It is evident from equations (8.17) and (8.18) that the Yukawa potential adds an exponential term $e^{-r/r_0}$ to the Coulomb potential that greatly shortens the range. For $r < r_0$ both the potentials vary in a similar manner. A look at the Yukawa potential

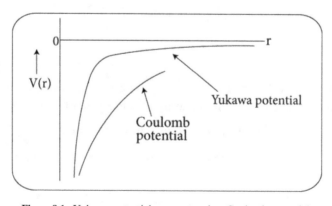

**Figure 8.1.** Yukawa potential as compared to Coulomb potential.

shows that apart from the exponentially decaying part $e^{-r/r_0}$ the rest portion viz. $-\frac{g^2}{r}$ resembles Coulombian potential. So with increase in $r$ Yukawa potential varies more rapidly—in fact, falls exponentially over and above the Coulomb potential due to the exponential factor.

Yukawa potential is also called screened Coulomb potential.

- Conclusions from Yukawa meson theory

☐ There are three types of interactions in a nucleus
  ✓ $nn$ interaction ✓ $pp$ interaction ✓ $np$ interaction.

  ☐ For all these interactions nuclear force is the same in magnitude because nuclear force is charge independent. We can say that a nuclear particle does not change its charge during the interaction.
  ☐ The exponentially decaying part of Yukawa potential suggests that nuclear force is short range. Range can be calculated from mass of virtual exchange particle and it follows that nuclear force is short range $r_0 \sim 1.4\,fm$.
  ☐ The value of coupling constant is large—taken to be unity. Large coupling constant suggests that more than one meson can be transferred simultaneously between two nucleons though we considered only one pion exchange and we obtained one-pion exchange potential.
  ☐ For very small separation nuclear force becomes repulsive, i.e. has a repulsive core. This follows from the presence of Dirac delta function used in this theory.

- Shortcomings of Yukawa's theory
  ☐ There are other mesons like $\omega$ meson with higher energy implying that $\Delta E$ is larger and this means larger violation is allowed by Heisenberg's uncertainty principle over a shorter time range, i.e. for $\Delta t =$ smaller. So these higher energy mesons will not take part in nuclear interactions having range $r_0 \geqslant 1.5\,fm$ but they can be exchanged at lower range $r_0 \leqslant 1.5\,fm$.
  ☐ Yukawa's theory in its simple form (as discussed) cannot explain many properties of nuclear force, for instance spin dependence of nuclear force, non-central nature of nuclear force or tensor force.

To explain them we have to make a rigorous approach, for instance incorporate the tensorial character and isobaric spin character in the meson field wave function.

## 8.5 Approaches to study nuclear potential

There are three independent approaches to study the nucleon–nucleon force and potential. We list them as follows.
  ✓ Phenomenological approach
    In this approach we focus on various scattering experiments at low and high energy like $n$–$p$ scattering, $p$–$p$ scattering, $n$–$n$ scattering etc. Based upon this experimental data and arguments potentials are constructed with

variable parameters. The parameters are adjusted to match experimental data.

For instance to explain nucleon–nucleon interaction we can start with a square well potential described by parameters like depth, width and these can be adjusted to match data. The potentials that are constructed and we work with are called phenomenological potentials.

✓ Yukawa's meson exchange theory

In this approach we focus on how the nuclear force is manifested or executed—its nature and cause. On the basis of pion hypothesis Yukawa explained it through a potential known as Yukawa potential

$$V(r) = -V_0 \frac{e^{-r/r_0}}{r/r_0}. \tag{8.19}$$

In this approach we start with a mechanism that interaction between nucleons occurs in a particular manner and based on that mechanism we derive a potential. Shape of the potential is not arbitrary here since it comes out from a definite mechanism. We discussed it in Yukawa's one-pion exchange potential denoted as OPEP in section 8.4.

✓ Quark theory (1964)

In this approach we focus on the origin or cause of nuclear force. We briefly outline this in the following.

## 8.6 Strong force as a colour force

Strong force operates in a nucleus between the nucleons.

Nucleons are built out of quarks. A proton has structure $p \equiv uud$ and a neutron has quark structure $n \equiv udd$. And quarks carry colour charge.

Each quark possesses one of three colour charges and the colour charges are red colour, green colour, blue colour.

We know that electrical charge carried by particles results in electromagnetic interaction. Similarly, colour charge carried by quarks results in strong interaction which is an interaction through colour force.

Each nucleon is colour neutral or white, i.e. total colour charge balances out to zero. For example

✓ a blue $u$ quark, red $u$ quark, green $d$ quark constitute a proton which is colourless or white.

✓ a blue $u$ quark, red $d$ quark, green $d$ quark constitute a neutron which is colourless or white.

The interaction between quarks inside nucleons is mediated, i.e. accomplished by eight force carriers called gluons. These gluons glue the coloured quarks inside the nucleons through strong force interaction referred to as colour force interaction.

The gluons are massless particles, also carrying colour charge and remaining confined within the nucleus—they cannot leak out.

## 8.7 Form of nuclear potential

Nuclear force binds nucleons in a nucleus. To study the nature of nuclear force, nuclear structure, binding energy and other properties of a nucleus and to carry on with requisite calculations and make theoretical predictions we need to assume some potential form that would represent nuclear force.

Theoretical predictions are of course subject to experimental verification in order to be accepted.

The explicit form of nucleon–nucleon potential is not exactly known. There is no formula which depicts the exact form of nuclear force since nuclear force is not determined completely by knowing the separation between two nucleons.

This means that nuclear force does not depend only on the inter-nucleon separation and there are other factors which are too many.

This is in contrast to the completely known case of Coulomb interaction or electromagnetic interaction, gravitational interaction where the form of the laws (the inverse square laws) are well tested and accepted—they simply depend on the separation between the interacting particles. But this is not so for two nucleons.

A nucleus has a complex structure and houses multiple nucleons in general. It is a many-body problem that is obviously difficult to analyse. This is in addition to the fact that the exact form of nucleon–nucleon potential is not known.

Nucleus is a quantum mechanical system and necessitates use of quantum mechanics for studying it.

Nucleon–nucleon force is a strong force and is of short range ($\leqslant fm$).

Nuclear force can be attractive as well as repulsive depending on inter-nucleon distance. Nuclear force is attractive in the range $\leqslant fm$.

But nuclear force is repulsive in the range $\leqslant 0.4\, fm$.

Some common potentials that are employed to investigate the nature of attractive nuclear force are as follows.

Here $r$ is the relative coordinate of the two nucleons, as shown in figure 8.2. $V_0$ is depth of potential, $r_0$ is the range of nuclear force.

✓ Square well potential

$$V(r) = \begin{cases} -\,V_0 \text{ for } r \leqslant r_0 \\ 0 \quad \text{ for } r > r_0 \end{cases}$$

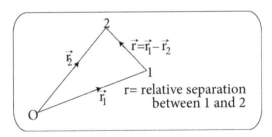

**Figure 8.2.** Defining inter-nucleon separation.

✓ Exponential potential.

$$V(r) = -V_0 e^{-r/r_0}$$

✓ Gaussian potential

$$V(r) = -V_0 e^{-r^2/r_0^2}$$

✓ Yukawa potential (figure 8.1)

$$V(r) = -V_0 \frac{e^{-r/r_0}}{r/r_0}$$

These potentials are space dependent, i.e. $V = V(r)$ and are collectively referred to as Wigner potentials.

As the potential is spherically symmetric the angular part will be spherical harmonic $Y_l m_l(\theta, \phi)$. Thus it is the radial part that we have to consider.

These potentials cannot explain the exchange property of nuclear force.

Nuclear force has a central part and also a non-central part. Nuclear force is also velocity dependent.

Dependence of potential on direction $\hat{r}$ as well as on nucleon spin $\vec{s} = \frac{\hbar}{2}\vec{\sigma}$ is also necessary to explain short-range character and exchange character of nuclear force. Here $\vec{\sigma}$ is Pauli spin matrix operator.

Obviously other potentials are needed for explanation of various nuclear properties.

The short-range property, exchange character, charge independence, spin dependence etc can be incorporated by using certain potentials, as follows, which are kind of mathematical tools required to probe into the details of nuclear force. We discuss them briefly in the following.

✓ Wigner potential

$$V_W = V_W(r)\rho_W \tag{8.20}$$

where $\rho_W$ is called Wigner operator.

This potential does not exchange the space or spin coordinates of two nucleons 1 and 2, namely $\vec{r}_1$, $\vec{\sigma}_1$ and $\vec{r}_2$, $\vec{\sigma}_2$ respectively.

The operator $\rho_M$ operates on two-nucleon wave function $\psi(\vec{r}_1\vec{r}_2, \vec{\sigma}_1, \vec{\sigma}_2)$ and keeps it unchanged.

$$\rho_W \psi(\vec{r}_1\vec{r}_2, \vec{\sigma}_1, \vec{\sigma}_2) = \psi(\vec{r}_1\vec{r}_2, \vec{\sigma}_1, \vec{\sigma}_2) = 1\psi(\vec{r}_1\vec{r}_2, \vec{\sigma}_1, \vec{\sigma}_2) \tag{8.21}$$

Clearly

$$\rho_W = 1 \tag{8.22}$$

i.e. $\rho_W$ is a unit operator or identity operator.

This is the ordinary potential.

✓ Majorana potential

$$V_M = V_M(r)\rho_M \tag{8.23}$$

where $\rho_M$ is called Majorana exchange operator or space exchange operator.

This potential describes the exchange of the space coordinates of two nucleons 1 and 2, namely

$$\vec{r}_1 \equiv x_1, y_1, z_1; \vec{r}_2 \equiv x_2, y_2, z_2$$

The operator $\rho_M$ operates on two-nucleon wave function and would exchange the space coordinates of the first and second nucleon.

If $\phi(\vec{r}_1, \vec{r}_2)$ is the spatial part of the quantum mechanical wave function of the two-nucleon system

$$\psi(\vec{r}_1 \vec{r}_2, \vec{\sigma}_1, \vec{\sigma}_2) = \phi(\vec{r}_1 \vec{r}_2)\chi(\vec{\sigma}_1, \vec{\sigma}_2)$$

then

$$\rho_M \phi(\vec{r}_1, \vec{r}_2) = \phi(\vec{r}_2, \vec{r}_1) \tag{8.24}$$

✓ Bartlett potential

$$V_B = V_B(r)\rho_B \tag{8.25}$$

where $\rho_B$ is called Bartlett exchange operator or the spin exchange operator.

This potential describes the exchange of the spin coordinates of two nucleons 1 and 2, namely

$\vec{\sigma}_1, \vec{\sigma}_2$ which are the Pauli spin operators of two nucleons.

It is of attractive nature. The operator $\rho_B$ operates on two-nucleon wave function and would exchange the spin coordinates of the first and second nucleon.

If $\chi(\vec{\sigma}_1, \vec{\sigma}_2)$ is the spin part of the quantum mechanical wave function of the two-nucleon system

$$\psi(\vec{r}_1 \vec{r}_2, \vec{\sigma}_1, \vec{\sigma}_2) = \phi(\vec{r}_1 \vec{r}_2)\chi(\vec{\sigma}_1, \vec{\sigma}_2) \text{ then}$$

$$\rho_B \chi(\vec{\sigma}_1, \vec{\sigma}_2) = \chi(\vec{\sigma}_2, \vec{\sigma}_1) \tag{8.26}$$

✓ Heisenberg potential

$$V_H = V_H(r)\rho_H \tag{8.27}$$

where $\rho_H$ is called Heisenberg exchange operator or space–spin exchange operator.

This potential describes the exchange of both space coordinates and spin coordinates of two nucleons 1 and 2, namely

$$\vec{r}_1, \vec{\sigma}_1 \equiv x_1, y_1, z_1, \vec{\sigma}_1; \vec{r}_2, \vec{\sigma}_2 \equiv x_2, y_2, z_2, \vec{\sigma}_2$$

The operator $\rho_H$ operates on nucleon wave function and would exchange the space–spin coordinates of the first and second nucleon.

If $\psi(\vec{r}_1, \vec{r}_2, \vec{\sigma}_1, \vec{\sigma}_2)$ is the quantum mechanical wave function of a two-nucleon system then

$$\rho_H \psi(\vec{r}_1, \vec{r}_2, \vec{\sigma}_1, \vec{\sigma}_2) = \psi(\vec{r}_2, \vec{r}_1, \vec{\sigma}_2, \vec{\sigma}_1) \tag{8.28}$$

Clearly the action of the exchange operators on wave function $\psi(\vec{r}_1, \vec{r}_2, \vec{\sigma}_1, \vec{\sigma}_2)$ is as follows

$$\rho_W \psi(\vec{r}_1, \vec{r}_2, \vec{\sigma}_1, \vec{\sigma}_2) = \psi(\vec{r}_1 \vec{r}_2, \vec{\sigma}_1, \vec{\sigma}_2) \tag{8.29}$$

$$\rho_M \psi(\vec{r}_1, \vec{r}_2, \vec{\sigma}_1, \vec{\sigma}_2) = \psi(\vec{r}_2 \vec{r}_1, \vec{\sigma}_1, \vec{\sigma}_2) \tag{8.30}$$

$$\rho_B \psi(\vec{r}_1, \vec{r}_2, \vec{\sigma}_1, \vec{\sigma}_2) = \psi(\vec{r}_1 \vec{r}_2, \vec{\sigma}_2, \vec{\sigma}_1) \tag{8.31}$$

$$\rho_H \psi(\vec{r}_1, \vec{r}_2, \vec{\sigma}_1, \vec{\sigma}_2) = \psi(\vec{r}_2 \vec{r}_1, \vec{\sigma}_2, \vec{\sigma}_1) \tag{8.32}$$

## 8.8 Saturation of nuclear forces

We can explain the saturation property in terms of exchange mechanism and also in terms of isotopic spin formalism.

Dealing with nuclei having mass number $A > 2$ leads to complications since we have to consider the following.

  ✓ Density of all nuclei are roughly equal $\rho \sim 10^{17}\ kg\ m^{-3}$ (saturation of density)
  ✓ Binding energy per nucleon of most nuclei is roughly equal $\frac{B}{A} = 8 \frac{MeV}{nucleon}$ (saturation of binding energy)
  • Force between two nucleons cannot be always attractive, for if it were attractive always then we can not explain the finite nuclear size and volume. The nucleons should not come too close as to shrink to a point but should keep a distance to maintain a finite volume. To explain nuclear size there should be a short-range repulsion.

   If $r_c \sim 0.4\ fm$ is the order of distance between nearest neighbour nucleons in nucleus then force between two nucleons will be

$$\begin{cases} 0 \leqslant r \leqslant r_c & \text{Strong repulsive} \\ r_c \leqslant r < R & \text{Strong attractive} \\ \quad r > R & \text{Strong force vanishes} \end{cases} \tag{8.33}$$

where $R$ is nuclear radius.

   Also, force between two nucleons whether attractive or repulsive depends on the state of two nucleons w.r.t. each other, i.e. w.r.t. coordinates (space, spin) characterizing the nucleons that are exchanged.
  • Let us now discuss what we mean by exchange property.

The purpose of Heisenberg was to explain saturation of nuclear force. He proposed that nuclear forces were exchange forces and depend explicitly on the symmetry of the wave function.

Exchange of $\pi$ meson takes place between the nucleons.

$$p \leftrightarrows p + \pi^0$$

$$n \leftrightarrows n + \pi^0$$

$$p \leftrightarrows n + \pi^+$$

$$n \leftrightarrows p + \pi^-$$

From the last two relations we state that exchange of pions $\pi^\pm$ is equivalent to exchange of charge and we can think that the nucleons are exchanging their space and spin coordinates.

Consider a two-nucleon system and as shown in figure 8.3 the space–spin coordinates of the first nucleon are $\vec{r}_1$, $\vec{\sigma}_1$ and that of the second nucleon are $\vec{r}_2$, $\vec{\sigma}_2$ respectively.

The Schrödinger wave equation in a centre of mass system is

$$H\psi(\vec{r}_1, \vec{r}_2, \vec{\sigma}_1, \vec{\sigma}_2) = E\psi(\vec{r}_1, \vec{r}_2, \vec{\sigma}_1, \vec{\sigma}_2)$$

$$\left(-\frac{\hbar^2}{m}\nabla^2 + V\right)\psi(\vec{r}_1, \vec{r}_2, \vec{\sigma}_1, \vec{\sigma}_2) = E\psi(\vec{r}_1, \vec{r}_2, \vec{\sigma}_1, \vec{\sigma}_2)$$

$$\left(\frac{\hbar^2}{m}\nabla^2 + E\right)\psi(\vec{r}_1, \vec{r}_2, \vec{\sigma}_1, \vec{\sigma}_2) = V\psi(\vec{r}_1, \vec{r}_2, \vec{\sigma}_1, \vec{\sigma}_2) \qquad (8.34)$$

where $\mu = \frac{m}{2}$ is the reduced mass of two-nucleon system and $m$ is average mass of a nucleon.

$\psi = \psi(\vec{r}_1, \vec{r}_2, \vec{\sigma}_1, \vec{\sigma}_2)$ is the wave function of a two-nucleon system.

- There are four types of exchange forces. They are
  - ☐ Wigner force
  - ☐ Majorana exchange force
  - ☐ Bartlet exchange force
  - ☐ Heisenberg exchange force

  These forces have different nature and different properties and effects.

  We discuss the wave equations for four types of forces corresponding to different types of exchanges and the corresponding wave equations.
- Wigner force or non-exchange force or ordinary central force

  Wigner force corresponds to no exchange of space and spin coordinates. This interaction is characterized by the Wigner operator $\rho_W$ [equations (8.21) and (8.22)]. For this interaction there is no change in wave function $\psi(\vec{r}_1, \vec{r}_2, \vec{\sigma}_1, \vec{\sigma}_2)$.

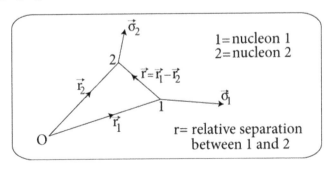

**Figure 8.3.** Nucleon 1 has space–spin coordinates $\vec{r}_1$, $\vec{\sigma}_1$ and nucleon 2 has space–spin coordinates $\vec{r}_2$, $\vec{\sigma}_2$.

The Schrödinger wave equation (8.34) using $V = V_W = V_W(r)\rho_W$ [equation (8.20)] becomes

$$\left(\frac{\hbar^2}{m}\nabla^2 + E\right)\psi(\vec{r}_1, \vec{r}_2, \vec{\sigma}_1, \vec{\sigma}_2) = V_W(r)\rho_W\psi(\vec{r}_1, \vec{r}_2, \vec{\sigma}_1, \vec{\sigma}_2)$$

With equations (8.22) and (8.29)

$$\left(\frac{\hbar^2}{m}\nabla^2 + E\right)\psi(\vec{r}_1, \vec{r}_2, \vec{\sigma}_1, \vec{\sigma}_2) = V_W(r)\psi(\vec{r}_1, \vec{r}_2, \vec{\sigma}_1, \vec{\sigma}_2) \tag{8.35}$$

- Majorana exchange force

If during the interaction there is exchange of space coordinates of the nucleons, i.e.

$$(\vec{r}_1, \vec{r}_2) \leftrightarrows (\vec{r}_2, \vec{r}_1)$$

but no exchange of spin coordiantes of nucleons $(\vec{\sigma}_1, \vec{\sigma}_2) \rightarrow (\vec{\sigma}_1, \vec{\sigma}_2)$ then the nuclear force is called Majorana force.

The Schrödinger wave equation (8.34) using $V_M = V_M(r)\rho_M$ [equation (8.23)] becomes

$$\left(\frac{\hbar^2}{m}\nabla^2 + E\right)\psi(\vec{r}_1, \vec{r}_2, \vec{\sigma}_1, \vec{\sigma}_2) V_M(r)\rho_M\psi(\vec{r}_1, \vec{r}_2, \vec{\sigma}_1, \vec{\sigma}_2) \tag{8.36}$$

where operation of $\rho_M$ is given by equation (8.24).

Since $\vec{r}_1$ and $\vec{r}_2$ are interchanged, this means the relative position vector of figure 8.3 viz.

$$\vec{r} = \vec{r}_1 - \vec{r}_2$$

changes sign. So $\vec{r}$ is replaced by $-\vec{r}$ in the wave function.

Clearly the effect of this interaction is equivalent to the parity eigenvalue $(-1)^l$ where $l$ is the orbital angular momentum quantum number (figure 8.5). We can thus write

$$\rho_M\psi(\vec{r}_1, \vec{r}_2, \vec{\sigma}_1, \vec{\sigma}_2) = (-1)^l\psi(\vec{r}_1, \vec{r}_2, \vec{\sigma}_1, \vec{\sigma}_2) \tag{8.37}$$

With equations (8.36) and (8.37)

$$\left(\frac{\hbar^2}{m}\nabla^2 + E\right)\psi(\vec{r}_1, \vec{r}_2, \vec{\sigma}_1, \vec{\sigma}_2) = V_M(r)(-1)^l\psi(\vec{r}_1, \vec{r}_2, \vec{\sigma}_1, \vec{\sigma}_2) \tag{8.38}$$

Obviously the effect of Majorana force or potential is decided by whether $\psi$ does not change sign for different $l$ values, i.e.

$$(-1)^l\psi = \psi(\text{then force is attractive})$$

or whether $\psi$ changes sign for different $l$ values, i.e.

$$(-1)^l\psi = -\psi \quad (\text{then force is repulsive}).$$

✓ If $l$ is even then

$$(-)^l = +ve$$

and $\psi(\vec{r})$ does not change sign (symmetric) and the force is attractive for the state of even $l$ say $l = 0, 2, 4...$ i.e. attractive for even parity states say $S, D, G, ...$

In other words the two nucleons attract each other if the wave function describing the system of nucleons does not change its sign when the space coordinates of the two particles are interchanged.

This corresponds to even parity $\rho_M = +1$

✓ If $l$ is odd then

$$(-)^l = -ve$$

and $\psi(\vec{r})$ changes sign (antisymmetric) and the force is repulsive for the state of odd $l$ say $l = 1, 3, ...$ i.e. repulsive for odd parity states say $P, F, ...$

In other words the two nucleons repel each other if the wave function describing the system of nucleons changes sign when the space coordinates of the two particles are interchanged.

This corresponds to odd parity $\rho_M = -1$

It is clear that Majorana force may be attractive or repulsive depending on orbital angular momentum quantum number $l$.

- Bartlett exchange force.

If during the interaction there is no exchange of space coordinates of the nucleons $(\vec{r}_1, \vec{r}_2) \rightarrow (\vec{r}_1, \vec{r}_2)$ but there is exchange of spin coordinates of nucleons, i.e.

$$(\vec{\sigma}_1, \vec{\sigma}_2) \leftrightarrows (\vec{\sigma}_2, \vec{\sigma}_1)$$

then the nuclear force is called Bartlett force.

The Schrödinger wave equation (8.34) using equation (8.25), i.e. $V_B = V_B(r)\rho_B$ becomes

$$\left(\frac{\hbar^2}{m}\nabla^2 + E\right)\psi(\vec{r}_1, \vec{r}_2, \vec{\sigma}_1, \vec{\sigma}_2) = V_B(r)\rho_B\psi(\vec{r}_1, \vec{r}_2, \vec{\sigma}_1, \vec{\sigma}_2) \qquad (8.39)$$

where operation of $\rho_B$ is given by equation (8.26).

Interchange of $\vec{\sigma}_1, \vec{\sigma}_2$ means the effect of this interaction is equivalent to the parity eigenvalue $(-1)^{l+s}$ where $s$ is the spin angular momentum quantum number (figure 8.5). We can thus write

$$\rho_B\psi(\vec{r}_1, \vec{r}_2, \vec{\sigma}_1, \vec{\sigma}_2) = (-1)^{l+s}\psi(\vec{r}_1, \vec{r}_2, \vec{\sigma}_1, \vec{\sigma}_2) \qquad (8.40)$$

With equations (8.39) and (8.40)

$$\left(\frac{\hbar^2}{m}\nabla^2 + E\right)\psi(\vec{r}_1, \vec{r}_2, \vec{\sigma}_1, \vec{\sigma}_2) = V_B(r)(-1)^{l+s}\psi(\vec{r}_1, \vec{r}_2, \vec{\sigma}_1, \vec{\sigma}_2) \qquad (8.41)$$

Obviously the effect of Bartlett force or potential is decided by whether $\psi$ does not change sign for different $s$ values, i.e.

$$(-1)^{l+s}\psi = \psi \quad \text{(then force is attractive)}$$

or whether $\psi$ changes sign for different $s$ values, i.e.

$$(-1)^{l+s}\psi = -\psi \quad \text{(then force is repulsive)}.$$

✓ If $s = 1$ (spin triplet state) then

$$(-)^{l+s} = (-)^{l+1} = +ve$$

and $\psi(\vec{r})$ does not change sign (symmetric) and the force is attractive for the spin triplet states say $^3S$, $^3P$, $^3D$, ....

In other words the two nucleons attract each other when their spins are parallel.

So for triplet $\rho_B = +1$

✓ If $s = 0$ (spin singlet state) then

$$(-)^{l+s} = (-)^{l+0} = -ve$$

and $\psi(\vec{r})$ changes sign (antisymmetric) and the force is repulsive for the spin singlet states, say $^1S$, $^1P$, $^1D$, ....

In other words, the two nucleons repel each other when their spins are antiparallel. So for singlet $\rho_B = -1$

Obviously Bartlett force explains why nuclear potential is different for $s = 1$ and $s = 0$, i.e. different for parallel and antiparallel orientation of nuclear spins.

- Heisenberg exchange force.

If during the interaction there is exchange of space coordinate of the nucleons, i.e.

$$(\vec{r}_1, \vec{r}_2) \leftrightarrows (\vec{r}_2, \vec{r}_1)$$

and there is also exchange of spin coordiantes of nucleons, i.e.

$$(\vec{\sigma}_1, \vec{\sigma}_2) \leftrightarrows (\vec{\sigma}_2, \vec{\sigma}_1)$$

then the nuclear force is called Heisenberg force.

The Schrödinger wave equation (8.34) using $V_H = V_H(r)\rho_H$ [equation (8.27)] becomes

$$\left(\frac{\hbar^2}{m}\nabla^2 + E\right)\psi(\vec{r}_1, \vec{r}_2, \vec{\sigma}_1, \vec{\sigma}_2) = V_H(r)\rho_H\psi(\vec{r}_1, \vec{r}_2, \vec{\sigma}_1, \vec{\sigma}_2) \tag{8.42}$$

where operation of $\rho_H$ is given by equation (8.28).

This is a general exchange force and can be treated as a combination of Majorana force and Bartlett force.

Accordingly the effect of this interaction is equivalent to the parity eigenvalue $(-1)^{l+1+s}$.

We can thus write

$$\rho_H\psi(\vec{r}_1, \vec{r}_2, \vec{\sigma}_1, \vec{\sigma}_2) = (-1)^{l+s+1}\psi(\vec{r}_1, \vec{r}_2, \vec{\sigma}_1, \vec{\sigma}_2) \tag{8.43}$$

With equations (8.42) and (8.43)

$$\left(\frac{\hbar^2}{m}\nabla^2 + E\right)\psi(\vec{r}_1, \vec{r}_2, \vec{\sigma}_1, \vec{\sigma}_2) = V_H(r)(-1)^{l+s+1}\psi(\vec{r}_1, \vec{r}_2, \vec{\sigma}_1, \vec{\sigma}_2) \qquad (8.44)$$

Obviously the effect of Heisenberg force or potential is decided by whether $\psi$ does not change sign for different $l,s$ values, i.e.

$$(-1)^{l+s+1}\psi = \psi \quad \text{(then force is attractive)}$$

or whether $\psi$ changes sign for different $s$ values, i.e.

$$(-1)^{l+s+1}\psi = -\psi \quad \text{(then force is repulsive)}.$$

✓ We discuss for $l + s = $ odd.
  If $l = 0, 2, \ldots = $ even and $s = 1$ (triplet) state, i.e. even $l$ triplet

$$(-)^{l+s+1} = (-)^{\text{even}+1+1} = +ve$$

and $\psi(\vec{r})$ does not change sign (symmetric) and the force is attractive for even $l$ triplet states, say $^3S$, $^3D$, .....
  If $l = 1, 3, \ldots = $ odd and $s = 0$ (singlet) state, i.e. odd $l$ singlet

$$(-)^{l+s+1} = (-)^{\text{odd}+0+1} = +ve$$

and $\psi(\vec{r})$ does not change sign (symmetric) and the force is attractive for odd $l$ singlet states, say $^1P$, $^1F$, .....
✓ We discuss for $l + s = $ even

If $l = 0, 2, \ldots = $ even and $s = 0$ (singlet) state, i.e. even $l$ singlet

$$(-)^{l+s+1} = (-)^{\text{even}+0+1} = -ve$$

and $\psi(\vec{r})$ changes sign (antisymmetric) and the force is repulsive for even $l$ singlet states say $^1S$, $^1D$, .....
  If $l = 1, 3, \ldots = $ odd and $s = 1$ (triplet) state, i.e. odd $l$ triplet

$$(-)^{l+s+1} = (-)^{\text{odd}+1+1} = -ve$$

and $\psi(\vec{r})$ changes sign (antisymmetric) and the force is repulsive for odd $l$ triplet states say $^3P$, $^3F$, ....
  We conclude that
  ☐ For even $l$ triplet state ($^3S$) and odd $l$ singlet state ($^1P$) the force will be attractive.
    For even parity triplet and for odd parity singlet $\rho_H = +1$.
  ☐ For odd $l$ triplet state ($^3P$) and even $l$ singlet state ($^1S$) the force will be repulsive.
    For even parity singlet and odd parity triplet $\rho_H = -1$.

- Each of three exchange operators $\rho_M$, $\rho_B$, $\rho_H$ have eigenvalues $\pm 1$ since parities are (figure 8.5)

$$\rho_M \to (-1)^l$$

$$\rho_B \to (-1)^{l+s}$$

$$\rho_H \to (-1)^{l+s+1}$$

Sign of exchange operators for various states as we have derived has been shown in figure 8.4. We mention that the value $+1$ means $\psi$ does not change sign and force is attractive while the value $-1$ means $\psi$ changes sign and force is repulsive.

The Majorana exchange operator is $+1$ for even parity states, i.e. for even $l$ states and $-1$ for odd parity states, i.e. for odd $l$ states. The Bartlett exchange operator is $+1$ for triplet states and $-1$ for singlet states. The Heisenberg exchange operator is $+1$ for even parity triplet and odd parity singlet states and $-1$ for odd parity triplet and even parity singlet states.

We note a few important properties.

✓ Consider

$$\rho_M \rho_B \psi(\vec{r}_1, \vec{r}_2, \vec{\sigma}_1, \vec{\sigma}_2) = \rho_M \psi(\vec{r}_1, \vec{r}_2, \vec{\sigma}_2, \vec{\sigma}_1) \quad \text{[using equation (8.31)]}$$

$$= \psi\left(\vec{r}_2, \vec{r}_1, \vec{\sigma}_2, \vec{\sigma}_1\right) \text{[using equation (8.30)]}$$

Again, by equation (8.32) $\rho_H \psi(\vec{r}_1, \vec{r}_2, \vec{\sigma}_1, \vec{\sigma}_2) = \psi(\vec{r}_2, \vec{r}_1, \vec{\sigma}_2, \vec{\sigma}_1)$.
Hence

$$\rho_M \rho_B \psi(\vec{r}_1, \vec{r}_2, \vec{\sigma}_1, \vec{\sigma}_2) = \rho_H \psi(\vec{r}_1, \vec{r}_2, \vec{\sigma}_1, \vec{\sigma}_2) \quad \text{[as both are equal to } \psi(\vec{r}_2, \vec{r}_1, \vec{\sigma}_2, \vec{\sigma}_1)\text{]}$$

This means

$$\rho_H = \rho_M \rho_B \qquad (8.45)$$

This is the relation between the exchange operators $\rho_M$, $\rho_B$, $\rho_H$.

✓ Consider

$$\rho_M^2 \psi(\vec{r}_1, \vec{r}_2, \vec{\sigma}_1, \vec{\sigma}_2) = \rho_M \psi(\vec{r}_2, \vec{r}_1, \vec{\sigma}_1, \vec{\sigma}_2) = \psi(\vec{r}_1, \vec{r}_2, \vec{\sigma}_1, \vec{\sigma}_2) \quad \text{[using equation (8.30)]}$$

$$\rho_M^2 = 1 \qquad (8.46)$$

| Exchange | Even parity states | | Odd parity states | |
|----------|---------|---------|---------|---------|
| Operator | Triplet | Singlet | Triplet | Singlet |
| $\rho_M$ | +1 | +1 | −1 | −1 |
| $\rho_B$ | +1 | −1 | +1 | −1 |
| $\rho_H$ | +1 | −1 | −1 | +1 |

**Figure 8.4.** Eigenvalues of exchange operators $\rho_M$, $\rho_B$, $\rho_H$.

✓ Consider

$$\rho_B^2 \psi(\vec{r}_1, \vec{r}_2, \vec{\sigma}_1, \vec{\sigma}_2) = \rho_B \psi(\vec{r}_1, \vec{r}_2, \vec{\sigma}_2, \vec{\sigma}_1) \quad \text{[using equation (8.31)]}$$

$$= \psi(\vec{r}_1, \vec{r}_2, \vec{\sigma}_1, \vec{\sigma}_2) \quad \text{[using equation (8.31)]}$$

$$\rho_B^2 = 1 \tag{8.47}$$

✓ Consider

$$\rho_H^2 \psi\left(\vec{r}_1, \vec{r}_2, \vec{\sigma}_1, \vec{\sigma}_2\right) = \rho_H \psi\left(\vec{r}_2, \vec{r}_1, \vec{\sigma}_2, \vec{\sigma}_1\right) \text{[using equation (8.32)]}$$

$$= \psi\left(\vec{r}_1, \vec{r}_2, \vec{\sigma}_1, \vec{\sigma}_2\right) \text{[using equation (8.32)]}$$

$$\rho_H^2 = 1 \tag{8.48}$$

Since

$$\rho_M^2 = \rho_B^2 = \rho_H^2 = 1$$

it follows that the exchange operators $\rho_M$, $\rho_B$, $\rho_H$ have only two eigenvalues viz. $\pm 1$.

Figure 8.5 summarises the properties of various exchange forces.

✓ We have shown that Bartlett force is attractive in $^3S$ and repulsive in $^1S$. This is contrary to what is actually observed in an $np$ scattering experiment.

So Bartlett spin exchange force cannot be the only exchange force in $np$ interaction.

| Wigner force<br>Non-exchange force<br>No exchange<br>$V_W = V_W(r) \, \rho_W$ | $\rho_W \psi(\vec{r}_1, \vec{r}_2, \vec{\sigma}_1, \vec{\sigma}_2) = \psi(\vec{r}_1, \vec{r}_2, \vec{\sigma}_1, \vec{\sigma}_2)$<br><br>$(\frac{\hbar^2}{m}\nabla^2 + E)\, \psi = V_W \psi$ |
|---|---|
| Majorana force<br>Exchange force<br>Exchanges space<br>$V_M = V_M(r) \, \rho_M$ | $\rho_M \psi(\vec{r}_1, \vec{r}_2, \vec{\sigma}_1, \vec{\sigma}_2) = \psi(\vec{r}_2, \vec{r}_1, \vec{\sigma}_1, \vec{\sigma}_2)$<br><br>$(-)^l \psi = \begin{cases} \psi \ \text{(attractive)} & l = \text{even} \quad \text{S, D, G,...} \\ -\psi \ \text{(repulsive)} & l = \text{odd} \quad \text{P,F,...} \end{cases}$ |
| Bartlett force<br>Exchange force<br>Exchanges spin<br>$V_B = V_B(r) \, \rho_B$ | $\rho_B \psi(\vec{r}_1, \vec{r}_2, \vec{\sigma}_1, \vec{\sigma}_2) = \psi(\vec{r}_1, \vec{r}_2, \vec{\sigma}_2, \vec{\sigma}_1)$<br><br>$(-)^{1+s}\psi = \begin{cases} \psi \ \text{(attractive)} \ \text{Triplet s=1} & ^3S, ^3P, ^3D \\ -\psi \ \text{(repulsive)} \ \text{Singlet s=0} & ^1S, ^1P, ^1D \end{cases}$ |
| Heisenberg force<br>Exchange force<br>Exchanges space<br>and spin<br>$V_H = V_H(r) \, \rho_H$ | $\rho_H \psi(\vec{r}_1, \vec{r}_2, \vec{\sigma}_1, \vec{\sigma}_2) = \psi(\vec{r}_2, \vec{r}_1, \vec{\sigma}_2, \vec{\sigma}_1)$<br><br>$(-)^{l+s+1}\psi = \begin{cases} \psi \ \text{(attractive)} & l = \text{even} \ \text{Triplet s=1} & ^3S, ^3D \\ \quad l+s=\text{odd} & l = \text{odd} \ \text{Singlet s=0} & ^1P, ^1F \\ -\psi \ \text{(repulsive)} & l = \text{even} \ \text{Triplet s=0} & ^1S, ^1D \\ \quad l+s=\text{even} & l = \text{odd} \ \text{Singlet s=1} & ^3P, ^3F \end{cases}$ |

**Figure 8.5.** Definition of various exchange forces.

✓ We have shown that Heisenberg force is attractive in $^3S$ and repulsive in $^1S$. This is contrary to what is actually observed in an $np$ scattering experiment.

So Heisenberg spin exchange force cannot be the only exchange force in $np$ interaction.

✓ Saturation character of nuclear force requires a mixture of Majorana and Wigner type potentials.

✓ Not all combinations give agreement with experimental result.

✓ The most general form of the nuclear potential of the exchange type is

$$V = V_W(\vec{r}) + V_M(\vec{r})\rho_M + V_B(\vec{r})\rho_B + V_H(\vec{r})\rho_H \qquad (8.49)$$

where

$V_W(\vec{r})$ is Wigner potential, $V_M(\vec{r})$ = Majorana potential, $V_B(\vec{r})$ = Bartlett potential, $V_H(\vec{r})$ = Heisenberg potential

## 8.9 Tensor force and exchange character

Tensor force may or may not possess exchange character.

• The tensor force operator $S_{12}$ [equation (6.116)] commutes with the Bartlett spin exchange operator $\rho_B$, i.e.

$$\left[ S_{12}, \rho_B \right] = 0$$

$$S_{12}\rho_B - \rho_B S_{12} = 0$$

$$S_{12}\rho_B = \rho_B S_{12} \qquad (8.50)$$

We say that there are two types of tensor forces.

✓ Ordinary tensor force, also called Wigner type tensor force having no exchange character.

✓ Tensor force having Majorana exchange character

If we include the exchange version of tensor force then the general form of the potential is given as

$$V = V_W(\vec{r}) + V_M(\vec{r})\rho_M + V_B(\vec{r})\rho_B + V_H(\vec{r})\rho_H + V_{TW}(\vec{r})S_{12} + V_{TM}(\vec{r})S_{12}\rho_M \quad (8.51)$$

$V_{TW}(\vec{r})S_{12}$ is a mixture of Wigner and tensor potential
$V_{TM}(\vec{r})S_{12}\rho_M$ is a mixture of tensor potential and Majorana potential

## 8.10 Saturation in terms of exchange force

Saturation of nuclear force is revealed by two characteristics of nuclei. They are

✓ Volume of nucleus

$$V = \frac{4}{3}\pi R^3 = \frac{4}{3}\pi(R_0 A^{1/3})^3 = \frac{4}{3}\pi R_0^3 A \propto A.$$

So $\frac{V}{A}$ = constant.

✓ Nuclear binding energy per nucleon

$$\frac{B}{A} = \text{constant, i.e. } B \propto A$$

- Bartlett spin exchange force does not lead to saturation of binding energy per nucleon. If nuclear force is of Bartlett type then heavy nuclei will exist with all spins aligned in the same direction as we do not consider tensor force in that case. Then the number of pairs of interacting nucleons will be

$$\frac{A(A-1)}{2}$$

where $A$ is mass number. So in this case

$$\frac{B}{A} \propto A, \text{ i.e.} \frac{B}{A} \text{ is } A \text{ dependent.}$$

If we consider the Bartlett spin exchange potential only then binding energy per nucleon should increase with mass number. For larger mass number $\frac{B}{A}$ should be larger and for small $A$, $\frac{B}{A}$ would be smaller. So we cannot explain the saturation property by Bartlett spin exchange property.

- Saturation of binding energy per nucleon is due to the space exchange of Majorana and Heisenberg forces. The condition for saturation of $\frac{B}{A}$ can be explained by the average potential calculated by Blatt and Weiskoff

$$\langle V \rangle = \left[ \frac{A}{2}\left( V_W - \frac{V_M}{4} \right) - \frac{T^2}{2A} V_M \right]\left( \frac{r_0}{R_0} \right)^3 \left[ 1 - \frac{9}{16}\frac{r_0}{R_0 A^{\frac{1}{3}}} + \frac{1}{32}\frac{r_0^3}{R_0^3 A} + \right] \quad (8.52)$$

where $r_0$ is the range of nuclear force, $R = R_0 A^{1/3}$ is radius of nucleus, $R_0 = 1.2\, fm$,

$$T = \frac{1}{2}(N - Z)$$

where $N$ = number of neutrons in the nucleus and $Z$ = number of protons in the nucleus. So $T$ = half the excess number of nucleons.

If small terms in the equation are neglected then we have the form (can be shown)

$$\langle V \rangle \cong -a_V A + a_S A^{2/3} + C\frac{T^2}{A} \quad (8.53)$$

The first term in equation (8.52) is the strongest and is

$$-a_V A = \frac{A}{2}\left( V_W - \frac{V_M}{4} \right)\left( \frac{r_0}{R_0} \right)^3$$

where

$$a_V = -\frac{1}{2}\left( V_W - \frac{V_M}{4} \right)\left( \frac{r_0}{R_0} \right)^3.$$

This first term is −ve since both $V_W$ and $V_M$ are −ve, i.e.

$$a_V = -\frac{1}{2}\left(-|V_W| - \frac{-|V_M|}{4}\right) = \frac{1}{2}\left(V_W| - \frac{|V_M|}{4}\right) = +ve \tag{8.54}$$

For positive value of $a_V$ it needs to be ensured that

$$|V_W| \geqslant \frac{1}{4}|V_M|$$

$$4|V_W| \geqslant |V_M|$$

$$|V_M| \leqslant 4|V_W| \tag{8.55}$$

Then potential energy $<V>$ is negative.

Saturation of binding energy occurs under this condition, i.e. when Majorana force contributes such that it is less than four times the Wigner force.

## 8.11 Isotopic spin formalism

Charge symmetry and charge independence of nuclear force and near equality of masses of neutron and proton suggest that neutron and proton are the same particle in two different charge states.

Mathematically it is possible to treat neutron and proton on an equal footing and it is done by mathematically introducing a 3D charge space or isotopic spin space or isospin space or simply isospace.

In isospace the neutron and proton are described by an isotopic spinor or iso spinor field.

There are two iso spinor components corresponding to the two charge states of nucleon viz. neutron and proton.

- The ordinary spin also has two states—up and down states.

    The formal development of the theory of isospin is the same as that of ordinary spin.

    For ordinary spin

$$\vec{s} = \frac{\hbar}{2}\vec{\sigma}$$

where $\vec{\sigma} = (\sigma_1, \sigma_2, \sigma_3)$ is Pauli spin vector and the $z$ component is

$$s_3 = \pm\frac{\hbar}{2} = \pm\frac{\hbar}{2}\sigma_3. \text{ So}$$

$$\sigma_3 = \pm 1.$$

In the field of charge space or isospace $n$ and $p$ are treated on a common footing. These two particles have a common name called nucleon. We define in isospace the isospin angular momentum which obeys the same algebra as the algebra obeyed by spin angular momentum.

A nucleon is a particle with isospin defined as

$$\vec{I} = \frac{\hbar}{2} \, \vec{\tau} \tag{8.56}$$

$\vec{I}$ is analogous to $\vec{s}$ and the vector $\vec{\tau}$ is analogous to the Pauli spin matrix

$$\vec{\sigma} = (\sigma_1, \sigma_2, \sigma_3).$$

Hence

$$\vec{\tau} = (\tau_1, \tau_2, \tau_3) \tag{8.57}$$

The $z$ compoment of isospin $\vec{I}$ in isospin space is

$$I_z \equiv I_3 = \pm\frac{\hbar}{2} = \frac{\hbar}{2}\tau_3 \tag{8.58}$$

$$\tau_3 = \pm 1 \tag{8.59}$$

Let us choose

$$\tau_3 = +1 \text{ for proton and}$$

$$\tau_3 = -1 \text{ for neutron as per popular convention.}$$

Since $\tau_1, \tau_2, \tau_3$ are analogous to $\sigma_1, \sigma_2, \sigma_3$ let us define the three components $\tau_1, \tau_2, \tau_3$ of isospin matrix $\vec{\tau}$ as follows

$$\tau_1 = \begin{pmatrix} 0 & 1 \\ 1 & 0 \end{pmatrix}, \quad \tau_2 = \begin{pmatrix} 0 & -i \\ i & 0 \end{pmatrix}, \quad \tau_3 = \begin{pmatrix} 1 & 0 \\ 0 & -1 \end{pmatrix} \tag{8.60}$$

From these values we can express the wave function of proton and neutron in isospin space.

Let the isospin wave function of the proton and neutron be $\eta(p)$ and $\eta(n)$ respectively defined in the isospin space such that

$$\eta(p) = \eta(\tau_3 = 1) = \begin{pmatrix} 1 \\ 0 \end{pmatrix} = |\, I = \frac{1}{2} \quad I_z = \frac{1}{2} > = |\, p > = |\uparrow > \tag{8.61}$$

$$\eta(n) = \eta(\tau_3 = -1) = \begin{pmatrix} 0 \\ 1 \end{pmatrix} = |\, I = \frac{1}{2} \quad I_z = -\frac{1}{2} > = |\, n > = |\downarrow > \tag{8.62}$$

- Charge operator $Q$

Let us define a charge operator $Q$ as

$$Q = \frac{1}{2}(I + \tau_3)$$

where $I = \begin{pmatrix} 1 & 0 \\ 0 & 1 \end{pmatrix}$ is 2 × 2 unit matrix.

✓ Let us find the matrix form of $Q$.

$$Q = \frac{1}{2}(I + \tau_3) = \frac{1}{2}\left[\begin{pmatrix} 1 & 0 \\ 0 & 1 \end{pmatrix} + \begin{pmatrix} 1 & 0 \\ 0 & -1 \end{pmatrix}\right] = \frac{1}{2}\begin{pmatrix} 2 & 0 \\ 0 & 0 \end{pmatrix}$$

$$Q = \begin{pmatrix} 1 & 0 \\ 0 & 0 \end{pmatrix} \tag{8.63}$$

✓ We now find the eigenvalue of charge operator $Q$.

The eigenvalue equation for the operator $Q$ is

$$Q\eta = \lambda\eta \Rightarrow Q\eta = \lambda I\eta \tag{8.64}$$

where $\lambda$ is the eigenvalue of this matrix $Q$ and $\eta = \begin{pmatrix} x_1 \\ x_2 \end{pmatrix}$ is the eigenstate, i.e. equation (8.64) is

$$\begin{pmatrix} 1 & 0 \\ 0 & 0 \end{pmatrix}\begin{pmatrix} x_1 \\ x_2 \end{pmatrix} = \lambda\begin{pmatrix} 1 & 0 \\ 0 & 0 \end{pmatrix}\begin{pmatrix} x_1 \\ x_2 \end{pmatrix}$$

$$\left[\begin{pmatrix} 1 & 0 \\ 0 & 0 \end{pmatrix} - \lambda\begin{pmatrix} 1 & 0 \\ 0 & 0 \end{pmatrix}\right]\begin{pmatrix} x_1 \\ x_2 \end{pmatrix} = 0$$

$$\begin{pmatrix} 1 - \lambda & 0 \\ 0 & -\lambda \end{pmatrix}\begin{pmatrix} x_1 \\ x_2 \end{pmatrix} = 0 \tag{8.65}$$

From equation (8.64)

$$(Q - \lambda I)\eta = 0$$

As $\eta \neq 0$

$$|Q - \lambda I| = 0$$

This is the characteristic equation which can be solved to obtain the eigenvalues.

$$\begin{vmatrix} 1 - \lambda & 0 \\ 0 & -\lambda \end{vmatrix} = 0$$

$$-\lambda(1 - \lambda) = 0$$

$$\lambda = 0, 1$$

Eigenvalue equation (8.65) for $\lambda = 0$ is

$$\begin{pmatrix} 1 - 0 & 0 \\ 0 & -0 \end{pmatrix}\begin{pmatrix} x_1 \\ x_2 \end{pmatrix} = \begin{pmatrix} 0 \\ 0 \end{pmatrix} \Rightarrow \begin{pmatrix} 1 & 0 \\ 0 & 0 \end{pmatrix}\begin{pmatrix} x_1 \\ x_2 \end{pmatrix} = \begin{pmatrix} 0 \\ 0 \end{pmatrix}$$

$$\begin{pmatrix} x_1 \\ 0 \end{pmatrix} = \begin{pmatrix} 0 \\ 0 \end{pmatrix}$$

$x_1 = 0$, $x_2$ = arbitrary = 1, say.

$$\begin{pmatrix} x_1 \\ x_2 \end{pmatrix} = \begin{pmatrix} 0 \\ 1 \end{pmatrix} \equiv \eta(n)$$

is eigenvector of charge operator $Q$ with eigenvalue 0. This is neutron state.
Eigenvalue equation (8.65) for $\lambda = 1$ is

$$\begin{pmatrix} 1-1 & 0 \\ 0 & -1 \end{pmatrix}\begin{pmatrix} x_1 \\ x_2 \end{pmatrix} = \begin{pmatrix} 0 \\ 0 \end{pmatrix} \Rightarrow \begin{pmatrix} 0 & 0 \\ 0 & -1 \end{pmatrix}\begin{pmatrix} x_1 \\ x_2 \end{pmatrix} = \begin{pmatrix} 0 \\ 0 \end{pmatrix}$$

$$\begin{pmatrix} 0 \\ -x_2 \end{pmatrix} = \begin{pmatrix} 0 \\ 0 \end{pmatrix}$$

$x_2 = 0$, $x_1$ = arbitrary = 1, say.

$$\begin{pmatrix} x_1 \\ x_2 \end{pmatrix} = \begin{pmatrix} 1 \\ 0 \end{pmatrix} \equiv \eta(p)$$

is eigenvector of charge operator $Q$ with eigenvalue 1. This is proton state.
Hence

$$Q\eta(p) = \begin{pmatrix} 1 & 0 \\ 0 & 0 \end{pmatrix}\begin{pmatrix} 1 \\ 0 \end{pmatrix} = +1\begin{pmatrix} 1 \\ 0 \end{pmatrix} = \eta(p) \text{ i.e.}$$

$$Q \mid p > = \mid p>$$

(8.66)

$$Q\eta(n) = \begin{pmatrix} 1 & 0 \\ 0 & 0 \end{pmatrix}\begin{pmatrix} 0 \\ 1 \end{pmatrix} = 0\begin{pmatrix} 0 \\ 0 \end{pmatrix} = 0, \text{ i.e.}$$

$$Q \mid n > = 0$$

(8.67)

- Neutron is one unit of charge less than the proton charge. Proton is one unit of charge more than the neutron charge.

  To build a mechanism of increase or decrease of charge of nucleon by one unit let us define raising and lowering operators.

  Let $\tau_+$ be the raising operator. It raises charge of nucleon (neutron) by one unit and converts a neutron to a proton.

  Let $\tau_-$ be the lowering operator. It lowers charge of nucleon (proton) by one unit and converts a proton to a neutron.

  Definition

$$\tau_+ = \frac{1}{2}(\tau_1 + i\tau_2)$$

(8.68)

$$\tau_+ = \frac{1}{2}\left[\begin{pmatrix} 0 & 1 \\ 1 & 0 \end{pmatrix} + i\begin{pmatrix} 0 & -i \\ i & 0 \end{pmatrix}\right] = \frac{1}{2}\begin{pmatrix} 0 & 2 \\ 0 & 0 \end{pmatrix} = \begin{pmatrix} 0 & 1 \\ 0 & 0 \end{pmatrix}$$

(8.69)

and

$$\tau_- = \frac{1}{2}(\tau_1 - i\tau_2) \tag{8.70}$$

$$\tau_- = \frac{1}{2}\left[\begin{pmatrix} 0 & 1 \\ 1 & 0 \end{pmatrix} - i\begin{pmatrix} 0 & -i \\ i & 0 \end{pmatrix}\right] = \frac{1}{2}\begin{pmatrix} 0 & 0 \\ 2 & 0 \end{pmatrix} = \begin{pmatrix} 0 & 0 \\ 1 & 0 \end{pmatrix} \tag{8.71}$$

We arrive at the following important relations

$$\checkmark \quad \tau_+\eta(p) = \begin{pmatrix} 0 & 1 \\ 0 & 0 \end{pmatrix}\begin{pmatrix} 1 \\ 0 \end{pmatrix} = 0 \tag{8.72}$$

$\tau_+$ cannot raise the charge of proton as there is no such nucleon carrying 2 units of charge.

$$\checkmark \quad \tau_-\eta(p) = \begin{pmatrix} 0 & 0 \\ 1 & 0 \end{pmatrix}\begin{pmatrix} 1 \\ 0 \end{pmatrix} = \begin{pmatrix} 0 \\ 1 \end{pmatrix} = \eta(n) \tag{8.73}$$

$\tau_-$ can lower the charge of proton and convert it to neutron.

$$\checkmark \quad \tau_+\eta(n) = \begin{pmatrix} 0 & 1 \\ 0 & 0 \end{pmatrix}\begin{pmatrix} 0 \\ 1 \end{pmatrix} = \begin{pmatrix} 1 \\ 0 \end{pmatrix} = \eta(p) \tag{8.74}$$

$\tau_+$ can raise the charge of neutron and convert it to proton.

$$\checkmark \quad \tau_-\eta(n) = \begin{pmatrix} 0 & 0 \\ 1 & 0 \end{pmatrix}\begin{pmatrix} 0 \\ 1 \end{pmatrix} = 0 \tag{8.75}$$

$\tau_-$ cannot lower the charge of neutron as there is no such nucleon carrying one negative unit of charge.

Evidently the operators $\tau_+$, $\tau_-$ changes the charge state of a nucleon.
- Two-nucleon wave function in isospin space

Total isospin $\vec{I}$ is the vector sum of the isospins $\vec{I_1}$, $\vec{I_2}$ of the two nucleons.

$$\vec{I} = \vec{I_1} + \vec{I_2} = \frac{\vec{1}}{2} + \frac{\vec{1}}{2}$$

$$I = \begin{cases} \dfrac{1}{2} - \dfrac{1}{2} = 0 \text{ for antiparallel isospins } |\uparrow\downarrow> \\ \dfrac{1}{2} + \dfrac{1}{2} = 1 \text{ for parallel isospins } |\uparrow\uparrow> \end{cases} \tag{8.76}$$

We now consider the two-nucleon wave function in isospin space.

Consider a composite system of two nucleons defined through their isospin quantum numbers. We note that

$$[I^2 \quad I_z \quad ] = 0$$

and so simultaneous eigenstates $|\,I\,m>$ can be constructed for the operators $I^2$ and $I_Z$ the eigen equations being

$$I^2\,|\,I\,m> = I(I+1)\hbar^2\,|\,I\,m> \tag{8.77}$$

$$I_z\,|\,I\,m> = m_I\hbar\,|\,I\,m> \tag{8.78}$$

We first develop the uncoupled basis.

For the first nucleon isospin quantum numbers are $I_1$, $m_1$ and we define the state $|\,I_1,\,m_1>$ where

$$I_1^2\,|\,I_1,\,m_1> = I_1(I_1+1)\hbar^2\,|\,I_1,\,m_1>,\ I_1 = \frac{1}{2} \tag{8.79}$$

$$I_{1z}\,|\,I_1,\,m_1> = m_1\hbar\,|\,I_1,\,m_1>,\ m_1 = \pm\frac{1}{2} \tag{8.80}$$

So for the first nucleon we define $2I_1 + 1 = 2.\frac{1}{2} + 1 = 2$ states

$$|\,I_1 m_1> \rightarrow |\,\frac{1}{2}\frac{1}{2}>_1 \equiv |\,p>_1 \equiv \eta^{(1)}(p),\ |\,\frac{1}{2}-\frac{1}{2}>_1 \equiv |\,n>_1 \equiv \eta^{(1)}(n) \tag{8.81}$$

For the second nucleon isospin, quantum numbers are $I_2$, $m_2$ and we define the state $|\,I_2,\,m_2>$ where

$$I_2^2\,|\,I_2,\,m_2> = I_2(I_2+1)\hbar^2\,|\,I_2,\,m_2>,\ I_2 = \frac{1}{2} \tag{8.82}$$

$$I_{2z}\,|\,I_2,\,m_2> = m_2\hbar\,|\,I_2,\,m_2>,\ m_2 = \pm\frac{1}{2} \tag{8.83}$$

So for the second nucleon we define $2I_2 + 1 = 2.\frac{1}{2} + 1 = 2$ states

$$|\,I_2 m_2> \rightarrow |\,\frac{1}{2}\frac{1}{2}>_2 \equiv |\,p>_2 \equiv \eta^{(2)}(p),\ |\,\frac{1}{2}-\frac{1}{2}>_2 \equiv |\,n>_2 \equiv \eta^{(2)}(n) \tag{8.84}$$

Operators $I_1^2$, $I_{1z}$, $I_2^2$, $I_{2z}$ commute as the two nucleons are different systems and for the two-nucleon system we can construct a complete set of $(2I_1 + 1)(2I_2 + 1) = 2.2 = 4$ uncoupled base states in the representation as follows

$$|\,I_1 m_1>_1 \otimes |\,I_2 m_2>_2 \equiv |\,I_1 m_1 I_2 m_2> \underrightarrow{\text{shorten notation}}\ |\,m_1 m_2> \tag{8.85}$$

$$|\,\frac{1}{2}\frac{1}{2}>_1 \otimes |\,\frac{1}{2}\frac{1}{2}>_2 \equiv |\,\frac{1}{2}\frac{1}{2}\frac{1}{2}\frac{1}{2}> \rightarrow |\,\frac{1}{2}\frac{1}{2}> \tag{8.86}$$

$$|\,\frac{1}{2}\frac{1}{2}>_1 \otimes |\,\frac{1}{2}-\frac{1}{2}>_2 \equiv |\,\frac{1}{2}\frac{1}{2}\frac{1}{2}-\frac{1}{2}> \rightarrow |\,\frac{1}{2}-\frac{1}{2}> \tag{8.87}$$

$$| \frac{1}{2} - \frac{1}{2} >_1 \otimes | \frac{1}{2}\frac{1}{2} >_2 \equiv | \frac{1}{2} - \frac{1}{2}\frac{1}{2}\frac{1}{2} > \rightarrow | - \frac{1}{2}\frac{1}{2} > \tag{8.88}$$

$$| \frac{1}{2} - \frac{1}{2} >_1 \otimes | \frac{1}{2} - \frac{1}{2} >_2 \equiv | \frac{1}{2} - \frac{1}{2}\frac{1}{2} - \frac{1}{2} > \rightarrow | - \frac{1}{2} - \frac{1}{2} > \tag{8.89}$$

Here $\otimes$ represent number of combinations. In view of equations (8.81) and (8.84) we can rewrite equations (8.86)–(8.89) through a different notation as follows.

$$\eta^{(1)}(p)\eta^{(2)}(p) = | \frac{1}{2}\frac{1}{2} > = | p >_1 | p >_2 = | pp > \tag{8.90}$$

$$\eta^{(1)}(p)\eta^{(2)}(n) = | \frac{1}{2} - \frac{1}{2} > = | p >_1 | n >_2 = | pn > \tag{8.91}$$

$$\eta^{(1)}(n)\eta^{(2)}(p) = | - \frac{1}{2}\frac{1}{2} > = | n >_1 | p >_2 = | np > \tag{8.92}$$

$$\eta^{(1)}(n)\eta^{(2)}(n) = | - \frac{1}{2} - \frac{1}{2} > = | n >_1 | n >_2 = | nn > \tag{8.93}$$

Now we develop the coupled basis.

Let isospin quantum numbers $I$, $m$ refer to the composite system of two nucleons having $I_1 = \frac{1}{2}$, $I_2 = \frac{1}{2}$ where

$$\vec{I} = \vec{I_1} + \vec{I_2} = \frac{1}{2} + \frac{1}{2} = \begin{cases} \frac{1}{2} - \frac{1}{2} = 0 \Rightarrow 1 \text{ projection} \Rightarrow 0 \text{ (singlet)} \\ \frac{1}{2} + \frac{1}{2} = 1 \Rightarrow 3 \text{ projections} \Rightarrow 1, 0, -1 \text{ (triplet)} \end{cases} \tag{8.94}$$

and

$$I_z = I_{1z} + I_{2z}$$

where the projections $m_1 = \pm\frac{1}{2}$ of the first nucleon and $m_2 = \pm\frac{1}{2}$ of the second nucleon are related to the projection of the coupled state $| Im >$ as

$$m = m_1 + m_2 \tag{8.95}$$

The state $| Im >$ is a simultaneous eigenstate of the operators $I^2$ and $I_z$ as

$$I^2 | Im > = I(I + 1)\hbar^2 | Im > \tag{8.96}$$

$$I_z | Im > = m\hbar | Im > \tag{8.97}$$

We note that for $I = 0$ there are $2I + 1 = 2.0 + 1 = 1$ state with $m = 0$, the state being

$$| Im > = | 00 > = \eta_{00} \tag{8.98}$$

And for $I = 1$ there are $2I + 1 = 2.1 + 1 = 3$ states with $m = 1, 0, -1$, the states being

$$| I m > \rightarrow | 11 > = \eta_{11}, | 10 > = \eta_{10}, | 1 - 1 > = \eta_{1-1} \qquad (8.99)$$

So altogether there are four states.

Operators $I_1^2, I_2^2, I^2, I_z$ commute and for the composite system of two nucleons we can construct a complete set of basis states, the number of which is

$$\sum_{I=0,1} (2I + 1) = (2.0 + 1) + (2.1 + 1) = 1 + 3 = 4$$

So there are four uncoupled base states. Let us use the representation as follows.

$$| I_1 I_2 I m > \xrightarrow{\text{shorten notation}} | I m > \qquad (8.100)$$

$$\left| \frac{1}{2}\frac{1}{2}00 > \rightarrow | 00> \right. \qquad (8.101)$$

$$\left| \frac{1}{2}\frac{1}{2}11 > \rightarrow | 11> \right. \qquad (8.102)$$

$$\left| \frac{1}{2}\frac{1}{2}10 > \rightarrow | 10> \right. \qquad (8.103)$$

$$\left| \frac{1}{2}\frac{1}{2}1 - 1 > \rightarrow | 1 - 1> \right. \qquad (8.104)$$

In view of equations (8.98) and (8.99) we can rewrite equations (8.101)–(8.104) through a different notation as follows.

$$\left| \frac{1}{2}\frac{1}{2}00 > \rightarrow | 00 > = \eta_{00} \right. \qquad (8.105)$$

$$\left| \frac{1}{2}\frac{1}{2}11 > \rightarrow | 11 > \quad = \eta_{11} \right. \qquad (8.106)$$

$$\left| \frac{1}{2}\frac{1}{2}10 > \rightarrow | 10 > = \eta_{10} \right. \qquad (8.107)$$

$$\left| \frac{1}{2}\frac{1}{2}1 - 1 > \rightarrow | 1 - 1 > = \eta_{1-1} \right. \qquad (8.108)$$

We thus have two complete sets of orthonormalized base states
(i) the uncoupled complete set of base states

$$\left| \frac{1}{2}\frac{1}{2} >, \left| \frac{1}{2} - \frac{1}{2} >, \left| - \frac{1}{2}\frac{1}{2} >, \left| - \frac{1}{2} - \frac{1}{2} > \Leftarrow | m_1 m_2 > \text{ and} \right.\right.\right.\right. \qquad (8.109)$$

(ii) the coupled complete set of base states

$$| 00 >, | 11 >, | 10 >, | 1 - 1 > \Leftarrow | Im>$$  (8.110)

We can inter-relate the two basis, i.e. express one set of base states in terms of the other set of base states. In other words we can linearly expand the basis of one set in terms of the basis of the other set.

✓ Consider $| Im> = | 11 >$ .

The value $m = 1$ can come only from the combination viz.

$$m_1 + m_2 = \frac{1}{2} + \frac{1}{2} = 1 = m.$$

This means $| 11>$ is same as the state $m_1 = \frac{1}{2}$, $m_2 = \frac{1}{2}$, i.e. $| \frac{1}{2}\frac{1}{2}>$. So

$$| 11 > = | \frac{1}{2}\frac{1}{2}>$$  (8.111)

Using equation (8.90) we can rewrite equation (8.111) as

$$| 11 > = \eta^{(1)}(p)\eta^{(2)}(p) = | pp > = | \frac{1}{2}\frac{1}{2}>$$  (8.112)

✓ Consider $| Im> = | 1 - 1 >$ .

The value $m = -1$ can come only from the combination viz.
$m_1 + m_2 = -\frac{1}{2} - \frac{1}{2} = -1 = m.$
This means $| 1 - 1>$ is same as the state $m_1 = -\frac{1}{2}$, $m_2 = -\frac{1}{2}$, i.e.
$| -\frac{1}{2} - \frac{1}{2}>$. So

$$| 1 - 1 > = | -\frac{1}{2} - \frac{1}{2}>$$  (8.113)

Using equation (8.93) we can rewrite equation (8.113) as

$$| 1 - 1 > = \eta^{(1)}(n)\eta^{(2)}(n) = | nn > = | -\frac{1}{2} - \frac{1}{2}>$$  (8.114)

✓ Consider $| Im> \rightarrow | 00>$ and $| 10>$

The value $m = 0$ can come from the combination viz.

$$m_1 + m_2 = \frac{1}{2} - \frac{1}{2} = 0 = m$$

or from the combination

$$m_1 + m_2 = -\frac{1}{2} + \frac{1}{2} = 0 = m.$$

This means $|00>$ and $|10>$ can be expressed as a linear combination of the states $|\frac{1}{2} - \frac{1}{2}>$ and $|-\frac{1}{2}\frac{1}{2}>$ as follows

$$| 00 > = a \left| -\frac{1}{2}\frac{1}{2} \right> + b \left| \frac{1}{2} - \frac{1}{2} \right> \tag{8.115}$$

$$| 10 > = c \left| -\frac{1}{2}\frac{1}{2} \right> + d \left| \frac{1}{2} - \frac{1}{2} \right> \tag{8.116}$$

where $a$, $b$, $c$, $d$ are constant coefficients to be determined.

The normalization condition is

$$a^2 + b^2 = 1, \; c^2 + d^2 = 1 \tag{8.117}$$

Let us use the relation (with $\hbar = 1$)

$$I_- | Im > = \sqrt{(I + m)(I - m + 1)} \; | I\, m - 1> \tag{8.118}$$

and we consider

$$I_- \left| 11 \right> = \sqrt{(1 + 1)(1 - 1 + 1)} \left| 11 - 1 \right>$$

$$I_- | 11 > = \sqrt{2} \; | 10> \tag{8.119}$$

Also

$$I_{1-} | m_1 m_2 > = I_{1-} | I_1 m_1 I_2 m_2 >$$

$$= \sqrt{(I_1 + m_1)(I_1 - m_1 + 1)} \; | I_1 m_1 - 1 I_2 m_2 > \tag{8.120}$$

and we consider

$$I_{1-} \left| \frac{1}{2}\frac{1}{2} \right> = I_{1-} \left| \frac{1}{2}\frac{1}{2}\frac{1}{2}\frac{1}{2} \right>$$

$$= \sqrt{\left(\frac{1}{2} + \frac{1}{2}\right)\left(\frac{1}{2} - \frac{1}{2} + 1\right)} \left| \frac{1}{2}\frac{1}{2} - 1\frac{1}{2}\frac{1}{2} \right>$$

$$= \left| \frac{1}{2} - \frac{1}{2} \; \frac{1}{2}\frac{1}{2} \right>$$

$$I_{1-} \left| \frac{1}{2}\frac{1}{2} \right> = \left| -\frac{1}{2} \; \frac{1}{2} \right> \tag{8.121}$$

Also

$$I_{2-} | m_1 m_2 > = I_{2-} | I_1 m_1 I_2 m_2 >$$

$$= \sqrt{(I_2 + m_2)(I_2 - m_2 + 1)} \; | I_1 m_1 I_2 m_2 - 1> \tag{8.122}$$

and we consider

$$I_{2-} \left| \frac{1}{2} \frac{1}{2} \right> = I_{2-} \left| \frac{1}{2} \frac{1}{2} \frac{1}{2} \frac{1}{2} \right>$$

$$= \sqrt{\left( \frac{1}{2} + \frac{1}{2} \right) \left( \frac{1}{2} - \frac{1}{2} + 1 \right)} \left| \frac{1}{2} \frac{1}{2} \frac{1}{2} \frac{1}{2} - 1 \right>$$

$$= \left| \frac{1}{2} \frac{1}{2} \frac{1}{2} - \frac{1}{2} \right>$$

$$I_{2-} \left| \frac{1}{2} \frac{1}{2} \right> = \left| \frac{1}{2} - \frac{1}{2} \right> \tag{8.123}$$

Apply $I_- = I_{1-} + I_{2-}$ on equation (8.111), namely $| 11 > = | \frac{1}{2}\frac{1}{2} >$ to get

$$I_- | 11 > = (I_{1-} + I_{2-}) \left| \frac{1}{2} \frac{1}{2} \right>$$

$$= I_{1-} \left| \frac{1}{2} \frac{1}{2} \right> + I_{2-} \left| \frac{1}{2} \frac{1}{2} \right> \tag{8.124}$$

Using equations (8.119), (8.121), and (8.123) we get

$$\sqrt{2} \, | 10 > = \left| -\frac{1}{2} \; \frac{1}{2} \right> + \left| \frac{1}{2} - \frac{1}{2} \right>$$

$$| 10 > = \frac{1}{\sqrt{2}} \left| -\frac{1}{2} \; \frac{1}{2} \right> + \frac{1}{\sqrt{2}} \left| \frac{1}{2} - \frac{1}{2} \right> \tag{8.125}$$

Using equations (8.92) and (8.91) we can rewrite equation (8.119) as

$$| 10 > = \frac{1}{\sqrt{2}} \eta^{(1)}(n) \eta^{(2)}(p) + \frac{1}{\sqrt{2}} \eta^{(1)}(p) \eta^{(2)}(n) \tag{8.126}$$

$$= \frac{1}{\sqrt{2}} | np > + \frac{1}{\sqrt{2}} | pn > = \frac{1}{\sqrt{2}} \left| -\frac{1}{2} \; \frac{1}{2} \right> + \frac{1}{\sqrt{2}} \left| \frac{1}{2} - \frac{1}{2} \right> \tag{8.127}$$

Compare equation (8.116) with equation (8.119) that gives

$$c = \frac{1}{\sqrt{2}}, \, d = \frac{1}{\sqrt{2}}$$

Hence

$$c^2 + d^2 = 1.$$

Again equation (8.115) viz. $| 00 > = a | -\frac{1}{2}\frac{1}{2} > + b | \frac{1}{2} - \frac{1}{2}>$ is orthogonal to equation (8.119) viz.

$$|10 > = \frac{1}{\sqrt{2}}| -\frac{1}{2}\frac{1}{2} > +\frac{1}{\sqrt{2}}|\frac{1}{2} - \frac{1}{2} > \text{ i.e. we demand that}$$

$$<00 | 10 > = 0 \tag{8.128}$$

This gives

$$\left[ a^*<-\frac{1}{2}\frac{1}{2} | + b^*<\frac{1}{2} - \frac{1}{2} | \right]\left[ \frac{1}{\sqrt{2}}| -\frac{1}{2} \ \frac{1}{2} > +\frac{1}{\sqrt{2}} |\frac{1}{2} - \frac{1}{2} > \right] = 0$$

$$\frac{a^*}{\sqrt{2}} <-\frac{1}{2}\frac{1}{2} |-\frac{1}{2} \ \frac{1}{2} > +\frac{a^*}{\sqrt{2}} < -\frac{1}{2}\frac{1}{2} |\frac{1}{2} - \frac{1}{2} > +$$

$$\frac{b^*}{\sqrt{2}} < \frac{1}{2} - \frac{1}{2} |-\frac{1}{2} \ \frac{1}{2} > +\frac{b^*}{\sqrt{2}} < \frac{1}{2} - \frac{1}{2} |\frac{1}{2} - \frac{1}{2} > = 0$$

$$a^* + b^* = 0$$

$$a = -b$$

Using equation (8.117) viz. $a^2 + b^2 = 1$

$$(-b)^2 + b^2 = 1$$

$$b = \frac{1}{\sqrt{2}}, a = -\frac{1}{\sqrt{2}} \tag{8.129}$$

Hence

$$| 00 > = -\frac{1}{\sqrt{2}} | -\frac{1}{2}\frac{1}{2} > +\frac{1}{\sqrt{2}} |\frac{1}{2} - \frac{1}{2}> \tag{8.130}$$

Using equations (8.92) and (8.91) we can rewrite equation (8.130) as

$$| 00 > = -\frac{1}{\sqrt{2}}\eta^{(1)}(n)\eta^{(2)}(p) + \frac{1}{\sqrt{2}}\eta^{(1)}(p)\eta^{(2)}(n) \tag{8.131}$$

$$= -\frac{1}{\sqrt{2}} | np > +\frac{1}{\sqrt{2}} | pn > = -\frac{1}{\sqrt{2}} | -\frac{1}{2}\frac{1}{2} > +\frac{1}{\sqrt{2}} |\frac{1}{2} - \frac{1}{2}> \tag{8.132}$$

We write the relation between the two complete orthonormalized sets as follows

$$| 00 > = -\frac{1}{\sqrt{2}} | -\frac{1}{2}\frac{1}{2} > +\frac{1}{\sqrt{2}} |\frac{1}{2} - \frac{1}{2}> \tag{8.133}$$

$$| 11 > = |\frac{1}{2}\frac{1}{2}> \tag{8.134}$$

$$| 10 > = \frac{1}{\sqrt{2}} | - \frac{1}{2} \ \frac{1}{2} > + \frac{1}{\sqrt{2}} | \frac{1}{2} - \frac{1}{2} > \qquad (8.135)$$

$$| 1 - 1 > = | - \frac{1}{2} - \frac{1}{2} > \qquad (8.136)$$

In matrix representation we write

$$\begin{pmatrix} | 00> \\ | 11> \\ | 10> \\ | 1-1> \end{pmatrix} = \begin{pmatrix} 0 & -\frac{1}{\sqrt{2}} & \frac{1}{\sqrt{2}} & 0 \\ 1 & 0 & 0 & 0 \\ 0 & \frac{1}{\sqrt{2}} & \frac{1}{\sqrt{2}} & 0 \\ 0 & 0 & 0 & 1 \end{pmatrix} \begin{pmatrix} | \frac{1}{2}\frac{1}{2}> \\ | -\frac{1}{2}\frac{1}{2}> \\ | \frac{1}{2}-\frac{1}{2}> \\ | -\frac{1}{2}-\frac{1}{2}> \end{pmatrix} \qquad (8.137)$$

and the matrix is called the Clebsch–Gordon matrix and the coefficients are called Clebsch–Gordon coefficients.

We rewrite the relation in terms of isospin wave functions as follows.
For isospin singlet
Equation (8.133) can be written using equations (8.105) and (8.131) as

$$\eta_{00} = -\frac{1}{\sqrt{2}} \ \eta^{(1)}(p)\eta^{(2)}(n) + \frac{1}{\sqrt{2}}\eta^{(1)}(n)\eta^{(2)}(p) \qquad (8.138)$$

For isospin triplet
Equation (8.134) can be written using equations (8.106) and (8.112) as

$$\eta_{11} = \eta^{(1)}(p)\eta^{(2)}(p) \qquad (8.139)$$

Equation (8.135) can be written using equations (8.107) and (8.126) as

$$\eta_{10} = \frac{1}{\sqrt{2}}\eta^{(1)}(p)\eta^{(2)}(n) + \frac{1}{\sqrt{2}}\eta^{(1)}(n)\eta^{(2)}(p) \qquad (8.140)$$

Equation (8.136) can be written using equations (8.108) and (8.114) as

$$\eta_{1-1} = \eta^{(1)}(n)\eta^{(2)}(n) \qquad (8.141)$$

We can also write the relations as [with equations (8.132), (8.112), (8.127), and (8.114)]

$$| 00 > = -\frac{1}{\sqrt{2}} | np > + \frac{1}{\sqrt{2}} | pn > \qquad (8.142)$$

$$| 11 > = | pp > \qquad (8.143)$$

$$| 10 > = \frac{1}{\sqrt{2}} | np > + \frac{1}{\sqrt{2}} | pn > \qquad (8.144)$$

$$| 1 - 1 > = | nn > \qquad (8.145)$$

## 8.12 Exchange operator in terms of isospin operator

We can express the exchange operator in terms of Pauli spin operator $\vec{\sigma}_1$, $\vec{\sigma}_2$ and the isotopic spin operator $\vec{\tau}_1$, $\vec{\tau}_2$ for the two nucleons.

For two nucleons each of spin $\frac{1}{2}$ we get from *exercise 6.22*

$$\vec{\sigma}_1 \cdot \vec{\sigma}_2 = \begin{cases} +1 \text{ for triplet } s = 1 \\ -3 \text{ for singlet } s = 0 \end{cases}$$

Since isopspin and spin obey the same algebra we can write by analogy

$$\vec{\tau}_1 \cdot \vec{\tau}_2 = \begin{cases} +1 \text{ for iso triplet } I = 1 \\ -3 \text{ for iso singlet } I = 0 \end{cases} \qquad (8.146)$$

The total wave function for the two-nucleon system is

$$\psi_{\text{total}}(1, 2) = \psi(\vec{r}_1, \vec{r}_2)\chi(\vec{\sigma}_1, \vec{\sigma}_2)\eta(\vec{\tau}_1, \vec{\tau}_2) \qquad (8.147)$$

where $\psi(\vec{r}_1, \vec{r}_2)$ is function of space coordinates $\vec{r}_1$, $\vec{r}_2$, $\chi(\vec{\sigma}_1, \vec{\sigma}_2)$ is function of spin coordinates $\vec{\sigma}_1$, $\vec{\sigma}_2$ and $\eta(\vec{\tau}_1, \vec{\tau}_2)$ is function of isospin coordinates $\vec{\tau}_1$, $\vec{\tau}_2$.

The two nucleons are fermions and the total wave function $\psi_{\text{total}}$ is antisymmetric under interchange of all the coordinates $\vec{r}_1$, $\vec{r}_2$ ; $\vec{\sigma}_1$, $\vec{\sigma}_2$ ; $\vec{\tau}_1$, $\vec{\tau}_2$.

We now mention the symmetry properties of the functions occurring on the RHS of equation (8.147).

$$\psi(\vec{r}_2, \vec{r}_1) = (-1)^l \psi(\vec{r}_1, \vec{r}_2)$$

$$\chi(\vec{\sigma}_2, \vec{\sigma}_1) = (-1)^{1+s}\chi(\vec{\sigma}_1, \vec{\sigma}_2)$$

$$\eta(\vec{\tau}_2, \vec{\tau}_1) = (-1)^{1+I}\eta(\vec{\tau}_1, \vec{\tau}_2) \text{ wriiten by analogy}$$

Hence the symmetry property of the two-nucleon wave function would be

$$\psi_{\text{total}}(2, 1) = \psi(\vec{r}_2, \vec{r}_1)\chi(\vec{\sigma}_2, \vec{\sigma}_1)\eta(\vec{\tau}_2, \vec{\tau}_1)$$

$$= (-1)^l \psi(\vec{r}_1, \vec{r}_2)(-1)^{1+s}\chi(\vec{\sigma}_1, \vec{\sigma}_2)(-1)^{1+I}\eta(\vec{\tau}_1, \vec{\tau}_2)$$

$$= (-1)^{l+s+I+2}\psi(\vec{r}_1, \vec{r}_2)\chi(\vec{\sigma}_1, \vec{\sigma}_2)\eta(\vec{\tau}_1, \vec{\tau}_2)$$

$$= (-1)^{l+s+I+2}\psi_{\text{total}}(1, 2) \qquad (8.148)$$

And for antisymmetry of $\psi_{\text{total}}$ the requirement is

$$\psi_{\text{total}}(2, 1) = -\psi_{\text{total}}(1, 2)$$

Hence

$$(-1)^{l+s+I+2} = -1$$

$$(-1)^{l+s+I} = -1 \tag{8.149}$$

For $l = 0$ space symmetric, $s = 0$ (spin singlet, antisymmetric) we should have $I = 1$, i.e. iso triplet, iso symmetric.

For $l = 0$ space symmetric, $s = 1$ (spin triplet, symmetric) we should have $I = 0$, i.e. iso singlet, iso antisymmetric wave function.

• Consider the operator

$$P_\sigma = \frac{1}{2}(1 + \vec{\sigma}_1 . \vec{\sigma}_2)$$

For triplet $s = 1$, $\vec{\sigma}_1 . \vec{\sigma}_2 = +1$ by *exercise 6.22*. Hence

$$P_\sigma = \frac{1}{2}(1 + 1) = +1$$

For singlet $s = 0$, $\vec{\sigma}_1 . \vec{\sigma}_2 - 3$ by *exercise 6.22*. Hence

$$P_\sigma = \frac{1}{2}(1 - 3) = -1$$

Combining we write

$$P_\sigma = \begin{cases} +1 \, for \, triplet \, s = 1 \\ -1 \, for \, singet \, s = 0 \end{cases} = (-1)^{1+s} \tag{8.150}$$

And using equation (8.40)

$$P_\sigma = (-1)^{1+s} = P_B = \frac{1}{2}(1 + \vec{\sigma}_1 . \vec{\sigma}_2) \tag{8.151}$$

• Consider the operator

$$P_\tau = \frac{1}{2}(1 + \vec{\tau}_1 . \vec{\tau}_2)$$

By analogy to equation (8.150) we can write

$$P_\tau = \begin{cases} +1 \text{ for iso triplet } I = 1 \\ -1 \text{ for iso singet } I = 0 \end{cases} = (-1)^{1+I} \tag{8.152}$$

This $P_\tau = (-1)^{1+I}$ is called charge exchange operator or isotopic spin operator.

This operator gives $+1$ when applied to symmetric isospin triplet state since for $I = 1$, $P_\tau = (-1)^{1+I} = (-1)^{1+1} = +ve$

This operator gives $-1$ when applied to antisymmetric isospin singlet state since for $I = 0$, $P_\tau = (-1)^{1+I} = (-1)^{1+0} = -ve$.

Hence $\rho_\tau$ is equivalent to exchange of the isotopic spin coordinates $\vec{\tau}_1$, $\vec{\tau}_2$ of the two nucleons (just as $\rho_\sigma = \rho_B$ changes spin coordinates $\vec{\sigma}_1$, $\vec{\sigma}_2$). Thus we can write

$$\rho_\tau \eta(\vec{\tau}_1, \vec{\tau}_2) = \eta(\vec{\tau}_2, \vec{\tau}_1) \tag{8.153}$$

• Consider

$$\rho_\tau^2 \eta(\vec{\tau}_1, \vec{\tau}_2) = \rho_\tau \rho_\tau \eta(\vec{\tau}_1, \vec{\tau}_2)$$

Using equation (8.153)

$$\rho_\tau^2 \eta(\vec{\tau}_1, \vec{\tau}_2) = \rho_\tau \eta(\vec{\tau}_2, \vec{\tau}_1)$$

$$= \eta(\vec{\tau}_1, \vec{\tau}_2)$$

$$= 1\eta(\vec{\tau}_1, \vec{\tau}_2)$$

$$\rho_\tau^2 = 1 \tag{8.154}$$

• Let us consider the combined effect of Heisenberg space–spin exchange operator $\rho_H$ and charge exchange operator $\rho_\tau$ on
$\psi_{\text{total}}(1, 2) = \psi(\vec{r}_1, \vec{r}_2)\chi(\vec{\sigma}_1, \vec{\sigma}_2)\,\eta(\vec{\tau}_1, \vec{\tau}_2)$ [equation (8.147)]

$$\rho_H \rho_\tau \psi_{\text{total}}(1, 2) = \rho_H \rho_\tau \psi(\vec{r}_1, \vec{r}_2)\chi(\vec{\sigma}_1, \vec{\sigma}_2)\eta(\vec{\tau}_1, \vec{\tau}_2)$$

$$= \rho_H[\psi(\vec{r}_1, \vec{r}_2)\chi(\vec{\sigma}_1, \vec{\sigma}_2)]\rho_\tau \eta(\vec{\tau}_1, \vec{\tau}_2)$$

$$= \psi(\vec{r}_2, \vec{r}_1)\chi(\vec{\sigma}_2, \vec{\sigma}_1)\eta(\vec{\tau}_2, \vec{\tau}_1)$$

$$= \psi_{\text{total}}(2, 1)$$

$$= -\psi_{\text{total}}(1, 2) \tag{8.155}$$

since two-nucleon wave function is antisymmetric w.r.t. interchange of coordinates.

Comparison gives

$$\rho_H \rho_\tau = -1 \tag{8.156}$$

Now operate both sides of equation (8.156) by $\rho_\tau$ from the right to get

$$\rho_H \rho_\tau \rho_\tau = -\rho_\tau$$

$$\rho_H \rho_\tau^2 = -\rho_\tau$$

Using $\rho_\tau^2 = 1$ [equation (8.154)]

$$\rho_H = -\rho_\tau = -\frac{1}{2}(1 + \vec{\tau}_1 \cdot \vec{\tau}_2) \tag{8.157}$$

Again from equation (8.45)

$$\rho_H = \rho_M \rho_B$$

Operate by $\rho_B$ from the right to get

$$\rho_H \rho_B = \rho_M \rho_B \rho_B = \rho_M \rho_B^2$$

Using $\rho_B^2 = 1$ [equation (8.47)] we have

$$\rho_M = \rho_H \rho_B \tag{8.158}$$

$$= -\rho_\tau \rho_\sigma \text{ [using equations (8.151) and (8.157)]}$$

As spin space and isospin spaces are independent and commute

$$\rho_M = -\rho_\sigma \rho_\tau$$

$$= \rho_\sigma(-\rho_\tau)$$

$$\rho_M = \rho_B \rho_H \text{ [using equation (8. 157)]} \tag{8.159}$$

The Majorana exchange operator is a combination of Bartlett and Heisenberg exchange operator. Again using equations (8.151) and (8.157)

$$\rho_M = \rho_B \rho_H = \frac{1}{2}(1 + \vec{\sigma}_1 . \vec{\sigma}_2)\left[-\frac{1}{2}(1 + \vec{\tau}_1 . \vec{\tau}_2)\right]$$

$$\rho_M = -\frac{1}{4}(1 + \vec{\sigma}_1 . \vec{\sigma}_2)(1 + \vec{\tau}_1 . \vec{\tau}_2) \tag{8.160}$$

This is the expression of Majorana exchange operator in terms of spin and charge operators.

## 8.13 Properties of nuclear force—a recapitulation

✓ Short-range behavior

Magnitude of nuclear force acting between two nucleons depends on the relative separation between the two nucleons.

Nuclear force is effective and very strong when nucleon–nucleon separation is small, i.e. $r \leqslant 1 fm = 10^{-15} m$, i.e. within the nucleus. So at short distances (within nucleus) nuclear force is stronger than Coulomb force. Nuclear force overshadows the Coulomb repulsion of protons in nucleus.

As the separation becomes $r \sim 10^{-14} m$ or more nuclear force becomes weaker than electromagnetic force. At long distance, i.e. outside nucleus— say at distance $\sim$ atomic size—the nuclear force is negligibly small. Interactions among nuclei in a molecule can be understood based only on Coulomb force.

Electrons do not feel nuclear force.

✓ Nuclear force is of exchange type and this explains the short range behavior.
Inside the nucleus there is exchange of a particle called pion of different
charge states between the nucleons.

☐ $n^0 \leftrightarrows p^{+1} + \pi^{-1}$

A neutron emits a negative pion and converts to proton. The reverse
process also occurs when a proton gains a negative pion and converts to
a neutron.

☐ $p^{+1} \leftrightarrows n^0 + \pi^{+1}$

A proton emits a positive pion and converts to a neutron. The reverse
process also occurs when a neutron gains a positive pion and converts
to a proton.

☐ $n^0 \leftrightarrows n^0 + \pi^0$

There is conversion of a neutron into a neutron through absorption or
emission of a neutral pion.

☐ $p^{+1} \leftrightarrows p^{+1} + \pi^0$

There is conversion of a proton into a proton through absorption or
emission of a neutral pion.

According to Yukawa's meson theory there is exchange of pions of
different charge states between the nucleons. This causes change in
momentum which appears in the form of nuclear force. So the
mechanism of generation of nuclear force is explained.

✓ Nuclear range from Yukawa meson exchange theory

Let $m$ be the rest mass of pion and $mc^2$ be the rest mass energy of pion

$$\Delta E = mc^2 \text{ say}$$

According to the Heisenberg's uncertainty principle, the time required by
nucleons to exchange the $\pi$ mesons cannot exceed the time

$$\Delta t = \frac{\hbar}{\Delta E} = \frac{\hbar}{mc^2}$$

otherwise it would get detected.

This is not the case since any energy violation beyond the time span $\Delta t$ is
not permitted. Energy violation may occur within $\Delta t$ so that it should not get
detected.

In this time interval $\Delta t$ the distance $r_0$ covered by pion assumed to be
moving at speed $c$ will be

$$r_0 = c\Delta t = c . \frac{\hbar}{mc^2} = \frac{\hbar c}{mc^2}$$

$$= \frac{197 MeV \, fm^2}{135 \, MeV} = 1.5 \, fm \text{ (taking } mc^2 = 135 \, MeV \text{ and using } exercise \text{ 1.2)}.$$

This $r_0$ is the range of nuclear force which shows that nuclear force is short
range. So Yukawa was able to explain the short-range behavior of nuclear force.

✓ Charge independence of nuclear force.

Gravitational force depends on mass. Electromagnetic force depends on charge. Nuclear force is completely independent of charge.

It is found from scattering experiments that in all nucleon–nucleon interaction nuclear force is the same. For instance in $pp$ interaction involving interaction between two charged particles, in $nn$ interaction involving inter-action between two neutral particles and in $np$ interaction involving interaction between one charged and one neutral particle the nuclear force is the same. So we conclude that nuclear force does not depend on the charge of the interacting particles—i.e. on whether they are neutrons or protons.

The proton is a charged particle and the neutron is a neutral particle. But w.r.t. nuclear force, proton and neutron are identical particles—hence they are collectively known as nucleons.

In fact, coupling coefficients are the same for the three interactions implying that the strengths of the interactions are same. Exchange of $\pi$ meson proves the charge independence of nuclear force. Neutron and proton are different charge states of nucleon.

✓ Spin dependence of nuclear force.

Nuclear force depends on spin state of nucleons—it depends strongly on the orientation or direction of nucleon spins—i.e. on whether spins of nucleons are parallel or antiparallel.

☐ Nuclear force is stronger when the spins of the two interacting nucleons are parallel

↑↑ i.e. both nucleons have up spins $m_{sp} = \frac{1}{2}$, $m_{sn} = \frac{1}{2}$ or

↓↓ i.e. both nucleons have down spins $m_{sp} = -\frac{1}{2}$, $m_{sn} = -\frac{1}{2}$.

☐ Nuclear force is relatively weaker if nucleons have antiparallel spin orientation

↑↓ i.e. one nucleon has $m_{sp} = \frac{1}{2}$ while the other has $m_{sn} = -\frac{1}{2}$ or

↓↑ i.e. one nucleon has $m_{sp} = -\frac{1}{2}$ while the other has $m_{sn} = \frac{1}{2}$.

Spin dependence of nuclear force has been confirmed from study of nuclear energy levels. In fact, the energy of a two-nucleon system corre-sponding to parallel spins is different from the energy of antiparallel spin. As energy is different the corresponding force is also different.

Also, a nucleon–nucleon system has a bound state called deuteron where the two nucleons have parallel spins $S = 1$.

There is no nucleon–nucleon bound state corresponding to antiparallel spins.

This means that bound state is possible only when two nucleons combine with parallel spins to $S = 1$ state. If they combine with antiparallel spins $S = 0$ no bound state results. Clearly the existence of deuteron is a remarkable proof of strong spin dependence of nuclear force.

✓ Non-central behaviour of nuclear force.

Central force is expressed as $\vec{F} = f(r)\hat{r}$. The factor $f(r)$ shows that central force depends only on $r$. Nuclear force has a predominant central force part.

But nuclear force also depends on spin orientation and so it is not a central force. Even in the simplest nucleus of deuteron the orbital angular momentum of the two nucleons relative to their centre of mass is not constant (in contrast to the central force field case where orbital angular momentum is conserved). Obviously nuclear force is not a central force—rather, it is a non-central force that does not conserve orbital angular momentum.

✓ Nuclear force is a tensor force

A part of nuclear force is tensor force which is velocity dependent. This is because nuclear force has a tensor force part that depends on interaction between nucleon spins and the angular momentum of the nucleons and so is velocity dependent.

✓ Nature of nuclear force—an attractive part and a repulsive part are present.

Whether nuclear force is attractive or not depends on the separation between the nucleons. If distance between any two nucleons is $r$ and $r_c \sim 0.4\,fm$ is a critical distance which is the average separation between two neighbouring nucleons then

☐ if $r > r_c$ nuclear force is attractive

☐ if $r < r_c$ nuclear force is repulsive

Nuclear force thus has a part that is of repulsive nature. This is needed to keep the nucleons at a certain average separation and thereby maintain nuclear size. In the case of only attraction, all nucleons would coalesce and nucleus would assume a point-like structure. In fact, as we add more nucleons the nucleus grows in such a way as to keep its density constant and this means something is keeping the nucleons from crowding too closely together.

✓ The nature of nucleon–nucleon interaction from the $V(r)$ versus $r$ plot in $n$–$p$ scattering and $p$–$p$ scattering

The plot of potential energy $V(r)$ against separation $r$ determines the nature of force or interaction between the particles. In other words, the nature of the curve of $V(r)$ determines the nature of interaction.

If $V(r)$ is $+ve$ interaction will be repulsive and if $V(r)$ is $-ve$ interaction will be attractive.

Different scattering experiments suggest a plot of $V(r)$ versus $r$, as shown in figure 8.6.

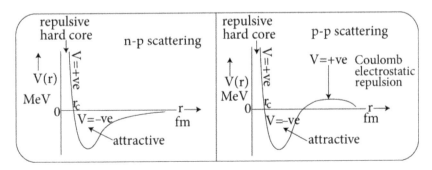

**Figure 8.6.** (a) $V(r)$ versus $r$ plot for $n$–$p$ scattering, (b) $V(r)$ versus $r$ plot for $p$–$p$ scattering.

Figure 8.6(a) shows an experimental plot of $V(r)$ versus $r$ for $np$ scattering.

Clear that in the region $0 \leqslant r \leqslant r_c$, $V(r) = +ve$, nuclear force is repulsive—there is a repulsive core.

Then there is an attractive part represented by $V(r) = -ve$.

As $r$ increases further, $V(r) \to 0$, nuclear force tends to zero signifying short-range behavior of nuclear force.

Figure 8.6(b) shows experimental plot of $V(r)$ versus $r$ for $pp$ scattering.

It is clear that in the region $0 \leqslant r \leqslant r_c$, $V(r) = +ve$, nuclear force is repulsive—there is a repulsive core.

Then there is an attractive part represented by $V(r) = -ve$.

As $r$ increases further $V(r)$ becomes $+ve$ representing Coulombian long range electrostatic repulsion $\frac{1}{4\pi\varepsilon_0} \frac{|e|^2}{r}$ between the two protons.

With further increase of $r$, $V(r) \to 0$, nuclear force tends to zero.

In view of the above discussion, we can express the nuclear potential as follows

$$V(r) = \begin{cases} +\infty, \ r < 0.4 \text{ fm} & \text{(repulsive core)} \\ -V_0, \ 0.4 \text{ fm} \leqslant r \leqslant r_0 & \text{(attractive, assumed to be square well)} \\ 0, \ r > r_0 & \text{(zero outside nucleus)} \end{cases}$$

The change from an attractive to a repulsive force is manifested by $s$ wave phase shift from positive (at $E < 300 \ MeV$) to negative (at $E > 300 \ MeV$) corresponding to change from attractive force to repulsive force, as shown in figure 8.7)

✓ Nuclear force has saturation property.

    Nuclear force saturates. Density of nucleus is constant $\sim 10^{17} \ kg \ m^{-3}$, i.e. all nuclei have the same density. Binding energy per nucleon is constant if mass number of nuclei $A > 40$. These facts imply that nuclear force saturates. This is because nucleons attract only those nucleons which are closely situated to it, i.e. lie in its immediate neighbourhood. It does not interact with a far away nucleon—this is a short-range phenomenon. This property of a nucleon interacting with a limited number of nucleons in close proximity is referred to as saturation property of nuclear force.

✓ Possible states of a two-nucleon system

    Figure 8.8 shows various possible states of a two-nucleon system viz deuteron, unbound $np$ system, $nn$ system, $pp$ system all corresponding to

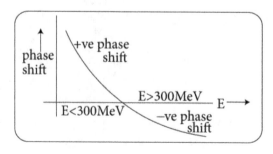

**Figure 8.7.** $S$ wave phase shift changes from positive to negative at energy about 300 $MeV$.

| System | Space wave function | Spin part | Isospin part | Total wave function | State | Scattering length |
|---|---|---|---|---|---|---|
| deuteron np | $l=0$ symmetric | $s=1$ symmetric Triplet | $I=0$ antisymmetric Singlet | antisymmetric | Bound | 5.4 fm |
| Unbound np | $l=0$ symmetric | $s=0$ antisymmetric Singlet | $I=1$ symmetric Triplet | antisymmetric | Unbound | −23.7 fm |
| di neutron nn | $l=0$ symmetric | $s=0$ antisymmetric Singlet | $I=1$ symmetric Triplet | antisymmetric | Unbound | −7.8 fm |
| di proton pp | $l=0$ symmetric | $s=0$ antisymmetric Singlet | $I=1$ symmetric Triplet | antisymmetric | Unbound | −16.6 fm |

**Figure 8.8.** Various possible states of a two-nucleon system.

$l = 0$ (i.e. symmetric space part). Only *np* system in spin triplet symmetric state corresponds to positive scattering length signifying a bound state called deuteron.

✓ Pion exchange mechanism is consistent with the nuclear range of force.

This is shown in *exercise 8.30*.

## 8.14 Exercises

**Exercise 8.1** *Find the range of strong, electromagnetic and gravitational interaction.*
    [Ans] In Yukawa's exchange theory based on quantum field concept range $R$ of an interaction is given by

$$R = v\Delta t = v\frac{\hbar}{\Delta E} = v\frac{\hbar}{\mathrm{mc}^2} \qquad (8.161)$$

where
    $v$ is speed of exchange particle $\sim c$ (take), $m =$ mass of exchange particle. Hence

$$R = c\frac{\hbar}{mc^2} = \frac{\hbar}{mc}$$

$$R = \frac{\hbar c}{mc^2} = \frac{197\,MeV.fm}{mc^2} \qquad (8.162)$$

where we have used the result of *exercise 1.2*
    ✓ For pion that mediates strong interaction

$$m_\pi c^2 = 135\,MeV. \text{ Hence range is}$$

$$R = \frac{197\,MeV.fm}{mc^2} = \frac{197\,MeV.fm}{135\,MeV} = 1.46\,fm.$$

So strong force is short range.

✓ For a photon that mediates electromagnetic interaction

$$m_\gamma c^2 = 0. \text{ Hence range is}$$

$$R = \frac{197 \, MeV. fm}{0} = \infty.$$

So electromagnetic force is long range.

✓ For a graviton that mediates gravitational interaction

$$m_{\text{graviton}} c^2 = 0. \text{ Hence range is}$$

$$R = \frac{197 \, MeV. fm}{0} = \infty.$$

So gravitational force is long range.

**Exercise 8.2** *Show that $\phi = -g\dfrac{e^{-kr}}{r}$ is a solution of the equation* $(\nabla^2 - k^2)\phi = 4\pi g\delta(\vec{r})$.

$\boxed{Ans}$ Let us put

$$\phi_1 = \frac{1}{r}, \, \phi_2 = e^{-kr} \text{ so that}$$

$$\phi = -g\frac{e^{-kr}}{r} = -g\phi_1\phi_2$$

By Green's theorem we have

$$\nabla^2(\phi_1\phi_2) = \phi_1\nabla^2\phi_2 + \phi_2\nabla^2\phi_1 + 2\vec{\nabla}\phi_1 \cdot \vec{\nabla}\phi_2$$

Hence

$$(\nabla^2 - k^2)\phi = (\nabla^2 - k^2)(-g\phi_1\phi_2) = -g(\nabla^2 - k^2)\phi_1\phi_2$$

$$= -g\nabla^2(\phi_1\phi_2) + gk^2\phi_1\phi_2$$

$$= -g\left[\phi_1\nabla^2\phi_2 + \phi_2\nabla^2\phi_1 + 2\vec{\nabla}\phi_1 \cdot \vec{\nabla}\phi_2\right] + gk^2\phi_1\phi_2$$

$$(\nabla^2 - k^2)\phi = -g\left[\frac{1}{r}\nabla^2 e^{-kr} + e^{-kr}\nabla^2\frac{1}{r} + 2\vec{\nabla}e^{-kr} \cdot \vec{\nabla}\frac{1}{r}\right] + gk^2\frac{e^{-kr}}{r} \qquad (8.163)$$

Now

$$\nabla^2 e^{-kr} = \frac{1}{r^2}\frac{d}{dr}r^2\frac{d}{dr}e^{-kr}$$

$$= \frac{-k}{r^2}\frac{d}{dr}r^2 e^{-kr} = -\frac{k}{r^2}\left(r^2\frac{d}{dr}e^{-kr} + e^{-kr}\frac{d}{dr}r^2\right)$$

$$= -\frac{k}{r^2}(-kr^2 e^{-kr} + e^{-kr}2r)$$

$$\nabla^2 e^{-kr} = ke^{-kr}(k - \frac{2}{r})$$

and

$$\nabla^2\frac{1}{r} = -4\pi\delta(r) \text{ [where } \delta(r) \text{ is Dirac delta function]}$$

$$\vec{\nabla}e^{-kr}.\,\vec{\nabla}\frac{1}{r} = \hat{r}\frac{d}{dr}e^{-kr}.\,\hat{r}\frac{d}{dr}\frac{1}{r} = \hat{r}(-k)e^{-kr}.\,\hat{r}\left(-\frac{\hat{r}}{r^2}\right)$$

$$= \frac{k}{r^2}e^{-kr}$$

From equation (8.163) we have

$$(\nabla^2 - k^2)\phi = -g\left[\frac{1}{r}\left(ke^{-kr}\left(k - \frac{2}{r}\right)\right) + e^{-kr}(-4\pi\delta(r)) + 2\left(\frac{k}{r^2}e^{-kr}\right)\right] + gk^2\frac{e^{-kr}}{r}$$

$$= -ge^{-kr}\left[\frac{k^2}{r} - \frac{2k}{r^2} - 4\pi\delta(r) + \frac{2k}{r^2} - \frac{k^2}{r}\right]$$

$$= -ge^{-kr}(-4\pi\delta(r)) = 4\pi ge^{-k.0}\delta(r)$$

$$= 4\pi g\delta(r)$$

So $\phi = -g\frac{e^{-kr}}{r}$ is the solution of the equation $(\nabla^2 - k^2)\phi = 4\pi g\delta(r)$.

**Exercise 8.3** *Calculate the range of nuclear force from meson theory if a nucleon emits a virtual pion of rest mass $270m_e$.*
 [Ans] From equation (8.161)

$$R = \frac{\hbar c}{m_\pi c^2} = \frac{197\, MeV.\,fm}{m_\pi c^2}$$

$$= \frac{197\, MeV.\,fm}{270m_e c^2} = \frac{197\, MeV.\,fm}{270 \times 0.511\, MeV} = 1.43\, fm$$

**Exercise 8.4** *Let $|n\rangle$ and $|p\rangle$ denote the isospin state with $I = \frac{1}{2}, I_3 = \frac{1}{2}$ and $I = \frac{1}{2}, I_3 = -\frac{1}{2}$ of a nucleon, respectively. Which one of the following two-nucleon states has $I = 0, I_3 = 0$ ?*

$(a) \frac{1}{\sqrt{2}}(| nn > -| pp >)$     $(b) \frac{1}{\sqrt{2}}(| nn > +| pp >)$

$(c) \frac{1}{\sqrt{2}}(| np > -| pn >)$     $(d) \frac{1}{\sqrt{2}}(| np > +| pn >)$

Hint: Given that $| n > = | \frac{1}{2}\frac{1}{2}>, | p > = | \frac{1}{2} - \frac{1}{2}>$. This choice is opposite to what we assumed in equations (8.61) and (8.62). This leads to

$| 00 > = -\frac{1}{\sqrt{2}} | pn > + \frac{1}{\sqrt{2}} | np>$

from equation (8.142).

**Exercise 8.5** *Nuclear forces are*
  *(a) Spin dependent and have no non-central part.*
  *(b) Spin dependent and have a non-central part.*
  *(c) Spin independent and have no non-central part.*
  *(d) Spin independent and have no non-central part.*

**Exercise 8.6** *With reference to nuclear forces which of the following is not true?*
  *(a) Short range, (b) charge independent, (c) velocity dependent, (d) spin independent.*

**Exercise 8.7** *Pick up the wrong statement.*
  *(a) The nuclear force is independent of electric charge.*
  *(b) The Yukawa potential is proportional to $r^{-1}e^{\frac{mc}{\hbar}r}$ where r is the distance separation between two nucleons.*
  *(c) The range of nuclear force is of the order of $10^{-15} - 10^{-14}$ m.*
  *(d) The nucleons interact among each other by the exchange of mesons.*

**Exercise 8.8** *Show that deuteron is an iso singlet.*
  $\boxed{Ans}$ Deuteron is a system of two nucleons and nucleons are fermions. So the total wave function has to be antisymmetric.

The total wave function is written as

$$\psi_d(\vec{r}, \vec{\sigma}, \vec{\tau}) = \psi(\vec{r})\chi(\vec{\sigma})\eta(\vec{\tau})$$

The spatial part $\psi(\vec{r})$ has parity $(-1)^l$ and for deuteron state $l \to 0, 2$

$$| d > = a_S | l = 0 > +a_D | l = 2>$$

So the space part $\psi(\vec{r})$ is $(-1)^l = (-1)^{even} = +ve$, i.e. symmetric under interchange of space coordinates of nucleons.

Deuteron is a spin triplet state with $s = 1$ (parallel spins) and so the spin part has parity $(-1)^{1+s} = (-1)^{1+1} = +ve$, i.e. symmetric w.r.t. interchange of spin coordinates of nucleons.

Clearly to make $\psi_d(\vec{r}, \vec{\sigma}, \vec{\tau})$ antisymmetric the isospin part $\eta(\vec{\tau})$ has to be antisymmetric since

$$\psi_d(\vec{r}, \vec{\sigma}, \vec{\tau}) = (-1)^l(-1)^{l+s}\eta(\vec{\tau})$$

Writing explicitly the parities we have

$$(-1) = (-1)^{\text{even}}(-1)^{l+1}\eta(\vec{\tau})$$

$$\eta(\vec{\tau}) = -ve$$

Again parity of isospin part $\eta(\vec{\tau})$ is decided by the factor $(-1)^{l+l}$ and so

$$(-1)^{l+l} = -ve$$

$$I = \text{even}$$

Again nucleons are iso doublets

$$|p> = |\frac{1}{2}\frac{1}{2}>, |n> = |\frac{1}{2} - \frac{1}{2}> \text{ in the } |Im_I> \text{ representation.}$$

In deuteron which is an $np$ system the possible isospin values are

$$\vec{I} = \vec{I}_n + \vec{I}_p = \frac{1}{2} + \frac{1}{2} = \begin{cases} \frac{1}{2} - \frac{1}{2} = 0 \Rightarrow I = 0 \text{ is iso singlet, } 2I + 1 = 2.0 + `1 = 1 \text{ state} \\ \frac{1}{2} + \frac{1}{2} = 1 \Rightarrow I = 1 \text{ is iso triplet, } 2I + 1 = 2.1 + `1 = 3 \text{ states} \end{cases}$$

Also, the parities are

$$(-1)^{l+l} = \begin{cases} (-1)^{l+l} = (-1)^{l+0} = -ve \text{ (antisymmetric) for } I = 0 \\ (-1)^{l+l} = (-1)^{l+1} = +ve \text{ (symmetric) for } I = 1 \end{cases}$$

Clear that deuteron corresponds to even $I$, i.e. $I = 0$ which is antisymmetric.

In other words, deuteron is an iso singlet with isospin part antisymmetric. Explicitly $\eta(\vec{\tau})$ can be expressed as [equation (8.142)]

$$|I = 0 \ m_I = 0> \equiv |00> = \frac{1}{\sqrt{2}}(|pn> - |np>)$$

**Exercise 8.9** *Why can we not choose l = 1 for deuteron?*

[Ans] Deuteron is a spin triplet with $s = 1$, iso singlet with $I = 0$. Deuteron total wave function is

$$\psi_d(\vec{r}, \vec{\sigma}, \vec{\tau}) = \psi(\vec{r})\chi(\vec{\sigma})\eta(\vec{\tau}) \tag{8.164}$$

Deuteron is a system of nucleons which are fermions that have antisymmetric total wave function. So $\psi_d(\vec{r}, \vec{\sigma}, \vec{\tau}) \rightarrow -ve$, i.e. antisymmetric. Also, the symmetry of spin part is $\chi(\vec{\sigma}) \rightarrow (-1)^{l+s} = (-1)^{l+1} = +ve$, and symmetry of isospin part is $\eta(\vec{\tau}) \rightarrow (-1)^{l+l} = (-1)^{l+0} = -ve$.

Again, parity of spatial part is $\psi(\vec{r}) \rightarrow (-1)^l$.

Equation (8.164) suggests that

$$(-) = (-1)^l(+)(-)$$

$$(-1)^l = +ve$$

$$l = \text{even}$$

As $l$ = even it follows that $l \neq 1$.

**Exercise 8.10** *Why is there no spin singlet state of deuteron?*
  [Ans] We can present the following arguments that establish that deuteron cannot have a singlet state.

✓ Deuteron is a spin triplet $s = 1$ state in which the spins of neutron and proton are parallel. The nuclear force is stronger for parallel spins so as to make it a stable bound system.

  The spin singlet $s = 0$ state corresponds to antiparallel orientation of nucleon spins. The nuclear force is not strong enough to bind antiparallel neutron and proton into a stable bound system.

✓ Deuteron is a system of two nucleons which are fermions and so its total wave function will be antisymmetric as per the Pauli exclusion principle.

  Also, the space part of deuteron is

$$| d > = \sqrt{0.96} \ | l = 0 > + \sqrt{0.04} \ | l = 2>$$

which means $l$ = even. So space part has symmetry

$$(-1)^l = (-1)^{\text{even}} = +ve \text{ (symmetric)}.$$

  Also, deuteron is an iso singlet, i.e. has isospin $I = 0$. So the isospin part has symmetry

$$(-1)^{1+I} = (-1)^{1+0} = -ve \text{ (antisymmetric)}.$$

Deuteron total wave function is

$$\psi_d(\vec{r}, \vec{\sigma}, \vec{\tau}) = \psi(\vec{r})\chi(\vec{\sigma})\eta(\vec{\tau})$$

Putting symmetries we have

$$(-ve) = (+ve)(-1)^{1+s}(-ve)$$

where $(-1)^{1+s}$ is the symmetry of the spin part.
  Clearly $(-1)^{1+s} = +ve$ (symmetric)
  i.e. $s$ = odd.
  This means $s = 1$.
  So deuteron is spin triplet symmetric under exchange of spin coordinates of nucleons as required by the Pauli exclusion principle.
  A spin singlet state $s = 0$ of deuteron will not be consistent with Pauli exclusion principle, i.e. it will not make the total wave function of deuteron

nucleus antisymmetric. So due to symmetry requirement the deuteron system cannot be a spin singlet.

✓ The scattering lengths of singlet states are negative implying unbound state, namely

$$a_s^{np} = -23.7\,fm,\; a^{pp} = -7.8\,fm,\; a^{nn} = -16.6\,fm$$

But the scattering length of triplet state is positive implying bound state, namely

$$a_t^{np} = 5.4\,fm$$

**Exercise 8.11** *Which of the following phrases is acceptable for $l = 0$?*
(a) *deuteron spin singlet,*
(b) *deuteron spin triplet,*
(c) *deuteron iso singlet,*
(d) *deuteron iso triplet.*
Hint. Figure 8.8.

**Exercise 8.12** *Which of the following phrases is acceptable for $l = 0$?*
(a) *np spin singlet bound state,*
(b) *np spin triplet bound state,*
(c) *np spin singlet unbound state,*
(d) *np spin triplet unbound state.*
Hint. Figure 8.8.

**Exercise 8.13** *Which of the following phrases is acceptable for $l = 0$?*
(a) *pp spin singlet bound state,*
(b) *pp spin triplet bound state,*
(c) *pp spin singlet unbound state,*
(d) *pp spin triplet unbound state.*
Hint. Figure 8.8.

**Exercise 8.14** *Which of the following phrases is acceptable for $l = 0$?*
(a) *nn spin singlet bound state,*
(b) *nn spin triplet bound state,*
(c) *nn spin singlet unbound state,*
(d) *nn spin triplet unbound state.*
Hint. Figure 8.8.

**Exercise 8.15** *Which of the following phrases is not acceptable regarding scattering length for $l = 0$?*
(a) *Negative for unbound np,*

(b) negative for nn,
(c) negative for deuteron,
(d) negative for pp.
Hint. Figure 8.8.

**Exercise 8.16** *Which of the following phrases is acceptable regarding scattering length of a two-nucleon system for l = 0?*
  (a) *Positive for triplet,*     (b) *Positive for deuteron,*
  (c) *Positive for singlet,*     (d) *Positive for pp.*
Hint. Figure 8.8.

**Exercise 8.17** *The scattering length and effective range for nn scattering are, respectively,   a, $R_0 \cong -16.6\,fm$, 2.66 fm   and   those   for   pp   scattering   are a, $R_0 \cong -7.8\,fm$, 2.8 fm. If we apply correction for Coulomb interaction, which of the following would be the correct parameters for pp scattering?*
  (a) *$-16.6\,fm$, 2.66 fm,*     (b) *$-7.8\,fm$, 2.8 fm,*
  (c) *$-12.2\,fm$, 2.73 fm,*     (d) *$-23.7\,fm$, 2.7 fm.*
Hint : The pp scattering parameters would be identical to the nn scattering parameters if we remove the Coulomb effect because nuclear force does not depend on Coulomb interaction and is charge independent.

**Exercise 8.18** *Discuss if pp scattering is possible from spin triplet state?*
  Ans For low energy for which $l = 0$, pp scattering can only be in the singlet state (figure 8.5). For higher $l$ value, pp scattering can be either in the singlet state or in the triplet state depending on whether $l =$ even or $l =$ odd, respectively.

**Exercise 8.19** *According to meson theory of nuclear force the potential energy of interaction between two nucleons is*
  (a) *$r^{-2}e^{-\mu r}$,*     (b) *$r^{-1}e^{-\mu r}$,*     (c) *$r^{-2}e^{\mu r}$,*     (d) *$r^{-1}e^{-\mu r^2}$.*

**Exercise 8.20** *According to the meson theory of nuclear force*
  (a) *a neutron emits a $\pi^-$ meson and is converted to a proton.*
  (b) *a neutron absorbs a $\pi^0$ meson and is converted to a proton.*
  (c) *a neutron emits a $\pi^+$ meson and is converted to a proton.*
  (d) *a neutron absorbs a $\pi^-$ meson and is converted to a proton.*

**Exercise 8.21** *Which is the best option for nuclear force?*
  (a) *Short range, strongly repulsive, attractive core.*
  (b) *Long range, strongly repulsive, attractive core.*
  (c) *Short range, strongly attractive, repulsive core.*
  (d) *Long range, strongly attractive, repulsive core.*

**Exercise 8.22** *Identify the incorrect statement.*
(a) *Nuclear force is the strongest force.*
(b) *Net nuclear force on the nucleons inside the nucleus is zero.*
(c) *Nuclear force between two protons is the same as that between two neutrons.*
(d) *Nuclear force between two protons over a distance is always less than the electrostatic force between them.*

**Exercise 8.23** *Shape of Yukawa potential is which of the following?*
(a) $V_0 e^{-r/r_0}$,   (b) $-V_0 e^{-r^2/r_0^2}$,   (c) $\frac{V_0 e^{-r/r_0}}{r/r_0}$,   (d) $-\frac{V_0 e^{-r^2/r_0^2}}{r^2/r_0^2}$.

**Exercise 8.24** *The saturation character of nuclear force can be explained by*
(a) *central force,*
(b) *exchange interaction,*
(c) *electromagnetic interaction,*
(d) *strong interaction.*

**Exercise 8.25** *Bartlett operator exchanges*
(a) *position of particles,*      (b) *spin of particles,*
(c) *both position and spin of particles,*   (d) *neither position nor spin of particles.*

**Exercise 8.26** *Wigner operator exchanges*
(a) *position of particles,*      (b) *spin of particles,*
(c) *both position and spin of particles,*   (d) *neither position nor spin of particles.*

**Exercise 8.27** *Majorana exchange operator exchanges*
(a) *position of particles,*      (b) *spin of particles,*
(c) *both position and spin of particles,*   (d) *neither position nor spin of particles.*

**Exercise 8.28** *Heisenberg exchange operator exchanges*
(a) *position of particles,*      (b) *spin of particles,*
(c) *both position and spin of particles,*   (d) *neither position nor spin of particles.*

**Exercise 8.29** *Which of the following statements is false?*
(a) *Nuclear forces are spin dependent.*
(b) *Nuclear forces are charge dependent.*
(c) *Nuclear forces have saturation character.*
(d) *Nuclear forces are partly central and partly non-central forces.*

**Exercise 8.30** *Find the time interval within which a proton can emit and absorb a pion without any observation of violation of conservation of energy. Hence find the range of nuclear potential.*

$\boxed{Ans}$ *From* equation (8.2)

$$\Delta t = \frac{\hbar}{\Delta E} = \frac{\hbar}{m_\pi c^2}$$

Since pion rest energy is $m_\pi c^2 = 140\ MeV$ we have

$$\Delta t = \frac{\hbar}{m_\pi c^2} = \frac{\hbar}{140\ MeV}$$

Multiplyimg numerator and denominator by $c = 3 \times 10^8\ m\ s^{-1}$ we have

$$\Delta t = \frac{\hbar}{140\ MeV} = \frac{\hbar c}{140\ MeV \left(3 \times 10^8\right) m\ s^{-1}} = \frac{197\ MeV.fm}{140\ MeV \left(3 \times 10^8\ m\ s^{-1}\right)} \text{ (using } exercise\ 1.2\text{)}$$

$$\Delta t = \frac{197 \times 10^{-15}\ m}{140 \times 3 \times 10^8\ m\ s^{-1}} = 4.7 \times 10^{-24}\ s$$

In a time interval shorter than $4.7 \times 10^{-24}\ s$ a proton can emit and absorb a pion without detection of any violation of energy conservation.

A pion will travel a distance equal to the range $r_0$ of nuclear force during this time interval.

$$r_0 = c\Delta t = (3 \times 10^8\ m\ s^{-1})(4.7 \times 10^{-24}\ s) = 1.4 \times 10^{-15}\ m = 1.4\,fm.$$

Obviously if the distance between two nucleons is $r \leqslant 1.4\,fm$ they would interact strongly through exchange of pions. But if the distance between the two nucleons is $r \geqslant 1.4\,fm$, pion exchange will not occur and they will not interact strongly.

It follows that the pion exchange mechanism is consistent with the nuclear range of force.

**Exercise 8.31** *Which of the following phrases is acceptable for $l = 0$?* (Hint. Figure 8.8)

    *(a) pp spin asymmetric isospin asymmetric state.*
    *(b) pp spin asymmetric isospin symmetric state.*
    *(c) pp spin symmetric isospin asymmetric state.*
    *(d) pp spin symmetric isospin symmetric state.*

**Exercise 8.32** *Which of the following phrases is acceptable for $l = 0$?* (Hint. Figure 8.8)

    *(a) np spin asymmetric isospin symmetric unbound state.*
    *(b) np spin symmetric isospin asymmetric unbound state.*
    *(c) np spin asymmetric isospin symmetric bound state.*
    *(d) np spin symmetric isospin asymmetric bound state.*

Ans to Multiple Choice Questions

8.4 c 8.5 b, 8.6 d, 8.7 b, 8.11 b, 8.12 b, c, 8.13 c, 8.14 c, 8.15 c, 8.16 a, b, 8.17 a, 8.19 b, 8.20 a, 8.21 c, 8.22 d, 8.23 c, 8.24 b, 8.25 b, 8.26 d, 8.27 a, 8.28 c, 8.29 b, 8.31 b, 8.32 a, d.

## 8.15 Question bank

Q8.1  Discuss the force generation mechanism within a nucleus.

Q8.2  Why is it that particle exchanges within a nucleus between the nucleons is virtual?

Q8.3  Which particles are said to be mediators of:
(a) strong interaction, (b) weak interaction, (c) electromagnetic interaction, (d) gravitational interaction?

Q8.4  Describe Yukawa's meson theory of nuclear force.

Q8.5  How do you explain the fact that even after absorbing a $\pi^0$ ($m_{\pi_0} \neq 0$) a neutron remains a neutron in the interaction $n \to n + \pi^0$ within a nucleus?

Q8.6  How many types of pions were proposed by Yukawa and how many were discovered later? What is their mass and charge?

Q8.7  Establish that the pion exchange mechanism is consistent with the nuclear range of force.

Q8.8  What is the role of Heisenberg's uncertainty principle in the pion exchange mechanism within a nucleus as proposed by Yukawa?

Q8.9  Find the range of nuclear force from pion exchange mechanism within a nucleus as proposed by Yukawa.

Q8.10  Find the mass of pions from the pion exchange mechanism within a nucleus as proposed by Yukawa.

Q8.11  What do we mean by exchange character of nuclear force?

Q8.12  Obtain one-pion exchange potential to characterize force exchanges within a nucleus.

Q8.13  Compare Yukawa potential and Coulomb potential through a diagram.

Q8.14  Why is Yukawa potential also known as screened Coulomb potential?

Q8.15  What are the conclusions of Yukawa's meson theory? What are the shortcomings of Yukawa theory?

Q8.16  What is colour force?

Q8.17  How do you explain that a nucleon is colour neutral or white?

Q8.18  How does nuclear interaction differ fron Coulomb interaction?

Q8.19  How do you explain that nuclear force has a repulsive part?

Q8.20  Define and explain (a) Wigner potential, (b) Majorana potential, (c) Bartlett potential, (d) Heisenberg potential. Write down the corresponding exchange operators and obtain their effect on the wave function $\psi(\vec{r}_1, \vec{r}_2, \vec{\sigma}_1, \vec{\sigma}_2)$.

Q8.21  Write down the nature of nuclear force as a function of distance within nucleus.

Q8.22 What are the four types of exchange forces?

Q8.23 What is Wigner force? Why is it called a non-exchange force?

Q8.24 What is Majorana exchange force? Show that Majorana force is decided by whether wave function changes sign or not for different $l$ values?

Q8.25 What is Bartlett exchange force? Show that Bartlett force is decided by whether wave function changes sign or not for different $s$ values?

Q8.26 What is Heisenberg exchange force? Show that effect of Heisenberg exchange force is decided by whether wave function changes sign for different $l$ and $s$ values.

Q8.27 What is the eigenvalue of Majorana, Bartlett and Heisenberg exchange operators? Why is it so?

Q8.28 Show that Majorana exchange force is attractive for the states $S$, $D$, $G$, ... but repulsive for the states $P$, $F$, ....

Q8.29 Establish that Bartlett exchange force is attractive for the states $^3S$, $^3P$, $^3D$... but repulsive for the states $^1S$, $^1P$, $^1D$...

Q8.30 Establish that Heisenberg exchange force can be treated as a combination of Majorana exchange force and Bartlett exchange force.

Q8.31 Establish that Heisenberg exchange force is attractive for the states $^3S$, $^3D$... $^1P$, $^1F$... but repulsive for the states $^1S$, $^1D$... $^3P$, $^3F$...

Q8.32 Show that the square of the Majorana exchange operator, Bartlett exchange operator and Heisenberg exchange operator is unity.

Q8.33 Write down the most general form of nuclear potential of exchange type.

Q8.34 What is tensor force?

Q8.35 Write down the most general form of nuclear potential incorporating exchange version of tensor force.

Q8.36 Outline the theory that shows that saturation of binding energy occurs when Majorana force contributes such that it is less than four times the Wigner force.

Q8.37 Define nucleons in terms of their isospin.

Q8.38 Define charge operator and find its eigenvalues and eigen vectors.

Q8.39 Neutron and proton are not differentiated so far as their strong interaction is concerned. Define an operator that converts proton to neutron and vice versa. Also show that this operator can neither increase the charge of proton nor implant negative charge to neutron.

Q8.40 Construct two-nucleon wave functions in isospin space using (a) coupled basis, (b) uncoupled basis. Obtain the Clebsch–Gordon coefficients connecting the two bases.

Q8.41 Define charge exchange operator or isotopic spin operator $\rho_\tau$ and show that (a) $\rho_\tau^2 = 1$, (b) $\rho_H = -\rho_\tau$, where $\rho_H$ is Heisenberg exchange operator.

Q8.42 Show that Majorana exchange operator is a combination of Bartlett and Heisenberg exchange operators.

Q8.43    Show that Heisenberg exchange operator can be expressed as
$\rho_H = -\frac{1}{4}(1 + \vec{\sigma}_1 . \vec{\sigma}_2)(1 + \vec{\tau}_1 . \vec{\tau}_1)$ where $\vec{\sigma}, \vec{\tau}$ are spin and isospin operators.

Q8.44    How can exchange type behavior of nuclear force explain short-range behavior of nucleus?

Q8.45    Find nuclear range from Yukawa's meson theory.

Q8.46    Discuss if exchange of $\pi$ mesons proves charge independence of nuclear force.

Q8.47    How is it that nuclear force is non-central?

Q8.48    Is tensor force velocity dependent?

Q8.49    Show the plot of nuclear potential aginst distance for (a) $pp$ scattering, (b) $np$ scattering. Explain the plot in different regions of distance.

Q8.50    Nuclear force is a short-range force. Is it purely attractive? Is there a repulsive part?

Q8.51    How do you account for the existence of attractive and repulsive part in a nucleus? How do you explain the change from attractive to repulsive force?

Q8.52    Define saturation property of nucleus.

Q8.53    Write down the various possible states of a two-nucleon system (bound and scattering states).

## Further reading

[1]  Krane S K 1988 *Introductory Nuclear Physics* (New York: Wiley)

[2]  Tayal D C 2009 *Nuclear Physics* (Mumbai: Himalaya Publishing House)

[3]  Satya P 2005 *Nuclear Physics and Particle Physics* (New Delhi: Sultan Chand & Sons)

[4]  Guha J 2019 *Quantum Mechanics: Theory, Problems and Solutions* 3rd edn (Kolkata: Books and Allied (P) Ltd))

[5]  Lim Y-K 2002 *Problems and Solutions on Atomic, Nuclear and Particle Physics* (Singapore: World Scientific)

Printed in the USA
CPSIA information can be obtained
at www.ICGtesting.com
LVHW080331110924
790375LV00003B/7

9 780750 350259